SAFETY AND HEALTH
FOR ENGINEERS

SECOND EDITION

SAFETY AND HEALTH
FOR ENGINEERS

ROGER L. BRAUER, Ph.D., CSP, PE
Tolono, Illinois

**WILEY-
INTERSCIENCE**

A JOHN WILEY & SONS, INC., PUBLICATION

Published by John Wiley & Sons, Inc., Hoboken, New Jersey
Published simultaneously in Canada

For general information on our other products and services or for technical support, please contact our Customer Care Department within the United States at (800) 762-2974, outside the United States at (317) 572-3993 or fax (317) 572-4002.

Wiley also publishes its books in a variety of electronic formats. Some content that appears in print may not be available in electronic formats. For more information about Wiley products, visit our web site at www.wiley.com.

Library of Congress Cataloging-in-Publication Data:

Brauer, Roger L.
    Safety and health for engineers / Roger L. Brauer.—2nd ed.
        p. cm.
    Includes index.
    ISBN-13: 978-0-471-29189-3 (cloth)
    ISBN-10: 0-471-29189-7 (cloth)
    1. Industrial safety.   2. Product safety.   3. Products liability—United States   I. Title.
    T55.B72   2005
    620.8′6—dc22   2005009403

Printed in the United States of America

10 9 8 7 6 5 4 3

CONTENTS

PREFACE

Since the first edition of this book, some things have not changed and others have. Today, engineers still have a moral, legal, and ethical responsibility to protect the public in professional practice and in design of products, buildings, processes, equipment, work, and workplaces. The importance of safety in engineering education remains a concern for most engineering degree programs. The need for safety specialists to understand basic technical fundamentals essential in hazard recognition, evaluation, and control continues. As a result, there is still a need for this book.

The laws, regulations, standards, and standard of practice in safety and health continue to change on a regular basis. As soon as a book is complete or updated, it is likely to be out of date in certain regulatory areas. The reader should recognize this type of change and consult government and voluntary standards to ensure compliance with current requirements.

Technology continues to change. Computer technology has changed the toolbox for nearly every professional field, and it impacts safety practice as well. Since the first edition was published, the Internet has become an integral part of professional practice, business and business transactions, and many other elements of daily life. Although the explosion in availability of information continues, one must be able to sort out valid, quality information and reliable information sources from those sources that are not. It is far easier today to find information as well as misinformation on a wide variety of safety issues.

The overall field of safety has changed. One significant trend is the continued growth in education of those practicing at the professional level. More individuals than ever who specialize in safety have advanced degrees. At the same time, many employers have achieved significant improvements in safety performance by moving safety knowledge and skills deeper into their organizations and workgroups. There seems to be a growing interest among people from other areas of work experience in finding a professional home in the broad safety field. Another trend is the rapid convergence of several related areas of practice. Two decades ago, safety, industrial hygiene, environmental science and engineering, environmental health, ergonomics, fire protection, and other areas of practice often were isolated from each other. Today, many of these have converged into a single organizational unit for an employer, and many individuals—regardless of their original backgrounds—have responsibility for many of these areas simultaneously. The overall impact is a change in what safety and health specialist do.

The original goal for this book was to help engineers and others gain a broad, quick overview of safety and health practices and to identify some of the detailed resources that may provide expanded help with applications. One of the most valued results of having written this book in the first place is having people who I have never met express appreciation for the assistance it provided them in their professional development. Many have told me that it helped them to understand what safety and health practice is about. It is rewarding to know that a personal project has assisted others professionally.

In completing the update, there are many to thank who may have contributed in some way to the insights offered among the revisions and who pressed me to keep working to complete the revision. I also want to thank my family for their continued support and for tolerating the time often stolen from family activities to make room for the revision effort after abnormally long but typical work weeks.

ROGER L. BRAUER

Tolono, IL

PART *1*

INTRODUCTION

THIS SECTION of the book identifies the technological foundation of safety engineering, summarizes its history, and outlines some fundamental concepts for safety.

CHAPTER 1

THE IMPORTANCE OF SAFETY AND HEALTH FOR ENGINEERS

1-1 INTRODUCTION

Technological Change

Engineers have played a major role in technological advancements that have created many changes for mankind. Some advancements have improved society, some have been detrimental. Some have aided life, others have created new economic, social, political, environmental, or safety and health problems.

One noteworthy change brought about by technology is faster and more efficient travel. Not long ago, people traveled approximately 8 km/hr or less either walking or via animal-powered conveyances. Automobiles made travel approximately 10 times faster than that, airplanes 100 times faster, and rockets more than 1,000 times faster. A horse-drawn wagon could carry a 1- or 2-ton load. Today, a 200-car freight train can carry 20,000 tons, and supertanker ships carry similar or larger loads.

Communication and electronics technologies continue to shrink the world and change lifestyles. The Pony Express moved only small pouches of information at one time. Today, there are many communication satellites in orbit, transmitting millions of bits of information every second. At least 95% of American homes have a television set. Nearly half have more than one DVD player. Children spend an average of three and one-half hours per day in front of a TV set; adults average more than 4 hours per day. One used to associate a telephone with a place, whereas today one associates a telephone with a person. The Internet and personal computers offer electronic mail and access to specific information sources around the globe at any time.

Technology not only has increased the flow of information, it has increased information density. A printed page in a book contains approximately 450 words. A 600-page book contains approximately 270,000 words and occupies approximately 70 cubic inches. A DVD can store nearly 1.5 million pages of text. A small memory stick can store the equivalent of 1,000 books in less than 1 cubic inch.

Because of technology, the number of materials and substances known to humanity has increased rapidly. Today there are approximately 5 million substances listed in the *Registry Handbook*.[1] Nearly 100,000 chemical substances are now in use, with several hundred new ones entering the marketplace each year.

Advances in medicine, supported by new technology, have extended human life. In the early stages of the industrial revolution, life expectancy for the working class in Manchester, England, was 17 years; for the gentry, it was roughly 35 years. Today the life expectancy of American males is more than 72 years; for females, it is nearly 80 years.

Safety and Health for Engineers, Second Edition, by Roger L. Brauer
Copyright © 2006 John Wiley & Sons, Inc.

Diseases that were once a major threat, such as smallpox, typhoid, cholera, bubonic plague, diphtheria, tuberculosis, and polio, are now well under control. Vaccination, improved treatment, wonder drugs, and sanitation made these advances possible. And now we are beginning the age of biological medicine, with diagnosis from DNA analysis and biological growth of substances, tissue, and perhaps even organs for treatment.

Aided by advances in medicine and improved standards of living, the world's population has risen from approximately 0.3 billion in 1 A.D. to 1.1 billion in 1850 and to more than 6 billion today. The increase is creating a new demand on available resources in the world. For example, the per capita energy consumption in the United States is more than 350×10^6 BTU annually.

Manual labor has given way to industrialization and automation. Production rates have increased rapidly as a result. The industrial production index, which represents the rate of industrial output (equal to 100 in 1967), grew from 42 in 1950 for transportation equipment to 140 by 1979. For chemicals, the index grew in the same period from 26 to 208.[2]

The Risks

Although life has improved and has been extended, citizens of the United States pay a high price for their high-technology lifestyle. Each year, there are more than 100,000 accidental deaths and nearly 10 million disabling injuries. The cost of all accidents in the United States is approximately $600 billion annually, excluding some indirect costs and the value resulting from pain and suffering. Accidents are the fifth leading cause of death. For those aged 65 or older, the accidental death rate is increasing. Only heart disease, cancer, stroke and chronic respiratory disease exceed it. For the total population, the two leading causes of accidental death are motor vehicles and falls. Nine times more workers die accidently off the job than at work. The accidental death rate in the United States has declined from approximately 85 to 90 deaths per 100,000 persons in 1910 to fewer than 35 today.[3]

Not only has technological change introduced new methods, materials, products, and equipment into use by society, but also new hazards. For example, electricity replaced gas and oil lighting. Electricity may be less hazardous than gas and oil lighting; however, it is identified as the cause of one of every seven fires and produces roughly 100 electrocution deaths each year.

Another example of a new hazard is asbestos. In the 1930s, asbestos became a widely used material for thermal insulation, roofing, brakes, and other applications. A 1978 estimate by the federal government said that 8 to 11 million workers had been exposed to asbestos. Of those, one million were significant to the point that half of these individuals could expect to die of cancer in the next 30 years. Some believe that this is an overestimate and does not explain the full story. It does illustrate that hazards associated with new technology are sometimes widely distributed in society.

The automobile arrived at the end of the nineteenth century. Today, there are approximately 1.5 motor vehicles per American household. The use of these vehicles now results in roughly 45,000 traffic deaths and 2 million disabling injuries each year in the United States.

Society's Response

Society has responded to the safety and health risks placed on them by technology, primarily through regulation and litigation. Federal, state, and local governments have passed many laws and regulations dealing with safety and health issues. More than 15,000 new laws are passed each year. Approximately 10% or more of these involve safety and health.

The 1960s and early 1970s saw the creation of several federal safety and health agencies and the emergence of others through restructuring of some existing federal organizations. Each of these created new regulations. Counterparts often have appeared at state and local levels and produced additional regulations and standards.

Society has turned to the courts to recover losses from injury and damages for pain and suffering. According to congressional estimates, there are between 60,000 and 140,000 product liability claims filed each year. In addition, legal interpretations place a greater burden on the manufacturers and sellers of products to minimize the risks to their users. As a result, product liability insurance rates have grown. Tort reform efforts seek to limit liability claims in size and frequency.

Although death and injury rates are holding steady or are on the decline, the public is not fully satisfied with the protection offered by government and industry. In one opinion survey,[4] public respondents rated the job being done by the federal government, the business community, and state and local government to make society acceptably safe. The differences among the ratings for the three groups were small. Overall, approximately 25% to 33% of the public said these groups did a very good job, 50% said they were doing only a fair job, and 15% to 23% reported they were doing a poor job.

The survey results also suggest that the public continues to look to government and society for protection from technological risks. One of every five public respondents believed that "no matter what risks an individual takes, there should be no personal economic penalty; society as a whole should bear the cost." In another survey,[5] 75% of the respondents wanted government to cut back in size. However, nearly 50% of the people surveyed believed that the government was doing less than it should to regulate major corporations in areas like product safety and other matters that have to do with protecting the public. Twenty-two percent of the respondents believed that the federal government was doing more than it should and 27% said the government was doing the right amount.

People said they want to exercise control and choice in the risks they face. The public does not always see eye to eye with industrial and government leaders regarding technological risks placed on them. In the first survey mentioned above, more than half of the respondents wanted a choice in making tradeoffs between risk and cost. One question asked whether the higher risk of fatal accidents with small cars was worth the savings from fuel and initial cost. Almost 50% of the public said it was not. In contrast, only 11% of the top corporate executives and 15% of the congressional representatives included in the study shared the same view. More recently, the public love affair with large cars has shifted to minivans, sports utility vehicles, and trucks.

A Closer Look

Technology has brought new things to modern life. We live better lives through chemistry, electricity, transportation, electronics, and communication. Society has accepted the benefits, but not all the risks. It has placed new demands on engineering and other professions to reduce safety and health problems.

1-2   OCCUPATIONAL SAFETY AND HEALTH

According to National Safety Council statistics, there are approximately 4,500 work-related deaths each year, with a death rate of more than 3 per 100,000 for all industries.

Annually, there are more than 3.5 million injuries involving one or more days away from work. The total cost in lost wages, medical expenses, insurance, fire losses, and other indirect costs associated with these work-related accidents is more than $150 billion annually. This figure does not include business interruption costs. Workplace injuries result in more than 100 million lost workdays each year. Each worker in the United States loses approximately two days each year from job-related accidents.

Since the 1930s, when such record keeping began, the highest number of work-related deaths occurred in 1937: 19,500. However, estimates for earlier years projected a peak of 35,000 deaths in 1913. In general, the trend in recent decades has been toward fewer worker deaths and a lower work-related death rate. At the same time, the number of workers has risen.

Death, injury rate, lost work days, and other statistics do not distinguish job-related injuries from job-related illnesses. It is often very difficult to establish that an illness is job related. Some illnesses have a long latency period between exposure and onset of disease. Workers may have had off-the-job exposures to health hazards, may have had exposure on different jobs, or may have changed jobs. Some employers are reluctant to report occupational illnesses, and many employees and physicians fail to recognize a disease as being job related. These factors suggest that the preceding statistics about worker deaths and injuries may be underestimated.[6]

Accurate estimates of the ratio of job illness to job injuries are hard to find. For federal employees, there are roughly four job illnesses reported for every 100 job injuries. A study cited in a government report on occupational diseases[7] listed the causes of occupational disabilities: approximately one third are caused by job injury and two thirds by job disease. Estimates say that lost earnings resulting from disabling occupational diseases cost more than $11 billion in 1978, and the cost is significantly higher today. Death, pain, suffering, and other intangibles are not included in the estimate.

There are other factors to consider about long-term trends in safety records. There are continual changes in the injuries and illness that are recognized under workers' compensation. These changes influence which incidents are included in records. For example, silicosis was not compensable until the 1940s to 1950s. Formerly, hernia injuries were recognized as job-related when the pain was so severe that workers could not work. Today, hernia symptoms do not have to be as obvious to achieve compensation. We now recognize cumulative trauma injuries as work related and compensable. In the early 1980s, many ergonomics-related injuries were not compensable. The shift in the definition of compensable and job-related injuries may account, in part, for the inability to reduce the work-related injury and illness statistics as much as we would like.

The source of accident, injury, and illness data from industries often is derived from the larger companies that have organized safety programs and organizations. It is not uncommon to find an order of magnitude difference in accident statistics within an industry when all types and sizes of companies are considered. When only a portion of an industry is the source of data, and if this portion is comprised of the better companies in terms of accident records, the actual record may be quite different. The real statistics may differ from published or reported statistics.

Although great progress has been made in occupational safety and health, the toll in terms of dollars, lives, injuries, and illnesses is still high. The statistics often overlook the personal impacts on the individuals and their families.

1-3   CONSUMER PRODUCTS AND HOME ACCIDENTS

Accidental death, injury, and illness at home and from consumer products is also a large problem. Many accidents in this group go unreported. The National Safety Council estimates there are roughly 12,000 deaths and 2.9 million disabling injuries annually caused by accidents at home. The death rate for home accidents, now approximately 1.5 per 100,000 persons, and the number of deaths annually have shown a slight decline over the years. The total cost of home accidents, lost wages, medical expenses, fire losses, and insurance administrative costs is roughly $135 billion per year. Some indirect costs are not included in this estimate.

Many home accidents involve consumer products, although all accidents involving consumer products do not occur at home. In 1970, the National Commission on Product Safety attempted to determine the scope of the safety problem associated with consumer products. In their final report,[8] the Commission estimated that there are approximately 20 million injuries at home associated with consumer products each year. Also, consumer products cause 110,000 permanent disabilities and 30,000 deaths annually. These data exclude injuries and deaths associated with foods, drugs, cosmetics, motor vehicles, firearms, tobacco products, radiological hazards, and certain flammable fabrics. The Consumer Product Safety Commission tracks product injuries in hospital emergency rooms through the National Electronic Injury Surveillance System. Data from 1973 suggested that more than 6 million product-associated injuries occur each year.[9]

Today, injury and death from firearms has become a public issue. Individuals and local governments use the courts to make firearm manufacturers liable even though the right to bear arms is protected by the Second Ammendmen.

1-4   TRANSPORTATION

Losses from transportation accidents are also very large. Transportation includes motor vehicles, aircraft, railroads, and waterways. By far the greatest cause of accidental death is motor vehicle accidents. Each year, nearly 50,000 people die in motor vehicle accidents and more than 2 million sustain disabling injuries. The overall death rate in the United States from motor vehicle accidents is presently approximately 15 per 100,000 persons and 1.6 deaths per 100 million miles traveled for the 240 million registered vehicles. For drivers in the 15- to 24-year old age group, the death rate is nearly double that of the total population. Although little attention has been given to the death rate from vehicles while on the job, some studies suggest that 25% to 33% of all job-related deaths involve motor vehicles.

The population death rate for air transportation is roughly 0.5 per 100,000 persons. There are some differences between general aviation and commercial aviation. The National Safety Council reports a death rate of approximately 16 per 100,000 persons for general aviation and 0.1 per 100,000 persons for commercial aviation. The National Transportation Safety Board estimated that general aviation had 3.3 fatalities per 100,000 hours of flight, whereas commercial aviation had 5.1 per 100,000 hours.

Over recent decades, there has been a decline in railroad passengers and railroad employees. Over the same period, there has been a decline in railroad deaths and injuries. Each year there are roughly 1500 deaths and 20,000 injuries associated with railroad accidents. Approximately 60% of the deaths and 15% of the injuries occur at rail–highway grade crossings. Other railroad accidents, such as derailments, result in explosions, fires, chemical releases, major property and environmental damage, and legal claims. For example, in the mid-1970s, a 40-car derailment occurred in Florida, apparently caused by vandalism. It resulted in a chlorine tank car leak that killed occupants of an automobile traveling on an adjacent highway and caused other injuries. Resulting liability claims totaled more than $200 million, whereas the small railroad company had assets of less than $7 million.

The U.S. Coast Guard reports that more than 1,500 boating accidents occur each year. Here, too, a major accident can result in large losses, not just death and injury. For example, in May 1980, a freighter rammed the Sunshine Skyway Bridge in St. Petersburg, Florida, ripping out a 1,400-ft section. Thirty-one people died as their vehicles plunged 140 ft to the water below. Authorities reopened the rebuilt bridge after seven years of diverted traffic that impacted businesses and added travel time and expenses for many thousands of people.

1-5   ENVIRONMENTAL PROBLEMS

It is difficult to assess the impact of air and water quality on human safety and health. Even when it is known that a substance affects humans, it is difficult to prove that a disease or illness is caused directly by exposure to it. The expenditures made to reduce air and water pollution are assessed more easily. The Environmental Protection Agency estimated that the annual cost for 17 major industries to comply with the Resource Conservation and Recovery Act of 1976 was $750 million.

Another aspect of the environmental problem is the scale and cost of cleanup. Estimates say that in 1980, industry generated 60 million tons of hazardous waste as acids, solvents, oils, caustics, explosives, and other forms. The Environmental Protection Agency estimated there are 30,000 to 50,000 hazardous waste sites in the United States. In 1980, Congress established a $1.6 billion Superfund for site cleanup. In late 1985, the Superfund was extended for five years and an additional $7.5 billion. This funding resulted in cleanup for only a small portion of the known sites.

The costs for claims and for cleanup of a particular site can be very large. Approximately 20,000 tons of waste, made up of more than 80 different substances, were buried at Love Canal in New York. By 1981, $36 million, or approximately $1.00 per pound, were spent in cleanup, relocation of residents, health and environmental testing, and other expenses. This does not include most health expenses, the cost of suffering, and much of the depreciation in real estate values. Nearly $3 billion in lawsuits were filed by 1980. Reported costs do not include most legal settlements.

1-6   SIGNIFICANCE FOR ENGINEERS

For a long time, society has sought to protect itself from risk. One means in recent times has been through laws requiring registration or licensing of professions, including engineers. The one justification for engineering registration laws is "protecting public health, safety and welfare." This concept assumes that those who appropriate education and experience and are able to sit for and pass an examination are qualified to provide the protection expected by the public. The public expects engineers to protect them against unnecessary and undesirable risks, particularly those brought on society through technological advancement and change.

Spectacular failures erode public confidence in engineers. Examples include the collapse of the Tacoma Narrows Bridge near Tacoma, Washington, in 1940; the March, 1979, nuclear accident at Three Mile Island near Middletown, Pennsylvania; the chemical waste tragedy at Love Canal in Niagara Falls, New York, in 1978 through 1980; the toxic chemical release in Bhopal, India, in December, 1984, that killed approximately 2,500 people and injured thousands more. A later example, witnessed on live television around the world, was the spectacular Challenger space shuttle accident at Cape Kennedy, Florida, on January 18, 1986, and more recently the Columbia space shuttle reentry accident on February 1, 2003.

The National Council of Examiners in Engineering and Surveying (formerly the National Council of Engineering Examiners) surveyed practicing engineers to find out what they do on their jobs. They found that nearly all engineering disciplines and all kinds of engineering jobs included significant responsibilities for safety and health.[10]

Some have claimed that engineers do not know what they are doing when it comes to health and safety. They point to the fact that professional engineering examinations in most states do not include, or include very few, questions dealing with safety and health. They also note that most engineering curriculums do not include safety and health courses. Many engineering programs incorporate safety and health issues into capstone design projects.

Engineering schools and the engineering profession are becoming more aware of the safety and health challenge. One group of design educators states: "Through the combined voices of society, the government and the courts the message to the industrial/technological community is clear. Consumer groups, regulatory agencies and the law of strict liability all demand that unreasonable risks be eliminated from the interaction of technology with society. The design engineering educator can no longer overlook the fact that the network of regulations, standards and litigation as it has evolved in recent years represents an important set of criteria for design that are added to the traditional constraints of function, cost, manufacturability and, marketability."[11]

At the 1986 National Congress on Engineering Education, representatives from engineering societies passed resolutions about educational requirements for engineers. The resolutions included recommending that training in safety and health be strengthened and that design constraints, such as safety, are essential in engineering courses.[12] These recommendations led to modified accreditation criteria in 1987 for all engineering programs in the United States. The criteria now require that knowledge about engineering practice must include an understanding of the engineer's responsibility to protect both occupational and public health and safety.[13]

Recently, the National Safety Council added a program called the Institute for Safety Through Design to affect engineering design. Its goal is to elevate safety in the design of products, processes, equipment, and vehicles.

Engineers do have an important role in reducing risks placed on society by modern technology, its products, and its wastes. Although engineers cannot bear the total blame for safety and health risks, engineers are able to help reduce them to levels acceptable to society.[14] In planning, design, operations, maintenance, or management activities, engineers should be able to recognize hazards and implement controls for them. Engineers should know how to eliminate, reduce, or control safety and health risks within their sphere of responsibility. Every engineer must know when and how to use other professions, including safety professionals, in analyzing and reviewing their procedures and design decisions. Every engineer needs to know when to say, "I don't know; I need other expertise." Every engineer should know how to manage risks while making tradeoffs with cost, convenience, and other factors. Creating a functional or economical product or system is not enough. It also must be safe. Engineers cannot set arbitrary standards to determine when something is safe enough. They must be knowledgeable about the thousands of standards society has established. Engineers must work with society and other professions in meeting the health and safety challenge of registration laws: "To protect public health, safety and welfare."

EXERCISES

1. Discuss who should set safety and health standards.
2. Discuss the public confidence in engineering and technology.
3. Discuss the effectiveness of registration laws in protecting public health and safety.
4. Find out what accreditation criteria exist to ensure that safety and health are incorporated into engineering and engineering technology education.
5. Contact the National Safety Council's Institute for Safety Through Design to find examples of designs which incorporate safety.

REVIEW QUESTIONS

1. How many chemical substances are registered? Are in use?
2. How many accidental deaths occur each year in the United States? How are these distributed among work, vehicles, and home?
3. How many disabling injuries occur each year in the United States?
4. What is the annual cost of accidents in the United States?
5. What means has society used to reduce safety and health risks resulting from technological advances?
6. Are job-related illnesses well represented in accident and death statistics? Explain.
7. Approximately how many hazardous waste sites are known to exist in the United States?
8. What is the one common reason for state engineering registration laws?
9. What measures help ensure that engineers have some competency in dealing with health and safety in practice?

NOTES

1 *Registry Handbook*, Chemistry Abstract Service, Columbus, OH, annual.

2 *Statistical Abstract of the United States*, U.S. Department of Commerce, Bureau of the Census, Washington, DC, annual.

3 *Incident Facts*, National Safety Council, Chicago, IL, annual.

4 *Risk in a Complex Society*, Marsh & McLennen Companies, Inc., Boston, MA, 1980.

5 *The Vital Consensus: American Attitudes on Economic Growth*, Union Carbide, New York, 1979.

6 H. J. Kilaski, "Understanding Statistics on Occupational Illnesses," *Monthly Labor Review*, 104: 3:25–29 (1981).

7 *An Interim Report to Congress on Occupational Disease*, Assistant Secretary for Policy Evaluation and Research, U.S. Department of Labor, Washington, DC, 1980.

8 *Final Report of the National Commission on Product Safety*, Washington, DC, June 1970.

9 *Handbook & Standards for Manufacturing Safer Consumer Products*, U.S. Consumer Product Safety Commission, Washington, DC, June 1975.

10 *A Task Analysis of Licensed Engineers*, National Council of Engineering Examiners, Clemson, SC, March 1981.

11 T. A. Jur, L. C. Peters, T. F. Talbot, A. S. Weinstein and R. M. Wolosewicz, "Engineering, the Law, and Design Education," *Engineering Education*, 71:271–274 (1981).

12 *National Congress on Engineering Education*, Accreditation Board for Engineering and Technology, New York, November 1986.

13 "Criteria for Accrediting Programs in Engineering in the United States," in *1986 Annual Report*, Accreditation Board for Engineering and Technology, New York, 1986.

14 *Public Safety—A Growing Factor in Modern Design*, National Academy of Engineering, Washington, DC, 1970.

BIBLIOGRAPHY

CULLEN, LISA, *A Job to Die For—Why So Many Americans are Killed, Injured or Made Ill at Work and What to Do About It*, American Industrial Hygiene Association, Fairfax, VA, 2002.

DOTTER, EARL, *The Quiet Sickness: A Photographic Chronicle of Hazardous Work in America*, American Industrial Hygiene Association, Fairfax, VA, 1998.

HAMILTON, ALICE, *Exploring the Dangerous Trades: The Autobiography of Alice Hamilton, M.D.*, American Industrial Hygiene Association, Fairfax, VA, 1995.

Injury Facts, National Safety Council, Itasca, IL, annual.

CHAPTER **2**

SAFETY AND HEALTH PROFESSIONS

2-1 INTRODUCTION

Keeping people safe involves the work of many disciplines. Engineers have made many contributions to safety and have helped resolve many safety problems; so have many other professions. Engineers need to know what role other professions have in safety and health and need to work with them. Often an interdisciplinary effort is required to identify hazards, to develop effective solutions to safety problems, and to achieve safe products, buildings, operations, and systems.

In today's high-technology society, no individual can be an expert in every aspect of safety and health. It is impossible to keep up with all the new laws and regulations at the federal, state, and local level. There are many changes and interpretations of them. To know everything about safety, one would have to be an expert in law, engineering, technical equipment, manufacturing processes, behavioral sciences, management, health sciences, finance, insurance, and other fields. Current safety professionals can be specialists in a particular area of safety. They also can function as generalists who coordinate and facilitate the actions of other knowledgeable professionals in applying safety principles to particular problems.

Sometimes people with limited safety training become involved in safety activities. There are many different reasons:

They may have a good knowledge of plant operations, a valuable characteristic.

They may have had an accident themselves and become a safety proponent.

They may know fellow workers well and have a good rapport with them, also important.

They may be good leaders and effective communicators.

These characteristics may make them good workers and advocates for safety, but do not give them the safety knowledge and skills necessary to deal effectively with complex and technical safety and health problems today. Achieving safety now involves professionals, often from a variety of backgrounds.

2-2 SAFETY PROFESSIONALS

Many individuals who moved into safety and health jobs from other fields learned the principles of safety and health on the job. Some learned safety through continuing edu-

Safety and Health for Engineers, Second Edition, by Roger L. Brauer
Copyright © 2006 John Wiley & Sons, Inc.

cation programs after joining into the field. Many began a safety and health career after study in specialized programs. More and more people entering the discipline today have baccalaureate and advanced degrees in safety and health.

There are many different specialists in the safety and health field. Many are safety professionals. A safety professional is "an individual who, by virtue of his specialized knowledge and skill and/or educational accomplishments, has achieved professional status in the safety field."[1] Without contributions from these different specialties, many of which are discussed herein, the safety field would not be where it is today.

Today, job analysis studies for certifications[2] define the tasks, knowledge, and skills required for safety professionals. Safety students need a solid foundation in mathematics and sciences (physics, chemistry, human physiology, and human behavior) and in business and technology. They receive specialized training in principles and practices of safety, industrial hygiene, ergonomics, and fire protection. They also receive some training in environmental matters and hazardous materials management.

Persons working in safety and health careers express a high level of job satisfaction. A 1990[3] salary survey showed that 90% of respondents were somewhat to highly satisfied with their careers in safety. Those holding nationally accredited certifications, such as the Certified Safety Professional (CSP),[4] earn significantly more than those with no certification.[5]

2-3 ENGINEERING

Every engineering discipline has important contributions to make to safety and health within its areas of specialization. Jobs in virtually every engineering discipline include a significant number of safety-related tasks.[6] At the risk of slighting some disciplines, contributions of certain disciplines are noted below. Engineers work mainly on the preventive side of safety. In this role, engineers must identify hazards during design and must eliminate or reduce them. They also prevent unsafe behavior by designing products, workplaces, and environments so that unsafe behavior cannot or is not likely to occur. They also mitigate the effects of unsafe behavior through design so that the effects are controlled or of limited scope.

Civil Engineering

Civil engineers have advanced many areas of safety and health. Civil engineers pursue structural integrity of buildings, bridges, and other constructed facilities. Civil engineers seek safe and sanitary handling, storage, treatment, and disposal of wastes. They study and develop controls for air and water pollution and contribute to transportation safety in design and construction of facilities for railroads, motor vehicles, ships, and aircraft.

Industrial Engineering

Being concerned with industrial processes and operations, industrial engineers try to fit jobs to people and make work methods and work environments safe. Many industrial engineers receive some training in occupational safety and health, safety engineering, ergonomics, or human factors engineering.

Mechanical Engineering

Mechanical engineers took the lead in establishing safety requirements for machines, boilers and pressure vessels, elevators, and other kinds of mechanized equipment and facilities. They started safety standards for some of these systems before 1900.

Electrical Engineering

Electrical engineers have contributed to safety through design of electrical safety devices, electrical interlocks, ground fault circuit interrupters, more compact electrical circuits, and other items. Today, electronics engineers and computer engineers must include software safety analysis in their designs to prevent injuries to system users.

Chemical Engineering

Through the design of less hazardous processes, chemical engineers have contributed to safety. They have applied system safety techniques to process design, have helped develop requirements for less hazardous chemicals, and have developed waste reclamation processes.

Safety Engineering

Safety engineering is devoted to the application of scientific and engineering principles and methods to the elimination and control of hazards. Safety engineers need to know a great deal about many different engineering fields. They specialize in recognition and control of hazards, and they work closely with other engineering and nonengineering disciplines.

Ergonomics and Human Factors Engineering

Ergonomics and human factors engineering are very similar. They specialize in the application of information from the biological and behavioral sciences to the design of systems and equipment. Their goal is to improve performance, safety, and satisfaction. They try to improve the fit between people and equipment, environments, systems, work-places or information. Specialists in this field try to improve performance and safety by reducing task errors and physical stresses involved in physical activity. Ergonomics has a strong emphasis on physiological and biomechanical aspects whereas human factors engineering emphasizes the behavioral and cognitive aspects of performance and safety.

Fire Protection Engineering

Fire protection engineering is the field of engineering concerned with safeguarding life and property against loss from fire, explosion, and related hazards. Fire protection engineers are specialists in prevention, protection, detection and alarms, and fire control and extinguishment for structures, equipment, processes, and systems. They design egress routes to allow for safe exiting from fires.

2-4 MANAGEMENT SCIENCES

People in business and personnel management also contribute to the advancement of the safety and health field. In some companies, a safety and health program is part of the personnel, human resources, or labor relations branch of the organization. Sometimes safety and health programs are part of a risk management or loss control unit. Other areas of management, such as advertising, marketing, sales, and procurement, can contribute to safety, too. For a number of years the National Institute of Occupational Safety and Health operated Project Minerva, which had a goal of advancing safety within the management sciences. Some management theories, such as those of Juran and Demming (see Chapter 34), offer constructs that aid management in reducing hazards and concurrently in improving the organization's quality and bottom line.

Risk Management

The field of risk management attempts to reduce all types of losses to an acceptable level at the lowest possible cost. Risk managers often administer accident prevention, risk assessment, and insurance programs. Part of risk management is transferring risk through insurance.

Loss Control Specialists

A loss control specialist is a person responsible for the development of programs to prevent or minimize business losses other than speculative losses. Losses include personal injury, damage to property, fire, explosion, theft, pilferage, vandalism, industrial espionage, air and water pollution, employee illness, and product defects.[7] Loss control and risk management are related. The casualty and property insurance industry employs loss control specialists who provide loss control assistance to policy holders. A job analysis study identifies the minimum tasks, knowledge, and skills for loss control specialists.[8]

2-5 HEALTH SCIENCES

The health sciences play an important role in safety. In the industrial workplace, it is common practice in large companies to see the safety professionals and the medical professionals working closely together. Although the task of physicians and nurses is to treat those who are ill and injured, much of their attention in safety is on prevention. In recent times, new specialties have emerged in the health field, including industrial hygienists, health physicists, toxicologists, environmental health specialists, public health specialists, and others.

Occupational Medicine and Nursing

Occupational medicine applies medical science to the prevention and treatment of occupational injuries, illnesses, and diseases. Because there are fewer specialists in this field than needed, a number of medical schools received federal grants beginning in the 1970s to train physicians in occupational medicine. Occupational nurses often fill the role of health specialists in a company or plant when an occupational physician is not available.

Industrial Hygiene

Industrial hygiene is the science and art devoted to the recognition, evaluation, and control of those environmental factors or stresses arising in or from work situations that may cause illness or impaired health.

Health Physics

Health physics is a branch of medical physics concerned with protecting humans and their environments from unwarranted radiation exposure. Health physicists engage in the study of radiation problems and methods to provide radiation protection as well as study the mechanisms of radiation damage. They also develop and implement methods necessary to evaluate radiation hazards and to provide and properly use radiation protective equipment.

2-6 BEHAVIORAL SCIENCES AND EDUCATION

Among the behavioral sciences, psychologists sometimes work in health and safety. Some psychologists work directly in safety and health programs, whereas others contribute through the fields of human factors, organizational behavior, industrial psychology, or personnel management. Some psychologists specialize in behavior modifications to reduce accidents. Education specialists often develop training programs and apply training methods for safety and health. More recently, some specialists have applied behavioral modification theories to safety practices. The literature for achieving safety now includes the term *behavior based safety*.

2-7 LEGAL PROFESSION

Legal issues have long been a part of the safety and health field. Even before worker compensation laws came into existence, lawyers helped resolve work-related injury and illness claims. During the nineteenth and twentieth centuries, lawyers helped frame laws and regulations to protect workers, consumers, and the public. They played an important role in seeking just compensation and protection and in preventing unjust claims. Some believe that the legal profession has inhibited American business through liability litigation. However, litigation often improves the safety of products and workplaces. There are continued attempts to limit liability through legislation.

2-8 OTHER PROFESSIONS

Undoubtedly, other professions and interdisciplinary specialties play important roles in advancing the safety and health field. Naming some runs the risk of overlooking others that have made significant contributions.

Architecture

Architects have contributed to safety and health in a number of ways. They are often responsible for the structural integrity of buildings and work closely with civil and structural engineers to prevent collapse. Architects work with fire protection engineers in developing and implementing life safety codes and other fire protection standards. Architects also can affect safety when selecting flooring and designing stairs and railings that prevent falls or when designing buildings to minimize risks during maintenance work. Their designs can affect safety for workstations, buildings, and sites.

Urban Planning

Urban planners have developed zoning ordinances to remove congestion from lots and streets, and they have participated in the development and implementation of building codes and environmental standards. Urban planners are an integral part of teams working to reduce air and water pollution, to provide community fire protection, to separate housing from noisy and hazardous activities, and to improve traffic flow.

2-9 CERTIFICATION, REGISTRATION, AND PROFESSIONALISM

All states have laws governing the registration or licensing of certain professionals, such as engineers, architects, lawyers, and physicians. The one justification found in each of these laws is protection of the safety, health, and welfare of the public. Many disciplines that contribute to safety and health are not licensed.

Safety engineering is recognized as an engineering specialty in very few states. Most engineers who specialize in safety become registered as engineers in a traditional engineering field. In some states, there is no distinction among engineering specialties for reasons of registration. Licenced engineers are simply registered professional engineers.

In most engineering registration examinations, very few questions deal with safety and health. As a result, engineering registration does not ensure that an individual can recognize and control hazards. The engineering registration process and examinations often are criticized because of this. In fact, some believe that the public is not adequately protected through engineering registration.

For many disciplines not covered by registration laws, including some in safety and health, it is difficult to tell whether individuals really have professional qualifications. For disciplines that do not have state licensing, it is becoming commonplace to establish certification programs. A respected panel of professionals in the field usually oversees the program. Typically, candidates for certification must meet education and experience standards and must pass one or more professional practice examination. The process for certification is very similar to that for government-operated registration, but is managed by professional peer groups.

Certified Safety Professional (CSP), Certified Industrial Hygienist (CIH), and Certified Health Physicist (CHP) programs began in the 1960s. These programs follow strict education, experience, and examination procedures in awarding such designations to applicants. Boards of directors who are recognized for their professional qualifications in these fields manage these programs. These certifications are the most notable in safety and health. Today, peer certifications use accreditation to assure the public, employers, and government organizations that the programs adhere to quality procedures. The National Commission for Certifying Agencies (NCCA),[9] the Council of Engineering and Scientific Specialty Boards (CESB),[10] and the American National Standards Institute[11] publish standards for peer certification boards and operate accreditation programs for peer certification programs.

Individuals with certification do not have the same status under state laws that registered or licensed professionals do. The certification process provides an orderly means to assure the public and others that specialists have achieved certain professional standards. Over time employers, government units, the public, and the profession rely on peer certification to assess minimum competency in particular disciplines. One frequently finds government agencies and companies requiring certification for certain positions or job functions in contracts and when recruiting.

In the future engineers will probably be required to demonstrate competence in safety and health more than they do now. Safety and health will probably receive greater attention in engineering courses and degree programs. Accreditation reviews for engineering programs will look for safety and health training. Engineering registration examinations undoubtedly will contain more questions dealing with particular safety and health issues. In the future, engineering registration and certification may involve a two-tier process: registration in engineering by states, followed by peer group certification in specialties.

EXERCISES

1. Develop a library of course catalogs and brochures for academic programs in safety and health. Visit the database of safety and related academic programs on the web site of the Board of Certified Safety Professionals: www.bcsp.org.

2. Talk to active safety and health professionals. Report on such things as how they entered their field, where they received their training, what their responsibilities are, what job challenges they have experienced, and how they achieve job satisfaction.

REVIEW QUESTIONS

1. Explain why safety and health specialists must work as a team, both within the safety and health field and with others professions outside the field.

2. Define the main responsibilities of the following specialties:

   (a) safety engineer

   (b) human factors engineer

   (c) fire protection engineer

   (d) risk manager

   (e) physician in occupational medicine

   (f) industrial hygienist

   (g) health physicist

3. Explain the major differences between registration and certification.

4. Explain how quality of peer certification programs can be verified.

5. What peer certifications are highly recognized in the safety and health field?

6. What is the one reason used to justify state laws that establish registration for engineers?

NOTES

1 *The Dictionary of Terms Used in the Safety Profession*, 3rd ed., American Society of Safety Engineers, Des Plaines, IL, 1988.

2 "Job Analysis Study for Certified Safety Professional Examinations," BCSP Technical Report 2001-1, Board of Certified Safety Professionals, Savoy, IL, February 2001.

3 *Member Salary Survey,* American Society of Safety Engineers, Des Plaines, IL, 1990.

4 Certified Safety Professional and CSP are certification marks issued by the U.S. Patent and Trademark Office to the Board of Certified Safety Professionals.

5 "Professional Certification: Its Value to SH&E Practitioners and the Profession," *Professional Safety*, 49:12:26–31 (2004).

6 *A Task Analysis of Licensed Engineers*, National Council of Engineering Examiners, Clemson, SC, updated periodically.

7 J. A. Fletcher, *The Industrial Environment—Total Loss Control*, National Profile Limited, Willowdale, Ontario, Canada, 1972.

8 *Role Delineation Study for Occupational Health and Safety Technologist and Loss Control Specialist Examinations*, CCHEST Technical Report 2003-1, Council on Certification of Health, Environmental and Safety Technologists, March 2003.

9 National Commission for Certifying Agencies, 1200 19th Street, NW, Suite 300, Washington, DC

10 Council of Engineering and Scientific Specialty Boards, 130 Holiday Court, Suite 100, Annapolis, MD 21401.

11 American National Standards Institute, 1819 L Street NW, 6th Floor, Washington, DC 20036, operates ISO/IEC 17024 (Certification of Persons) within the United States.

BIBLIOGRAPHY

ADAMS, PAUL S, BRAUER, ROGER L, KARAS, BRUCE, BRESNAHAN, THOMAS F., and MURPHY, HEATHER, "Professional Certification: Its Value to SH&E Practitioners and the Profession," *Professional Safety*, 49:12:26–31 (2004).

Dictionary of Terms Used in the Safety Profession, American Society of Safety Engineers, Des Plaines, IL, 1981.

MANUELE, FRED A., *On the Practice of Safety*, 3rd ed., John Wiley & Sons, New York, 2003.

RUSSELL, J. E., "Board of Certified Safety Professionals: How It All Began," *Professional Safety*, 34:6:38–43 (1989).

WEIS, W. J. III, PURCELL, T. C., STREET, M. H., and KENDRICK, P. A., *Directory of Academic Programs in Occupational Safety and Health*, U.S. Department of Health, Education and Welfare, Public Health Service, Center for Disease Control, National Institute for Occupational Safety and Health, Cincinnati, OH, January 1979 (DHW-NIOSH Publication No. 79–126).

CHAPTER 3

FUNDAMENTAL CONCEPTS AND TERMS

As people applied their skills to making our technological world safer, their ideas and concepts became the tools for others in the safety field. They developed new approaches to understanding complex, often confusing, events and conditions. Some of the concepts have endured and become part of the vocabulary of safety professionals. Some concepts are helpful to many, inadequate for others. A student of safety and health should know some of these concepts. They will help when talking with others in the field and in solving safety problems.

3-1 WHY SAFETY?

Why is safety important? Why bother with it? There are several major reasons for safety. Our society places high value on human life and welfare. This fact provides the first and overriding reason for safety—*humanitarianism*. This is the moral basis for safety.

Each person has a different degree of regard for others and uses different standards for right and wrong. To minimize these differences, society formalizes standards of conduct among people. This body of formalized standards, the *law*, provides a second reason for safety. It is derived from the first.

Society's standards recognize that life and the ability to live it fully has worth. Property, too, has worth. As part of an economic system, at times society must determine the actual value of property, human capabilities, and life itself. The third reason for safety is *cost*. Cost is measured in actual outlays, in avoidance of expenditures, or in the value of lost abilities and property.

Although each of these three reasons, humanitarianism, the law, and cost, forms a basis for further discussion, each is not treated with the same detail in this book. Part Two of this book deals with the law as it applies to safety and health. It is particularly important for engineers to recognize the wide range of standards society expects them to follow and apply. Part Five deals with cost and techniques for dealing with cost.

3-2 ACCIDENTS, INJURIES, AND LOSSES

Accidents Defined

The dictionary defines an accident as "a happening or event that is not expected, foreseen, or intended." An event itself is the key element of this definition. Other definitions include the effects of an event: "An accident is an unexpected, unforeseen, or unintended event that causes injury, loss or damage." The term *accident* usually evokes thoughts about undesirable effects or consequences. The term suggests to most people an immediacy between event and effect. We tend to think of an accident as a sudden event and of short duration. The term *accident* often suggests that the event occurred by chance—it just happened.

These definitions and many of the commonly held ideas associated with the term accident create problems for the safety and health field. Three difficulties are:

1. The idea of chance occurrence
2. The relationships between accident events and consequences
3. The duration of events

To the safety specialist, every accident has one or more identifiable causes. Chance may play a role in bringing causes together. There are two fundamental types of accident causes: unsafe acts and unsafe conditions. Accidents involve either of these two causes or both. Recognizing that accidents are caused and are not just functions of chance allows one to pursue accident prevention. To avoid the connotation of chance, a number of organizations no longer use the term *accident*. Instead, they use the term *incident*.

A frequent error is assuming that relationships about accident events and consequences are related. We often assume that an accident includes adverse consequences. For example, if we hear that close friends had an accident, we immediately ask, "Are they alright?" We assume there is an injury when we hear the word *accident*. It is incorrect to assume a relationship between accident events and consequences. Most accidents do not include injury or significant loss.

Another kind of accident–consequence relationship that can be in error is immediacy. We commonly think of results of accidents appearing right away, immediately after the event. For some injuries and many kinds of losses, results appear right after the event. However, in safety and health, one must also deal with illness and disease. For these effects, there is usually a delay or latency period between an event and the results. For example, the symptoms for sunburn are most intense several hours after exposure. Some cancers have a latency period of 20 to 40 years after exposure. There are also injuries from continuous or repeated activities that may extend over days, weeks, or months. The idea of immediate results implied in the term *accident* makes it difficult to include illness, diseases, or cumulative injury as accident effects. The term *accident* does not seem to fit these situations well.

Another difficulty with the term *accident* is the idea that the event itself is of short duration. An event that produces an injury or illness may occur over a period of hours, days, weeks, or even years. Many diseases will not occur after only a short exposure. Chronic and long-term events may be necessary to cause some effects. For example, carpenter's elbow, an inflammation of the elbow, follows a long period of stressful hammering activity.

One must recognize the limitations of the term *accident* and its common definition. The English language does not have simple terms that fully resolve these difficulties. Therefore, a modification to the common definition may make the term accident more useful: An accident is an unintended, unplanned single or multiple event sequence that is caused by unsafe acts, unsafe conditions, or both and may result in immediate or delayed undesirable effects.

Incidents and Accidents[1]

Because of the limitations the word *accident* poses, many organizations use the term *incident* to refer to any unplanned event or event sequence, whether it results in loss, injury, illness, disease, death, or none of these. The word *incident* does not carry the connotation that the event or event sequence cannot be prevented, which is often implied in the term *accident*. The term *accident* has a long use history and is not easily replaced.

Types of Losses

Losses from incidents can take many forms. Besides injury, illness, disease, and death, there are damage to property, equipment, materials, and the environment and the cost of repair or replacement. Losses can include loss of time, production, and sales. Incidents can result in the need to complete and submit forms. Incidents may result in travel, record keeping, investigations, cleanup, legal and medical services, hospitalization, rehabilitation, and recovery of public image. All these cost money.

Direct Versus Indirect Costs   One way of classifying costs associated with incidents is to group them into direct costs and indirect (or hidden) costs. Direct costs are those expenses incurred because of an incident and ascribed to it. Direct costs typically include medical expenses and compensation paid to an injured employee for time away from work and costs for repair or replacement of damaged items.

Indirect costs are real expenses associated with incidents, but difficult to assess for an individual case. Table 3-1 lists eleven categories of indirect costs, which H. W. Heinrich developed to point managers' attention toward prevention of accidents. He wanted to convince them that medical costs and compensation of worker time are not the only costs related to accidents. Based on his own investigation in 1926, he introduced the "4:1 ratio," which suggests that the total cost associated with accidents is much higher than the obvious, direct expenses. Although the ratio varies for different companies and different types of operations, the basic idea is sound.

Insured Versus Uninsured Costs   It is often difficult to establish which costs are direct and which are indirect. Insurance covers many losses. As a result, many people clas-

---

TABLE 3-1   Hidden Costs Associated with Incidents

A. Lost time of injured employee
B. Time lost by other employees to assist injured coworker, to see what is going on, and to discuss events
C. Time lost by a supervisor to assist injured worker, investigate incident, prepare reports, and make adjustments in work and staffing
D. Time spent by company first aid, medical, and safety staff on case
E. Damage to tools, equipment, materials, or property
F. Losses due to late or unfilled orders, loss of bonuses, or payment of penalties
G. Payments made to injured employee under benefit programs
H. Losses resulting from less than full productivity of injured workers on return to work
I. Loss of profit because of lost work time and idle machines
J. Losses due to reductions in productivity of coworkers because of concern or reduced morale
K. Overhead costs that continue during lost work

Adapted from: H. W. Heinrich, *Industrial Accident Prevention*, 4th ed., McGraw-Hill, New York, 1959.

sify incident-related losses as insured or uninsured. Insured costs are paid through insurance claims. Uninsured costs are paid directly from other sources. The distinction between insured and uninsured losses is confounded by large companies using self-insurance or a combination of purchased insurance and self-insurance.

Unsafe Acts and Unsafe Conditions

A few decades ago, some people tried to identify the portion of incidents caused by unsafe acts compared with unsafe conditions. Heinrich analyzed 75,000 accidents and found that 88% were caused by unsafe acts, 10% by unsafe conditions, and 2% by unpreventable causes. This is Heinrich's 88:10:2 ratio. A study in 1960 by the Pennsylvania Department of Labor and Industry found that *both* unsafe acts and unsafe conditions were contributing factors in more than 98% of the 80,000 industrial accidents analyzed.

The lesson is that both unsafe acts and unsafe conditions do contribute to incidents. The relative significance of each will probably always be debated. Engineers have many opportunities to eliminate or reduce unsafe conditions. This book emphasizes that role of engineers. Engineers also have many opportunities to minimize unsafe acts. Designs that reflect an understanding of human error and behavior can limit the range of human behavior that leads to or causes incidents. In the early half of the twentieth century, many used the Heinrich data to blame employees for incidents. Today, some continue to cite the Heinrich data to emphasize the importance of controlling employee unsafe behavior. However, effective safety programs work to eliminate both unsafe conditions and unsafe acts.

Incident–Injury Relationships

Heinrich introduced another important concept. He said that preventive actions should focus primarily on accidents and their causes (unsafe acts and unsafe conditions). Less attention should be placed on effects, like injuries and their immediate causes. To demonstrate this point, he developed the 300:29:1 ratio from a study of accident cases. For every group of 330 accidents of the same kind, 300 result in no injuries, 29 produce minor injuries, and 1 results in a major, lost-time injury. Thus, there are many opportunities to implement preventive actions before minor or serious injuries occur.

Others have tried to duplicate Heinrich's ratio. In another study, Bird and Germain[2] included prevention of damage-causing accidents, not just injury-causing accidents. It showed a 500:100:1 relationship among property-damage accidents, minor-injury accidents, and disabling-injury accidents. Fletcher[3] reported a ratio of 175:19:1 for no-injury accidents, minor-injury accidents, and serious-injury accidents.

The exact ratio among incidents and various kinds of injuries or results is not the important outcome of these studies. One key lesson is that serious injuries occur less frequently than minor injuries and minor injuries occur less frequently than no-injury incidents. Another key lesson is that even information about those incidents that do not produce injury can be useful in formulating preventive actions. Often people do not recognize the events as accidents, but may view them as incidents.

Incident–Cost Relationships

There is also an important relationship between the frequency of injury accidents and direct costs. A concept termed *the vital few*, originally introduced by Gordon Lembke of Wausau Insurance Companies, recognizes that costs are unequally distributed for similar accidents.

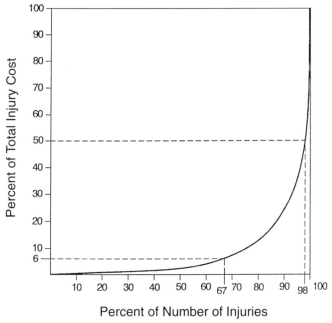

Figure 3-1.   The vital few concept shows that a few accidents account for the majority of the insured costs. (From Wausau Insurance Companies; used with permission.)

The distribution was first applied to economic data by Vilfredo Paretois (often spelled Pareto in English literature), an Italian economist from the nineteenth century. He noted that significant items in a given group normally are a relatively small portion of the total. For a group of similar incidents resulting in injuries and direct costs (insurance claims), only a small percentage of the injuries account for most of the total costs of the group, and most of the injuries account for merely a small portion of the total injury cost. Figure 3-1 illustrates an unequal cost distribution based on many similar incident cases.

Other Terms

A few other terms that are important are now introduced and defined.[4]

Safety is the state of being relatively free from harm, danger, injury or damage.

Risk is a measure of both the likelihood and the consequences of all hazards of an activity or condition. It is a subjective evaluation of relative failure potential. It is the chance of injury, damage, or loss.

A *hazard* is the potential for an activity, condition, circumstance, or changing conditions, or circumstances to produce harmful effects. A hazard is an unsafe condition.

Safety engineering is the application of engineering principles to the recognition and control of hazards.

Safety practice involves the recognition (and sometimes anticipation), evaluation, and control (engineering or administrative) of hazards and risk and management of these activities.

3-3 INCIDENT AND ACCIDENT THEORIES

There are a number of theories about incidents and accidents. The theories give us insight into preventive actions. None is totally adequate, either at describing all the factors that contribute to the occurrence of incidents or at predicting with reasonable accuracy the likelihood that an incident will take place. People will find some theories more helpful than others in preventing incidents.

Domino Theory

An early theory (still in use by some) is the domino theory of W. F. Heinrich. For many it is a helpful concept. The theory states that an incident sequence is like a series of five dominos standing on end. One can knock others over. The five dominos in reverse sequence are (1) an *injury* caused by (2) an *incident*, which, in turn, is caused by (3) *unsafe acts or conditions*. The latter are caused by (4) *undesirable traits* (such as recklessness, nervousness, violent temper, lack of knowledge, or unsafe practices) that are inherited or developed through one's (5) *social environment*. The incident sequence can be stopped by removing or controlling contributing factors. The theory places strong emphasis for incident prevention on the middle domino: unsafe acts and unsafe conditions.

As noted earlier, Heinrich believed that unsafe acts are more frequently involved in incidents than unsafe conditions. Therefore, his philosophy of incident prevention emphasized unsafe acts and person-related factors leading up to them. Individuals involved in prevention of incidents may find some value in this theory. For engineers who do not have control over unsafe acts as much as unsafe conditions, portions of this theory are of limited value.

Multiple Factor Theories

There are other theories for incidents in which incidents are deemed to be caused by many factors acting together. The immediate cause may be an unsafe act or an unsafe condition acting alone. In multiple causation theories, factors combine in random or other fashion and cause incidents. Grose,[5] for example, proposed a multiple factor model referred to as the four Ms: *man*, *machine*, *media*, and *management* (see Figure 3-2). Obviously, *man* refers to people. *Machine* refers to any kind of equipment or vehicle. *Media* includes such

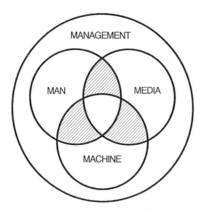

Figure 3-2. Four system safety factors: the four Ms. (From V. L. Grose, "System Safety in Rapid Rail Transit," from the August 1972 issue of the *ASSE Journal*, official publication of the American Society of Safety Engineers.)

things as environments, roadways, and weather. *Management* is the human context in which the other three Ms exist and operate.

The factors included in each multiple factor theory vary. In each multiple factor theory, characteristics of the factors that may be involved in a particular incident are identified. For example, characteristics of man are age, height, gender, skill level, amount of training, strength, posture, motivation, emotional state, and so on. Characteristics of media could include thermal conditions in buildings, water or snow on a roadway, fresh water compared with salt water or a contaminant in air. Characteristics of management might be management style, organizational structure, communication flow, policies, and procedures. Characteristics of machines might include size, weight, shape, energy source, type of action or motion, availability or placement of controls, and materials of construction.

Multiple factor theories are useful in incident prevention. They help identify which characteristics or factors are involved in a given operation or activity. Characteristics can be analyzed to see which combinations are most likely to cause an incident or result in losses. Statistical techniques, such as factor analysis, multiple regression analysis, and other multivariate methods, may be used in analyzing characteristics. Fault tree analysis, similar branched event-chain analysis, and other methods are also used to establish associations among characteristics and their relationships to damage, injuries, illnesses, and death. Both quantitative and qualitative methods can be helpful in multiple factor theories. Many of the methods used in multiple factor theories do not establish cause and effect, but rather relationships.

Energy Theory

More recently, William Haddon[6] proposed the idea that many accidents and injuries involve the transfer of energy. Objects, events, or environments interacting with people illustrate this idea: fires, hurricanes, projectiles, motor vehicles, various forms of radiation, and other items produce injuries and illnesses of various sorts. The energy theory suggests that quantities of energy, means of energy transfer, and rates of transfer are related to the kind and severity of injuries. Sometimes the theory is called the *energy release theory*, because the rate of release is an important component. This theory is attractive for many safety engineering problems and suggests ideas for controlling many unsafe conditions.

Using energy transfer as the accident-injury model, Hadden suggests 10 strategies for preventing or reducing losses. The order for these strategies follows the accident sequence.

1. *Prevent the marshalling of energy.* In this strategy, the goal is not producing energy or changing it to a form that cannot cause an accident or injury. Examples are not producing gun powder, substituting a safe substance for a dangerous one, preventing the accumulation of snow where avalanches are possible, removing snow where slips and falls can occur, not letting small children climb to levels above the floor, and not setting a vehicle in motion.

2. *Reducing the amount of energy marshalled.* Examples are keeping vehicle speeds down, reducing the quantities or concentration of high energy or toxic materials, limiting the height to which objects are raised, and reducing machine speed to the minimum needed when a machine is unguarded for cleaning or maintenance.

3. *Prevent the release of energy.* Examples are using various means or devices to prevent elevators from falling, flammables from igniting, or foundations from being undercut by erosion.

4. *Modify the rate at which energy is released from its source or modify the spatial distribution of the released energy.* Slowing the burning rate of a substance or using an inhibitor and reducing the slope of roads are examples.

5. *Separate in space or time the energy being released from the structure that can be damaged or the human who can be injured.* Examples include separate paths for vehicular and pedestrian traffic, placing electric power out of reach, using traffic signals to phase pedestrian and vehicular traffic, and using energy-absorbing materials.

6. *Separate the energy being released from a structure or person that can suffer loss by interposing a barrier.* Safety glasses, barrier guards, radiation filters or shields, median barriers on roadways, thermal insulation, and explosive barricades are examples of barriers.

7. *Modify the surfaces of structures that come into contact with people or other structures.* Rounded corners, blunt objects, dull edges, and larger surface areas for tool handles are examples.

8. *Strengthen the structure or person susceptible to damage.* Examples for this strategy are fire- and earthquake-resistant construction of buildings, training of personnel, and vaccination for disease.

9. *Detect damage quickly and counter its continuation or extension.* Sprinklers that detect heat and spray water to prevent the spread of a fire and wear indicators built into the treads of vehicle tires are examples of this strategy.

10. *During the period after damage and the return to normal conditions, take measures to restore a stable condition.* Examples are rehabilitating an injured worker and repairing a damaged vehicle.

Unlike Heinrich, who advocated a serial model, Haddon argues for a parallel model of preventive action. A parallel model includes multiple actions working at the same time. A serial model has actions working one at a time. Haddon notes there is no reason to select one preventive strategy over another or to prioritize countermeasures according to the accident sequence. Any measure that prevents the damage or undesired result is satisfactory. There is one exception to this parallel model, the quantity of energy involved. As the amount of energy increases, countermeasure higher in the list are more desirable.

Errors in Management Systems

As part of their approach to management through quality, Juran[7] and Demming[8] focused on work processes and the role management has in establishing the processes provided for workers to follow. Both focus on errors by workers as attributes of poor management processes. Deming claims that 85% of errors are the result of poor processes, and no matter how hard someone tries to improve within a given process, it is not possible unless there is a change in the process itself. The focus is on management getting the process right, reducing errors in poor processes, and avoiding the need to correct things after they have resulted in errors. Errors are a management issue, not a worker issue. Incidents and accidents are simply a form of error. They interrupt processes and reduce quality. However, by engaging both workers and managers together in helping to get the processes right, all participants work together to achieve quality (and safety).

Juran notes that critical processes, those which present serious dangers to human life, health, and the environment or which create losses of very large sums of money, require planning and design to reduce the opportunity for human error to a minimum.

Therefore, emphasis is on continuous, incremental improvement. An extension of this concept is that of "six sigma."[9]

Similarly, reengineering concepts of Hammer and Champy[10] focus on improving business processes, but rely on major redesign principles and technological change to achieve operational improvements and reduced errors.

Quality improvement, six sigma, and reengineering approaches create an opportunity to reduce incidents and losses by improvements to the processes used to accomplish work. These concepts are expanded in Chapter 34.

Single-Factor Theories

Many individuals, particularly those not trained in incident prevention and investigation, have the idea that there is a single cause for an incident. A single factor theory assumes that when one finds a cause, there is nothing more to find out. Single-factor theories have limited use in prevention, because contributing factors and corresponding corrective actions will be overlooked. The single-factor theory is a very weak tool in the arsenal of incident prevention and safety management. In fact, it is often a hindrance.

3-4   PREVENTIVE STRATEGIES

Figure 3-3 outlines a general approach for using data from incidents to prevent them from occurring in the future. Regardless of the theory and methods used, the causes of incidents are identified and corrective actions are taken to prevent future incidents of the same type. Different strategies are possible for this approach. The strategies are based on frequency, severity, and cost. Each has merit, depending on preventive goals.

The *reactive approach* in Figure 3-3 requires that at least one incident must occur to identify preventive actions. In Part Five, other approaches, which have the goal of keeping incidents from occurring the first time, are discussed. This approach, illustrated in Figure 3-4, is called a *proactive approach*.

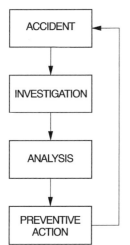

Figure 3-3.   A reactive approach for deriving preventive actions from accidents.

Figure 3-4.   A proactive approach for developing preventive actions before accidents occur.

Frequency

Frequency strategies try to prevent as many incidents as possible. Therefore, investigation, analysis, and preventive actions are directed toward incidents that occur frequently. Preventive actions attempt to reduce the frequency of occurrence.

Recognition of these related factors will help direct preventive efforts where they will be most effective. For example, nearly 50% of injuries occur to workers in their first year on the job. Half of these occur in the first three months. Centering corrective actions (such as proper training) on new employees and their work environments should reduce incident frequency more than would applying the effort with equal intensity to all workers.

Severity

Another approach is directed at serious cases: those cases involving long-term disability, long or serious illnesses, death, large numbers of people, or large property loss. One study[11] reported that serious injuries occur most frequently in four kinds of work activities: construction, nonproductive activities, rarely performed and unusual nonroutine work and work involving high health risks. Data like these can help formulate strategies to prevent serious injury and illness.

Cost

Another strategy is to prevent high-cost incidents. This strategy, based on the principle of Pareto's law, uses cost as the basis for measuring seriousness of incident consequences, not the injury or illness itself. Although cost and severity strategies are much the same, a cost strategy includes losses other than human ones.

Combinations

Another strategy is to use a combination of frequency, severity, and cost. To establish priorities for preventive actions, one can use a number of risk analyses and related techniques (see Chapters 35 and 36). They rely on the probability that an event will occur or the frequency of its occurrence, the seriousness of the event if it does occur, the cost of losses that could be avoided, and the cost to implement corrections.

The Three Es of Safety

Another concept for selecting preventive actions can be structured around the "three Es of safety": *engineering*, *education*, and *enforcement*. *Engineering* includes such actions as substituting less hazardous materials, reducing the inventory of hazardous materials, modify processes, designing out hazards, incorporating fail-safe devices, using warning devices and prescribing protective equipment.

Education includes:

- training people in safe procedures and practices
- teaching people how to perform a job correctly and safely
- teaching users how to use a product safely
- teaching people what hazards exist in a product, process, or task and how to take appropriate protective actions
- training engineers about hazard recognition, hazard evaluation, compliance with safety standards, and legal responsibilities

Enforcement is achieving compliance with federal, state, and local laws and regulations, with consensus standards and with company rules and procedures.

Sometimes a fourth E is part of the paradigm: *enthusiasm*. It refers to motivating people in an organization to cooperate with safety programs through participation and other means. It is motivating users to follow safe practices. Behavior based safety (Chapter 31) expands principally on education, enforcement, and enthusiasm.

3-5 HOW SAFE IS SAFE ENOUGH?

What is accepted as safe is neither constant nor absolute. Each person and society establishes what level of safety and health is acceptable. Not everyone agrees on whether things are safe enough. People would like to be free from risks. However, every activity has some risk. The level of risk that society finds acceptable is a moral issue, not just a technical, economic, political, or legal one. Society participates in deciding what risk is acceptable and at what price. The standards are not constant. They change over time, may vary by location, and are also affected by who is paying for the risk reduction.

As illustrated in Figure 3-5, there is a region of uncertainty between that which is acceptably safe and that which is unacceptably dangerous. Engineers face a dilemma in dealing with this middle region because they cannot depend on their own intuition to decide what is safe enough. To achieve acceptably safe products and environments, engineers must be able to recognize hazards and apply current standards of society found in laws, regulations, judicial interpretation, and public expectation. There is a trend toward lowering levels of acceptable risk, requiring engineers to anticipate tighter standards than exist at the time they design something. There will never be a final answer to the ques-

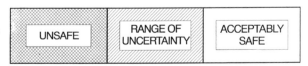

Figure 3-5. Many things are not clearly safe or unsafe. Some things fall in a region of uncertainty.

tion "How safe is safe enough?" Society, through political, economic, and legal processes, will define the price it wants to pay for acceptable levels of risk.

EXERCISES

Obtain copies of completed incident reports. Use data found in each to complete the following:

1. Identify unsafe acts and unsafe conditions in each report.
2. Identify preventive actions that were possible for the cases from a domino theory and energy theory perspective.
3. Identify factors involved in each case using a multiple factor theory, such as the four Ms.
4. Use the history of automobile safety (or some other product) to show the shift in public acceptance of and preference for safety features on products.
5. Review the management principles of Juran and Deming and identify how safety can be incorporated into continuous improvement and quality.

REVIEW QUESTIONS

1. What are the three major reasons for safety?
2. Give a definition of the term *accident*.
3. What are the deficiencies of the term *accident*? What term is an alternate for accident?
4. Why is the term *incident* preferred today?
5. What are the differences between the following pairs of terms:
   (a) incidents and injuries
   (b) injuries and illnesses
   (c) direct and indirect costs of incidents
   (d) insured and uninsured incident costs
   (e) unsafe acts and unsafe conditions
6. What is Heinrich's 4:1 ratio? What is important about it? What are its limitations?
7. What is Heinrich's 300:29:1 ratio? What is important about it? What are its limitations?
8. What is the vital few concept? What is significant about it?
9. Define
   (a) safety
   (b) hazard
   (c) control
   (d) risk
10. Distinguish between hazard and risk.
11. Explain the domino theory.

12. What is the energy theory? Identify at least 5 strategies for incident prevention based on the energy theory.

13. How are the management theories of Juran and Deming helpful in developing strategies for safety?

14. What is a single factor theory? What is its main limitation?

15. What are multiple factor theories? How are they helpful in incident prevention?

16. Explain the four Ms.

17. Name 4 strategies for incident prevention. Briefly explain each.

18. What is a reactive approach to safety?

19. What is a proactive approach to safety?

20. What are the three Es of safety? What is the fourth E?

21. Can absolute safety be achieved? Why not?

NOTES

1 This book uses both *incident* and *accident*, often interchangeably.

2 F. E. Bird, Jr., and G. L. Germain, *Damage Control*, American Management Association, New York, 1966.

3 J. A. Fletcher, *The Industrial Environment—Total Loss Control*, National Profile Limited, Willowdale, Ontario, Canada, 1972.

4 *Dictionary of Terms Used in the Safety Profession*, American Society of Safety Engineers, Des Plaines, IL, 1981.

5 V. L. Grose, "System Safety in Rapid Rail Transit," *ASSE Journal*, August: 18–26 (1972).

6 W. Haddon, Jr., "On the Escape of Tigers: An Ecological Note," *Technology Review*, May: 45–47 (1970).

7 J. M. Juran, *Juran on Quality by Design*, The Free Press, New York, 1992.

8 Mary Walton, *The Deming Management Method*, Perigree Books, New York, 1986.

9 Mikel Harry and Richard Schroeder, *Six Sigma: The Breakthrough Management Strategy Revolutionizing the World's Top Corporations*. Doubleday, New York, 2000.

10 Michael Hammer and James Champy, *Reengineering the Corporation*, HarperCollins, New York, 1993.

11 C. W. Ross, "Serious Injuries Are Predictable," *Professional Safety*, December: 22–27 (1981).

BIBLIOGRAPHY

DeReamer, R., *Modern Safety and Health Technology*, Wiley, New York, 1990.

DiNardi, Salvatore R., and Luttrell, William E., *Glossary of Occupational Hygiene Terms*, American Industrial Hygiene Association, Fairfax, VA, 2000.

Haddon, W. J., Jr., "On the Escape of Tigers: An Ecological Note," *Technology Review*, May: 45–47 (1970).

Lack, Richard W., ed., *The Dictionary of Terms Used in the Safety Profession*, 4th Ed., American Society of Safety Engineers, Des Plaines, IL, 2001.

Lowrence, W. W., *Of Acceptable Risk*, William Kaufman, Los Altos, CA, 1976.

Manuele, Fred A., *Heinrich Revisited: Truisms or Myths*, National Safety Council, Itasca, IL, 2002.

Safety and Health Classics, National Safety Council, Itasca, IL, 2000.

PART II

LAWS, REGULATIONS, AND STANDARDS

THIS SECTION of the book discusses legal concepts, laws and regulations, and standards of government and private organizations that are important for safety engineering. It is essential that someone practicing safety engineering know the legal context and basic legal theories. For many pursuing safety, the first concern is compliance with the hundreds of federal, state, and local laws, regulations, and standards. Today, international companies face many legal contexts for their products or operations. The main focus of this book is on United States legal considerations.

CHAPTER 4

FEDERAL AGENCIES, LAWS, AND REGULATIONS

In a discovery deposition, attorneys ask the chief engineer for a manufacturer of warehouse storage racks what standards his company uses to test the structural integrity of its products. The engineer replies, "I am not aware of any standards for that." At issue is a claim of a fork lift driver who became a paraplegic when he apparently was struck by rolls of paper falling from a storage rack he backed into. Imagine later in court when the attorney for the driver is presenting evidence for his client. He introduces information about a standard published by an association of manufacturers of warehouse equipment. The standard includes procedures for evaluating the structural integrity of storage racks under a variety of loading and use conditions. The evidence also shows that the chief engineer's company is a member of the association. Then, with the chief engineer on the witness stand, the attorney quotes the discovery question and answer for the jury. Imagine the credibility of the chief engineer and manufacturer.

4-1 FEDERAL LAWS AND REGULATIONS

The Congress of the United States enacts laws and appropriates money. Members of both the House of Representatives and the Senate propose bills and act on them. When approved by Congress and signed by the President, the bills become the law of the land. Many of the laws intend to protect the public and provide for their health and safety.

To implement and enforce the laws, Congress assigns responsibility for particular acts to organizations within the executive branch or to independent agencies. Most organizations in the executive branch are part of a department headed by a cabinet-level secretary. These organizations may issue regulations that establish how acts are to be implemented and enforced. Regulations created in support of an act have the authority of law.

To make laws and regulations apply to executive branch organizations the president signs executive orders requiring compliance. To make them apply to Congress, an act must include provisions that include Congress within the scope of the law, or Congress must pass a separate act assigning responsibility to its members and agencies.

Interpretation of federal laws and regulations is not done with absolute authority by an implementing and enforcing agency. The system of justice in the United States gives individuals and organizations the right to due process. Citizens may protest laws and regulations that adversely affect them by filing a complaint in a federal court and arguing for their position. One can appeal citations for violation of regulations.

Safety and Health for Engineers, Second Edition, by Roger L. Brauer
Copyright © 2006 John Wiley & Sons, Inc.

Civil and Criminal Law

There are two kinds of laws in the United States: civil law and criminal law. Civil laws deal with the private rights of individuals. Under civil laws, an individual (a person or organization) seeks to obtain compensation for a loss or to prevent a loss from occurring. Criminal laws deal with harmful acts or crimes against individuals, society, or the government. Crimes are prosecuted by the state or the federal government, depending on which has jurisdiction. On conviction, one faces fines and imprisonment. Violation of safety and health laws and regulations most often involve civil law. Some involve criminal actions and penalties.

United States Code   The laws enacted by Congress are codified and logically grouped into the body of laws called the *United States Code* (USC). The USC is published and bound in volumes. The published version contains the full text of congressional acts. It is updated periodically to reflect additions, changes, and deletions resulting from each session of Congress. Federal laws and data about them appear in a multivolume, annual publication called *U.S. Statutes at Large*.

There are several methods for labeling congressional acts. Each public law has a number, such as Public Law 91-596. In this example, the act is the 596th law enacted by the 91st Congress. Each congress sits for two years. An act may also have a name as part of the provisions of the act. An example is the Occupational Safety and Health Act of 1970. Some acts also are cited by the name of the two individuals who sponsored it as a bill in Congress (one person from each legislative body—the House of Representatives and the Senate). For example, the OSHAct of 1970 also is known as the Williams-Steiger Act.

Note that two acts are needed to set a government organization into action: an authorization act and an appropriation act. An authorization act assigns responsibility to a government agency, empowering it to perform certain functions. An appropriation act provides the money for a fixed period to pay for the activities. A federal agency cannot function unless Congress passes both acts. By limiting appropriations, Congress can control the effectiveness of a government agency.

Code of Federal Regulations

Federal agencies propose and adopt regulations and standards. The *Code of Federal Regulations* (CFR) contains adopted final rules. Regulations are organized into 50 topics or titles. A title is normally assigned to a particular agency. The CFR is updated annually, with certain portions appearing each quarter. The CFR has an index and other aids for locating particular regulations. Today, the current CFR also appears in electronic media, such as CD-ROM or on the Internet.

Often the provisions of an act itself have little direct impact on engineers and others who must follow them. More often, one must comply with regulations issued by an agency in response to an act.

Each federal regulation is indexed by an alphanumeric code. For example, OSHA Safety and Health Standards for General Industry have the identifier: 29 CFR 1910. 29 CFR refers to Title 29 of the Code of Federal Regulations (Title 29 is assigned to OSHA) and 1910 refers to that part within Title 29 dealing with general industry. Each part is divided into subparts or sections, paragraphs, and subparagraphs. Part 1910 is divided into 26 subparts, A through Z. Part 1910 is also divided into sections, paragraphs, and several levels of subparagraphs. For example, Section 1910.1049 might have a sub-

sub-subparagraph labeled (d)(6)(iii)(e). Often particular portions have a descriptive name. For example, 29CRF1910.146 is called the OSHA Permit-Required Confined Spaces Standard.

Federal Register

Each agency adopts or modifies federal regulations through an orderly process. After publication of proposed additions or changes, there is a period in which interested parties submit comment in writing or at public hearings. A proposed regulation may then be modified, left as proposed originally, or withdrawn. After public comment, a proposed regulation becomes an official government policy and procedure after publication as a final regulation. Proposed and final regulations, together with supporting data and arguments, appear in the *Federal Register.*

The *Federal Register* is a daily publication for communicating regulations and other legal documents of federal agencies to the public. It typically runs 50,000 to 80,000 pages per year. In presenting proposed or final changes in regulations, agencies may include supporting data, summaries of research, hearing dates, procedures for submitting comments, arguments for and against the regulation, projected impacts for the private sector, cost–benefit analysis, and effective implementation dates.

The date of publication forms the basis for citing items published in the *Federal Register.* A table of contents at the front of each daily issue identifies the sections included, the pages on which they appear, and the issuing agency.

Other Federal Publications

In addition to developing regulations, most government agencies prepare publications for the public or make them available to the public. Many of these are helpful in complying with federal regulations. Others may be research, statistical, or other kinds of reports prepared by an agency or its contractors. All publications prepared for public distribution are indexed and listed in the *Monthly Catalog of U.S. Government Publications.* The National Technical Information Service in Springfield, Virginia, indexes and catalogs research and other reports. Many agencies publish a listing or catalog of their own documents. The Superintendent of Documents prints and sells many government publications.

A number of organizations distribute searchable versions of government regulations and standards and regularly update them as part of the service. These publications may be in electronic form, such as CD-ROM or distributed via the Internet. The *Congressional Quarterly, Federal Register*, and proposed federal legislation can be accessed on various government and private services on the Internet.

Private Publications

Keeping track of changes in federal laws and regulations is a time-consuming task. A number of private publishing companies offer current-awareness publications, some of which are available in electronic form for computers. These publishers monitor what is going on in the federal government for various special fields of interest. They summarize actions of federal agencies. The publications often contain copies of proposed and final changes in federal and state laws and regulations. Information is organized in a convenient manner for readers. These current-awareness publications are available for occupational safety and health, product liability, environment, consumer products, food and drugs, workers' compensation, nuclear energy, mine safety and health, chemicals, hazardous materials transportation, noise, insurance and loss control, and other topics important in safety engineering.

Other publishers simply keep track of federal and state agencies and their responsibilities. Some contain names, addresses, phone numbers, and similar information for contacting the agencies and their employees.

4-2 LEGISLATIVE BRANCH

The legislative branch enacts the laws. In addition, there are two agencies under Congressional control that are of interest to the safety and health field. These are the Government Printing Office and the General Accounting Office.

The Government Printing Office (GPO) is the printing service for the federal government. Copies of most government publications, including safety and health topics, are available through the Superintendent of Documents at GPO.

Among other duties, the General Accounting Office (GAO) audits federal agencies. Safety and health agencies are audited to determine that they are properly using the powers assigned to them and performing their duties efficiently and effectively. The GAO reviews the regulations of each agency to be sure that they reflect the intent of Congress written into public law. Congress may propose changes to laws based on GAO audits.

4-3 JUDICIAL BRANCH

Complaints and appeals regarding safety and health laws and regulations of the federal government are under the authority of the Supreme Court. Initially, district courts within each of the 10 judicial circuits of the United States hear cases. Appeals move to circuit courts of appeal and ultimately may reach the Supreme Court. Special organizations, such as the Occupational Safety and Health Review Commission, hear complaints regarding enforcement of particular federal regulations. Some complaints may proceed to district or circuit courts.

4-4 EXECUTIVE BRANCH

The President heads the executive branch of the federal government. Special offices and commissions and 14 departments (this number may change from time to time), headed by secretaries, report directly to the President. There are also a number of standing and specially appointed offices, councils, and commissions that report to the President. One example was the Kemmeny Commission that investigated the Three Mile Island nuclear accident in 1979. Another is the Rogers Commission that investigated the *Challenger* Space Shuttle accident in 1986.

Figure 4-1 diagrams the key safety and health organizations within the executive branch. Many functions of these organizations are summarized in the following text. No summary can be kept complete and fully up to date, because there are frequent reorganizations and changes in programs and funding. Refer to current directories of government organizations found in most local libraries or on the Internet.

```
┌─────────────────────────────────────────┐
│        EXECUTIVE BRANCH AGENCIES         │
│    WITH MAJOR SAFETY RESPONSIBILITIES    │
└─────────────────────────────────────────┘
```

Department of Agriculture (USDA)

Department of Commerce
- National Institute for Standards and Technology (NIST)

Department of Defense (DOD)

Department of Health and Human Services (HHS)
- Center for Disease Control (CDC)
  - National Institute for Occupational Safety and Health (NIOSH)
  - Food and Drug Administration (FDA)

Department of Homeland Security (DHS)
- United States Coast Guard (USCG)
- Federal Emergency Management Agency (FEMA)

Department of Labor (DOL)
- Occupational Safety and Health Administration (OSHA)
- Mine Safety and Health Administration (MSHA)
- Bureau of Labor Statistics (BLS)
-Employment Standards Administration
   - Office of Workers' Compensation Programs (OWCP)

Department of Transportation (DOT)
- Federal Aviation Administration (FAA)
- Federal Railroad Administration (FRA)
- Federal Highway Administration (FHWA)
- National Highway Traffic Safety Administration (NHTSA)
- Urban Mass Transportation Administration
- Research and Special Programs Administration

Figure 4-1. Organizational structure for agencies within the executive branch that have major safety responsibilities.

Department of Agriculture (USDA)

Animal and Plant Health Inspection Service   This organization is responsible for protecting and improving animal and plant health for the benefit of humans and their environment. Also, it works to control and eradicate pests and diseases and to insure that drugs for animal use are pure and safe.

Food Safety and Quality Service   This organization ensures that foods for human consumption are safe, wholesome, nutritious, and of good quality. It sets standards for and inspects meat, poultry, eggs, dairy products, and fresh and processed fruit and vegetables. As a result of some bacteria-related deaths from several incidents that received national publicity in the early 1990s, the inspection methods of this agency were changed in the late 1990s. Previously, the inspection procedures used had been much the same as that introduced in the early 1900s.

Department of Commerce

National Institute of Standards and Technology (NIST)   This agency was formerly the National Bureau of Standards. As part of its broad mission, NIST conducts research and develops codes and standards in fire protection and prevention, fire equipment, fire behavior, and safety of consumer and building products.

Department of Defense (DOD)

This large agency has several million military and civilian employees. The agency has a safety office that addresses safety of peacetime, training, and combat military affairs. There are also safety organizations and safety schools within each of the services (Army, Navy, Air Force, and Marines). These organizations deal with special hazards associated with the manufacture, distribution, use, and disposal of weapons and weapon materials. They also oversee the safety of construction and maintenance of military facilities and installations.

Department of Health and Human Services (HHS)

Public Health Service (PHS) PHS is responsible for promoting and assuring the highest level of health for Americans. The operating agencies within PHS have direct and indirect significance for safety and health professionals. Key agencies are the Centers for Disease Control (CDC; which also operates the National Institute for Occupational Safety and Health) and the Food and Drug Administration (FDA). Programs in the Alcohol, Drug Abuse, and Mental Health Administration (ADAMHA) and the National Institutes of Health (NIH), particularly the National Institute of Environmental Health Sciences and the National Library of Medicine, may provide help for safety and health professionals.

CDC Within the CDC, the most important safety organization is the National Institute for Occupational Safety and Health (NIOSH). The mission of NIOSH is to assure safe and healthful working conditions for all working people. It recommends occupational safety and health standards, conducts research, and performs related activities in occupational safety and health.

FDA The FDA protects people against impure and unsafe foods, drugs, and cosmetics and against other potential hazards. The Bureau of Biologics regulates biological products. The Bureau of Drugs regulates drugs, including drug safety, effectiveness, and labeling. The Bureau of Foods is responsible for the composition, quality, nutrition, and safety of foods, food additives, colors, and cosmetics. The Bureau of Radiological Health carries out programs concerned with hazards of and human exposure to ionizing and nonionizing radiation. The Bureau of Medical Devices is charged with the safety, efficacy, and labeling of medical devices. Study of the toxic effects of chemical substances is the responsibility of the National Center for Toxicological Research.

Department of Homeland Security (DHS)

This is a relatively new department that consolidated a number of government agencies. Although the emphasis is on security of the nation, some agencies have safety repsonsibilities.

United States Coast Guard The Coast Guard directs many of its functions at safety and health. It conducts search-and-rescue operations to protect life and property at sea and to remove navigational hazards. It enforces safety standards for the design, construction, equipping, and maintenance of commercial vessels and offshore structures. It investigates marine accidents, is responsible for protecting the marine environment from pollution, and enforces rules and regulations governing the safety and security of ports and the anchorage and movement of vessels in U.S. waters. The Coast Guard operates and maintains a system of aids to navigation. It develops and directs national boating safety programs for small craft and creates uniform safety standards for recreational boats and equipment. The Marine Safety Council reviews proposed Coast Guard regulations.

Federal Emergency Management Agency (FEMA)   FEMA is responsible for preparedness, mitigation, relief, and response activities for natural, artificial, and nuclear emergencies. FEMA supports training, education, and research for many kinds of emergencies, develops emergency plans and policies, and provides response and recovery assistance to state and local governments or other organizations when disasters occur. FEMA's United States Fire Administration works to reduce the national fire loss through training at the U.S. Fire Academy.

Department of Housing and Urban Development (HUD)

Some of the programs operated by HUD intend to eliminate conditions detrimental to health, safety, and welfare in housing and community development. HUD develops standards, including structural and building sewer codes, for conventional and manufactured homes.

Department of the Interior

The Department of the Interior protects and preserves public natural resources. This includes activities regarding water quality. The Bureau of Mines is a research and factfinding agency. Areas of research include mine safety, health, and pollution abatement. The department also operates the National Mine Health and Safety Academy, which trains inspectors, managers, and other specialists for various safety and health positions in the mining industry.

Department of Labor (DOL)

The DOL has many activities to foster and promote the safety and health of workers. The Women's Bureau is devoted to improving women's working conditions. Other important agencies are the Occupational Safety and Health Administration (OSHA), the Bureau of Labor Statistics, and the Office of Workers' Compensation Programs.

OSHA   OSHA develops and implements standards and regulations and conducts inspections and investigations to ensure compliance, issues citations, and proposes penalties for violations. It also provides assistance to employers in complying with standards and regulations through consultations, training programs, and publications.

Mine Safety and Health Administration (MSHA)   MSHA is responsible for safety and health in surface and underground mines in the United States. It develops, promulgates, and enforces standards, investigates accidents, and conducts training.

Bureau of Labor Statistics   This agency conducts economic and statistical research. As part of its activities, it collects injury and illness data from employers and compiles national and regional statistics regarding worker safety and health.

Office of Workers' Compensation Programs This agency develops and recommends standards for state workers' compensation laws and provides technical assistance to states. It also administers three workers' compensation programs:

1. federal employees workers' compensation
2. workers' compensation for longshoremen and harbor workers
3. the "black lung" benefit program for coal miners and their survivors

Department of Transportation (DOT)

DOT, divided into eight administrations, conducts programs concerned with all forms of transportation. Responsibilities include the safety of air, water, highway, rail, and pipeline transportation.

Federal Aviation Administration (FAA) As part of its functions, FAA fosters aviation safety through a number of activities. It issues and enforces rules, regulations, and standards for the manufacture, use, and maintenance of aircraft. It certifies pilots, other flight personnel, and airports; operates and maintains air navigation systems; manages air traffic; and conducts research in systems, procedures, facilities, and devices to ensure aviation safety.

Federal Railroad Administration (FRA) One of the responsibilities of the FRA is to administer and enforce rail safety laws and regulations concerned with locomotives, signals, safety appliances, brakes, hours of service, transportation of hazardous material, and the reporting and investigation of railroad accidents.

Federal Highway Administration (FHWA) In carrying out highway transportation programs, one of FHWA's duties is to make highways safe. FHWA develops and implements standards for highway design, construction, and maintenance; promotes the correction of street and highway hazards for vehicles and pedestrians; and conducts research in highway safety and traffic. It enforces safety regulations for motor carriers (trucking), seeks noise abatement, and performs activities relating to the transport of hazardous materials on highways.

National Highway Traffic Safety Administration (NHTSA) NHTSA conducts programs to reduce the frequency of motor vehicle crashes, the severity of injuries, and economic losses that result. It issues Federal Motor Vehicle Safety Standards that establish safety features and characteristics for motor vehicles. It tests vehicles for damage susceptibility, crashworthiness, and ease of repair, and it tests motor vehicles and equipment for compliance with standards. It conducts research and development projects to improve the safety of motor vehicles and related equipment and to make motor vehicles safe for operators, occupants, and pedestrians. It also operates programs to assist state and local motor vehicle safety programs, to set motor vehicle fuel economy standards, and to measure fuel efficiency of vehicles.

Urban Mass Transportation Administration This agency promotes and tries to improve urban mass transportation, including safety of mass transit equipment.

Research and Special Programs Administration In this branch of DOT, the Materials Transportation Bureau develops standards, monitors compliance, conducts research, and coordinates the activities of other agencies for transportation of hazardous materials by air, water, rail, highway, and pipeline. In the Transportation Program Bureau, the Transportation Safety Institute promotes safety and security management through training programs for government and industry.

4-5 INDEPENDENT AGENCIES

Independent agencies operate under their own administration, not falling directly under any of the three main branches of the federal government. One reason these agencies are independent is to minimize the influence by related agencies that promote a technology or by general policies of a current presidential administration.

Consumer Products Safety Commission (CPSC)

The CPSC protects the public against unreasonable risk of injury from consumer products. It assists consumers in evaluating the safety of products, develops standards for consumer product safety, and supports research in the causes and prevention of injury, illness, and death from consumer products. It also operates the National Injury Information Clearinghouse, which compiles data on consumer product injuries from a sampling of hospital emergency room cases across the country.

Environmental Protection Agency (EPA)

The EPA is responsible for protecting and enhancing the environment. It develops and enforces standards, assists state and local governments, and conducts research in prevention and control of air and water pollution. Its responsibility governs pollution from solid waste, noise, radiation, and toxic substances.

National Transportation Safety Board (NTSB)

The NTSB helps assure that all forms of transportation are operated safely. It investigates transportation accidents (all civil aviation and serious rail, pipeline, marine, selected highway, and other catastrophic accidents) and develops recommendations for other government agencies and transportation industries regarding transportation safety, transport of hazardous materials, accident investigation methods, regulations, and reporting of accidents.

Nuclear Regulatory Commission (NRC)

The NRC protects the public health and safety and the environment by licensing and regulating the use of nuclear energy. It also develops and enforces regulations concerning nuclear safety, and it inspects licensed activities, sponsors research, and publishes reports related to its mission.

Occupational Safety and Health Review Commission (OSHRC)

The OSHRC adjudicates disagreements resulting from citations issued to employers for noncompliance with OSHA standards. Decisions by OSHRC judges may be appealed to the U.S. courts.

4-8 OTHER DEPARTMENTS, AGENCIES, AND SAFETY PROGRAMS

Many departments and agencies that have not been listed have safety programs as well, at least for their own employees or contractors. For example, the U.S. Army Corps of Engineers has detailed safety rules and regulations that Corps construction contractors must follow. By executive order of the President, all agencies within the executive branch are required to comply with safety laws and regulations.

Under the OHSAct of 1970, states may choose to operate programs to protect the safety of workers under federal guidelines or to allow the federal OSHA administration to operate such programs with the states. Similarly, states may choose to operate their own Environment Protection Agency.

States often establish laws, regulations, and standards and operate enforcement agencies to protect the safety of its citizens with regard to many kinds of products, operations, and services. Some estimate that states generate far more safety and health laws and standards than does the federal government.

EXERCISES

1. Find a safety or health regulation from the CFR on
   (a) ladders for construction
   (b) ladders and walking surfaces affixed to truck trailers
   (c) elevators in mines
   (d) hazardous waste disposal
   (e) windshields in automobiles
2. What is the public law number for
   (a) The OSHAct of 1970?
   (b) The Resource Conservation and Recovery Act of 1976?
3. Find announcements of proposed changes to safety regulations, schedules for public hearings, or final rule adoption in recent issues of the *Federal Register*.
4. Make a literature search on some topic in safety and health using the Internet. Identify if the sources of information are reliable.
5. Find out what publications the Consumer Products Safety Commission has available by contacting a regional or area office or by looking in the *Monthly Catalog of U.S. Government Publications*.
6. Find rulings of the Occupational Safety and Health Review Commission in commercial legal review publications.
7. Find out whether your state operates a state plan for occupational safety and health or for environmental protection. Compare the state regulations to those issued by the corresponding federal agencies.

REVIEW QUESTIONS

1. Describe differences and similarities between federal laws and federal regulations.

2. By what means are federal executive branch agencies required to comply with safety and health laws and regulations?

3. What is the main difference between civil and criminal law?

4. What is the *U.S. Code*?

5. Describe the methods used to label acts of Congress.

6. What is the *Code of Federal Regulations*? How is it indexed?

7. What is the *Federal Register*? What is its significance for safety and health information?

8. What is the name for each federal agency identified below by acronym? Is it an independent agency? What safety and health responsibility does it have?

   (a) GAO

   (b) NIST

   (c) PHS

   (d) NIOSH

   (e) FDA

   (f) OSHA

   (g) MSHA

   (h) BLS

   (i) FAA

   (j) FRA

   (k) FHWA

   (l) NHTSA

   (m) MTB

   (n) EPA

   (o) FEMA

   (p) NTSB

   (q) NRC

   (r) OSHRC

BIBLIOGRAPHY

Federal Regulatory Directory, Congressional Quarterly Inc., Washington, DC, revised periodically.

United States Government Manual, Office of the Federal Register, National Archives and Records Service, Washington, DC, biennial.

TOMPKINS, NEVILLE C., *A Manager's Guide to OSHA*, Crisp Publications, Menlo Park, CA, 1993.

CHAPTER 5

OTHER LAWS, REGULATIONS, STANDARDS, AND CODES

The federal government is not the only organization producing various forms of safety and health rules. State and local governments issue many such rules and standards. Companies produce rules for their own operation and products. Professional societies, associations, and laboratories develop rules and standards for adoption and use by others. Some work within consensus or voluntary standard-setting bodies. In addition, foreign governments and international organizations create safety and health rules and standards.

It is impossible to list all rule- and standard-making organizations and to keep up with their changes. This chapter includes only major organizations.

5-1   STATE GOVERNMENTS

State governments and their agencies issue many laws and regulations and have agencies assigned to enforce them. States may have agencies that enforce federal regulations. In fact, the 50 state legislatures passed roughly 250,000 laws during the 1970s, whereas the U.S. Congress enacted 3,359 laws during the same period.[1] Perhaps 10% of these at each level had to do with safety and health of the public, at least to some extent.

Federal Programs Administered by States

In an attempt to keep the federal bureaucracy from growing too large and to ensure local control, a number of federal laws encourage states to administer federal laws and regulations. Federal funds often defray administrative expenses. Examples are state-operated environmental protection agencies and occupational safety and health agencies. In many cases, states have not elected to establish agencies and have left enforcement with the federal government for their states.

State Laws and Regulations

States have their own laws and administering agencies for many aspects of safety and health. Some state laws and enforcing agencies were in effect before federal safety and health laws were created. Others appeared after federal laws were enacted. In some cases, federal laws and regulations supercede state laws and regulations, but not always. What laws and regulations apply can become quite complicated.

Safety and Health for Engineers, Second Edition, by Roger L. Brauer
Copyright © 2006 John Wiley & Sons, Inc.

To complicate matters, local governments may adopt ordinances that conflict or differ with state and federal laws and regulations. All may be applicable or those of higher governments may supercede local ones. Not only are the laws and regulations confusing, but the methods and procedures for compliance may be as well.

State Agencies and Regulations

Safety and health regulations commonly issued and administered by states are listed in Table 5-1. Most states have regulations dealing with life safety and structural safety of buildings and with safety in construction and industrial operations. Most regulate transportation, including vehicles, highways, and waterways. All states have regulations governing the licensing of occupations that can affect the public safety and health. Most have standards or codes for sanitary systems and fire protection.

Directories of state governments that list agencies and general responsibilities help identify sources for regulations or assistance. These directories are often called red books or blue books.

TABLE 5-1 An Incomplete List of Safety Laws and Regulations Commonly Issued or Adopted by State Governments

Building	Personal protective clothing and equipment
Building code	Proximity to high voltage lines
Guarding of floor and wall openings	Tank truck vehicles
Separation distances between structures	Welding and cutting equipment
Gasoline stations	Woodworking machines
Institutions, hospitals, schools	Fire safety regulations
Public assembly places	Blasting and explosives
Residences, hotels, apartments	Flammable liquids
Restaurants, dance halls	Hazardous materials
Theaters, movie houses	Housekeeping and maintenance of work areas
Fire-resistant construction	Health regulations
Emergency lighting	Air and water pollution control
Exits	Employee toilet, washroom, and eating facilities
Fire alarm systems	Lighting of work areas
Fire extinguishers	Radiation control
Sprinklers and other fire protection equipment	Exposure to chemical and physical agents
Flame retardant finishes and materials	Ventilation and dust control
Electrical code	Right-to-know/hazards communication
Access for the disabled	Industry safety codes
Construction regulations	Mining of coal, metals, and other materials
Asbestos removal	Dry cleaning and dying
Demolition work	Liquified petroleum gas
Excavation work	Petroleum refining, handling, storage, and
Material hoists	transport
Steel erection	Railroads and grade crossings
Storage of construction materials	Licensing and qualifications of occupations
Temporary electrical wiring	Boiler inspectors
Equipment and machinery regulations	Engineers
Boilers	Health-related professions
Elevators, dumbwaiters, escalators	Mine inspectors
Ladders	Field safety representatives for workers'
Mechanical power transmission apparatus	compensation insurance companies
Painting and spraying equipment	Safety professionals and industrial hygienists

Local Governments

Most villages, cities, and counties have safety and health laws of some kind. Frequently, local governments adopt national standards or portions of them as part of local ordinances. Typical laws and codes at the local level that address safety and health issues include zoning codes, building codes, fire codes, plumbing and sewer codes, and traffic codes. Major cities commonly have regulations and codes that are unique.

5-2 PRIVATE COMPANIES

Most companies have rules about safety and health for employees, customers, products, and use of equipment. These may take several forms: policy statements, rule books, operating procedures and manuals, assembly or maintenance manuals, agreements with unions, contracts, or agreements with suppliers and buyers. These rules may deal with employee activities or they may deal with procedures for certain kinds of work, such as procurement, selection and training of workers, settling of grievances, or operation of particular equipment. There may be handbooks or reference manuals for design that include specific safety information. Special rules may exist for fire, transportation, weather, and other emergencies. Publications may be guides for customers or users.

5-3 VOLUNTARY AND CONSENSUS STANDARDS

There are many nongovernment organizations, like professional societies, trade associations, and others, that develop and publish standards for their field of interest. A few organizations specialize in creation and publication of standards.

Committees of individuals create or update standards that are of interest to companies or organizations who send committee members. Sponsoring organizations are usually members of the organization that will publish a standard. Several organizations may publish the same standard. There have been some challenges to voluntary standards, particularly when the participants on the committees have the interest of their own companies or products in mind and there is no open participation by the public. Challenges also relate to prescribing requirements in the standards that only participating product manufacturers can meet.

Because membership in the organizations that set the standards is voluntary and because compliance is often voluntary, standards created or published by most standards organizations are called *voluntary standards*. Because the standards include those elements that at least a majority of committee members can agree on, the standards are also called *consensus standards*. Compliance with voluntary and consensus standards is required when they are adopted by local, state, or federal governments or are incorporated into government agency regulations or contracts.

The Internet, computer data banks, index services, CD-ROMs, and printed directories help locate voluntary and consensus standards.

Standard and Code Organizations

Two of the largest and best-known voluntary standards organizations are the American National Standards Institute (ANSI) and the American Society for Testing and Materials (ASTM). (ANSI was originally called the American Standards Association. Before its current name, ANSI was named the United States of America Standards Institute and, for a brief period, the American Standards Institute.) Both ANSI and ASTM publish standards on a wide range of topics, including safety and health. ANSI does not endorse the content, but merely provides a format, development and administrative procedures, and publishing services. Volunteer committees establish the contents.

Professional Societies

Many professional societies have developed standards on matters related to their fields. Some of these are listed and distributed by ANSI and ASTM. Others, like the American Society of Mechanical Engineers and the Society of Automotive Engineers publish their own standards. Some societies serve as secretariats for certain standards that are published by organizations like ANSI. Table 5-2 lists many professional and technical societies that develop voluntary safety and health standards.

Associations

Associations generally promote the common interest of members. Many associations exist for a wide range of fields and interests. Some associations develop standards for products or operating procedures, and some of these standards address safety and health topics. For example, the National Fire Protection Association (NFPA) publishes the *National Fire Code*. The Association of Truck Trailer Manufacturers publishes standards on the design of ladders and climbing devices for tank trailers. Directories list associations and data about them. The directories help locate possible sources of standards but do not identify which associations write safety standards.

TABLE 5-2   An Incomplete List of Professional and Technical Societies That Have Developed Voluntary Standards and Codes

ACI	American Concrete Institute
ACGIH	American Conference of Government Industrial Hygienists
AIHA	American Industrial Hygiene Association
AISI	American Iron and Steel Institute
ANS	American Nuclear Society
API	American Petroleum Institute
ARI	Air Conditioning and Refrigeration Institute
ASA	Acoustical Society of America
ASAE	American Society of Agricultural Engineers
ASHRAE	American Society of Heating, Refrigerating and Air Conditioning Engineers
ASME	American Society of Mechanical Engineers
ASQC	American Society of Quality Control
ASSE	American Society of Safety Engineers
AWS	American Welding Society
IEEE	Institute of Electrical and Electronics Engineers
IES	Illuminating Engineering Society
ISA	The Instrumentation, Systems, and Automation Society
ITE	Institute of Traffic Engineers
SAE	Society of Automotive Engineers
SOLE	Society of Logistics Engineering

5-4 PRIVATE LABORATORIES

A number of private laboratories exist to provide independent testing, certification, and other technical services to customers for a fee. Some laboratories were created to support the needs of the insurance industry. Two of the most well-known independent laboratories are the Underwriters Laboratory (UL) and the Factory Mutual System (FM). Both have written some safety standards relating to testing procedures and products tested.

Underwriters Laboratory Incorporated is a nonprofit organization that conducts scientific investigations, studies, experiments, and tests related to hazards of life and property. As part of its function, it publishes standards, classifications, and specifications aimed at reducing hazards.

Factory Mutual System is devoted to control of losses from industrial fires, explosions, and related calamities. It provides inspection services for clients, conducts studies and tests, and produces some standards related to fire protection systems. It tests fire protection devices against its standards for the manufacturers of the devices.

5-5 FOREIGN AND INTERNATIONAL LAWS AND REGULATIONS

Foreign governments and organizations issue laws, regulations, and standards for safety and health. They may impact companies doing business or selling products where they have jurisdiction. At least for some European countries, one can locate regulations and publications related to them through computer data banks and the Internet.

One international organization that has a high rate of growth in standards is the International Organization for Standardization (ISO). It has member organizations throughout the world. Its member organization from the United States is ANSI. ISO may adopt standards proposed by member organizations.

With the implementation of the European Community (EC) during the 1990s, standards for the EC have emerged. They apply to member countries and companies doing business within the EC. For example, companies manufacturing and selling production machines in the EC must follow EC standards for machine safety and must complete risk analyses on the machines being sold. Sellers must inform buyers of risks that remain with the machines and what protection is provided or is left for users.

EXERCISES

1. Determine if ANSI or ASTM has safety standards for
   (a) stepladders
   (b) floor slipperiness
   (c) sports equipment
   (d) glass for doors and windows
2. Determine if your state has any of the following:
   (a) fire code
   (b) ventilation code
   (c) plumbing code
   (d) construction safety regulations

(e) regulations for asbestos removal projects

(f) regulations for cleanup of contaminated soil

3. Find out what agency in your state is responsible for each of the items in Exercise 2.

4. Determine if your local government has a building code, fire code, zoning ordinance, or waste disposal ordinance. Obtain a copy of each and identify which provisions are safety related. Find out how the ordinances and codes are enforced.

5. Skylights in roofs allow daylight to enter interior portions of buildings. When workers are on a roof, a skylight can become a working surface on which people may stand, walk, or set things. Find organizations that may produce standards for skylights and determine what safety considerations are included in skylight design, placement, installation, or maintenance.

6. Identify organizations that write standards for indoor air quality.

7. Determine the difference between the Committee Method and the Canvas Method when developing an ANSI standard.

8. Locate major sources of internationals standards for occupational safety and health.

REVIEW QUESTIONS

1. Where would one look to determine what agencies in a state are responsible for promulgating and/or enforcing fire codes, occupational safety and health standards, and traffic codes?

2. What forms do safety rules and regulations usually take in a company?

3. How would one find associations that may have developed safety and health standards?

4. Name two major organizations that publish voluntary standards, including safety standards.

5. Name two major safety testing laboratories.

6. Describe the process usually used to develop voluntary standards.

NOTE

1 John Naisbitt, *Megatrends*, Warner Books, New York, 1982.

BIBLIOGRAPHY

AKEY, D. S., ed., *Encyclopedia of Associations*, Gale Research Company, Detroit, MI, annual.

Directory and Index of Safety and Health Laws and Codes, U.S. Department of Labor, Wage and Labor Standards Administration, Bureau of Labor Standards, 1969.

National Trade and Professional Associations of the United States and Canada and Labor Unions, Columbia Books, Inc., Washington, DC, annual.

The National Directory of State Agencies, Information Resources Press, Arlington, VA, biennial.

Yearbook of International Organizations, International Chamber of Commerce, New York, annual.

CHAPTER **6**

WORKERS' COMPENSATION

6-1 THE DEMAND FOR COMPENSATION

With the growth of the industrial revolution, the toll in human lives, injuries, medical expenses, and lost income rose rapidly for the men, women, and children employed in factories. Society found these results unacceptable and pushed for reform that would make jobs safer. They also sought to place at least some burden on employers to pay for the losses workers experienced. However, efforts were thwarted, because common law defenses gave employers a great deal of protection. If a worker wanted to obtain compensation or indemnity under common law, the worker had to sue the employer and prove that the employer's negligence was the sole cause of injury. The employee carried virtually all the risks in employment. Furthermore, an attempt to obtain compensation through a lawsuit was likely to result in loss of employment and ill will.

Common Law Defenses

In compensation lawsuits, employers could claim there was no negligence on their part. Three other common law defenses could also be used against an injured worker:

1. assumption of risk
2. contributory negligence
3. the fellow servant rule

Assumption of Risk   The principle of tort law called *assumption of risk* says that if a person voluntarily assumes a risk and is injured as a result, he cannot be indemnified for the losses. This principle provided the employer near absolute protection against claims for work-related injuries of employees. By accepting a job, an employee assumed all the risks the job entailed.

Contributory Negligence   If a plaintiff were able to prove negligence on the part of an employer and establish assumption of risk as an inadequate defense, an employer could claim contributory negligence. For example, assume an employee was caught in a machine and injured. The employer could claim that the employee acted carelessly (was negligent), and therefore had no reason to bring action against the employer. At worst, the employer might have to pay some compensation if both parties were negligent.

Fellow Servant Rule   When assumption of risk and contributory negligence were not sufficient, employers often used a third line of defense. Because servants (employees) had certain duties toward each other, an employer could attempt to show that a fellow employee was negligent and caused the injury of the worker. For example, suppose one worker fed material into a machine and another worker removed the material after the machine completed some action on it. Suppose also that the first worker accidentally started the machine and thereby injured the hands of the second worker. The first worker was negligent. The employer was not responsible for the injury.

Early Workers' Compensation Laws

After the Industrial Revolution, society found the stout defenses of the employer unacceptable. As a result, compensation claims were awarded more frequently, and awards grew larger.

Society in the industrialized nations of Europe and in the United States sought better ways to resolve job-related injury compensation. Near the dawn of the twentieth century, employers were ready for a change. A means for providing workers' compensation emerged. The United States followed the lead of Germany and England.

Early legislation tried to increase employer responsibility by removing some of the common law defenses: assumption of risk and the fellow servant rule. Some liability laws also changed contributory negligence to comparative negligence and allowed juries to determine whether the employer or employee was more negligent. Under employer liability acts, the injured worker had to take his claim to court, find fellow workers who would risk their jobs to testify for him, and avoid being coerced by the employer to sign a release from liability for an inadequate payment. The employers began to lose cases and pay larger awards. The employer liability acts, though an improvement, were still not fully adequate.

Workers' compensation laws followed. Several states and the federal government passed them. Initial laws were declared unconstitutional over issues of due process and mandatory participation by employers. Subsequent state laws were primarily elective, allowing employers to elect to come under the law. Since the first constitutionally acceptable workers' compensation law passed in 1911, all states have implemented such laws. They continue to change to include more workers, to broaden and modify benefits, to change administrative procedures, and to restructure benefit methods.

6-2 WORKERS' COMPENSATION LAWS

No-Fault Concept

In workers' compensation laws, employers and employees struck a balance in rights. Workers gave up the right to sue employers for compensation for injuries arising out of and in the course of employment. Employers agreed to provide compensation for work-related injuries as a cost of producing a product or service. Employers were no longer liable for negligence resulting in worker injury. Legal battles were no longer required to determine who was at fault.

Proliferation of Laws

There are at least 53 separate workers' compensation laws in the United States. Attempts to standardize compensation laws or create federal standards for them have not progressed very far. Each of the 50 states has its own workers' compensation law. The federal government has three compensation programs, each covering a different group of employees. The three acts are the Federal Employees Compensation Act, the Longshoremen's and Harbor Workers Act, and the District of Columbia Workmen's Compensation Act. There are many differences among these laws. Changes occur continuously. The provisions, benefits, and changes are summarized in an annual report.[1]

Types of Laws

Today there are two types of workers' compensation laws—compulsory and elective. A compulsory law requires each employer that is under its jurisdiction to accept its provisions and to provide for benefits as specified. Under an elective law, an employer has the right to accept or reject participation. If an employer rejects compliance with the law, he loses the three common-law defenses and is rendered virtually defenseless. In effect, elective laws are compulsory. Most early workers' compensation laws found constitutional were elective. Nearly all are now compulsory.

Objectives of Workers' Compensation Laws

There are at least six objectives for workers' compensation programs. They are:

1. Replace lost income and provide medical treatment promptly
2. Provide a single remedy without costly litigation and delays
3. Relieve public and private charities of financial drains
4. Encourage employer interest in accident reduction and prevention
5. Restore earning capacity and work capability of workers through rehabilitation
6. Encourage open investigation of accidents to prevent similar occurrences in the future (not to find fault)

One could debate whether these objectives are achieved by existing compensation laws. For example, some thought that employers would become more interested in safety by becoming responsible for indemnification of injured workers, but the competition among insurance companies for employers' business may have done as much for increased employer interest in safety. Insurance companies provide loss control services to employers. Preventing work-related accidents helps employers reduce claims and lower insurance premiums.

Workmen's Versus Workers' Compensation

Until the 1970s, *workmen's compensation* was the accepted term. *Workers' compensation* is now the accepted term because it does not infer gender.

6-3   WORKERS COVERED

Today workers' compensation laws cover approximately 90% of all wage and salary employees. However, several categories of workers are commonly excluded from protection. The exceptions vary among the different state and federal laws. Most common exceptions are domestic servants, casual (short-term, temporary) laborers, agricultural or seasonal farm laborers, volunteer workers, and workers who are covered by other laws (railroad and maritime workers). Recently, professional athletes were excluded. They often have injury compensation in their contracts. In many states, employers with fewer than two to five employees are also exempt. Under most laws, excluded employees may be covered through voluntary action of the employer. In some states, exempted workers must concur with an employer who elects coverage voluntarily.

In the past, states have avoided jurisdictional problems by not requiring public employees of local government units to be covered by compensation laws. Now most state laws require all public employees, whether career, elected, or appointed, to be covered. Here again, there are exceptions.

Under most workers' compensation laws, minors are covered. The definition of a minor varies slightly. For some states, minors who are illegally employed (below minimum age) and become eligible for compensation receive maximum benefits at double or triple the standard rates. This provides a penalty for the employer and accounts for lost future earning capacity of the minor. An employer may be subject to additional penalties under the law if an illegally employed minor is injured on the job.

6-4   BENEFITS

Eligibility Criteria

The main goal of workers' compensation laws is to compensate workers for injuries caused by accidents arising out of and in the course of employment. This goal gives rise to a number of issues regarding eligibility: What is an accident? What is an injury? What does "out of and in the course of employment" include? There are many interpretations to these questions.

Accident and Injury   As noted in Chapter 3, the term *accident* suggests an event of very short duration. This was the meaning for early interpretations under workers' compensation claims. For most workers' compensation laws today, *accident* may refer to extended exposures and may recognize other factors. In the early 1980s, claims increased significantly for cumulative trauma injuries. These disorders result from repeated trauma to the part of the body affected, such as the arm of a carpenter swinging a hammer. More recently, claims for various forms of "job stress" have been on the rise.

The term *injury* was limited originally to physical damage to the body, such as cuts, punctures, fractures, and burns. Today most laws recognize a variety of job-related illnesses as a form of injury, but not all job-related illnesses are covered.

To avoid these language problems, different terminology is now being used. For example, the Federal Employees Compensation Act states that compensation will be paid for "the disability or death of an employee resulting from personal injury sustained while in the performance of his duty." It defines injury to include "in addition to injury by accident, a disease proximately caused by the employment."

Employment   There are many legal questions regarding the definition of employment. Self-inflicted, intentional injuries are excluded, as are injuries resulting from willful misconduct (often including those resulting from intoxication), most injuries resulting from personal conflict with a fellow worker, and injuries occurring off the job. Many difficulties remain. The courts must answer these questions on the merits of individual cases. For example, are workers covered while going to and from work? Are they covered during lunch hours? Are they covered when intoxicated while performing job-related tasks, like a salesman wining and dining a customer? Is a heart attack at work covered? Is a worker covered when injured in a boating accident at a company picnic?

It is difficult to establish whether certain kinds of injuries occur during employment. For example, hernias, back injuries, and diseases with a latency period between exposure and observable symptoms all create problems in eligibility. A worker may file a claim stating that the injury was job-related and occurred on the job. Diagnostic procedures may not be able to establish the time or place of injury to verify whether it was job related. Many of the laws have special provisions to deal with these problem cases.

Types of Disability

Most workers' compensation laws recognize four classes of disability: temporary total, permanent partial, permanent total, and death. Some states recognize an additional class: temporary partial. Definitions for and interpretations of each class vary by compensation law.

Temporary Total Disability   Temporary total disability applies to a worker who is completely unable to work for a time because of a job-related injury. Eventually, the person recovers fully and returns to full job duties. No disability or reduction in work capacity remains after recovery. Most disability cases are temporary total cases.

Temporary Partial Disability   This classification applies to injured workers who are unable to perform their regular job duties during the recovery period, but are able to work at a job requiring lesser capabilities. After recovery, the worker returns to work with full capability.

Permanent Partial Disability   This classification refers to a worker who endures some permanent reduction in work capability but is still able to retain gainful employment. Examples of permanent partial disability include the *loss* of a body member, such as a hand, eye, or finger, or the *loss of use* of a body member, such as an eye, or permanent reduction in the movement or functionality of an elbow or other joint.

Permanent Total Disability   This refers to a worker injured on the job and no longer able to work, even after medical and rehabilitative treatment. In many states, certain disabilities are classified as permanent total disability by definition. Defined impairments typically include loss of both eyes, loss of both legs, and loss of both an arm and a leg.

Benefits

Workers' compensation laws provide payments for medical expenses, burial expenses, loss of wages, and impairments. Most provide payment for physical and vocational rehabilitation. Some provide for mental rehabilitation.

Loss of Wages   Injured employees receive compensation for their loss of earnings, which can occur under all the types of disability. Most laws provide a percentage of the average weekly earnings of the injured employee. Payment schedules usually have upper and lower limits. Because disability income is not usually subject to income tax, a claimant receives only a portion of regular earnings. The percentage (commonly $66^2/_3\%$) may vary by type of disability, number and ages of dependents, and other criteria. Some states limit loss-of-wage payments to a maximum length of time (usually for temporary total disability). A few pay the difference between preinjury wages and postinjury wages when the injury reduces the earning capacity, but not the ability to be gainfully employed. Payments are made for life to a worker with permanent total disability. In the event of a job-related death, the dependents of the worker usually receive benefits for loss of income until a spouse remarries or dies and minor children reach adult age or complete school.

All workers' compensation laws require a waiting period before loss of wage payments begin. This waiting period ranges from one to seven days. However, if the disability extends long enough (usually two weeks), then compensation starts on the first day of lost wages. The purposes for this waiting period are to reduce administrative costs for minor disabilities and to discourage malingering by workers.

Medical Expenses Workers' compensation payments normally cover unlimited medical expenses deemed necessary in the treatment of the injured worker. These include physician charges, hospital costs, physical therapy, cost of prosthetic devices, and many other medical costs. There is no waiting period before payment of medical expenses.

Burial Expenses All compensation laws provide an allowance or fixed payment for burial expenses. The allowance varies. Some laws provide an additional allowance for transportation of the deceased if the death occurred away from home.

Rehabilitation Expenses Physical rehabilitation is typically covered as a medical expense. Provisions vary considerably for vocational rehabilitation. Some states require the employer to pay for vocational rehabilitation. Some laws have maximum payments, limit the period allowed for training, or limit total expenses per case. Under the Federal Vocational Rehabilitation Act, states receive federal funds to help cover the cost of retraining persons disabled in industrial accidents.

Payments for Impairments Workers who sustain permanent partial disabilities receive compensation for the loss of a body member or the loss of its function (loss of use). The fundamental idea is that an individual's ability to work and earn an income is impaired by the disability. As a result, he will earn less over the rest of the working years. In most states, payments for impairments are in addition to payments for loss of earnings during the period of healing.

There are a number of theories for determining the amount of compensation. Three major ones are the whole-man theory, the lost wages theory, and loss of earning capacity. One or more of the theories may apply under a particular law.

Whole-Man Theory The whole-man theory considers only the functional effect of the loss—its impact on normal functions and abilities. The disability is rated as a percentage of a whole, fully functional person. A formula that relates degree of disability to income potential establishes disability payments. For example, in Nevada, compensation is $1/2\%$ of a person's average monthly earnings for each 1% of disability.

Lost Wages Theory The lost wages theory considers the actual loss in wages relative to a standard that estimates what the individual would have earned. When actual earnings are less than the standard and the reduction in earnings is the result of the impairment, the actual compensation will maintain the income at or near the standard.

Loss of Potential Earnings Theory The loss of potential earnings theory is by far the most common approach for paying compensation for impairments. Future earning capacity is estimated from such factors as impairment, age, occupation, gender, and education. The benefits are the difference between preinjury earnings continued into the future and estimated future capacity after injury.

Schedule Payments The administrative problem of evaluating each permanent partial disability has given way to the widespread practice of schedules. Schedules establish in advance the value of each kind of disability. Units for disability are weeks of lost earnings. For example, under the Federal Employees Compensation Act, the loss of a thumb is worth 75 weeks. The value of the loss in weeks is multiplied by a percentage of the normal weekly wage of the person before injury. Practices in using schedules vary by state and the value of a scheduled loss can be quite different.

Functional impairments or loss of use are normally expressed as a percentage of total loss of the member or function. An impairment is the schedule value multiplied by the percent of impairment. For example, a 20% loss of use of a thumb in the preceding example would be worth 15 weeks (20% of 75 weeks).

Duration of Disability Most compensation laws use calendar days to establish the period of disability. Not counted are the day of the injury and the day an injured worker returns to work. All days between the injury and the return to work are counted as calendar days of disability. This avoids the problem of establishing the schedule that a person would have worked. Swing shifts, variable work schedules, flexible hours, holidays, plant vacations, layoffs, and the other work schedules do not create difficulties in computing benefits.

Loss of wages and payments for impairments usually are based on average weekly earnings. Sometimes monthly earnings determine death benefits. Many individuals have biweekly, monthly, or annual pay rates and conversions to weekly rates could affect actual payments. Each compensation law has its own procedure for computing time and rate conversions.

6-5 FINANCING

Types of Insurance

Depending on state regulations, one or more methods of providing workers' compensation insurance is available to employers: state-operated insurance, private insurance policies, or self-insured benefits. As of 1980, only six states required employers to participate in the state-operated insurance. Twelve states operated a state insurance fund, but permitted employers to purchase private policies from commercial insurance companies. Most states do not operate an insurance fund. At least forty-seven states allow employers to be self-insured, if they qualify.

Large corporations may reduce administrative costs by becoming self-insured. Group self-insurance arrangements also may be possible and allow smaller companies to benefit from self-insurance. To become self-insured, a company must create a large reserve fund to ensure that claims will be paid. Self-insurance programs often include a wide variety of employment types to avoid concentrating risks. Many companies cannot afford to establish the required reserve fund because the funds might be used better elsewhere in the company. Also, reserve funds are not always deductible for tax purposes, whereas insurance premiums usually are. In addition, self-insurers must maintain medical, legal, and safety staffs to administer the program, resolve problems, and work to reduce claims.

Cost of Workers' Compensation

U.S. employers spend approximately $100 billion per year for workers' compensation insurance of all types. Although costs will vary, approximately one fourth of the expenditures are for medical care, nearly half for compensation payments, and less than one third for administrative costs and expenses for safety and health and legal services provided by insurers.

Premiums

Employee payroll forms the basis for workers' compensation insurance premiums: units for premiums are dollars per $100 of payroll. Average costs are roughly $2.00 per $100 of payroll, but vary widely with employment type.

The National Council on Compensation Insurance, an actuarial organization, sets basic premium rates for most states. Rates are adjusted to keep up with changes in compensation laws. Each state has its own rate table or book. Tables include premium rates for many kinds of operations or work activities.[2] Rates for each state reflect different risks and claim histories that are accounted for in setting rates. The system for classification of operations or work activity used to be the Standard Industrial Classification (SIC) system. However, with many new kinds of work and international commerce, a new system is now in use called the North American Industry Classification System (NAICS). Some kinds of work had major changes in classifications.

It is somewhat complicated to determine the total premium paid by an employer. If an employer has one kind of operation, the premiums are based on the rate for that operation. If there are two or more kinds of operations, premiums will usually be based on the operation with the largest amount of payroll. If employees participate in several operations, the premium for those employees usually is based on the highest rated activity. For large, complex companies, combinations of rates usually determine the premiums.

Kinds of Rates and Discounts

Depending on provisions in applicable compensation laws, a number of methods may establish premium rates for an insurance customer. The key methods are manual rating, schedule rating, experience rating (prospective and retrospective), fixed rates, and premium discounts.

Manual Rates   In manual rates, one applies premiums directly from the rate book for the applicable state. The premiums will be the same from all insurance companies. For example, a company engaged in sheet metal work has a payroll for the year of $853,200. Assuming all employees are sheet metal workers and the manual rate is $4.48 per $100 of payroll, the annual premium would be $853,200 × $4.48/$100 = $38,223.36.

Schedule Rates   In the earlier days of workers' compensation, employers could receive a percentage reduction in the premium rates by engaging in certain hazard reduction activities that were listed in a schedule. This technique is no longer used, one major reason being that it was difficult and expensive to monitor compliance.

Experience Rating-Prospective   Under this method, the accident experience record of a policy holder can influence future premiums. To avoid excessive fluctuation in the premiums, the experience of three years is used. The results of an immediate past year will affect the premiums three years later. Each state determines the average losses by employment classification (such as meat packing, carpentry, etc.). The average rate times the payroll for that category in a company determines the expected losses. If the actual losses for an employer exceed that expected based on state average loss rates, a surcharge will be added to the manual rate. If the actual losses are less than expected, a credit will be applied to the manual rate. The surcharge or credit is called the *experience multiplier*, *experience modification*, or *experience rating modifier*. This method provides an incentive to control losses.

Suppose the sheet metal firm above has experience rating modifiers during the three previous years of 1.32, 1.04, and 0.88, respectively. It would pay $38,223.36 × 1.32 = $50,454.84 for its premiums next year, and $38,223.36 × 0.88 = $33,636.56, two years later.

Experience Rating-Retrospective   In a very similar method, employers with sufficiently large policies can affect their rates while the policy is in force, rather than waiting for three years. Before a policy is put into force, the employer and the insurer agree to a set of adjustments in premiums within upper and lower limits. Claim experience will affect premiums during the life of the policy (normally one year).

Fixed Rate Premiums   For small companies that cannot qualify for experience rating modifiers, the manual rate in effect at the inception of the policy applies. The premium will change from year to year, depending on the losses of all businesses within the state for that employment classification.

Premium Discounts   For large policies, administrative costs are relatively less than for small policies. As a result, states allow discounts for premiums in graduated steps based on total premiums paid. For example, there may be no discount for the first $1,000 of premiums, 3% or more for the next $4,000, and larger discounts for higher steps.

Competitive Premium Rates   Until recently, workers' compensation premiums were fixed for each program. All insurers quoted rates from the same manual rate book. Competition among insurance companies was based on supporting services for clients. Recently, some states have initiated competitive premium rates in which insurance companies can set premiums on their own. Programs operated this way expect to produce lower rates, but often produce reduced loss control services.

Other Strategies to Reduce Workers' Compensation Costs   A variety of methods are now in use to reduce workers' compensation claims and to put injured people back to work. The employer, employee, and insurer all come out ahead. One approach is dealing with the psychological and behavioral aspects of injured workers. Being removed from work because of injury, even if temporary, can create fears and stress for injured workers and their families. Supervisors, coworkers, and company staff often treat injured workers differently after a compensation claim is filed. The goal is reducing supervisors' negative feelings and employers' lack of concern. This method attempts to rebuild strained relationships and to make workers want to return to work. It seeks to build worker confidence, particularly when some job capabilities are lost.

Another approach involves systematic and objective evaluation of worker capabilities and job requirements. Special programs then rebuild physical strength and endurance through work hardening, modifying the workplace, or developing new job skills. Many hospitals now have worker rehabilitation programs that apply interdisciplinary evaluation and treatment to workers' compensation cases.

A number of states now require safety committees with participation by both management and labor. Building a cooperative environment and a team effort to reduce hazards and risks often lowers incidents and claims.

In some large, multicontractor construction projects, the project management firm or owner may use reductions in worker compensation claims to reward those contractors who meet project safety goals.

6-6 ADMINISTRATION

Efficient administration of workers' compensation programs keeps cost down. The fact that such programs are "no fault" relieves many of the delays in making compensation available to injured workers. Employers must notify employees of workers' compensation benefits and claim procedures and must keep records of claim-causing or potentially claim-causing injuries (usually other than first aid cases).

To initiate action, the worker must provide the employer with notice (usually written on a standard form) that he was injured on the job. Because the injuries happen on company premises, employers are aware of most injuries and they may assist with some formalities. As soon as an employee files a notice, the employer must file a claim with the insurance carrier and with the state agency (if it is not the carrier). In many cases, employer-maintained reports of on-the-job injuries are submitted with claims. After review and approval of a claim, payments are authorized and made.

Most payments are made by direct settlement. The insurer pays benefits at the prescribed rates. In some cases, the employer and employee reach an agreement on the benefits (subject to state approval) before funds are disbursed. Usually there is no dispute between employee and employer. In a third method, a commission or its representative reviews each claim to determine benefits. When employees believe that the compensation offered is inadequate, under most programs they may file an appeal within a certain time period (normally 1 to 3 years). Only 5% to 10% of the five million or more cases each year are contested. Each program has established procedures for reviewing cases and proceeding toward final resolution. There may be several levels of appeal, and an employee may engage an attorney in claim and appeal procedures. Many states have established approved fee structures for legal work in workers' compensation cases.

6-7 THIRD-PARTY LAWSUITS

As noted earlier, employees cannot file suit against their employers for job-related injuries. However, an employee may sue the manufacturer of a machine or product that caused injury. An employee may sue another employer on a multiemployer job site or another organization or individual involved in the injury-causing accident. In a few states, an employee can sue a fellow worker. After the theory of strict liability for products appeared, the frequency of third-party suits increased. Most often the suit is against a manufacturer of a product causing the injury or another organization contributing to the accident and injury.

Defendant manufacturers or other employers may initiate a third-party action against the injured worker's employer. Ultimately, the worker's employer may have to pay part of the settlement.

If an injured worker wins such a lawsuit and receives an award that is larger than that obtained through workers' compensation, the worker may have to repay the compensation obtained through workers' compensation. The employer may be able to place a lien against the third-party award to ensure repayment of workers' compensation benefits. If the third-party award is less than that obtained through workers' compensation, the employer may have to pay only the difference between the third-party award and what would have been paid by workers' compensation alone. All such adjustments would occur after payment of legal and other direct expenses for the suit. If the worker fails to win a third-party award, there is probably no loss in workers' compensation benefits.

Third-party lawsuits by injured workers are not the only means for achieving payment other than workers' compensation for job-related injuries. Under certain conditions, the employer may file suit on its own behalf or that of the employee against a third party. If the suit is on behalf of the employee, any award in excess of workers' compensation benefits and expenses necessary to bring the suit pass to the employee.

EXERCISES

1. Find out what the manual rate is for your state for
   (a) paint manufacturing
   (b) grocery store workers
   (c) roofing work
   (d) traveling carnival workers

2. For the occupations in Exercise 1, try to find out what the rates are for one or more neighboring states.

3. Obtain a copy of the workers' compensation regulations for your state and a neighboring state. Compare such factors as benefits paid for different disabilities. Compare procedures for submitting, processing, and appealing claims.

4. Discuss fairness of benefits and cost of workers' compensation premiums with
   (a) a local attorney who deals in workers' compensation
   (b) a local business executive
   (c) a workers' compensation insurance broker or agent

5. Find out what the job duties are of an engineer who is a loss control representative for an insurance company.

6. Visit a rehabilitation facility at a local hospital or clinic that helps get injured workers back on the job. Find out how they approach minimizing workers' compensation claim costs.

7. A grain elevator is considering a location for a new plant. A site is to be selected in your state or one or more neighboring states. All employees fall into two job classifications, listed in the following table with annual payroll for each classification. Find out the current manual rates in order to complete the table below.

Job Classification	Total Annual Payroll ($)	Your State	First Adjacent State	Second Adjacent State
Grain elevator operator	2,500,000	—	—	—
Truckman	850,000	—	—	—

   (a) If the company will pay manual rates for the first three years, what is the total cost of premiums over three years for each of the possible sites?
   (b) Compared with the site with the highest premium rates, how much is saved over three years at each of the other sites?

8. The company in Exercise 7 had the following experience rating for all job classifications:

Year	Rating
1st	0.92
2nd	0.87
3rd	1.21
4th	1.02

If there is no change in the manual rates over the years, what workers' compensation premiums will a company in your state pay for each of the four years after the initial policy?

REVIEW QUESTIONS

1. What are the three common law defenses that protect employers from legal claims for compensation resulting from on-the-job injuries?
2. When were constitutionally acceptable workers' compensation laws first passed in the United States?
3. What agreement was reached between employers and employees under the no-fault concept of workers' compensation?
4. What are the two types of workers' compensation laws?
5. How many workers' compensation laws are there in the United States?
6. What was the original term for workers' compensation?
7. What employees are often exempt from workers' compensation benefits?
8. What injuries does workers' compensation typically cover?
9. What are the four most commonly used classifications for disabilities? Define each.
10. What benefits are normally provided by workers' compensation?
11. What are schedule payments?
12. Describe theories used to establish payments for impairments.
13. Name seven methods for establishing workers' compensation premiums. Briefly explain each.
14. How can an employer reduce workers' compensation claims?
15. How can an employer reduce workers' compensation premiums?
16. What is a third-party lawsuit? How can it result from a workers' compensation case?
17. How is NAICS used in pricing workers' compensation premiums?

BIBLIOGRAPHY

Analysis of Workers' Compensation Laws, The United States Chamber of Commerce, Washington, DC, annual.

CHEIT, E. F., *Injury and Recovery in the Course of Employment*, Wiley, New York, 1961.

HANES, D. G., *The First British Workmen's Compensation Act*, Yale University Press, New Haven, CT, 1968.

MARTIN, R. A., *Occupational Disability*, Charles C. Thomas, Springfield, IL, 1975.

Right Off the Docket, Penton Educational Division, Penton Publishing Inc., Cleveland, OH, 1986.

Supplemental Studies for the National Commission on State Workmen's Compensation Laws, Washington, DC, 1973 (three volumes).

CHAPTER 7

PRODUCT LIABILITY

7-1 INTRODUCTION

Industrial, commercial, and consumer products are a significant source of injuries and death. Injured parties frequently sue manufacturers and those in the distribution chain for compensation. Estimates of the number of product liability lawsuits in courts throughout the United States range from 100,000 to 1,000,000 each year. Over the last few decades, there has been a major increase in product liability lawsuits. Along with this increase in the number of suits, there were many changes in product liability laws and legal interpretations of them. There is growing pressure for many forms of liability reform to reduce the legal burden on business in the United States.

Product liability litigation is one means for society to cope with the technological risks imposed on it. Not all product liability litigation is initiated for this reason. Decisions and actions of engineers, managers, and others during planning, design, manufacturing, distribution, and marketing of products can impact their safety. Because of this, engineers need to know the fundamentals of product liability. Knowledge of the legal concepts and processes for seeking remedies is important for engineers so they can act prudently, professionally, and ethically at an early stage to keep unnecessary risks associated with products out of the marketplace.

7-2 THEORIES OF LIABILITY

A manufacturer or seller of a product is not liable for all injuries that may result from a product. That would be absolute liability. However, in most states, three theories of liability apply to products and establish the duties of a manufacturer or seller toward a user or consumer. The three theories are (1) warranty, (2) negligence, and (3) strict liability. Warranty addresses performance of a product regarding implied or explicit claims made for it by the manufacturer or seller. Negligence involves the conduct or behavior of a person or corporate body regarding something they did or failed to do. Strict liability deals with characteristics of products that are unreasonably dangerous. More than one theory may apply in a legal case.

The theories of negligence and strict liability are part of tort law. Torts are wrongful acts, injuries, or damages for which civil (as opposed to criminal) action can be brought. Warranty is part of contract law and the relationships between buyers and sellers.

Product liability developed from English common law. As the industrial revolution of the late 1800s placed new products on the market, the social and legal climate at that time gave them an esteemed position. The legal concept was *caveat emptor*—let the buyer

Safety and Health for Engineers, Second Edition, by Roger L. Brauer
Copyright © 2006 John Wiley & Sons, Inc.

beware. Complaints about a product usually were virtually ignored. The law held that a buyer was negligent for not examining a product for defects at the time of purchase.

A manufacturer was further protected by "privity of contract," or the doctrine of privity. It limits the parties involved in a negligence case to those directly involved in a transaction—the buyer and seller. As long as a manufacturer was not part of the direct selling of its product, there was no need for concern over suits from buyers. There was little need to worry about defective and unsafe products. In 1916, the decision in *MacPherson v. Buick Motor Company*[1] ended the privity doctrine for negligence cases and opened the door to changes in product liability law. The court ruled negligence occurred on the part of a remote (from the sales transaction) manufacturer of an automobile for a defectively made wheel that broke and injured the plaintiff. The court's opinion noted: "Without regard to a contract between buyer and seller and when a buyer is not likely to check a product for defects, the manufacturer of a thing of danger has a duty to make it carefully."

Similarly, a 1960 decision removed the doctrine of privity as a barrier in implied warranty cases.[2] The court held that a buyer is not capable of determining the fitness of an automobile for use. It also recognized that under modern market conditions, a manufacturer who places a product on the market and promotes its sale becomes a party to the sale through implied warranty.

In 1962, the theory of strict liability emerged. It removed the need to show breach of express warranty on the part of a plaintiff.[3] The court ruled: "A manufacturer is strictly liable in tort when an article he places on the market, knowing that it is to be used without inspection for defects, proves to have a defect that causes injury to a human being." In 1965, the American Law Institute published the Second Restatement of Torts (Section 402A). Most courts accept it as the rules for strict tort liability.

As a result of the changes in liability law, approximately 95% of all liability suits are now handled under the theory of strict liability. With these shifts in the law, society has recognized that users and consumers should receive compensation in many cases for injuries resulting from defective products. The legal pendulum has swung from manufacturers, who had been virtually immune from liability, toward users and consumers. Adjustments in product liability continue as the courts determine if the pendulum has swung too far in favor of product users or not far enough. More recently, the use of negligence has increased and there is a growing effort to limit liability and to minimize frivolous product liability suits.

7-3   PRODUCT LIABILITY EVIDENCE

The plaintiff in a product liability lawsuit must bring certain evidence in support of his claim. Except in expressed warranty cases, the plaintiff must prove

1. that the product was defective
2. that the defect existed at the time it left the defendant's hands
3. that the defect caused the injury or harm and was proximate to the injury

In strict liability, cases no other evidence is required to establish the basis for a case. However, under negligence, additional evidence is needed. The plaintiff must show that the defendant was negligent in some duty toward the plaintiff. In warranty cases, the plaintiff must merely show that a product failed to meet implied or expressed warranty or represented claims for the product.

The defendant may use a number of defenses for the three kinds of evidence. The questions surrounding the existence of a defect in a product can be complex. The defendant may try to show that although the product is dangerous, the danger by itself is not a defect. The defendant may try to show that the plaintiff altered the product or unreasonably misused it. The defendant may claim that the product met accepted standards of government, industry, or self-imposed standards related to the product, to the claimed defects, and to the use of the product. In addition, the defendant may try to show that the product did not cause the injury or was not the proximal cause.

7-4 NEGLIGENCE

Besides the three elements of evidence just noted, a plaintiff acting in a negligence case must show that the defendant had a duty toward the plaintiff in providing a product free of the claimed defect and was negligent in performing that duty. Negligence includes acts of omission (failure to act) or commission (performing an act). Because negligence has to do with the behavior of an individual or organization, it is often very difficult for the plaintiff to gather sufficient information about the behavior of the defendant to prove negligence. It would be difficult, for example, to show what decisions a defendant made in the process of designing a product. It may be hard to find out how or why they were made. Such records may not exist. Similarly, without the defendant's records it would be difficult to portray a quality control program in manufacturing that was not being implemented according to policy and standards for the batch containing the injury-causing product. Through discovery procedures, a plaintiff can seek to obtain such information about the defendant if it exists. A plaintiff may attempt to demonstrate that a manufacturer did not use technology available at the time the product was made.

A defendant may claim that he had no duty toward the plaintiff or that the duty was performed without negligence. The defendant may argue that he met government, industry, consensus, or even self-imposed standards and standards of professional practice applicable to the product or the defect. The defendant may try to show that the plaintiff was negligent in the use of the product (contributory negligence), which led to the injury. The defendant may also try to show that the plaintiff was fully aware of the defect and voluntarily accepted the risks associated with the defect in using the product.

In judging behavior on the part of a defendant or plaintiff, actions are compared with the "reasonable person." Negligent conduct occurs only when an act is less than that which a reasonable person would have performed under similar circumstances. Creating the reasonable person standard opens the door for many legal arguments. Included are arguments about the probability of preventing harm, the likelihood that injury will occur, how serious a resulting injury would be, and the cost of preventing injury from occurring.

7-5 WARRANTY

There are two types of warranty: implied and express. Through the Uniform Commercial Code, adopted by nearly all states, the user or consumer of a product receives some guarantee regarding the quality of a product. This is implied warranty. Implied warranty is divided into (1) merchantability and (2) fitness for a particular purpose. Merchantability means that a product is fit for the ordinary purposes for which such goods are used. Merchantability applies only to the sellers who normally deal in particular goods. Buyers assume that such sellers have knowledge about the products they sell. Buyers do not expect the same kind of expertise about a product with a one-time seller.

The other type of implied warranty is fitness for a particular purpose. Before purchasing a product, a buyer may wish to know whether a product will perform for a particular application, not just in general. The buyer may ask the seller for advice, a recommendation, or to select a suitable product. If the product purchased on the basis of the seller's assistance does not perform, the implied warranty of fitness for a particular purpose is breached.

Implied warranty is a branch of contract law rather than a tort. If injury results to the buyer from the intended use of the product, the buyer can act against the seller. The buyer and members of the buyer's household are the only persons who can bring a case against the seller. However, the buyer cannot act against the producer of the product under this theory.

Express warranty occurs when a seller makes expressed claims or representations for a product that become a basis for the bargain. The plaintiff must establish only that the product failed to meet the seller's warranty or representations and that an injury resulted from the failure. The plaintiff does not have to prove that a defect or unreasonable danger existed in the product.

Advertising frequently creates express warranty. Overselling a product and making claims for characteristics it does not have can lead to product liability lawsuits. In an early case of this nature, the purchaser of a new automobile relied on the manufacturer's claim that the windshield was shatterproof.[4] While driving the car, a stone struck the windshield and a fragment of the glass lodged in the plaintiff's eye, causing injury. The plaintiff received compensation in the case. The court ruled:

[It would] be unjust . . . to permit manufacturers . . . to create a demand for their products by representing that they possess qualities which they, in fact, do not possess, and then, because there is no privity of contract existing between the consumer and the manufacturer, deny the consumer the right to recover if damages result from the absence of those qualities when such absence is not readily noticeable.

One problem associated with express warranty is trying to differentiate actual misrepresentations from overstatements of a product's qualities (called puffing) that buyers typically expect salespeople to make. In express warranty cases, a jury must decide if there is misrepresentation.

7-6 STRICT LIABILITY

Negligence is difficult to prove. Warranty often restricts the parties involved in a case to buyer and seller. As a result, the theory of strict liability emerged in the early 1960s. Operating under the Second Restatement of Torts, Section 402A,[5] a plaintiff in a strict liability lawsuit does not have to prove negligence. The behavior of the defendant is irrelevant. The defendant cannot show how well his quality control or product safety program was operated to prevent defects. Neither must breach of warranty be proven. Strict liability focuses on the qualities of the product that caused injury. The plaintiff must present the three fundamental elements of evidence:

1. that the product was defective
2. that the defect existed at the time it left the defendant's hands
3. that the defect caused the injury or harm and was proximate to the injury

7-7   DEFECTS

Defects in a product may arise from design, from manufacturing, or from inadequate warnings and instructions. Defects are conditions that are not compensated by the ultimate consumer and that are unreasonably dangerous to him or her.

Design Defects

Design defects are unreasonably dangerous characteristics of a product resulting from decisions, calculations, drawings, or specification of the design process. Design defects occur in all products of a particular make or model.

There are many factors in design from which defects may result. One factor is selection of materials. For a particular product, selection of materials is based on such considerations as cost, durability, function, maintenance, appearance, and strength. In one case involving selection of materials, the use of soft pine that was not acceptable for ladders according to a consensus standard resulted in a plaintiff winning a negligence case.[6]

Another design factor involves management of energy. A baseball pitching machine depended on a spring to energize the arm and cause it to throw a ball. Even when the machine was unplugged, the spring could be storing energy that could be released suddenly. A boy's face was injured by such a machine. He recovered damages when vibration caused the catch holding the spring and arm to release.[7]

Providing functional features in a product is another important factor in design. Reasonable safety in arrangement of features is needed. For example, an outdoor lounge was designed to adjust to different positions. However, the court found it to be unreasonably dangerous when a plaintiff severed a finger in the part of the chair's arm that moved for adjustment.[8]

A design must include safety features. The court found the design of an earth-moving machine defective because it did not have a rearview mirror as a safety feature. A mirror would allow the driver to see a blind area behind the machine when backing up. A worker, standing in the blind zone, was injured and recovered damages from the manufacturer when the machine backed over him.[9]

An important factor to consider in design is the use environment. Use environment refers to the context in which a product is used. What may otherwise seem safe could become unreasonably dangerous when one understands the physical, social, and behavioral context for the product's use. For example, it is likely that a storm door will face the impact of a rolled-up newspaper thrown by a delivery boy. The use environment includes such behavior. Another example is the load a kitchen drawer must withstand when a child uses it as a step to climb to the countertop.

In product design, it is important to comply with government and consensus standards. Lack of compliance may prove that a design defect exists. Standards are minimums. Even complying with them will not ensure that a design is adequate. The best protection is designing out the hazard. One should note that standards may go beyond published standards; they can include standards of practice. Standards of practice may be principles or practices appearing in textbooks or taught in courses or practices typically used in a discipline or a company.

Besides complying with standards, it is important for designers to stay abreast of technology, even that outside their specialty field. Failure to use available technology in a design may place unnecessary liability on a product.

Manufacturing Defects

Manufacturing defects occur in a limited number of products of the same make. A manufacturing defect in a product can be identified easily by comparing a good product from the same manufacturer with the defective one. Manufacturing defects usually result from inadequate quality control, testing, and inspection or from errors in assembly. One example of a manufacturing defect is a poor weld that fails at a later time.

The legal doctrine of *res ipsa loquitur*—the thing speaks for itself—frequently applies to negligence cases involving manufacturing defects. Classic cases are exploding soft drink bottles or food products containing foreign material, such as metal or glass.

Defects in Instructions and Warnings

A product may meet all necessary standards of design and contain no production flaws, yet it may be unreasonably dangerous, because instructions for use or warnings about dangers during use or misuse are inadequate or absent. Under both the theories of negligence and strict liability, a supplier has a duty to warn of dangers that remain in a product or occur during its use. See Chapter 35 for a discussion of some standards requiring risk analysis, hazard reduction, and protection for hazards that remain.

One must make a clear distinction between instructions (or directions) and warnings. Warnings identify dangers inherent to the product or dangers that may result from its use or misuse. Instructions explain how to use a product effectively or safely. Instructions explain what actions one must take to eliminate or reduce the likelihood of injury from a product's dangers.

Instructions and warnings must have many characteristics that are based on good writing skills, knowledge of use environments, ergonomic principles, and other factors. Table 7-1 lists 15 important characteristics of warnings.

A common error in writing instructions is representing them as descriptions of what a product does, not as imperative statements or what steps must be followed and in what order. A review of warnings by legal experts, human factors specialists, users, and others may be helpful in making them effective. Also important is the education, reading skills, and ability of the ultimate user and the language of the warning or instructions. Warnings and labels also are discussed with several other topics.

7-8 MISUSE AND FORESEEABILITY

In some product liability cases, the supplier of a product may be liable even when a product is used for some purpose or in some manner other than intended. In cases of misuse, the courts use a test of "foreseeability." This test determines whether a misuse reasonably could have been anticipated on the part of the supplier. A classic case involving foreseeability is that of a child standing on the open door of a kitchen range to reach something in the cupboard and having the range tip over on him. A manufacturer must allow for abuses and misapplication of a product and minimize the liability by designing the product for or providing warnings and instructions that address foreseeable misuses.

7-9 MODIFICATIONS AND SUBSTANTIAL CHANGE

A defect must have existed at the time the product left the defendant for liability to exist. Sometimes a user or owner modifies or alters a product in some way during its life. A sup-

TABLE 7-1   Characteristics of Warnings

READABILITY. The ability to read or receive a message. Multiple languages, pictorials or symbolics, and braille are all methods to ensure that a message is received.

UNDERSTANDING. The ability to understand individual components of a message. Some words are beyond the vocabulary of certain readers. Not all symbols are recognized or understood by every viewer.

COMPREHENSIBILITY. The ability to understand the overall message. Messages must be simply stated, must require little technical or specialized knowledge, and must be precise.

PRACTICALITY. The ability to heed or comply with a warning in light of behavior that is normally expected or given a normal context for the warning.

EFFECTIVENESS. Having valid and reliable test data to establish whether a warning does, in fact, communicate its message and is not just assumed to do so by its writer or designer.

BEHAVIOR MODIFICATION. Achieving the behavior desired by the warning, that is, preventing unsafe or injury-causing acts that might otherwise occur.

COMPATIBILITY. Suitable for and consistent with expectations of individual applications. Warnings should agree with local customs and practices, should be consistent in similar situations (standardized), should meet requirements of consensus and local standards, and should be appropriate for a particular application situation.

CONSPICUOUS. Provide a reasonable certainty of perception, without search and in a short time. This characteristic includes size, color contrast, stimulus novelty, brightness level, and other characteristics.

DURABILITY. The ability to resist environmental conditions, such as abrasion, wear, wetness, chemicals, sunlight, and so forth.

RELIABILITY. Must be present when needed. This property is particularly applicable to visual and audio warning devices that must act when a danger is present.

REINFORCEMENT. Giving people additional or more detailed data about a warning or its importance through training sessions, operating manuals or other means. The goal is to influence the receiver's sensitivity toward the warning.

DANGER SIGNAL. Attention-getting enhancements, such as underlined or boxed text, bright colors, signal words like danger or warning, special auditory tones, and so forth.

PLACEMENT. Locating warnings where they are likely to be seen or heard and where the danger is; proximity in distance and time.

NOVELTY. Use of attention-getting features like animation, voice synthesized messages, color, and so forth.

TYPE. Classification of purpose or function. For example, one might classify a warning as (a) advisory, (b) explaining what to do, (c) reminder, and so forth.

Derived from G. A. Peters, "15 Cardinal Principles to Ensure Effectiveness of Warning Systems," *Occupational Health and Safety*, May:76–79 (1984).

plier is responsible for those risks that he introduced. He may be liable for some modifications introduced by a user, but generally, the one who modifies a product is liable for modifications. Failure to include an important feature, which then necessitates a user modification, may shift the liability to a manufacturer.

7-10   STATUTE OF LIMITATIONS

Another problem for a manufacturer is that of expected product life and its role in liability. Many states have statutes of limitations that limit the period during which product liability claims can be filed. The time allowed under statutes of limitations varies considerably, but usually involves a fixed number of years from the date of sale or a time limit for claim after injury. A typical design problem is whether the product and its components will fail within the statute of limitations period and whether the failure may lead to injury.

7-11   THE LAWSUIT PROCESS

The procedures for a liability suit involve three main steps: complaint, discovery, and trial. Variations from this simplified model occur in particular cases. Within each step, a number of activities may occur and the entire process can end at any point. A number of factors can impact conclusion of a case. A defendant may find that a plaintiff has a good case. Parties may want to avoid legal costs and reach a settlement. A defendant may petition the judge overseeing the case for a summary judgment that removes the defendant from the case. A defendant may not want the arguments to become general knowledge through case law.

Complaint

In the first step, the attorney for the plaintiff files a complaint with the court that has jurisdiction. Before filing a complaint, significant investigation may be needed to establish that a lawsuit has a reasonable chance of success. After the defendants receive a copy of the complaint, defense attorneys usually deny the accusations. In suits naming several defendants, each defendant may file a petition stating why they should not be named in the suit. One defendant may bring additional defendants into the case by filing additional complaints against the additional parties.

There are several reasons for naming a person or organization as a defendant in a complaint: the potential defendants have a duty toward the plaintiff and may have a role in a defect causing injury to the plaintiff. Another consideration is the ability of the defendant to pay damages. A defendant with the capability (through assets or insurance) to pay is commonly called a deep pocket.

Discovery

In the discovery step, the plaintiff sends written interrogatories to the defendant, who may have to answer them in a certain number of days. The defendant may not have to answer them if they are unreasonable or cause unreasonable expense to prepare an answer.

Based on the complaint and written interrogatories, each party begins to develop its case by identifying witnesses who will testify in the case. Each party may question the opponent's witnesses under oath in discovery depositions. A legal reporter makes a record of the questions and answers. The plaintiff and others who may have witnessed the injury events are deposed. Expert witnesses—persons with specialized knowledge, like doctors, engineers, and others—may be deposed about their knowledge and opinions of the facts in the case. Each side develops a sense of whether they can win the case. If both believe they have solid arguments, the process continues into the trial step. If there is a good case but the issue revolves around which parties must pay, the case may also continue. If the plaintiff has a weak case, one or more defendants may petition the court for dismissal.

Trial

In a jury trial, each side presents its arguments. Witnesses are questioned once again under oath. Not all witnesses in the deposition step may appear at trial. An attorney may question a witness about statements made during a deposition. After each side completes arguments, the jury must decide whether the plaintiff should receive compensation and how much to award. If the case involves the theory of comparative negligence (allowed in some states), the jury must decide the portion of negligence attributable to each party and apportion the total award accordingly. For example, a manufacturer might be assigned 20% of the total dollar value of an award, a user 50%, and the employer 30%.

At any time before a case goes to the jury, the parties may negotiate a settlement. If a settlement is reached before the case goes to a jury for a decision, the evidence presented does not go on the court record. As soon as the jury is given the case for decision, the evidence presented is public record. Similar cases by others may use information in the court records.

7-12   EXPERT WITNESSES

If the facts in a legal case involve specialized and technical subject areas, expert witnesses may testify in the case. In product liability cases, engineers often are needed to testify about a product, existence of defects, use of the product, design alternatives, negligence, compliance with published standards or standards of practice, the state of the art, and other matters. A case may require the expertise of engineers, safety professionals, and other specialists.

Besides giving testimony, an engineer may serve other functions in a product liability case. An engineer may help the attorney understand the technology involved in the case; may help establish whether a defect existed through testing and evaluation of products, literature searches, or other means; may help reconstruct the incident and help the attorney prepare interrogatories; and may locate standards, gather facts, and perform tests.

Before an engineer serves as an expert witness, the attorney doing the hiring will determine whether the potential witness is qualified in the area of specialization needed. The attorney will examine the candidate's training, experience, and professional credentials. Later, in depositions or at trial, the opponents may challenge the qualifications of the expert to testify on the subject matter in question.

Ultimately, the attorney will seek the technical opinions of the expert on issues in the case. Often sought are opinions "with a reasonable degree of scientific and engineering certainty." In a legal sense, this infers a certainty of 51% or more. The question is whether the expert is more sure than not sure on an issue. It is not to be confused with certainty in a statistical sense, where one uses a 95% or similar confidence level in drawing inferences or conclusions from data.

7-13   REDUCING LIABILITY RISKS

There are risks in any product. A manufacturer or seller of a product must face those risks in putting a product on the market. A manufacturer or seller cannot prevent a user from initiating a lawsuit after being injured by a product. However, liability does not mean absolute liability. A manufacturer or seller can minimize liability in a number of ways. Attorneys will defend a manufacturer in the courts. Engineers can prevent many lawsuits by defending the manufacturer in design, manufacturing, packaging, and the marketplace.

For product liability, the primary role of an engineer is to remove unreasonable dangers from products and environments and to prevent defects from reaching the marketplace. Products with few defects will produce few product injuries and even fewer liability claims. Engineers must account for the use environment, foreseeable misuses, product life, possible product modifications, hazards, potential injury, seriousness of injury, compliance with standards (as a minimum), state-of-the-art practices, quality control, packaging and handling, advertising, and claims for products. They must face concerns like cost, function, maintenance, maintainability, and durability of a product. Engineers must see that warnings identify remaining hazards and instructions necessary for user protection. There are detailed programs and guides for managing these items in a systematic way.

A good technique for reducing hazards in a product is thorough design review. A review team not involved in the design, and thus independent and with limited bias, can analyze a product for hazards and acceptable controls. The team may include engineers, attorneys, safety professionals, and others. The collective knowledge and experience of the team can provide a broad foundation of experience and expertise. The review team may work closely with the designers throughout the design process, rather than coming in after a design is completed. Sometimes this review team is called an audit team, particularly when the team is reviewing for compliance with laws, regulations, standards, and practices.

EXERCISES

1. **(a)** Select a product. Identify its primary use.
   (b) Try to identify possible use environments for it.
   (c) Try to identify foreseeable misuses and the hazards involved.
   (d) Evaluate the product for product safety. Consider alternatives for design and manufacture that would reduce or eliminate its hazards.
   (e) Compare design alternatives in terms of risk, cost, function, product life, and other factors.
   (f) Prepare a set of instruction for use of the product.
   (g) Prepare a set of warnings for the product and its hazards and draft instruction for its safe assembly, installation or use.

2. **(a)** Obtain the warnings and instructions accompanying some product. Identify uses and misuses for the product.
   (b) Determine whether the warnings and instructions adequately identify the risks for a user and whether instructions adequately tell users how to protect themselves from the risks.

3. Arrange with an attorney working on an actual product liability case or a law school holding mock proceedings to monitor the deposition of an expert witness or the conduct of a trial.

REVIEW QUESTIONS

1. What are the three theories of product liability?
2. Explain major differences among the three theories.
3. What evidence must the plaintiff provide for each of the three theories?
4. Under which theory are most product liability lawsuits argued today?

5. What is absolute liability?

6. What is privity of contract or the privity doctrine?

7. What is contributory negligence?

8. What is a defect?

9. What are the three types of defects? Give an example of each.

10. What is the difference between warnings and instructions?

11. Name at least five characteristics of warnings.

12. What is the doctrine of proximate cause?

13. What is *res ipsa loquitur*?

14. What does *caveat emptor* mean?

15. What is the statute of limitations?

16. Explain the role of an engineer as an expert witness.

17. How can engineers reduce liability for a product?

18. What is the reasonable person test?

19. What is merchantability?

20. What is comparative negligence?

21. Explain the difference between implied and express warranty.

22. What does "reasonable scientific and engineering certainty" mean?

NOTES

1 217 New York 382, 111 Northeastern 1050 (1916).

2 *Henningsen v. Bloomfield Motors*, 32 New Jersey 358, 161 Atlantic 2d 69 (1960).

3 *Greeman v. Yuba Power Products, Inc.*, 59 California 2d 57, 27 California Reporter 697, 377 Pacific 2d 897 (1962).

4 168 Washington 456, 12 Pacific 2d 409 (1932).

5 *Restatement (Second) of the Law: Torts*, American Law Institute, St. Paul, MN, 1965.

6 *Wilson v. Loe's Asheboro Hardware, Inc.*, 259 North Carolina 660, 131 Southeastern 2d 501 (1963).

7 *Dudley Sports Co. v. Schmitt*, 279 Northeastern 2d 266 (Indiana App.) (1972).

8 *Mathews v. Lawnlite Co.*, 88 Southern 2d 299 (Florida) (1956).

9 *Pike v. Frank G. Hough Co.*, 2 California 3d 465, 85 California Reporter 629, 467 Pacific 2d 229 (1970).

BIBLIOGRAPHY

BASS, L., *Products Liability: Design and Manufacturing Defects*, McGraw-Hill, New York, 1986.

BRESNAHAN, THOMAS F., LHOTKA, DONALD C., and WINCHELL, HARRY, *The Sign Maze—Approaches to the Development of Signs, Labels, Markings and Instruction Manuals*, American Society of Safety Engineers, Des Plaines, IL, 2000.

CASTRO, CANDIDA, and HORBERRY, TIM, *The Human Factors of Transport Signs*, CRC Press, Boca Raton, FL, 2004.

GOODDEN, RANDALL L., *Product Liability Prevention: A Strategy Guide*, American Society for Quality, Milwaukee, WI, 2000.

GRAY, I., *Product Liability—A Management Response*, AMACOM (Division of American Management Associations), New York, 1975.

HAMMER, WILLIE, *Product Safety Management and Engineering*, 2nd Ed., American Society of Safety Engineers, Des Plaines, IL, 2001.

Handbook and Standard for Manufacturing Safer Consumer

Products, U.S. Consumer Product Safety Commission, Washington, DC, June 1975.

KOLB, J., and ROSS, S. S., *Product Safety and Liability*, McGraw-Hill, New York, 1974.

LAUGHERY, KENNETH R., Sr., WOGALTER, MICHAEL S., and YOUNG, STEPHEN L., *Human Factors Perspectives on Warnings: Selections from Human Factors and Ergonomics Society Annual Meeting Proceedings, 1980–1993*, Human Factors and Ergonomics Society, Santa Monica, CA, 1994.

ROSEN, STEPHEN I., *The Duty to Warn Handbook*, Hanrow Press, Rancho Santa Fe, CA, 1996.

PETERS, G. A., *Product Liability and Safety*, Coiner Publications, Ltd., Washington, DC, 1971.

SEIDEN, R. M., *Product Safety Engineering for Managers*, Prentice-Hall, Englewood Cliffs, NJ, 1984.

SHERMAN, P., *Products Liability*, McGraw-Hill, New York, 1981.

SCHOFF, GRETCHEN HOLSTEIN, and ROBINSON, PATRICIA A., *Writing and Designing Manuals*, 2nd Ed., Lewis Publishers, Boca Raton, FL, 1991.

WEINSTEIN, A. S., TWERSKI, A. D., PIEHLER, H. R., and DONAHER, W. A., *Products Liability and the Reasonably Safe Product*, Wiley-Interscience, New York, 1978.

CHAPTER 8

RECORD KEEPING AND REPORTING

8-1 WHY KEEP RECORDS AND FILE REPORTS?

Very early in life people discover that a good way to learn is through experience. This idea is carried into safety and health. Understanding what happened in an incident and why it occurred can lead to preventive actions in a similar situation. The idea of developing lessons learned from incidents that have happened and using those ideas for preventive actions in the future is depicted in Figure 3-3. After an incident occurs, it is investigated and data are compiled in a report. Data from the report, and possibly related ones, are analyzed. Preventive actions are taken so that future incidents of the same type will not occur. The idea of learning from past events and making changes is a reactive approach.

Learning from incident experience is one reason for compiling records and reports. Making use of that process and information derived from it is a management function. The process in Figure 3-3 is discussed in the last part of this book, along with other techniques. The fact that laws and regulations require record keeping and reporting is another major reason for such activity. That is why this topic is included at this point in the book.

Beside being required by the law and providing a basis for correcting safety problems, there are many other reasons for maintaining records and reports about incidents and other safety and health matters. Records and reports often are needed to protect the legal rights of employers and employees. Records and reports form the basis for measuring safety performance. They can help identify hazards, they are used to establish or adjust insurance rates, and they may be used to assign legal penalties.

Requirements of the Law

Federal, state, and local governments require that certain safety records be maintained and that certain reports be submitted. For example, employers must keep records of job-related incidents. Automobile drivers must complete a police incident report after a vehicle incident. Building owners must maintain records on maintenance and inspection of elevators. In the following, some record-keeping requirements are explored in more detail.

Protecting Legal Rights

If there were no records of an on-the-job injury or illness, an employee would have no way to validate a claim for workers' compensation benefits. Records about design decisions, production, testing, and sales of products may be used by a manufacturer in defending claims in a product liability lawsuit.

Measures of Performance

Many companies have award programs based on the number of work hours completed without an incident. Without records and reports, these programs would be impossible. Statistics based on data compiled from records can be used by managers to develop quantitative indicators of safety performance. A number of frequency and severity statistics are used for decision making in safety.

Making Contract Awards

For contract work, some government organizations and private companies require examination of bidders' safety records and safety plans. Safety performance and plans are one factor in deciding which bidder will receive a contract.

Hazard Recognition

By collecting data on incidents and studying them, one can often establish that particular hazards are involved. Knowing what contributing factors and hazards are recurring provides the basis for specific corrective actions.

Corrective Actions

Corrective actions can be implemented only by those in authority. Therefore, information from records must be provided to those in charge. By communicating to those in authority through reports in units of measure that they understand, appropriate actions can be initiated. Data from incident records and reports may be used to make decisions about evacuations in an emergency. Reports may form the basis for budget requests or a manager's performance rating. Incident data can be used in safety promotion programs.

Managing Safety

Information from safety and incident records and reports may form the basis for ranking problems, ranking corrective actions, and assigning limited resources to achieve the greatest risk reduction.

Insurance Rates

In Chapter 6, the methods by which premium rates are set for workers' compensation insurance were discussed. In several methods, the cost of claims made against a policy are used to determine future rates. Insurance rates for liability or other protection also are based on claims. Incident records and reports are often used to establish or record the value of a loss from injury or damage.

Legal Penalties

The severity of legal penalties, fines, or terms of imprisonment are sometimes based on record keeping, the lack of it, or data contained in records. Recurring injuries may form the basis for claims of negligence or willful wrongdoing. Failure to maintain Occupational Safety and Health Administration (OSHA) records properly has resulted in companies being fined more than one million dollars at a single facility.

8-2 KINDS OF RECORDS AND REPORTS

There are many kinds of records and reports that safety professionals must complete. The task of record keeping for safety professionals grew exponentially for a time. Employers must keep records and file reports on incidents. They must also keep records on training, exposures, issue of safety equipment, conditions, and tests of certain kinds of equipment and many other health and environmental matters. The requirements are different for different kinds of businesses, operations, activities, or equipment.

A review of many of the federal requirements gives one an idea of the magnitude and complexity of the record-keeping and reporting tasks. It would be impossible to detail all current federal reporting requirements. The requirements discussed below are organized into four groups.

Accident and Incident Reporting

The major types of records and reports required by the federal government for safety and health include work-related incidents, transportation incidents, and incidents arising out of the use of radioactive materials.

Work-Related Incidents   At least three government organizations require that work-related incidents and injuries or illnesses be reported: OSHA, the Mine Safety and Health Administration (MSHA), and the Nuclear Regulatory Commission (NRC).

OSHA requires that each employer having more than 10 employees must maintain a log of recordable occupational injuries and illnesses and a summary by calendar year. In addition, a more detailed supplementary record must be made of each recordable occupational injury or illness. Data to be included on the log, summary, and supplementary record are specified on OSHA forms. These records must be available for inspection and, when requested, submitted to the Bureau of Labor Statistics (BLS). If an employee is killed on the job or if five or more employees are hospitalized or killed in an incident, the employer must report the incident to OSHA either orally or in writing within 48 hours. The OSHA record-keeping requirements are described in detail in Section 8-3. Incidents of excessive radiation exposure or release must be reported to OSHA if workers are not covered by the NRC.

The MSHA requires that each mine operator submit a report for each mine incident, injury, or illness. One form must be submitted for each injured or ill person. In addition, for purposes of computing incident and injury statistics, the total employee hours worked must be submitted quarterly.

Depending on severity, licensed operators of nuclear facilities and operations must report incidents to the NRC. Reportable incidents are excessive exposures of workers to ionizing radiation, excessive release of radioactive material, loss of operation, and property damage. The report must be made immediately (by phone or other media) within 24 hours or within 30 days, depending on the severity of the incident. Oral reports must be followed by written ones within 30 days.

The NRC also requires that organizations under its jurisdiction notify the NRC immediately in some instances. Immediate reports are required when one or more workers receives exposures exceeding the allowable quarterly dose by 20 times, when release of radioactive material 5000 times greater than allowed occurs, when loss of one working week of operation of a facility occurs, or when $200,000 property damage occurs. When worker exposures are above the quarterly standard and releases of radioactive material are 10 times greater than allowed, reports must be made within 30 days.

Transportation Incidents The federal government requires reports for a variety of transportation incidents and incidents.

Aircraft Operators of aircraft must notify the National Transportation Safety Board of all incidents and those incidents in which a flight control system malfunctions, the flight crew is injured or ill, there is an in-flight fire, or there is a structural failure of a turbine engine. The NTSB must approve moving wreckage, contents, and records.

Railroads All railroad incidents in which there is a fatality, five or more people are injured, or there is more than $150,000 in damage must be reported to the NTSB and to the Federal Railroad Administration. Railroads must also file a telephone report (with written follow-up) of any signal system failure. Each railroad must report and maintain a log of all incidents arising out of railroad operations. There are three classes:

1. rail-highway grade crossing cases
2. rail equipment cases
3. death, injury, or occupational illness

Boats and Ships For all vessels in U.S. waters and for vessels owned in the United States and operated on the high seas, a casualty or incident report must be filed with the Coast Guard when a person dies, an injury requiring medical treatment occurs, or there is $200 damage to a vessel or a vessel is lost in a boating incident. Operators must notify the Coast Guard immediately if a person dies or disappears from a vessel because of an incident.

Trucks When a trucking incident results in a death, injury requiring medical treatment, or $2,000 property damage, a motor carrier must file a telephone report (and written follow-up) with the Motor Carrier Safety Office of the Federal Highway Administration (FHA). Motor carriers also must maintain a register of incidents. In addition, state traffic laws may require reporting of incidents other than those required by the federal government.

Motor Vehicles There is no requirement for owners and drivers of motor vehicles to report incidents to the federal government. However, the FHWA strongly encourages state compliance with reporting standards so that incident data are consistent for compiling national statistics.

Pipelines Carriers who transport liquids by pipeline must report incidents to the Office of Pipeline Safety. They must report by telephone when there is an explosion or fire; 50 barrels or more are lost; there is an evaporative loss of five gallons or more per day; a death results; there is bodily harm resulting in loss of consciousness, medical treatment, or disability; or there is $1,000 property damage. The operators of pipelines carrying natural and other gases must report leaks by telephone with written follow-up when leaks cause death or injury requiring hospitalization. They must also report when they remove a pipeline segment from service, gas is ignited, or there is $5,000 property damage. They must also report smaller leaks and submit an annual report of leaks.

Hazardous Material   Certain incidents in transporting hazardous material must be reported. The incident may include loading, unloading, or temporary storage. A carrier must report immediately to the Centers for Disease Control if the incident results in the death of a person, an injury requiring hospitalization, $50,000 in damage or fire, or spill or leakage of radioactive material or etiological agents. The carrier must send a written follow-up report to the Department of Transportation.

Defects and Noncompliance with Federal Standards

Manufacturers of products, those constructing or operating certain facilities, and owners and users of certain equipment must maintain records of inspection and repair or design and testing. These regulations intend to insure that facilities and products placed into use are safe. The regulations also intend to keep equipment and products that are in use in a safe condition. Owners or designers often maintain similar records. In some cases, annual reports are required from manufacturers, whereas in other cases, manufacturers must report defects or safety problems when they are known.

Equipment in Use   The MSHA requires that companies keep records for inspection, testing, and maintenance of person-hoisting equipment, shafts, boilers, compressed air equipment, ventilation equipment, emergency escapeways and facilities, fire doors, smokers' articles, hazardous conditions, methane, roof bolt torque, and electrical equipment.

OSHA requires that employers maintain records for the maintenance and inspection of equipment such as powered platforms, cranes and derricks, fire extinguishers, forging machines, manlifts, presses, respirators, and safety valves for pressure vessels.

The NRC requires that firms constructing, owning, operating, or supplying components to regulated facilities report defects and noncompliance with regulations and licenses.

Railroads must maintain records of track inspections, operational tests, and inspections of equipment and tests involving the repair of signal equipment.

Cargo containers used in international shipping must be examined periodically for safety. The Coast Guard requires records of such examinations.

Trucking firms must keep records of inspections, repairs, and maintenance of motor vehicles for the Motor Carrier Safety Office of the FHA.

Aircraft owners must log maintenance, alterations, and rebuilding of aircraft, engines, propellers, and appliances.

Products and Facilities   The Coast Guard certifies boats and associated equipment to ensure that manufacturers comply with safety regulations and produce safe vessels.

Through a certification process, the Federal Aviation Administration (FAA) ensures that aircraft, engines, propellers, and related products and parts are safe and airworthy. Manufacturers and owners participate in the certification process.

Every manufacturer of motor vehicles or items of equipment for motor vehicles must report defects related to motor vehicle safety and noncompliance with Federal Motor Vehicle Safety Standards. After the National Highway Traffic Safety Administration (NHTSA) receives notice of such safety problems, the manufacturer must implement a program to remedy the defects and report progress quarterly in the implementation of that program.

The Consumer Products Safety Commission (CPSC) requires that manufacturers, importers, distributors, and retailers must report product defects that could create a substantial hazard. They are also required to report failure to comply with the CPSC standards or bans of products.

For manufacturers of electronic products that emit radiation (ionizing or nonionizing), the Food and Drug Administration (FDA) requires the reporting of information about such products and data about their design, quality control, and testing. Manufacturers also must maintain quality control and test records for those products, distribution data, and sales data about purchasers. The FDA requires reports of suspected accidental radiation occurrences, defects in products, and failure to comply with federal standards, together with plans to repair, repurchase, or replace such products. The FDA depends on voluntary reporting of ingredients in cosmetics and of unusual experiences with cosmetic products.

A number of federal agencies ensure that facilities meet safety standards by reviewing and approving designs. For example, the FAA must approve the construction, modification, or abandonment of any airport. The NRC closely monitors the planning, design, and construction of licensed nuclear facilities. The FHA uses an approval process for highways funded with federal money.

Hazardous Materials

Regulations of several federal agencies require records and reports about hazardous materials. For example, a manufacturer of explosives must report all sales of explosives to the Bureau of Alcohol, Tobacco and Fire Arms. The report must identify the quantity, date, and other data regarding each transaction.

Under the regulations issued by the Environmental Protection Agency (EPA) in response to the Resource Conservation and Recovery Act of 1976, a manifest system helps manage hazardous waste materials. Generators of hazardous (ignitable, corrosive, reactive, or toxic) waste must prepare manifests for all hazardous material that they dispose. The manifest moves with the material during transport, treatment, storage, or disposal. To track the materials, copies of the manifest are filed with state or federal EPA offices, or both, at each point in the disposal process. The Toxic Substances Control Act of 1976 requires manufacturers of potentially toxic substances to notify the EPA about such substances.

The NRC similarly keeps track of all fissile material. Various organizations participate in creating or managing records of packaging, transport, transfer, and disposal of licensed material. In addition, the NRC requires records of inspections and tests of materials, facilities for use or storage, radiation monitoring, and other equipment. There are also requirements for security records for the transport, storage, and use of fissile material. Reports of lost, unaccounted for, or stolen material must be filed with the NRC.

Other Records and Reports

The federal government requires many other records and reports to ensure the safety of workers and the public. The FDA, for example, requires that manufacturers involved in the preparation, compounding, assembly, or processing of medical devices for human use must register with the agency.

OSHA requires employers to keep records of workers' exposures to asbestos, ionizing radiation, noise, and hazardous chemicals. Under right-to-know regulations, employers must maintain information about the hazards of materials and substances. Workers with particular exposures (such as asbestos workers) must undergo periodic medical examinations. OSHA also requires that employers keep an inventory of Class I flammable liquids and records regarding the issuance, inspection, and maintenance of respirators. Employers must have records of safety training that employees have completed.

The MSHA requires that employers submit a training plan and records of training of miners regarding hazard recognition, emergency procedures, safety rules, and the use of safety and rescue equipment. Other regulations require records of exposure of miners to radon daughters, dusts, and noise and the submission of plans for mine ventilation, escape and evacuation, and roof control.

Organizations involved with fissile materials must keep records of exposure of workers to radiation and report data periodically to the NRC. Data must be available to workers.

Truck drivers must complete daily logs and submit them to their employers, who retain them under rules of the Motor Carrier Safety Office.

Under provision of the NHTSA, manufacturers of tires must maintain a list of first purchasers so customers can be notified of recalls. Similarly, motor vehicle manufacturers must maintain a list of registered owners and must compile complaints, reports, and other records concerning motor vehicle malfunctions.

This list goes on, not only for federal agencies, but also for state and local governments, insurance companies, and good safety management within individual companies. Record keeping and reporting are an essential part of safety regulations and safety programs.

8-3   OSHA METHOD FOR INJURY AND ILLNESS RECORD KEEPING

Although there are many kinds of incident and injury record-keeping requirements and forms, the most commonly known system is that of OSHA and the BLS. It provides an example of incident record keeping. Details of the OSHA record keeping requirements appear in 29 CFR 1904.

The OSHA system requires that employers keep an injury and illness log (OSHA Form 300). The log must be available for OSHA inspectors. Employers must submit summary data annually on OSHA Form 300-A. The log must be retained for 5 years. OSHA also requires a supplemental record for each recordable case (OSHA Form 301 or equivalent).

Recordable Cases

Not every occupational injury and illness is reported, only those that are "recordable." Recordable cases include every occupational death, every occupational illness, and every occupational injury involving days away from work, restricted work or transfer to another job, medical treatment beyond first aid, loss of consciousness, or a significant injury or illness diagnosed by a physician or other licensed health care professional. In addition, work-related cases involving cancer, chronic irreversible disease, a fracture or cracked bone, or a punctured eardrum are recordable. Also recordable are needlestick injuries, cuts from a sharp object that are contaminated with another person's blood or potentially other infectious materials, tuberculosis infection after exposure to a known case of active tuber-

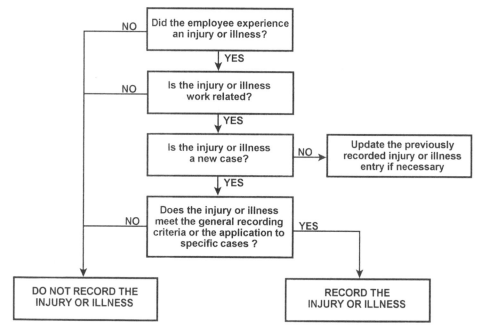

Figure 8-1. Decision chart for OSHA recordable cases.

culosis, musculoskeletal disorders (MSDs), and certain other cases. Figure 8-1 is a decision chart for determining whether a case is recordable.

Occupational Injury

An occupational injury is any wound or damage to the body resulting from an event in the work environment. Examples of injuries are cuts, punctures, lacerations, abrasions, fractures, bruises, contusions, chipped tooth, amputations, animal or insect bites, electrocutions or a thermal, chemical, electrical, or radiation burn. Injuries include sprains and strains that result from a slip, trip, fall, or other similar accident.

Occupational Illness

Occupational illness involve skin diseases or disorders caused by work exposure to chemicals, plants, or other substances; respiratory conditions associated with breathing hazardous biological agents, chemicals, dust, gases, vapors, or fumes at work; poisonings; noise-induced hearing loss; or all other occupational illnesses.

Musculoskeletal Disorders

Musculoskeletal disorders are disorders of the muscles, nerves, tendons, ligaments, joints, cartilage, and spinal discs that are not caused by slips, trips, falls, motor vehicle accidents, or other similar accidents. Examples include many described in Chapter 13 and Table 8-1.

TABLE 8-1   OSHA Definitions Affecting Recordable Cases

First aid case

Incidents requiring *only* the following types of treatment are not recordable:

Using nonprescription medications at nonprescription strength

Administering tetanus immunization

Cleaning, flushing, or soaking wounds on the skin surface

Using sound coverings, such as bandages, gauze pads, or butterfly bandages

Using hot or cold therapy

Using any totally nonrigid means of support, such as elastic bandages, wraps, or back belts

Using temporary immobilization devised while transporting an accident victim

Drilling a fingernail or toenail to relieve pressure or draining liquid from blisters

Using eye patches

Using simple irrigation or a cotton swab to remove foreign bodies not embedded in or adhered to the eye

Using irrigation, tweezers, cotton swab, or other simple means to remove splinters or foreign material from areas other than the eye

Using finger guards

Using massages

Drinking fluids to relieve heat stress

Medical treatment

Medical treatment is managing and caring for a patient for the purpose of combating disease or disorder, other than the following:

Visits to a doctor or health care professional solely for observation or counseling

Diagnostic procedures, including administering prescription medications that are used solely for diagnostic purposes

Any procedure that can be labeled first aid

Restricted work

Restricted work activity occurs when, as the result of a work-related injury or illness, an employer or health care professional keeps, or recommends keeping, an employee from doing the routine functions of his or her job or from working the full workday that the employee would have been scheduled to work before the injury or illness occurred.

Counting restricted work activity

Restricted work is counted as the number of calendar days the employee was on restricted work activity or was away from work as a result of the recordable injury or illness. The day of the injury or illness is not counted. Counting includes both the sum of the days away from work and the days of restricted work activity. Counting ends when either or both reach 180 days.

Skin diseases or disorders

Skin diseases or disorders are illnesses involving the worker's skin that are caused by work exposure to chemicals, plants, or other substances. Examples are contact dermatitis, eczema, or rash caused by primary irritants and sensitizers or poisonous plants; oil acne; friction blisters; chrome ulcers; and inflammation of the skin.

Respiratory conditions

Respiratory conditions are illnesses associated with breathing hazardous biological agents, chemicals, dust, gases, vapors, or fumes at work. Examples are silicosis, asbestosis, pneumonitis, pharyngitis, rhinitis or acute congestion; farmer's lung; beryllium disease; tuberculosis; occupational asthma; reactive airways dysfunction syndrome (RADS); chronic obstructive pulmonary disease (COPD); hypersensitivity pneumonitis; toxic inhalation injury, such as metal fume fever; chronic obstructive bronchitis; and other pneumoconioses.

Poisoning

Poisoning includes disorders evidenced by abnormal concentrations of toxic substances in blood, other tissues, other bodily fluids, or the breath that are caused by the ingestion or absorption of toxic substances into the body. Examples are poisoning by lead, mercury, cadmium, arsenic or

(continued)

TABLE 8-1 continued

other metals; poisoning by carbon monoxide, hydrogen sulfide, or other gases; poisoning by benzene, benzol, carbon tetrachloride, or other organic solvents; poisoning by insecticide sprays such as parathion or lead arsenate; and poisoning by other chemicals such as formaldehyde.

Hearing loss

Noise-induced hearing loss is defined for record-keeping purposes as a change in hearing threshold relative to the baseline audiogram of an average of 10 dB or more in either ear at 2,000, 3,000, and 4,000 hertz, and the employee's total hearing level is 25 dB or more above audiometric zero (also averaged at 2,000, 3,000, and 4,000 hertz) in the same ear(s).

All other illnesses

These include all other occupational illnesses. Examples are heatstroke, sunstroke, heat exhaustion, heat stress, and other effects of environmental heat; freezing, frostbite, and other effects of exposure to low temperatures; decompression sickness; effects of ionizing radiation (isotopes, x-rays, radium); effects of nonionizing radiation (welding flash, ultraviolet rays, lasers); antrhrax; bloodborne pathogenic diseases, such as AIDS, HIV, hepatitis B, or hepatitis C; bruccellosis; malignant or benign tumors; and histoplasmosis and coccidioidomycosis.

Musculoskeletal disorders

Musculoskeletal disorders are disorders of the muscles, nerves, tendons, ligaments, joints, cartilage, and spinal discs that are not caused by slips, trips, falls, motor vehicle accidents, or other similar accidents. Examples are carpal tunnel syndrome, rotator cuff syndrome, De Quervain's disease, trigger finger, tarsal tunnel syndrome, sciatica, epicondylitis, tendinitis, Raynaud's phenomenon, carpet layers' knee, herniated spinal disc, and low back pain.

Case Classification

Under the OSHA system, there are four classes of recordable cases. One is any work-related death. A second is any case resulting in days away from work (not counting the day for the onset of the injury or illness). The other two classes involve cases in which a worker remains at work. One of these includes those cases in which a worker is transferred to a different job because of the injury or illness or is restricted in job duties. The other involves any other recordable case in which the person remains at work.

In calculating the days away from work, OSHA counts calendar days. Weekend days, holidays, vacation days, or other days off are included in the total number of days recorded if the employee would not have been able to work on those days because of a work-related injury or illness. OSHA sets the maximum number of days for a case at 180 days.

Incident Rate

A statistic used to measure safety performance and compare the performance of different groups or employers is the incident rate (IR). The BLS computes and publishes incident rates for different industries and sizes of companies each year. The OSHA incident rate is defined as

$$\text{OSHA incident rate} = \frac{\text{number of injuries and illnesses} \times 200,000}{\text{number of employee hours worked}}. \quad (8\text{-}1)$$

The 200,000 hours in the formula represents 100 employees working 40 hours per week and 50 weeks per year. This number keeps the value that results from the formula small. The number of employee hours comes from company records. They represent hours worked. They do not include hours paid but not worked, such as vacation, sick, or holiday time.

Example 8-1   Assume the XYZ Machine Company had 192 employees who worked a total of 385,728 hours during a calendar year. The company experienced 10 recordable injuries and illnesses among employees. The incident rate would be

$$IR = \frac{10 \times 200,000}{385,728} = 5.19.$$

Examples of industry composite incident rates reported by the BLS for 2002 are as follows:

Private industry	5.0
Agriculture, forestry, and fishing	6.0
Mining	3.8
Construction	6.9
Manufacturing	6.4
Transportation and public utilities	5.8
Wholesale and retail trade	5.1
Finance, insurance, and real estate	5.1
Services	4.3

In reporting industry composite incident rates, the BLS sometimes uses a baseline of 10,000 employees instead of 100, which increases the incident rates by a factor of 100.

Severity Measure

OSHA does not use a measure of severity of injury and illness cases. The BLS reports a severity measure that is the median days away from work.

Some use a similar severity statistic, often called the severity measure (SM). It is computed for composite or particular injury categories from

$$SM = \frac{\text{sum of days} \times 200,000}{\text{hours of employee exposure}}. \tag{8-2}$$

The sum of days may represent both days away from work and restricted work days.

Example 8-2   Suppose a mining company had several recordable cases that produced a total of 95 days away from work and restricted work. There were 389,295 hours worked at the mine. The severity measure is then

$$SM = \frac{95 \times 200,000}{389,295} = 48.8.$$

8-4   OTHER RECORD KEEPING STANDARDS AND RECORDS

For a long time, the record keeping standards for measuring work injury and similar accident experience were published by the American National Standards Institute (ANSI). They provided uniform record-keeping methods that employers and others used to assess performance of safety programs. The National Safety Council (NSC) has published annual accident and incident statistics for a long time. Many of the NSC records were derived from ANSI record-keeping methods. When OSHA was established in 1970, some of the methods changed in the OSHA record-keeping requirements.

For certain workers' compensation systems falling under state jurisdictions, the method for tracking and reporting injuries, illness, and deaths may also differ from the above examples. Today, safety records may differ also by country and standards organization. One should investigate the current applicable standards that might apply and be promulgated by the government of a country or by government organizations or other standard setting bodies.

EXERCISES

1. Obtain a copy of OSHA Form 300 and determine if the following cases should be logged on the form. If so, enter the appropriate data on the form. When logging days away from work or days of restricted work, assume that all workers work a 5-day schedule (Monday through Friday) and have Christmas, New Year's Day, Memorial Day, Labor Day, Independence Day, and Thanksgiving Day as holidays. Use a calendar for the current year. If dates shown in problems fall on a weekend or holiday for the current year, move the injury date back to the last workday before the date in the problem. Move the date a person returns to work (restricted or full duties) forward to the next normal workday.

   (a) John W. is employed in the press department as a press operator's assistant. On February 4, he cut his right hand on a sheet metal scrap. The company doctor treated him and gave him a tetanus shot because he had not had one recently. To avoid contaminating the wound with the grease used in the operation, he remained working in the department, but not at his regular job. He returned to his regular duties on February 18.

   (b) Mary J. is employed in the packing department as a packer. On March 6, she dropped a carton on her foot, injuring the third toe on her left foot. The toe was so badly crushed that amputation of the entire toe was necessary. She returned on a full-time basis on May 7, but worked a job requiring less time on her feet. She returned to her regular job duties on May 17.

   (c) Gary P. is employed as a paint sprayer in the paint department. On April 15, the exhaust system failed in the spray booth where he worked. He became ill after inhaling fumes, experiencing breathing difficulty and a headache. He stayed at home 2 workdays and then returned to work at his regular job.

   (d) William O. works in the maintenance department as an electrician. While working on a high-voltage line supplying some heavy equipment, he was electrocuted, because he failed to lock out the power. The incident occurred on April 19.

   (e) Sylvia P. is also employed in the maintenance department as a gardener for the summer. The week of June 29 was a scorcher. On June 30, she collapsed from heat exhaustion and did not return to work until July 10.

   (f) Sylvia's sister, Sally P., also does gardening work during the summer. On July 18, she was stung by a hornet and developed a respiratory reaction that caused her to be out of work until July 24.

   (g) Joseph C. works as a bookkeeper in the accounting department. On August 11 when he was checking some files, he turned around to return to his desk to answer the phone, tripped over an open file drawer, and broke his right wrist. He returned to work on September 11, but could not operate the accounting computers (part of his regular duties) until the cast was removed on October 16.

(h) Jerry D. works as a press operator in the press department. On November 5, he received a cut on his left arm that required medical attention (10 stitches). He returned to his regular job in 1 hour.

(i) Marilyn G. works in the metal coatings department as a coatings specialist II. On October 5, while preparing some liquids used to finish products for a special order, she splattered some acid on both arms and received chemical burns to the skin. She remained off the job, receiving much special treatment, and returned to her regular job on November 19.

(j) On May 12, Marvin K., a forklift driver in the shipping department, was seriously injured when the vehicle he was driving tipped over on him. He suffered several fractures and some nerve damage in one leg that prevented him from returning to gainful employment. (For this exercise, compute this case to the end of the year only.)

(k) On September 22, Elmer F., a carpenter in the maintenance department, was helping stack lumber and ran a large splinter into his right hand. It was removed at the first aid station and treated. He returned directly to his job.

2. Compute the (OSHA) incident rate for the company in Problem 1 for the year. Assume employees worked 123,413 hours during the year.

3. Obtain a copy of the standards used by a different state or country or published by a standards organization. Then evaluate each of the cases in Exercise 1 and determine whether the cases would fall under the record-keeping rules and how the recording keeping may differ from those applicable to OSHA Form 300.

4. For the company and cases in Exercise 1 and based on 2,000 hours per employee year, what is the average time (work hours) between recordable incidents? Assume the company has 65 employees.

5. Locate a company whose records have been inspected recently by an OSHA representative. Discuss experiences in trying to comply with OSHA record keeping with a company representative.

6. Locate a representative of a federal agency that requires safety record keeping and reporting. Have the representative discuss the effectiveness of their procedures and the value of such records for the agency in attempting to meet the laws and regulations that require such records and reports.

REVIEW QUESTIONS

1. List eight reasons why records and reports are important for safety.
2. What federal agencies require incident records and reports for employees?
3. For what kinds of transportation must incident records or reports be made?
4. Describe records or reports required for equipment in use.
5. Describe records or reports required for products and facilities.
6. Under the OSHA record keeping requirements for injuries and illnesses, define the following:

    (a) Recordable case

    (b) Occupational injury

    (c) Occupational illness

    (d) Musculoskeletal disorder

    (e) Lost workdays

    (f) Days of restricted work

7. What is the OSHA incident rate? For what is this statistic useful?

8. What is a severity rate? What does this statistic measure?

BIBLIOGRAPHY

American National Standards Institute, Inc., New York, NY 10018: D16.1 and D16.1a, Manual on Classification of Motor Vehicle Traffic Accidents.

Code of Federal Regulations. See applicable sections for record keeping requirements of particular agencies.

Roughton, James E., *OSHA 2002 Recordkeeping Simplified*, Butterworth-Heinemann, Burlington, MA, 2002.

PART III

HAZARDS AND
THEIR CONTROL

THIS SECTION of the book, the largest, deals with hazards and their control. When seeking to achieve safety, a major role for engineers is prevention. Prevention requires that engineers be able to recognize hazards, to know available controls, and to apply them. All too often, engineers do not recognize hazards and factors that contribute to incidents. Therefore, appropriate controls that are available to engineers are not applied at all or as fully as needed.

Sometimes engineers assume that when they apply their skills and knowledge to the best of their ability, things are safe enough. They assume that the products, equipment, workplaces, processes, and environments that they design, implement, and manage are safe. They see themselves as professionals, people who know what they are doing. But incidents do happen, contributing factors are overlooked, errors are made, and things do go wrong. Too often engineers do not have the knowledge and skill to prevent such problems. They do not make things as safe as they could be, as safe as society expects, or as safe as the law requires.

The goal of this section is to develop the reader's general knowledge of hazard recognition and hazard control. Because of practical limits, this book cannot include every hazard or every control for each topic or application.

CHAPTER 9

GENERAL PRINCIPLES OF HAZARD CONTROL

9-1 INTRODUCTION

In this chapter, basic concepts for controlling hazards are developed. Hazard control begins with recognition. It ends with implementation of a control for a hazard selected from one or more options. In the steps from recognition to control, one must apply several principles that are important.

This chapter presents several approaches for recognizing hazards and selecting controls. There are helpful constructs for thinking through hazard recognition and considering the use environment in which they occur. These aids are useful to envision a use environment and other factors that can contribute to an incident or its severity.

9-2 MURPHY'S LAW

Yes, things do go wrong. Despite one's best efforts to prevent undesired events, errors, misunderstandings, and incidents do occur. Murphy's law captures the idea "whatever can possibly go wrong, will."

The origin of Murphy's law is ascribed to an Air Force engineer, Captain Ed Murphy, and his colleagues, who were conducting crash tests in 1949. Finding a strain gage bridge wired incorrectly, Captain Murphy declared, "If there is any way the technician can do it wrong, he will." To this a colleague ascribed the name Murphy's law.

Captain Murphy and his colleagues achieved an excellent safety record. During several years of crash testing, they ascribed their results to a firm belief in Murphy's law and a concerted effort to prevent its fulfillment. When this claim was announced at a press conference by Colonel Stapp, the project director, Murphy's law quickly became a part of our vocabulary.[1] Variations and corollaries have been added as people applied Murphy's law to different fields. Table 9-1 lists a few applicable to safety engineering.

One goal in safety engineering is to prevent fulfillment of Murphy's law. For many engineers who have a role in products, equipment, processes, and environments, the goal is to reduce hazards. Through planning, design, and analysis of production and operations, factors that contribute to incidents can be eliminated or reduced.

TABLE 9-1   Safety Engineering Corollaries of Murphy's Law

A car and truck approaching each other on an otherwise deserted road will meet at the narrow bridge.

Most projects require three hands.

Hindsight is an exact science.

Only God can make a random selection.

When all else fails, read the instructions.

Any system that depends on human reliability is unreliable.

If a test installation functions perfectly, all subsequent systems will malfunction.

In any calculation, any error which can creep in will do so. Any error in any calculation will be in the direction of most harm.

A fail-safe circuit will destroy others.

A failure will not appear until a unit has passed final inspection.

From A. Block, *Murphy's Law and Other Reasons Why Things Go Wrong* and *Murphy's Law Book Two*, Price/Stern/Sloan Publishers. Inc., Los Angeles. CA, 1977, 1980.

9-3   HAZARDS AND HAZARD CONTROL DEFINED

A hazard is "a condition or changing set of circumstances that presents a potential for injury, illness or property damage." It is the "potential or inherent characteristics of an activity, condition or circumstance which can produce adverse or harmful consequences."[2] Hazard control is any means of eliminating or reducing the risk resulting from a hazard. Hazard recognition is perceiving or being aware that a hazard does or can exist.

9-4   SOURCES OF HAZARDS

There are many sources for hazards. Some hazards are introduced by people. All too often hazards arise from engineering activities, such as planning, design, production, operations, and maintenance. Hazards are seldom introduced by engineers or others deliberately; more likely, they are created inadvertently, unknowingly, or unintentionally. Many factors may contribute to the introduction of hazards: pressure to meet design or production schedules, job stress, poor communication, and lack of knowledge may influence hazard recognition and control. Also important are lack of instruction, personnel, funds, management concern, and assistance from safety and health specialists.

Planning and Design

Planning is the process of developing a method for achieving something, formulating a program of action, or structuring an orderly arrangement of parts. Designing is an extension of planning. More detail and specific information is incorporated into a method, program of action, or physical object. In planning and design activities, engineers may create hazards in sites, buildings, facilities, equipment, operations, and environments. A hazard may result from a computational error, failure to envision the use environment, making poor assumptions, or not envisioning how things will actually work.

There are many examples of planning and design errors. A few will suffice. A common computational problem for engineers is converting units of measure. For example, failure to convert square inches to square feet will produce a large error in a load calculation. Failure to include a factor of safety in a structural calculation can be disastrous. Using the wrong factor of safety can introduce a hazard.

Failure to envision the use environment can introduce hazards. For example, the force required by an operator to push or pull an object may be adequate when a floor is dry. The task may be hazardous when the floor is wet or shoes are muddy. The visibility of a display may be excellent for the designer, but obscured for an operator who is taller or shorter. An opening or access for servicing equipment may be large enough for a bare arm, but inadequate when a mechanic wears heavy clothing in cold weather. A skylight on a roof may not be strong enough to stand on or its strength may diminish with continued exposure to sunlight. It becomes a dangerous stepping stool when placed adjacent to a refrigeration unit that must be serviced.

Making inadequate assumptions is another way hazards are introduced. Assuming that a load is static when it is really dynamic may result in failure. Football stands may not be capable of rhythmic loading as the crowd sways and stomps to the music of the band. We may make bad assumptions when we fail to obtain the best possible data from literature, user testing, or input from specialists. One may assume that a product will be used one way for a function whereas in practice, there may be other ways in which a product is used. There may also be misuses that are not envisioned.

Selection of materials can introduce hazards during design. A material may be attractive, but may produce toxic substances if it catches on fire. A material may have adequate strength, but may have other properties, like creep or brittleness, that can lead to disaster. A material may quickly lose its strength when exposed to sunlight or dampness found in some use environments.

Failing to consider the life of a product can introduce hazards during design and planning. A product may be safe when new, but may become dangerous during use. Use factors, such as heat, chemicals, weather, vibration, freezing, wear, abrasion, or other adverse conditions, can shorten product life.

Production and Distribution

Hazards also can result from production and distribution activities that engineers plan or manage. It is not always possible to construct or produce items the way they are drawn or described on paper. Changing fasteners or connectors because those specified are not available could weaken a structural joint. Replacing one chemical with another may introduce toxic or flammable hazards. Poor packaging design may contribute to the introduction of hazards during handling and shipping. Inadequate packaging could result in a release of hazardous materials to handlers, distributors, or buyers.

Maintenance and Repair

Hazards may come from insufficient, delayed, and improper maintenance and repair. Controlling hazards related to normal use is not sufficient. Many designs fail to recognize hazards during setup, maintenance, and cleaning activities. For example, poor access to service points or the need to carry out servicing with high levels of energy present can be dangerous. Hazards during or resulting from maintenance, repair, or cleaning, not just normal operation or use, must be recognized.

Failure to provide manual power or inching controls for powered equipment may make service and setup activities dangerous. Failure to tighten a bolt or tightening it too much may create a hazard. Failure to lock out or provide lockout capabilities for electrical, steam, or mechanical power or fuel sources during maintenance creates hazards. Failure to clean up work areas before, during, or after servicing and repair can introduce hazards. Errors in maintenance procedures or poorly written procedures can cause hazards.

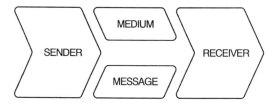

Figure 9-1. The four components of communication.

Failure to block areas undergoing maintenance activities may allow unqualified or unaware individuals into dangerous areas.

Communication

Poor communication or failures in communications can introduce hazards. Hazards can be introduced when changes in design, operations, and procedures are not communicated adequately to those impacted by them. The way information is communicated and the knowledge and understanding of receivers is important. Instructions and user manuals need the knowledge of the designer and others. Too often, instructions are descriptions of how an item works, rather than a series of actions one must take to make something work correctly. Poor communication leads to errors, incidents, and losses.

The four components of communication are essential in safety engineering. The four components are sender, receiver, media, and message (see Figure 9-1).

Designers, safety engineers, and other specialists have important roles in communications. They need to communicate designs, specifications, and procedures involving safe operations, use, maintenance, setup, and cleaning. They should even participate in preparation of advertising materials. If hazards or controls are not communicated to users or if protection is not illustrated in advertising, results may be disastrous.

9-5   PRINCIPLES OF HAZARD CONTROL

To minimize hazards, one must be able to

1. recognize them
2. define and select preventive actions
3. assign responsibility for implementing preventive actions
4. provide a means for measuring effectiveness

Together, these four steps achieve hazard control. A number of methods are available to accomplish these steps systematically. Part Five of this book details several methods.

Knowledge and Recognition of Hazards

As noted earlier, no one individual can be fully knowledgeable about all hazards. Several disciplines and specialists may need to work together. Safety engineering requires a knowledge of hazards in many different topics. Safety engineering also requires a broad knowledge of engineering and systems. In contrast, many engineering disciplines provide in-depth knowledge of particular topics. Thus, the specialty of safety engineering requires knowledge of hazards and potential controls across many engineering disciplines.

TABLE 9-2 Most Frequently Cited OSHA Violations (2003)

Rank	Topic	No. of Citations
1	Scaffolding	8,682
2	Hazard communication	7,318
3	Fall protection	5,680
4	Lockout/tagout	4,304
5	Respiratory protection	4,302
6	Electrical-wiring	3,337
7	Machine guarding	3,245
8	Powered industrial trucks	3,130
9	Electrical systems	2,399
10	Mechanical power	2,321

After one has developed a knowledge of hazards, there is a need to develop skill at recognizing and understanding hazards. Sometimes one must anticipate hazards by knowing that bringing certain materials, activities, or conditions together produces hazards that otherwise are not present. One must consider the use environment and many different contexts. Only after hazards are recognized can one identify and select suitable controls.

Historical data often helps in identifying or anticipating hazards that may exist or potentially exist. For example, OSHA publishes annual statistics based on the frequency of citations of OSHA standards. Table 9-2 provides example data for 2003. The rate of citations may help identify hazards to look for and resolve. Internal company data from workers' compensation claims, OSHA or other logs of incidents, or company accident reports can help identify hazards that require attention.

Priorities

There is a set of priorities that many find helpful for selecting controls for hazards. Some refer to this list as "design order of precedence." The priorities, in order of importance, are:

1. eliminate the hazard
2. reduce the hazard level
3. provide safety devices
4. provide warnings
5. provide safety procedures (and protective equipment)

Many factors must be considered when selecting and implementing controls for hazards. Risk, cost, kind or severity of loss, practicality, and not introducing additional hazards are all important. For kind of loss, the first priority is to protect people and human life. Protection of property, environments, and operations follows. Haddon's energy release theory (Chapter 3) provides ideas for dealing with these priorities.

Eliminate the Hazard

The highest priority in hazard control is to eliminate or avoid the hazard. As soon as it is eliminated, the potential for harm or loss is gone. Hazards can be eliminated by making process or design changes or by substituting a nonhazardous material for a hazardous one. For example, elimination of manual handling steps in an operation will eliminate lifting hazards. A noncombustible material can replace a combustible one. Sharp corners can be rounded. Wastes can be removed.

Reduce the Hazard

If one cannot remove a hazard, the degree of hazard often can be reduced. Two approaches are reducing the degree of severity or reducing the probability of occurrence.

Reductions in degree of severity lead to less injury, illness, or damage. For example, moving a fire hazard where it is distant from people is a reduction in degree of severity. Fewer are likely to be injured. Placing hazards where there are few people reduces hazard severity. Using smaller quantities of flammable or toxic material or reducing energy levels at an occupied location is also a severity reduction. A sprinkler system does not prevent fires. It simply minimizes their severity.

Reducing the probability of occurrence means that a hazard is less likely to result in an incident. One means to accomplish this is to use parts that have a longer life. Designing for lower failure rates or using redundancy are others. Avoiding single point failures is another.

Redundancy The probability of error or failure can be reduced by providing redundancy in an operation or system. Redundancy means providing more than one means to accomplish something, where each means is independent of the other. There are several kinds of redundancy and ways to implement redundancy. One is to provide two or more parallel subsystems or components. For example, Figure 9-2(a) illustrates a circuit that will not operate if the single actuating switch is open or fails in an open position. The circuit in Figure 9-2(b) will operate when either switch A or switch B is closed, or when both A and B are closed. A failure of either switch alone will not disable the ability to energize the circuit.

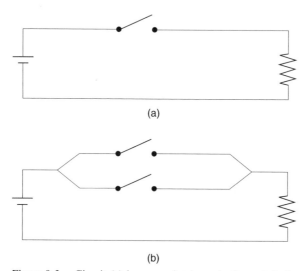

(a)

(b)

Figure 9-2.   Circuit (a) has no redundancy in the switch that activates the circuit, whereas circuit (b) does by having two independent switches that can close the circuit.

Another way to provide redundancy is to use a backup system. For example, on some aircraft, the aileron control is activated by hydraulic devices. Some aircraft have as many as four separate hydraulic systems to minimize the chances of control failure. Failure in one system will actuate a second system. The second is not normally in operation until the first one fails. For example, having candles or lanterns available for use when the electricity goes off and lights are out is a form of backup system.

In some systems there is partial redundancy. Suppose there is only one pump supplying two sets of hydraulic lines to an actuating cylinder. The pump and cylinder have no redundancy, whereas the lines do. The system has only partial redundancy. A failure in the pump or cylinder would produce a failure of the system. A blockage failure in one of the lines would protect the system from failure because there is a second line. Running the hydraulic lines through the same location where damage to both is likely reduces the value of the redundant lines.

Redundancy can involve both human operators and automatic equipment. The cruise control on an automobile is an automatic device that keeps a car moving at a steady speed. The driver can also control the speed of the car by depressing the accelerator pedal. The driver and the speed control are redundant. The driver can fully override the speed control by disengaging it with a switch on the brake pedal or throwing a dashboard switch to deactivate it.

Redundancy also can be accomplished through the use of more than one person. In aircraft, a pilot and a copilot can perform the same function. If one is incapacitated, the other can take over (parallel redundancy). Another example (series redundancy) is a two-person press. Both operators face the hazard of getting caught in the machine when it is in motion. If each operator has a two-hand control, all four hands of the operators must be depressing a switch before the machine will operate. Another example is the use of two people for cleaning or repairing a closed container. After checking the container for hazards, only one person enters the enclosure. The other watches and provides help in case the first encounters some difficulty, but does not enter the enclosure without a replacement at the backup position.

Single Point Failure   A single point failure is a failure of a component or subsystem that results in failure of the entire system. A broken starter switch or a dead battery in a car renders it inoperable. Single point failures must be avoided if a failure of the system can produce dangerous conditions.

Safety Devices

Safety devices can reduce hazards in many cases. Safety devices are features or controls that prevent people from being exposed to a hazard that exists. As soon as a safety device is in place, operating correctly, and properly maintained, it requires no action on the part of people. Safety devices are automatic devices. One must remember that safety devices do not remove a hazard. A major difficulty with safety devices is that they often are removed or are rendered inoperative, exposing someone to a hazard.

Machine guards are examples of safety devices. They prevent operators from entering a hazardous area. Fences, interlocks, shielding, and enclosures are all forms of safety devices.

Fail-safe devices are safety devices designed to prevent exposure to hazards. They also prevent injury or damage when a system or machine fails. Examples of fail-safe devices are automatic fire doors, air brakes on truck trailers, a dead-man switch on a powered hand tool, and safety cans with a spring-closing lid for flammable liquids. Fail-safe devices can be classified as fail-passive, fail-active, or fail-operational. A fail-passive device, such as an electrical circuit breaker or fuse, renders a system inoperative or deenergized until corrective action is taken. A fail-active device keeps a system energized but in a safe mode until corrective action is taken. A fail-operational device allows a system to function safely, even when the device fails.

Warning Devices

Another way to reduce hazards is to warn people. Warnings notify people of a hazard or danger. Warnings depend on people to take some action that will prevent them from being exposed to or injured by a hazard. Warnings do not remove a hazard. Warnings depend on human action to implement protection and are effective only when humans perceive and understand them and act correctly in response to them. Warning devices often rely on sensors to establish that a hazard exists for which a warning must be given.

Most warnings signal people through visual or auditory senses. Some common examples are signs, symbols, and visual or auditory alarms. Flags, labels, signs, flashing or changing lights, sirens, whistles, horns, and other means are used to notify people that a hazard exists. Because communication is a complex process, select and use warnings with care. Warnings can fail or be ineffective because of the complexities involved in their use. The following sequence is typically involved:

1. A hazard must be recognized during design by a designer or sensed by some sensor device.
2. The hazard must be differentiated from other hazards.
3. A warning must be actuated or presented.
4. The warning must operate.
5. The warning must be sensed by a receiving person.
6. The warning must be perceived as a warning relative to the background and its meaning understood.
7. The receiver must know what protective action should be taken.
8. The receiving person must take the appropriate protective action.
9. The correct action must be completed in a timely manner.

A warning is useless if any one of these steps is not completed. Table 7-1 identified 15 characteristics that warnings should have.

Warnings that seem similar can result in the incorrect action. For example, a fire horn in a school has a long continuous sound, whereas a tornado warning on the same system produces a sound that alternates between high- and low-pitch sounds. The appropriate actions in each case are opposite. For a fire, children must exit the building. For a tornado, they are to get down along the wall in a central corridor. An error in action can be deadly, as shown by the events in a Midwest grade school. The children exited when there was actually a tornado.

When several warnings are present at one time, they can be confusing, particularly if priorities among competing warnings are not clear. During the major loss of coolant incident at the Three Mile Island nuclear power plant, 500 or more audio and visual warnings went off during the first minute of the incident sequence; more than 800 went off by the end of the second minute.[3] Operators had a sensory and decision-making overload, which contributed to the overall severity of the incident.

Procedures

Another way to reduce the danger from hazards is by using procedures. Procedures are sets of actions that must be executed. People must learn to use safe procedures. Procedures must be developed and understood before they are used, must be safe, and must accomplish the desired goal in an efficient manner. One can establish procedures for efficiency, management control, and many other purposes beside safety. There are a number of methods (see Part Five) available for analyzing procedures to determine whether they are safe, sufficient, and effective.

One needs to design procedures to minimize danger to anyone using them. Procedures should not introduce unsafe practices and should not put someone in danger. People must be taught and develop skills in following safe procedures. People should learn why safe procedures exist and what hazards the procedures attempt to help them avoid.

People need to recognize hazards that may occur during the use of procedures and how to act if such contingencies occur. For example, people are often taught how to operate a machine. Then they start to use the machine and something unexpected occurs that their training did not include. Because the procedures did not cover such an event, the operator must use individual judgment to take the correct action. Too often the wrong action is taken. Because new and inexperienced operators are not familiar with the unexpected and how to protect themselves, the incident and injury rate for new employees is very high. Often new employees are not taught how to deal with nonroutine conditions.

Procedures are the lowest control on the priority list because they depend totally on human behavior to recognize the hazard and take appropriate corrective action. The hazard is still present. A person must be able to recognize the situation calling for a procedure, to know what procedure is correct for that situation, to recall the procedure, and to execute it correctly. The correct situation and procedure must be differentiated from all other similar ones. Skill is required in completing the procedure, and frequent practice may be necessary to retain the proper skill. The person must have the physical capabilities to perform it. All actions must occur in a timely manner. Failure in any one of these steps can result in inadequate protection.

Personal Protective Equipment

Personal protective equipment is sometimes needed if controls that are higher in the priority list cannot be implemented, but one must recognize that personal protective equipment is an element of a procedure. Wearing special equipment depends on human behavior and cooperation. Even if good fit and proper selection are accomplished, the use of equipment is not ensured. The hazard against which it provides protection is still present or likely to be present.

9-6   ENVIRONMENTAL HAZARDS

When dealing with hazards of environments, additional factors are important. Environments include such things as heat, light, noise, vibration, pressure, chemicals, and radiation (nonionizing and ionizing). One must consider the effects on people, how they occur, and how they are observed. We cannot observe most environmental hazards or assess them accurately without instrumentation and reference to standards. Therefore, procedures for determining whether a hazard exists are important.

Effects

Exposures to environments produce few traumatic injuries. Most often there are health effects, nontraumatic injuries, or cumulative effects. Thermal environments can cause burns. Exposure to extremely loud noise can cause injury to the eardrum. Exposure to high-intensity ultraviolet radiation can injure receptor cells in the eye. More often, environmental exposures lead to health disorders. Exposures to hot environments can produce various illnesses and physiological disorders. Exposures to noise can produce forms of stress and lower tolerance for others. Exposure to high levels of ionizing radiation can result in acute illness and death and exposures to low levels may lead to delayed illnesses, such as cancer.

Some effects of environmental exposures are delayed. The delay in manifestation of illness may be hours, days, or even years. The time between exposure and onset of symptoms is the latency period. Some cancers associated with exposures to certain materials and environmental conditions may not appear for years. The most extreme latency period is on the order of 30 to 40 years.

Some effects of environmental exposures appear as behavioral effects. A person changes the way he behaves. Some behavioral changes are easy to recognize. For example, consider the parents who are irritated by the constant blare of their teenager's stereo. They may feel tense and yell at their child as a result. Some chemicals affect nerve transmission or muscle action. A person exposed to such materials may exhibit noticeable reduction in motor skills. Other materials can cause loss of memory that affects a person's job skills. Often these behavioral changes are not associated immediately with some environmental exposure. The symptoms may result from many other causes as well. Sometimes treatment is initiated for the behavior problem and not the real cause. An example is "mad hatter's disease" or "Danbury shakes." Employees in the hat-making industry around Danbury, Connecticut, were exposed to mercury and became nervous and irritable and exhibited shaking.

There are significant differences among people in their physical, emotional, and behavioral response to environmental exposures. For example, some people burn easily in sunlight; others do not. Some people may experience a skin rash from contact with certain solvents, whereas others may not for the same exposure. In some cases, people become sensitized. For a long time they do not exhibit any effect when exposed to an environmental agent; then they do. After the first response is initiated, further exposures at even low levels will initiate the response.

Information Requirements

Exposures to environmental conditions and materials do not always produce effects. Not all exposures are harmful. Some are beneficial. For example, exposure to sunlight provides a means for acquiring vitamin D. Excessive exposure can lead to burns and skin cancer.

To determine if an environmental condition is hazardous, several items of information are needed. The information must be estimated for design purposes. When actual exposures are the concern, one needs to make measurements.

First, one must know the agent. Whereas it is easy to distinguish thermal conditions from noise, it is not so easy to tell what chemicals are present in the air and whether they are airborne solids or gases and vapors. For ionizing radiation, one must know what kind of radiation is present. For ultraviolet radiation, one needs to know the wavelengths present.

Second, one must know the values for attributes of an environmental condition. For thermal environments, for example, one needs the dry bulb air temperature, humidity, air velocity, and radiant heat load. For nonionizing radiation, one must know the intensity and wavelengths. For airborne chemicals, one must know the contaminants present and their concentrations.

Third, one must often know how long a person could be or has been exposed. The degree of hazard for many environmental agents is a function of the dose, determined in most cases from length of exposure and concentration.

Hazard Recognition

From knowledge about the presence of an agent, its form and intensity, and the potential or actual duration of exposure, one cannot establish if there is a hazard. For some agents, computations are needed to convert this information into some index value. In addition, the indices or measurements themselves must be compared with exposure standards, which establish what environmental conditions constitute a minimally acceptable exposure.

Instrumentation and Measurement

Special instruments are needed to determine the agents and their form and intensity present in an environment. Instruments may be grouped into two classes: laboratory instruments and field instruments.

Many times it is impractical or impossible to bring specialized instruments to the location where there is concern over an exposure. Laboratory instruments may not be portable or may require support systems that cannot be provided in field settings. Laboratory instruments may not be rugged enough to take the physical abuse and conditions found in the field. Laboratory instruments may be difficult to set up and calibrate when they are moved. Some instrumentation is difficult to read correctly, and an untrained user is likely to make errors.

There are two approaches for resolving these problems. One can use field instruments if they are available for the agents of concern. Field instruments overcome many limitations of laboratory instruments, although accuracy may be compromised in doing so. However, they may provide sufficiently accurate information so that decisions about exposures can be made in the field.

The second approach requires collecting samples and bringing them to a laboratory for analysis. Samples cannot be collected for all environmental agents. Some agents, like radioactive materials, decay with time. A delay from a sampling point to a laboratory may reduce the accuracy of readings.

For each kind of chemical and physical exposure, there are accepted procedures and instruments for making measurements. Because this book cannot provide full details on instruments and measurement procedures, one should refer to current publications and regulations for accepted practices or seek assistance from occupational safety and health or industrial hygiene professionals.

Health Standards

Standards for environmental agents (physical and chemical) are updated regularly. The updates are necessary to incorporate new knowledge and new agents into standards. Often there is incomplete information about the exposures themselves, effects of exposures, and the mechanisms for illness or injury. Information about hazards of agents are derived primarily from incidental exposures and from testing. Testing most often involves animal or other studies. Occasionally, human volunteers participating in studies provide direct information on effects of environmental agents on humans. In general, standards are based on past events. The trend is to make them more restrictive because new experience indicates that past standards do not provide adequate protection.

The main sources of environmental standards are OSHA and EPA standards. OSHA sets workplace standards; the EPA sets standards for air and water quality for the general public. Other agencies, like the CPSC and the FDA, also set certain environmental standards as they pertain to products. For work environments, the American Conference of Governmental Industrial Hygienists publishes recommended standards of exposure for chemical and physical agents.[4]

9-7   HAZARD CONTROL MODELS

The complex relationships among people, machines, environments, and organizations can make hazard control difficult. Using only one means for control may not be sufficient. Consider the problem of protecting people from falling into an excavation. Barricades may be placed around a trench or hole. However, at night someone may not see the barricade, so a flashing light is mounted on the barricade for visibility. For blind people, the flashing light is useless. When appropriate, a beeper is added to the flashing unit. Children may ignore the warning devices and their features and crawl under or over the barricade and fall in. A strong wind could knock the barricades over. The battery for the light and beeper may fail. A warning sign in English may be installed, but someone may not be able to read or understand English. The complexities of a seemingly simple problem often make it difficult to eliminate or control a hazard.

In the process of hazard recognition and control, one must identify the complexities of contributing elements. One must consider the hazards in their use environment. A number of conceptual models have been proposed to help one think of the many elements that are involved in incidents. Individually, people, machines, environments, materials, and other factors may not create hazardous conditions. Taken together in certain situations, a hazardous condition may be created or a danger increased. The appropriateness of a control method can only be determined in light of the complex array of elements potentially present.

Four Ms

One conceptual model, illustrated in Figure 3-2, is the four Ms: man, media, machine, and management.[5] Media can be thought of as environment. The model helps one think of the many factors and their interrelationships that contribute to potential incidents.

Goal Accomplishment Model

Another conceptual model, the goal accomplishment model, is illustrated in Figure 9-3. It assumes that people and organizations are goal oriented. The model includes nine factors that are typically involved in accomplishing a goal. *People* (1) perform *activities* (2) and use *equipment* (3) to help them. People perform the activities in some *place* or *facility* (4) under constraints of *physical* (5), *social* (6), and *regulatory* (7) environments. There are *time* (8) and *cost* (9) limits for the activities. Each of these elements has many characteristics that can affect the achievement of the goal (see Table 9-3). One

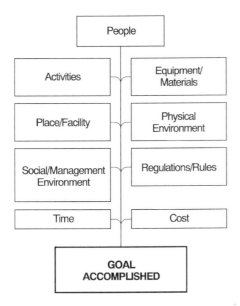

Figure 9-3. A goal accomplishment model for identifying and controlling hazards.

TABLE 9-3   Factors in the Goal Accomplishment Model

Factor	Typical Characteristics
People	Age, gender, size, strength, training, knowledge, emotion, state of health, culture, attitudes
Activities	Sensory and motor skills, actions taken
Equipment	Machines, vehicles, systems, materials, supplies, containers
Place	Facility, building, land area, road, air space, waterways, and characteristics of them
Environment	Thermal, electrical, sound, chemical, illumination, radiological, biological
Social/management environment	Organizational and work climate, interpersonal relationships, communication, language
Regulatory/procedural environment	Laws, regulations, procedures, policies, work rules and practices, rules of the road, etc. (both written and unwritten)
Time	Time available, rates, shifts, work hours, changes in shifts
Cost	Initial cost, operating cost, rent, losses, medical cost, repair cost, replacement costs, demolition or decommissioning costs, etc.

can analyze situations for these elements to help identify what can go wrong in reaching the goal.

9-8   SOME BASICS

Housekeeping and sanitation are fundamental in preventing injuries and illnesses. When incidents do occur, first aid or emergency action can reduce the severity of losses. People often overlook these basics. Industrial workers fought hard to achieve some of these and some workers are still fighting for them. A few comments about them are needed. These fundamentals must not be overlooked.

Housecleaning and Housekeeping

One way to control hazards is through housecleaning and housekeeping. Housecleaning involves involves picking up, wiping up, and sweeping up. It includes removal of scrap and waste. Housekeeping reflects the adage "having a place for everything and everything in its place." Not having proper storage places and storage equipment often is the problem. Some would delegate housecleaning and housekeeping to janitorial services, but everyone should share the responsibility for them.

Lack of housecleaning and housekeeping creates hazards. It is a symptom of unorganized, unplanned, and sloppy work and work management methods. In fact, many companies find that good planning and organization of work solves many housecleaning and housekeeping problems. At the same time, the planning creates profit. One can often tell how well an activity is planned and managed and how profitable it is by simply observing the housecleaning and housekeeping.

Sanitation

Sanitation is another important concept related to safety and health. Control of health hazards requires sanitation. Disease transmission and ingestion of toxic or hazardous materials are controlled through a variety of sanitation practices:

1. proper design and operation of sanitary and storm sewers
2. availability of safe drinking water and sanitary dispensing equipment
3. clean, operable toilet facilities
4. frequent garbage, scrap, and waste removal
5. sanitary food preparation, service, handling, and eating areas
6. insect and rodent control
7. sufficient and sanitary cleanup areas, locker rooms, and showers
8. use of appropriate personal protective equipment and clothing

First Aid and Emergency Action

Treating injuries immediately can reduce their severity and prevent further injury. Trained personnel, who know correct treatment, should administer first aid and maintain records of treatment. Adequate supplies and equipment should be available, and special equipment, such as deluge showers and eyewash fountains, should be provided at points where chemical hazards require them. Maintaining first aid supplies, equipment, and training is also important.

Emergency actions help mitigate the severity of an incident by limiting exposures of people, property, and the environment. Emergency actions may take several forms, such as evacuation, emergency communications, treatment, and recovery and may require the use of specially trained teams (fire brigades, spill response teams, etc.) and special equipment (fire protection systems, spill containment equipment, flood control equipment, communication systems, etc.). Chapter 29 discusses emergency actions in more detail.

EXERCISES

1. Identify hazards in your place of work or residence, applying the four Ms model and the goal accomplishment model.

2. Discuss the importance of communication for safety and cases of communication errors or failures with
   (a) a communication specialist
   (b) a safety professional
   (c) an attorney

REVIEW QUESTIONS

1. What is Murphy's law?

2. What is a hazard?

3. What is hazard control?

4. What are the sources of hazards?

5. What are the four components of communication?

6. What are the priorities for hazard control?

7. What are the general effects of exposures to hazardous environmental conditions?

8. What is a latency period?

9. What range in time can a latency period cover for various exposures to hazardous environments?

10. Do all people exhibit the same response to exposures to environmental hazards? If not, why?

11. What three items of information does one need to evaluate an exposure to an environmental condition?

12. How does one know if an exposure is hazardous?

13. How does one acquire information about an exposure?

14. What are the elements in the four Ms model?

15. What are the elements in the goal accomplishment model?

16. How are housecleaning and housekeeping related to hazards?

17. Explain the following terms:
    (a) redundancy
    (b) single point failure
    (c) safety device
    (d) safety warning

18. Why do procedures have the lowest priority in the list of hazard controls?

19. Why is personal protective equipment included with procedures?

NOTES

1 A. Block, *Murphy's Law and Other Reasons Why Things Go Wrong*, Price/Stern/Sloan Publishers, Inc., Los Angeles, CA, 1977.

2 *The Dictionary of Terms Used in the Safety Profession*, American Society of Safety Engineers, Des Plaines, IL, 1981.

3 Sheridan, T. B., "Human Error in Nuclear Power Plants" *Technology Review*, February: 23–33 (1980).

4 *Threshold Limit Values and Biological Exposure Indices*, American Conference of Governmental Industrial Hygienists, Cincinnati, OH, annual.

5 Grose, V. L., "System Safety in Rapid Rail Transit," *ASSE Journal*, 17: 18–26 (1972).

BIBLIOGRAPHY

Accident Prevention Manual: Administration & Programs, 12th ed., 2001, *Engineering & Technology*, 12th ed., 2001, *Environmental Management*, 2nd ed., 2000, *Security Management*, 1997, National Safety Council, Itasca, IL.

Best's Loss Control Engineering Manual, A. M. Best Co., Inc., Oldwick, NJ, annual.

BISESI, MICHAEL S., *Bisesi and Kohn's Industrial Hygiene Evaluation Methods*, 2nd ed., CRC Press, Boca Raton, FL, 2004.

BURGESS, W. A., *Recognition of Health Hazards in Industry*, John Wiley & Sons, New York, 1981.

CHRISTENSEN, WAYNE C., and MANUELE, FRED A., *Safety Through Design*, National Safety Council, Itasca, IL, 2000.

CONFER, ROBERT G., *Workplace Health Protection*, Lewis Publishers, Boca Raton, FL, 1994.

DeBERARDINIS, LOUIS J., *Handbook of Occupational Safety and Health*, 2nd ed., John Wiley & Sons, New York, 1999.

DiNARDI, SALVATORE R., *The Occupational Environment: Its Evaluation, Control, and Management*, 2nd ed., American Industrial Hygiene Association, Fairfax, VA, 2003.

Engineering Reference Manual, 2nd ed., American Industrial Hygiene Association, Fairfax, VA, 1999.

KOREN, HERMAN, *Illustrated Dictionary and Resource Directory of Environmental and Occupational Health*, CRC Press, Boca Raton, FL, 2004.

LACK, RICHARD W., ed., *Safety, Health, and Asset Protection—Management Essentials*, 2nd ed., Lewis Publishers, Boca Raton, FL, 2002.

PLOG, BARBARA A., and QUINLAN, PATRICIA J., *Fundamentals of Industrial Hygiene*, 5th ed., National Safety Council, Itasca, IL, 2002.

SCOTT, RONALD M., *Introduction to Industrial Hygiene*, Lewis Publishers, Boca Raton, FL, 1995.

STELLMAN, JEANNE MAGER, editor-in-chief, *Encycolpaedia of Occupational Health and Safety*, 4 vol., 4th ed., International Labour Organization, Geneve, Switzerland, 1998.

SWARTZ, GEORGE, *Job Hazard Analysis: Guide to Identifying Risks in the Workplace*, Government Institutes, Rockville, MD, 2001.

CHAPTER 10

MECHANICS AND STRUCTURES

10-1 INTRODUCTION

April 27, 1978: In West Virginia, 51 construction workers fell 170 feet to their deaths as the scaffold and form work system peeled from the top of a cooling tower under construction. The lack of some required bolts connecting the scaffold to the tower and inadequately cured, insufficient strength concrete contributed to the accident.

May 30, 1979: A DC-10 crashed in Chicago, killing 271 people. A $^3/_8$-inch diameter bolt supporting the engine pylon failed, causing the engine to break away from the wing. As it broke away, it ripped through three redundant hydraulic flight control lines.

May 12, 1982: A report to Congress stated that more than 212,000 of the nation's 525,600 highway bridges (40.5%) were structurally deficient or functionally obsolete. A structurally deficient bridge is one that has a reduced load, is closed, or must be rehabilitated immediately. A functionally obsolete bridge can no longer safely serve its current traffic load because of lane width, load carrying capacity, clearance, or approach alignment.

June, 1979: The driver of an off-highway dump truck was crushed to death in the cab when the loaded truck's chassis collapsed. Although the exact cause is not known, some speculated that metal fatigue caused the collapse.

August, 1989: It was found that bolts that did not meet standards for strength and other properties were marketed for use in aircraft, trucks, and many other applications without the knowledge of the using companies. The bolts had been certified to meet standards, when in fact, they were manufactured and imported as inferior. Their lower production cost provided a price advantage to their marketing companies. The Federal government became heavily involved in investigating the distribution of inferior bolts throughout the United States. Some companies that imported and sold inferior bolts were criminally charged. The problem of knowing which bolts are inferior and where they are in use is virtually impossible to solve.

1995: As a worker stepped on a plastic skylight cover to gain access to an air conditioning unit on a plant roof, the plastic cover failed and the worker fell to the concrete floor 20 feet below the skylight. After being exposed to ultraviolet light for many years, the plastic skylight cover had lost much of its strength.

1998 and 1999: After six deaths and many more injuries to auto racing spectators at racing events, designers re-evaluated standards for separating speeding vehicles and crash debris from fans.

Safety and Health for Engineers, Second Edition, by Roger L. Brauer
Copyright © 2006 John Wiley & Sons, Inc.

December 26, 2004: One of the largest earthquakes on record, registering 9.0 on the Richter scale with an epicenter off the northwest coast of Samatra, created a tsunami that extended throughout the Indian Ocean as far as 1,000 miles or more from the epicenter. The tsunami changed tide patterns half way around the globe. Within a few days, the death toll in the region exceeded 170,000. The damage to buildings, vehicles, and other elements from the wall of water that was more than 25 feet in some locations and rushed inland resulted in more than two million people without homes and five million in need of assistance. The disaster created the largest international relief effort on record.

Many accidents and injuries are caused by forces that have too great a magnitude for a structure or a material. An important part of engineering is the study of forces and their actions: the field of mechanics. To make systems, devices, or products safe, engineers must account for the forces that act or might act on buildings, vehicles, toys, bottles, or other devices. In addition, engineers must account for the forces from objects that may act on the human body and its tissues. The strength of some body tissue may be the limiting factor.

In engineering mechanics, there are many specialized fields. This chapter cannot review them all. The goal is to look at some of the fundamentals and their relationships to safety.

Forces, Distribution, and Materials

The magnitude of a force acting on a body is obviously important. As a rule, large forces are more likely to cause failure or damage than small ones.

How a force acts on a body is also important. The direction of a force, its location or point of application, and the area over which it acts are also important in safety. A 50-lb force applied to the edge of a sheet of glass and parallel to it may not break it. If a hammer strikes the center of the sheet with the same force, the glass will probably break. A wood panel of the same size undergoing the same force will not break.

When evaluating the strength of a material, it is essential to evaluate the distribution or concentration of forces as they act on bodies. Figure 10-1 gives some examples of distributed and concentrated loads.

Experience tells us that different materials have different strength properties. Striking a glass panel will cause it to shatter, whereas striking a wood panel will cause a dent. The effect of a force is related to the strength of a material and its ability to deform. Important properties of materials include strength, brittleness, ductility (ability to bend or deform), thermal expansion and contraction, shape, age, exposure to environmental conditions, and exposures to chemicals. Even strength can vary, depending on whether forces are pulling, crushing, twisting, or cutting.

A key relationship between a force F and a body on which it acts is

$$F = sA, \tag{10-1}$$

where

s = force per unit area or stress (such as pounds per square inch) and

A = area (such as square inches) over which a force acts.

The stress that a material can withstand is a function of the material's strength properties and the type of loading.

If the material and the area over which the load acts are given, the designer must determine what forces the object can withstand safely. In other cases, one estimates the expected force first and then selects the material and designs for the load area.

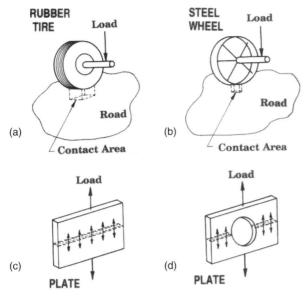

Figure 10-1. Examples of distributed and concentrated forces. In (a) tire flexion distributes the load over a larger road area than does the steel wheel in (b). The hole in the plate in (d) concentrates the load over a smaller internal area compared with the plate without a hole in (c).

A designer must envision the use environment. For example, building designers must determine the weight of building components and potential loads from building contents, wind, snow, rain, ice, and earthquakes. The designer of a wrench must consider how hard a user can pull on it. The designer of a toy must estimate how hard a child (young or old) can push or pull on it and how the toy's surfaces interface with human tissue. The toy designer should even consider the impact forces of someone falling on the toy.

The forces that an object can encounter are often different from the forces that an object should be able to withstand. For example, designers of breakaway sign posts and light poles along highways want the structure to fail at loads much lower than they could possibly encounter. The designer of a toy may want the toy to fail and fail safely rather than damaging body tissue when a child falls on it. In other cases, the designer may want a structure to withstand the greatest possible load.

Safety Factor

In applying Equation 10-1, a safety factor or factor of safety is often introduced. A factor of safety makes an allowance for many unknowns related to materials, assembly, or use. Unknowns may be inaccurate estimates of real loads or differences between actual materials and those tested in laboratories. They may be changes in area resulting from corrosion, wear, manufacturing, assembly, or use. They may be irregularities or nonhomogeneity in materials. The unknowns may include suddenly applied, dynamic loads. Technically, a safety factor (SF) refers to the ratio of a failure-producing load to the maximum safe stress a material may carry. The maximum safe stress is often called the allowable stress. Failure may not be by rupture or fracture. A failure could be a change in area or properties of the material that affect the load-carrying capacity and its safety. For structural steel, the allowable stress is derived at the yield point in a stress-strain (load per unit area-unit elongation or deformation) diagram from laboratory tests. For other materials, the allowable stress is based on the ultimate strength from similar tests. Refer to references on strength of materials for more details about test methods and stress-strain diagrams. There are many ways to determine a safety factor SF. A common way is

$$\text{SF} = \frac{\text{failure producing load}}{\text{allowable stress}}. \tag{10-2}$$

Safety factors are often based on experience with the material in question and many of its properties and applications. Safety factors should include analysis of risk and potential consequences of failure. Different safety factors may be appropriate for different applications and use conditions for the same material. Usually, the safety factor will be higher for materials with less homogeneity. Safety factors are higher for sudden, dynamic loads. Designs that anticipate reductions in a cross-sectional area of a component through wear or some other change in properties may incorporate higher safety factors. Some safety factors are specified in regulations and standards.

In safety engineering, one must be very careful in using data from tables dealing with strength of materials. Some tables include a factor of safety, whereas others do not. Using strength data in error from a table for which a factor of safety is not included poses a significant risk. The factor of safety incorporated in a table also must be applied carefully to ensure that the assumed safety factor is suitable for the actual use conditions.

When load and strength tables are intended for field use, they should incorporate appropriate safety factors. Field personnel who have to perform computations and complex interpretations of data tables are more likely to make errors as the number of steps in using a table increases. Field tables should reflect decision tasks and situations expected.

Kinds of Forces and Stresses

If one were to slice an object that is under external load, one can describe the kind of stress acting on the object. The key is the direction in which the stress acts relative to the plane of the section. Figure 10-2 illustrates several examples. Stresses acting perpendicular to the plane are normal stresses. They can be tension or compression stresses. Stresses acting parallel to the plane are shear stresses.

Forces on an object are classified by the way they act on a body. Forces that pull an object apart are called tensile forces; those that squeeze an object are compression forces; those that cut an object are shear forces; those that twist an object are torsional forces; those that cause an object to bend are called bending or flexural loads. When one object acts on, presses against, or bears on another, the force of one on the other is a bearing force or load.

10-2 MODES OF STRUCTURAL FAILURE

Materials and structures can fail in a number of ways. The main modes of static failure are shearing, tension, compression, bearing (crushing or deforming), bending, and buckling. Names for most modes of failure come from the kinds of forces applied.

Beside static loads, dynamic loads can cause materials to fail. Impact failure and fatigue failure are dynamic failures. The ability of a material to withstand an impact load gives rise to a property called toughness. Dynamic loads, that is, continually changing loads, can change the strength, ductility, and other properties of materials. Dynamic loading itself or the changes in material properties that result can cause fatigue failures.

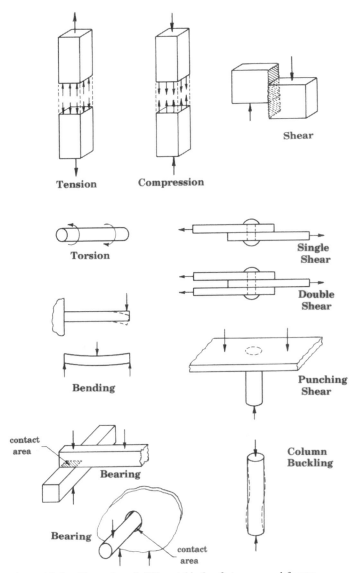

Figure 10-2.   Examples of different kinds of stresses and forces.

Instability is a form of failure for an object, rather than a failure of some material it contains. Tipping over is a common instability failure. If the resultant forces on an object act outside the support area or exceed the capability of anchors, the object will fall over. Failures of structural components can shift loads so that instability results. Because their joints tend to act as hinged joints, rectangular frames are less stable than triangular ones. Rectangular bases are more stable than triangular ones, because the base extends over a larger area. Gravity, friction, inertia of some moving mass, or externally applied loads may contribute to the resultant force. To help maintain stability, football players will spread their feet apart to increase the support area or obtain support by placing their feet in line with the resultant force. If both feet are close together, the player would get knocked over easily. A crane may tip over when an excessive load is too far from the support area or a load is swinging and creating an inertial load. That occurs when the resultant force acts outside the support envelope of wheels, tracks, or outrigger pads.

There are other forms of instability worth noting. During construction of buildings, it is important to add cross bracing for rectangularly arranged structural components. Torsion and lateral loads created by wind or the materials of construction can lead to collapse, even when the load-carrying members of a building are in place and adequate. A step ladder depends on spreaders and cross bracing to keep the legs of the ladder in position relative to one another. Because of the slope of the legs, a torsional load is applied when someone climbs the ladder. The ladder has a tendency to twist and buckle without adequate bracing. It is also important that spreaders be fully down and locked. If they fold up or are not fully in place to start with, the ladder can collapse.

Another form of failure for some materials is creep. Creep is a very slow but permanent deformation of a material under load. Some plastic materials are subject to creep failures. The cross-sectional area of a part may change and weaken the part as a result of creep. Another example is electrical aluminum wire. During a shortage of copper in the late 1970s, solid aluminum was substituted for solid copper in some applications. Tight connector screws became loose later as the local load of the connector on the aluminum wire caused creep in the aluminum material, something that did not occur for copper wires. The loose connections may eventually lead to arcing and fire.

Other changes in properties of a material can lead to failures. For example, exposing some metals to caustics will make the metals more brittle. Brittleness increases the likelihood of fractures or other failures. Exposing materials to ionizing radiation may reduce strength and increase brittleness. Exposing some plastics to ultraviolet radiation, such as from sunlight, will change the strength properties. Dynamic loading of some ductile materials will make them more brittle. Freezing may make some materials more brittle; heating may reduce the strength of others. Making some materials, like cardboard or paper products, wet may significantly change their properties.

Several methods of failure are possible for an object. One must analyze what kinds of loads and what methods of failure are possible. One must analyze each method to determine what method of failure is most likely for each condition.

10-3 CAUSES OF STRUCTURAL FAILURE

There are many different causes for failures. One scheme for classifying structural failures is the following: design errors, faulty materials, physical damage, overloading, and poor workmanship and poor maintenance and inspection practices.

Design Errors

One form of design error is incorrect or poorly made assumptions. For example, one may assume some load or a maximum load as the basis for a design. The actual load may be much different in normal, adverse, or misuse conditions. In selecting a load for design, one may make tradeoffs with cost and other factors. Here are some typical issues for a designer in estimating the load on a lever:

How hard can someone push or pull on a lever?

Is the 95th percentile male strength data found in a design handbook a good choice?

Should one use a value for two people pulling on the lever?

Will using a "cheater pipe" or extension on the handle of a ratchet wrench overload it?

Another form of design error is assuming a static condition or load, even though a dynamic one is a more representative of the real conditions. The collapse of the Kemper Arena roof in Kansas City in 1979 gives us an example. The roof tended to swing a little from its suspension during windy conditions. The hanger bolts supporting the roof from an external space frame were high-strength steel. After this bolt material was selected during design, later test data on similar bolts of the same material showed a rapid reduction in strength each time a nut was tightened and induced a load. Some engineers believe that the dynamic loading of the roof bolts reduced the strength of the bolts to a point where they could no longer support the roof.[1]

Designs that are difficult to fabricate or build is another kind of design error. The error can result from lack of practical experience on the part of a designer or from improper implementation of a design in the field. For example, a welder may not have experience with special welds called for on a drawing.

Computational errors are another form of design error. Manual calculations or errors in computer programs can lead to structural failure if computations are not checked or validated.

Another form of design error can be material selection. A selection error may result from lack of knowledge or data about particular materials. Similar materials may have different properties that are critical. A selection error may result from lack of knowledge or from lack of field data or test data about a use environment. Selection of incompatible materials may induce or accelerate corrosion, fatigue, embrittlement, or other effects and reduce strength of the material.

Another form of design error is specification of materials. A designer may have selected the right material, but the specification used by others may lack precise information for purchase and application. For example, a lubricating oil may be selected for a particular flammability property to minimize the danger of fire. Similar oils, although matching other requirements, may not have that required property.

Faulty Materials

Two factors that can affect the safety of materials are lack of homogeneity and changes in properties over time. Homogeneity refers to the uniformity of a material or the similarity among several samples. Wood, for example, has knots and grain variations that affect strength across a sample. Cast and molded materials often have voids. Some materials, like glass, may have internal stresses that result from uneven temperatures during manufacturing. Composite materials may not be thoroughly mixed and have uneven distribution of components. For example, the United Airlines crash of a DC-10 in Iowa in 1989 may have been caused by a tiny flaw in the material used in the turbine wheel. The wheel flew apart and ripped out hydraulic lines that controlled flight of the aircraft.

One way to control homogeneity is through testing. Another way is grading of materials, like wood. In some cases, the cost to ensure homogeneity is too high. A proper factor of safety or accurately estimated operational loads can help compensate for nonhomogeneity.

Changes in properties of a material over time take many forms. Changes that affect strength are of great importance, but other properties, like ductility, brittleness, or toughness, are also important.

The changes may result from corrosion, dynamic loading or vibration and noise, rotting or decay, wear and exposure to sunlight or other radiation, salt air, chemicals, water, or dissimilar materials. The changes may be minimized to extend safe use by anticipating the use environment, proper selection and use, maintenance, inspection, and special treatments.

Physical Damage

Objects and structures may be damaged through use, abuse, and unplanned events so that strength and dimensions are modified. The damage to an element may not cause failure by itself. However, when a load shifts to other elements of a structure, elements may not be able to withstand the load change.

One control that may minimize physical damage is placement. A house built very close to a railroad track is likely to be hit should a train derail at that location. A mailbox placed right next to the pavement of a highway is much more likely to be struck than one set back. Someone is likely to run into or trip over objects protruding into an aisle of a storage area.

Another control is the use of barriers. Placement of wires in conduits will reduce the likelihood of damage to the wires. Bulbs in trouble lights have a protective metal cage. Islands and concrete-filled steel columns protect gas pumps in service stations so cars will not strike them. Shields in automobile engine compartments protect some components from thermal damage.

Another control is structural design that allows for some damage. A standard for warehouse storage racks, for example, requires that damage to one leg of a four-legged structure not cause collapse of a rack. The rack must stand even when one leg does not support a load.

Overloading and Inadequate Support

Conditions change in the use environment. When not foreseen by a designer or user, the changes may result in overloading or inadequate support. For example, a warehouse in Florida was converted to offices. Because there was inadequate parking for employees (not a problem for the prior use), the roof was converted to a parking deck. When the roof collapsed, it became clear that the roof was inadequate for the weight of vehicles. In another example, a flatbed truck trailer was designed to carry uniformly distributed loads of bagged material. When used to haul an earthmover with concentrated loads on the outer edges, the sides collapsed.

Inadequate support refers to an object or structure not having enough load carrying capacity. If designers or users do not foresee these problems, failure can result. There are many examples. If an operator sets supporting outriggers of cranes or backhoes in mud or disturbed soil, the soil may compress and allow the machine to tip over. The legs of tubular scaffolds are fine when they rest on concrete. When they rest on soil, they tend to sink in. A bearing plate placed under them on soils will prevent sinking. Soil with a certain moisture content provided a firm foundation for the Winchester Cathedral in England when it was built. When a nearby stream was diverted away, the soil compacted as it dried and caused a corner of the Cathedral to sink. The foundation had to be shored up to prevent the cathedral from collapsing. The vibrations from tracked earthmoving equipment can travel through the soil and cause the walls of nearby excavations to collapse. Nearly every rainy season in southern California, homes slide down hills because wet and saturated soils can no longer support the loads. Stacking cartons too high may cause the carton at the bottom to collapse and tip the stack over. Many facilities under construction are adequate when completed, but have significant weak points during construction.

Poor or Faulty Workmanship

Another cause of structural failure is improper assembly and maintenance. Some failures may result from human error; some may be the result of designs that are difficult, impractical, or impossible to implement; some may simply result from careless work and poor decisions on the part of workers and management. Sometimes these are interrelated. Lack of communication, skill, knowledge, training, procedures, and management commitment can all contribute to faulty workmanship.

One theory for the cause of the collapse of the Hartford (Connecticut) Coliseum roof in 1978 is that workers did not assemble some joints as specified in the design. The original design allowed a 160,000-lb load through the center of the connecting plate and a moment of 0 ft-lb. As actually fabricated, the joint created a 15,440-lb load and a 9,490 ft-lb moment.[2]

In 1981, a walkway collapsed in the lobby of the Hyatt Regency Hotel in Kansas City. The walkway hung from rods that protruded through box beams in the walkway. The design required supporting nuts to be threaded several feet along the rod. Because that task was difficult to complete, the design was changed on site. The change doubled the shear load at the lower supporting nut on the box beam (see Figure 10-3).[3]

Another form of faulty workmanship is a change in procedures, particularly when its consequences are not fully considered. One example is the DC-10 crash in Chicago in 1979. The manufacturer's procedure for maintenance called for removing the engine first, then the pylon that attached the engine to the wing. To save time, workers suggested changing the procedure so they could remove both engine and pylon at the same time. Some believe this practice may have placed excessive loads on the pylon–wing connection and caused cracking of components and ultimate failure.

Poor Maintenance, Use, and Inspection

Materials, products, structures, and buildings do not stay the way they are at the time of manufacture, assembly, or construction. Exposures to various conditions during their life will change them. It is important that proper maintenance be applied to prevent corrosion or damage. Improper use can affect the likelihood of structural failure. Normal use can

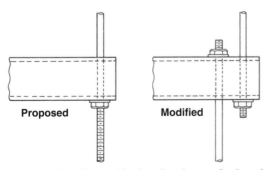

Proposed   **Modified**

Figure 10-3.   Changed load on box beams. In the original design, the load on one skywalk bears on one supporting nut below the box beam. A modification places the load of a lower skywalk on the box beam supporting the upper one. As a result, both loads bear on the nut under the upper box beam.

affect the structural integrity. Worn components need to be identified and replaced. Corroded elements may need strengthening or replacement. Inspections are an important method for identifying the change in properties. A wide range of inspection methods is possible, depending on the potential changes on the product or facility. In some cases, very specialized and sophisticated equipment may be needed to inspect the condition of structures and their components.

10-4   EARTHQUAKES

Earthquakes result from the movement of the subterranean plates forming the earth's surface. The movements between plates typically occur along fault lines. Earthquakes occur suddenly and typically are over in less than 1 minute, with smaller tremors occurring thereafter for a period of time. The Richter scale, a logarithmic scale, is a measure of earthquake intensity, or energy released during the plate movement and surface wave magnitude. An earthquake of magnitude 6 on the Richter scale is 10 times greater than one of 5, and an earthquake of magnitude 9 is 1,000 times greater than one of 6. An earthquake of 8 is an annual occurrence somewhere in the world and one of 7 is weekly. Those earthquakes originating under the sea will create ripples on the surface and the ripple may travel at rates of 300 to 400 miles per hour over great distances. A large ripple is called a tsunami.

An earthquake will cause the ground to vibrate at low frequencies. Any structure that has flexibility and can stretch to some extent or bend through connected joints has a greater chance of sustaining the vibration of the earth than a structure that has little joint strength. Mortar joints found in many structures are more brittle and are not likely to withstand the structural flexion resulting from an earthquake.

Some additional mechanics of soils can come into play during an earthquake. Because of the moisture content and the makeup of some soils, vibration from an earthquake will make them behave much like a liquid during the vibration rather than like a solid under normal conditions. This is called liquification. Some of the soils are man-made fills, while others are ancient lake bed sediments or simply soft soils. Much of Mexico City and towns and cities in the area of the New Madrid Fault in the area between Memphis, Tennessee, and southeastern Missouri are likely to exhibit the change in soil strength during an earthquake. The result is significantly greater damage to structures because foundation designs are based on normal soil properties rather than the "liquified" properties.

Another earthquake-related phenomenon affecting structures occurs when the frequency of vibration in an earthquake is at or very near the resonant frequency of the structure. The amplitude of the vibration becomes amplified and the degree of damage is significantly greater than expected from the earth's movement from the earthquake itself. A number of elevated highway structures have exhibited unexpected damage from earthquakes because of their resonant frequencies.

When an undersea earthquake occurs, it can cause a tsunami, a large wave effect. The normal water elevation changes and large amounts of water can wash into built up areas, causing severe damage from the energy produced by the moving water or from the flooding. The earthquakes cause a surge through the water that results in excessively large waves as the energy in the surge approaches the shore. In shallow areas, a tsunami can wipe out the entire built up area and most of the population located there. Usually the water's action occurs much faster than anyone can react. The force of the moving water can knock down structures and move people and vehicles uncontrollably. In recent years, a tsunami warning system has been put into place at a few locations subject to undersea earthquakes.

Many locations are defined by seismic zones that denote the likelihood and severity of potential earthquakes. It is important to know the seismic zone for any location and to follow the latest designs for buildings and structures for such zones to ensure the greatest degree of structural stability and to achieve the minimum amount of damage from an earthquake. For locations subject to structural property changes in soils or subject to potential tsunamis, other design considerations should be made.

10-5 CONTROLLING STRUCTURAL HAZARDS

There is no simple prescription for the elimination and control of structural hazards. Knowledge of the technology involved is essential. So is knowledge of materials and their behavior. One must complete calculations correctly and check them. Careful communication between designers and builders is needed. Attention to the use environment is necessary. Skill and care in assembly are needed. Designers must consider the consequences of failure. Not all structural failures cause injury, death, or major damage. In some cases, a structural failure may be desirable to control the point of failure and ensure that there are no catastrophic results. In some designs, the point of failure is controlled to minimize adverse effects.

10-6 APPLICATIONS

A safety engineer must have a good understanding of the principles of mechanics. This will help in recognizing hazards and selecting and implementing appropriate controls. A safety engineer must work with other engineers, metallurgists, architects and other structural specialists to ensure safety.

Static Mechanics

The field of static mechanics deals with forces acting on a body. Static mechanics involves bodies at rest or in equilibrium. Forces acting on them do not create motion. Common applications are bolts, rivets, welds, load-carrying components such as ropes and chains, and other structural elements. Equations 10-1 and 10-2, discussed earlier, apply to many static situations.

Example 10-1 Consider the bolt in Figure 10-4. It is loaded in tension and holds two elements together. One force acting on it is the load on the lower element (100-lb load plus 10-lb suspending elements). Another force is that caused by the tightened nut (20 lb). The total effective load on the bolt is 130 lb (100 + 10 + 20).

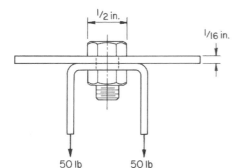

Figure 10-4.   Example of tensile strength. The U-shaped member places a tensile load on the bolt.

For a mild steel bolt, one can determine its ultimate tensile strength from tables ($60,000\,\text{lb/in}^2$). For a $1/4$-inch diameter bolt, the cross sectional area is $0.196\,\text{in}^2$. The actual stress in the bolt, using Equation 10-1, is $130\,\text{lb}/0.196\,\text{in}^2 = 663\,\text{lb/in}^2$.

Assume that for this application, a reasonable factor of safety is 3. By applying Equation 10-2, the actual factor of safety is $(60,000\,\text{lb/in}^2)/663\,\text{lb/in}^2 = 90.5$. Because 90.5 is much greater than 3, the bolt will easily carry the load.

Example 10-2 The plate in Figure 10-4 will fail in shear if the head of the bolt pulls through the plate. To determine the safe load capacity of the plate, one uses Equations 10-1 and 10-2. The bolt carries a 100-lb load. The outside diameter of its head is $1/2\,\text{in}$. The thickness of the plate is $1/16\,\text{in}$. The shear area in the plate is $\pi \times 0.5 \times 0.0625 = 0.098\,\text{in}^2$. The actual shear stress is $100\,\text{lb}/0.098\,\text{in}^2 = 1,020\,\text{lb/in}^2$. If the plate is aluminum, the ultimate shear strength is approximately $35,000\,\text{lb/in}^2$ from tables. It is obvious that the plate will not fail in shear for the assumed load; $35,000\,\text{lb/in}^2$ is much greater than $1,020\,\text{lb/in}^2$.

Welds

Figure 10-5 shows some forms of common weld connections. The strength, P, of a butt weld is

$$P = LtS_a, \tag{10-3}$$

where

L = the length of the weld,

t = the thickness of the thinner plate of the joint, and

S_a = the allowable stress of the weld.

The strength of a fillet weld is usually given as strength per linear inch of weld for a certain size of fillet. Because fillet welds are not often the full thickness of a plate, the size of a fillet is taken as something less than the thickness of a plate. Data on weld strength are available from the American Welding Society.

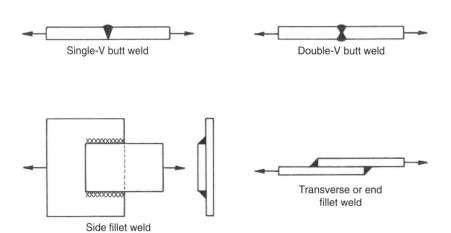

Figure 10-5.   Examples of welds.

Dynamics

Dynamic mechanics deals with the forces acting on a body to cause acceleration. The motion may be linear and angular. Impulse, momentum and kinetic energy are part of the field of dynamics. Table 10-1 gives a summary of key equations for dynamics.

Many dynamic loading conditions are important for safety engineers. Some examples are deciding if rotating equipment will fly apart and whether objects striking the body will cause injury. A forklift turning a corner too sharply may cause it to tip over. The distance needed to stop a vehicle in motion is a dynamic problem. Later chapters discuss some of these in more detail.

Friction

Friction deals with one body in contact with another that is on the verge of sliding or is sliding. Friction allows us to walk, drive vehicles, and power equipment. The force tangent to the contact surface that resists motion is the friction force. When no motion occurs, the resistance is the result of static friction. If motion occurs, the resistance is due to kinetic friction. Kinetic friction values are generally lower than those for static friction. The coefficient of friction, μ, is the ratio of the frictional force F_f to the normal force N between the two bodies:

$$\mu = F_f/N. \tag{10-4}$$

Friction has limits, however. Friction will prevent motion until the coefficient of friction is exceeded. Because friction causes wear, lubricants are used to reduce friction. Some substances become lubricants. Water, snow and ice, oils, greases, soaps, and plastics may reduce friction in locations where high friction is desirable.

Example 10-3   Assume someone is about to push a large box. It may slide. It could also tip over. Which will occur, the sliding or the tipping? Assume the coefficient of friction between the box and the floor is 0.6.

Referring to Figure 10-6, one can determine what force will tip the box over by computing the moments about corner B of the box:

$$\Sigma M_B = 0, \quad F(4) - 500(^3/_2) = 0, \quad F = 188\,\text{lb}.$$

By summing forces in the horizontal direction, one can determine what force would cause the box to slide. Solving for the frictional force F_f using Equation 10-4,

$$F_f = \mu N = 0.6(500) = 300\,\text{lb},$$

$$\Sigma F_x = 0, \quad F - F_f = 0 = F - 300, \quad F = 300\,\text{lb}.$$

Because the force required to overcome friction (300 lb) is greater than the force required for tipping (188 lb), the box will tip.

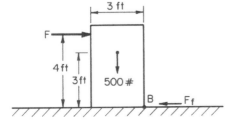

Figure 10-6.   Diagram for Example 10-3.

TABLE 10-1   Summary of Mechanics Equations for Dynamics

Property	Linear Motion		Angular Motion	
	Mathematical Expressions and Formulas	Typical Units	Mathematical Expressions and Formulas	Typical Units
Displacement	s, dv, dy, dt, $s = v/2g$ $s = v_0 t + 1/2\, at^2$	in, ft, m	$d\theta$ $s = r\theta$	deg, rad ft, m
Velocity	v, dx/dt, dy/dt, dz/dt $v = v_0 + at$	ft/s, mi/hr, km/hr	ω, $d\theta/dt$ $\omega = \omega_0 + \alpha t$	rad/s, deg/s
Acceleration	a, dv/dt, d^2x/dt^2	ft/s^2, m/s^2	α, $d\omega/dt$, $d^2\theta/dt^2$ Tangential: $a_t = r\alpha$ Radial: $a_n = r\omega^2 = v^2/r$	rad/s^2, deg/s^2
Newton's second law of motion	$F = ma$ $m = W/g$ $W = $ weight $g = $ gravitational constant slugs, kg	lb$_f$, newtons, lbm	$T = I_\alpha$ $I = \Sigma mr^2 = mk^2$ $k = $ radius of gyration	in-lb, ft-lb, kg-m^2, n-m
Momentum	$M = mv$	kg-m/s, slug-ft/s	$L = I\omega$	kg-m^2/s, slug-ft^2/s
Kinetic energy	$KE = 1/2\, mv^2$	ft-lb, kg-m	$KE = 1/2\, I\omega^2$	ft-lb, kg-m
Potential energy	$PE = Wh = mgh$	ft-lb, kg-m		
Work	$\text{Work} = 1/2\, m(v_2^2 - v_1^2) = \Delta KE$	ft-lb, kg-m	$\text{Work} = 1/2\, I(\omega_2^2 - \omega_1^2) = \Delta KE$	ft-lb, kg-m
Power	Rate of work, $P = Fv$	watts, hp, BTU/hr	$P = T\omega$	watts, hp, BTU/hr

Fluid Mechanics

Fluid mechanics is the study of forces on fluids. The field is sometimes called hydraulics when only liquids are involved and not gases. An understanding of fluid mechanics is necessary to predict and control the behavior of fluids. Safety engineers encounter many fluid mechanics problems and applications of fluid mechanics.

A major difference between mechanics of solids and fluids is that fluids have very little shear strength. Other important properties of fluids are density, specific weight, compressibility, viscosity, surface tension, and vapor pressure.

Pascal's law states that at any level, a fluid exerts an equal force in all directions. For a contained column of fluid, the pressure will vary with the vertical location. For incompressible fluids (such as water), the pressure p along the vertical column is given by

$$p = \gamma h, \tag{10-5}$$

where h is the vertical distance from the top surface to the point under consideration and γ is the specific weight.

Example 10-4   A tank contains oil to a depth of 25 ft (see Figure 10-7). The oil has a specific gravity of 0.9. What is the pressure at a point 8 ft from the surface? At the bottom of the tank?

Using Equation 10-5, one can solve for pressure at the two locations. The specific weight of water is assumed to be 62.4 lb/ft³. The specific gravity of water is 1.0. The specific weight and specific gravity for a fluid have a constant ratio, the force of gravity. The specific weight of the oil can be determined: $\gamma_0 = 0.9(62.4) = 56.2$ lb/ft³. The pressure at a depth of 8 ft is then $p = 56.2(8)/144 = 3.12$ lb/in². Similarly, at a depth of 25 ft, the pressure would be $56.2(25)/144 = 9.75$ lb/in².

Pressure increases linearly with depth in a fluid. Knowing this, one can develop a simple expression for the total pressure on a plane surface submerged in a fluid. Because the mean or average pressure p_m acting on the surface occurs at a depth located at the midpoint between the highest and lowest submerged point of the surface, the total force F is

$$F = p_m A, \tag{10-6}$$

where A is the area of the submerged plane.

Example 10-5   A 10-ft wide rectangular gate holds back water as shown in Figure 10-8. What is the force on the gate?

The midpoint of the submerged gate is $8(\sin 45°)/2 = 2.83$ ft. The mean pressure is $p_m = 62.4(2.83)/144 = 1.23$ lb/in². The total force on the gate acts perpendicular to it (the force of the fluid is exerted equally in all directions) and is $F = 1.23(8)(10)(144) = 14{,}170$ lb.

The volume flow of a fluid Q, or discharge, through some cross section (pipe, duct or channel) is given by

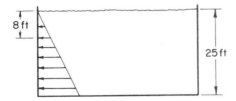

8 ft

25 ft

Figure 10-7.   Diagram for Example 10-4.

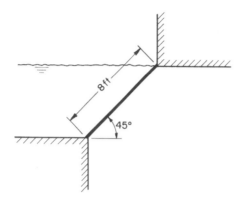

Figure 10-8.   Diagram for Example 10-5.

$$Q = VA, \tag{10-7}$$

where V is the average velocity (the flow of a fluid is not uniform over its cross section) and A is the cross sectional area. Equation 10-7 is called the continuity equation. Sometimes correction factors are used with Equation 10-7 for flow through orifices of various shapes. An example for ventilation is found in Chapter 25.

In fluid dynamics, the energy of a flowing fluid remains constant (conservation of energy). The form of the energy changes. A relationship that brings these energy terms together is the Bernoulli equation, Equation 10-8. The units are the equivalent column of water or head represented. The three main components in the Bernoulli equation are pressure head (p/γ), the elevation head (z), and the velocity head ($V^2/2g$). The elevation head is measured against some vertical reference point. The sum of the elevation head and the pressure head is called the piezometric head h. The Bernoulli equation is written

$$\frac{V_1^2}{2g} + \frac{p_1}{\gamma} + z_1 = \frac{V_2^2}{2g} + \frac{p_2}{\gamma} + z_2 = C \tag{10-8}$$

where subscripts refer to locations selected for particular applications and C is a constant for a particular application.

When fluid flows through pipes, energy may change form. For example, there are "losses" resulting from surface roughness, turns, valves, and other pipe components. These are called shear losses and form losses. The velocity head is reduced as a result. The losses for each component are added and form the total loss H_L. To maintain the energy conservation in the Bernoulli equation, H_L is included in one side of the equation

$$\frac{V_1^2}{2g} + \frac{p_1}{\gamma} + z_1 = \frac{V_2^2}{2g} + \frac{p_2}{\gamma} + z_2 + H_L \tag{10-9}$$

Example 10-6   A fire truck (see Figure 10-9) pumps water to the third floor (25 ft from ground level) of a building. Water for the pump is in an open tank. The flow rate at the nozzle must be 50 gal/min. The nozzle has a 2-in diameter opening. The pressure loss resulting from friction in the hose between the pump and the nozzle is equivalent to 3 ft of water. What pressure must the pump produce? Assume that a gallon of water occupies 0.1337 ft³.

First, one must determine the fluid velocity v at the nozzle. This is determined from the continuity equation. The cross sectional area, A, at the nozzle is $\pi d^2/4 = \pi(4)/4(144)$ $= 0.0218\,\text{ft}^2$. The velocity is 50 gal/min (0.1337 ft³/gal)/0.0218 ft² = 306.7 ft/min = 5.11 ft/s.

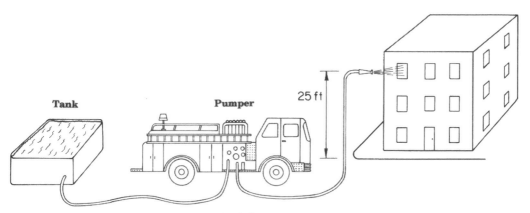

Figure 10-9.   Diagram for Example 10-6.

The velocity head at the nozzle is then $v^2/2g = 5.11^2/2(32.2) = 0.406$ ft. The velocity head at the tank is zero. From the data given, the elevation head at the nozzle is 25 ft relative to the pump. The pressure head at the pump and at discharge are both zero. The friction component H_L is included in the Bernoulli equation.

Then, the resulting equation for this situation is

$$C = v^2/2g + p/\gamma + h + H_L = 0.406 \text{ ft} + 0 + 25 + 3 = 28.4 \text{ ft water} \quad \text{or}$$

$$28.4(62.4 \text{ lb/ft}^3)(1/144 \text{ ft}^2/\text{in}^2) = 12.3 \text{ lb/in}^2$$

Soils

The branch of engineering that deals with action of forces on soils is called soil mechanics or soil engineering. Almost all structures ultimately rest on soil. Media over which vehicles travel (roads and rails) depend on sufficient soil strength for support.

There are many kinds of soils with different properties. Sand, for example, behaves much like a fluid. Clay behaves more like a solid. Soils engineering uses many empirical equations, because soils and their properties vary considerably. A thorough knowledge of the field and much experience is needed to apply soil engineering practices skillfully.

Properties of Soils   Many soils properties are documented. These properties help classify soils and apply soil engineering practices. Important properties include weight, density, modulus of elasticity, internal resistance, internal friction, cohesion, and volume changes resulting from various causes.

The weight of a given soil depends on the moisture content or the amount of water it contains. The amount of solid material for a unit volume is the dry weight. Density increases by the processes of settling and compaction and decreases by disturbing soil through excavation, tillage, and other actions. The moisture content of many soils is constantly changing from climatic conditions, natural or induced drainage, and compaction.

Internal resistance, which may vary for a soil, is a combination of frictional and cohesive forces acting on a soil. Several methods help determine this property. Results are quite dependent on the method. Internal resistance is an index of shear resistance. Internal friction is another index of shear resistance of soils and can never exceed the value of internal resistance.

The particles of some soils tend to adhere together, whereas others (sand, silt, gravel) do not. The fact that some soils tend to hold together even when well saturated results in the term cohesion. Cohesion refers to the internal tensile strength of a soil.

Volume changes in soil result from several factors. When some soils dry out, they shrink. When moisture increases, they expand. When compressed by external loads, soils will reduce in volume. The voids between particles become smaller in size from the loads. Water is squeezed out. Any process that reduces the water content of a bed of saturated soil is called consolidation.

Bearing Foundations must transfer the load of a structure to the soil. The load that a soil can support is sometimes simplified to Equation 10-1. The actual design of footings is much more complicated. Not only must the footings and soil carry the weight of the building and its contents, but the loads caused by wind and other imposed loads. Soils must carry the bearing load as well as moments that may be present. Borings help determine actual soil conditions. Building codes specify the maximum bearing loads for different soils, usually in tons per square foot. These allowable values often contain a sizeable safety factor, typically from 2 to 5.

In most foundation failures, the footings seldom fail. Failures frequently involve soil compression, unequal soil compression or movement, and changes in soil conditions (water content, volume, chemical content).

Piles Piles are slender underground columns used to support loads at their top. Loads transfer to soils by the friction and adhesion along the sides of the piles and by bearing at the bottom end. Designers establish the number, spacing, size, type, and angle of piles necessary to meet the capacities of local soil and anticipated loads.

Retaining Walls Soils exert lateral pressure on retaining walls, much like a fluid (see Example 10-5). Soils can exert one of two kinds of lateral pressure: active pressure or passive pressure. Active pressure exists when a wall resists the tendency of a soil to slide into the wall. For example, a pile of cohesionless sand will want to form a natural slope or angle of repose. A wall that restrains this action must overcome active pressure. Active pressure includes vertical force components. Active pressure varies with soil type, geometric characteristics of the wall, and the soil restrained.

The horizontal component of active and passive lateral pressure are both a function of the unit weight of soil, the square of the height of soil restrained, and the internal resistance of the soil.

Another force that can add to the pressure on a retaining wall stems from poor drainage that may cause the soil behind the wall to act like a fluid. Drainage of soils behind a wall will reduce the design load on the wall.

The design of sheeting and bracing for excavations can be complicated. Many pertinent factors must be analyzed. A qualified person must perform the design to meet acceptable engineering standards. Sheeting can be flat or corrugated and made of wood, steel, or other materials. Sheeting may be anchored or braced in a variety of ways. Sheeting itself may be embedded without braces and act as a cantilevered restraint. Poles or uprights can extend in front of the sheeting and be embedded below the sheeting. Braces can be placed in the excavation or anchors extended into the soil behind the sheeting.

Shoring for trenches is often constructed from tables like Table 10-2. Major components are illustrated in Figure 10-10. Depending on the source of the law or regulations, shoring is required in trenches more than 4 or 5 ft in depth. For trenches that are not open very long and not of great depth, a sliding trench shield (see Figure 10-11) or portable

TABLE 10-2 Minimum OSHA Requirements for Trench Shoring[a,b]

Depth of Trench (ft)	Kind or Condition of Earth	Uprights Minimum Dimension (in)	Uprights Maximum Spacing (ft)	Stringers Minimum Dimension (in)	Stringers Maximum Spacing (ft)	Cross Braces; Width of Trench Up to 3 ft (in)	3–6 ft (in)	6–9 ft (in)	9–12 ft (in)	12–15 ft (in)	Maximum Spacing Vertical (ft)	Maximum Spacing Horizontal (ft)
5–10	Hard, compact	3 × 4 or 2 × 6	6			2 × 6	4 × 4	4 × 6	6 × 6	6 × 8	4	6
	Likely to crack	3 × 4 or 2 × 6	3	4 × 6	4	2 × 6	4 × 4	4 × 6	6 × 6	6 × 8	4	6
	Soft, sandy, or filled	3 × 4 or 2 × 6	Close sheeting	4 × 6	4	4 × 4	4 × 6	6 × 6	6 × 8	8 × 8	4	6
	Hydrostatic pressure	3 × 4 or 2 × 6	Close sheeting	6 × 8	4	4 × 4	4 × 6	6 × 6	6 × 8	8 × 8	4	6
10–15	Hard	3 × 4 or 2 × 6	4	4 × 6	4	4 × 4	4 × 6	6 × 6	6 × 8	8 × 8	4	6
	Likely to crack	3 × 4 or 2 × 6	2	4 × 6	4	4 × 4	4 × 6	6 × 6	6 × 8	8 × 8	—	6
	Soft, sandy or filled	3 × 4 or 2 × 6	Close sheeting	4 × 6	4	4 × 6	6 × 6	6 × 8	8 × 8	8 × 10	4	6
	Hydrostatic pressure	3 × 6	Close sheeting	8 × 10	4	4 × 6	6 × 6	6 × 8	8 × 8	8 × 10	4	6
15–20	All kinds or conditions	3 × 6	Close sheeting	4 × 12	4	4 × 12	6 × 8	8 × 8	8 × 10	10 × 10	4	6
Over 20	All kinds or conditions	3 × 6	Close sheeting	6 × 8	4	4 × 12	8 × 8	8 × 10	10 × 10	10 × 12	4	6

[a] 29 CFR 1926.652 (OSHA Table P-2).

[b] Braces and diagonal shores in a wood shoring system shall not be subjected to compressive stress in excess of values given by the formula $S = 13 - (20\, L/D)$, where L = length, unsupported, in inches; D = least side of the timber, in inches; S = allowable stress in pounds per square inch of cross section; and the maximum ratio of $L/D = 50$.

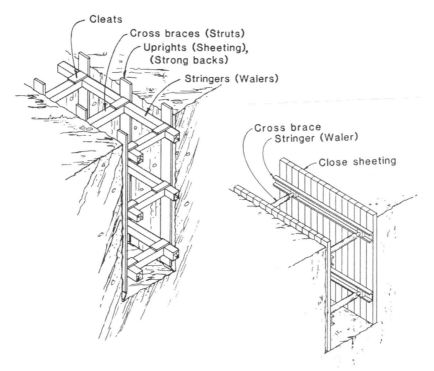

Figure 10-10. Examples of trench shoring and the components involved.

Figure 10-11. A trench shield in use.

trench box can be used. It is towed along as the trench is dug and provides a safe area for workers. In addition to shoring, there are many other requirements for safe trenching and excavation work.

Angle of Repose   When soil is excavated, unrestrained walls will tend to collapse at some point in time. When that will occur is not always predictable. The remaining soil will form some angle relative to horizontal, called the angle of repose. The angle formed varies with type of soil, moisture content, presence of loose materials, and other factors. The same is true of soil or other bulk material that is piled up. The sides slide out and form some angle. In excavations, the walls can be cut back or stepped back to an angle less than the angle of repose to reduce the danger from cave-in. Figure 10-12 illustrates typical angles of repose for some soils.

Dewatering   Changing the moisture content of soils can have significant effects. One effect is the change in load-bearing properties of the soil. Another is the change in volume of the soil. Pumping water from soils for construction of one facility may cause dewatering in adjacent areas and may induce damage on existing foundations and buildings.

Beams

Loading of beams is another important aspect of structural safety. A load on a beam induces stresses in its material. The strength of the beam material and the kind of loading determine the size of load that it can carry. Bending or deflection can create hazards even before total failure occurs. For example, a flat roof that deflects can cause water to accumulate or pond. The more water that accumulates, the more the roof deflects. This cycle could lead to collapse. A water buildup can be started by buildup of ice, leaves, or debris around a roof drain inlet.

As a beam bends, part of its cross section is in compression, part in tension. Figure 10-13 is a diagram of the stress distribution. The neutral axis is defined at the point where the stress is zero.

Properties of the beam cross section are important in determining the load that can be carried. One property is the moment of inertia, I. The moment of inertia is the sum of differential areas multiplied by the square of the distance from a reference plane (often the neutral axis) to each differential area. Because the distance is squared, the strength of a beam increases rapidly as its cross section is moved farther from the neutral axis. A rectangular beam will be much stronger when it is loaded along its thin dimension than along its flat dimension.

Another property used in beam load computations is the section modulus, Z. It is the moment of inertia divided by the distance from the neutral axis to the outside of the beam cross section.

The maximum bending stress s_b in a beam under a bending or flexural load is

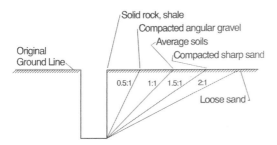

Figure 10-12.   Slopes of sides of excavations recommended by OSHA.

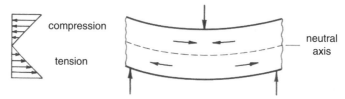

Figure 10-13. Distribution of stress in a beam cross section during bending.

$$s_b = Mc/I = M/Z, \qquad (10\text{-}10)$$

where

M is the maximum moment created on the beam by the load and

c is the distance from the neutral axis to the remotest element of the beam.

A problem faced by a designer is to minimize the size and cost of a beam while providing a safe load capacity. Similar to the procedures for evaluating other stresses, computed loads are compared with allowable loads. Allowable loads differ from maximum loads that produce failure by some appropriate factor of safety. Beams are usually selected from standard types, materials, and shapes. Section properties and other data about beams can be found in engineering tables.

The deflection exhibited by a beam under load is a function of material, section properties, length, means of support or attachment, and loading. Formulas for maximum deflections and slopes created are found in engineering tables.

Floors

Determining the safe load on a floor is a commonly encountered structural issue. Loads placed on the floor are transferred to joists. The joists transfer loads to a wall or to beams. Figure 10-14 illustrates a typical assembly for a floor in a building.

Designers find floor load values in handbooks or building codes. The task is to provide an economical, attractive, and functional floor system that will safely carry expected loads. There are usually two load components. Dead loads include the weight of the building and its components. Live loads are the loads that are placed on the floor. One would expect different live loads for a warehouse, an office, and a parking garage. In an office, file cabinets may be distributed among work areas or concentrated in one location. The designer must consider such use conditions in a floor design. Some building codes require that floor loads that were used in a design be posted, at least in certain kinds of buildings.

Example 10-7 A floor has a uniformly distributed load of 150 lb/ft². Floor joists are 18 ft long and spaced 2 ft on center. What is the load on one joist? Ignore dead loads. If the joists are simply supported at each end by a beam, what load is transferred to each beam by each joist?

The floor area acting on each joist is 2 ft × 18 ft = 36 ft². The total load on one joist is 36 × 150 = 5,400 lb, evenly distributed over its length. The load transferred to each beam is 5,400/2 = 2,700 lb.

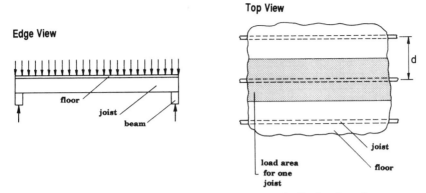

Figure 10-14. Typical structural components and load distribution for a floor.

Columns

Columns are structural members loaded in compression that have an unsupported length 10 times greater than the smallest lateral dimension. There are long and intermediate columns. Long columns fail by buckling or excessive lateral bending. Intermediate columns fail by a combination of crushing and buckling.

For a long column, the critical load is defined as the maximum possible axial load while still remaining straight. At the point of critical loading, the column is unstable and would bow easily if a slight lateral load were imposed. At greater axial loads, the column will buckle. The equation for computing the critical load P is

$$P = NEI\pi^2/L^2, \tag{10-11}$$

where

      N is an adjustment factor for end conditions,

      E is the modulus of elasticity for the material, and

      L is the length of the column.

For fixed ends, $N = 4$. For one end fixed and the other hinged, $N = 2$. When both ends are hinged, $N = 1$. When one end is fixed and the other is free, $N = {}^1\!/_4$. I is the smallest moment of inertia for the column cross section. Equation 10-11 is called Euler's formula for long columns.

Multiple Modes of Failure

For each assembly of structural components, there are several ways it can fail. A designer must analyze the assembly and identify all modes of failure. To determine which mode of failure is most likely to occur, each must be analyzed. Even very simple structures are complex.

Consider anchoring a shelf to a wall by means of a bracket and screws, one near the top of each bracket and one near the bottom (see Figure 10-15). The shelf must support some books. It may also have to support a child hanging from it, if other foreseeable conditions are considered. Several modes of failure are possible. In one mode, the screws could fail in shear at the wall–bracket interface. In a second mode, the screw at the top of a bracket could fail in tension. The moments on the bracket will be pulling the top screw away from the wall. The top screw could pull out of the wall. The head of the top screw could shear through the bracket material. The bracket could shear the head off the top

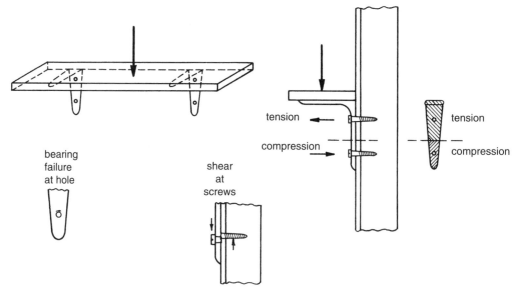

Figure 10-15.   A shelf supported by brackets attached to a wall, the load distribution on the brackets, and some modes of failure.

screw. The lower end of the bracket, which is in compression, could crush the wall material. The bracket could bend, because the top leg is, in essence, a cantilevered beam. The wall material could fail in bearing where the screws bear down at the holes in the wall. The threads on the screws could cause shear failure in the wall material. Any of these modes of failure could occur. In a comprehensive evaluation, all would have to be analyzed to determine which is most likely to occur.

EXERCISES

1. In using a water slide, a number of youth piled up in one section to form as long a human chain as possible. In doing so, they overloaded the joint between sections, causing the joint to fail. The 12 young people fell about 40 ft to the ground.

    There are two possible designs for the connection between sections of the slide. In each case, the joint has three bolts connecting the two sections (upper and lower) together. Assume the three bolts carry the entire load at the joint.

    In the first design, a shear load is on the connecting bolts, whereas in the second design a tension load is on the bolts when the lower section carries a load.

    What force, F, acting on one side of the joint (assume the other side or section if rigidly supported) is required to cause failure when the design data at the end of the exercise are used and the mode of failure is

    (a) bolt shear (for first design)

    (b) plate shear in the upper section when the slide users placed a load on the lower section (for first design)

    (c) bolt tension failure (for second design)

    Data: factor of safety = 3, bolts are 1/2 in diameter, breaking tensile stress = 45,000 lb/in^2, breaking shear stress = 62,000 lb/in^2, holes are 1/2 in diameter, plates are 1/10 in thick, breaking shear stress = 40,000 lb/in^2.

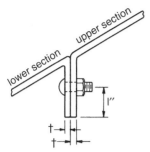

Design (1) for Exercise 1.

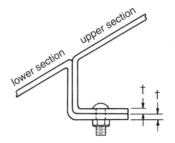

Design (2) for Exercise 1.

Bolt shear diagram (a) for Exercise 1.

Plate shear diagram (b) for Exercise 1.

2. A floor is supported by joists that transfer their load to beams at each end of the joist. Joists are 2 ft on center (O.C.) and 30 ft long.

   (a) If w, the load on the floor, is uniformly distributed and is 30 lb/ft^2, what is the load on one joist?

   (b) If the maximum flexural stress S in a joist is given by $S = M/Z$, where M is the maximum bending moment (pound-foot) and Z is the section modulus (cubic inches), and the maximum bending moment for a simply loaded joist is given by $M = wL^2/8$, what is the stress in a joist that has a section modulus of 50 in^3 and is 30 ft long?

   (c) If the joists are made of pine, which has an allowable bending stress of 9,300 lb/in^2, will it carry the load?

3. In an office it was decided to centralize files. All file cabinets in one department are to be placed in a row. The depth of the row is centered over a joist. Each file cabinet is 18 in wide by 30 in deep and weighs 300 lb. Determine

   (a) the total load on the one joist (neglect the weight of the floor itself)

   (b) the load transferred to the beam located at each end of the joist (assume the cabinets are also centered on the joist length)

   (c) the maximum bending moment in the joist

   (d) the maximum flexural stress in the joist

4. A 5-ft wide trench will be dug 12 ft deep in sandy soil. In considering shoring, determine the following: (a) upright dimensions and spacing, (b) size and spacing of stringers, and (c) size of cross braces and their maximum horizontal and vertical spacing.

5. A home swimming pool recirculates the water in the pool through a filter system. A pump moves the water from the drain(s) in the bottom or sides of the pool through

the filters and returns it to the pool. It is known from experience that children have sat on the single drain port in some designs after the drain cover was removed and not replaced and had their intestines sucked into the recirculation system, causing serious medical problems for the children.

Consider design options for reducing or eliminating the hazard of injury to body parts caused by the suction at the drain port(s) for pool recirculation systems. What is the likelihood of occurrence and the severity of potential injury for each option? What legal cases have resulted from various recirculation designs for pools?

6. Select a product. Analyze one or more structural components and identify modes of failure for each component.

REVIEW QUESTIONS

1. List three characteristics of forces related to failure or damage.
2. Explain the concept of a structural safety factor.
3. Why is a safety factor used in structural analysis?
4. Why should a table of material strength or load capacity have a factor of safety incorporated into it?
5. Why should a field table have a factor of safety incorporated into it?
6. Define the following:
   (a) tensile force
   (b) compression force
   (c) shear force
   (d) torsional force
   (e) bending force
   (f) bearing force
7. Name eight possible methods of failure for structures.
8. Name five causes for structural failure and give an example of each.
9. Give an example of an application for safety of each of the following areas of mechanics:
   (a) static mechanics
   (b) dynamics
   (c) friction
   (d) fluids or hydraulics
   (e) soils
   (f) strength of beams
   (g) strength of columns
10. Locate an article on a significant earthquake event and identify the causes of failure for structures that resulted. Identify what changes in designs resulted from a study of the effects or could have been made to reduce the degree of damage in one or more of the damaged structures.

NOTES

1 "Rocking That Fatigued Bolts Felled Arena Roof," *Engineering News Record*, August 16, 1979, pp. 10–12.

2 "Design Flaws Collapsed Steel Space Frame Roof," *Engineering News Record*, April 6, 1978, pp. 10–12.

3 "Altered Design Probed in Hyatt Collapse," *Building Design and Construction*, September: 17–18 (1981).

BIBLIOGRAPHY

American National Standards Institute, New York:
ANSI/ASCE 7, *Minimum Design Loads for Buildings and Other Structures*,
ANSI A10.21, *Safety Requirements for Excavations*.
Forging Safety Manual, National Safety Council, Itasca, IL, 1991.

LEVY, MATTHYS, and SALVADORI, MARIO, *Why Buildings Fall Down—How Structures Fail,* W. W. Norton & Company, New York, 1987.

MERRITT, F. S., ed., *Standard Handbook for Civil Engineers,* 3rd ed., McGraw-Hill, New York, 1983.

CHAPTER

CHAPTER *11*

WALKING AND WORKING SURFACES

The surfaces and devices on which people stand, walk, work, and climb contribute to many accidents, injuries, and deaths. Falls result in 20% of all accidental deaths. Slips and falls are the leading cause of accidents and deaths in the home. A study of California workers' compensation claims found that work surfaces are the most common agent for job-related injuries (21%). Falling objects also cause many on-the-job injuries.[1]

11-1 TRIPPING AND SLIPPING

Tripping

Most everyone has caught the toe of their shoe on a protruding or irregular surface of a floor, carpet, or sidewalk. In tripping, the motion of the foot is interrupted during a step. If the interruption of motion is sufficient, a fall will result.

Hazards Conditions that lead to tripping are irregular surfaces, objects protruding from the floor or walking surface, objects left lying where someone walks, and objects extending into a walking zone from the side and near the floor. Warped floor boards, missing floor tile, uneven tile or brick, carpet edges, loose carpet or rugs, protruding nails and screws, and chipped and cracked concrete are all examples of irregular surfaces or protruding objects. Other common tripping hazards are electrical cords, pipes, boards, and toys.

People do not always monitor the detailed condition of the floor or surface they are walking on. The normal line of sight is approximately 10° to 15° below horizontal relative to the eyes. Most of the time, people do not walk around looking down at their feet. As a result, even small changes in surface elevation are not always seen. Also, if someone is looking down at the surface, irregularities may not be perceived. Studies have shown that color, texture, low light levels, and glare may obscure changes in walking surfaces.

Not all tripping incidents result in falls, and not all falls lead to serious injury. Surrounding conditions contribute to the severity of tripping incidents. On an elevated surface, tripping may lead to a long fall. Even on a flat surface, the fall may be against a protruding object, or one may land in a manner that causes serious injury.

Controls Tripping hazards can be controlled; most often, good housekeeping is all that is needed. Tools, scrap, and waste should be picked up and objects like pipes, lumber, pallets, and file drawers that protrude into the walking zone should be moved out of the

Safety and Health for Engineers, Second Edition, by Roger L. Brauer
Copyright © 2006 John Wiley & Sons, Inc.

way. One or two step changes in elevation should be avoided, and the intersection of different floor finishes should be at the same level.

Inspection and maintenance can help remove tripping hazards. Protruding nails and screws should be removed or set even with the floor surface. Damaged tile, floor boards, or carpet should be repaired. Curled or wrinkled mats or flooring should be removed and electrical cords or similar objects that extend across walking zones should be recessed. (When there are temporary runs of electrical or communication cables across a walking zone, the cords should be taped down to minimize tripping hazards or should be routed overhead.)

Changes in elevation often are hard to see, but they can be made more visible by making different levels different colors. Avoid textured patterns that tend to hide changes in elevation. Changing levels should be well lit and warning signs should be posted at locations where there are tripping hazards. Direct or reflected glare that can interfere with the ability to see changes should be avoided.

Slipping

A slip is the sliding of one or both feet on a surface, and if it is unexpected, it can lead to a fall. Even without a fall, a slip can cause strains to muscles and joints. In a fall resulting from a slip, the feet slide out from under the body, producing an unstable condition. People expect to encounter a certain resistance between the floor and their shoes or feet and if that resistance is not there or it changes suddenly, a slip will probably result. A slip occurs when the lateral force applied at the foot–surface interface is greater than the frictional resistance available. Although this principle seems simple, it is complicated by continually changing forces during walking. To a great extent, the possible resistance is a function of the combination of shoe and flooring materials at their interface. Activity and gait or walking style affect the force created by the body. The resistance may be altered by wet, dry, or oily surface conditions, the presence of foreign material, and the roughness or polish of interface materials. Differences between static and dynamic friction coefficients further complicate the resistance possible.

It is difficult to predict when a slip will occur. However, activities like pushing, pulling, accelerating, turning corners, and throwing will produce higher horizontal forces at the foot–surface interface than normal walking. Horizontal forces increase as the angle formed by the leg with vertical increases. When people know that a surface is slippery, they will walk with a short stride to prevent slipping. Sloped surfaces add to horizontal forces and may increase the likelihood of slipping.

Measurement of Floor and Shoe Slipperiness   There are a number of methods for measuring floor slipperiness. Typically, each produces a reading on a scale from 0 to 1.

Different devices produce different resulting values. One type, a slip meter, is a small instrument that is pulled along the floor. A dial gives a reading of force created by the device on the string used to pull it. There are several patented slip meters, such as the horizontal pull slipmeter (HPS).[2]

Another type of slipperiness testing device is a swinging pendulum. It gives a reading of drag as a shoe at the end of the swinging pendulum slides for a short time across a surface. The British pendulum tester uses the pendulum principle.[3]

A third type of slipperiness measuring device is an articulating arm device. It is usually a bench-top instrument that places a load on prepared samples of shoe material and floor material. A static load starts directly over the sample. A hinged bar holds the load above the material sample at the other end. The machine moves the load slowly to one side of vertical. The bar begins to form an angle from vertical. The angle of offset is increased until the shoe sample pad slides against the floor sample. Two machines of this type are the James machine[4] and the NBS-Brungraber slip-resistance tester.[5] There is a portable version of the NBS-Brungraber slip-resistance tester.

A more recent device is the English XL Variable Incidence Tribometer (VIT).[6] It is designed to provide reliable testing for wet surfaces.

Some measurement standards suggest that a criterion of 0.5 defines whether a shoe and flooring combination is safe. Higher values are defined as slip resistant; lower values are slippery. However, it is difficult to relate a reading from one of these instruments to a qualitative description of slippery or safe. It is also difficult to relate test values to actual conditions or to predict when someone will slip. For example, a real floor may have wax buildup, small amounts of sand, mud, water, oil, or other material present or may be highly polished. Such conditions are difficult to incorporate into test procedures, and specially prepared test specimens may not replicate actual shoe–floor conditions. Test data provide valuable information for design and for material and finish selection but may not determine why a slip occurred or predict accurately when a slip will occur.

Hazards   One hazard related to slips is having a combination of shoe and floor materials and finishes that may cause slipping. Polished shoe and floor materials are more likely to be slippery than rough ones. Repeated mopping of some floor materials may increase slipperiness. The pores of the flooring material that originally were slip-resistant become filled with oily material after the substances dissolved by detergent water dry.

Another hazard is a sudden change in floor conditions. For example, when one moves from a dry surface to one that is wet, muddy, icy or oily there is an increased chance for slipping unless one adjusts the style of walking and movement. A sloped surface can add to a slip hazard.

A rapid change from a low slip resistance surface (slippery) to a high one (not slippery) can be hazardous. In this situation, people may stumble, rather than slip. In a stumbling fall, the body moves faster than the feet to an unstable condition. Some standards require surfaces to have consistent slip resistance.

An additional hazard is the risk associated with a fall resulting from a slip. For example, surfaces with a potential for a fall from an elevated surface to one below may require a higher degree of slip resistance than for those potentially having a fall to the same surface.

Controls   One control for preventing slips is housekeeping. As much as possible, walking and working surfaces must be kept free of foreign materials that can result in slipping. Water, mud, snow, ice, oil, grease, loose materials, scrap, and waste must be wiped up or picked up. In some cases, oversprays and foreign material can be prevented from getting on surfaces where people will walk and stand.

In areas where wet processes are expected, surfaces should be well drained to minimize standing liquid. In certain situations, raised floor surfaces may be an option so that workers do not have to stand in accumulated wet, oily, or scrap material. As a temporary solution, absorbent materials may be used to clean up spilled oils.

Where a change in surface conditions occurs, warnings should be provided. For example during mopping activities, workers should mark areas being mopped with warning signs. Procedures like mopping half the width of a hall at a time may help so that people do not have to walk through a wet area. At locations where ice, snow, or water are tracked in, warnings can help. Mats and rugs also can help reduce hazards for such conditions by providing a transition zone and by reducing tracking of foreign materials.

However, protective runners that become slippery when they are wet should be avoided. If there is heavy tracking, cleanup is essential.

Another control is selection of shoe and surface materials and floor finishes. Employers should help employees select appropriate footwear for their jobs. Purchasers should avoid shoe materials that are slippery when dry or wet. There is a wide range of flooring materials that designers can select from. Many manufacturers have slip test data for their flooring and surface products and technical references contain representative data on slipperiness properties of various materials. Designers should avoid sudden changes in slip-resistance properties in flooring, stairs, and other walking and working surfaces.

Selection, application, and maintenance of surface treatments and finishes also are important. Many manufacturers and suppliers have test data on floor finish products, and independent testing and evaluation of samples may be worthwhile before final selection. To lessen slipperiness, abrasive strips can be placed in strategic places, such as on stair nosing or wet areas, or fine silica sand can be mixed with flooring paints. Waxes and other finish materials must be applied and maintained properly because they have slip properties that can be affected by maintenance procedures. For example, excessive buildup of finish material or excessive buffing can alter slip properties from those obtained in test conditions. Locations where slip hazards are high may need different materials and finishes than other locations.

11-2 FALLS

Falls often cause injury. They may result from slipping, tripping, or stumbling and they include falling from one surface to another or on the same surface where standing or walking occurs. Falling objects that may strike people or things below are also included in this category.

Physics

In a moment of humor people say, "Its not the fall that is so bad, it's the sudden stop when you hit the ground." One must understand the physics of falls to understand the potential severity of a potential fall and associated impact and the hazard reduction resulting from controls. Three important aspects of falls are (1) the displacement and motion of a body, (2) the impact, and (3) the ability to withstand impact.

Displacement and Motion One characteristic of a fall is how far a body moves vertically during the fall. Knowing the distance that an object falls, s, allows computation of the velocity, v, at any point in the fall:

$$v = (v_o^2 + 2gs)^{1/2}, \tag{11-1}$$

where

v_o is the initial velocity and

g is the acceleration of gravity.

If the weight of a body, W, is known, one can compute the kinetic energy at the point where the body reaches a velocity, v:

$$KE = \frac{mv^2}{2} = \frac{Wv^2}{2g} \tag{11-2}$$

One can estimate the KE at any point in a fall by combining Equations 11-1 and 11-2.

Impact   When one body strikes another, the two bodies absorb much or all of the stored energy. Much of the energy is absorbed by the deformation of the two bodies. The energy in deformation may not be distributed equally between the two bodies. The energy not absorbed by deformation is transferred into motion of the bodies. Very often a falling object strikes the earth, a floor, or other structure that does not deform or deforms very little.

The injuries that result from a fall of a person onto a surface are, in part, a function of the rate of deceleration. From an estimate of the stopping distance (the distance the center of mass moves after initial impact), one can determine the rate of deceleration, a:

$$a = V^2/2s, \tag{11-3}$$

where

V is the velocity at the point of impact and

s is the stopping distance.

Some surfaces are hard or massive and deflect very little.

Often the rate of deceleration is compared with the acceleration of gravity, G. The number of Gs is determined from

$$G = \frac{V^2}{2gs} = \frac{a}{g} \tag{11-4}$$

Another important factor affecting the severity of injury is A, the contact area between two impacting bodies. If the force of a falling object striking a body is distributed over a large area, the severity of injury will be less than if the same force were applied to a small area. For example, a pointed object or sharp edge is more likely to cause injury than a flat object or rounded edge. In general, a similar relationship exists for a person landing following a fall. The force of impact F_i is

$$F_i = \frac{Wa}{g}. \tag{11-5}$$

This force must be resisted by the material, which may be human tissue, to which the force applies. The ability of the material to withstand the impact force can be determined from

$$F = sA, \tag{11-6}$$

where

s is the stress in the material and

A is the area over which the force is applied.

One must compare the induced stress with the tensile, compression, shear, or bending stress the material can withstand.

Injury to tissues other than those receiving the initial impact occur because the force of impact is transferred to other elements of the body, such as muscles, ligaments, bones, and joints.

Impact Limits of the Human Body

Data about the strength properties of human tissue and structure, often from cadaver or animal studies, can be used to estimate the likelihood of injury or severe injury in some situations. The data may be helpful in reconstructing certain accidents. However, because the body and actual conditions of an accident are complex, it is difficult to describe analytically what happened and what caused the resulting injuries. Figure 11-1 provides some data about human tolerance to impacts.

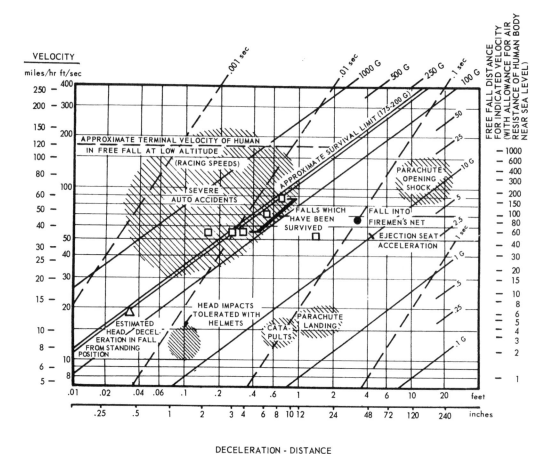

Figure 11-1. The effects of human impacts from falls. (From Webb, P., *Bioastronautics Data Book*, NASA SP3006, Washington, DC, 1964.)

11-3 PREVENTING FALLS AND INJURIES

There are four objectives in fall protection: (1) prevent people from falling, (2) prevent objects from falling, (3) reduce energy levels if falls do occur, and (4) reduce injury at impact. The latter two are not needed if the first two are met. In the following, these objectives are discussed, and they are summarized in Table 11-1.

Preventing People from Falling

Remove Slipping and Tripping Hazards Controls for slipping and tripping hazards were discussed in the preceding text and apply to any surface where people may be present.

Warnings and Barriers Particularly where there are changes in level between surfaces, warnings and barriers are needed. A barrier is a restraint that prevents a fall from an upper to a lower level. It must withstand the force of people running into it, leaning on it, or sometimes standing on it. Common barriers are guardrails, covers over openings, and cages on fixed ladders. OSHA requires that these devices withstand a load of 200 lb at any

TABLE 11-1   Summary of Fall Protection Methods

Objective	Method
A. Prevent falls of people	1. Remove tripping and slipping hazards
	2. Protect edges and openings
	a. Provide barriers (guardrails, covers, cage, etc.)
	b. Proved visual and auditory warnings
	3. Provide grab bars, handrails, and handholds
	4. Provide fall-limiting equipment
B. Prevent objects from falling on people	1. Housekeeping (remove objects that could fall)
	2. Barrier (toe boards, guardrail, infill, covers, etc.)
	3. Proper stacking and placement
	4. Fall zone
	5. Overhead protection
C. Reduce energy levels	1. Reduce fall distances
	2. Reduce weight of falling objects
	3. Control fall deceleration
D. Reduce injuries from falls and impact	1. Increase area of impact force
	2. Increase energy absorption distance

point, but because the weight of many people exceeds this, higher design loads may apply. Designs for covers should not introduce tripping hazards.

Covers over openings should be attached rather than be unsecured or, if they are temporary, should have warnings on them to indicate their purpose. It is not uncommon to have a sheet of plywood covering an opening in a roof or floor during construction. It is also not unusual to have two people pick up the plywood sheet to remove it and have the person carrying the rear of the sheet fall through the opening.

Most warning devices do not fully restrain someone. A barricade placed around a temporary excavation and a rope placed around the perimeter of elevated floors during building construction are warning devices. Flags and bright colors make warnings easier to see, and visual and auditory signals, such as flashing lights and beepers, help people recognize warning devices.

The Standard Guardrail   A commonly specified barrier is the standard guardrail, illustrated in Figure 11-2. It is comprised of vertical supports, which are typically 10 ft or less apart, and three horizontal components: the top rail, middle rail, and toe board. Infill between these components is important, too.

The function of the top rail is to prevent someone from falling. The height of a top rail is related to its effectiveness. It should be at least 42 in from a floor; a shorter dimension will protect fewer people. OSHA requires a 42-in height.[7]

Consider the mechanics of someone falling or leaning against a horizontal rail. We know that the center of gravity for a human body is approximately 3 in above the midpoint of one's height. If the center of gravity acts above the rail, a person falling against the rail would rotate over the top of the rail; if it acts below the rail, a body would rotate under the rail. For a person 6 ft tall, the center of gravity acts at a height of approximately 39 in. Therefore, if 99% of the population is less than 6 ft 6 in tall, a 42-in high top rail will prevent rotation over the rail for all but very few people.

The middle rail will keep someone from falling when their body rotates under the top rail. As a precaution for children, infill may be needed.

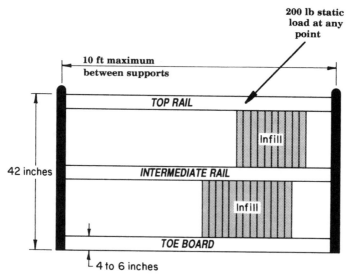

Figure 11-2.   Features of the standard guardrail.

A toe board is normally a 4- to 6-in high barrier along the walking surface. Its has two purposes: to prevent someone from placing their foot over the edge of an elevated surface and to prevent objects from sliding or rolling over the edge onto someone below.

Infill also prevents objects from falling from an elevated surface. The size of objects that could fall between the rails and toe board determines the size of the opening in infill material. For architectural handrails, the space between balusters should be small enough to prevent children from falling through the openings or getting their heads caught.

Handholds   When people move up or down between two different levels, it is important to provide a capability for three-point support, which means having two hands and one foot or two feet and one hand supported. Steps or ladder rungs provide support points for the feet and grab bars; handrails and handholds provide support for the hands. There are times during climbing activity when a foot or hand must be repositioned to a new support. Should the foot in contact slip, the only way to prevent a fall is with a firm grip with the hands. Handholds must be available at all points of climbing until a person is standing firmly with both feet on the new upper or lower level. There should be enough space in handholds or behind handrails and grab bars for fingers, even when wearing gloves. The cross-sectional shape should permit near maximum grip strength, which occurs when the fingers are well curled, and the grab bars should be able to carry a person's weight. The Department of Transportation has specifications for access to the rear of trucks and truck trailers.[8] Several OSHA regulations address grab bars and handholds.[9]

An important design consideration for handholds, particularly those that extend along lengths of ladders or stairs, is a touch indicator at the end. For people hanging on, but focusing their attention on something other than where they place their hand, some cue, such as change in texture or shape, would warn them that they are at the end of the handhold.

Other Barriers   Designs should include barriers where falls may occur. For example, in multistory buildings, people sometimes fall out of windows. Architectural components, like windows and skylights, can be dangerous openings or adequate barriers. The size of the opening, its placement, and the strength of its frame and infill all determine whether such elements are adequate barriers against falls. Openings in doors and walls should be placed properly to prevent falls through them or should be protected with adequate guard rails, covers, or screens. A retaining wall that holds back soil in a landscaped setting should be away from walkways and stairs to discourage people from getting near the edge. Railings and shrubs are also effective. Even Moses recognized the need for barriers: "When you build a new house, you shall make a parapet for your roof, that you may not bring the guilt of blood upon your house, if anyone fall from it."[10]

Fall-Limiting Devices If a fall does occur, intercepting it can prevent injuries. There are several kinds of patented fall-limiting devices, some of which attach to fixed ladders or to other climbing and elevated work equipment. A person using a fall-limiting device wears a harness that attaches to the device with a connection, normally a short, fixed-length rope, called a lanyard. The lanyard connects to the harness through a D-ring. The short length minimizes fall distance and deceleration forces that result from stopping the fall. The longer the fall, the greater the chances of injury when the rope becomes fully extended and the body stops quickly. A supporting rope, called a lifeline, must attach with a minimum of slack to an independent support, not to scaffolding or other equipment. Some patented devices are available that connect a lanyard to the life-line and will control the rate of deceleration in a fall so that injury is less likely. A user must position some devices along the life-line, whereas others move freely, but lock automatically during a fall.

The goal in fall arresting equipment is to minimize the force imposed on a falling person when the fall limit is reached. OSHA requires all safety belt and lanyard hardware to withstand a tensile load of 4,000 lb and requires the anchor and lifeline to withstand 5,400 lb. Body belts are seldom used for fall protection because they can cause injury when the fall is arrested and someone can slip out of them, particularly if the body belt does not fit well. A full harness is highly preferred, because it distributes the arresting load and is not likely to separate from the user if properly attached. Requirements for full harnesses and other components of fall arresting systems appear in ANSI Z359.1.

An empirical formula for estimating the maximum arrest force,[11] *MAF*, is

$$MAF = \frac{\left(W + 1.45(kfW)^{1/2}\right)abs}{c},$$ (11-7)

where

W is the weight of the falling person,

f is a fall factor $(0.1 \leq f \leq 2.0) = h/L$ (h is the fall distance in feet permitted by the lanyard, L = total lanyard length in feet),

k is the rope modulus (pound-force; see Figure 11-3),

a is the factor from Table 11-2,

b is the factor from Table 11-3,

s is the factor from Table 11-4, and

c is the conversion factor (see Figure 11-4).

Equation 11-7, which assumes a rigid anchorage point, was tested for accuracy by the developer and shown to produce results within ±5% of actual test values. This formula permits one to analyze work situations by showing that arresting forces are significantly reduced by the use of shock-absorbing mechanisms. In actuality, however, reduction varies with the kind of shock-absorbing device used and the arresting force is lower than that calculated when the anchor point is not rigid, but has some deflection. Also, the fall

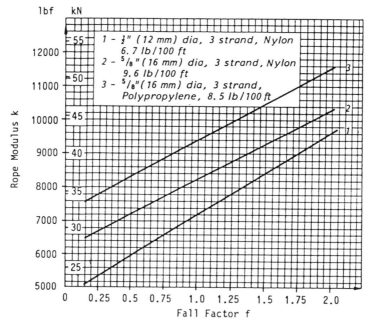

Figure 11-3.   Rope modulus (*k*) versus fall factor (*f*). (Reprinted with permission from the April 1981 issue of the *National Safety News*, a publication of the National Safety Council. Note: Although the information and recommendations contained in this publication have been compiled from sources believed to be reliable, the National Safety Council makes no guarantee as to and assumes no responsibility for the correctness, sufficiency, or completeness of such information or recommendations. Other or additional safety measures may be required under particular circumstances.)

TABLE 11-2   Fall Arrester Reduction Factor

| Type of Fall Arrester | *a* | | Comments |
	Range	Recommended[a]	
Inertia type, wire rope	0.5–0.7	0.7	for $0 < f < 2$
Inertia type, synthetic	0.75–0.9	0.9	for $0 < f < 2$
Friction type	0.5–0.75	0.7	for $0 < f < 2$
Mechanical lever	0.9–1.0	1.0	for $0 < f < 2$
No fall arrester used in fall arrest system	N/A	1.0	

[a] Recommended when exact value of fall arrester reduction factor, *a*, is not available.

TABLE 11-3   Body Gripping Device Reduction Factor

| Belt or Harness | *b* | | Comments |
	Range	Recommended[a]	
Belt (nylon, 2 in wide)	0.8–0.9	0.9	for $0 < f < 2$
Harness (full body, parachute style)	0.5–0.8	0.8	for $0 < f < 2$
No belt or harness	N/A	1.0	

[a] Recommended when exact value of fall arrester reduction factor, *a*, is not available.

TABLE 11-4   Shock Absorber Factor

Type of Shock Absorber	Range	Recommended[a]	Comments
		s	
Tear-the-stitches (TS)	0.2–0.6	0.6	for $0.1 < f < 2$
Tear-the-fabric (TF)	0.3–0.7	0.7	for $0.5 < f < 2$ (synthetic lifeline)
Tear-the-fabric (TF)	0.2–0.6	0.6	for $0.2 < f < 2$ (5/16-in diameter wire rope)
No shock absorber used N/A		1.0	

[a] Recommended when exact value of fall arrester reduction factor, *a*, is not available.

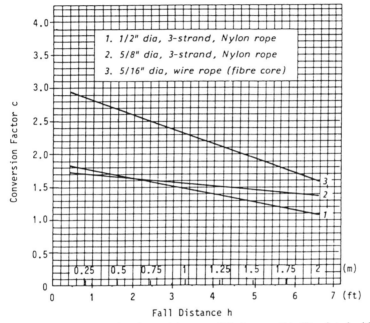

Figure 11-4. Conversion factor (*c*) versus fall distance (*h*). (Reprinted with permission from the April 1981 issue of the *National Safety News*, a publication of the National Safety Council. Note: Although the information and recommendations contained in this publication have been compiled from sources believed to be reliable, the National Safety Council makes no guarantee as to and assumes no responsibility for the correctness, sufficiency, or completeness of such information or recommendations. Other or additional safety measures may be required under particular circumstances.)

distance does not include any clearance required for elongation of the fall arrest system (see Table 11-5) or the extension of the arms or legs of a supported person.

Safety Nets   Where other fall protection is impractical, such as construction of bridges, towers, and other structures, safety nets can stop a worker's fall. The nets and anchorages must meet standards and pass tests. OSHA requires a minimum breaking strength of 5,000 lb. Nets are placed to minimize the fall distance. Again, the longer the fall, the greater the likelihood of injury. A net intended for stopping a falling person is a personnel net and one intended for stopping falling materials is a debris net. Debris nets must be strong enough to stop the size and weight of objects that could fall. Their mesh may be as small

TABLE 11-5   Approximate Additional Elongation of Fall Arrest System Resulting from Shock Absorber (ft)

Type of Shock Absorber	Starting Force (lb)	Additional Elongation (ft)
Tear-the-stitches	550	3.6
Tear-the-fabric	1,000	2.9

as $^1/_4$ in, whereas the maximum mesh for a personnel net is 6 in. A safety net may serve both purposes. Clearing debris regularly from the net will prevent it from injuring a falling worker.

Catch Platforms   Catch platforms, which are normally 2 ft wide, are placed at the edge of a sloped roof to prevent someone who slides down the roof from falling to the ground. They may not be necessary if workers use lifelines.

Preventing Objects from Falling

Housekeeping   If objects such as tools, waste, and other items are left lying around on an elevated surface, there is a chance that they will fall onto the surface below. Overflowing trash containers create the same hazard.

Standard Guardrail   As previously explained, the standard guardrail has a toe board and sometimes infill to prevent objects from falling to lower levels.

Fall Zone   A fall zone is a fenced or guarded area where people at lower levels cannot enter so that falling debris will not strike them. The guarding consists of overhead and side protection of the occupied area. The protecting material must have enough strength to provide protection for the size and weight of objects that could fall. Side protection protects occupants from splattering or scattering material.

Overhead Protection   In some operations, materials travel above areas where people are. Elevated conveyors and parts lines in assembly operations are common examples. Overhead protection can prevent materials from falling on those below. From time to time, fallen materials must be removed from the overhead screen or barrier, so the barrier must be strong enough to support workers during the removal activities.

Covers   Pipes, ducts, conveyors, and other equipment pass through holes in floors and walls. The holes may be oversized for installation and maintenance activities. Openings should be covered to prevent someone from falling through them or items from falling on anyone below. When covers are not in place, other forms of protection may be needed.

Stacking and Storage   Proper stacking and storage of materials can prevent objects from falling. Keeping stored items away from openings and not stacking them above barriers will help. Adequate work space and traffic aisles are needed to minimize the danger of running into stacked items. Impact barriers for material handling equipment can reduce the danger of knocking stacked materials over and damaging storage racks. For many kinds of stacked materials, stepping the materials back with each higher tear increases stability of the stacked materials. Cross ties also may be necessary to improve stability. Other storage requirements for fire protection purposes are discussed in Chapter 16.

Reducing Energy Levels

In some cases, one can control energy in a falling body or a body that could fall. One method is minimizing the vertical distance. The other is minimizing the weight of elevated objects.

Reducing Injuries

Cushion the Impact   Preventing falls is always the primary objective. However, a variety of things can be done to prevent injury when people do fall. One is increasing the stopping distance. Padding can help. In sports facilities, pads are applied to walls, poles, and other vertical surfaces. In some sports, like gymnastics, floors surfaces are padded. In motor vehicles, the dash, visors, and other interior surfaces are padded with energy-absorbing materials. For building fires, one type of rescue device is a large inflated cube that people can leap to. Fall-arresting devices cushion the impact and extend the distance over which deceleration occurs.

Distribute the Forces   If forces of impact are distributed over a larger area, the likelihood of injury is reduced. Pads and cushions not only extend the stopping distance, but distribute the forces over a larger area. The function of a suspension system inside a hard hat or helmet is to distribute the impact force over a larger area. Knee pads for flooring installers serve this same function. Knobs and controls that one can strike can be flat or recessed to enlarge the impact area. Protruding elements on toys made from deformable materials help distribute loads if a child falls on them. A safety harness usually will distribute the force of a fall over a larger area than a life belt.

11-4   APPLICATIONS

There are many kinds of walking and working surfaces and equipment involved with them. The principles discussed in the preceding text apply to them. This section looks at a number of them in more detail.

Stairs

Stairs are the most common device for helping people move from one elevation to another. Normally built on structural supports called stringers, the main components of stairs are the treads (the surfaces stepped on) and the risers (the vertical faces between the treads). Most treads extend beyond the face of the risers; this leading edge is called the nosing. As early as the ancient Egyptians,[12] some stair standards existed. Currently, there are a number of standards for stair design that can be found in OSHA regulations, building codes, life safety codes, books on carpentry, and other sources.

There are a number of factors that affect stair safety. They include uniformity of dimensions, slip-resistant treads, overall slope, visibility, structural strength, width, tread depth, placement, and other features.

Uniformity   Stairs should have uniform dimensions for all steps in a flight. A person walking up or down a set of stairs intuitively establishes a measure of what the stair dimensions are and expects the dimensions found in the first step to occur for the others. A sudden change in dimension can cause stumbling similar to when a person misjudges the number of steps.

Slip Resistance   Stair treads must have the same slip-resistance characteristics that floors do. When people walk on stairs, they expect the resistance to be the same or very similar to adjacent surfaces. Sudden changes may cause slipping or stumbling.

Slope   The slope of a stairs is the ratio of riser height to tread depth. The preferred slope for stairs is approximately 30° to 35° from horizontal, although standards differ on actual values, and slopes may fall between 20° and 50°. Tread and riser dimensions should be easy to establish with a tape or rule. Otherwise, it is difficult to lay them out or buy them. Table 11-6 lists some combinations of tread and riser dimensions that result in slopes between 30° and 40°. The Life Safety Code (NFPA 101) limits riser heights to 7 in, minimum height to 4 in, and minimum tread depth to 11 in.

Visibility   Visibility on stairs is very important. It includes having enough light to see steps easily and avoiding glare that obscures the ability to see steps. For example, a window or door placed at the base of a stairs can create glare that makes it difficult to see steps. When one enters a building from bright outdoor conditions, minimum lighting levels for stairs (often 1 to 5 footcandles) are inadequate because the eyes do not have time to adjust to low light levels. Visibility also includes clear definition of tread nosing and avoidance of surface finishes and textures that make one step blend in with another.

Structure   Stairs must be able to carry an anticipated load. Standards for stairs include minimum load capacity. OSHA requires a stairs to carry five times the normal live load and not less than a moving concentrated load of 1,000 lbs. In fire exiting, one often plans exit stairs to carry the maximum concentration of people for which there is space.

TABLE 11-6   Some Acceptable Combinations of Stair Riser and Tread Dimension[a]

Angle to Horizontal	Riser (in)	Tread Depth (in)
30°35′	$6\frac{1}{2}$	11
32°08′	$6\frac{3}{4}$	$10\frac{3}{4}$
33°41′	7	$10\frac{1}{2}$
35°16′	$7\frac{1}{4}$	$10\frac{1}{4}$
36°52′	$7\frac{1}{2}$	10
38°29′	$7\frac{3}{4}$	$9\frac{3}{4}$
40°08′	8	$9\frac{1}{2}$
41°44′	$8\frac{1}{4}$	$9\frac{1}{4}$
43°22′	$8\frac{1}{2}$	9
45°00′	$8\frac{3}{4}$	$8\frac{3}{4}$
46°38′	9	$8\frac{1}{2}$
48°16′	$9\frac{1}{4}$	$8\frac{1}{4}$
49°54′	$9\frac{1}{2}$	8

[a] 29 CFR 1910.24 (e).

Width   Life safety codes define stair widths for exiting during fires. NFPA 101 generally requires stairs to be at least 44 in wide, unless there are fewer than 50 occupants at all stories served by the stairs, which can then be at least 36 in wide. Chapter 16 gives more information on life safety for exits, including stairs, doorways, and corridors.

Distractions   One study[13] filmed people walking down stairs. The study found that the people who stumbled or fell usually did so at the same location on the stairs, even though there were no changes in the physical characteristics of the stairs. The conclusion was that as their heads moved below the floor opening, the people were distracted by the sudden view of the large room. Many retailers have experienced similar effects when displays or pictures are placed on landings or in view of people using stairs. Enclosing a stairs or otherwise preventing such a visual distraction can help minimize this problem.

Other Features   The number of steps included in a staircase can create hazards. A one- or two-step change often is not seen and causes people to fall. Consequently, for small changes in elevation, ramps are better than steps. Because climbing stairs is hard work and too many steps can be tiring, multistoried buildings and long flights of stairs should have landings to give people a chance to recover.

   The base of a stairway should be free of hazards. For example, if someone fell down a stairs into a glass wall or door, a severe injury could result. The base of a stairway also should be free of objects that could cause injury. An example of this is people who use stairs for storage because they do not want to carry items all the way up or down. Stairs should be free of waste, like paper or other objects, that can create tripping or slipping hazards.

   Stairs must have handrails that will support a person's weight. Handrails provide at least two-point support during ascending or descending activities. Handrail cross sections should be easy to grasp and should be placed according to life safety codes and other standard specifications. In addition to handrails along the sides of stairs, wide stairs should have one or more handrails located at intermediate locations across the width of the stairs.

Ramps

Ramps have a hazard related to their slope because a force component runs parallel to their surface and could contribute to slipping. To minimize slipping, ramps must have a limited slope and a slip-resistant finish. The preferred angle for a ramp is 15° or less; the maximum for access by the handicapped is 11°. The Life Safety Code (NFPA 101) limits ramps to a 1 in 12 slope. For slopes of more than 15°, cleats are needed to help provide traction. Ramps may have one or both sides open to the lower level. Unless protected by guardrails, someone could fall from such ramps. Handrails are needed to assist those negotiating them. Long ramps need landings similar to long stair runs.

Dock Plates and Gangplanks

Dock plates and gangplanks bridge gaps between a platform or surface and a vehicle or ship. They often serve both pedestrian traffic and material-handling vehicles. Major hazards for dock plates and gangplanks are structural failure, sliding out of position, falling off supporting edges, and difficulty in placing them into position. Slipping and tripping hazards and controls previously discussed also apply.

   To reduce the possibility of structural failure, dock plates and gangplanks must carry anticipated loads, and they should be labeled to show load capacity. They should be inspected regularly for defects or failures and they should be repaired or replaced when defective or damaged.

To help prevent vehicles from running off their edge, dock plates should have a curb or lip along their sides. Dock plates often are made from lightweight materials, like aluminum or magnesium, to make them easier to handle; some even have handholds. Some have features that facilitate material-handling equipment or forklift trucks. Some dock plates are integral parts of loading platforms and are equipped with powered devices to adjust them to different positions without manual lifting.

Dock plates will slide if they are not locked into position. Some of the devices for locking them into position are stop pins, cleats, or built-in flanges and blocks. Anchoring methods should not create tripping hazards.

Ladders

Types There are many types of ladders. They are classified by material of construction (wood, metal), load capacity, function, and design. Some ladders are classified as type I, II, or III, depending on the standard loads (200, 225, and 250 lb) used for certain tests. Most ladders are manufactured and sold for use. However, ladders called job-made ladders are made on job sites. Ladders can be portable or fixed in place. Common types of portable ladders are step, platform, straight, and extension ladders. Some ladders, such as those on truck trailers, have custom designs to fit around other design elements. There are many standards for ladder construction.

Hazards and Controls Ladder rungs or steps must have slip resistance to prevent slips and falls. Especially for ladders used around wet, oily, and muddy conditions, rungs or steps must have side protection, so a foot cannot slip sideways and off them. There must be enough space between a rung and a wall behind it so the arch of the foot, not just the toes, can fit on the rung. Flat rungs or steps should be very near horizontal when the ladder is in position for use. If not, they create sliding forces parallel to the surfaces.

Ladders must withstand reasonable forces applied to them without collapsing or having other structural failures. Inspections will help locate defects and ensure integrity. Defective ladders should be repaired or thrown away. Misuse and improper use can create structural failure. For example, an extension ladder must have a minimum amount of overlap between sections. If not, the bending load at the connection will cause failure. The exact amount depends on the ladder length.

Ladders can tip over sideways, tip over backwards, or slip out at the bottom, depending on ladder type. If the user of a straight or extension ladder leans out to reach something and the resultant force for the ladder and occupant falls outside the support area, the ladder will tip over. An outrigger attachment at the top will reduce this hazard. Anchoring or tying a ladder to a support structure will also prevent sliding. All ladders should rest on a solid base because on soft soil the feet may sink in, causing tipping.

Step or platform ladders may collapse or tip if the spreaders are not fully opened and locked. A load on the ladder may cause it to "walk," twist, and close up when spreaders are not fully extended and locked.

As one climbs up a ladder, the center of mass goes higher. The higher the center of mass, the easier it is to tip a ladder over. Generally, users should never stand on upper rungs or steps, although the type and length of ladder determines which rungs should not be used.

Proper positioning will prevent straight or extension ladders from tipping over backward. The preferred angle is 75° from horizontal. A rule of thumb is to place the feet of these ladders so that the horizontal distance from the support point at the top to the ladder feet is one quarter of the working length of the ladder. If the angle formed by the base with vertical is too large, these ladders will slide out at the bottom. To prevent sliding, the top should be tied off and the bottom restrained. Special feet are available on some ladders to reduce the chances of sliding.

When a straight or extension ladder provides a route to a higher work surface, the top of the ladder should extend 3 ft above the upper surface. This provides a handhold when getting on or off the ladder.

Metal ladders can conduct electricity. Therefore, they should not be used around electrical conductors or equipment.

Inspect ladders regularly and before use for cracks, wear, sway, breaks, bends, and other kinds of damage. Examine extension locks, bolts, fasteners, and feet for defects or looseness. Remove defective ladders from service until they are properly repaired. Unrepairable ladders should be destroyed and replaced.

People must learn how to use ladders correctly. They need to know how to select them for the job, place them, carry them, and store them. Ladders are not intended for transporting materials. If someone carries an object and cannot grasp onto the ladder, the three-point support principle is violated. A slip is very likely to lead to a fall. Users should clean their shoes and ladder steps or rungs to reduce the chances of slipping.

Scaffolds

Types   There are many kinds of scaffolds: wood, metal, fixed, moveable. Many scaffolds have patented designs and features. There are tubular, suspended, and special classes of scaffolds. There are standards for most of these. Large scaffold systems require structural design and analysis by people trained and qualified for such work.

Tubular Scaffolds   There are two main types of tubular scaffolds: frame and tube. Both are modular and are erected from components to fit particular job requirements. Frame types have fewer individual parts and connections and have standard frame components used to assemble and lock parts together.

Suspended Scaffolds   Hanging scaffolds are classified by number of suspension points, type of anchoring, and suspension system. Single-point suspension scaffolds include a boatswain's chair and a single person work cage. Suspended scaffolds usually hang by ropes from cornice hooks, outrigger beams, or other anchors. Work platforms assembled at a job site include the platform, guardrails, and suspension. Manufactured stages have preassembled components. Powered or manually operated pulleys, winches or other devices move platforms up and down. Proper rigging is very important.

Special Scaffolds   There are many kinds of specialty scaffolds. Ladder jacks attach to straight or extension ladders. Two ladders with ladder jacks will support planks between them. Roofing brackets, anchored to a roof with steep slopes, support planks between them. The brackets are nailed to a roof or are suspended with ropes and are removed after shingles are in placed. Some masonry scaffold systems allow for incrementally jacking the working surface upward as the courses of masonry are completed.

Another special type of scaffolding, called slip form scaffolding, combines concrete forms with scaffolding. The scaffolding is raised as the concrete cures. In some cases, these systems allow for continuous pouring of concrete.

Planking   Planking for scaffold work surfaces may be wood or metal. Wood planking is a special grade of lumber free of knots and other defects that would reduce load capacity. Cleats may lock planks in place. Many scaffolding accidents are the result of unsecured and loose planks.

Loads   A major hazard for scaffolding is overloading and structural failure. Often scaffolding is load tested before use. Scaffolding is rated light, medium, and heavy duty, depending on designed working loads of 25, 50, and 75 lb/ft², respectively. All load-carrying components must meet design standards. Scaffolds should be inspected before use. Complete assembly of all fastening bolts, connectors, and bracing is important. Ropes, suspension fittings, wear on ropes, counterweight adequacy, outrigger beams, and clamp tightness should all be checked.

Tipping   For ground-supported scaffolds, a hazard is tipping over. Placing each leg on a solid base will prevent tipping. Tying a scaffold to an adjacent building or structure also will prevent tipping.

Falls   Falls of people and objects from scaffolds is another hazard. Controls previously discussed are appropriate.

Use   Users of scaffolds must learn proper assembly, testing, inspection, and use. A qualified scaffold designer should oversee many scaffold applications, particularly large and complex scaffolding systems.

Elevated Work Platforms

Some work sites where assembly of scaffolding would be expensive require a temporary work platform at some height. A number of mobile devices provide vertically adjustable work platforms; some are self-propelled. One type of mobile device is an aerial basket, or "cherry picker," which can move vertically and horizontally. Other types may telescope vertically. Platforms used around electrical equipment must be insulated from electrical hazards. Properly positioned outriggers on some work platform devices provide stability and ensure that the telescoping section is plumb. The use site may need preparation to be sure that it is level. The surface also should be solid to prevent sinking of supports and tipping of equipment.

Some powered platforms have controls at the platform or basket. A second set of operator controls are located at ground level for general and emergency use. The platform should lock at the level desired and should not be subject to free fall if there is a power failure. Users must learn how to set up and operate elevated work platforms. Lifelines are worn and connected to the platform to prevent workers from falling. Controls should be fail safe and have features to prevent inadvertent actuation. Some controls are interlocked to level indicator switches to prevent tipping the equipment over.

EXERCISES

1. A fixed stairs is to be constructed between two floors of a building. The distance between floors (surface to surface), is 140 in. Using OSHA stairs criteria, determine

(a) how many steps are needed

(b) the depth of each tread and the height of each riser

(c) the stairs slope for your solution

2. A straight ladder that has a length, L, of 35 ft is placed against a building that is 50 ft tall. The ladder forms the desired angle with the wall, the base being $L/4$ ft from the wall. Assume the ladder weighs 90 lb and the person climbing on it weighs 200 lb. Assume that the person's center of gravity acts 3 ft vertically from the point of contact on the ladder rungs. The coefficient of friction between the ladder feet and surface it rests on is 0.5.

(a) How far up the ladder can the worker go before it slides out at the bottom?

(b) If the worker stands 3 ft from the top of the ladder, how much force is required to tip the ladder away from the wall, assuming the force is acting horizontally against the top of the ladder? Neglect vertical forces at the wall.

3. A worker is struck by a 30-in long piece of pipe weighing 1.92 lb/ft. The pipe fell out of an overflowing trash can from a platform at a level 70 ft above the worker's head.

(a) What was the velocity of the pipe when it struck the worker?

(b) What is the kinetic energy of the pipe at the point of impact?

4. While walking on a level surface, a person exerts a force, F, of 100 lb from the hip joint to the shoe of the forward foot just at the time the forward foot is planted fully. Assume the leg forms an angle of 15° with the vertical. What coefficient of friction is required to prevent the shoe from slipping on the floor?

5. A hemispherical plastic skylight is located on a roof directly adjacent to a work surface for air conditioning equipment. The skylight is not designed to carry the weight of a person. The work surface is about 44 in above the skylight and there are no access steps to it. What forms of protection would protect a worker who might

(a) fall from the platform onto the skylight?

(b) step on the skylight as a means to get onto the platform?

What standards should one apply to this situation?

REVIEW QUESTIONS

1. Define tripping.
2. Identify at least three tripping hazards.
3. What factors contribute to lack of visibility of tripping hazards?
4. What controls will help remove tripping hazards?
5. Define slipping.
6. Identify at least three hazards that may cause slipping.
7. What activities and environmental conditions contribute to slipping?
8. Name three types of instruments for measuring slipperiness.
9. Name five controls to prevent slipping.
10. What factors contribute to injury from falls?

11. What are four objectives of fall protection?

12. What kinds of controls prevent people from falling?

13. Identify the elements of the standard guardrail and the function of each.

14. Explain why guardrails should be at least 42 in high.

15. Explain the components of fall-limiting devices and the function of each.

16. Name seven controls for preventing objects from falling.

17. What are two ways to reduce injuries from falls?

18. Identify hazards and hazard controls for each of the following:

   (a) stairs

   (b) ramps

   (c) gangplanks

   (d) dock plates

   (e) ladders

   (f) scaffolds

   (g) elevated work platforms

NOTES

1 Szymusiak, S. M., and Ryan, J. P., "Prevention of Slip and Fall Injuries," Part 1, *Professional Safety*, June: 11–15; Part 2, July: 30–35 (1982).

2 Refer to ASTM F 609, Standard Test Method for Using a Horizontal Pull Slipmeter (HPS).

3 See ASTM E 303, Standard Test Method for Measuring Surface Frictional Properties Using the British Pendulum Tester.

4 See ASTM F 489, Standard Test Method for Using a James Machine; D 2047, Standard Test Method for Static Coefficient of Friction of Polish-Coated Floor Surfaces as Measured by the James Machine and D 6205, Standard Practice for Calibration of the James Static Coefficient of Friction Machine.

5 See ASTM F 1677, Standard Test Method for Using a Portable Inclineable Articulated Strut Slip Tester (PIAST) and ASTM F 1678, Standard Test Method for Using a Portable Articulated Strut Slip Tester (PAST).

6 See ASTM F1679, Standard Test Method for Using a Variable Incidence Tribometer (VIT) and ASTM D 5859, Standard Test Method for Determining the Traction of Footwear on Painted Surfaces Using the Variable Incidence Tester.

7 29 CFR 1910.23(e) and 29 CFR 1926.500(f).

8 49 CFR 399 subpart L.

9 29 CFR 1910 and 1926.

10 Deuteronomy 22:8, *Revised Standard Version Bible,* Thomas Nelson and Sons, New York, 1952.

11 Sulowski, A. C., "Assessment of Maximum Arrest Force," *National safety News*, April: 50–53 (1981).

12 Szyumusiak, S. M., and Ryan, J. P., "Prevention of Slip and Fall Injuries," Part 2, *Professional Safety*, July: 30–35 (1982).

13 Archea, J., Collins, B. L., and Stahl, F., *Guidelines for Stair Safety*, Building Sciences Series 120, National Bureau of Standards, Washington, DC, May 1979.

BIBLIOGRAPHY

American National Standards Institute, New York:
   A10.4 Safety Requirements for Personnel Hoists and Employee Elevators.
   A10.8 Safety Requirements for Scaffolding.

A10.11 Safety Requirements for Safety Nets.
A10.18 Construction Safety Requirements for Temporary Floor and Wall Openings, FlatRoofs, Stairs, Railings, and Toeboards.`

A10.28 Safety Requirements for Work Platforms Suspended from Cranes or Derricks.

A10.32 Personal Fall Protection Safety Requirements for Construction Demolition Operations.

A14.1 Safety Requirements for Portable Wood Ladders.

A14.2 Safety Requirements for Portable Metal Ladders.

A14.3 Safety Requirements for Fixed Ladders.

A14.4 Safety Requirements for Job-Made Wooden Ladders.

A14.5 Safety Requirements for Portable Reinforced Plastic Ladders.

A14.7 Safety Requirements for Mobile Ladder Stands and Mobile Work Platforms.

A14.9 Safety Requirements for Ceiling Mounted Disappearing Climbing Systems.

A14.10 Special Duty Ladders.

A90.1 Safety Standard for Belt Manlifts.

A92.2 Vehicle-Mounted Elevating and Rotating Aerial Devices.

A1264.1 Safety Requirements for Workplace Floor and Wall Openings, Stairs, and Railing Systems.

A1264.2 Provision of Slip Resistance on Walking/Working Surfaces.

Z359.1 Requirements for Personal Fall Arrest Systems, Subsystems, and Components.

ANSI/ALI ALCTV, Safety Requirements for Construction, Testing and Validation of Automotive Lifts.

ANSI/ALI ALIS, Safety Requirements for Installation and Service of Automotive Lifts.

ANSI/ALI ALOIM, Safety Requirements for Operation, Inspection and Maintenance of Automotive Lifts.

American Society of Mechanical Engineers, New York:

A17.1 Safety Code for Elevators and Escalators.

A17.2 Guide for Inspection of Elevators, Escalators, and Moving Walks.

A17.3 Safety Code for Existing Elevators and Escalators.

A17.4 Guide for Emergency Evacuation of Passengers from Elevators.

A17.5 Elevator and Escalator Electrical Equipment.

A18.1 Safety Standard for Platform Lifts and Stairway Chairlifts.

A120.1 Safety Requirements for Powered Platforms for Building Maintenance.

B56.10 Safety Standard for Manually Propelled High Lift Industrial Trucks.

American Society for Testing and Materials, West Conshohocken, PA:

C1028, Standard Test Method for Determining the Static Coefficient of Friction of Ceramic Tile and Other Like Surfaces by the Horizontal Dynamometer Pull-Meter Method.

D5859 Standard Test Method for Determining the Traction of Footwear on Painted Surfaces Using the Variable Incidence Tribometer.

D6205 Standard Practice for Calibration of the James Static Coefficient of Friction Machine.

F462 Consumer Safety Specification for Slip-Resistant Bathing.

F469 Standard Test Method for Using a James Machine.

F609 Test Methods for Using a Horizontal Pull Slipmeter (HPS).

F802 Guide for the Selection of Certain Walkway Surfaces when Considering Footwear Traction.

F1240 Guide for Categorizing Results of Footwear Slip Resistance Measurements on Walkway Surfaces with an Interface of Various Foreign Substances.

F1637 Practice for Safe Walking Surfaces.

F1646 Standard Terminology Relating to Safety and Traction for Footwear.

F1677 Standard Test Method for Using a Portable Inclineable Articulated Strut Slip Tester (PIAST).

F1678 Standard Test Method for Using a Portable Articulated Strut Slip Tester (PAST).

F1679 Standard Test Method for Using a Variable Incidence Tribometer (VIT).

F1694 Standard Guide for Composing Walkway Surface Evaluation and Incident Report Forms for Slips, Stumbles, Trips and Falls.

F2047 Standard Test Method for Static Coefficient of Friction of Polich-Coated Floor Surfaces as Measured by the James Machine.

F2048 Standard Practice for Reporting Slip Resistance Test Results.

G115 Standard Guide for Measuring and Reporting Friction Coefficients.

Institute of Electrical and Electronic Engineers, New York: IEEE 1307 Standard for Fall Protection for Utility Work.

ARCHEA, J., COLLINS, B. L., and STAHL, F. I., *Guidelines for Stair Safety*, Building Sciences Series 120, National Bureau of Standards, Washington, DC, May 1979.

BRUNGRABER, R. J., *An Overview of Floor Slip-Resistance Research*, Technical Note 895, National Bureau of Standards, January 1976.

BURKHART, MATHEW J., MCCANN, MICHAEL, and PAINE, DANIEL M., *Elevated Work Platforms and Scaffolding*, McGraw-Hill, New York, 2000.

CARSON, D. H., ARCHEA, J. C., MARGULIS, S. T., and CARSON, F. E., *Safety on Stairs*, Building Sciences Series 108, National Bureau of Standards, Washington, DC, November 1978.

CHANG, WEN-RUEY, COURTNEY, THEODORE K., GRONGVIST, RAOUL, and RDFERN, MARK, eds., *Measuring Slipperiness: Human Locomotion and Surface Factors*, CRC Press, Boca Raton, FL, 2003.

COHEN, H. H., and COMPTON. D. M. J., "Fall Accident Patterns," *Professional Safety*, June: 16–22 (1982).

DI PILLA, STEVEN, *Slip and Fall Prevention: A Practical Handbook*, CRC Press, Boca Raton, Fl, 2003.

ELLIS, J. NIGEL, *Introduction to Fall Protection*, 3rd ed., American Society of Safety Engineers, Des Plaines, IL, 2001.

ENGLISH, WILLIAM, *Slips, Trips and Falls: Safety Egnineering Guidelines for the Prevention of Falling Accidents*, Hanrow Press, Rancho Santa Fee, CA, 1989.

Injuries Resulting From Falls on Stairs, Bulletin 2214, Bureau of Labor Statistics, U.S. Department of Labor, Washington, DC, August 1984.

IRVINE, C. H., "Measurement of Pedestrian Slip Resistance," *Professional Safety*, December: 30–33 (1984).

KING, R., and HUDSON, R., *Construction Hazard and Safety Handbook*, Butterworths, London, 1985.

NFPA 101 Life Safety Codes, National Fire Protection Association, Boston, MA.

ROSEN, S. I., *The Slip and Fall Handbook*, Hanrow Press, Rancho Santa Fee, CA 1983.

SZYMUSIAK, S. M., and RYAN, J. P., "Prevention of Slip and Fall Injuries," Part 1, *Professional Safety*, June: 11–15; Part 2, July: 30–35 (1982).

CHAPTER 12

ELECTRICAL SAFETY

The use of electricity as a source of power has become extremely commonplace. There are very few homes in the United States and other developed countries that do not have electrical service. We are in the electronic and information age, in which electrical and electronic equipment makes things cheaper and more convenient and provides new communication and information capabilities.

As with all forms of energy, electricity has certain hazards associated with it. The goal is to eliminate or control these hazards.

12-1 FUNDAMENTALS OF ELECTRICITY

An understanding of the hazards and safeguards for electricity begins with an understanding of basic electrical phenomena.

Ohm's Law

The most important principle of electricity is Ohm's law, which defines the flow of electrical energy. It states that electron flow, I, called current, is a function of electrical potential, V, between two points and the resistance, R, between them. Ohm's law is stated as

$$I = \frac{V}{R}, \tag{12-1}$$

where

I is in amperes,

V is in volts, and

R is in ohms.

Resistance is a function of the material over which electrons move. If electrical energy or charge is to move from one point to another, there must be a difference in energy level between the two points. There must also be some conductive material that connects the two points.

If there is more than one path between two points that differ in electrical energy level, the electrons will flow primarily through the path of least resistance.

Current Density

Current density is the amount of current flowing through a conductor per unit of cross sectional area. If the area is large, the current density is low. Current and current density are important for safety.

Safety and Health for Engineers, Second Edition, by Roger L. Brauer
Copyright © 2006 John Wiley & Sons, Inc.

Resistance

All materials exhibit some resistance to the flow of electricity. Materials that allow electrons to flow easily are conductors. Some good conductors are copper, other metals, water, and electrolytic fluids. Materials that do not allow electrons to flow easily are called insulators. Examples of insulators are rubber, glass, wood, air and other gases, and most plastics.

The total resistance to the flow of electrons is the sum of all resistances presented by the flow path. The flow path can be created by one or more materials. For a single material of length l and a given cross-sectional area, the resistance, R, is given by

$$R = \rho l = \frac{l}{\sigma}, \qquad (12\text{-}2)$$

where

ρ is resistivity in ohms per unit length and

σ is conductivity in unit length per ohm.

Resistivity and conductivity of various materials are found in reference tables. Equation 12-2 shows that resistivity and conductivity are reciprocals, with conductivity the more commonly used property. Table 12-1 gives resistivity for selected copper wire sizes.

Heating

The fact that a material creates a resistance to electron flow gives rise to another important phenomenon for safety engineering. The temperature of a conductor will rise as the current flow increases. The energy lost to resistance changes to heat energy, a process called Joule heating. The amount of heat produced can be determined from Joules law,

$$P = I^2 R, \qquad (12\text{-}3)$$

where P is power in watts.

The increase in temperature depends on the amount of heat produced during Joule heating and how well heat transfers to the surrounding environment through convection, conduction, and radiation.

Another kind of electrical heating is inductive heating, which occurs when metal is placed or located inside a magnetic field. Inductive heating is highest inside an inductive coil. When high levels of alternating current pass through the coil, the magnetic reluctance of the metal causes it to heat up. Any metal in the field, including jewelry worn by someone, will heat up.

TABLE 12-1 Electrical Resistivity of Selected Copper Wire

Gauge	Resistivity ρ (ohms per 1000 ft at 20°C)	Cross Section (in²)
0	0.09827	0.08289
4	0.2485	0.03278
10	0.9989	0.008155
12	1.588	0.005129
14	2.525	0.003225
18	6.385	0.001276

Arcing

Arcing occurs when current flows through air between two conductors that are not in direct contact. Arcing produces light as electrons move across the gap between the conductors. Because dry air is a poor conductor, the distance over which an arc will travel between conductors is small. The distance over which arcing will travel in dry air increases with an increase in voltage between the conductors. We observe arcing in the form of lightning; we see it when a switch or other electrical contact is opened or closed; we see it in the contact between brushes and commutator in many electric motors. Arching is a function of voltage between conductors, the conductivity of the medium between them, and the distance between conductors.

12-2 ELECTRICAL HAZARDS

Electricity and electrical equipment create or contribute to a number of hazards. The most common ones are electric shock, heat, fire, and explosion. Electricity may produce other hazards indirectly. For example, when electricity energizes equipment, mechanical hazards may result. Some electrically powered devices produce harmful levels of X-rays, microwaves, or laser light. Certain equipment may create dangers from magnetic fields. Haddon's energy theory (see Chapter 9) helps people analyze electrical hazards.

There are many other kinds of indirect hazards that electrical and electronic equipment create or to which they contribute. For example, failure of electrical power can make building interiors dark and can make exiting dangerous or impossible. Failures of computer equipment or electronic sensors can contribute to hazards in processes or control systems in aircraft, industrial plants, or other places. Radio frequencies, field-induced currents, or static buildup can interfere with critical electronic equipment and can cause failures unless adequately shielded or insulated. This book does not detail these kinds of electrical and electronic hazards and controls for them.

Electric Shock

Electric shock refers to current passing over or through a human body or its members and to the injuries that result. For electric shock to occur, a person must become part of an electric circuit; that is, a person must become a conductor between two points that differ in electrical potential.

Electric shock effects are mainly a function of the amount of current that flows through the body. Besides current, other properties of electricity that affect the severity of shock include voltage, type of current (direct current [DC] or alternating current [AC]), and frequency of alternating current. Length of exposure and the part of the body through which the current passes are also important determiners of the probability and severity of injury. The effects are summarized in Table 12-2. Figure 12-1 presents shock effects from electricity of various forms. In general, alternating current is more dangerous than direct current and 60 Hz is more dangerous than high-frequency current.

As little as 35 mA of current has produced heart fibrillation when the current is applied directly to heart tissue. Currents as low as 50 mA at 120 V and 60 Hz have caused death. There is a threshold current at which a tingling feeling occurs; at higher current levels, there is pain; at even higher currents, muscles contract involuntarily. Normally, the brain sends electrical impulses to muscles, causing them to contract. An externally applied current can produce the same result, and at some level, the external current dominates brain impulses. People who have experienced involuntary contraction say they "can't let

TABLE 12-2 Effects of Electricity on the Human Body

| | Current (mA) | | | | | |
| | Direct | | 60 Hz | | 10,000 Hz | |
Effect	Men	Women	Men	Women	Men	Women
Slight sensation on hand	1	0.6	0.4	0.3	7	5
Perception threshold	5.2	3.5	1.1	0.7	12	8
Shock: not painful, muscle control not lost	9	6	1.8	1.2	17	11
Shock: painful, muscle control not lost	62	41	9	6	55	37
Shock: painful, let-go threshold	76	51	16	10.5	75	50
Shock: painful and severe, muscle contractions, breathing difficult	90	60	23	15	95	63
Shock: possible ventricular fibrillation from 3 s duration	500	500	100	100	—	—
Short shocks lasting t seconds	—	—	$165(t)^{1/2}$	$165(t)^{1/2}$	—	—
High-voltage surge	50[a]	50[a]	13.6[a]	13.6[a]	—	—

[a] Watt-seconds or joules.

go." The effect may vary, depending on what muscles are affected. If chest muscles are contracted by an external current, the normal contraction–relaxation cycle for breathing cannot occur. This can produce asphyxiation. A shock that produces muscle contractions often will cause sudden contractions at many locations in the body. People who "jump" from a ladder when shocked or "jump" away from a source of shock are often experiencing rapid involuntary muscle contraction.

Certain currents cause fibrillation, which is a disruption of the normal, cyclical contraction of heart muscle. Compared with other muscle tissue, heart muscle tissue has the unique ability to generate an electrical pulse. One location in the heart muscle dominates and starts the synchronous contraction process of heart muscle. An externally applied current can disrupt the synchronous pulse and can cause fibrillation. Then each local element of heart tissue generates its own current, which causes local contraction. During this random contraction of local tissue, the heart loses coordinated pumping action.

The current that is likely to cause fibrillation is called a probability current. For a 60-Hz current, the fibrillation threshold probability current I, for an arm-to-arm or arm-to-leg connection are given by Lee[1] as

$$I_{0.5\%} = \frac{W}{150} + 165(t)^{1/2} \tag{12-4a}$$

and

$$I_{99.5\%} = \frac{W}{150} + 495(t)^{1/2}, \tag{12-4b}$$

where

I_p is the current in microamperes at which p percent of the population is affected,

W is body weight in pounds, and

t is time of exposure in seconds (5 s is maximum).

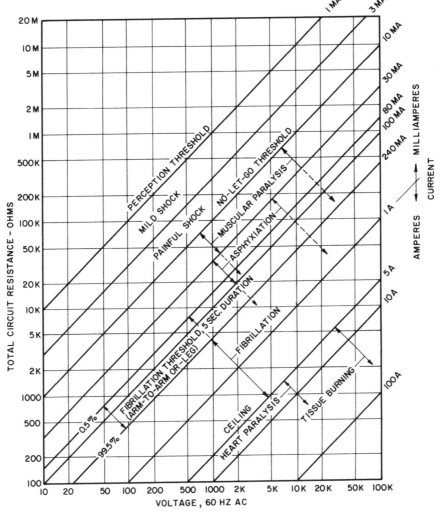

Figure 12-1.   Effects of 60 Hz AC electric current on a 150-lb man. (Reprinted by permission. Copyright © by The Instrumentation, Systems and Automation Society, 1965. From *Electrical Safety Practices.*)

For leg-to-leg connections, a current 10 times greater is needed to achieve the same probabilities.

Electric shock effects are a function of the resistance of the tissue involved in conduction. The resistance of the skin varies widely, depending on the moisture present from sweat or water. The resistance of wet skin is approximately 1,000 ohms. Below the skin the resistance of tissue is very low because it is essentially water or electrolyte.

High currents can burn tissue. Heating and arcing cause serious tissue damage. The areas of the skin that make contact with a conductor or where current enters or exits the body exhibit more serious tissue and burn damage than some internal tissue.

Heating and Fire

One of the leading causes of fires is electricity and electrical equipment. Many fires are caused when more current flows through conductors than their designed capacity, causing excessive heating that can ignite surrounding materials. Poor connections are another cause. As the contact areas between conductors become smaller, the current increases at the connections. This may increase temperatures at connections enough to cause components to glow. Poor maintenance, vibration, abuse, physical damage, and other factors may cause electrical "shorts"—current passing through routes it is not intended or designed to travel. When there is very poor contact, arcing may result and may ignite nearby combustible materials.

Explosions

Arcing in the presence of an atmosphere containing combustible dust or flammable vapors may cause an explosion. Even low-energy discharges of static electricity can initiate major disasters. Later chapters cover the principles of fire, combustion, and explosions.

12-3   CONTROL OF ELECTRICAL HAZARDS

There are a variety of controls that can reduce or eliminate electrical hazards. Groups of controls are physical controls, switching devices, grounding and bonding, ground fault circuit interrupters, and procedures. Electrical hazard controls also may eliminate or reduce other hazards. The National Electric Code and other codes, regulations, and standards, provide detailed specifications and procedures for safeguarding electrical equipment and systems.

Physical Controls

Physical controls refer to materials used, design of components, and placement of electrical equipment. Shielding, enclosing, and positioning of electrical devices can reduce contact with humans, other equipment, or hazardous materials and environments.

Wire Size and Length   Equation 12-2 suggests that the longer a wire of a given cross section or size, the greater the resistance. Also, the higher the resistance, the more the wire will heat when current flows. Each gauge and type of wire has a recommended maximum length to limit its temperature and safe use.

Location   When possible, electrical equipment should be placed where people and other equipment cannot come into contact with it. For example, poles keep power distribution lines out of people's reach and above most vehicles and equipment. Buried power lines reduce the likelihood of contact even further.

When distribution lines pass through or are located in the "people" zone, shields, conduit, and barriers should protect them. For example, covers protect power lines extending from atop a pole to below ground. Locked gates and fences keep people, animals, vehicles, and other things out of power distribution substations.

Conduit and Protective Coverings   One of the reasons for placing electrical lines in metal conduits is to prevent physical damage to them. Another reason is to reduce the chance that people will contact energized conductors. The National Electric Code specifies the number of conductors, capacities, and other factors that determine the size of conduits and related fittings that are part of the conduit system.

Nonconductive materials cover most electrical wires. There are a variety of protective materials and types of coverings. Protective materials have ratings for specific kinds of environments and conditions. Coverings may protect individual conductors, groups of them, or both. A common example is extension cords that have light-, medium-, or heavy-duty ratings. The rating does not indicate electrical capacity. The rating depends on the thickness and type of covering and indicates how well the cord can withstand physical abuse.

Sealed Equipment   When switches are turned on or off, an arc may be generated as the electrical contacts approach or separate. Similarly, electric motors arc as the brushes contact the commutator or slip rings. If an incandescent light bulb breaks, the filament will glow momentarily until it burns and breaks. When events such as these occur in an atmosphere containing a flammable mixture of air and gas or dust, a fire or explosion may result. To prevent this, special electrical equipment is installed in hazardous environments. This electrical equipment (switches, motors, lighting fixtures, conduits, etc.) is sealed to separate heat and sparks from the hazardous environment and to reduce the chance of physical damage. There are three classes of hazardous environments and divisions within the classes. Special electrical equipment is rated on its ability to comply with standards for each class and division. Table 12-3 gives a summary of hazardous location classifications.

Proper Connections   There are many ways to connect electrical conductors: plugs, receptacles, screw terminals, wire nuts, and other special fasteners. For assembled connections, the screws or other fasteners are tightened to be sure that the conductors make good contact with each other. In some cases, codes specify the force or torque required for a connection because later vibration, corrosion, creep of materials (see Chapter 10), damage, and other factors may reduce the connecting force, thereby reducing the contact area between conductors or between a conductor and a fastener. Unless the connections are tight and stay that way, contact areas get smaller and current density increases, resulting in a connection heating up and even glowing. In some applications, one should inspect connections from time to time to ensure that they are tight and in good condition.

As noted in Chapter 10, one problem with aluminum wire that was used in the late 1970s was creep. An additional problem concerns corrosion. In contrast to copper oxide, aluminum oxide is not a good conductor. As aluminum wire corrodes, aluminum oxide forms on the outside of the wire, and resistance increases, producing more heat and accelerating corrosion. This problem may lead to fire.

Isolation and Double Insulation   Another form of physical control is separating energized portions of electrical equipment from those components that people can contact (iso-

TABLE 12-3 Major Hazardous Location Classifications[a]

Class	Description
I	Locations where flammable gases or vapors are or may be present in the air in quantities sufficient to produce explosive or ignitable mixtures
II	Locations that are hazardous because of the presence of combustible dust
III	Locations that are hazardous because of the presence of easily ignitable fibers or flyings that are not likely to be in suspension in air in quantities sufficient to produce ignitable mixtures

[a]Classifications are further divided into divisions. See NFPA 70.

lation). Several means can accomplish this. Conductors can be separated from contact by covering them with nonconductive materials (insulation). Another method is to provide two layers of enclosure for energized components. At least one of the layers must be non-conductive, and the nonconductive layer must separate a user from possible contact with any energized component (double insulation). Most portable power tools (drills, saws, etc.) have internal components that cannot energize any portion of the external surface.

A person who works on electrical distribution lines sometimes stands on a rubber pad rated for the work being done. The pad is a form of isolation. It insulates the worker from other conductors so that current will not pass through the worker's body.

Other equipment relies on isolation. Line worker's tools have nonconductive handles; aerial baskets and their controls use nonconductive material to connect them to the lifting arm so that current cannot pass through the basket. For cranes, devices are available that prevent current from flowing to the cab and chassis if the boom gets close to or contacts an overhead power line.

Overcurrent Devices

Overcurrent devices limit the current that can flow through a circuit or electrical device. If current exceeds a given limit, the device shuts off power. Fuses and circuit breakers are two common overcurrent devices.

Fuses When placed in a circuit and current in the circuit exceeds some limiting value, the material in a fuse (usually lead or a lead alloy) heats above its melting point and separates, thereby stopping the flow of current. If the overcurrent is very large, components of the circuit or equipment connected to it may be damaged, because the fuse heats and melts too slowly. Fuses come in various sizes and shapes for different purposes. They may protect very low current electrical circuits or large distribution lines. There are fast-acting and slow-acting fuses. Codes specify what types are to be used in specific applications.

Circuit Breakers Circuit breakers are a form of switch that opens when current passing through them exceeds some designed limit. There are two kinds of breakers, each with a different principle of operation. One type opens when the temperature of the breaker reaches a predetermined level. The temperature of the environment around the breaker can affect its response. The second type is magnetic, which opens when a predetermined current level is reached. Environmental temperature has less affect on this type of breaker. There are many different breaker designs. Codes specify what types to use in certain applications.

Switching Devices

In addition to overcurrent devices, other switching devices can reduce or eliminate electrical hazards. They include lockouts, interlocks, and thermal or overspeed switches.

Lockouts Some switching devices use lockout devices and procedures. A lockout procedure involves placing a lock on a switch or other device to prevent the switch or equipment from being turned on or energized. As illustrated in Figure 12-2, one kind of lockout device has holes for several locks. Each person who works on equipment that can be energized by the switch places a lock through the lockout device. No one can open another person's lock and the switch will not operate until all locks are removed. A lockout pro-

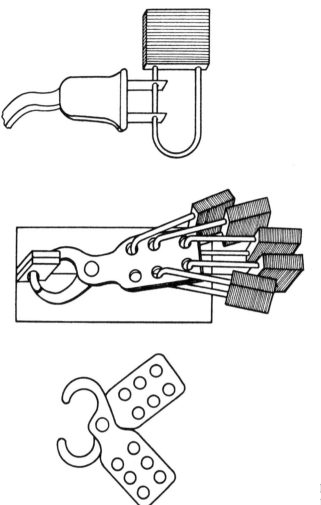

Figure 12-2. Examples of lockout devices.

cedure prevents power from being applied by someone who does not realize that another person is working on a circuit or its equipment.

Interlocks An interlock is a switch intended to prevent access to an energized or dangerous location. Interlocks are often attached to access doors, panels, and gates. When a door opens or a panel is removed, power to equipment is shut off by the interlock switch. Operator seats often have interlocks. When the operator leaves the seat, power is cut off. In some applications, interlocks that fail to work may not be detected, which will leave equipment energized and in a dangerous condition.

Thermal and Overspeed Cutouts The temperature of an electric motor will rise during use. If the temperature exceeds a certain value, a dangerous condition may exist. A temperature-sensitive switch with a preset temperature limit can interrupt power. Some electrical motors, equipment with resistance heaters (such as hair dryers), and other electrical equipment have thermal cutout switches.

There are many kinds of thermal switches. Some are normally open, some closed. Different switches have different preset action temperatures. Some operate one time, whereas others reset automatically after the temperature returns to a preset value. Still others have a manual reset button.

Overspeed switches sense when a motor or other device operates too fast. Excessive speed may create dangerous conditions or may indicate failure in the equipment. If a motor reaches excessive speed, the switch interrupts power to the equipment.

Grounding and Bonding

Grounding and bonding control the electrical potential between two bodies. If there is a difference in potential between two bodies, a conductor between them will allow charge or current to flow. That flow may be dangerous, particularly as a source of ignition.

Bonding   In bonding, two bodies have a conductor between them. As illustrated in Figure 12-3(b), bonding equalizes charge between the two bodies; it does not remove charge from them. Bonding often controls static charge buildup. Bonding is not a protection for electric shock, because a person can still become a conductor between a charged body and a ground.

Grounding   In grounding, one or more charged bodies have a conductor between them which is also connected to an electrical ground. Grounding removes charge from the bodies, as shown in Figure 12-3(c).

Grounding is usually accomplished by driving a conductive rod (usually copper) into the ground and attaching ground connections to it. Electrical codes specify size and other requirements for ground rods and ground conductors.

Grounding may protect people from electric shock. In 120-V electrical circuits, an extra conductor connects electrical equipment that people may contact with a ground. Although energized parts could shock someone who contacts them, the current most likely will flow through the ground wire, not through the person, if the parts connect to the ground wire.

Grounding of electrical equipment may not provide full protection. The ground connection from receptacle to ground rod may not exist or may have a break, rendering the ground connection to the equipment useless. If a device has a three-prong plug and someone has cut off the ground tab, the device is not grounded. Testing will determine the integrity of a grounding system. One type of low-cost ground testing device plugs into a receptacle. Colored display lights tell if the ground works properly and if the polarity of the circuit is correct. Another type of instrument measures leakage current. The ground is not adequate if there is too much current leaking from the circuit.

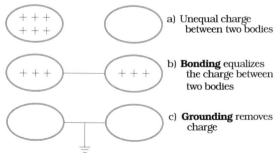

a) Unequal charge between two bodies

b) **Bonding** equalizes the charge between two bodies

c) **Grounding** removes charge

Figure 12-3.   Grounding compared with bonding.

Capacitors, chokes, and transformers retain electrical charge after external power is cut off. This charge will dissipate slowly, and one cannot tell by inspection if charge remains. In such cases, a grounding or shorting stick or hook can be used to discharge the energy. Because the tools will not remove all the charge at once, the tool may have to be applied several times to remove the charge. Some people leave a tool in place during work on equipment with devices storing electrical charge.

Ground Fault Circuit Interrupter

Another means to protect people from electric shock is a ground fault circuit interrupter (GFCI). A GFCI is a fast-acting circuit breaker that quickly senses very low current levels. Some GFCIs sense as little as 2 mA and shut off current in as little as 0.02 s. A GFCI compares current normally flowing through the power distribution wire and the grounded neutral wire of a circuit. The current flowing through one must pass through the other for the circuit to work. If current is not equal, some electrical energy is flowing to ground through other than the normal route, perhaps through a person. When the current is not equal, the GFCI detects this current differential and shuts off the current.

GFCIs protect 110/120 V circuits where users can form a ground with energized equipment. GFCIs do not work on line-to-line connections found in distribution of 220 V and higher. The National Electric Code requires GFCIs for outdoor receptacles or circuits and for bathrooms and other locations.

Low Voltage

In confined spaces and wet areas, it may be difficult to achieve grounding protection for normal power. "Safety" low-voltage equipment can reduce the electrical hazard. These special tools, lighting fixtures, power cords, and other electrical equipment are operated at less than 24 V. The safety low voltage power supply and its primary source at higher voltages are isolated from each other. This special equipment may include additional safety features for other than electrical hazards.

Smart Power Integrated Circuits

An emerging technology involves combining microelectronics and power controls. Smart-power integrated circuits (PIC) can help reduce electrical hazards. These devices are connected to a circuit and will have their own identification. The PIC will not permit current to flow to a device that it does not recognize. For example, this concept may prevent electrocution of a child who inserts a metal object into a receptacle.

Warnings

When there is a possibility that someone can gain access to energized electrical conductors and equipment, there should be warnings to indicate the dangers. The warning may take several forms, such as signs and visual or auditory warnings. Chapters 7 and 33 address warnings.

Procedures

Work procedures for installation, use of, and maintenance of electrical systems and equipment can reduce risk. Procedures differ for high-voltage power distribution, low voltage, and safety low voltage. Workers must learn these procedures. Electrical equipment for consumers must have instructions that explain electrical hazards and how to prevent electrical injury.

First Aid   Anyone working with electrical circuits and equipment should know rescue methods and first aid for electrical accidents. Too often, attempts at rescue in electrical accidents result in a rescuer becoming part of the circuit and an additional victim.

Because respiratory arrest and fibrillation are common effects, knowledge of cardiopulmonary resuscitation (CPR) is essential for those who work with electrical circuits and equipment. Without immediate treatment for these injuries, chances of survival are minimal.

12-4   STATIC ELECTRICITY

Physics

Electrical charge will build up when there is motion or friction between two insulated or partially insulated objects. The motion does not require rubbing or sliding. The likelihood of charge being created is usually greater when the two interfacing materials are different. Many activities produce static electricity. A common experience is static buildup on one's body from walking on carpet when the air is dry. A small arc occurs when a charged person touches another person or something at a different electrical potential. Belts moving over supporting rollers, sliding bulk material, the flow of fluids through pipes and hoses, and vehicle tires on pavements produce static charge. Static charge buildup is greater when the air has low moisture content. The amount of charge buildup and release of charge at undesirable times and locations are the main concerns.

The amount of energy stored or discharged, E, through a spark or arc is

$$E = \frac{1}{2}CV^2, \tag{12-5}$$

where

   E is energy in joules,

   C is capacitance in farads, and

   V is potential in volts.

Capacitance is a property of a material. Capacitance for a person is approximately 100 pF. A walk over a carpet in dry air may produce as much as 50,000 V. A resulting spark would release 0.125 J.

Hazards

The main hazard of static electricity is creation of an arc and ignition of certain vapor or dust mixtures in air. Reference tables give minimum ignition energy for various mixtures. Chapter 17 discusses this hazard further.

Controls

One control is minimizing the buildup of charge. Using materials that do not generate or store as much charge as others can help. Bonding and grounding are the simplest ways to minimize charge buildup. Grounding is preferred because it removes charge. Humidification of air in closed environments may help, but is usually more costly and less effective in reducing risks.

Grounding or bonding wires reduce hazards during fueling operations, where someone transfers fuel from one container to another. Conductive floors reduce risks in hospital operating rooms, where anesthetic gases may be explosive under the right conditions. In locations where static charge from clothing could be dangerous, workers wear conductive clothing, particularly shoes. Clothing made from fabric that resists charge buildup or is treated with antistatic chemicals can reduce risk. Conductive clothing, shoes, and flooring used to control static charge may increase the hazard of electric shock from sources other than static charge.

Training also can help reduce the hazards of static electricity. For example, some retail gasoline stations post information to reduce the likelihood of static buildup when customers slide in and out of their vehicle while fueling it or use cellular phones during the refueling activity. Employers will want to make employees aware of activities and operations that are likely to generate static charge.

There are instruments for measuring the presence and amount of static electricity. The instruments measure electrical potential of charged objects or surfaces.

12-5 HOSPITAL PATIENTS

Medical patients may have equipment and instruments attached inside or outside their bodies. Small currents leaking from one instrument to another may be sufficiently large to cause injury or death. There are many ways a patient can become part of a circuit: a patient could reach out and directly contact equipment; hospital staff could be in contact with electric equipment and touch a patient; even a person cleaning the floor with a vacuum cleaner could create a leakage current hazard. A serious hazard exists when a patient has a catheter in or near the heart or has an electrical connection attached to or near the heart during treatment, monitoring, or surgery.

A variety of techniques reduce the risk to a patient. Grounding and double insulation help. Isolation of circuits and sensor leads, minimal current for equipment operation, low voltages, and turning off unused equipment can all help. Shielding reduces magnetically induced currents. Some patient situations can become very complex, requiring special analysis to determine a safe solution. There are safe current limits and standards for the safe use of electrical equipment in hospitals and medicine.[2]

12-6 LIGHTNING

Lightning is the sudden release of static buildup in clouds, particularly during thunderstorms, which can produce very large currents. Lightning has occurred at nearly every location on our globe. However, there is a wide range in frequency of occurrence. Each year in the United States, lightning kills approximately 150 people and property damage reaches millions of dollars. Lightning can cause external damage to property, such as structural damage, heating, and fire, as well as internal damage to electrical distribution and communication and data systems within a building.

A system of lightning rods or air terminals connected to a special ground rod is the normal method for providing external protection. Air terminals are placed strategically along roof lines, on protruding building elements (chimneys, dormers, etc.), or in the form of a protective grid. The air terminals intercept lightning discharges in their vicinity and conduct the current to the ground.

Internal protection is achieved most commonly by providing common grounding points for all systems in a building. Other means include the use of surge-diverting or protection devices on electrical equipment, placing electrical equipment distant from lightning protection equipment, shielding of equipment and wires, and use of stranded and twisted overcurrent protection.

12-7   BATTERY CHARGING

Automobiles, trucks, industrial vehicles, and other equipment depend on lead-acid batteries. These batteries can explode during charging operations, causing battery acid and particles from the case to injure the eyes or skin. Two types of explosions are possible: one related to flammability of hydrogen gas and the other electrical in nature.

During charging, lead-acid batteries produce hydrogen gas from the electrolytic fluid. If the hydrogen gas reaches a flammable concentration (4%–75% by volume of air) in the air near the battery, a spark or flame can ignite it and can produce an explosion. The arc may come from attaching or removing charging cables. An external heat or flame can come from a variety of sources.

The second type of explosion can occur when making connections. An explosion can occur if two batteries being connected are of different voltage or when joining terminals of differing polarities.

To prevent the first type of explosion, dilute the air around the battery with uncontaminated air to keep hydrogen gas from reaching an explosive concentration. In a closed battery charging room, an exhaust system is needed. Charging rooms must have charging racks, and special coatings on walls and floors are desirable to prevent acid damage. Charging rooms should have an emergency eye wash fountain and emergency shower. Workers must wear protective eyewear and other protective clothing.

To prevent the second type of explosion, make sure batteries being connected together have the same voltage. Connect negative terminals to a ground last (for cars that have a negative ground) and disconnected them first. Also, one should wear protective eyewear during this operation.

EXERCISES

1. A worker was using a metal rod to unclog a spout at a grain elevator. The rod accidently contacted a 7,200-V power line. The worker suffered burns and other injuries that resulted from a subsequent fall. Estimate how much current was flowing through the worker's body. Assume a skin resistance of 80,000 ohms.

2. An operator of a boom crane became part of a circuit when the boom touched a 12,000-V power line. Assuming a skin resistance of 18,000 ohms, how much current flowed through the crane operator's body?

3. Soy protein dust requires 0.06 J for ignition. A belt moving the material has a capacitance of 300 pF. How many volts of electrical potential must accumulate on the belt to reach sufficient energy for ignition?

4. Each leg of a 150-lb hospital patient is attached to a different monitoring device. Both devices have 120 V, 60 Hz power supplies. How much current must pass through the patient to produce fibrillation at a 99.5% probability level if the exposure time is 0.1 sec?

5. A copper wire will supply current to a 2,500-W electric space heater. The supply is 110 V, 60 Hz. What is the resistance of the heating device?

6. A copper wire supplying the heater in Problem 5 will be 6 ft long. What gauge is required if the resistivity must stay below 1.5 ohms/1,000 ft to prevent excessive heating of the wire?

REVIEW QUESTIONS

1. State Ohm's law.
2. What is
   (a) current density?
   (b) resistivity?
   (c) conductivity?
3. How are resistivity and conductivity related?
4. State Joule's law.
5. Explain what may cause an electrical conductor to heat up.
6. What causes arcing?
7. What are the four main hazards of electricity?
8. What electrical characteristics contribute to the danger of electrical shock?
9. Which electrical parameter is most associated with the likelihood of shock?
10. How does electricity cause fibrillation?
11. What characteristic of skin most affects its electrical resistance?
12. What causes the "can't let go" phenomenon in electric shock?
13. How does electricity cause fires? Explosions?
14. What are the main types of controls for electrical hazards?
15. Name five physical controls for electrical hazards.
16. What are the two kinds of overcurrent devices?
17. Name three kinds of switching devices. What does each protect against?
18. How do grounding and bonding differ?
19. Explain how a GFCI works.
20. Explain one way a PIC reduces electrical hazards?
21. What is the hazard of static electricity? How can it be controlled?
22. What special electrical hazards do medical patients face?
23. How is external protection from lightning achieved? Internal protection?
24. What is the electrical hazard in charging of lead-acid batteries? What are some controls?

NOTES

1 Lee, R. H., "Human Electrical Safety," in *Electrical Safety Practice*, Monograph #110, Instrument Society of America, Research Triangle Park, NC, 1965.

2 Association for the Advancement of Medical Instrumentation, Arlington, VA.

BIBLIOGRAPHY

American National Standards Institute, New York:
    Z244.1 *Safety Requirements for the Lock Out/Tag Out of Energy Sources*
    AAMI D80 Medical Electrical Equipment
    AAMI HF18 Electrosurgical Devices
    AAMI NS14 Implantable Spinal Cord Stimulators
    AAMI NS15 Implantable Peripheral Nerve Stimulators
    UL 101 Leakage Current for Appliances
    UL 817 *Cord Sets and Power-Supply Cords*
    UL 859 *Household Electric Personal Grooming Appliances*
National Fire Protection Association Standards:
    70 *National Electrical Code*
    70B Electrical Equipment Maintenance
    70E Electrical Safety Requirements for Employee Workplaces
    77 *Static Electricity*

79 Electrical Standards for Industrial Machinery
780 Standard for the Installation of Lightning Protection Systems
Electrical Inspection Illustrated, 3rd ed., National Safety Council, Itasca, IL, 1993.
FORDHAM-COOPER, W., *Electrical Safety Engineering*, 2nd ed., Butterworths, London, 1986.
GOLDE, R. H., ed., *Lightning*, Volume 2, *Lightning Protection*, Academic, New York, 1977.
GRUND, EDWARD V., *Lockout/Tagout: The Process of Controlling Hazardous Energy*, National Safety Council, Itasca, IL, 1995.
KOVACIC, THOMAS M., ed., *An Illustrated Guide to Electrical Safety*, 5th ed., American Society of Safety Engineers, Des Plaines, IL, 2003.
U.S. Air Force Regulation 160-3, *Prevention of Electrical Shock Hazards in Hospitals*.

CHAPTER 13

TOOLS AND MACHINES

13-1 TOOL AND MACHINE HAZARDS

Since the early days humans have always used tools to extend human capabilities, and tools have gradually become more sophisticated. The industrial revolution combined different tools into machines, and a variety of power sources have improved the efficiency of tools and machines. Tools and machines using manual and other sources of power are more common than ever. There are many special kinds of tools that are often similar in function or action.

Tools and machines are a major source of injuries. Hand tools cause approximately 8% of lost-time occupational accidents; machines cause an even greater share. Because powered tools and machines involve much greater energy and power, injuries from them are likely to be more severe than for hand-operated ones.

Hazards

There are a variety of hazards associated with tools and machines. Some hazards are unique to particular tools and machines; others are related to the material acted on or the use environment. Certain hazards are common for many kinds of tools and machines.

One hazard is being struck by a tool, moving machine, or machine part. For example, most everyone has struck a finger or thumb when learning to drive a nail with a hammer. Motor vehicles, one type of machine, strike people every day.

Another hazard of certain tools and machines is being struck by materials acted on. When using a cold chisel or star drill and hammer to break a hole in concrete, particles of concrete fly from the tip of the tool with each blow. Particles fly from the cutting tool as a lathe removes material. The particles may strike the tool or machine user or someone else. Some parts of the body, such as the eyes, have a greater risk of injury than others when struck by flying materials.

Another kind of hazard is getting caught in a machine or tool. Many people have experienced the pinching action of a pair of pliers. One can get caught in many ways: in power transmission elements, such as belts, chains, gears, linkages, and other components, and in the portions of the machine that act on something else.

Repeated motion by users in operating a tool or machine can lead to a group of injuries called cumulative trauma disorders or repeated motion disorders.

The power source may add to the inherent hazards of a tool or machine itself. For example, electrical power brings certain hazards (discussed in Chapter 12). Combustible fuels, explosives, hydraulics, and pneumatics each bring an additional hazard to the tools and machines they power. Machines and tools may create noise hazards, produce air contaminants, become hot, or create other hazards.

Injuries

There are many kinds of injuries related to tools and machines. Tools and machines designed to cut materials also will cut people. Knives, hatchets, and axes are examples of hand tools. Hedge trimmers, paper cutting machines, and shears are examples of powered tools and machines.

Abrasion injuries result from contact with machines that use friction or abrasion, such as grinding and sanding equipment. Puncture wounds can result from contact with pointed tools, like drills, punches, and awls. Tissue tears may result from contact with sharp edges on equipment.

A variety of crushing injuries, fractures, and other injuries can result from getting caught in the compression action of a tool or machine.

Flying particles may cause injuries to surface tissue on impact. The likelihood of injury depends on the tissue struck, the strength of the tissue, and the ability of the tissue to absorb and distribute forces. The larger the material in motion, the greater the chances of injury. Initial impact may not be the only concern. The materials may have other hazards associated with them. For example, they may be caustic or they may be rough and cause injury if they contact skin or other tissue.

Objects may fall from machines and crush hands, feet, or bodies. Materials being operated on by tools and machines may be poorly anchored and subsequently may fall, producing crushing or related injuries.

Cumulative Trauma or Repeated Motion Injuries

Cumulative trauma injuries, cumulative trauma disorders, or repeated motion injuries are a family of injuries that result from repeated motion or repeated use of a tool or other equipment. They are one form of musculoskeletal disorder. Most involve inflammation of or damage to various tissues. Several may occur at the same time. The appearance of symptoms will vary from person to person, with frequency of activity, forces, and movements involved in the activity and other factors. Factors other than repeated motion, such as age and personal differences, may contribute to the incidence of these disorders.

Trigger Finger Trigger finger is characterized by irritation and soreness of the hand resulting from repeated use of individual fingers for operation of switches and buttons. Operation of spray guns, video games, power tools, or machine controls with one or more fingers may result in trigger finger.

Carpal Tunnel Syndrome This is an affliction caused by the compression of the median nerve, median artery, and tendons passing through the carpal tunnel of the wrist (see Figure 13-1). The medial side of the wrist has a ligamentous band that serves as a "pulley" of sorts for the tendons extending from muscles in the forearm to the fingers. The size of the passage under this band may decrease as a result of inflammation, aging, and other factors. Extreme bending of the wrist, particularly ulnar deviation, causes tissues in and surrounding the carpal tunnel to become irritated and to swell. Symptoms of carpal tunnel syndrome include tingling, pain, or numbness in the thumb and first three fingers and reduced manipulative skills in the hand.

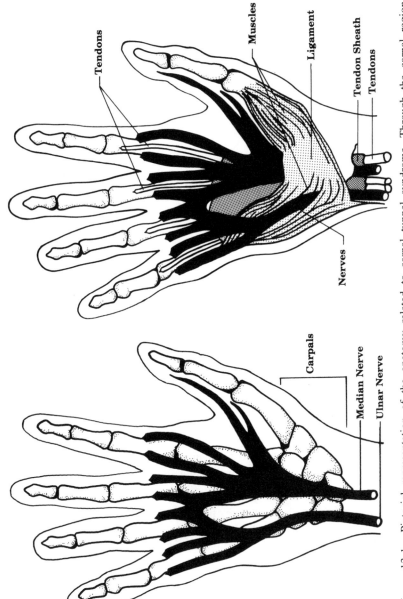

Figure 13-1. Pictorial representation of the anatomy related to carpal tunnel syndrome. Through the carpal region of the wrist pass nerves, blood vessels, tendons, and tendon sheaths. The large ligament over the carpal region helps keep tissues in place during wrist articulation. Various forms of inflammation and disorders result from repetitive motion.

Raynaud's Syndrome This disorder, also called white finger, is caused by blood vessels in the hand constricting from cold temperature, vibration, emotion, or unknown causes. Signs and symptoms include both hands becoming simultaneously cold, blue, or white and numb and loss of manipulation of fine objects. On recovery, the hands become red, accompanied by a burning sensation. Of particular concern in design of tools and controls for machines is vibration that induces Raynaud's syndrome. The handles on some machines and tools may produce very high compression forces on the tissues of the palm and may lead to this disorder. Examples of activities that lead to Raynaud's syndrom are use of impact tools and vibrating sanders.

Dequervain's Disease This disorder results from narrowing of the tendon sheath for both long and short abductor muscles of the thumb. This disease results from manual tasks involving radial or inward hand motions and firm grips.

Bursitis Each joint of the body has an encapsulating tissue called the bursa. Bursitis is the inflammation of the bursa.

Tendinitis Tendinitis is inflammation of a tendon.

Perimyotendinitis This disorder is the inflammation of tissues surrounding a tendon or its sheath.

Tenosynovitis, Stenosing Tenosynovitis, or Tenovaginitis These terms refer to inflammation involving the compression or narrowing of a tendon sheath.

Epicondylitis This disorder is caused by the combined action of pronation of the hand and ulnar deviation. For example, swinging a hammer involves these combined motions. Symptoms include pain in the elbow, forearm, and hand. Epicondylitis also has activity-related names, such as carpenter's elbow or tennis elbow.

Ganglion Cysts This disorder is the enlargement of a nerve cell or ganglion. The disorder is often manifested with other cumulative trauma disorders.

Rotator Cuff This disorder stems from a tearing of a particular ligament in the shoulder joint. As a result, the mechanics of the shoulder change. One has difficulty beginning to raise the arms from its position along the side of the body or raising it above shoulder height.

Thoracic Outlet Syndrome This disorder stems from restriction of the blood supply or compression of the nerves passing from the neck to the shoulder. Symptoms may include lateral arm pain, tingling of the ring and small finger, and hand swelling or weakness. Activities contributing to this disorder include pulling the shoulders back and down, carrying heavy loads in the hand (such as carrying suitcases or similar objects), working overhead with repeated arm abduction and adduction, and holding something between the shoulder and neck.

13-2 SAFEGUARDING MACHINES

History

Machines became prominent during the industrial revolution. With the growth of machine use, injuries resulting from machines increased. Principles of guarding closely paralleled the industrial revolution. Whereas machines, designs, and materials have changed, the principles of guarding have been around for a long time. Over the years, guarding principles and standards have appeared in many publications on machines. New technologies have provided some new capabilities for machines, but the basic ideas for guarding have changed little.

The first interlock, a mechanical arrangement, was patented in 1868. The first electrical interlock on a power transmission guard was patented in 1923. In 1899, a patent was issued for a press guard that prevented the machine from acting if the guard was not fully closed. The earliest set of standards for machine guarding appeared in 1914. The current series of ANSI machine guarding standards (B11 series) began in 1922 with adoption of the first standard for power presses and foot and hand presses. The first edition of the standard for guarding power transmission apparatus (B15) appeared in 1927.

Guarding Principles

Guards, one type of machine safeguard, are preferred over other types. Guards on machines are intended to keep people and their clothing from coming into contact with hazardous parts of machines and equipment. They also prevent flying particles from an operation and broken machine parts from coming into contact with or striking people. Guards may serve other functions, such as enclosing noise or dust and forming part of an exhaust system for contaminants.

Guards must have certain characteristics. They must be a permanent part of the machine or equipment, must prevent access to the danger zone during operation, and must be durable and constructed strongly enough to resist the wear and abuse expected in the environment where machines are used. Guards must not create hazards. They should not interfere with the operation of the machine. Even though some designers incorporated guards that protect machine users only during normal operations, where possible they should be designed so routine inspection, adjusting, lubricating, cleaning, and repairing can be performed without removing them. Because the equipment likely will pose the same hazards during setup, maintenance, and repair, guards should protect these activities as well.

Guard Openings   Guards may have openings for several reasons: to insert materials into a machine for processing; to allow access for inspection or lubrication; to monitor machine action. The larger an opening, the farther one can reach through it. In approximately 1950, the insurance industry studied the opening-reach distance relationship. Since then, people have used the resulting data, found in Table 13-1 and Figure 13-2, to define how far a barrier must be placed from a danger point for certain size openings. A danger point or line is where the dangerous action of a machine occurs or is a machine action location narrowing to $^3/_8$ in, which forms a pinch point. Three-eighths of an inch is the approximate thickness of a finger. Figure 13-3 illustrates the danger line for openings between rollers and the application of distance-opening data. For rollers that have openings larger than $^3/_8$ in, the fingers and hands may not sustain severe crushing. Not only are straight line distances important, but so is the potential of reaching around or over some

TABLE 13-1   Distances a Guard Must Be from a Danger Line for Various Sizes of Openings

Distance of Opening from the Danger Line (in)	Maximum Width of Opening (in)
$1/2-1^1/2$	$1/4$
$1^1/2-2^1/2$	$3/8$
$2^1/2-3^1/2$	$1/2$
$3^1/2-5^1/2$	$5/8$
$5^1/2-6^1/2$	$3/4$
$6^1/2-7^1/2$	$7/8$
$7^1/2-12^1/2$	$1^1/4$
$12^1/2-15^1/2$	$1^1/2$
$15^1/2-17^1/2$	$1^7/8$
$17^1/2-31^1/2$	$2^1/8$

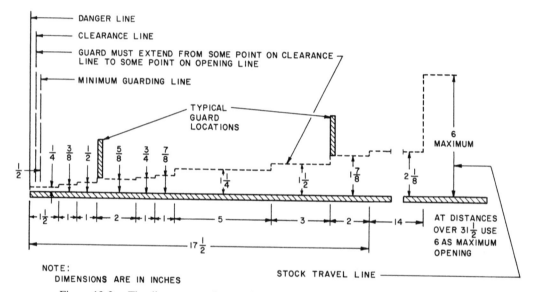

Figure 13-2.   The distance guards must be located from a danger point depends on the size of the opening in the guard.

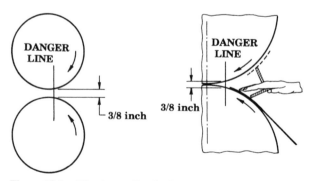

Figure 13-3.   The danger line for in-running rolls. Guard placement is determined from the defined location.

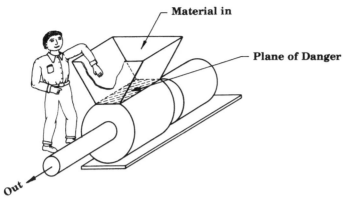

Figure 13-4. Example of using distance to keep someone from reaching into dangerous parts of a machine. The application and various standards determine whether the use of distance provides sufficient protection.

obstacle. Figure 13-4 gives an example. In some cases, one also must use human anthropometric data to determine if people can reach into the hazardous part of a machine.

Guard Construction   Guards must be substantial. To make them substantial, designers must consider the forces and events they are intended to protect and the use environment for the machines and guards. The environment may be dusty. Vehicles dumping material may strike guards. In some cases, the structural criterion is the force of a person leaning or falling on a guard. In other cases, the force and energy of flying machine parts or materials will establish what is substantial. Environmental factors, like a constant shower of particulates, a corrosive mist, or excessive heat, will determine what is substantial. Some standards give recommended materials of construction.

Types of Machine Motion

There are five types of machine motion (see Figure 13-5) that create machine hazards: rotation, reciprocating or transverse motion, in-running nip points or pinch points, cutting actions, and punching, shearing, and bending.

Rotation   One example of rotating motion is the spindle, chuck, and bit of a drill. Others are shafts, splines, couplings, feed stock, flywheels, and handwheels. Burrs, set screws, keys, rough surfaces, and rotary or eccentric motion all tend to grab clothing or hair. Even smooth, rotating surfaces can catch clothing, hair, or other loose material and wind it up. After being caught, clothing or hair quickly pulls a person into a machine. Skin contact with these moving surfaces also can produce injury.

Reciprocating or Transverse Motion   Back-and-forth motion and motion in a straight line are dangerous because someone may be in the line of motion, resulting in their being struck by the moving part of the machine. Someone also may become sheared or pinched between fixed and moving parts of the machine or between the moving part of the machine and the surrounding fixed object. Even the body of a crawler-type backhoe can create a shear or pinch point when operated near a concrete wall. When the body is facing forward, the rear of the body may have clearance from a wall located along side. When the body

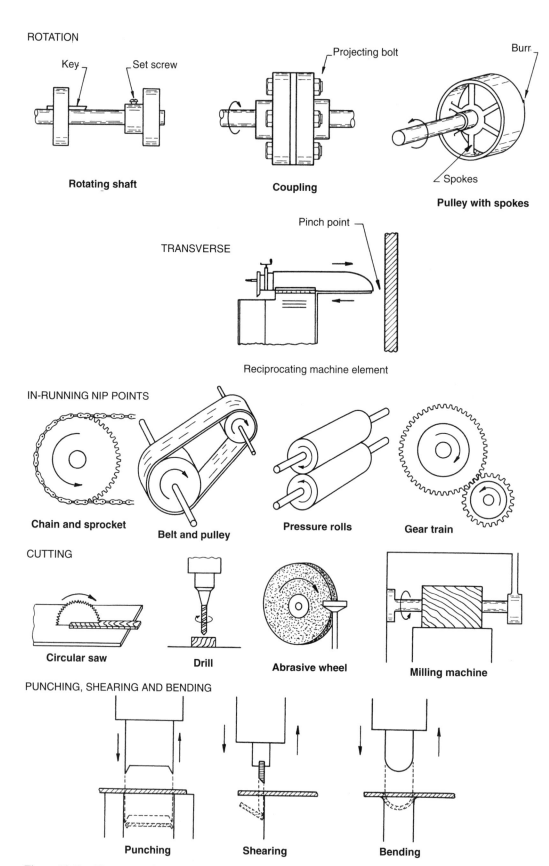

ROTATION

Key — Set screw

Rotating shaft

Projecting bolt

Coupling

Burr

Spokes

Pulley with spokes

Pinch point

TRANSVERSE

Reciprocating machine element

IN-RUNNING NIP POINTS

Chain and sprocket

Belt and pulley

Pressure rolls

Gear train

CUTTING

Circular saw

Drill

Abrasive wheel

Milling machine

PUNCHING, SHEARING AND BENDING

Punching

Shearing

Bending

Figure 13-5.   Five types of machine motions and examples of each.

rotates to the side, the rear of the body may extend over the top of the wall or may crush someone against the wall.

In-Running Nip Points Machine parts that rotate toward each other or rotate toward a fixed component create in-running nip points or pinch points. If these pinch or nip points catch someone, severe injury can result. Examples of in-running nip points are belts and pulleys, chains and sprockets, gears, conveyor belts and rollers, and rolls used for various forming, mixing, and processing functions.

Cutting Actions Machines that cut or remove material have cutting actions. Examples are saws, milling machines, shaping and planing machines, lathes, grinders, and boring mills. If the machine can cut wood, leather, metal, or paper, it can cut human tissue. The location of a machine's cutting action is the point of operation.

Punching, Shearing, and Bending This type of motion describes two machine components that come together. Both may move or one may be fixed. Examples of this motion are punching, shearing, bending, stamping, drawing, and trimming. If the machine can perform these functions on metal, paper, wood, or other material fed into it, it can also damage human tissue. Again, the danger is at the point of operation.

Types of Safeguards

Safeguards include guards, devices, distance, or location. Safeguards for machines fall into two major groups: mechanical power transmission safeguards and point-of-operation safeguards.

Guards Guards are barriers that prevent the entry of a person's body and clothing into a hazardous part of a machine. They also prevent materials from striking and injuring someone.

Devices Devices are controls or attachments that inhibit normal operation of a machine if any portion of a person's body is within a hazardous area.

Distance Distance places the hazardous portion of a machine vertically or horizontally out of reach to prevent inadvertent contact with or access to a dangerous part on motion.

Location Location means placing a hazardous machine or component where people will not normally be.

13-3 POWER TRANSMISSION SAFEGUARDS

Most machines use power transmission components to transfer power from a motor, engine, or other prime mover to an element of a machine where a useful function occurs. Some common power transmission components are belts, pulleys, ropes, sheaves, chains, sprockets, gears, and friction rollers. Pinch, nip, or shear points are created as these components come together.

Rotating shafts and couplings are also power transmission components that pose the danger of rotary motion. Some power transmissions include flywheels, which store energy during the power transmission process. At excessive speeds or after extended stress, they may fly apart. Unless these components are fully enclosed within the machine, people can come into contact with them.

Guards

If possible, power transmission guards (see Figure 13-6) should totally enclose the hazardous components. They should prevent fingers, hands, or other parts of the body from coming into contact with them. Designs should prevent personnel from reaching around, over, under, or through the guard to the danger point.

Guards should be fastened to the machine to prevent access to the hazardous areas and they should have an interlock that will de-energize the machine when the guard is removed. When possible, guard designs should allow lubrication, adjustment, and inspection to be performed from outside the guard, without removing it.

Devices

If the guard has an access panel, the panel should have an interlock or presence-sensing device that de-energizes the machine or enclosed power transmission equipment. There should be an emergency shutoff control within reach of the hazardous components for each and every operator. This is necessary in case someone does get caught when the guard is removed or a panel is opened while the machine is, for some reason, energized.

Distance and Location

Mechanical power transmission equipment that is out of reach of people may not require guards. The minimum distance required from a floor or walking surface to mechanical power transmission apparatus varies with different standards, although 7 or 8 ft is often used. Most people would not be likely to contact or reach above this distance. Anthropometric reach data also can establish safe distances from people to danger points. Distances within reach require guarding, that is, enclosed screens, walls, or fences that provide safe locations and are permanent and substantial. They should be at least 8 ft high or have functionally equivalent features. Only trained workers should have access to the enclosed locations.

Warnings

Guards and access points to hazardous areas of power transmission components should have warnings. The warnings should tell people to de-energize and lockout power before opening the access or removing the guard.

Openings

Openings in power transmission guards, partitions, or screens must prevent people from coming into contact with hazardous elements. Such openings must comply with data in Table 13-1 and Figure 13-2.

13-4 POINT-OF-OPERATION SAFEGUARDS

A point of operation is the location where a machine performs work. Although there are many kinds of machines, each performing a different function, the basic kinds of machine

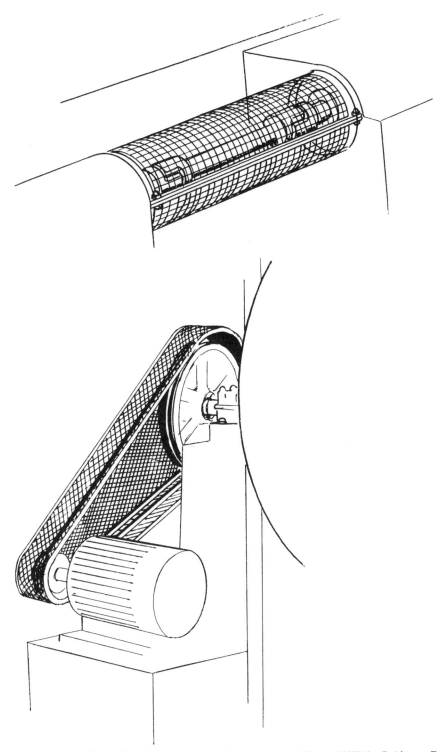

Figure 13-6.   Examples of power transmission guards. (From *MSHA's Guide to Equipment Guarding*.)

motions previously described cover them all. The safeguards for points of operation are guards, devices, tools, awareness barriers and signals, and emergency stop controls. Not all of these safeguards apply to every machine; references about particular machines will give more details and will identify suitable safeguards.

Guards

Enclosure Guards   Point-of-operation guards (see Figure 13-7) should prevent fingers, hands, other body parts, and clothing from reaching through, under, over, or around the guard to the point of operation. Criteria in Table 13-1 and Figure 13-2 apply. A guard should be fixed to the machine so that it does not create additional hazards, such as a pinch point. Construct and attach guards with fasteners that cannot be removed without tools to help prevent unauthorized removal. Most guards should permit viewing the point of operation.

Interlocked Guards   In some cases, a guard needs a hinged or moveable section for setup, adjustment, or maintenance, whereas in other cases, a guard is removed for such activities. The section or entire guard must be interlocked to the machine motion, so the machine will not operate while the section is open or the guard is removed. An example of a guard interlock is shown in Figure 13-8.

Adjustable Guards   Some guards must be adjusted for different operations (see Figure 13-9). If sections are adjustable, they should meet the preceding requirements.

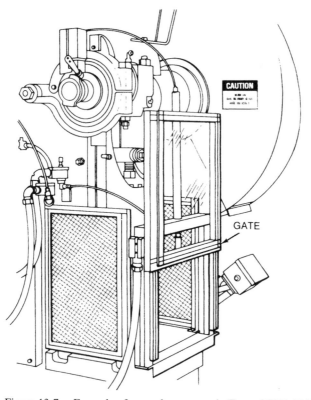

Figure 13-7.   Example of an enclosure guard. (From OSHA 3067.)

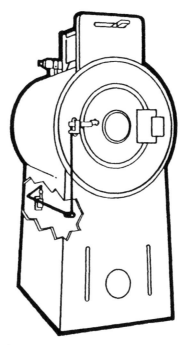

Interlock guard for drying tumbler. When door is opened the circuit is broken and the revolving cylinder stops.

Figure 13-8. An example of an interlock.

Special Guards There are many special guards (see Figure 13-10) that are unique to particular machines or machine actions. A few examples illustrate some of these special problems.

Ring Guard A rotating cutter on a shaper may be exposed on all sides. A ring guard adjusts to allow material to pass under it and to provide protection from the remaining exposed portion of the cutter.

Leg of Mutton Guard A leg of mutton guard covers the cutting head of a jointer. It gets its name from its shape. It swings horizontally out of the way when wood material moves across the jointer table. Its position varies with adjustments made in the fence that guides the material. The part of the cutting head behind the fence needs to be guarded as well. Because a leg of mutton guard can move only horizontally, it will not protect all jointer cuts. Other guards that float vertically give protection when material moves over the cutting head and extends over the side of the jointer table.

Hood Guard Circular table saws typically have a hood guard over the blade. The guard "floats" vertically as material is moved into the blade.

Grinding Wheel Guards Guards for grinding wheels perform several functions. They help keep operators from coming into contact with the wheel itself and they help contain fragments should a wheel fracture, disintegrate, and fly apart. As a result, grinding wheel guards are made of much heavier material than other guards and the opening provided for the grinding operation is held to a minimum. In addition, these guards often enclose the spindle end, nut, outer flange, and side of the abrasive wheel. The guard may

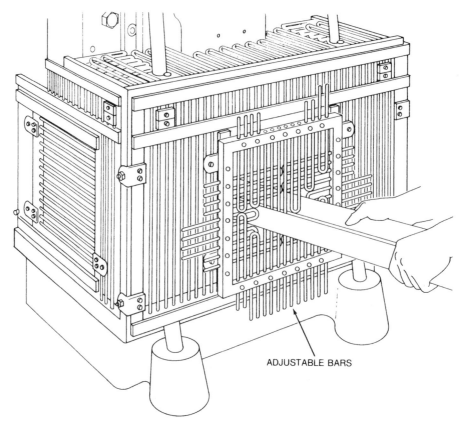

Figure 13-9. Example of an adjustable guard. (From OSHA 3067.)

form part of an exhaust system that collects particles from the operation. It may have a work rest attached to it and have enough structural strength for that purpose. Because there are many kinds of grinding operations and grinding wheels, there are many variations and additional design features to fit each variation.

Devices

When it is not possible to provide an adequate guard, point-of-operation devices are used. Point-of-operation devices ensure that the machine cannot operate or create a hazard while an operator has fingers or hands in the point of operation. For example, it may be necessary to open a guard to insert or remove parts the machine must act on, and hands may reach the hazardous area of the machine through such openings.

There are many kinds of point-of-operation devices. They include gates or moveable barrier devices, presence sensing devices, pull-out devices, sweep devices, hold-out or restraint devices, two-hand controls, and hand-feed tools.

Automatic or Semiautomatic Feed and Ejection One means for preventing the need to place fingers and hands on the point-of-operation is the use of automatic or semiautomatic feed and ejection equipment. Automatic devices require no action by an operator; semiautomatic devices require a control action by an operator. Material is fed and removed through openings in an enclosure guard (Figure 13-11). Mechanical actions, air pressure,

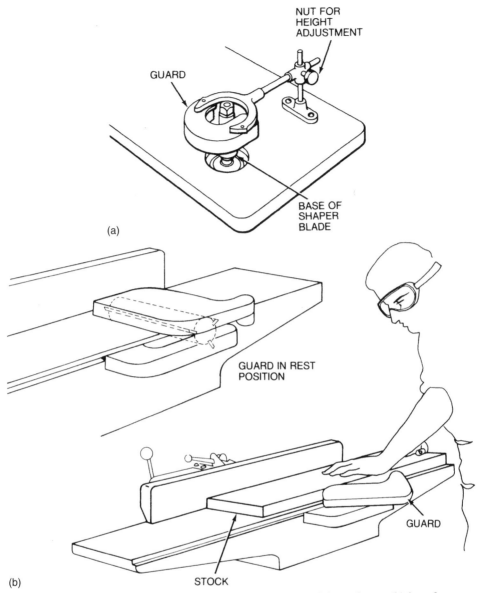

Figure 13-10. Examples of some special guards: (a) ring guard for a shaper; (b) leg of mutton guard for a jointer. (From OSHA 3067.)

and gravity are used for feeding and ejecting materials. The guard openings should meet the criteria in Table 13-1 and Figure 13-2. The means to accomplish this depend on the machine, its function, the material acted on, and other factors.

Gates or Moveable Barrier Devices   These are panels or barriers that open to allow materials to be inserted into or removed from a machine. These devices must enclose the point of operation before the machine can start and they stop the machine if the gate or barrier is opened.

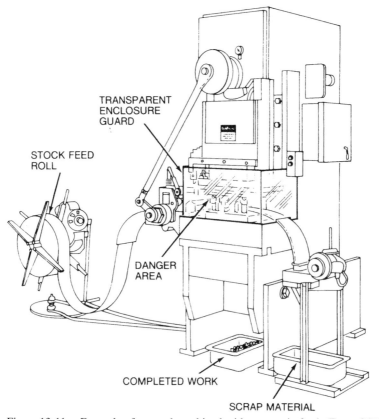

Figure 13-11.   Example of a guard combined with automatic feed. (From OSHA 3067.)

Presence-Sensing Devices   Presence-sensing devices detect whether the operator's fingers or hands are or could be in the point of operation. If they are, the machine will not operate or will stop quickly enough to prevent injury. Presence-sensing devices may not be a suitable alternative for all machines. Each type of device has particular limitations. Guards should protect all areas not protected by presence-sensing devices.

The sensing device must be far enough from the danger point to ensure that someone cannot reach into the point of operation faster than the machine can be stopped. OSHA uses the following formula to establish the safe distance, D, between sensing point and point of operation:

$$D = 63 \, \text{in/s} \times T_s, \tag{13-1}$$

where

$\qquad D$ is in inches from the point of operation,

$\qquad 63 \, \text{in/s}$ is an assumed hand speed, and

$\qquad T_s$ is the stopping time of the machine in seconds.

One must check presence-sensing devices from time to time to be sure they are working properly. Sometimes an indicator light combined with the control circuit notifies the user of a failure. Assuming a device is working properly when it is not could result in an injury.

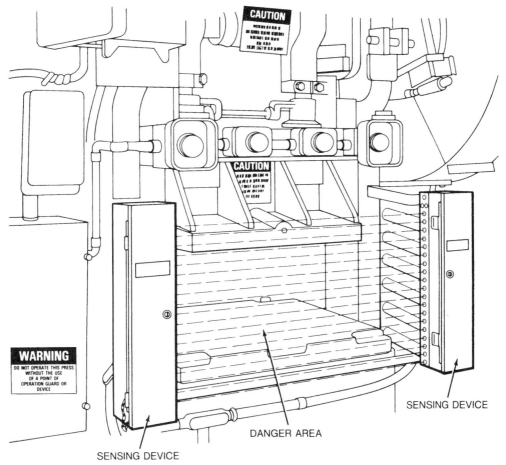

Figure 13-12. Example of a photoelectric presence sensing device on a machine. (From OSHA 3067.)

Mats One kind of presence-sensing device is a mat that detects whether someone is standing on it. These mats are placed in locations near a machine where someone could reach into the point of operation. Such a mat can be used to shut off a machine.

Photoelectric Another type of presence-sensing device uses light beams and photoelectric sensors (Figure 13-12). An array of beams forms a horizontal, vertical, or combined pattern. If something is inserted into the array that interrupts one or more of the beams, the machine stops or will not operate. Because the array is not easily seen, one must depend on a display to know if it is functioning correctly.

Radio Frequency Field Another kind of device uses a radio frequency field. Electronic circuitry that creates the field detects the presence of a human body. The field is positioned around the point of operation by an antenna, which may be a wire or conduit. The sensitivity of the field is adjustable. Grounding characteristics for an operator may change with position and other factors, such as humidity, conductivity of shoes, and so forth, and may affect the reliability of some models.

Mechanical   Some machines include a mechanical sensing device. For example on a riveting machine, a ring surrounding the point of operation moves down first. If a finger or hand obstructs the ring's full motion, the machine will not cycle the riveting action.

Pull-Out Devices   Pull-out devices (Figure 13-13) are mechanisms that attach to an operator's hands and to the moving part of the machine, usually by means of a lightweight cable. This setup couples the motion of the machine with the operator's hands in such a way that the machine action will pull the hands out of the point of operation before they can become injured. To ensure their effectiveness, these devices must be properly fitted and adjusted for each worker. The rigging must be checked each time it is attached to ensure that the device will be effective.

Hold-Out or Restraint Devices   A hold-out device is a mechanism that connects the operator's hands to some point of restraint. When properly fitted and adjusted, hands and fingers cannot reach the point of operation.

Sweep Devices   A sweep device is a mechanism that is connected to the point-of-operation action of a machine. The device sweeps an operator's hands or body away from the point of operation if they are present when the machine action begins. Sweep devices

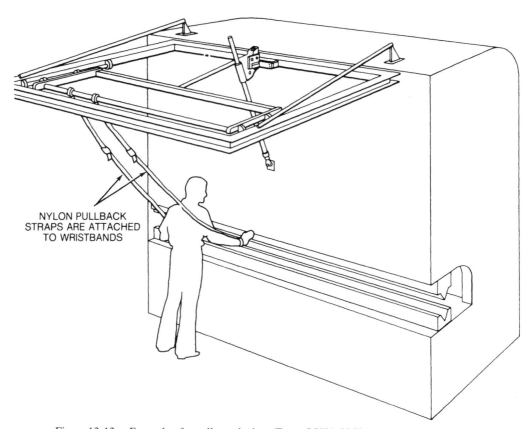

NYLON PULLBACK
STRAPS ARE ATTACHED
TO WRISTBANDS

Figure 13-13.   Example of a pull-out device. (From OSHA 3067.)

are not allowed under some regulations. In some applications, they may create a shear hazard between the sweep and some other machine component.

Two-Hand Controls A two-hand control device (Figure 13-14) is an electronic or pneumatic control assembly that requires an operator to place pressure on a control with each hand during all or most of the machine operation. The device is designed so that hands cannot be in the point of operation when they are on the controls. The controls must be designed so that they cannot be operated with one hand or arm or by one hand and another part of the body. Some two-hand controls have a time delay limit between activation of two controls. This prevents someone from pressing one control, waiting or doing something else and then pressing the second control with the hand off the first control.

Distance and Time The two-hand control must be far enough from the machine so one cannot reach into the point of operation after the machine starts in motion. Equation 13-1 would apply here as well.

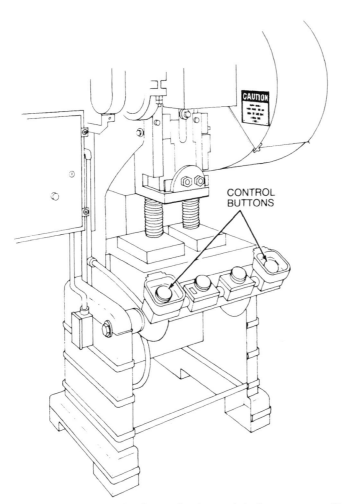

Figure 13-14. Example of a two-hand control device on a press. (From OSHA 3067.)

Multiple Operators If there is more than one operator for a machine, each should have a separate set of controls. Controls must be interlocked. All operators must have their hands free of the point of operation before the machine can start its motion.

Hand-Feed Tools Hand-feed tools (Figure 13-15) are special tools that allow an operator to place materials into the point of operation or to remove them without inserting

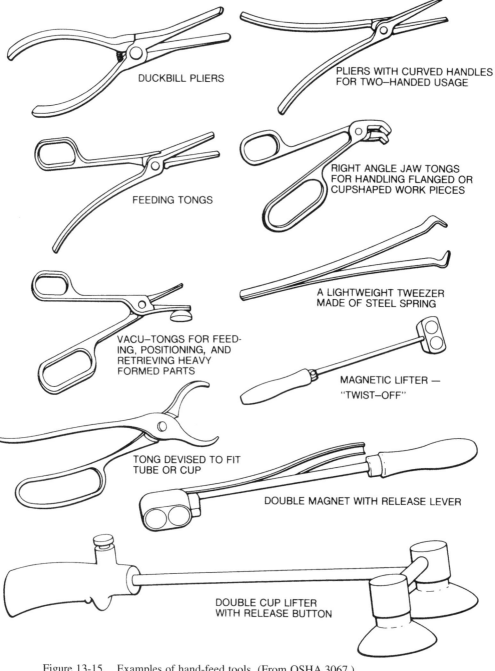

DUCKBILL PLIERS

PLIERS WITH CURVED HANDLES
FOR TWO–HANDED USAGE

FEEDING TONGS

RIGHT ANGLE JAW TONGS
FOR HANDLING FLANGED OR
CUPSHAPED WORK PIECES

A LIGHTWEIGHT TWEEZER
MADE OF STEEL SPRING

VACU–TONGS FOR FEED-
ING, POSITIONING, AND
RETRIEVING HEAVY
FORMED PARTS

MAGNETIC LIFTER —
"TWIST–OFF"

TONG DEVISED TO FIT
TUBE OR CUP

DOUBLE MAGNET WITH RELEASE LEVER

DOUBLE CUP LIFTER
WITH RELEASE BUTTON

Figure 13-15. Examples of hand-feed tools. (From OSHA 3067.)

hands or fingers. On some machines, such as table saws, jointers, and belt sanders, push blocks and push sticks have a similar use. They keep hands and fingers away from the point of operation.

Awareness Barriers   An awareness barrier does not prevent access to the point of operation, but it does alert people to a hazardous area or operation. It may work either of two ways. It may be a guard rail, gate, or similar barrier that prevents easy entry or access to the hazard area, or it may contact any part of the body of a person exposed to the point of operation to let them know that they are near or approaching a hazard. In conjunction with an awareness barrier, a warning should explain the machine's hazard and the purpose of the barrier.

Awareness Signals   These are audio or visual signals used with point-of-operation safeguards. The signals alert an operator or others that a hazard exists or is approaching.

Emergency Stop Controls   An emergency stop control is an electrical, mechanical, pneumatic, or other control used to stop or de-energize a machine when an emergency occurs. An emergency may include someone getting caught in the machine or point of operation. An emergency control overrides all other controls and requires a separate control to restart the machine. Emergency stop controls should be at locations where an emergency can arise and where operators and others can reach them easily. Stop controls are red, clearly labeled, and require only momentary contact to activate them. They are usually larger than other machine controls, so they are easy to locate and operate. In a machine where hands are occupied when an emergency occurs, there should be a foot-operated emergency stop. Multiple emergency stop controls may be needed if a machine is large or if one emergency stop control is difficult to reach or to access by someone other than the operator. If there are multiple operators, each should have easy access to an emergency stop control. If there are multiple operator positions, there should be an emergency stop control within easy reach of each position. An emergency stop that cannot be reached by a worker who gets caught is not of value to the operator. Emergency stop controls may be combined with braking devices to stop the machine quickly or may reverse the motion to render it safe.

On some machines, a pressure-sensitive body bar, safety trip rod or trip wire, cable, or cord is used. Requiring a worker to reach over a pressure-sensitive body bar to perform some function near a point of operation will allow any excessive leaning to trip the control and stop the machine. Pushing or pulling on a trip rod or trip wire also will stop machines in an emergency.

Other Machine Safeguards

Depending on the kind of machine, other safeguards may be needed.

Antirepeat   Many machines operate one cycle at a time. An operator inserts a part into the machine and then activates the machine, which operates on that part and stops. Then the operator removes the completed part. The operator becomes used to the rhythm of the machine. If, on occasion, the machine cycles more than once, the operator may have reached into the point of operation at the end of the first cycle and injury can occur. Where this situation might occur, the design for a machine action must incorporate an antirepeat mechanism as a safeguard to prevent an inadvertent extra cycle.

Brakes Some machines, like certain presses, have elevated components that may fall because of gravity. Therefore, a mechanical brake is built into them to keep the components in position until a control cycles the machine. Some machines have brakes that can interrupt a machine at any point in its cycle.

Brake monitors determine whether a mechanical brake is performing correctly on each cycle. The monitor displays a position reading. If the brake does not stop the machine component at a desired position within some tolerance, the operator will know that maintenance on the brake is needed.

Other machines, like rubber mills and calenders, use reversing current in the drive motor to stop the rolls when an emergency trip or switch is activated. According to some standards, the rolls must stop within 1.5% of the peripheral no-load surface speed of the roll.

Circular saws and other rotary machine actions tend to rotate for some time after they are shut off. Electronic motor brakes can stop the rotary motion within 1 to 2 s after the power is off.

Foot Controls Foot controls actuate some machines or a machine cycle. They may be inadvertently tripped if something falls on them or if someone inadvertently steps on them. However, a guard over the control and recessing the control inside the guard will prevent most of these occurrences (Figure 13-16). There are a variety of designs for foot controls and foot control guards. Some have a pin that locks them when not in normal use to prevent inadvertent operation of the machine.

When an operator keeps a foot on a foot control between cycles, it is possible to activate it at the wrong time. To help prevent inadvertent operation, foot controls should have a significant resistant force and should require an amount of movement consistent with intentional actions. Human factors literature and other sources give criteria for minimum switch movement.

Jog Control During cleaning, maintenance, or setup operations, it is often necessary to move machine components to certain positions. One means for accomplishing this is with an inching or jogging control. Jog controls allow the machine to be turned on and off

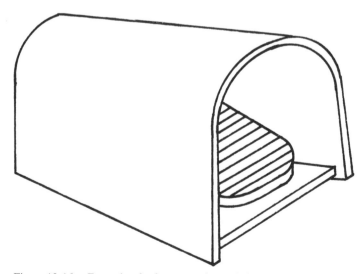

Figure 13-16. Example of a foot control guarded on the top and sides.

quickly where the danger is in turning a machine on for too long and becoming caught or creating some other hazard. When pressed or pushed, a jog control runs the machine; when released, the control stops the machine without any additional action on the part of the user.

Low-Energy Operation and Inch Controls   A low-energy mode and low speed (inching) controls can reduce the hazards of machine operation during setup, maintenance, or cleaning. The control may be a mode control switch that has a normal speed or normal energy position and a slow speed or low energy position. In some cases, manual power, such as a hand wheel or crank, may provide the low energy. Because the machine is significantly slowed from its normal operation, an operator has a much better chance to react to dangers and to avoid injury. Fully stopped and de-energized machines are preferred unless there is a clear need for low speed or inching for setup, maintenance, or cleaning.

Mode Selection   Some machines operate in different modes. Modes may include continuous operation, single cycle, low energy, or de-energized. A keyed control selects the desired mode. A key is placed into the control and is turned to the desired mode. The operator or supervisor removes the key to lock the selected mode. A key procedure is used to ensure that someone does not change the mode without appropriate safeguards in place. A mode selection switch does not replace lockout and tagout procedures.

Run Controls   Run controls turn on a machine and its action or activate the action only. If the machine is turned on inadvertently by someone or something pressing or falling against the power switch, a hazardous condition could result. Putting guards on or recessing run controls prevents accidental starting of a machine. There are various methods for guarding switches and run controls; see Chapter 33.

Blocks and Stops   During maintenance and setup activities, the ram on a press could fall and cause injury. To prevent this, a die block or safety block is used to prop it open (Figure 13-17). Some blocks set in a rack attached to the press when not in use, and removal of the block from the rack operates a switch or interlock that de-energizes the machine.

Antikickback Devices   Circular table saws have the capability of throwing or "kicking back" stock that binds or catches on the blade. A 2-in × 2-in piece of lumber thrown by a large saw can have enough energy to penetrate an operator's chest. An antikickback dog is a device that allows material to move freely into the saw blade, but pinches the material against the saw table when the material is pulled or pushed backward. A similar device, called antikickback fingers, protects radial saw operators during ripping operations.

Warnings

One or more warnings are needed on a machine to communicate hazards that may be present. If there are guards for a machine, the warnings should include a notice to keep guards in place and not operate the machine without them. When there are guard devices, the warnings should state the hazards or dangers, any limitation the device may have, and protective actions the operator must take. See also Table 7-1.

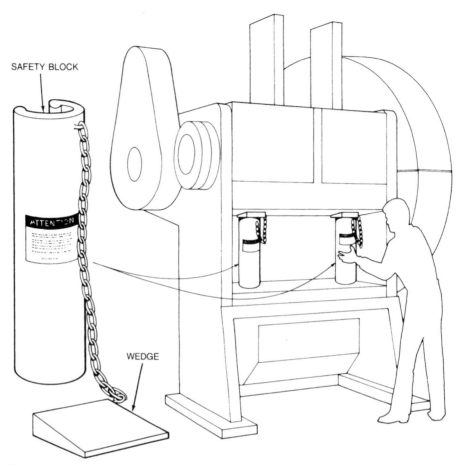

SAFETY BLOCK

ATTENTION

WEDGE

Figure 13-17.   Example of a safety block for presses. (From OSHA 3067.)

Robots

Industrial robots are now quite common. A robot is "taught" or programmed to perform certain movements and functions. Robots have many of the same hazards that machines do, and in addition, their movements are less predictable than simpler machines. The volume of space enclosing the maximum designed reach of the robot or objects it manipulates is the work envelope. One or more safeguard devices protect the work envelope for normal operations. These safeguard devices include presence sensing devices, barriers, interlocked barriers, perimeter guards, awareness barriers, or awareness signals. If an operator must enter the work envelope to maintain or train the robot, additional safeguards are needed. Lockout and tagout procedures, reduced operating speed (10 in/s or less), blocks or stops, emergency shutoff controls, keeping a second person at the robot control panel, and other safeguards may be applied. A pendant control also may help safeguard robots.

Pendant Control   A pendant control allows an operator to control a robot from within the work envelope. The pendant control must have certain features. It must be a single point of control; that is, no other controls can operate the robot when it is in the pendant control mode. The robot motions being controlled should be only at slow speed, and when buttons or other controls on the pendant are released, they must stop the robot motion. The pendant control should not have the capability to place the robot in automatic mode and it must have an emergency stop.

Safeguarding Procedures

Guards and guard devices may not provide enough protection for all the operations and activities involved in the use, maintenance, and repair of machines and tools. A number of safeguarding procedures will help prevent accidents from occurring.

Training   An operator of a machine must learn to perform all tasks associated with the machine as well as learning what hazards the machine has and what safeguards protect users from each hazard. Operators should learn where every control is and what it is for, particularly all emergency stop controls. They should know how to make adjustments and to perform setup, maintenance, repair, and cleaning tasks in a safe manner. Learning how to run a machine is not enough. Operators must learn to manage contingencies. They should learn what kinds of things can go wrong, what hazards such situations present, and how to deal safely with them when they do occur.

Enforcement   Employers must ensure that workers follow correct and safe operating procedures. Employers and workers must be sure that all guards and guard devices are in place and we working properly. Feedback from workers can provide ideas for making equipment and related operations safer.

Inspections   Machines should be inspected regularly to detect potential problems early. Included should be regular checks and testing of guards and guard devices. Any hazards or inadequate safeguards should be corrected before using a machine further.

Clothing and Jewelry   Loose clothing, jewelry (particularly rings), and long hair can catch in many machine motions and can pull parts of a body into the machine, causing injury. During machine use, it is good practice not to wear jewelry or loose clothing and to keep long hair covered.

Some machines and operations create hazardous environments for which protective clothing is needed to reduce dangers. One must analyze the hazards of each machine and use appropriate personal protective clothing or equipment.

Lockout and Tagout Procedures   During setup, maintenance, or cleaning, a machine should be locked out and tagged out of service. This will prevent anyone from activating it inadvertently or while someone else expects it to be de-energized.

Zero Mechanical State   Zero mechanical state (ZMS) recognizes that locking out the main power sources of a machine or system may not remove all sources of energy. Pneumatic, hydraulic, or other fluid lines or components may still be pressurized and may need to be relieved or isolated to make them safe. Valves from other energy sources may not be closed. Springs may have stored energy and need to be blocked or tied. Suspended or loose components may fall or cause movement in the machine and need to be restrained. The ZMS concept recognizes that detailed procedures will help ensure that a machine or system is safe for maintenance, setup, or cleaning operations.

13-5   CONTROLS FOR HAND TOOL HAZARDS

Two groups of controls for hand tool hazards are proper practices and safeguards.

Safe Practices

A number of safe practices that apply to many hand tools are detailed in the succeeding text. In addition, there are individual practices that apply to particular tools. References on hand tools and manufacturers' publications give particular practices.

Select the Right Tool for the Job   All too often, people try to make do with an available tool, rather than obtaining the correct one. One example of a wrong use is opening a paint can with a screwdriver, which could slip from the lid and puncture the hand. Other examples of wrong uses are using the wrong-sized screwdriver for a fastener, using a screwdriver for a chisel, striking hard objects or hardened tools with a carpenter's hammer that has a hardened face, or using a pliers for a wrench. Every hand tool has a purpose and is designed with certain features for its purpose. Proper selection will prevent misuse.

Know the Hazards of the Tool   Closely associated with the function of each tool are particular capabilities, limitations, and hazards. For example, a carpenter's hammer, like certain other kinds of hammers, has a hardened face. If it strikes another hardened tool or some hard object, the face is likely to spall, causing fragments that can fly into one's eye. Sledge hammers have a softer face that makes them more appropriate for such tasks. The tip of a screwdriver is also hardened, and when it is used for chiselling or prying, it can fragment easily, whereas the struck end of a chisel is intentionally soft so that it will not fragment easily. An ax has a long handle to reach wood objects farther away, whereas a hatchet has a short handle for short swings and nearby wood objects. Adjustable and open-end wrenches do not fit nuts as well as box-end and socket wrenches. Therefore, they should not be used to loosen frozen nuts or to perform final tightening because they are apt to slip.

Use Tools Correctly   Socket wrenches have certain torque limits. Using handle extensions or adapters for smaller wrench sizes can cause sockets or wrenches to break. Sockets for hand wrenches are not strong enough for power or impact wrenches. Poor footing and excessive pulling on a wrench may cause one to slip and fall. Wearing eye protection when using striking tools will prevent eye injuries. Using woven metal gloves for cutting operations with knives will prevent hand injuries.

Maintain Tools   Inspect tools regularly to be sure that they are in good condition and repair or discard broken, worn or damaged tools. Broken handles on hammers, sledges, axes, and hatchets may cause the head to fly off and strike someone. Sprung or damaged jaws on wrenches can slip, and worn screwdriver tips can slip from the head of a screw and cause injury. Sharpen cutting edges to make cutting tools easier to use and grind the heads of impact tools, like chisels or drift pins, to prevent mushroomed heads, which can fragment.

Store Tools Properly   When tools are left lying around, someone may bump into them and become injured; they also create tripping hazards. Pointed tools left in pockets can cause injury if a person falls on them. Tool belts, boxes, chests, and cabinets can help keep tools in order and can keep them safe when not in use. Sharp tools should be stored in protective sheaths when not in use.

Tool Safeguards

There are a number of safety features found on tools that can help prevent injuries. Certain kinds of tools have particular features.

Tool Guards   When using cold chisels, star drills, and other tools intended to be struck, there is the danger of missing the tool head and striking the hand that holds the tool. A round pad of foam with a small hole can slip over the tool and on top of the hand. If one misses the tool, the pad will absorb some of the shock.

Some knives have blade guards that keep the user's hand from sliding down the handle onto the blade (Figure 13-18). Such guards are particularly useful when hands are wet or greasy, such as in meat cutting.

Handle Design   Handles on tools come in a variety of shapes (Figure 13-19). Not only do lengths vary, but so do cross sections, which normally vary over the length. Some tools (hatchets, for example) have enlarged ends to minimize the chances of the handle slipping from the hands; others are smaller. Long handles on tools like axes and sledge hammers keep the tool at some distance from the user, allowing for full swings, and keep the tool from striking the legs and feet. Tools with short handles are intended for work closer to the body, which requires partial swings.

The rationale for some handle characteristics is not clear. Recently, however, handle characteristics have received some attention. In the 1950s, Damon and others at Western Electric in Kansas City developed bent handles for needle-nose pliers used in electronics assembly. In the 1970s, without knowledge of previous work, Bennett patented handles for tools and sporting goods. His patent has a bend of $19° \pm 5°$ and a particular aspect ratio for the handle cross section. He noted that when one curls the fingers to grasp something, the opening formed by the fingers is not round, but is oval. He also noted that when one grips some tools, the user's wrist bends to its limit. He chose to bend the handle instead. Examples of Bennett's designs are shown in Figure 13-19(b,c).

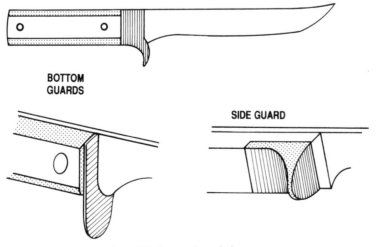

Figure 13-18.   Examples of blade guards on knives.

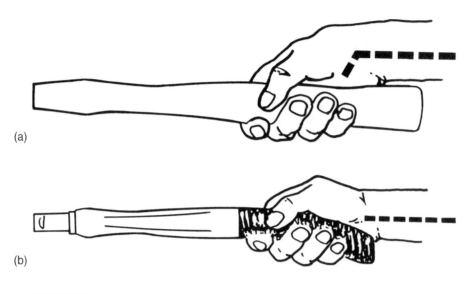

(a)

(b)

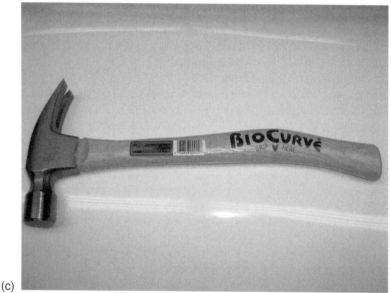

(c)

Figure 13-19.   Examples of bent handles compared with a straight handle. (a) A conventional hammer handle that requires signficant ulnar deviation of the wrist. (b) A curved handle reduces ulnar deviation. (c) A rip hammer with a BioCurve handle. (d) A mallet with a Bio-Curve handle. (Photos of BioCurve tools provided by and used with permission of BARCO Industries Inc., Reading, PA.)

Handle designs affect the biomechanics of arm and wrist motion. Carpal tunnel syndrome and some related hand and elbow disabilities seem to be relieved by adjustments in the shape or length of handles. For many tools, bending the handle reduces the bending of the wrists. With some handle designs, forces are transferred from the wrist and elbow to more powerful shoulder and back muscles. Bent handles for such tools as pliers, compression cutting tools, hammers, and lopping shears seem to reduce repeated trauma disorders and enhance use.

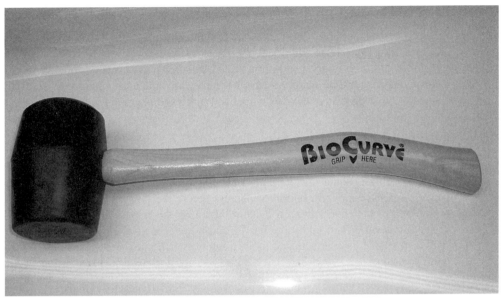

(d)

Figure 13-19. *continued*

Muscles generate the greatest force when they are about midway through their range of motion. Thus, fingers generate the strongest grip when they are approximately half flexed. The flexion of the wrist joint also affects the ability of the fingers to grip something. For example, when the wrist is fully flexed, it is difficult to grasp something tightly with the fingers.

During use, some tools create pressure in the hand. Examples are pushing on the end of a broom handle, gripping something with a pliers or cutter, and pushing on a paint scraper or putty knife. Special handle shapes can distribute the load against the hand over a greater area. This reduces the likelihood of tissue compression disorders.

High-friction plastics can increase the coefficient of friction between the hand and a tool it holds. Many tools have such plastic grips. There is a potential disadvantage for this material. If there is a tendency for the handle to slip against the skin, one can develop abrasion injuries.

This high-friction plastic material also comes in a moldable form. This permits custom shaping of handles for individual users. Such handles conform to the natural shape of a user's hand and reduce the forces needed for a good grip.

13-6 PORTABLE POWER TOOL CONTROLS

Training

Users of portable power tools need training to use them safely and properly. Training should include how to operate the tools. Users need to understand the hazards of the tool, its action, and its power supply. They need to know what protections are built in and how to protect themselves from hazards during use, and they need to know how to maintain the tools.

Proper Use and Condition

Like other tools, portable power tools are designed for particular tasks. If used for other purposes, they may create additional hazards.

Tools must be in good condition to perform well and to produce quality work. For example, blades must be sharp, because when they are dull, a user must apply extra force. Guards or other protective devices must be in place and operational.

Start Switch Lockout

Some power tools have a lockout for the start switch. The idea is to prevent inadvertent operation. A tool cannot operate until a keyed switch selects the operating mode. Some power tools have an extra button to depress before the power switch will operate. For example, a power saw without a lockout button on the start switch could be turned on accidently when picked up by the handle and a finger inadvertently activates the power switch. On some powder-actuated tools, a safety switch must be released before the tool will operate.

Interlocks

Some tools have interlocks that protect the operator or others. For example, a riding mower has a switch under the operator's seat that shuts off the blades or engine when the operator stands up. This protects the operator from being run over or getting a foot under the mower deck. Riding mowers have an interlock switch that stops the blades or engine when the operator selects reverse gear. This prevents the mower from backing over someone with the blades engaged.

Power-actuated tools have a guard over the driver end. The user presses the guard against the surface into which a fastener is to be installed by the tool. The guard is interlocked to the activating switch or trigger. If the guard is not pressed against a surface, the trigger cannot activate the tools and drive a fastener.

"Dead Man" Switch

Many power tools have activating switches that shut off power to the tool when the switch is released. Drills, saws, mowers, hedge trimmers, grinders, and other power tools have such controls. The switch that activates the tool action must be depressed at all times during use. If the operator leaves the operator position or lets go of the control, the tool stops. There is little chance of injury to the operator. For some tools, like mowers, standards require that the action of the tool stop within a certain time. This prevents the user from moving to a position of danger while the dangerous part of the machine is still in motion or coasting to a stop.

Vibration

Continuous use of vibrating tools can result in temporary or permanent disorders, so the design of the tool should minimize vibration. For some tools, vibration is an integral part of the tool and its action, so limiting the duration of use will minimize potential disability from vibration. It may be possible to isolate the vibration within the machine from the operator handles and controls or to reduce its amplitude by design. One option is incorporating isolation pads within the machine or between handles and operator.

Guarding Tool Action

Many portable power tools have the same dangers as fixed power tools. When any of the five kinds of dangerous machine actions are present, they may need guarding. The goal is to provide guards where possible. For some equipment, guarding may not be possible. For example, one cannot cover the cutting action of a chain saw. Other features, such as blade brake and antikickback designs, reduce the potential for contact with the chain.

Personal Protective Equipment

The dangers of each kind of power tool and the tool's use will dictate the need for and type of personal protective equipment. Chapter 28 provides more detail about personal protective equipment.

Some personal protective equipment is used in conjunction with certain power tools. For example, there are coveralls designed for use with chain saws. The outer fabric is tough and the chain does not easily cut through it. Another layer of material prevents the chain from contacting the skin.

Safeguarding Energy Sources

The energy sources for powered tools have hazards particular to them. One must consider these hazards and controls to safeguard them. For example, electrically powered equipment should have the safeguards discussed for electricity. Controls for hazards of gasoline or other flammable fuels used in mowers, trimmers, and other tools are important. Controls for air and fluids under pressure apply to tools with these power sources.

EXERCISES

1. A point-of-operation guard will be installed on a machine with a feed table. An operator will feed material into the machine. The material will be $1\frac{1}{8}$ in thick and require a $\frac{1}{8}$-in clearance in an opening in the guard. How far from the danger line must the guard be located?

2. Two in-running rolls (with no feed table) compress parts fed into them. The rolls are each 4 in in diameter, there is a $\frac{1}{8}$-in gap between them, and the material is $\frac{5}{8}$ in thick before compression in the rolls. A guard will protect the pinch point. The opening for the material will have a $\frac{1}{8}$-in clearance. Determine

    (a) where the danger line is located relative to the center line between the rolls

    (b) the location for the guard relative to the danger line

3. A calender with 18-in diameter rolls processes rubber. The rolls rotate at 60 rev/min.

    (a) What is the surface speed for the rolls?

    (b) If the rolls must be stopped in an emergency, what surface distance must not be exceeded? [Refer to OSHA regulations 29 CFR 1910.216(f)(3).]

4. A machine will have a presence-sensing device and an interlocking barrier gate. How far from a point of operation must the two-hand trip controls for the operator be located?

5. The doors on a food storage refrigerator or freezer in a supermarket are placed close together. When the hinge side rotates, it can create a pinch point for someone other

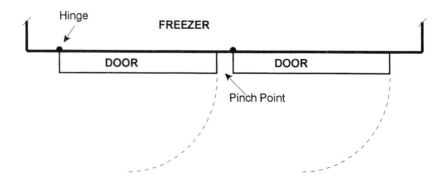

than the door operator, such as a child accompanying a parent. The design above is a cross-sectional view of two adjacent doors. What are alternate designs for the door structure that will reduce or eliminate the pinch point between doors?

REVIEW QUESTIONS

1. What are four kinds of hazards of tools and machines?

2. What additional hazards do power sources (such as gasoline, electricity, compressed air, or hydraulic fluid) add to tools and machines beside the hazard of the tool and machine actions?

3. What kinds of injuries can tools and machines inflict?

4. What are cumulative trauma or repeated motion disorders?

5. Describe the symptoms for each of the following:

   (a) trigger finger

   (b) carpal tunnel syndrome

   (c) Raynaud's syndrome

   (d) Dequervain's disease

   (e) bursitis

   (f) perimyotendinitis

   (g) tenosynovitis

   (h) epicondylitis

   (i) ganglion cysts

   (i) rotator cuff

   (j) thorasic outlet syndrome

6. When was the first patent issued for a machine interlock?

7. About when were standards for machine guarding formalized?

8. What functions can machine guards serve?

9. What safeguards are necessary for maintenance, cleaning, and setup of machines?

10. If openings are required in guards, how is protection provided?

11. What criteria are important in guard construction?

12. What are the five kinds of machine motions and the hazards of each?

13. Explain the following safeguards:

    (a) guards

    (b) devices

    (c) distance

    (d) location

14. What safeguards may protect power transmission equipment?

15. Describe each of the following point-of-operation guards:

    (a) enclosure guard

    (b) interlocked guard

    (c) adjustable guard

    (d) special guard

16. Describe the following point-of-operation devices:

    (a) automatic or semiautomatic feed and ejection devices

    (b) gates or moveable barrier devices

    (e) presence sensing devices

    (d) pull-out devices

    (e) hold-out or restraint devices

    (f) sweep devices

    (g) two-hand controls

    (h) awareness barriers

    (i) awareness signals

    (j) emergency stop controls, including trip rod, trip wire, and body bar

17. Describe the following machine safeguards:

    (a) antirepeat mechanism

    (b) brake

    (c) brake monitor

    (d) foot controls

    (e) jog control

    (f) low-energy control

    (g) mode selection

    (h) run control

    (i) block or stop

    (j) antikickback device

18. What roles do warnings have in safeguarding machines?

19. What safeguards are used for robots?

20. What safeguarding procedures are important for machines?

21. Describe the following:

    (a) lockout and tagout procedure

    (b) zero mechanical state

    (c) key procedure

22. What safeguards are important for hand tools?
23. What safeguards are important for portable power tools?
24. Describe the following power tool safeguards:

    (a) start switch lockout

    (b) interlocks

    (c) dead-man switch

BIBLIOGRAPHY

American National Standards Institute, New York:

A10.3 Safety Requirements for Powder Actuated Fastening Systems

B7.7 Safety Requirements for Abrading Materials with Coated Abrasives Systems

B11.1 Safety Requirements for Construction, Care, and Use of Mechanical Power Presses

B11.2 Safety Requirements for Construction, Care, and Use of Hydraulic Power Presses

B11.3 Safety Requirements for the Construction, Care, and Use of Power Press Brakes

B11.4 Safety Requirements for Shears

B11.5 Safety Requirements for Construction, Care, and Use of Iron Workers Machine Tools

B11.6 Safety Requirements for Construction, Care, and Use of Lathes

B11.7 Safety Requirements for Construction, Care, and Use of Cold Headers and Cold Formers

B11.8 Safety Requirements for the Construction, Care, and Use of Drilling, Milling, and Boring Machines

B11.9 Safety Requirements for the Construction, Care, and Use of Grinding Machines

B11.10 Safety Requirements for Construction, Care, and Use of Metal Sawing Machines

B11.11 Safety Requirements for the Construction, Care, and Use of Gear Cutting Machines

B11.12 Safety Requirements for the Construction, Care, and Use of Roll-Forming and Roll-Bending Machines

B11.13 Safety Requirements for Construction, Care, and Use of Single- and Multiple-Spindle Automatic Bar and Chucking Machines

B11.14 Safety Requirements for Construction, Care, and Use of Coil-Splitting Machines

B11.15 Safety Requirements for Construction, Care, and Use of Pipe, Tube, and Shape Bending Machines

B11.16 Safety Requirements for Powder/Metal Compacting Presses

B11.17 Safety Requirements for Construction, Care, and Use of Horizontal Hydraulic Extrusion Presses

B11.18 Safety Requirements for Construction, Care, and Use of Machines and Machinery Systems for Processing Strip, Sheet, or Plat from Coiled Configuration

B11.19 Performance Criteria for the Design, Construction, Care, and Operation of Safeguarding When Referenced by the Other B11 Machine Tool Safety Standards

B11.20 Safety Requirements for the Construction, Care, and Use of Manufacturing Systems/Cells

B11.21 Safety Requirements for the Design, Construction, Care, and Use of Machine Tools Using Lasers for Processing Materials

B11.22 Safety Requirements for Turning Centers and Automatic Numerically Controlled Turning Machines

B11.23 Safety Requirements for Machining Centers and Automatic Numerically Controlled Milling, Drilling, and Boring Machines

B11.24 Safety Requirements for Transfer Machines

B28.1 Safety Specifications for Mills and Calenders in the Rubber Industry

B65.1 Safety Standard for Printing Press Systems

B65.2 Binding and Finishing Systems

B65.3 Safety Standard for Guillotine Paper Cutters

B65.4 Safety Standard for Stand-Alone Bindery Trimmers

B65.5 Safety Standard for Stand-Alone Platen Presses

OPEI B71.1 Safety Specifications for Consumer Turf Care Equipment—Walk-Behind Mowers and Ride-On Machines with Mowers

OPEI B71.3 Safety Specifications for Snow Throwers

B71.4 Safety Specifications for Commercial Turf Care Equipment

OPEI B71.3 Safety Specifications for Powered Shredder/Grinders and Shredder/Baggers

OPEI B71.8 Safety Specifications for Outdoor Power Equipment—Walk-Behind Powered Rotary Tillers and Hand Supported Cultivators

SNT 101 Safety Requirements for Portable, Compressed Air Actuated, Fastener Driving Tools

SPI B151.1 Safety Requirements for Manufacture, Care, and Use of Horizontal Injection Molding Machines

SPI B151.2 Construction, Care, and Use of Film Casting Machines

SPI B151.4 Construction, Care, and Use of Blown Film Take-Off and Auxiliary Equipment

SPI B151.5 Manufacture, Care, and Use of Plastic Film and Sheet Winding Machinery

SPI B151.7 Requirements for the Manufacture, Care, and Use of Plastics Extrusion Machines

SPI B151.20 Manufacture, Care, and Use of Plastic Sheet Production Machinery

SPI B151.21 Safety Requirements for the Construction, Care, and Use of Injection Blow Molding Machines

SPI B151.27 Safety Requirements for the Integration, Care, and Use of Robots Used with Horizontal Injection Molding Machines

SPI B151.29 Requirements for the Manufacture, Care, and Use of Vertical Clamp Injection Molding Machines

B175.1 Safety Requirements for Gasoline Powered Chain Saws

B175.3 Safety Requirements for Grass Trimmers and Bruchcutters

B177.1 Safety Requirements for Three-Roller Printing Ink Mills

B177.2 Safety Requirements for Printing Ink Vertical Post Mixers

B154.1 Safety Requirements for Construction, Care, and Use of Rivet Setting Equipment

PMMI B155.1 Safety Requirements for Construction, Care, and Use of Packaging and Packaging-Related Converting Machinery

ABMA B165.1 Safety Requirements for Design, Care, and Use of Power-Driven Bruching Tools

RIA R15.06 Safety Requirements for Industrial Robots and Robot Systems

ASAE S493.1 Guarding for Agricultural Equipment

UL 1740 Standard for Safety for Robots and Robotic Equipment

Z50.1 Safety Requirements for Bakery Equipment

Z245.1 Safety Requirements for Refuse Collection, Processing, and Disposal Equipment—Mobile Refuse Collection and Compaction Equipment

Z245.2 Safety Requirements for Stationary Compactors—Equipment Technology and Operations for Wastes and Recyclable Materials

Z245.5 Safety Requirements for Baling Equipment—Equipment Technology and Operations for Wastes and Recyclable Materials

Z245.30 Safety Requirements for Refuse Collection, Processing, and Disposal Equipment—Waste Containers

Z245.41 Safety Requirements for Facilities for the Processing of Commingled Recyclable Materials—Equipment Technology and Operations for Wastes and Recyclable Materials

American Society of Mechanical Engineers, New York:

B15.1 Safety Standards for Mechanical Power Transmission Apparatus

B107.5M Hand Socket Wrenches (Metric Series)

B107.6 Combination Wrenches (Inch and Metric Series)

B107.8M Adjustable Wrenches

B107.9 Wrenches, Box, Open End, Combination, and Flare Nut (Metric Series)

B107.13M Pliers—Long Nose, Long Reach

B107.20M Pliers (Lineman's, Iron Worker's, Gas, Glass, and Fence)

B107.23 Pliers, Multiple Position, Adjustable

B107.27 Pliers, Multi-Position Electrical Connector

B107.34M Socket Wrenches for Spark Plugs

B107.37M Pliers—Wire Cutters/Strippers

B107.39 Open End Wrenches (inch and Metric Series)

B107.40 Flare Nut Wrenches (Inch and Metric Series)

B107.41M Safety Requirements for Nail Hammers

B107.42M Safety Requirements for Hatchets

B107.43 Electronic Tester, Hand Torque Tools

B107.43M Safety Requirements for Wood Splitting Wedges

B107.44M Safety Requirements for Glaziers' Chisels and Wood Chisels

B107.45M Safety Requirements for Ripping Chisels and Flooring/Electricians' Chisels

B107.46M Safety Requirements for Stud, Screw and Pipe Extractors

B107.47M Metal Chisels

B107.48M Safety Requirements for Metal Punches and Drift Pins

B107.49M Safety Requirements for Nail Sets

B107.50M Safety Requirements for Brick Chisels and Brick Sets

B107.51 Safety Requirements for Star Drills

B107.52M Safety Requirements for Nail-Puller Bars

B107.53M Safety Requirements for Ball Peen Hammers

B107.54 Safety Requirements for Heavy Striking Tools

B107.55M Safety Requirements for Axes

B107.56 Safety Requirements for Body Repair Hammers and Dolly Blocks

B107.57 Safety Requirements for Bricklayers' Hammers and Prospecting Picks

B107.58M Safety Requirements for Riveting, Scaling, and Tinners' Setting Hammers

B107.59 Slugging and Striking Wrenches

B107.60 Pry Bars

BERNARD, BRUCE P., *Musculoskeletal Disorders and Workplace Factors*, U.S. Department of Health and Human Services, Public Health Service, Centers for Disease Control and Prevention, National Institute for Occupational Safety and Health, DHHS (NIOSH) Publication No. 97–141, July 1997.

BLUNDELL, J. K., *Safety Engineering—Machine Guarding Accidents*, Hanrow Press, Del Mar, CA, 1987.

BRAMMER, A. J., and TAYLOR, W., *Vibration Effects on the Hand and Arm in Industry*, Wiley, New York, 1983.

FREIVALDS, ANDRIS, *Biomechanics of the Upper Limbs—Mechanics, Modeling, and Musculoskeletal Injuries*, CRC Press, Boca Raton, FL, 2004.

Guidelines for Controlling Hazardous Energy During Maintenance and Servicing, DHHS (NIOSH) Publication No. 83–125, National Institute for Occupational Safety and Health, Division of Safety Research, Morgantown, WV, 1983.

Hand Tool Safety-Guide to Selection and Proper Use, Hand Tool Institute, White Plains, NY, 1976.

OSBORNE, D. M., *Robots—An Introduction to Basic Concepts and Applications*, Midwest SciTech Publishers, Inc., Detroit, MI, 1983.

PARKER, KATHRYN G., and IMBUS, HAROLD R., *Cumulative Trauma Disorders*, Lewis Publishers, Boca Raton, FL, 1992.

Power Press Safety Manual, 5th ed., National Safety Council, Chicago, IL, 2002.

PUTZ-ANDERSON, VERN, *Cumulative Trauma Disorders: A Manual for Musculoskeletal Diseases of the Upper Limbs*, Taylor & Francis, New York, 1988.

RIA Robot Safety Seminar Proceedings, Robotic Industries Association, Ann Arbor, MI, 1985.

ROBERTS, V. L., *Machine Guarding—Historical Perspective*, Institute for Product Safety, Durham, NC, 1980.

Safe Chain Saw Design, Institute for Product Safety, Durham, NC, 1986.

Safeguarding Concepts Illustrated, 7th ed., National Safety Council, Chicago, IL, 2002.

Safe Maintenance Guide for Robotic Workstations, Publication No. 88-108, National Institute for Occupational Safety and Health, Division of Health and Human Services, Cincinnati, OH, March 1988.

SEVART, J. B., and HULL, R. L., *Power Lawn Mowers: An Unreasonably Dangerous Product*, Institute for Product Safety, Durham, NC, 1986.

STUBHAR, P. M., ed., *Working Safely with Industrial Robots*, Robotic Industries Association, Ann Arbor, MI, 1986.

ZENZ, CARL, editor-in-chief, *Occupational Medicine*, 3rd ed., Mosby, St. Louis, MO, 1994.

CHAPTER 14

TRANSPORTATION

14-1 TRANSPORTATION ACCIDENTS

The Problem

Each year in the United States, there are nearly 45,000 motor vehicle-related deaths and more than 2 million disabling injuries, many of which result in some degree of permanent impairment. Motor vehicles accidents are the leading cause of death between ages 1 and 44 years, and motor vehicle deaths account for more than half of all accidental deaths each year. The cost of motor vehicle accidents now exceeds $250 billion per year. The death rate from motor vehicle accidents is approximately 16 per 100,000 persons.

In addition, studies have revealed a number of other facts about motor vehicle accidents, injuries, and deaths. The data may change over time. Motor vehicle accidents are the leading cause of occupational injuries (35%) and work-related deaths (approximately 40%). For example, in many police departments, more police die as a result of squad car accidents than of gun shots.

Thirty-seven percent of motor vehicle deaths occur between 10:00 P.M. and 4:00 A.M. The death rate per crash generally increases with age. However, it is highest for people between ages 17 and 30 years, and it also increases dramatically with age for people older than 65 years. The death rate is lower for high-income individuals and is higher for males than for females, even after there are adjustments for the greater travel of males.

Approximately 50% of accidents involve frontal crashes; approximately 2% are rear-end collisions. Of the 50,000 motor vehicle accidents each year, roughly 8,000 involved rollover of the vehicle, and more than 5,000 involved impact with trees and utility poles.

Estimates suggest that 50% of drivers involved in fatal crashes have a blood alcohol content (BAC) of 0.10 or more. Some studies suggest that this percentage is significantly higher. Impairment of driver skills begins at a BAC of 0.05 or less, and all states now set driving-under-the-influence (DUI) or driving while intoxicated (DWI) limits at 0.08 percent.

There are many factors that contribute to vehicle accidents and statistics, such as gender and age. Women have a lower accident rate than men do, and the female death rate from motor vehicle accidents is nearly three times less than that for males. The accident rate is by far the highest for people in their late teenage years and early twenties.

Driving for long periods of time leads to higher frequency of crashes. Truck drivers who drive for 10 hr have nearly twice the risk of being in a crash compared with drivers on the road for less than 2 hr.

Passenger restraints contribute to lowering injury and death rates in vehicles accidents. The National Highway Traffic Safety Administration (NHTSA) estimates that more than 8,000 lives were saved between 1983 and 1987 by the use of seat belts. The addition

Safety and Health for Engineers, Second Edition, by Roger L. Brauer
Copyright © 2006 John Wiley & Sons, Inc.

of airbags also has helped to reduce injury and death rates at the same time that the number of vehicles and total miles driven continues to increase.

Vehicle size is an important factor in vehicle injuries and deaths. The frequency of injuries among drivers of small cars involved in crashes is more than twice that for large cars. There is also a tendency for small cars to be involved more frequently than large cars in crashes resulting in damage claims.

Event Sequence

The dynamics of vehicle crashes are quite complicated. There are many factors that contribute to the actual crash sequence, depending on vehicle speed, design, and motion. However, a simplified discussion explains some of the basic phenomena of vehicle crashes and impacts.

Frontal Impact When a vehicle runs into a fixed object or another vehicle in a frontal collision, there is a sequence of events that takes place within a fraction of a second. At very low speeds, the bumper and its mounting will absorb the energy of impact. The NHTSA Bumper Standard establishes that a passenger vehicle bumper should absorb the energy from a 2.5 mi/hr impact of the vehicle. For a time, the standard was a 5 mi/hr impact, which some manufacturers still meet. At higher speeds, the structure of the vehicle will absorb the energy of impact by crushing. Today's vehicles are designed to absorb a great deal of energy; older vehicles were not. Because of the absorption of energy by the structure, the rate of deceleration falls quickly with distance from the point of impact. Figure 14-1 illustrates this concept by plotting G load (the ratio of impact deceleration to the pull of gravity). At some distance from the point of impact, there is little deformation of the vehicle structure, and from that point on, the deceleration rate is fairly constant over the rest of the vehicle length.

When a vehicle strikes another object or barrier, it stops quickly. Objects within or occupants continue to move at the original vehicle speed until they impact on something to slow them down. If passengers are restrained to the vehicle structure or its components, they will stop with the vehicle.

Rear-End Collision In a rear-end collision, one vehicle strikes another vehicle in front of it. The front vehicle is usually at rest or may be moving at some speed less than the

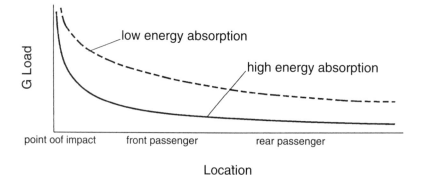

Figure 14-1. Approximate G loads from frontal impact as a function of distance from the point of impact.

striking vehicle. At slow speeds, the bumpers will absorb some energy. At higher speeds the structure of the two vehicles will crush and absorb much of the energy. The front vehicle will accelerate forward unless restrained in some way.

Within a few hundredths of a second after impact, the vehicle occupants in the forward vehicle will press into the seat backs. The seat cushioning and structure will absorb much of the energy related to the acceleration of the occupants. Consider a person sitting in the front seat during a rear-end collision. Unless the head is well restrained by the headrest, it will move rearward over the top of the seat in a phenomenon called whiplash. There are two components to the head movement: translation away from the body trunk and rotation about the neck. If the seat back remains rigid and quite vertical, the rotational component will likely be much greater than if the seat back structure yields. If the seat back yields, it absorbs some of the energy and reduces the rotational component. The resultant force on the head from the two components can be quite sizeable even at fairly low-impact speeds. Depending on the speed at impact, the body may slide backward up the slope of the seat and impact on the rear structure of the passenger compartment.

Rollover If there is enough lateral impact and movement, a vehicle may roll over. The tires may slide or contact another object. If sliding friction or restraining forces are large enough compared with the forces of lateral motion, the vehicle will roll over. The chance of rollover increases for vehicles with high centers of gravity. The vehicle structure may crush during the rollover, causing reduction of the occupant space. Because occupants may be thrown around, use of restraints will lessen the chance for injury. Some vehicles incorporate side impact structural elements to minimize reduction of occupant space. An increasing number of models include side impact air bags to cushion impacts on occupants.

The Second Crash

When a vehicle crashes, the occupants continue in motion until they impact on the interior of the passenger compartment. The term *second crash* refers to this crash of occupants against the interior surfaces of the vehicle after vehicle impact. Very often, the second crash is the primary source of passenger injuries, not the vehicle crash itself.

Vehicles designed to absorb energy and to reduce the transfer of forces to the occupants will reduce the second crash. Passenger restraints and air bags reduce the likelihood of the second impact occurring. Well-anchored seats help minimize the second crash. A seat that lets loose can add mass behind the passenger in a frontal crash. Interiors have many design features to reduce the sources of injury. Padded dashboards that distribute forces, laminated windshields that prevent glass from being imbedded in tissue, and elimination of protruding knobs and controls all reduce the probability of injuries. Energy-absorbing, collapsible steering columns significantly reduced the rate of injury to drivers during the second crash.

Crashworthiness

Crashworthiness refers to the ability of a vehicle to withstand an accident without intrusion of or reduction in the integrity of the passenger compartment. When a crash occurs, not only is it important to minimize the second crash, it is also important to ensure that the passenger compartment stays closed and retains it shape. The idea is to keep occupants inside a protected zone during a crash. Passengers thrown from a vehicle face a much greater risk of injury and death than those who stay inside.

It is important that door latches and locks perform during crashes. Doors must meet structural standards that are intended to provide compartment integrity in side impacts. The energy-absorbing steering column also reduces the injury caused by the steering column being driven into the driver's chest in a frontal crash. Compartment integrity also involves adequate structural strength to prevent collapse during rollover. The main energy absorption components of the vehicle are placed in elements outside the passenger compartment. Passenger restraints not only reduce the second crash, but keep people inside the protected zone. It is also important that fuel systems retain their integrity. If a fuel tank ruptures, spills fuel in the passenger compartment, and ignites, the opportunities for escape and rescue often are lost.

Crashworthiness and integrity of a passenger compartment permit race car drivers to survive crashes at very high speeds. Crashworthiness has improved in most cars over the years. Improved design features have contributed to the reduction in crash injuries and occupant deaths.

14-2 CONTROLLING TRANSPORTATION HAZARDS

There are many factors that can and do contribute to the reduction of motor vehicle accidents and the resulting injuries and losses. In discussing controls, one can apply the models discussed in Chapters 3 and 9. For example, in discussing motor vehicles, the four Ms representing man, machine, media, and management can refer to drivers, vehicles, roadways, weather conditions, regulations, enforcement, and cost.

Vehicles

There are many safety features built into motor vehicles today. The NHTSA has safety standards for motor vehicles: the Federal Motor Vehicle Safety Standards (FMVSS). Table 14-1 lists their titles and identifying numbers. The Society of Automotive Engineers (SAE) also publishes numerous standards for vehicles, including many safety standards. Other Department of Transportation (DOT) agencies also establish vehicle safety standards. The Urban Mass Transportation Administration has safety standards for buses and rail vehicles, and the Federal Highway Administration has safety standards for trucks.

An important design feature for trucks and cars is antilock brakes. For years, large aircraft had antilock brakes. Later, this technology moved to trucks and automobiles. The idea is to sense the moment wheels lock up as static friction between the tires and roadway changes to sliding friction. Because the coefficient of sliding friction is generally less than that for static friction, the sensors interrupt braking for a fraction of a second, allowing a return to static friction, and then braking continues. This sequence repeats frequently until the vehicle stops.

As for back as 1908, air bags were proposed for automobiles. Today, passenger cars have driver side and passenger side air bags as standard equipment. After the introduction of air bags in certain models, it took more than 20 years to make them standard equipment. Some adjustments in inflation rates were made after performance data suggested that some adults were seriously injured during inflation. There also has been controversy about the use of infant and child seats in front seats of passenger cars, because some air bags have caused some deaths. More recently, air bags have appeared in doors of some automobile models to help minimize injury during side impacts.

Some states require annual vehicle inspections to help ensure that vehicles and safety devices are in proper working order. The assumption is that vehicles with working head-

TABLE 14-1 Federal Motor Vehicle Safety Standards (FMVSS, from 49 CFR 571)

Number	Topic	Car[a]	Truck	Bus[b]	Multipurpose Passenger Vehicle[c]	Equipment[d]
100 Series: Accident prevention						
101	Controls and displays	•	•	•	•	
102	Transmission shift lever sequence starter interlock and transmission braking effect	•	•	•	•	
103	Windshield defrosting and defogging systems	•	•	•	•	
104	Windshield wiping and washing systems	•	•	•	•	
105	Hydraulic and electric brake systems	•	•	•		
106	Brake hoses	•	•	•	•	•
108	Lamps, reflective devices, and associated equipment	•	•	•	•	•
109	New pneumatic bias ply and certain specialty tires	•				•
110	Tire selection and rims for motor vehicles	•				•
111	Rearview mirrors	•	•	•	•	
113	Hood latch system	•	•	•		
114	Theft protection	•	•		•	
116	Motor vehicle brake fluids	•	•	•		•
117	Retreaded pneumatic tires	•				•
118	Power-operated window, partition, and roof panel systems	•	•		•	•
119	New pneumatic tires for vehicles other than passenger cars					•
120	Tire selection and rims for motor vehicles other than passenger cars		•	•	•	•
121	Air brake systems		•	•		
122	Motorcycle brake systems					
123	Motorcycle controls and displays					
124	Accelerator control systems	•	•	•	•	
125	Warning devices					•
129	New nonpneumatic tires for passenger cars: new temporary spare nonpneumatic tires for use on passenger cars	•				•
131	School bus pedestrian safety devices					•
135	Light vehicle brake systems					
138	Tire pressure monitoring systems	•	•	•	•	
139	New pneumatic radial tires for light vehicles	•	•	•	•	
200 Series: Injury protection						
201	Occupant protection in interior impact	•	•	•	•	
202	Head restraints	•	•	•	•	
203	Impact protection for driver from the steering control system	•	•	•	•	
204	Steering control rearward displacement	•	•	•	•	
205	Glazing materials	•	•	•	•	•
206	Door locks and door retention components	•	•		•	
207	Seating systems	•	•	•	•	
208	Occupant crash protection	•	•	•	•	
209	Seat belt assemblies	•	•	•	•	
210	Seat belt assembly anchorages	•	•	•	•	

(continued)

TABLE 14-1   continued

Number	Topic	Car[a]	Truck	Bus[b]	Multipurpose Passenger Vehicle[c]	Equipment[d]
212	Windshield mounting	•	•	•	•	
213	Child restraint systems	•	•	•	•	
214	Side impact protection	•	•	•	•	
216	Roof crush resistance	•	•	•	•	
217	Bus emergency exits and window retention and release			•		
218	Motorcycle helmets					•
219	Windshield zone intrusion	•	•	•	•	
220	School bus rollover protection			•		
221	School bus body joint strength			•		
222	School bus passenger seating and crash protection					
223	Rear impact guards					•
224	Rear impact protection			•		
225	Child restraint anchorage systems	•	•	•	•	
300 Series: Postaccident protection						
301	Fuel system integrity	•	•	•	•	
302	Flammability of interior materials	•	•	•	•	•
303	Fuel system integrity of compressed natural gas vehicles	•	•	•	•	
304	Compressed natural gas fuel container integrity					•
305	Electric-powered vehicles: electrolyte spillage and electric shock protection	•	•	•	•	
400/500 Series: Other						
401	Interior trunk release	•			•	
403	Platform lift systems for motor vehicles					•
404	Platform lift installations in motor vehicles	•	•	•	•	
500	Low speed vehicles					

[a] Passenger cars.
[b] Busses other than school busses.
[c] Multipurpose passenger vehicles.
[d] Item of motor vehicle equipment.

lights, brakes, and other features will improve highway safety. One study found no detectable impact on accident rates between states that required periodic vehicle inspections and those that did not. The main problem is that equipment failures may occur at anytime, whereas inspection occurs only once each year. Random inspections are reported to be as effective as periodic inspections.

Operators

There are many characteristics of drivers that affect the likelihood of an accident. Drivers must be licensed to operate a motor vehicle, and they must pass written and road tests to receive a license.

The Federal Highway Administration (FHWA) sets minimum qualifications for drivers of motor carriers (commercial motor vehicles). Some of the qualifications include being 21 years old, meeting physical qualifications, having knowledge of safe methods for securing cargo, passing road and written tests, and not being disqualified for certain criminal or driving offenses (including driving under the influence of alcohol or drugs). The FHWA tightened driver qualification standards in response to the Commercial Motor Vehicle Safety Act of 1986.

It is assumed that well-trained drivers drive more safely than those with little or no training. Many states require high school students to take a driver education course before a license is issued. The value of high school driver education is somewhat controversial. For example, one state dropped mandatory driver training in high schools and found that the death rate among teenage drivers dropped. Much of the decline was the result of a delay in the onset of driving. Instead of all students receiving a license at the earliest possible date, many did not begin to drive until after high school.

Another study compared three driver programs for 16,000 participants. The three programs included a control group (no training), a normal training course, and a 70-hr intensive course. Subsequent driving records were monitored for accidents and violations. There were no significant differences in the overall mean rate of accident or overall violation rates.

Recently, drunken driving has received more attention than ever before. Public groups across the nation have formed to get tough on drunk driving. Mothers Against Drunk Driving (MADD) and Students Against Driving Drunk (SADD) are two with national prominence. An accident in Kentucky in May, 1988, stirred national attention and illustrates the public pressure. A busload of school children returning home from an outing was struck by a pickup truck. As a result, two gasoline tanks on the bus ruptured and the spilled fuel burst into flames. Twenty-seven children died from the ensuing fire and many others were severely burned. The driver of the pickup truck reportedly had a BAC two and one-half times more than the legal limit. Some people believe that the driver should have been prosecuted as a murderer rather than a violator of motor vehicle laws. Some believe that design standards for school buses should be improved.

There has been a national effort to change many laws governing alcohol and driving. Many states have raised their legal drinking age to 21 years. In 1987, the Supreme Court upheld the federal law that restricted federal funds to states that did not comply with a set legal age for purchase of alcoholic beverages. Where states have raised the drinking age to 21 years, one study found a reduction of 13% in nighttime motor vehicle deaths. Many employers are becoming stricter with truck drivers who drink. One study showed that in the early 1980s, 15% to 16% of fatally injured tractor trailer drivers tested had BACs of 0.10 or more. By 1986, only 3% had BACs that high. All states have lowered BACs to 0.08 for DUI and DWI cases.

Facilities

The FHWA specifies highway design standards and traffic control devices. The *Manual on Uniform Traffic Control Devices*[1] (MUTCD) provides details on signage, traffic lights, lane markings, traffic and separation devices, and other items. The FHWA approves many standards as acceptable for design of highways and roads; 23 CFR 625 contains these standards. Most are standards of the American Association of State Highway and Transportation Officials (AASHTO).[2]

A few examples of design features that have improved the safety of highways include break-away light poles, break-away signs, energy absorbing guardrail ends, divider barriers, and protected bridge supports.

In the early days of interstate highway design, the bases for light poles were heavy concrete structures that simply stopped a vehicle on impact. Today, they are much smaller and are anchored to small bases by bolts that will shear and separate the pole from the base when struck by a vehicle (see Figure 14-2). Similar break-away technology also has been applied to highway signs (see Figure 14-3).

Guardrails have become a common design feature for major highways, replacing wooden poles with steel wire rope strung between them. Initially, guardrails had an end at the same height as the rest of the guardrail. When a vehicle struck the end of the guardrail just right, the guardrail entered the windshield and pierced through the passenger compartment. Today, the ends of some guardrails extend below ground level. In some locations where the end cannot be buried, a telescoping, energy-absorbing segment is used to terminate the guardrail. Figure 14-4 illustrates some guardrail features.

Early interstate highway designs had a variety of protective barriers for bridges and other locations. Subsequent experience and research led to better barrier designs that prevent vehicles of various sizes from overriding them. Today, there is a standard barrier with a unique cross-sectional shape (see Figure 14-5). These barriers are used in perma-

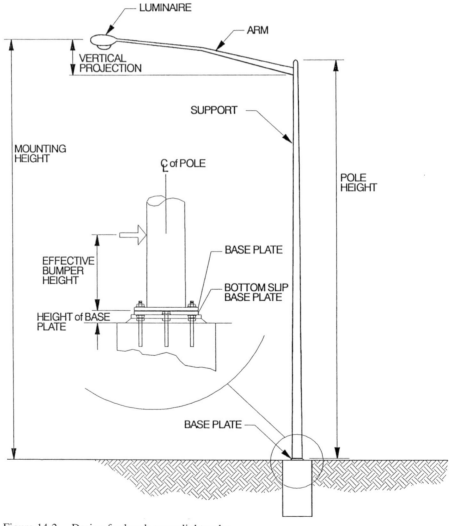

Figure 14-2.   Design for break-away light poles.

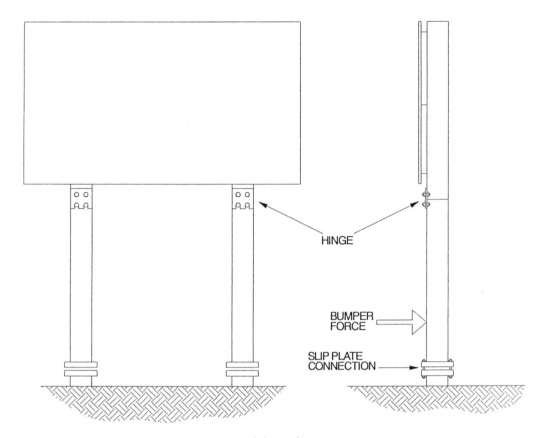

Figure 14-3.   Design for break-away highway signs.

nent bridge construction and for temporary and permanent lane separation when there is not enough distance between opposing lanes.

Early interstate highway designs left unprotected those bridge supports and columns located at some distance from the pavement. The columns were immoveable when struck. Occasionally, people used them to commit suicide by driving into them at high speeds. Today, energy-absorbing barriers containing sand or water protect some of them; guardrails and the standard concrete barriers protect others. Figure 14-6 illustrates protection of bridge piers.

Environments

Traffic safety environments are comprised of such factors as time of day, weather conditions, pavement conditions, and other factors.

Accidents occur more frequently at night. Not only is visibility a problem, but the proportion of drivers who have been drinking is higher at night.

Other factors contribute to the increase in nighttime accidents. Lighting of highways and intersections in particular reduce accident frequency. Light from headlights diminishes with the square of the distance from the vehicle. Although objects are visible at some distance, the driver has a limited amount of time to react and stop the vehicle. The distance required for perception, reaction, and stopping must be less than the effective

Figure 14-4.   Example of an energy-absorbing guardrail (the QuadGuard system). Telescoping sections collapse when impacted and crushable, foam-filled cartridges absorb energy. (Photo provided by Quixote Corporation, Chicago, IL.)

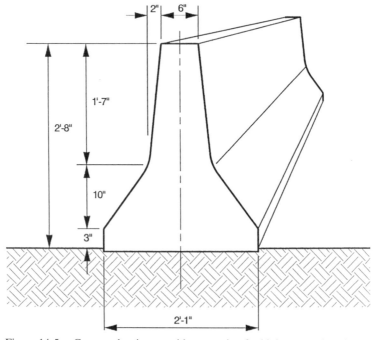

Figure 14-5.   Concrete barriers provide protection for highway workers in construction zones, separate opposing traffic lanes and form bridge guardrails.

Figure 14-6. Example of sand-filled plastic barrels (the Energite III inertial barrier system) that protect vehicles from impacting bridge columns and other objects near highways. (Photo provided by Quixote Corporation, Chicago, IL.)

distance for headlights. Wearing sunglasses, having tinted windows, or having certain vision impairments will reduce the effective distance.

Falling rain and snow can hinder visibility, and wet roads and roads covered by ice and snow reduce friction between the tires and the pavement. Consequently, cars cannot stop as quickly nor maneuver as well. Grooves ground into the pavement will improve traction on wet roads.

Pavements normally have some amount of crown to allow water to drain so that pools do not form. If water does accumulate on pavement, hydroplaning can occur. In hydroplaning, a film of water separates the tires from the pavement, reducing friction even further. To enhance traction on wet pavement, tires have grooves in the tread pattern and sipes (cuts or narrow slots in the tread design that allow the tire to flex in small elements). Both tire grooves and sipes allow water to move on the tire surface and help keep tires in contact with the pavement.

Governments in northern climates spend large amounts of money to keep highways clear in winter weather. Snow fences keep drifting snow from accumulating on highways, and roads are plowed and salted. The salt lowers the melting point of snow and ice, allowing them to be removed or to evaporate. Special tread designs on tires can increase traction, particularly for snow conditions, but have a limited effect on ice. Studded tires can improve traction, but they damage pavements, which resulted in their being banned not long after they were introduced. In severe snow and ice conditions, chains provide good traction, but destroy vehicle ride, create vibration and noise, and damage pavements. Some mountan roads require chains at certain times.

Other highway conditions might be affected by shoulder width. Where there is little room between the pavement and a ditch, there is little room to maneuver away from a dangerous situation. Traffic density, merging traffic, and visual distractions are factors of the environment that are related to vehicle accidents. Trees, bushes, and buildings that obscure intersections and people or vehicles emerging from driveways and yards are envi-

TABLE 14-2 Results of One Study of 4,800 Drivers and Driving Distractions

Distraction	Percent
Rubbernecking (viewing a crash, vehicle, roadside incident, or traffic)	16
Driver fatigue	12
Looking at scenery or landmarks	10
Passenger or child distraction	9
Adjusting radio or changing CD or tape	7
Cell phone	5
Eyes not on road	4.5
Not paying attention or daydreaming	4
Eating or drinking	4
Adjusting vehicle controls	4
Weather conditions	2
Unknown	2
Insect, animal, or object entering or striking vehicle	2
Document, book, map, directions, or newspaper	2
Medical or emotional impairment	2

ronmental conditions, too. Community zoning ordinances often control setback of buildings, vegetation, fencing, and other obstruction from roadways.

Recently, distractions of various kinds received increasing attention. Cell phones, global positioning displays, video and DVD players, and other built-in or driver-carried equipment may contribute to visual, auditory, or operational distractions. Table 14-2 presents the results of one study of 4,800 drivers and driving distractions.

Additional study may help understand the effects of a growing number of in-vehicle equipment items. Eating, drinking, smoking, reading, putting on make-up, and similar activities while driving also impact driver attention, control, errors, reaction times, and stopping times. Some government entities have banned cell phone use and other distracting activities in cars. Although the exact relationships between use of various equipment and driver performance is not clear, it is clear that distractions of all kinds can affect driver error rates and performance.

Social, Political, and Managerial Context

In the social, political, and managerial context there are many laws and regulations governing drivers, pedestrians, driving employees, and others. Social pressures to drink or take drugs and drive are part of the context. Traffic problems increase in rapidly growing communities with insufficient funding to develop roadways to meet traffic loads. Lack of funds to maintain roadways and lack of standard laws, signage, and traffic procedures can contribute to accidents.

14-3 ACCIDENT RECONSTRUCTION

Accident reconstruction is the process of collecting evidence from a vehicle accident to determine what happened and what may have caused the accident. Over the years, engineers have developed a number of analytical techniques. Today, computer programs are used to store, compile, and analyze data about particular accidents.

In accident reconstruction, one first collects data at the accident scene. One records the type and model of vehicle, takes photos and records other data about the damage to the vehicle, and records or plots skid marks, impact points, locations of vehicles and vehicle components, and type of pavements or surfaces on a drawing of the scene. Because length of skid marks is critical, police often record this data as part of their on-scene reports.

From the compiled data, equations applying conservation of energy and conservation of momentum allow one to estimate the speed of vehicles before the crash, what vehicle was in what position at the point of impact, and other information. Vehicle damage and examination of remaining components also suggest whether equipment, such as lights and brakes, operated properly. Other observations may suggest defects in the vehicle.

There is a growing number and level of sophistication of software programs for modeling potential crashes or reconstructing crashes for accident site data. Enhancements include data files of vehicle mass and damage data by vehicle make and model. Some programs prepare graphic plots of the vehicle positions before, during, and after a crash overlaid on skid and other site data. Enhancements in analysis software include aerodynamic drag and disturbances, engine drag, tire rolling drag and lateral force, steering maneuvers, braking effects, impact models, and interrelationships among some of these. Computer modeling of crashes is reasonably accurate for certain scenarios. By validating the models against staged accidents, researchers have achieved accuracies within 5% to 10% of real effects.

Some simple examples of accident reconstruction analysis illustrate the approach. One can compute stopping distance, s, in feet, for full, four-wheel braking of an automobile, from

$$s = \frac{V^2}{30\mu}, \tag{14-1}$$

where

V is the initial velocity (miles per hour) and

μ is the coefficient of friction for the tires on the pavement.

Using Equation 14-1, one can estimate the initial speed of a vehicle from skid marks, where the vehicle skidded to a full stop.

If only two or three wheels are skidding, an additional factor, n, is introduced into Equation 14-1:

$$s = \frac{V^2}{30\mu n}, \tag{14-2}$$

where

n is 0.75 for three wheels braking and

n is 0.50 for two wheels braking.

The total stopping distance is

$$S_p = \text{prebraking distance} + \text{stopping distance.} \tag{14-3}$$

The prebraking distance is composed of perception time plus reaction time. Perception time is the time for recognizing that one should start braking. Reaction time is the time necessary to move the foot onto the brake pedal and begin braking.

In front and rear collisions, one can estimate vehicle velocities by applying conservation of momentum. The momentum for two colliding vehicles before a crash must be equal to that after a collision:

$$(W_1 V_1) + (W_2 V_2) = (W_3 V_3) + (W_4 V_4), \tag{14-2}$$

where

W is the weight (pounds),

V is the velocity (miles per hour),

subscripts 1 and 2 represent the two vehicles before the crash, and

subscripts 3 and 4 represent the two vehicles after the crash.

One must know vehicle weights and vehicle speed before or after the crash to solve this equation.

Example 14-1 Consider a rear-end collision between two vehicles. Vehicle 1 weighs 3,500 lb and vehicle 2 weighs 2,800 lb. Vehicle 1 is standing still at a stop light and the driver has the brakes fully engaged. Vehicle 2 crashes into it. Just at the point of impact, the second vehicle begins to skid. The two vehicles skid 25 ft beyond the point of impact, with all wheels locked. Assume the coefficient of friction between the dry pavement and tires is 0.85. How fast was the second vehicle going before the accident?

The velocity of vehicle 1 is 0 before the crash and there is no change in the weight of the vehicles from before to after the crash. Using Equation 14-4,

$$(3,500 \times 0) + (2,800 \times V_2) = (3500 + 2800)V_3,$$

V_3 is determined from Equation 14-1:

$$V_3 = (30 \times 0.85 \times 25)^{1/2} = 25.25 \text{ mi/hr.}$$

Then

$$V_2 = \frac{6,300(25.25)}{2,800} = 56.8 \text{ mi/hr}.$$

14-4 OTHER FORMS OF TRANSPORTATION

Railroads

Since their emergence in the mid 1800s, railroads have faced safety problems. In the rush to complete the transcontinental railroad in 1867, safety received limited attention. In 1888, 315 passengers were killed and 2,138 were injured. In the same year, 2,070 employees were killed and 20,148 were injured. One citizen, a farmer from Iowa named Lorenzo Coffin, was so irate that he launched his own campaign for railroad safety. After being elected to Congress, he was responsible for enactment of the Rail Safety Appliance Act of 1893. The act required the use of two inventions: the Westinghouse air brake, which replaced manual brakes on each car, and the Janney coupling, which replaced the oval links and pins. After several years, railroad employee accidents were reduced by 60%. Soon the railroads learned that safety of employees and passengers paid dividends in cash and good will.[3]

Another significant improvement was an automatic block signal system that kept trains from colliding on the same track. Also, boiler overpressure valves prevented steam engines from exploding.

Hazards The record of railroads has improved and the hazards have shifted. Today, derailments pose significant hazards from the transported materials, including explosions, fires, and releases of toxic materials. Most rail deaths and injuries now result from grade-crossing accidents with motor vehicles.

Controls Today's railroad cars incorporate many safety features. Ladders have rungs that prevent feet from sliding off. There are grab bars at climbing transition points and slip-resistant walkways and handrails. Thermally insulated tank cars prevent heat buildup and explosions from fires in adjacent cars. Cars carrying hazardous materials have interlocking couplings. Locomotives have dead-man controls to prevent runaway trains should an engineer become incapacitated. Cars containing hazardous materials have placards indicating contents and type of hazard. Automatic sensors placed strategically along main lines detect overheated bearings.

Based on traffic at crossings, more and more grade crossings are being protected by barrier gates and signals. Sensors that detect the speed of a train activate some gates and ensure adequate lead time for gates to close. Major highways avoid grade crossings with underpasses and bridges. Some cities are diverting main line rail traffic from the inner cities where dangerous traffic conflicts exist.

Aviation

The aviation industry in the United States has grown during this century to the point where millions of passengers travel each year by air. Most propeller-driven aircraft have given way to much larger jet-powered aircraft. Airways have grown more crowded. In the past decade, the number of takeoffs and landings has nearly doubled.

Hazards Aircraft must take off and land without incident. Because airspace in some areas is very crowded, sophisticated electronic gear helps fly the aircraft and control traffic in the highways of the air. Aviation hazards have changed with aircraft scale, speed, and altitude. Structural loads are greater, requiring exotic metals, like fiber-reinforced titanium. Structures face expansion and contraction from repeated loading, pressurization and depressurization, and thermal expansion and shrinkage.

A large portion of an aircraft's takeoff weight is fuel. Leaking fuel tanks and ignition can be disastrous. Icing of wings and loss of lift is still a danger in severe weather. Designing emergency exiting for as many as 450 passengers is a challenge. Detecting and avoiding wind shear, particularly in clear weather, also remains a challenge.

Controls The newest aircraft have on-board computers and instruments for navigation, flight control and management, fuel management, fire detection and extinguishment, collision avoidance, pressurization control, and many other functions. Some even have the capability for automated landing. Air traffic control systems are upgraded continually to handle increased traffic. Without these systems, aviation would not be possible or as safe as it is.

Federal Aviation Administration regulations set design standards and control air worthiness of the aircraft and its engines. Flight and maintenance logs help ensure that equipment is in good condition. Regulation of pilot training and certification place qualified crews in charge of flight. Standards for air traffic controllers are essential, too. Airlines and aviation employee unions set standards for employee qualifications. Strict management and enforcement of flight and operation regulations and procedures contribute to aviation safety.

Pipelines

Pipelines transport gases and liquids. Most materials transported by pipeline are fuels, including natural gas, liquified natural gas, and other petroleum products, although other hazardous materials, like anhydrous ammonia, move through pipelines, too. Department of Transportation regulations govern pipelines.

Other piping systems transport water, storm runoff, sanitary waste, and other wastes. Codes for water and waste fall under Environmental Protection Agency (EPA) regulations and state and local codes.

Hazards   Because most pipelines are located underground, there are hazards related to materials handling and excavation during construction. For pipelines carrying fuels and hazardous materials, pumping systems move contents from sources to processing and use locations. Fire and explosion are major concerns for fuels and flammables. A leak in a pipe, fitting, valve, pump, or other component of the pipeline could produce disastrous results. During maintenance, segments being repaired must be depressurized, blocked off, and purged so that any residual fuel does not ignite. Valves must be closed and locked before flanges are worked on. If the material in the pipeline is corrosive or toxic, there are dangers of contact and inhalation. Gaskets and packings in valves are common failure points.

Most special facilities, like pumping stations, storage tanks, collection stations, and transfer stations, are above ground and part of the pipeline systems. Physical damage, corrosion, and component failures can lead to major disasters. Opening the wrong valve and not knowing what is in a system are common errors.

A failure in a pressurized water system will not have nearly the same results as piping systems for flammable or toxic materials. Storm and sanitary systems depend on gravity to move the contents. Leaks may contaminate underground water sources and leaking material may leach to the surface or into water supplies and create health hazards. The EPA has regulations for underground storage tanks.

Controls   Pipelines and related components and facilities for fuels and hazardous materials that are under pressure must meet strict design standards of the DOT. Standards vary for different pipe materials and different materials transported. Pipeline and piping components are pressure tested following strict procedures before use. DOT also specifies detailed requirements for operation, maintenance, inspection, and reporting procedures. DOT standards related to pipelines incorporate the reference standards of many other organizations, including the American Society of Mechanical Engineers (ASME), the American Petroleum Institute (API), and the American Society for Testing and Materials (ASTM).

Piping and components should have labels to prevent errors in contents and what each component does. There are standards for pipe labeling. Maintenance of pipelines, components, and facilities requires detailed planning, communication among workers, and application of many safety procedures. For example, failure to remove the pressure from a pipeline and to purge residual fuels can lead to explosion, severe injury, and fire. Welding in the presence of residual fuels is dangerous. A box or hood placed on valve stems can prevent injury when a valve packing blows. Workers at some distance from a compressor used to pressure test a line must be in close communication with the compressor operator because they need to know what the pressure is at all times during the test procedure.

14-5   TRANSPORTATION OF HAZARDOUS MATERIALS

The Transportation Safety Act of 1974 recognized the dangers to life and property that can result from transportation of a variety of hazardous materials. In the event of an accident, a spill of materials could create significant harm for passengers, other vehicles nearby, and people living near the accident. The act applies to transportation, shipping, packaging, and labeling of materials defined as hazardous, and it provides for penalties for violations. Estimates suggest that there are more than 250,000 hazardous material shipments every day in the United States. Estimates also suggest that approximately 400 shipments per year are involved in injury accidents.

Hazards

The Transportation Safety Act defines several classes of hazardous materials, including explosives, radioactive material, flammable liquids or solids, combustible liquids or solids, oxidizing or corrosive materials, compressed gases, poisons, etiologic agents (hazardous biological materials), irritating materials, and other regulated materials (ORM). The act excludes firearms and ammunition. Other chapters in this book discuss hazards associated with some of these materials.

Controls

To minimize the danger to life and property, controls for hazards fall into several categories. Controls include defining and recognizing hazardous materials, excluding certain materials from particular transportation modes, limiting quantities, controlling placement, design and selection of packaging, labelling of containers, restricting transportation routes, and using shipping manifests, incident reporting, and training.

Definition and Recognition The DOT defines what materials are hazardous. They work with other agencies to maintain a list of hazardous materials. Table 14-3 lists DOT definitions for hazardous materials. Hazardous materials and information about them are listed in 49 CFR Part 172. Changes to the list are published in the *Federal Register*.

Those who transport or use hazardous materials must know which materials are hazardous. Training and labeling help convey this information.

Excluding and Isolating Materials No one is allowed to carry or ship radioactive material by aircraft. Certain materials cannot be transported on passenger aircraft or passenger railroad cars. Certain hazardous materials cannot be transported in the same shipment with certain other hazardous materials.

Limiting Quantities Another way to limit the dangers of hazardous materials is to limit the quantity present in containers or in transport. For example, one may carry very small quantities of certain materials on passenger aircraft. Regulations permit larger amounts on cargo aircraft and even greater amounts by rail, truck, or water vessels. Fines for violations can be large. Consider a case in which an employee of a chemical company carried a small container of nitric acid in his luggage. The acid spilled and damaged the aircraft baggage compartment. The company was fined $15,000.

Storage Rules Hazards of some materials increase if they come into contact with other materials. For example, a fire involving flammable materials will greatly intensify in the presence of an oxidizer other than air. Some chemicals will react violently when mixed. Because of this, hazardous materials tables give guidance on placement of materials. Some should not be placed in locations where people are normally present; some should always

TABLE 14-3 Department of Transportation Hazardous Materials Definitions[a]

Hazard Class	Definition
Explosive	Any chemical compound, mixture, or device, the primary or common purpose of which is to function by explosion.
Explosive A	Detonation or otherwise of maximum hazard.
Explosive B	In general, function by rapid combustion rather than detonation. *Flammable hazard.*
Explosive C	Certain types of manufactured articles containing class A or class B explosives in restricted quantities. *Minimum hazard.*
Blasting agents	A material designed for blasting that has been tested and found to be so insensitive that there is very little probability of accidental initiation of explosion or of transition from deflagration to detonation.
Combustible liquids	Any liquid having a flash point of more than 100°F and less than 200°F.
Corrosive material	Any liquid or solid that causes visible destruction of human skin tissue or a liquid that has a severe corrosion rate on steel.
Flammable liquid	Any liquid having a flash point less than 100°F. *Pyroforic liquid*—Any liquid that ignites spontaneously in dry or moist air at or less than 130°F. *Compressed gas*—Any materials or mixture having in the container a pressure exceeding 40 lb/in^2 absolute at 70°F or a pressure exceeding 104 lb/in^2 absolute at 130°F, or any liquid flammable material having a vapor pressure exceeding 40 lb/in^2 absolute at 100°F.
Flammable gas	Any compressed gas meeting the requirements for lower flammability limit, flammability limit range, flame projection, or flame propagation criteria of DOT.
Nonflammable gas	Any compressed gas other than a flammable compressed gas.
Flammable solid	Any solid material, other than an explosive, that is liable to cause fires through friction or retained heat from manufacturing or processing or that can be ignited readily and when ignited burns so vigorously and persistently as to create a serious transportation hazard.
Organic peroxide	An organic compound containing the bivalent –O–O– structure and that may be considered a derivative of hydrogen peroxide where one or more of the hydrogen atoms has been replaced by organic radicals. (Some exceptions)
Oxidizer	A substance such as chlorate, permanganate, inorganic peroxide, or a nitrate that yields oxygen readily to stimulate the combustion of organic matter.
Poison A	*Extremely dangerous poisons*—Poisonous gases or liquids of such nature that a very small amount of the gas, or vapor of the liquid, mixed with air is *dangerous to life.*
Poison B	*Less dangerous poisons*—Substances, liquids, or solids (including pastes and semisolids), other than class A or irritating materials that are known to be so toxic to humans as to afford a hazard to health during transportation, or that, in the absence of adequate data on human toxicity, are presumed to be *toxic to humans.*
Irritating material	A liquid or solid substance that on contact with fire or when exposed to air, gives off dangerous or intensely irritating fumes, but not including any poisonous (class A) material.
Etiologic agent	A viable microorganism, or its toxin, that causes or may cause human disease.
Radioactive material	Any material or combination of materials that spontaneously emits ionizing radiation and has a specific activity greater than 0.002 μCi/g.

TABLE 14-3   continued

Hazard Class	Definition
ORM (Other regulated materials)	Any material that does not meet the definition of a hazardous material, other than a combustible liquid in packaging having a capacity of 110 gal or less, and is specified as an ORM material or that possesses one or more of the characteristics of specific ORM classes of materials.
ORM A	A material that has an anesthetic, irritating, noxious, toxic, or other similar property and that can cause extreme annoyance or discomfort to passengers and crew in the event of leakage during transport.
ORM B	A material (including a solid when wet with water) capable of causing significant damage to a transport vehicle or vessel from leakage during transportation. Materials meeting one or both of the following criteria are ORM B materials: (1) a liquid substance that has a corrosion rate exceeding 0.250 in/yr on aluminum (nonclad 7075-T6) at a test temperature of 130°F; (2) specifically designated by name by DOT.
ORM C	A material that has other inherent characteristics not described as an ORM A or ORM B but that make it unsuitable for shipment, unless properly identified and prepared for transportation. ORM C materials are specifically named by DOT.
ORM D	A material such as a consumer commodity (packaged or distributed in a form intended and suitable for retail sale for consumption by individuals for personal care or household use, including drugs and medicines) that, although otherwise subject to the regulations of this subchapter, presents a limited hazard during transportation because of its form, quantity, and packaging. They must be materials for which exceptions in DOT regulations are provided.
ORM E	A material that is not included in any other hazard class, but is subject to DOT regulations. Included are hazardous waste and hazardous substance named by DOT in this class.

[a] See also 49 CFR 171–173.

be separated from other materials during storage and transit; others must be kept dry, cool, away from sunlight or heat, or meet other restrictions.

Packaging Design and Selection   DOT has many standards for the design of packaging. Regulations specify strength, dimensions, and other properties for fiberboard boxes and drums, glass carboys, steel drums and liners, tank trucks, rail tank cars, and many other containers. Regulations also specify what type of container to use for each hazardous materials. The major purpose of these design standards is to minimize the release of hazardous materials, even if the packaging is in an accident. To dramatize the value of hazardous material packaging, the British government placed a shipping container for nuclear fuel rods on a railroad track and crashed a train into it at 100 mi/hr. The container withstood the crash without a leak.

Labeling   Serious errors in the transport and handling of hazardous materials could occur when one does not know what is in containers. DOT has many regulations regarding labeling of particular materials. Typically, the label indicates what material is inside the container and gives warnings for handling and transport, DOT hazard class, and special information. In many cases, empty containers must have an "EMPTY" label. For some materials, there are dangers for an empty container that do not exist when it is full.

As part of the labeling, DOT requires a standard warning placard. Motor vehicles, freight containers, and rail cars must have placards listing the class of hazardous material by name and symbol. For some materials, a placard also must list the name of the material and a DOT identifying number. A United Nations Hazard Class Number[4] also may be displayed. Figure 14-7 gives examples of DOT warning placards.

Restricted Transportation Routes   Some states have laws that restrict the highways over which certain hazardous materials can travel. Restrictions are typical for areas where there are high densities of people and highly traveled or critical tunnels and bridges. In addition, there may be local ordinances that restrict transportation of hazardous materials. Some of these restrictions resulted from public reaction to disasters. For example, a railroad tank car containing LP gas caught fire and exploded in Crescent City, Illinois, in 1970. This disaster occurred as the train was passing through the town and the explosion destroyed most of the businesses.

Shipping Papers   DOT regulations require that a shipping order, bill of lading, manifest, or other document used to initiate a shipment must describe any hazardous material offered for shipment. The description must include the name of the material, the DOT hazard class, the amount being shipped, and the number and type of containers. The document must separate hazardous materials from other materials. The shipping papers must certify that hazardous materials are packaged and labeled properly. There are additional requirements for hazardous waste.

Incident Reports   During transportation, a carrier must report to DOT any unintentional release of a hazardous material. When an incident involves death, serious injury, major property damage, or certain releases of radioactive and etiologic agents, the carrier must make an immediate telephone report. DOT regulations detail these reporting procedures.

Training   To help minimize the release of hazardous materials, people in the entire chain must receive training. Designers of packaging, preparers of shipments, managers of ship-

Figure 14-7.   Examples of DOT warning placards.

ping and handling, drivers, operators, and handlers need to know the dangers of the materials and the procedures to ensure safety. DOT requires that each person who offers hazardous materials for transportation must instruct officers, agents, and employees about applicable regulations.

EXERCISES

1. A car skids to a stop. At the start of the skid, the car was traveling at 50 mi/hr. What was the stopping distance if the coefficient of friction between the tires and the pavement was 0.71?

2. An investigator finds skid marks 90 ft long. If the skid brought the car to a stop, what was the initial velocity of the car? Assume a coefficient of friction of 0.65 between the tires and the pavement.

3. For the situation in Problem 2, what was the initial velocity if only two wheels were sliding and the braking efficiency was 0.5?

4. For the situation in Problem 2, what was the total stopping distance if the driver's combined perception and reaction time was 0.65 s? Assume the vehicle weights 1,000 lb.

5. Car 1 is stopped with brakes locked. Car 2 skids and strikes car 1 in the rear. If, after the collision, both cars skid 60 ft to a stop and the coefficient of friction is 0.45 for a wet pavement, what was the velocity of car 2 at the time of impact? Car 1 weighs 2,500 lb and car 2 weighs 3,800 lb.

6. A forklift industrial truck turns a corner with a 1,000-lb load on the forks. The forks are raised 3 ft above the floor. The center of gravity for the forklift with driver is 38 in above the floor with empty forks at the floor. The center of gravity of the load is 22 in above the floor when it sets on the floor. The forklift weighs 2,800 lb. Its wheels are 48 in between lateral outside edges. If the radius of turn is 25 ft, the speed of the forklift is 8 mi/hr, and the coefficient of friction between tires and floor is 0.67, will the forklift turn over, skid, or negotiate the turn?

7. Locate consensus or government safety standards for each of the following. Select one design feature that contributes to safety for each and describe its function.
   (a) motor vehicles
   (b) school buses
   (c) truck tractors
   (d) truck trailers
   (e) highways
   (f) railroad cars
   (g) railroad traffic controls
   (h) aircraft
   (i) pipelines

8. Review the current regulations related to transportation of hazardous materials issued by the Department of Transportation. Identify key requirements for each mode of transportation.

9. Conduct a review of the safety standards for the pipeline industry issued by the American Petroleum Institute.

REVIEW QUESTIONS

1. Approximately how many people die each year on United States highways?

2. What is the leading cause of occupational injuries and work-related deaths?

3. The motor vehicle death rate is highest for what age groups?

4. What portion of drivers involved in fatal crashes have blood alcohol content (BAC) of 0.10 or more?

5. Describe the crash sequence for vehicle and passenger in

    (a) a frontal crash

    (b) in a rear-end collision

6. What is the "second crash"?

7. Describe the concept of crashworthiness.

8. What is meant by integrity of the passenger compartment?

9. What organizations publish standards for motor vehicles?

10. Describe a safety feature built into a motor vehicle.

11. Describe a driver characteristic important for motor vehicle safety.

12. Describe a design standard for highways that reduces injury or death.

13. Describe an environmental factor that contributes to motor vehicle safety.

14. What is accident reconstruction?

15. Name two hazards in rail transportation today.

16. Name four safety design features in railroad cars.

17. Name four factors that contribute to aviation safety.

18. Identify four controls for hazards of pipelines and related facilities.

19. What types of controls reduce the dangers associated with transportation of hazardous materials?

NOTES

1 Available from Superintendent of Documents, U.S. Government Printing Office, Washington, DC, 20402.

2 AASHTO, Suite 249, 444 North Capitol Street NW, Washington, DC 20001.

3 Holbrook, S. H., *The Story of the American Railroads,* Crown Publishers, New York, 1947.

4 "Transport of Hazardous Goods," United Nations Recommendation, United Nations, New York, 1970.

BIBLIOGRAPHY

Asa, D., *The Trucking and Truck Accident Handbook*, Hanrow Press, Del Mar, CA, 1984.

Aviation Ground Operation Safety Handbook, National Safety Council, Itasca, IL, 2000.

Baker, J. S., *Traffic Accident Investigation Manual*, The Traffic Institute, Northwestern University, Evanston, IL, 1975.

Bierlein, L. W., *Red Book on Transportation of Hazardous Materials*, 2nd ed., Van Nostrand Reinhold, New York, 1987.

Campbell, R. D., *Flight Safety in General Aviation*, Collins Professional Books, Glasgow, 1987.

Collins, J. C., *Accident Reconstruction*, Charles C. Thomas, Springfield, IL, 1979.

DUNLAP, E. SCOTT, *Motor Carrier Safety: A Guide to Regulatory Compliance*, Lewis Publishers, Boca Raton, FL, 1999.

General Aviation Ground Operation Handbook, 4th ed., National Safety Council, Chicago, IL, 1988.

LIMPERT, R., *Vehicle System Components: Design & Safety*, Wiley-Interscience, New York, 1982.

LIMPERT, R., *Motor Vehicle Accident Reconstruction & Cause Analysis*, 2nd ed., Michie Co., Charlottesville, VA, 1984.

McGREW, D. R., *Traffic Accident Investigation and Physical Evidence*, Charles C. Thomas, Springfield, IL, 1984.

Motor Fleet Safety Manual, 4th ed., National Safety Council, Chicago, IL, 1996.

MUTCD 2000: Manual on Uniform Traffic Control Devices, Federal Highway Administration, Washington, DC, 2000.

PETERS, G. A., and PETERS, B. J., eds., *Automotive Engineering and Litigation*, Vols. 1 and 2, Garland Publishers, New York, 1994 and 1988.

PLATT, F. N., *The Traffic Accident Handbook*, rev. ed., Hanrow Press, Del Mar, CA, 1986.

SWARTZ, GEORGE, *Forklift Safety: A Practical Guide to Preventing Powered Industrial Truck Indicents and Injuries*, National Safety Council, Itasca, IL, 1999.

CHAPTER 15

MATERIALS HANDLING

15-1 INTRODUCTION

Materials handling is the lifting, moving, and placing of items in various forms. It may be done manually or with equipment. Materials handling is one of the leading causes of disabling occupational injuries. According to the National Safety Council, 20% to 25% of all disabling occupational injuries result from materials handling.

Materials handling includes the use of many kinds of equipment designed to help in the tasks. Manipulators, jacks, hoists, derricks, industrial trucks, cranes, backhoes, conveyors, rigging, escalators, elevators, and other equipment are part of the materials handling arsenal. There are many kinds of objects and materials to handle, each posing different hazards. There may be individual objects or groups of objects in boxes, bins, totes, or on pallets. We use buckets and scoops of various types to handle bulk materials, like grain, gravel, earth, and loose parts.

This chapter discusses many of these activities and types of equipment used in them. Included will be a discussion of storage of materials and excavation and trenching.

Hazards

There are many kinds of hazards for materials handling activities and equipment. Some are unique to particular activities, equipment, or kinds of materials. Manual materials handling poses dangers that may be different from the use of cranes or hoists. Electrically powered equipment has some hazards resulting from electricity that are different from those powered by other energy sources. Mobile equipment has hazards different from fixed equipment. Lifting and moving a coil of steel has different hazards from loading grain into a bin. Materials may be flammable or toxic.

Environments may contribute to hazards in materials handling. Good lighting, sufficiently wide aisles, good ventilation, traffic controls and visibility, and uncongested and unobstructed pathways are important. So is keeping lift zones clear of people. Proper maintenance of materials handling equipment is essential. Failure of structural elements, brakes, controls, and other components can lead to accidents. Training is also important. Workers must learn how to lift items to minimize the chances of injury. Operators must learn how to operate materials handling equipment, to properly plan a safe lift, to understand what can go wrong and how to protect themselves, others, and property. Other participants in materials handling operations must know procedures, such as hand signals, staying out from under loads and away from elevated loads, and use of proper rigging. It is also important to plan materials handling jobs and instruct participants in the steps that will be taken.

One major class of hazard in materials handling is failure of the lifting equipment. The failures are often the result of overloads for certain lifting conditions. For example in humans, we see sprains and strains of backs, arms, and legs. A crane boom may buckle, a chain or wire rope that is part of the lifting device may break, rigging that restrains load may fail, or a conveyor support may collapse.

Another class of hazards is falling loads. Materials may fall on people and cause injury or they may fall on property and cause damage. A load may shift and tip over. A load may be inadequately rigged, restrained, or anchored.

Another class of hazard is material in motion. The speed and mass of materials and equipment are important considerations. Objects may strike something else and cause damage or strike a person and cause injury. One may operate equipment too fast, tip it over, and be out of control or unable to stop it quickly. People may be run over or have their hands or bodies caught, crushed, or pinched. The rate of flow is important, particularly when materials are handled in different ways in an overall process. Unless there is balanced flow, materials will pile up at certain points, possibly causing workers to rush and do things in an unsafe manner.

Controls

There are many different controls for preventing materials handling accidents. Controls are related to kind of activity, kind of equipment, and kind of material. There are also controls for the environments where material handling occurs.

Eliminate Handling   Analysis of operations may identify ways to eliminate material handling tasks. If material handling steps are eliminated, there are fewer opportunities for handling hazards, which makes sense from a safety point of view as well as making good economic sense. Material handling takes time, costs money, and increases the likelihood of damage to items handled.

Planning   If materials handling is needed, one should plan the details. Handling locations should be clear of hazards. Planning should include selection of correct equipment, identification and analysis of steps that may go wrong, and establishment of procedures for dealing with contingency problems. Hand signals, two-way radio systems, and other means of communication must be arranged, and participants must understand the plans. Even a seemingly simple, two-person lift requires planning. Participants should, as a minimum, go over how they will proceed from start to finish, what they will do if something starts to go wrong, and how they will communicate during the process. Mechanical handling is generally preferred to manual handling. Manual handling is usually more expensive than mechanical handling.

Design and Selection   Materials handling tools, devices, and equipment require proper design. Standards from various sources may be applicable. Design considerations must include structural strength, operational features, control systems, visibility, failure modes, incorporation of safety features, and other factors. Even permanently installed materials handling equipment must have safety features. For example, conveyors that move above workers in a factory must have overhead protection for the people below to prevent objects and materials from falling on them. Some materials handling equipment must have access ways and guardrails for maintenance and lubrication tasks. There may be a need for exhaust ventilation or sprayers to isolate or control dust. Power equipment may need emergency shutoff controls and guards may be required. Each design requires analysis of uses and use environments. Selection of equipment must match use requirements to availability of necessary features to ensure safe use.

Selection of the right handling equipment for a job also is important. Specific jobs require particular handling equipment. Special features may be needed for certain uses. Whoever makes the selection must know the task, the equipment, and the use environment.

Use   People must use equipment correctly. Many examples of proper and safe use can be cited. Loads on materials handling equipment must not exceed safe load limits. Operators must drive mobile equipment safely. Cranes should not be operated within certain distances of power lines.

Training   The use of each kind of materials handling equipment requires particular knowledge and skill. Operators and those involved in the area of use must learn what hazards equipment and its use impose and how to control the hazards. They need to develop skill in operating controls, to develop skill in recognizing when things could and do go wrong, and to be knowledgeable of the suitable action to take. They must develop skill in the procedures and judgments related to planning and executing materials handling tasks. They must know what conditions in the use environment add to the hazards of the material handling task, and they must know when stopping the activity is more important than loss of equipment or materials or more important than injury or loss of life.

Environments   There are many different and important use environment factors. Lighting, visibility, weather, terrain, properties of materials (weight, toxicity, stability, etc.), and location of people on or near a site must be evaluated. Proper controls must be in place before handling tasks start. Even communication means have to be worked out on loading docks where workers do not speak the same language.

15-2   MANUAL MATERIALS HANDLING

Manual materials handling accidents result in a variety of injuries. Objects and loads may fall and injure hands, feet, and legs. Lifting may cause muscle strains and joint injuries. By far the most common injuries from manual materials handling are back injuries. According to several studies, low back injuries account for approximately one quarter of all workers' compensation claims.

Back claims and complaints are widespread among people and occupations. They are not limited to industrial or construction activities. They are common among hospital employees, often resulting from lifting of patients. Back complaints are even prevalent among office workers. Results of one national survey estimated that more than half of all office workers have back complaints at some time. Another study notes that four of every five Americans will experience at least one episode of lower back pain between the ages of 20 and 60 years.

Hazards

Many things contribute to manual materials handling injuries. Included are materials handling techniques, job design, and physical condition and characteristics of individuals.

A biomechanical analysis of lifting gives us insight into some of the problems. When a person lifts and carries an object, the load must be counteracted by the back muscles. The spine is the fulcrum (see Figure 15-1) and the back muscles are a fixed, short distance from the spine. The load in front of the body is much farther from the spine, at minimum nearly the thickness of the trunk. The moment created by the load is greater when a load is held far from the body compared with holding it close to the body, whether standing,

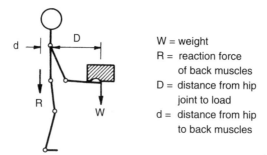

Figure 15-1. Biomechanics of manual lifting. The moment ($W \times D$) created by the load being lifted must be counteracted by the muscles of the back ($R \times d$).

sitting, or stooping. The moment created by the load must be counteracted by the back muscles.

Stooping to raise a load creates even greater moments because of the trunk length. To keep the moment small, the load must be held close to the body. In general, women have a slightly longer torso length relative to their body height than men do. As a result, a woman will experience a greater moment for a lifting task than will an man of equal height. Because there is considerable variability in body dimensions, this generalization may not apply to every woman.

The size of a load can contribute to the moment. A large object cannot be held as close to the body as a small object. Depending on its distance away from the body, a large but relatively light object may produce a greater moment than a small heavy object. The inertia created by acceleration during lifting can add to the static load and can increase the moment.

The length of a lift (vertical distance) can increase the potential for injury. Lifting overhead involves other muscle groups that may have less capacity than the back. Reaching while picking up an object or putting it down is more likely to result in dropped loads and to produce greater moments.

The weight of an object being lifted is also important. One study of 550 workers over a 2-yr period found that few injuries resulted when loads were kept to less than 45 lb. Other factors beside the weight of the object affect lifting stress. There are software programs, for example, that estimate the compressive load on the lumbar area of the back during lifting activities.

Frequency of lift is also important. Continuous lifting activity may exceed the physical work capacity of an individual and lead to fatigue, error, and injury. Because the body is not well suited to asymmetrical loads or rotation, lifting with one hand or twisting during a lift add to the likelihood of injury.

People vary in size, weight, strength, physical condition in general, physical condition of muscles, condition of joints, and other factors. Back muscles reacting to a lifted load compress the vertebrae of the spine. Some studies have estimated the compressive load limits of spinal elements, but the capacity for an individual and particular spinal locations varies. It is difficult to predict where and under what conditions an individual will experience pain, a strain of a muscle, or other form of injury.

Controls

Administrative Controls   Administrative controls include selection and training of workers. Selection includes physical assessment, strength testing, and testing for aerobic work capacity. Training involves recognition of dangers in manual materials handling, how to avoid unnecessary stress, and what a person can handle safely. Table 15-1 lists recommendations for lifting techniques compiled from various sources.

Engineering Controls   Engineering controls are divided into (1) mechanical, visual, and thermal environments, (2) alternatives for materials handling systems, and (3) potential safety and ergonomic problems. The mechanical environment includes unit size of load, container design, handle and handhold designs, and floor–worker interfaces. The visual environment refers to lighting, color, and labeling. Materials handling system alternatives involve materials handling equipment and job aids, like hooks, bars, rollers, and other devices. Other chapters discuss many environmental controls further.

The Revised National Institute for Occupational Safety and Health Lifting Equations

The National Institute for Occupational Safety and Health (NIOSH) studied many of the factors just discussed, and it combined much of the information into a guide called the *Work Practices Guide for Manual Lifting*.[1] The guide reviewed epidemiological, biomechanical, physiological, and psychophysical literature and recommended controls for minimizing lifting injuries. Later, NIOSH updated the study[2] and revised the lifting recommendations to incorporate additional lifting factors: asymmetrical lifting tasks and lifts of objects with less than optimal couplings between the object and the worker's hands.[3] The revised lifting equations compute two values: recommended weight limit (RWL) and lifting index (LI), based on seven lifting factors.

The RWL aids in decisions to separate an acceptable lifting condition from a hazardous lifting condition for which some redesign of the condition is required. If the weight of an object to be lifted is greater than the RWL, engineering or administrative controls are needed to reduce the weight or to increase the RWL.

Because the LI is simply a ratio of the weight of an object and its RWL, the LI provides an estimate of the hazard of overexertion injury (or degree of stress) for a manual lifting job.

There are limits for the applicability of the revised lifting equations. In summary, they do not apply if any of the following lifting/lowering conditions occur:

- with one hand
- for more than 8 hours

TABLE 15-1   Frequently Recommended Lifting Procedures

Get a firm footing. Make sure the floor is not slippery.
Size up the load. Determine what it weighs.
Spread your feet for a stable stance.
Get a firm grip. Use handles, gripping, or other lifting tools that will help.
Make sure the load is free, not locked down or stuck.
Keep your back straight. Keeping your chin tucked in will help keep your back straight.
Lift with your legs.
Tighten your stomach muscles.
Accelerate the load slowly. Don't jerk.
Hold the load close to your body. Position a load close to your body before lifting.
Watch out for your fingers and hands when carrying a load so you don't strike them against something.
Don't twist during lifting. Turn with your feet, not with your back.
Set the load down gently. Use your legs. Keep the back straight.
Watch your fingers so you don't pinch them.

- while seated or kneeling
- in a restricted work space
- unstable objects
- while carrying, pushing, or pulling
- with wheelbarrows or shovels
- with high speed motion (faster than approximately 30 in/s)
- with unreasonable foot/floor coupling (<0.4 coefficient of friction between sole and floor)
- in an unfavorable environment (temperature outside the ranges 66°–79°F [19°–26°C] and 35%–50% relative humidity)

For these conditions, a more comprehensive ergonomic evaluation of the activity is recommended.

The seven lifting task multipliers involved in the computations are presented in Table 15-2. The load constant (LC) is the maximum weight that can be lifted safely for a lift in which the lifting conditions are optimal (10 in horizontally from the body and 30 in vertically from the floor). The multipliers (which range between 0 and 1) reduce the load constant, depending on lifting conditions.

Associated with a lifting task are the load weight (weight of object lifted), L, and the following task variables:

H Horizontal location: distance of the hands away from the midpoint between the ankles, in inches or centimeters, measured at the origin and destination of a lift.

V Vertical location: distance of the hands above the floor, in inches or centimeters, measured at the origin and destination of a lift.

D Vertical travel distance: absolute value of the difference between the vertical heights at the destination and origin of the lift, in inches or centimeters.

A Asymmetry angle: angular measure of how far the object is displaced from the front (midsagittal plane) of the worker's body at the beginning or ending of the lift in degrees measured at the origin and destination of a lift. The asymmetry angle is defined by the location of the load relative to the worker's midsagittal plane (front-rear plane separating left and right) as defined by the neutral body posture, rather than the position of the feet or the extent of body twist.

F Lifting frequency: average number of lifts per minute over a 15-minute period.

C Coupling classification: coupling quality ratings are good, fair, or poor, depending on the quality of the hand-to-object coupling (see Table 15-4). The classification is necessary to use Table 15-5.

TABLE 15-2 The Seven Lifting Task Multipliers

Abbreviation	Term	English Units	Metric Units
LC	Load constant	51 lb	23 kg
HM	Horizontal multiplier	10/H	25/H
VM	Vertical multiplier	$1 - (0.0075|V - 30|)$	$1 - (0.003|V - 75|)$
DM	Distance multiplier	$0.82 + (1.8/D)$	$0.82 + (4.5/D)$
AM	Asymmetric multiplier	$1 - (0.0032A)$	$1 - (0.0032A)$
FM	Frequency multiplier	From Table 15-3	From Table 15-3
CM	Coupling multiplier	From Table 15-5	From Table 15-5

TABLE 15-3   Frequency Multiplier

Frequency Lifts/min $(F)^a$	≤1 hour		<1 but ≤2 hours		>2 but ≤8 hours	
	$V < 30$ in	$V \geq 30$ in	$V < 30$ in	$V \geq 30$ in	$V < 30$ in	$V \geq 30$ in
≤0.2	1.00	1.00	0.95	0.95	0.85	0.85
0.5	0.97	0.97	0.92	0.92	0.81	0.81
1	0.94	0.94	0.88	0.88	0.75	0.75
2	0.91	0.91	0.84	0.84	0.65	0.65
3	0.88	0.88	0.79	0.79	0.55	0.55
4	0.84	0.84	0.72	0.72	0.45	0.45
5	0.80	0.80	0.60	0.60	0.35	0.35
6	0.75	0.75	0.50	0.50	0.27	0.27
7	0.70	0.70	0.42	0.42	0.22	0.22
8	0.60	0.60	0.35	0.35	0.18	0.18
9	0.52	0.52	0.30	0.30	0.00	0.15
10	0.45	0.45	0.26	0.26	0.00	0.13
11	0.41	0.41	0.00	0.23	0.00	0.00
12	0.37	0.37	0.00	0.21	0.00	0.00
13	0.00	0.34	0.00	0.00	0.00	0.00
14	0.00	0.31	0.00	0.00	0.00	0.00
15	0.00	0.28	0.00	0.00	0.00	0.00
>15	0.00	0.00	0.00	0.00	0.00	0.00

To enter the table, first measure the number of lifts in a sample 15-minute period and divide by 15 to obtain the lifting frequency, F. Then select the applicable FM value from this table based on the length of the lifting task.
[a] For lifting less frequently than once per 5 minutes, set $F = 0.2$ lifts/min.

Recommended Weight Limit (RWL)   The RWL for a specific set of task conditions is the weight of the load that nearly all healthy workers (free of adverse health conditions that would increase the risk of musculoskeletal injury) could perform over a substantial period of time for up to 8 hours without an increased risk of developing lifting-related lower back pain. The RWL (lb) is defined as

$$RWL = LC \times HM \times VM \times DM \times AM \times FM \times CM. \qquad (15\text{-}1)$$

Lifting Index   The LI provides a relative estimate of the physical stress associated with a manual lifting job. As LI increases, the level of risk for a given worker increases, and a greater percentage of workers are likely to be at risk for developing lifting-related lower back pain. LI is computed as

$$LI = \frac{\text{load weight}}{RWL}. \qquad (15\text{-}2)$$

Procedures   The first step is to analyze a lifting task to define the load weight, L, and the task variables for both the origin and destination where applicable. The next step is to determine the task multipliers and then use them to compute RWL and LI. Then LI is used to make decisions about a lifting task and its design. RWL and LI guide the ergonomic design of a lifting task:

- Individual multipliers help to identify specific job-related problems
- RWL can help guide the redesign of an existing or new lifting task

TABLE 15-4 Hand-to-Container Coupling Classification

Good	Fair	Poor
For containers of optimal design, such as some boxes, crates, etc., a "good" hand-to-object coupling would be defined as handles or hand-hold cutouts of optimal design (see notes 1, 2, and 3).	For containers of optimal design, a "fair" hand-to-object coupling is defined as handles or hand-hold cutouts of less than optimal design (see notes 1, 2, 3, and 4).	Containers of less than optimal design or loose parts or irregular objects that are bulky, hard to handle, or have sharp edges (see note 5).
For loose parts or irregular objects, which are not usually containerized, such as castings, stock, and supply materials, a "good" hand-to-object coupling is defined as a comfortable grip in which the hand can be easily wrapped around the object (see note 6).	For containers of optimal design with no handles or hand-hold cutouts or for loose parts or irregular objects, a "fair" hand-to-object coupling is defined as a grip in which the hand can be flexed approximately 90° (see note 4).	Lifting nonrigid bags (i.e., bags that sag in the middle).

Select the classification that best fits the lifting task to enter Table 15-5.

Notes:

1. An optimal handle design has 0.75 to 1.5 in (1.9–3.8 cm) diameter, ≥4.5 in (11.5 cm) length, 2 in (5 cm) clearance, cylindrical shape, and a smooth, nonslip surface.

2. An optimal hand-hold cutout has the following approximate characteristics: ≥1.5 in (3.8 cm) height, 4.5 in (11.5 cm) length, semioval shape, ≥2 in (5 cm) clearance, smooth, nonslip surface, and ≥0.25 in (0.06 cm) container thickness (e.g., double thickness cardboard).

3. An optimal container design has ≤16 in (40 cm) frontal length, ≤12 in (30 cm) height, and a smooth, nonslip surface.

4. A worker should be capable of clamping the fingers at nearly 90° under the container, such as required when lifting a cardboard box from the floor.

5. A container is considered less than optimal if it has a frontal length >16 in (40 cm), height <12 in (30 cm), rough or slippery surfaces, sharp edges, asymmetric center of mass, unstable contents, or requires the use of gloves. A loose object is considered bulky if the load cannot easily be balanced between the hand grasps.

6. A worker should be able to wrap the hand comfortably around the object without causing excessive wrist deviations or awkward postures, and the grip should not require excessive force.

TABLE 15-5 Coupling Multiplier

Coupling Classification	Coupling Multiplier	
	$V < 30$ in (75 cm)	$V \geq 30$ in (75 cm)
Good	1.00	1.00
Fair	0.95	1.00
Poor	0.90	0.90

Select the coupling multiplier for the coupling classification and vertical location.

- LI helps estimate the degree of physical stress for a lifting task
- LI can help prioritize redesign of various lifting tasks based on rank order of LI values

Figure 15-2 provides a sample form for managing the analysis for a single task.

References provide examples of various lifting tasks, both for single tasks and for multiple tasks. They also explain how to deal with special circumstances.

JOB ANALYSIS WORKSHEET
Single Task - NIOSH Revised Lifting Equations

Department			Description of Job						
Job Title									
Analyst		**Date**							

Step 1. Measure and record task variables

Object Weight (lb)		Hand Location (in)				Vertical Distance (in)	Asymmetric Angle (degrees)		Freq Rate lifts/min	Duration	Object Coupling
		Origin		Destination			Origin	Destin.			
L (Avg)	L (Max)	H	V	H	V	D	A	A	F	Hrs	C

Step 2. Determine the mulitpliers and compute the RWL's
   RWL = LC x HM x VM x DM x AM x FM x CM

Multiplier	LC	HM	VM	DM	AM	FM	CM	RWL
ORIGIN								
DESTINATION								

Step 3. Compute the Lifting Index (LI)
   LI = Object Weight (lb) / RWL

ORIGIN	Object Weight =	RWL =	LI =
DESTINATION	Object Weight =	RWL =	LI =

Figure 15-2.   Job analysis worksheet for the revised NIOSH lifting equations.

Example 15-1   A worker parks a hand truck near a mixing hopper to be able to stand somewhat between the hand truck and the hopper. The task is to lift each of the eight bags from the hand truck and place the bag on the rim of the hopper to open it and let the contents spill into the hopper. During the lift, the worker twists from the hand truck position to the hopper position. The task occurs roughly 10 times per 8-hour shift. Because the main lifting problem is the bottom bag, it will be analyzed. Compute the RWL and LI for the task and recommend any revisions to the lifting task.

In analyzing the lifting task, the following are established:

The hands are at a vertical location of 15 in for the origin and 36 in at the destination.

The hands are at a horizontal location of 18 in at the origin and 10 in at the destination.

The asymmetric angle is 45° to the left from the origin to the destination.

The frequency is fewer than 0.2 lifts/min for less than 1 hour.

The worker can flex the fingers approximately 90° and the bags are fairly rigid (do not sag in the middle).

Each bag weighs 40 lb.

The following multipliers are determined:

LC = 51

HM = 10/H = 0.56

VM = 1 − (0.0075|V − 30|) = 1 − (0.0075|15 − 30|) = 1 − 0.11 = 0.89

DM = 0.82 + (1.8/D) = 0.82 + 1.8/21 = 0.91

AM = 1 − (0.0032A) = 1 − (0.0032[45]) = 1 − 0.14 = 0.86

FM = 1.0 from Table 15-3

CM = 0.95 (from Table 15-5), because the coupling classification is "fair" (Table 15-4) and V < 30 in.

Then, RWL = 18.9 lb and LI = 40/18.9 = 2.1.

Possibilities for improving the tasks are bringing the load closer to increase HM, reducing the asymmetry to increase AM by repositioning the hand truck relative to the hopper, and raising the height of the bags at the origin to increase VM. For example, by decreasing H to 10 in, RWL would increase to nearly 34 lb and LI would decrease to approximately 1.2.

15-3 JACKS

There are many kinds of jacks, including hydraulic, mechanical, and pneumatic ones. There are bumper jacks for cars, scissors jacks, jacks with ratchet mechanisms, and jack screws. Each jack has a maximum design load. Although jacks are very common, users should understand how to use them safely and what their load limits are before using them.

Hazards

Major hazards with jacks are overloading their capacity and that of related elements involved in the lift, improper placement and their inherent instability, and having a load slip from them.

A jack may have enough capacity to support a load, but the surface or structure that it sets on or bears against may not have enough capacity. For example, a jack placed on the ground may sink when fully loaded. An object being lifted may not withstand the load and may fail by bending, breaking, or puncture.

Because they only support a single point, jacks are rather unstable, and the instability may increase because of a small bearing support for the jack. Because of their instability, loaded jacks may tip easily. If not properly positioned, they may slip at the top or the base. The object being lifted may shift and make the jack unstable.

Jacks are most often made of steel and are very often used to lift steel objects. The coefficient of friction for steel against steel is very low, and a small lateral load may cause a load to slip.

Controls

A jack should have enough capacity for the load to be lifted and should have a solid bearing support so that it will not sink or slip. Jacks that are used to lift things should be plumb throughout their use. Sometimes jacks or jack-type devices are used to pry things apart. In such cases, they should be positioned so that the ends are well anchored and will not slip. The load being lifted should be stabilized at locations away from the jack. Blocking or anchoring the load may be needed. Avoid metal on metal when using a jack. A wood block placed between the jack and the load can reduce the chances of slipping because it will deform at compression points and help keep the jack and load aligned. People should stand clear of the load being raised. A jack should never be used to support a load. Some other suitable blocking device or support should be used.

15-4 HAND-OPERATED MATERIALS HANDLING VEHICLES

There are many kinds of hand-operated materials handling vehicles. Hand trucks, dollies, carts, and wheelbarrows are hand powered. There are also hand-operated vehicles that use batteries or other power sources.

Hazards

Like all other load-carrying devices, one hazard is a load shifting or tipping and falling. When maneuvering hand-powered vehicles in tight spaces, an operator may strike the handles against a wall or objects and injure the hands. A hand-operated vehicle loaded too high will obscure the visibility of an operator and may cause accidents.

A variety of traffic problems can occur when vehicles are in motion. Examples are blind corners and passage of vehicles moving in different directions. When pulling a load, the operator may slip. The momentum of the vehicle and load may cause the load to run over the operator. An example is a hospital crash cart loaded with oxygen tanks, medicines, and other emergency supplies. The array of materials, often weighing 200 to 300 lb, is rushed to a patient room to treat life-threatening conditions. Because of the weight and the difficulty of maneuvering the cart around turns in a corridor, two people often rush it to an emergency. In one case, such a cart ran over the heel of the person pulling on it and severed an Achilles tendon.

Controls

Loads should be stable and well secured, should be limited in height and provide good visibility for an operator, and should not exceed weights that operators can handle or the capacity of the vehicle. Most loads should be pushed rather than pulled. If a load on a powered, hand-operated vehicle can be raised overhead, the vehicle should have overhead protection for the operator and possibly outrigger devices for stability.

If a vehicle is designed to be pulled, the tongue and handle should extend far enough from the vehicle so the operator's feet are not run over. Hand-powered vehicles should be operated at speeds that allow an operator to easily maneuver and stop the vehicle. Design features, like large diameter wheels or pneumatic tires, can make maneuvering and operation easier on rough surfaces and over bumps. Handles should be designed with a recessed location for hands so that the handle structure will strike a wall or object, not the hands. Knuckle guards are available for some handles to provide this protection. Incorporating soft rubber bumpers on the handles or other protruding elements of a cart can reduce injury and damage if the cart runs into something.

There are special hand operated vehicles for particular handling tasks. For example, there are hand trucks designed for handling gas cylinders. They have restraints that fit around a cylinder and are part of the design. There are special carts for handling carboys of acids or cryogenic liquids. Operators should learn to operate hand-powered vehicles safely.

15-5   POWERED VEHICLES

Equipment

There are many different powered vehicles for materials handling. Many are included in the term *industrial trucks*. The most common industrial truck is a forklift truck, fitted with two forks or tines. Other devices may be installed on an industrial truck for special lifting tasks. A single, long fork is used for lifting carpet rolls and similar rolled material. A clamp device is used to pick up rolls of paper.

Some industrial vehicles pull carts or push objects. At an airline terminal, one can see some of these vehicles in use. There are backhoes, end loaders, scrapers, bulldozers and earthmovers, and other powered vehicles for handling earth and bulk materials.

Hazards

Powered materials-handling vehicles have some hazards in common. Other hazards are unique to particular vehicles and their use. There are also hazards related to the kind of power.

All powered materials-handling vehicles may have visibility problems for an operator, although the problems may differ in type and degree. The vehicles can drive or back over someone or something. The operator may not be able to see the load or how well it is positioned. All have hazards related to traffic and the movement of several vehicles in the same area, all have minimum space requirements for safe operation, and all have hazards related to proper design and functioning of controls, lifting devices, brakes, and steering. A failure of these components could lead to accidents. All vehicles have load limits. Exceeding safe load limits can lead to structural failure and accidents. All have hazards related to proper loading and load stability. A falling load could injure the operator or someone nearby or could damage the vehicle, materials, or other nearby objects.

All have hazards associated with improper operation and use. For example, driving with a load in an elevated position may result in striking an overhead, protruding object or door header.

There are hazards associated with the source of power. Most battery-powered vehicles use lead-acid batteries. There are dangers related to electricity and to battery charging. Gasoline and propane fuels, used to power some vehicles, are flammable. Engines and exhaust are hot. In poorly ventilated or confined spaces, the exhaust could create hazardous conditions from carbon monoxide and other products of combustion.

Many powered materials-handling vehicles have a high center of gravity. When operated too fast in a turn, they can roll over. The problem is even greater when a load is elevated during a turn. Because the loads on forklifts and similar vehicles are cantilevered from the vehicle, counterweights help ensure that tipping over from loads is minimized. The potential for rollover in a turn can be analyzed by evaluating moments about the lateral boundary of support formed by the tires. The centrifugal force is compared with the pull of gravity about the lateral support line. From this analysis, the maximum forward or rearward velocity, V_{max}, a vehicle can handle in a turn without tipping over is

$$V_{max} = \left[gr \left(\frac{d_{cg}}{2h_{ce}} \right) \right]^{1/2}, \qquad (15\text{-}3)$$

where

V_{max} is in feet per second,

g is the gravitational constant ($32.2\,\text{ft/s}^2$),

r is the turn radius (feet),

d is the distance from the composite center of gravity for the vehicle and load to the lateral support line (feet), and

h is the height of the composite center of gravity from the ground (feet).

Example 15-2　(a) A loaded, four-wheeled cart weighs 500 lb. The wheels on each axle are 24 in apart and the axles are 48 in apart. Its center of gravity is located at the center of the cart in plan view and 30 in from the floor. The cart is pulled around a corner. The turning radius is 6 ft. What is the maximum velocity for the cart that will not cause it to tip over? Applying Equation 15-3,

$$V_{max} = \left[32.2 \left(6 \left\{ \frac{\left[\frac{12}{12} \right]}{2 \frac{30}{12}} \right\} \right) \right]^{1/2} = 6.22 \, \text{ft/s} = 4.24 \, \text{mi/hr}.$$

(b) If the center of gravity were raised to 36 in and the turning radius was extended to 10 ft, what is the maximum velocity without tipping over?

$$V_{max} = 7.33 \, \text{ft/s} = 4.99 \, \text{mi/hr}.$$

A vehicle may skid in a turn. This potential can be determined by comparing forces acting horizontally on the vehicle, that of the centrifugal force from the turn at V_{max} and frictional force on the tires. If the centrifugal force at V_{max} is more than the resistance force of friction, the vehicle will skid before it will tip.

Example 15-3　Consider the vehicle in Example 15-2(a). Assume the coefficient of friction, μ, between wheels and the floor is 0.65. Will the vehicle skid before it will tip?

Assuming all wheels are in contact with the supporting surface, the frictional force acting at the wheels is $F_f = \mu N$. The centrifugal force $F_c = mV^2/r$. The centrifugal force is $F_c = (500/32.2)(4.9)^2/6 = 62.13$ lb. The frictional force is $F_f = 0.65(500) = 325$ lb. Because $F_f > F_c$, the cart will not skid, but may tip.

Operating materials-handling vehicles on rough, irregular, and sloping surfaces can add to instability. The momentum of a jostled and tilted vehicle and load could cause the load to fall or the vehicle to tip. A load or vehicle may tip, even on a smooth, sloped surface. If the sum of forces acting through the center of gravity falls outside the support zone created by the tires, the vehicle will tip over. An analysis of static conditions will reveal when forward, rearward, or lateral tilt will cause tipping for various load elevations. Dynamic conditions are more difficult to analyze.

Some vehicles, like forklifts, can raise a load overhead. A load or parts of it that fall from an elevated level can injure the operator or anyone nearby.

Another hazard for some vehicles is catching on protruding objects during operation. For example, forklifts often carry loads into or out of truck trailers. Some rollup doors on truck trailers have short, flat straps for pulling the door down. If a rope is used instead of the flat strap or a knot is placed in the rope or strap, the knot may catch on the structure of the forklift and pull the door down on the operator. Obstructions, like items protruding from storage racks, door frames, and suspended light fixtures, can extend into the potential operating zone of a materials-handling vehicle and create catch points or introduce other hazards.

Controls

To protect operators from vehicle rollover injuries, some vehicles have a rollover protection system (ROPS). The concept is the same as that for motor vehicles. It consists of a structure surrounding the operator and forming an operator compartment. During an accident, it is essential to retain the integrity of the operator compartment. ROPS may be formed by a rollover bar on some vehicles, like tractors. Other elements of the vehicle structure help establish the envelope for the compartment. For other vehicles, four columns and connecting crossbars or a cab that meets ROPS standards create the safety envelope for the operator. Seat belts prevent the operator from being thrown from the compartment or crushed by ROPS or the vehicle itself during a rollover. One study estimated that 80% of operators of farm tractors who were killed in rollover accidents had vehicles without ROPS. With ROPS, the fatality rate was very low.

To protect operators from falling objects, overhead protection is needed. Materials-handling vehicles that have this hazard need a falling object protection system (FOPS). The size of opening should be small enough to prevent objects in a load from penetrating into the operator compartment. The operator also must be able to see the load through or around the FOPS. If there is a hazard from materials falling on an operator when a vehicle backs into something, FOPS should extend to the rear of the operator. Often ROPS and FOPS are incorporated into the same protective system.

The Society of Automotive Engineers (SAE) has several ROPS and FOPS standards. OSHA incorporates many of these standards into its regulations.

The integrity of the operator compartment provides protection from other hazards as well. For example, the operator stands in some industrial trucks. If feet or other parts of the body extend outside the vehicle, they can be crushed if the vehicle runs into something or passes close to an object, wall, or column. Protective enclosures or doors can reduce this hazard.

Vehicles and their components must be properly maintained to ensure their safe use. Operators must be trained in safe use of the vehicle. In most cases, operators have primary responsibility for safe materials-handling operations for their vehicles. Where exhaust may create a hazard, there should be adequate ventilation. Electric vehicles may be safer in enclosed areas.

Other controls include keeping areas clear of obstructions where materials-handling vehicles are operated, having horizontal and vertical pathways large enough for safe maneuver of the vehicle and the load, and providing protective barriers for racks and other objects, which could pose dangers when struck by powered vehicles that operate nearby. Trailers being loaded by materials-handling vehicles should have wheel chocks, devices to lock trailers to loading docks or other means to lock them in place.

Where visibility is a problem, visual assistance, which may be as simple as a mirror, must be provided for the operator. A classic products liability case involved the lack of a rearview mirror on a materials-handling vehicle. Assistance may be in the form of an assistant who gives standard hand signals or in the form of audio and visual warning devices on the vehicle to indicate when the vehicle is moving backward or in a direction where the operators visibility is inhibited.

15-6   HOISTING APPARATUS

Equipment

There are many kinds of equipment for hoisting and positioning materials. Not all can be discussed here. Most involve vertical movement of materials. Some include lateral movement or a combination of motions. There are hoists, which may be hand operated or powered, blocks and tackle, derricks and winches of various types, fixed and mobile cranes, tower cranes, gantry cranes, overhead cranes and boom cranes, aerial baskets (for positioning people), and elevated baskets (for positioning supplies; not to be used by people). These devices may be independent or may be installed in buildings, on trucks, or on railroad cars, or may be incorporated into the design of other items. Most involve rigging of some kind. Each may have special features and fixtures for handling particular kinds of materials or objects.

Hazards

Hoisting devices have some of the same hazards as other materials-handling equipment. Elevated materials can fall on people or items below. Structural failure can occur when load limits are exceeded. There are also load limits for stability. When they are exceeded, a hoisting apparatus will tip over and operators or assistants may become caught in the machine.

Visibility may be limited during operation. Supporting surfaces for wheels, outriggers, or other supports may not be able to carry the load, and the equipment may then tip over. Operation near power lines can result in electrocution. Operations in windy conditions may cause overloads and structural failure.

There are also hazards associated with the kind of power, the type of operation, the materials being handled, and the condition of the vehicle and its components.

Controls

One control for hoisting apparatus is proper setup and planning. This can include placement on a level site, staying away from power lines, and proper assembly of tower, boom, jib, and other elements. Chains and ropes should be free of kinks and twists, and the hoisting equipment should be inspected regularly, preferably before use. Many hoisting devices must be certified regularly for safe use. Soil conditions and other factors in the work area should be checked. For some crane operations, a lift diagram or plan is written and approved by all participating parties. A crane and its load must have stability. Stability may be achieved in various ways, from outriggers for boom cranes to anchoring cables for tower cranes.

Another control is a safe load. A hoisting apparatus should have load limits clearly marked. The ratings assume a static load. Loads that are jerked or dropped some distance and place inertial loads on the rigging can overload the apparatus or its rigging. Cranes, for example, should have a load chart affixed to the operator's cab and should have an operator's manual in the cab. The weight of the load to be picked up must be known, and the capacity of the equipment should not be exceeded.

There must be a trained operator. Some hoisting equipment has simple controls, whereas other types have complex controls. The operator must know more than how to make the crane and its movements work. The operator must know what can go wrong and how to deal with such situations; must understand the changing conditions during an operation; must evaluate the site for other workers, potentially changing site conditions and other factors that could affect safe operation; must know if the load is properly rigged; and must be able to read a load chart, if one is applicable.

Operating hoisting equipment safely can be a very complex process. For example, to make a safe lift with a mobile crane, at least seven items of information are needed:

1. Is the vehicle level?
2. Are all outriggers extended or retracted?
3. Are extended outriggers supported by stable ground?
4. Are tires fully inflated?
5. What is the angle of the boom?
6. What are the boom and jib lengths?
7. What positions will the boom be in during the lift?
8. How much does the load weigh?

If these things are known, the operator can make readings on the load chart (see Figure 15-2) and determine if the operation will be within all structural and stability limits. The limits assume a level machine, and they are different when tires are not fully inflated and when the outriggers are not in place. The limits change suddenly when a load is moved from the front to the side and to the rear of the vehicle because of the changing distance of frontal, lateral, and rearward supports from the pivot point. The boom angle and length also are needed to determine the safe loads. Precise measurements may be needed when operations take place in tight spaces. The use of the crane can begin after evaluating the site and the environment around the crane and after comparing conditions with the load chart and deciding that a load can be lifted safely. Figure 15-2 illustrates how complex the use of a load chart can become while it helps ensure that lifts are completed safely.

With the wide variety of configurations of a mobile crane, using a load chart to determine a safe lift can be quite complicated. Today, some manufacturers build computers into the cab, require the operator to input a sequence of data in response to prompts, and combine the data with sensory data from crane components to handle the information sequence in deciding if a lift can be accomplished safely.

Excessive wind, potential contact with electrical lines and equipment, job site congestion, or poor soil and bearing under outriggers may require scrapping an operation. The load must be properly rigged, and all participants must understand the lifting plan. It is then lifted and moved, most likely with the assistance of others observing the load and the site, to make sure that no one is under the load or that other dangerous conditions develop. An assistant may use standard hand signals to guide the operator, who cannot see

LOAD CHARTS
RT530E

85% STABILITY
ON OUTRIGGERS
75% STABILITY
ON RUBBER

Figure 15-3.   Representative load chart for a mobile crane (Grove Mobile Hydraulic Crane, Model RT530E). Data show lifting capabilities for various crane operations. Capacities below bold lines are based on failure by tipping; capacities above bold lines are based on structural strength. Notes and diagrams identify many additional factors and assumptions that must be considered in deciding if a load can be lifted safely. (Reprinted with the permission of Grove U.S., LLC.)

NOTES FOR LIFTING CAPACITIES

GENERAL:

1. Rated loads as shown on lift chart pertain to this machine as originally manufactured and equipped. Modifications to the machine or use of optional equipment other than that specified can result in a reduction of capacity.
2. Construction equipment can be hazardous if improperly operated or maintained. Operation and maintenance of this machine shall be in compliance with the information in the Operator's and Safety Handbook, Service Manual and Parts Manual supplied with this machine. If these manuals are missing, order replacements from the manufacturer through the distributor.
3. The operator and other personnel associated with machine shall fully acquaint themselves with the latest American National Safety Standards (ASME/ANSI) for cranes.

SETUP:

1. The machine shall be level and on a firm supporting surface. Depending on the nature of the supporting surface, it may be necessary to have structural supports under the outrigger floats or tires to spread the load to a larger bearing surface.
2. For outrigger operation, all outriggers shall be properly extended with tires raised free of crane weight before operating the boom or lifting loads.
3. When machine is equipped with center front stabilizer, the front stabilizer shall be set in accordance with instructions in Operator's and Safety Handbook.
4. When equipped with removable and/or extendible counterweight, the proper counterweight shall be installed and fully extended before and during operation.
5. Tires shall be inflated to the recommended pressure before lifting on rubber.
6. With certain boom and hoist tackle combinations, maximum capacities may not be obtainable with standard cable lengths.
7. Unless approved by the crane manufacturer, do not travel with boom extension or jib erected unless otherwise noted. Refer to Operator's and Safety Handbook for job-site travel information.

OPERATION:

1. Rated loads at rated radius shall not be exceeded. Do not attempt to tip the machine to determine allowable loads. For clamshell or concrete bucket operation, weight of bucket and load must not exceed 80% of rated lifting capacities.
2. All rated loads have been tested to and meet the requirements of SAE J1063 - Cantilevered Boom Crane Structures - Method of Test, and do not exceed 85% of the tipping load on outriggers fully extended and SAE J1289 - Mobile Crane Stability Ratings [$1.25P < (T-0.1A)$] on outriggers 50% and 0% extended (fully retracted) as determined by SAE J765 - Crane Stability Test Code.
3. Rated loads include the weight of hookblock, slings and auxiliary lifting devices and their weights shall be subtracted from the listed rating to obtain the net load to be lifted. When more than the minimum required parts of line needed to pick the load are used, the additional rope weight as measured from the lower sheaves of the main boom nose shall be considered part of the load to be lifted. When both the hook block and headache ball are reeved, the lifting device that is NOT in use, including the line as measured from the lower sheave(s) of the nose supporting the unused device shall be considered part of the load.
4. Load ratings are based on freely suspended loads. No attempt shall be made to move a load horizontally on the ground in any direction.
5. The maximum in-service wind speed is 20 m.p.h. It is recommended when wind velocity is above 20 m.p.h., rated loads and boom lengths shall be appropriately reduced. For machines not in-service, the main boom should be retracted and lowered with the swing brake set in wind velocities over 30 m.p.h.
6. Rated loads are for lift crane service only.
7. **Do not** operate at a radius or boom length where capacities are not listed. At these positions, the machine may overturn without any load on the hook.
8. The maximum load which can be telescoped is not definable because of variations in loadings and crane maintenance, but it is safe to attempt retraction and extension of the boom within the limits of the capacity chart.
9. When the boom length or lift radius or both are between values listed, the smallest load shown at either the next larger radius or next longer or shorter boom length shall be used.
10. For safe operation, the user shall make due allowances for his particular job conditions, such as: soft or uneven ground, out of level conditions, high winds, side loads, pendulum action, jerking or sudden stopping of loads, experience of personnel, two machine (tandem) lifts, traveling with loads, electric wires, obstacles, hazardous conditions, etc. Side pull on boom or jib is extremely dangerous.
11. If machine is equipped with individually controlled powered boom sections, the boom sections must be extended equally at all times.
12. Never handle personnel with this machine unless the requirements of the applicable national, state, and local regulations and safety codes are met.
13. Keep load handling devices a minimum of 42 inches below boom head at all times.
14. The boom angle before loading should be greater than the loaded boom angle to account for deflection.
15. Capacities appearing above the bold line are based on structural strength and tipping should not be relied upon as a capacity limitation.
16. Capacities for the 29 ft. boom length shall be lifted with boom fully retracted. If boom is not fully retracted, capacities shall not exceed those shown for the 40 ft. boom length.
17. When operating the machine in the "On Outriggers 50% Extended (14' spread)" mode, the outrigger beam pins must be engaged. When operating in the "On Outriggers 0% Extended (7' 10" spread)" mode, the outrigger beams must be fully retracted. Failure to follow these precautions could result in structural damage or loss of stability of the machine.
18. Regardless of counterweight and outrigger spread configuration, no deduct is required from the main boom charts for a stowed boom extension.
19. **Do not** lift loads when boom is fully lowered. The Load Moment Indicator (LMI) senses pressure and will not provide warnings or lockout. The crane can become overloaded if lift cylinder(s) is fully retracted.
20. The maximum outrigger pad load is 48,900 lb.

DEFINITIONS:

1. Operating Radius: Horizontal distance from a projection of the axis of rotation to the supporting surface before loading to the center of the vertical hoist line or tackle with load applied.
2. Loaded Boom Angle (Shown in Parenthesis on Main Boom Capacity Chart): is the angle between the boom base section and the horizontal, after lifting the rated load at the rated radius with the rated boom length.
3. Working Area: Areas measured in a circular arc about the center line of rotation as shown on the working area diagram.
4. Freely Suspended Load: Load hanging free with no direct external force applied except by the lift cable.
5. Side Load: Horizontal force applied to the lifted load either on the ground or in the air.

RT530E - S/N

Figure 15-3.   *continued*

WEIGHT REDUCTIONS FOR LOAD HANDLING DEVICES

26 FT. OFFSETTABLE BOOM EXTENSION	
*Erected -	2,960 lbs.

26 FT. - 45 FT. TELE. BOOM EXTENSION	
*Erected (Retracted) -	4,220 lbs.
*Erected (Extended) -	5,780 lbs.

*Reduction of main boom capacities

AUXILIARY BOOM NOSE	142 lbs.
HOOKBLOCKS and HEADACHE BALLS:	
30 Ton, 3 Sheave	580 lbs.+
15 Ton, 2 Sheave	425 lbs.+
7.5 Ton Overhaul Ball (top swivel)	354 lbs.+
7.5 Ton Headache Ball	338 lbs.+

+Refer to rating plate for actual weight.

When lifting over swingaway and/or jib combinations, deduct total weight of all load handling devices reeved over main boom nose directly from swingaway or jib capacity.

NOTE: All load handling devices and boom attachments are considered part of the load and suitable allowances MUST BE MADE for their combined weights. Weights are for Grove furnished equipment.

LINE PULLS AND REEVING INFORMATION

HOISTS	CABLE SPECS.	PERMISSIBLE LINE PULLS	NOMINAL CABLE LENGTH
Main	5/8" (16 mm) 6x37 Class, EIPS, IWRC Special Flexible Min. Breaking Strength 41,200 lb.	11,640 lb.	450 ft.
Main & Aux.	5/8" (16 mm) Flex-X 35 Rotation Resistant (Non-rotating) Min. Breaking Strength 61,200 lb.	11,640 lb.	450 ft.

The approximate weight of 5/8" wire rope is 1.0 lb./ft.

HOIST PERFORMANCE

Wire Rope Layer	Hoist Line Pulls Available lb.*	Drum Rope Capacity (ft.)	
		Layer	Total
1	11,640	77	77
2	10,480	85	162
3	9,530	94	256
4	8,730	102	358
5	8,060	111	469
6	7,490	119	588

*Max. lifting capacity: 6x37 class = 11,640 lb.
                                     35x7 class = 11,640 lb.

Figure 15-3. *continued*

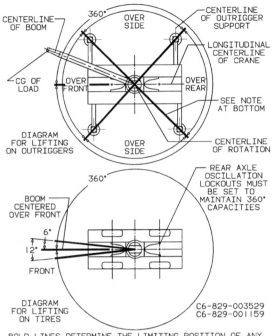

CENTERLINE OF BOOM
360°
OVER SIDE
CENTERLINE OF OUTRIGGER SUPPORT
LONGITUDINAL CENTERLINE OF CRANE
CG OF LOAD
OVER FRONT
OVER REAR
SEE NOTE AT BOTTOM
DIAGRAM FOR LIFTING ON OUTRIGGERS
OVER SIDE
CENTERLINE OF ROTATION

360°
REAR AXLE OSCILLATION LOCKOUTS MUST BE SET TO MAINTAIN 360° CAPACITIES
BOOM CENTERED OVER FRONT
6°
12°
FRONT
DIAGRAM FOR LIFTING ON TIRES
C6-829-003529
C6-829-001159

BOLD LINES DETERMINE THE LIMITING POSITION OF ANY LOAD FOR OPERATION WITHIN WORKING AREAS INDICATED
WORKING AREA DIAGRAM

WORKING RANGE DIAGRAM

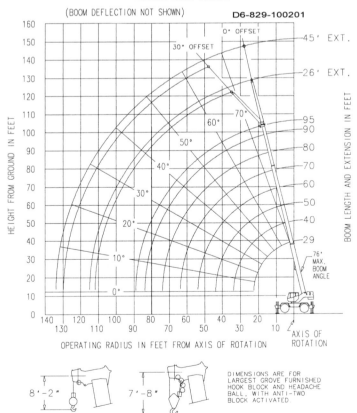

(BOOM DEFLECTION NOT SHOWN)   D6-829-100201

DIMENSIONS ARE FOR LARGEST GROVE FURNISHED HOOK BLOCK AND HEADACHE BALL, WITH ANTI-TWO BLOCK ACTIVATED.

Figure 15-3. *continued*

RATED LIFTING CAPACITIES IN POUNDS
29 FT. - 95 FT. BOOM
ON OUTRIGGERS FULLY EXTENDED - 360°

Radius in Feet	#0001							
	Main Boom Length in Feet							
	29	40	50	60	70	80	90	95
10	60,000 (60.5)	50,100 (69.5)	46,950 (74.5)					
12	54,650 (56)	50,100 (66.5)	44,950 (72)	*38,850 (76)				
15	42,850 (47.5)	43,800 (61.5)	41,050 (68)	36,000 (72)	*29,450 (76)	*22,450 (76)		
20	30,700 (30)	31,650 (53)	32,100 (61.5)	29,500 (67)	27,400 (71)	22,450 (73.5)	*18,550 (76)	*15,500 (76)
25		24,050 (42.5)	24,500 (54.5)	24,800 (61.5)	23,100 (66.5)	19,250 (70)	16,500 (72.5)	15,300 (74)
30		18,800 (29)	19,250 (47)	19,550 (56)	19,600 (61.5)	16,850 (66)	14,400 (69)	13,200 (70.5)
35			15,550 (38)	15,850 (49.5)	16,000 (56.5)	14,850 (61.5)	12,700 (65.5)	11,500 (67.5)
40			12,800 (26)	12,950 (42.5)	13,000 (51.5)	13,050 (57.5)	11,000 (62)	10,000 (64)
45	See Note 16			10,450 (34.5)	10,500 (46)	10,550 (53)	9,630 (58.5)	9,060 (60.5)
50				8,610 (23.5)	8,630 (39.5)	8,670 (48)	8,720 (54.5)	7,990 (57)
55					7,170 (32)	7,200 (43)	7,250 (50)	7,100 (53)
60					6,000 (22)	6,030 (37)	6,100 (45.5)	6,110 (49)
65						5,080 (30)	5,120 (40.5)	5,150 (44.5)
70						4,270 (20.5)	4,330 (35)	4,350 (40)
75							3,650 (28.5)	3,700 (34.5)
80							3,100 (20)	3,100 (28)
85								2,600 (20)
Minimum boom angle (°) for indicated length (no load)								0
Maximum boom length (ft.) at 0° boom angle (no load)								95

NOTE: ( ) Boom angles are in degrees.
#LMI operating code. Refer to LMI manual for operating instructions.
*This capacity is based on maximum boom angle.

Lifting Capacities at Zero Degree Boom Angle On Outriggers Fully Extended - 360°								
Boom Angle	Main Boom Length in Feet							
	29	40	50	60	70	80	90	95.2
0 deg.	26,100 (22.8)	15,800 (33.8)	11,000 (43.8)	7,430 (53.8)	5,220 (63.8)	3,730 (73.8)	2,660 (83.8)	2,220 (89)

NOTE: ( ) Reference radii in feet.

A6-829-101755

Figure 15-3. *continued*

26 FT. - 45 FT. TELE OFFSETTABLE BOOM EXTENSION
ON OUTRIGGERS FULLY EXTENDED - 360°

Radius in Feet	**26 ft. LENGTH		45 ft. LENGTH	
	#0021	#0023	#0041	#0043
	0° OFFSET	30° OFFSET	0° OFFSET	30° OFFSET
30	*8,200 (76)			
35	8,200 (73.5)		*5,250 (76)	
40	8,200 (71)	*5,780 (76)	5,250 (75)	
45	8,120 (68.5)	5,780 (73.5)	4,940 (73)	
50	7,350 (66)	5,360 (71)	4,540 (71)	
55	6,370 (63)	4,750 (68)	4,150 (68.5)	*2,730 (76)
60	5,670 (60.5)	4,290 (65)	3,890 (66)	2,730 (74.5)
65	4,820 (57.5)	3,870 (62)	3,740 (64)	2,730 (72)
70	4,200 (54.5)	3,530 (59)	3,600 (61.5)	2,580 (69.5)
75	3,680 (51.5)	3,230 (56)	3,470 (59)	2,520 (67)
80	3,080 (48.5)	3,000 (52.5)	3,240 (56.5)	2,460 (64)
85	2,520 (45)	2,780 (49)	3,050 (54)	2,420 (61.5)
90	2,050 (41)	2,410 (45)	2,820 (51)	2,390 (58.5)
95	1,670 (37)	1,970 (40.5)	2,480 (48.5)	2,370 (55.5)
100	1,370 (32.5)	1,580 (35.5)	2,090 (45.5)	2,310 (52)
105	1,020 (27.5)		1,740 (42)	2,000 (49)
110			1,430 (38.5)	1,580 (45)
115			1,150 (35)	1,260 (40.5)
120			900 (30.5)	
Min. boom angle for indicated length (no load)	24°	30°	30°	30°
Max. boom length at 0° boom angle (no load)	80 ft.		80 ft.	

NOTE: ( ) Boom angles are in degrees.   A6-829-100272A
#LMI operating code. Refer to LMI manual for instructions.
*This capacity based on maximum boom angle.
**26 ft. capacities are also applicable to fixed offsettable ext. However, the LMI codes will change to #0051 and #0053 for 0° and 30° offset, respectively.

Figure 15-3. *continued*

BOOM EXTENSION CAPACITY NOTES:

1. All capacities above the bold line are based on structural strength of boom extension.

2. 26 ft. and 45 ft. boom extension lengths may be used for single line lifting service.

3. Radii listed are for a fully extended boom with the boom extension erected. For main boom lengths less than fully extended, the rated loads are determined by boom angle. Use only the column which corresponds to the boom extension length and offset for which the machine is configured. For boom angles not shown, use the rating of the next lower boom angle.

   WARNING: Operation of this machine with heavier loads than the capacities listed is strictly prohibited. Machine tipping with boom extension occurs rapidly and without advance warning.

4. Boom angle is the angle above or below horizontal of the longitudinal axis of the boom base section after lifting rated load.

5. Capacities listed are with outriggers fully extended and vertical jacks set only.

ON RUBBER CAPACITIES

STATIONARY CAPACITIES
360°

Radius in Feet	#9005			
	Main Boom Length in Feet			
	29	40	50	60
10	25,550 (60.5)	25,550 (70)	*16,450 (76)	
12	20,600 (56)	20,600 (66.5)	16,450 (72)	
15	14,350 (47.5)	14,350 (62)	14,350 (68)	14,350 (72.5)
20	8,280 (30)	8,280 (53)	8,280 (61.5)	8,280 (67)
25		5,330 (42.5)	5,330 (54.5)	5,330 (61.5)
30		3,630 (29)	3,630 (47)	3,630 (56)
35			2,500 (38)	2,500 (49.5)
40			1,690 (26)	1,690 (42.5)
45				1,090 (34.5)
Min. boom angle for indicated length (no load)	34°			
Max. boom length at 0° boom angle (no load)	50 ft.			

NOTE: ( ) Boom angles are in degrees.
#LMI operating code. Refer to LMI manual for instructions.
*This capacity is based upon maximum boom angle.

Lifting Capacity at Zero Degree On Rubber - 360°			
Boom Angle	Main Boom Length in Feet		
	29	40	50
0°	6,110 (22.8)	2,730 (33.8)	1,210 (43.8)

NOTE: Reference radii in feet.   A6-829-100274C

Figure 15-3. *continued*

STATIONARY CAPACITIES
DEFINED ARC OVER FRONT (See Note 3)

Radius in Feet	#9005			
	Main Boom Length in Feet			
	29	40	50	60
10	30,100 (60.5)	26,550 (70)	16,450 (74.5)	
12	26,550 (56)	22,100 (66.5)	16,450 (72)	
15	22,100 (47.5)	22,100 (62)	16,450 (68)	16,450 (72.5)
20	16,050 (30)	16,050 (53)	16,050 (61.5)	16,050 (67)
25		11,005 (42.5)	11,005 (54.5)	11,005 (61.5)
30		8,060 (29)	8,060 (47)	8,060 (56)
35			6,110 (38)	6,110 (49.5)
40			4,720 (26)	4,720 (42.5)
45				3,680 (34.5)
50				2,870 (23.5)
Min. boom angle for indicated length (no load)	0°			
Max. boom length at 0° boom angle (no load)	60 ft.			

NOTE: ( ) Boom angles are in degrees.
#LMI operating code. Refer to LMI manual for instructions.

Lifting Capacity at Zero Degree On Rubber Stationary- Defined Arc Boom Centered Over Front			
Boom Angle	Main Boom Length in Feet		
	29	40	50	60
0°	12,700 (22.8)	6,500 (33.8)	3,890 (43.8)	2,360 (53.8)

NOTE: Reference radii in feet.   A6-829-100275B

ON RUBBER CAPACITIES (cont'd.)

PICK & CARRY CAPACITIES (UP TO 2.5 MPH) -
BOOM CENTERED OVER FRONT (See note 7)

Radius in Feet	#9006			
	Main Boom Length in Feet			
	29	40	50	60
10	25,900 (60.5)	25,900 (70)	18,250 (74.5)	
12	22,350 (56)	22,350 (66.5)	18,250 (72)	
15	18,250 (47.5)	18,250 (62)	18,250 (68)	13,350 (72.5)
20	13,350 (30)	13,350 (53)	13,350 (61.5)	13,350 (67)
25		10,350 (42.5)	10,350 (54.5)	10,350 (61.5)
30		8,060 (29)	8,060 (47)	8,060 (56)
35			4,810 (38)	4,810 (49.5)
40			3,770 (26)	3,770 (42.5)
45				2,930 (34.5)
50				2,240 (23.5)
Min. boom angle for indicated length (no load)				0°
Max. boom length at 0° boom angle (no load)				60 ft.

NOTE: ( ) Boom angles are in degrees.
#LMI operating code. Refer to LMI manual for instructions.

Lifting Capacity at Zero Degree On Rubber Pick & Carry - Boom Centered Over Front				
Boom Angle	Main Boom Length in Feet			
	29	40	50	60
0°	11,400 (22.8)	5,090 (33.8)	3,110 (43.8)	1,800 (53.8)

NOTE: Reference radii in feet.   A6-829-100276B

Figure 15-3. *continued*

NOTES TO ALL RUBBER CAPACITY CHARTS:

1. Capacities are in pounds and do not exceed 75% of tipping loads as determined by test in accordance with SAE J765.

2. Capacities are applicable to machines equipped with 20.5x25 (24 ply) tires at 75 psi cold inflation pressure, and 16.00x25 (28 ply) tires at 100 psi cold inflation pressure.

3. Defined Arc - Over front includes 6° on either side of longitudinal centerline of machine (ref. drawing C6-829-003529).

4. Capacities appearing above the bold line are based on structural strength and tipping should not be relied upon as a capacity limitation.

5. Capacities are applicable only with machine on firm level surface.

6. On rubber lifting with boom extensions not permitted.

7. For pick and carry operation, boom must be centered over front of machine, mechanical swing lock engaged and load restrained from swinging. When handling loads in the structural range with capacities close to maximum ratings, travel should be reduced to creep speeds.

8. Axle lockouts must be functioning when lifting on rubber.

9. All lifting depends on proper tire inflation, capacity and condition. Capacities must be reduced for lower tire inflation pressures. See lifting capacity chart for tire used. Damaged tires are hazardous to safe operation of crane.

10. Creep - not over 200 ft. of movement in any 30 minute period and not exceeding 1 mph.

RATED LIFTING CAPACITIES IN POUNDS
29 FT. - 95 FT. BOOM
ON OUTRIGGERS 50% EXTENDED (14.0 FT. SPREAD) - 360°

Radius in Feet	#4001							
	Main Boom Length in Feet							
	29	40	50	60	70	80	90	95
10	60,000 (60.5)	48,000 (69.5)	45,000 (74.5)					
12	53,300 (56)	48,000 (66.5)	44,950 (72)	*37,000 (76)				
15	42,100 (47.5)	40,500 (61.5)	38,350 (68)	36,000 (72)	*27,400 (76)	*21,000 (76)		
20	23,950 (30)	23,850 (53)	23,900 (61.5)	24,050 (67)	23,200 (71)	21,000 (73.5)	*17,000 (76)	*15,500 (76)
25		15,850 (42.5)	15,950 (54.5)	16,150 (61.5)	16,350 (66.5)	16,400 (70)	15,950 (72.5)	15,300 (74)
30		11,350 (29)	11,500 (47)	11,650 (56)	11,800 (61.5)	12,000 (66)	12,150 (69)	12,100 (70.5)
35	See Note 16		8,620 (38)	8,820 (49.5)	8,930 (56.5)	9,050 (61.5)	9,190 (65.5)	9,260 (67.5)
40			6,610 (26)	6,820 (42.5)	6,900 (51.5)	6,990 (57.5)	7,100 (62)	7,150 (64)
45				5,350 (34.5)	5,400 (46)	5,470 (53)	5,550 (58.5)	5,600 (60.5)
50				4,220 (23.5)	4,260 (39.5)	4,310 (48)	4,370 (54.5)	4,410 (57)
55					3,350 (32)	3,390 (43)	3,430 (50)	3,460 (53)
60					2,600 (22)	2,640 (37)	2,670 (45.5)	2,700 (49)
65						2,020 (30)	2,050 (40.5)	2,060 (44.5)
70						1,490 (20.5)	1,520 (35)	1,530 (40)
75							1,070 (28.5)	1,080 (34.5)
0.1A (lb.)	660	610	580	560	550	540	540	530
Minimum boom angle (°) for indicated length (no load)							15	20
Maximum boom length (ft.) at 0° boom angle (no load)							80	

NOTE: ( ) Boom angles are in degrees.
#LMI operating code. Refer to LMI manual for operating instructions.
*This capacity is based on maximum boom angle.

Lifting Capacities at Zero Degree Boom Angle On Outriggers 50% Extended - 360°								
Boom Angle	Main Boom Length in Feet							
	29	40	50	60	70	80		
0 deg.	18,800 (22.8)	9,000 (33.8)	5,400 (43.8)	3,480 (53.8)	2,100 (63.8)	1,130 (73.8)		

NOTE: ( ) Reference radii in feet.

A6-829-100270A

Figure 15-3. *continued*

26 FT. - 45 FT. TELE BOOM EXTENSION

ON OUTRIGGERS 50% EXTENDED (14.0 FT. SPREAD) - 360°

Radius in Feet	**26 ft. LENGTH		45 ft. LENGTH	
	#4021	#4023	#4041	#4043
	0° OFFSET	30° OFFSET	0° OFFSET	30° OFFSET
30	*8,200 (76)			
35	8,200 (73.5)		*5,250 (76)	
40	6,940 (71)	*5,780 (76)	5,250 (75)	
45	5,580 (68.5)	5,780 (73.5)	4,940 (73)	
50	4,490 (66)	5,360 (71)	4,540 (71)	
55	3,600 (63)	4,350 (68)	4,150 (68.5)	*2,730 (76)
60	2,860 (60.5)	3,430 (65)	3,490 (66)	2,730 (74.5)
65	2,190 (57.5)	2,670 (62)	2,870 (64)	2,730 (72)
70	1,610 (54.5)	2,030 (59)	2,340 (61.5)	2,580 (69.5)
75	1,120 (51.5)	1,490 (56)	1,840 (59)	2,520 (67)
80		1,020 (52.5)	1,400 (56.5)	2,260 (64)
85			1,020 (54)	1,760 (61.5)
90				1,310 (58.5)
0.1A (lb.)	570	540	500	460
Min. boom angle for indicated length (no load)	44°	46°	48°	49°
Max. boom length at 0° boom angle (no load)	60 ft.		60 ft.	

NOTE: ( ) Boom angles are in degrees.   A6-829-100273B
#LMI operating code. Refer to LMI manual for instructions.
*This capacity based on maximum boom angle.
26 ft. capacities are also applicable to fixed offsettable ext. However, the LMI codes will change to **#4051 and **#4053** for 0° and 30° offset, respectively.

BOOM EXTENSION CAPACITY NOTES:

1. All capacities above the bold line are based on structural strength of boom extension.

2. 26 ft. and 45 ft. boom extension lengths may be used for single line lifting service.

3. Radii listed are for a fully extended boom with the boom extension erected. For main boom lengths less than fully extended, the rated loads are determined by boom angle. Use only the column which corresponds to the boom extension length and offset for which the machine is configured. For boom angles not shown, use the rating of the next lower boom angle.

    WARNING: Operation of this machine with heavier loads than the capacities listed is strictly prohibited. Machine tipping with boom extension occurs rapidly and without advance warning.

4. Boom angle is the angle above or below horizontal of the longitudinal axis of the boom base section after lifting rated load.

5. Capacities listed are with outriggers properly extended and vertical jacks set only.

Figure 15-3. *continued*

RATED LIFTING CAPACITIES IN POUNDS
29 FT. - 95 FT. BOOM

ON OUTRIGGERS 0% EXTENDED (7 FT. 10 IN. SPREAD) - 360°

Radius in Feet	#8001							
	Main Boom Length in Feet							
	29	40	50	60	70	80	90	95
10	34,700 (60.5)	32,400 (69.5)	30,400 (74.5)					
12	26,200 (56)	25,400 (66.5)	24,100 (72)	*22,900 (76)				
15	17,750 (47.5)	17,550 (61.5)	17,550 (68)	17,250 (72)	*16,550 (76)	*10,900 (76)		
20	10,650 (30)	10,600 (53)	10,650 (61.5)	10,750 (67)	11,000 (71)	10,900 (73.5)	*10,500 (76)	*10,350 (76)
25		6,930 (42.5)	7,020 (54.5)	7,170 (61.5)	7,350 (66.5)	7,560 (70)	7,610 (72.5)	7,490 (74)
30		4,670 (29)	4,780 (47)	4,950 (56)	5,080 (61.5)	5,240 (66)	5,390 (69)	5,480 (70.5)
35			3,270 (38)	3,450 (49.5)	3,550 (56.5)	3,660 (61.5)	3,780 (65.5)	3,850 (67.5)
40			2,170 (26)	2,370 (42.5)	2,440 (51.5)	2,520 (57.5)	2,620 (62)	2,670 (64)
45				1,550 (34.5)	1,600 (46)	1,660 (53)	1,740 (58.5)	1,780 (60.5)
50							1,050 (54.5)	1,080 (57)
0.1A (lb)	660	610	580	560	550	540	540	530
Minimum boom angle (°) for indicated length (no load)				33	43	51	53	55
Maximum boom length (ft.) at 0° boom angle (no load)				50				

NOTE: ( ) Boom angles are in degrees.
#LMI operating code. Refer to LMI manual for operating instructions.
*This capacity is based on maximum boom angle.

Lifting Capacities at Zero Degree Boom Angle On Outriggers 0% Extended - 360°							
Boom Angle	Main Boom Length in Feet						
	29	40	50				
0 deg.	8,310 (22.8)	3,390 (33.8)	1,480 (43.8)				

NOTE: ( ) Reference radii in feet.   A6-829-100271A

TIRE INFLATION - PSI (BAR)				
SIZE (FRONT & REAR)	LOAD RANGE	TRA CODE	LIFTING SERVICE AND GENERAL TRAVEL STATIC, CREEP & 2.5 MPH (4.0 km/h)	EXTENDED TRAVEL
20.5x25	24 PR	E-3	75 (5.2)	70 (4.8)
16.00x25	28 PR	E-3	100 (6.9)	95 (6.6)

Figure 15-3.   *continued*

the load during parts of the operation. There may also be someone who is clear of the load, but guides it with a tag line.

There are many design features that can reduce risks in the use of hoisting apparatus. Not all safety features in cranes are fail-safe devices; each has its limitations. Not all can be discussed here. A few examples will suffice. An overhead crane often is built into a building's structure. Overhead cranes operate on rails attached to building columns. This allows the crane bridge to move through a building elevated well above the floor. A trolley attached to the bridge of the crane provides movement perpendicular to the movement of the bridge. The rails should be equipped with limit switches and bumpers to prevent the crane from running off the end of the rails or into the building structure.

Cranes may have built-in safety devices or warning devices. One type of device, load indicating devices, measure the load and provide the operator with a reading of load or an overload warning signal. There are limit switches to prevent "double blocking," which refers to the load block or hook reaching the top of its travel. Often the cable to which the block is attached runs into or tries to pass over another sheave. When double blocking occurs, the hoisting mechanism is likely to become overloaded and fail.

Boom cranes have special safety devices for their unique hazards. One device is a boom angle indicator or boom radius indicator. A reading from this device provides the operator with information to help determine if a load is within safe limits. There are also devices, called load-moment indicators, that combine boom angle and load. If the moment necessary to overturn a crane is approached or exceeded, an alarm sounds for the operator.

When crane booms operate near overhead power lines, there is a danger of contacting them and possibly electrocuting the operator or someone in contact with the crane. There are power line proximity warning devices that notify the operator if the boom becomes too close to an energized powerline. Regulations require that booms be kept certain distances from such energized power lines, and the distance is related to the voltage of the power line. There are devices for insulating cranes and booms from electrical power if contact is made with an overhead line. It is also important to ground cranes and other materials handling equipment that could come into contact with electrical power sources.

There are limit switches for hoists and cranes. On a crane, for example, the limit switch prevents the load block or hook from overrunning the sheave at the end of a boom. If the hook begins to pass over the sheave, the mouth of the hook may reach a position where the end loops of a sling fall out of the mouth. If this occurs, a load will fall from the hook.

15-7   ROPES, CHAINS, AND SLINGS

Equipment

Ropes, chains, and slings are the rigging used to lift loads. They form the interface between the hoisting equipment and the load. There are many kinds of rigging materials.

Ropes normally are twisted strands of material. They may be made from natural fibers like manila, from synthetic fibers like nylon and polypropylene, or from metal, usually steel. The nomenclature for wire rope usually includes an outer diameter. Wire rope also has a numbering system that indicates the numbers of wire bundles in the rope surrounding the core and the number of continuous wires in a bundle. For example, in a 6×19 wire rope, there are six bundles composed of 19 wires each for a total of 114 wires.

A common rigging component is a hook on the end of a hoist or crane rope. Like other elements of rigging, the hook has a load limit.

There are many kinds of slings. Slings normally have a fixed length and they may be made from various materials and have the form of rope, belts, mesh, or fabric. To prevent damage to an object being lifted, a sling may be a wide band of fabric or a belt to distribute the load. This prevents the sling from cutting into or scratching the object. Some slings have a loop on each end. One or both ends may be placed into the mouth of a hook. Slings are suspended from a hook in different ways. Figure 15-4 illustrates three common methods: vertical, choker, and basket hitch. Some slings have more than two legs; three- and four-legged slings are quite common. Some slings have a hook on the end of each leg instead of a loop. In some rigging arrangements, a spreader is placed between legs of a sling above the load to help the sling legs stay vertical rather than compress around the sides of the load. A four-legged sling and the use of a spreader are illustrated in Figure 15-5.

Hazards

The main hazards of rigging are failures resulting from overloading, deterioration or wear, exposures, and improper rigging or abuse.

Each type of rigging can carry some load. A rope of a certain diameter has a rated capacity, based on ultimate strength and a factor of safety. (A factor of safety for materials was discussed in Chapter 10.) Fittings on the rope, the way it is used, and its physical condition will only reduce its capacity. The amount of reduction can vary considerably. A rope, chain, or sling is no stronger than its weakest element. A rigging with full capacity is rated as 100% efficient. Fittings, splices, bends around sheaves or objects, and damaged segments or links all can reduce the load capacity to some efficiency level less than 100%. Tables and charts in several references give efficiencies for fittings and splices.

One factor that affects the load that can be lifted with a sling is related to the type of hitch applied in the rigging. The type of hitch determines the number of active legs. For a choker hitch, one live element or leg of the sling ends up carrying the entire load (see Figure 15-4). In a vertical hitch, the entire load is carried by one live leg. In a basket hitch, there are two live legs, each carrying half the load.

Another factor that determines the load that a sling can carry is the angle formed by the legs. Refer to Figure 15-6 for a two-legged sling. The greater the angle from vertical that a leg has, the greater the load on the leg.

HITCHES:

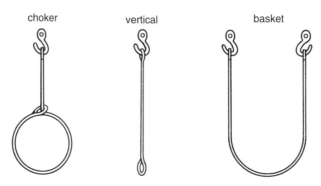

Figure 15-4.   Kinds of slings.

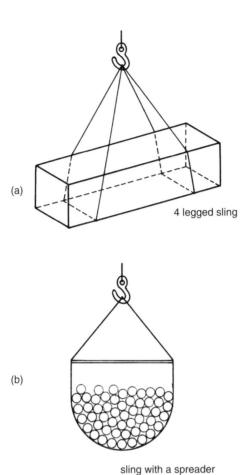

(a)

4 legged sling

(b)

sling with a spreader

Figure 15-5.   (a) A four-legged sling and (b) a sling with a spreader.

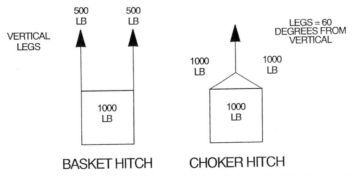

Figure 15-6.   Comparison of basket hitch and choker hitch load capacities.

A load may slip from rigging if the arrangement for the rigging is not properly attached. For example, it is not uncommon for a load to slip from a two-legged sling.

Hooks are usually made from ductile steel because if they were brittle, they would not be as durable under repeated loading. One hazard occurs when rigging becomes disengaged from the mouth of a hook, causing the load to fall. Another hazard is the widening of the mouth opening of a hook resulting from repeated loading. This can increase the chances of rigging slipping from the hook.

Improper storage and exposure to weather, sunlight, high temperatures, chemicals, moisture, or other elements can reduce the capacity of rigging. Sunlight affects the strength of certain synthetic materials. Moisture can induce rotting in natural materials and corrosion in metals. Physical damage occurs when rigging is run over, loads are dropped on rigging components, or when rigging is dragged around. When ropes bend around small radius corners or sheaves, the outer fibers are overstressed and may fail. As a result, some of the load capacity is lost. Some wires or fibers are made smaller simply by wear, thereby reducing the capacity.

Another hazard of rigging is injury to users or others nearby. If the rigging fails, the load can fall and possibly injure someone nearby or below it. Loaded rigging stretches and stores some energy, much like a rubber band (some rigging materials stretch more than others), and if there is a failure, the line can snap back in each direction from the break. The line, a clevis used for a connection or other component, may fly rapidly and strike someone. Even a rope used in a tug-of-war has been known to rip participants' hands apart when it breaks. The greater the load elongation and energy stored, the greater the danger of injury after a failure.

Controls

Important controls to minimize rigging hazards are proper selection of rigging materials, proper use, not overloading the rigging, regular inspection and testing of rigging components, training of riggers, and effective presentation of reference materials.

The *Handbook of Rigging* is one of the most complete references on proper selection, use, inspection, and testing of rigging. A number of standards define the frequency and method for testing of certain rigging components.

The *Handbook* also illustrates what to look for during inspections. For example, there are approximately 25 kinds of failures in rope that may be discerned. Some failures, like broken outer wires or fibers, may be easy to detect. Others are more difficult to find visually. Hooks must be inspected for cracks and the amount of growth in the jaw opening; when the angle of the jaw increases by 15°, the hook is removed from service. One reason is that as the jaw of a hook grows, the chances of rigging slipping from it increase. When chain is overloaded, the links will grow, causing the inside radius of one loop to wrap around the next loop. An inspector often can tell quite easily if a chain has been overloaded by observing how easily the links articulate. Any stiffness between links suggests that wrapping has taken place and capacity is reduced.

Hooks should be fitted with a retainer device that prevents rigging from slipping out of the jaw. There are a number of patented designs to accomplish this. The retainer allows rigging to slip into the jaw easily, but prevents it from slipping off the jaw. The retainer devices are not designed to carry a load, merely to keep rigging in the jaw.

There are load capacity charts for materials handling equipment and rigging. When these charts are for field use, they should avoid the need for computations before reading them. The need to make a computation creates a potential source of error. The charts or tables should include accepted safety factors and should indicate clearly what safety factors and other considerations are included in each table. Engineers may find tables with basic properties more valuable because they want to use them in many applications that require special computations. Tables 15-6 through 15-11 are examples of field tables.

Some examples of rigging consideration illustrate some important rigging principles.

Example 15-4 The angle formed by the legs of a sling affect the actual load carried by them. Figure 15-6 illustrates this. A basket hitch sling supports a 1,000-lb load; each vertical leg carries 500 lb. A choker hitch sling carries the same load, but the 1,000-lb load is carried by the single vertical element. The load in each of the legs that forms an angle of 60° from vertical between the choker and the object itself is quite different. The load in each leg, F_1, is solved by summing forces in the vertical direction: $F_y = 0 = 1,000 \; 2F_1 \cos 60$; $F_y = 1,000$ lb.

Example 15-5 A hoist with a 10,000-lb capacity will be used to lift a load of 7,500 lb. A wire rope sling will be used to lift the load. If a vertical hitch is used, what size rope is needed to lift the load safely, if the rope is

(a) 6×19 FC (fiber core) wire rope with a mechanical splice?

(b) nylon rope?

(c) polypropylene rope?

What factor of safety is included in each of these ropes?

From Tables 15-7, 15-9, and 15-11, respectively, the sizes are

(a) $^3/_4$ in diameter

(b) $1^3/_4$ in diameter

(c) 2 in diameter

TABLE 15-6 Rated Capacity (Working Load Limit; Pounds) for Alloy Steel Chain Slings[a]

Chain Size (in)	Single Branch Sling 90° Loading	Double Sling (Angle from Vertical)			Triple and Quadruple Sling (Angle from Vertical)		
		30°	45°	60°	30°	45°	60°
$^1/_4$	3,250	5,560	4,550	3,250	8,400	6,800	4,900
$^3/_8$	6,600	11,400	9,300	6,600	17,000	14,000	9,900
$^1/_2$	11,250	19,500	15,900	11,250	29,000	24,000	17,000
$^5/_8$	16,500	28,500	23,300	16,500	43,000	35,000	24,500
$^3/_4$	23,000	39,800	32,500	23,000	59,500	48,500	34,500
$^7/_8$	28,750	49,800	40,600	28,750	74,500	61,000	43,000
1	38,750	67,100	54,800	38,750	101,000	82,000	58,000
$1^1/_8$	44,500	77,000	63,000	44,500	115,500	94,500	66,500
$1^1/_4$	57,500	99,500	81,000	57,500	149,000	121,500	96,000
$1^3/_8$	67,000	116,000	94,000	67,000	174,000	141,000	100,500
$1^1/_2$	80,000	138,000	112,500	80,000	207,000	169,000	119,500
$1^3/_4$	100,000	172,000	140,000	100,000	258,000	210,000	150,000

[a] From 29 CFR 1926.251, Table H-1. Other grades of proof tested steel chain include proof coil, BBB coil, and hi-test chain. These grades are not recommended for overhead lifting and therefore are not covered by this table. Worn links may require removal of the assembly from service. See 29 CFR 1926.251, Table H-2.

TABLE 15-7   Rated Capacities for Single Leg Slings (6 × 19 and 6 × 37) Classification Improved Plow Steel Grade Wire Rope with Fiber Core[a]

| Rope Diameter (in) | Construction | Rated Capacities (tons, 2,000 lb) | | | | | | | | |
| | | Vertical[b] | | | Choker[b] | | | Vertical Basket[b,c] | | |
		HT	MS	S	HT	MS	S	HT	MS	S
$1/4$	6 × 19	0.49	0.51	0.55	0.37	0.38	0.41	0.99	1.0	1.1
$5/16$	6 × 19	0.76	0.79	0.85	0.57	0.59	0.64	1.5	1.6	1.7
$3/8$	6 × 19	1.1	1.1	1.2	0.80	0.85	0.91	2.1	2.2	2.4
$7/16$	6 × 19	1.4	1.5	1.6	1.1	1.1	1.2	2.9	3.0	3.3
$1/2$	6 × 19	1.8	2.0	2.1	1.4	1.5	1.6	3.7	3.9	4.3
$9/16$	6 × 19	2.3	2.5	2.7	1.7	1.9	2.0	4.6	5.0	5.4
$5/8$	6 × 19	2.8	3.1	3.3	2.1	2.3	2.5	5.6	6.2	6.7
$3/4$	6 × 19	3.9	4.4	4.8	2.9	3.3	3.6	7.8	8.8	9.5
$7/8$	6 × 19	5.1	5.9	6.4	3.9	4.5	4.8	10.0	12.0	13.0
1	6 × 19	6.7	7.7	8.4	5.0	5.8	6.3	13.0	15.0	17.0
$1 1/8$	6 × 19	8.4	9.5	10.0	6.3	7.1	7.9	17.0	19.0	21.0
$1 1/4$	6 × 37	9.8	11.0	12.0	7.4	8.3	9.2	20.0	22.0	25.0
$1 3/8$	6 × 37	12.0	13.0	15.0	8.9	10.0	11.0	24.0	27.0	30.0
$1 1/2$	6 × 37	14.0	16.0	17.0	10.0	12.0	13.0	28.0	32.0	35.0
$1 5/8$	6 × 37	16.0	18.0	21.0	12.0	14.0	15.0	33.0	37.0	41.0
$1 3/4$	6 × 37	19.0	20.0	24.0	14.0	16.0	18.0	38.0	43.0	48.0
2	6 × 37	25.0	28.0	31.0	18.0	21.0	23.0	49.0	55.0	62.0

[a] From 29 CFR 1926.251, Table H-3. See 29 CFR 1926.251 for additional regulations for wire rope.
[b] HT = hand tucked splice and hidden tuck splice. For hidden tuck splice, IWRC (Independent Wire Rope Core) use values in HT columns. MS = mechanical splice. S = swaged or zinc socket.
[c] Where values only apply when the D/d ratio for HT slings is 10 or more, and for MS and S slings is 20 or more, where D is the diameter of curvature around which the body of the sling is bent and d is the diameter of rope.

Because the tables are designed for field use, a factor of safety is already incorporated in each table value. Table 15-8 shows a factor of safety of 5, and the factors of safety are 9 for the nylon rope and 6 for the polypropylene rope, shown at the top of the tables.

Example 15-6   A sling will be used to lift a load. The sling is made from $1/2$-in 6 × 19 FC wire rope. If a vertical basket hitch is used, how large of a load can be lifted if the sling has a mechanical splice? What can be lifted if the splice is a hand-tucked splice?
     The capacity of a sling is determined from its weakest component, and the efficiency of a sling is affected by the kind of splice used to form the loops on the ends. From Table 15-7, the maximum load for the mechanical splice is 3.9 tons. For the hand-tucked splice, the maximum load is 3.7 tons.

15-8   CONVEYORS

Equipment

There are many kinds of conveyors used to move materials. There are roller, belt, screw, bucket, chain, and overhead conveyors; gravity chutes; wheeled carts and hoppers; and others. Some are powered, whereas others depend on people or gravity to move items along them.

TABLE 15-8 Manila Rope Slings[a]

		Rated Capacity (lb; Safety Factor = 5)													
		Eye and Eye Sling						Endless Sling						
				Basket Hitch (Angle from Vertical)						Basket Hitch (Angle from Vertical)			
Rope Diameter (in)	Maximum Breaking Strength (lb)	Vertical Hitch	Choker Hitch	0°	30°	45°	60°	Vertical Hitch	Choker Hitch	0°	30°	45°	60°
1/2	2,650	550	250	1,100	900	750	550	950	500	1,900	1,700	1,400	950
9/16	3,450	700	350	1,400	1,200	1,000	700	1,200	600	2,500	2,200	1,800	1,200
5/8	4,400	900	450	1,800	1,500	1,200	900	1,600	800	3,200	2,700	2,200	1,600
3/4	5,400	1,100	550	2,200	1,900	1,500	1,100	2,000	950	3,900	3,400	2,800	2,000
13/16	6,500	1,300	650	2,600	2,300	1,800	1,300	2,300	1,200	4,700	4,100	3,300	2,300
7/8	7,700	1,500	750	3,100	2,700	2,200	1,500	2,800	1,400	5,600	4,800	3,900	2,800
1	9,000	1,800	900	3,600	3,100	2,600	1,800	3,200	1,600	6,500	5,600	4,600	3,200
1 1/16	10,500	2,100	1,100	4,200	3,600	3,000	2,100	3,800	1,900	7,600	6,600	5,400	3,800
1 1/8	12,000	2,400	1,200	4,800	4,200	3,400	2,400	4,300	2,200	8,600	7,500	6,100	4,300
1 1/4	13,500	2,700	1,400	5,400	4,700	3,800	2,700	4,900	2,400	9,700	8,400	6,900	4,900
1 5/16	15,000	3,000	1,500	6,000	5,200	4,300	3,000	5,400	2,700	11,000	9,400	7,700	5,400
1 1/2	18,500	3,700	1,850	7,400	6,400	5,200	3,700	6,700	3,300	13,500	11,500	9,400	6,700
1 5/8	22,500	4,500	2,300	9,000	7,800	6,400	4,500	8,100	4,100	16,000	14,000	11,500	8,000
1 3/4	26,500	5,300	2,700	10,500	9,200	7,500	5,300	9,500	4,800	19,000	16,500	13,500	9,500
2	31,000	6,200	3,100	12,500	10,500	8,800	6,200	11,000	5,600	22,500	19,000	16,000	11,000
2 1/8	36,000	7,200	3,600	14,500	12,500	10,000	7,200	13,000	6,500	26,000	22,500	18,500	13,000
2 1/4	41,000	8,200	4,100	16,500	14,000	11,500	8,200	15,000	7,400	29,500	25,500	21,000	15,000
2 1/2	46,500	9,300	4,700	18,500	16,000	13,000	9,300	16,500	8,400	33,500	29,000	23,000	16,500
2 5/8	52,000	10,500	5,200	21,000	18,000	14,000	10,500	18,500	9,000	37,500	32,500	26,500	18,500

[a]From 29 CFR 1926.251, Table H-15.

TABLE 15-9 Nylon Rope Slings[a]

Rated Capacity (lb; Safety Factor = 9)

Rope Diameter (in)	Maximum Breaking Strength (lb)	Eye and Eye Sling						Endless Sling					
		Vertical Hitch	Choker Hitch	Basket Hitch (Angle from Vertical)				Vertical Hitch	Choker Hitch	Basket Hitch (Angle from Vertical)			
				0°	30°	45°	60°			0°	30°	45°	60°
1/2	6,080	700	350	1,400	1,200	950	700	1,200	600	2,400	2,100	1,700	1,200
9/16	7,600	850	400	1,700	1,500	1,200	850	1,500	750	3,000	2,600	2,200	1,500
5/8	9,880	1,100	550	2,200	1,900	1,600	1,100	2,000	1,000	4,000	3,400	2,800	2,000
3/4	13,490	1,500	750	3,000	2,600	2,100	1,500	2,700	1,400	5,400	4,700	3,800	2,700
13/16	16,150	1,800	900	3,600	3,100	2,600	1,800	3,200	1,600	6,400	5,600	4,600	3,200
7/8	19,000	2,100	1,100	4,200	3,700	3,000	2,100	3,800	1,900	7,600	6,600	5,400	3,800
1	23,750	2,600	1,300	5,300	4,600	3,700	2,600	4,800	2,400	9,500	8,200	6,700	4,800
1 1/16	27,360	3,000	1,500	6,100	5,300	4,300	3,600	5,500	2,700	11,000	9,500	7,700	5,500
1 1/8	31,350	3,500	1,700	7,000	6,000	5,000	3,500	6,300	3,100	12,500	11,000	8,000	6,300
1 1/4	35,625	4,000	2,000	7,900	6,900	5,600	4,000	7,100	3,600	14,500	12,500	10,000	7,100
1 5/16	40,850	4,500	2,300	9,100	7,900	6,400	4,500	8,200	4,100	16,500	14,000	12,000	8,200
1 1/2	50,350	5,600	2,800	11,000	9,700	7,900	5,600	10,000	5,000	20,000	17,500	14,000	10,000
1 5/8	61,750	6,900	3,400	13,500	12,000	9,700	6,900	12,500	6,200	24,500	21,500	17,500	12,500
1 3/4	74,100	8,200	4,100	16,500	14,600	11,500	8,200	15,000	7,400	29,500	27,500	21,000	15,000
2	87,400	9,700	4,900	19,500	17,000	13,500	9,700	17,000	8,700	35,000	30,000	24,500	17,500
2 1/8	100,700	11,000	5,600	22,500	19,500	16,000	11,000	20,000	10,000	40,500	36,000	28,500	20,000
2 1/4	118,750	13,000	6,600	26,500	23,000	18,500	13,000	24,000	12,000	47,500	41,000	33,500	24,000
2 1/2	133,000	15,000	7,400	29,500	25,500	21,000	15,000	26,500	13,500	53,000	46,000	37,500	26,500
2 5/8	153,900	17,100	8,600	34,000	29,500	24,000	17,000	31,000	15,500	61,500	53,500	43,500	31,000

[a] From 29 CFR 1926.251, Table H-16.

TABLE 15-10 Polyester Rope Slings[a]

		Rated Capacity (lb; Safety Factor = 9)											
		Eye and Eye Sling								Endless Sling			
				Basket Hitch (Angle from Vertical)							Basket Hitch (Angle from Vertical)		
Rope Diameter (in)	Maximum Breaking Strength (lb)	Vertical Hitch	Choker Hitch	0°	30°	45°	60°	Vertical Hitch	Choker Hitch	0°	30°	45°	60°
$1/2$	6,080	700	350	1,400	1,200	950	700	1,200	600	2,400	2,100	1,700	1,200
$9/16$	7,600	850	400	1,700	1,500	1,200	850	1,500	750	3,000	2,600	2,200	1,500
$5/8$	9,500	1,100	550	2,100	1,800	1,500	1,100	1,900	950	3,800	3,300	2,700	1,900
$3/4$	11,875	1,300	650	2,600	2,300	1,900	1,300	2,400	1,200	4,800	4,100	3,400	2,400
$13/16$	14,725	1,600	800	3,300	2,800	2,300	1,600	2,900	1,500	5,900	5,100	4,200	2,900
$7/8$	17,100	1,900	950	3,900	3,300	2,700	1,900	3,400	1,700	6,800	5,900	4,800	3,400
1	20,900	2,300	1,200	4,600	4,000	3,300	2,300	4,200	2,100	8,400	7,200	5,900	4,200
$1^{1}/_{16}$	24,225	2,700	1,300	5,400	4,700	3,800	2,700	4,800	2,400	9,700	8,400	6,900	4,800
$1^{1}/_{8}$	28,025	3,100	1,600	6,200	5,400	4,400	3,100	5,600	2,800	11,000	9,700	7,900	5,600
$1^{1}/_{4}$	31,540	3,500	1,800	7,000	6,100	5,000	3,500	6,300	3,200	12,500	11,000	8,900	6,300
$1^{5}/_{16}$	35,600	4,000	2,000	7,900	6,900	5,600	4,000	7,100	3,600	14,500	12,500	10,000	7,100
$1^{1}/_{2}$	44,460	4,900	2,500	9,900	8,600	7,000	4,900	8,900	4,400	18,000	15,500	12,500	8,900
$1^{5}/_{8}$	54,150	6,000	3,000	12,000	10,400	8,500	6,000	11,000	5,400	21,500	19,000	15,500	11,000
$1^{3}/_{4}$	64,410	7,200	3,600	14,500	12,500	10,000	7,200	13,000	6,400	26,000	22,500	18,000	13,000
2	76,000	8,400	4,200	17,000	14,500	12,000	8,400	15,000	7,600	30,500	26,500	21,500	15,000
$2^{1}/_{8}$	87,400	9,700	4,900	19,500	17,000	13,500	9,700	17,500	8,700	35,000	30,500	24,500	17,500
$2^{1}/_{4}$	101,650	11,500	5,700	22,500	19,500	16,000	11,500	20,500	10,000	40,500	35,000	29,000	20,000
$2^{1}/_{2}$	115,900	13,000	6,400	26,000	22,500	18,000	13,000	23,000	11,500	46,500	40,000	33,000	23,000
$2^{5}/_{8}$	130,150	14,500	7,200	29,000	25,000	20,500	14,500	26,000	13,000	52,000	45,000	37,000	26,000

[a] From 29 CFR 1926.251, Table H-17.

TABLE 15-11 Polypropylene Rope Slings[a]

Rated Capacity (lb; Safety Factor = 6)

Rope Diameter (in)	Maximum Breaking Strength (lb)	Eye and Eye Sling						Endless Sling					
		Vertical Hitch	Choker Hitch	Basket Hitch (Angle from Vertical)				Vertical Hitch	Choker Hitch	Basket Hitch (Angle from Vertical)			
				0°	30°	45°	60°			0°	30°	45°	60°
1/2	3,990	650	350	1,300	1,200	950	650	650	600	2,400	2,100	1,700	1,200
9/16	4,845	800	400	1,600	1,400	1,100	800	800	750	2,900	2,500	2,100	1,500
5/8	5,890	1,000	500	2,000	1,700	1,400	1,000	1,000	900	3,500	3,100	2,500	1,800
3/4	8,075	1,300	700	2,700	2,300	1,900	1,300	1,300	1,200	4,900	4,200	3,400	2,400
13/16	9,405	1,000	800	3,100	2,700	2,200	1,600	1,600	1,400	5,600	4,900	4,000	2,800
7/8	10,925	1,800	900	3,600	3,200	2,600	1,800	1,800	1,600	6,600	5,700	4,600	3,300
1	13,300	2,200	1,100	4,400	3,800	3,100	2,200	2,200	2,000	8,000	6,900	5,600	4,000
1 1/16	15,200	2,500	1,300	5,100	4,400	3,600	2,500	2,500	2,300	9,100	7,900	6,500	4,600
1 1/8	17,385	2,900	1,500	5,800	5,000	4,100	2,900	2,900	2,600	10,500	9,000	7,400	5,200
1 1/4	19,950	3,300	1,700	6,700	5,800	4,700	3,300	3,300	3,000	12,000	10,500	8,500	6,000
1 5/16	22,325	3,700	1,900	7,400	6,400	5,300	3,700	3,700	3,400	13,500	11,500	9,500	6,700
1 1/2	28,215	4,700	2,400	9,400	8,100	6,700	4,700	4,700	4,200	17,000	14,500	12,000	8,500
1 5/8	34,200	5,700	2,900	11,500	9,900	8,100	5,700	5,700	5,100	20,500	18,000	14,500	10,500
1 3/4	40,850	6,800	3,400	13,500	12,000	9,600	6,800	6,800	6,100	24,500	21,000	17,500	12,500
2	49,400	8,200	4,100	16,500	14,500	11,500	8,200	8,200	7,400	29,500	25,000	21,000	15,000
2 1/8	57,950	9,700	4,800	19,500	16,500	13,500	9,700	9,000	8,700	35,500	30,100	24,500	17,500
2 1/4	65,550	11,000	5,500	22,000	19,000	15,500	11,000	11,000	9,900	39,000	34,000	28,000	19,500
2 1/2	76,000	12,500	6,300	25,500	22,000	18,000	12,500	12,500	11,500	45,500	39,500	32,500	23,000
2 5/8	85,500	14,500	7,100	28,500	14,500	20,000	14,500	14,500	13,000	51,500	44,500	36,500	25,500

[a] From 29 CFR 1926.251, Table H-18.

Hazards

Some conveyors present hazards of material falling from them onto people or items below. Conveyors and the structures that support them may fail. For powered conveyors, there is the danger of people becoming caught in moving parts, such as belts and chains. Some conveyors present pinch points and places where workers hands and arms may become caught between materials and fixed parts of the conveyor or other fixed objects. Some conveyors, like belt conveyors, can extend for miles on elevated structures. To lubricate and maintain them, workers must have a walkway with a guardrail and access points to the conveyor components. The guardrail reduces the danger of falling.

Controls

Conveyors that pass overhead should have overhead protection or some enclosure to prevent materials from falling on people below. From time to time, someone must remove fallen materials and parts from the overhead structure. If workers are to remove the material, the overhead netting, screen, or other structure must be substantial enough to carry the load of materials and people. In the classic case that tested the imminent danger rule of OSHA, the Supreme Court ruled in 1980 that, after one worker who was cleaning fallen appliance parts from a wire mesh overhead protection system fell to his death when the screen failed, a second worker could not be suspended from work for refusing to perform the same task without reasonable safeguards.

The moving parts of conveyors, particularly powered conveyors, that pose machine dangers and that are within reach of people must be guarded. Pinch points created by belts and rollers, screw conveyors, chains and sprockets, and other components must meet the machine guarding principles discussed in Chapter 13. Included are emergency shutoff controls and warnings along the conveyor. In some cases, the edges of belts need to be guarded so fingers, hands, and other objects cannot become caught under them. Protection should include emergency shutoff switches at convenient locations, interlock switches, and low speed or jog controls for servicing and maintenance when appropriate.

Where objects suddenly come into view along a conveyor, their arrival may not be anticipated easily by workers nearby. If the objects pose pinch point dangers, audio and visual warning devices can alert people to the pending danger. This problem sometimes occurs on enclosed chute and roller conveyors.

Where access to conveyors requires elevated walking and working surfaces, there must be guardrails. Where conveyors have floor and wall openings through which people and material could fall, the openings must have guardrails with toeboards.

If workers must unjam materials from chutes or other conveyors, they should have tools, such as long poles, for chute conveyors. If a person must enter a chute or similar conveyor opening, power must be locked and tagged out. A lifeline may be necessary for the worker, and an observer/assistant should be available for securing emergency help or rescue.

15-9 ELEVATORS, ESCALATORS, AND MANLIFTS

Equipment

A number of devices move people and goods vertically. Elevators are the most common. An elevator has a car, moved by cables or a hydraulic system. There are also elevators for materials only. One type of elevator that moves goods only is a dumbwaiter. People enter an elevator, or goods are placed on them, through an opening or door. Escalators are moving belts with components that form steps.

Manlifts are belts that move vertically. Along the length of its belt are small platforms that one person can stand on. Handholds are also attached at appropriate heights above each platform. The belts move continuously at a fairly slow speed between two or more vertical locations. One can quickly step on a platform, grab the handhold, and move to the other level and step off.

Hazards

Major hazards for elevators include becoming caught in the opening to a car when the car moves, becoming caught in the doors, a runaway car, falling into the elevator shaft, and being trapped in a stalled car. Others include tripping or falling on entry or egress if the car does not line up with the floor opening and being crushed at the top or bottom of the shaft during service work.

For escalators, there is the danger of becoming caught in the steps as they fold into a fixed end or between the edges of the steps and the side walls. An example is a draw string on the collar or hood of a child's clothes. A free end can get caught in the folds of an escalator step, tighten around the child's neck, and strangle the child.

Hazards for manlifts include becoming caught as one moves through a floor opening or falling from the platform.

Controls

Standards for elevators have existed for a long time. Today, elevators are governed by many laws and codes at national, state, and local levels. Standards apply to design, construction, maintenance, and inspection.

Elevators, escalators, and manlifts must be maintained regularly and must be inspected by trained people. Many states and municipalities require elevator inspectors to be licensed. At regular intervals, inspectors check cables, sheaves, brakes, and other mechanical components. Inspectors sign forms that attest to the fact that all components meet standards.

Design standards incorporate safety features like brake systems, automatic braking if a cable fails, double doors for the car and openings, interlocks on all doors to stop the car if any doors are open, and interlocks on landing doors to keep them closed until the car arrives. Access doors have detector switches to prevent their closing on someone. Each size of car has a load capacity that affects the design and selection of various components. Elevators must contain emergency alarms and means of egress.

Escalators have emergency shutoff switches at each end. They have microswitches located at various points along and behind the side wall panels to stop the escalator if something becomes caught between the steps and the sidewalls.

Because elevators, escalators, and manlifts create vertical openings in buildings, they provide paths for heat and smoke to pass from floor to floor. There are many fire regulations governing elevators, escalators, and manlifts and the openings for them.

The importance of safeguards for elevators is illustrated by an accident involving a dumbwaiter in a restaurant. The dumbwaiter was used to move food and dishes from the kitchen to the dining room above. Employees locked out the interlock switches for the access doors. The doors were left open. One employee stuck his head in the opening to see if the dumbwaiter was coming and was fatally injured when his head was crushed between the door opening and the moving car. There have been many similar accidents involving elevators.

15-10 BULK MATERIALS, EXCAVATION, AND TRENCHING

Properties of Bulk Materials

Bulk materials, like grains, granular materials, sand, and soil, will form a slope, called an angle of repose, along their sides when they are piled up. Under certain conditions, bulk materials may hold together for a time, but when free to form their own shape, the angle of repose will form for loose, dry, bulk materials. Moisture, vibration, particle size, and other factors affect when the angle of repose will form and what the angle will be. Chapter 10 discussed angle of repose.

   Bulk materials have some adhesion properties that allow them to support some load. However, the adhesion properties may not be uniform throughout a given pile of materials. Moisture, temperature, pressure, frost, vibration, and other factors can affect the adhesion and load-bearing capabilities of a material.

Hazards

When bulk materials are stored, the components may adhere to each other. However, when material is removed from a pile, the disturbed edge of a pile may not take on the natural angle of repose immediately, and the wall of material could collapse suddenly and bury anyone working below the top level of the pile and near the unrestrained edge.

   A person may walk on a pile of some material and be supported by it and then may encounter a location that does not have the same load-bearing capacity and sink into the material. Someone may also fall into a pile of bulk material and sink into it because it cannot support the load presented by a person. Attempts to work free may simply sink the person deeper into the material, because the material acts more like a fluid for larger or moving loads than it does when only the particles of the material are involved. Not only is there a danger of suffocation if breathing air is sealed off from the face, but the bulk material can compress around the chest and severely limit chest movement necessary for inhalation, causing suffocation.

Controls

To prevent bulk materials from caving in on someone, the materials either must be restrained by shoring, must be sloped back to an angle more shallow than the natural angle of repose, or must be restrained by other suitable means. According to some standards, this protection is required when depths are more than 4 ft; others have these requirements for depths of 5 ft or more. Shoring or other restraints must meet accepted engineering standards.

   Workers who must work above bulk materials that cannot support the weight of a person should be protected by guardrails and lifelines. Two people should work together so that one can perform emergency rescue tasks if the other falls into the material below. The second person should have a backup if there is a need to approach the entrapped person. Emergency rescue equipment should be nearby.

   For work involving bulk materials in enclosed or confined spaces, there may be additional dangers related to breathable atmospheres and requiring necessary supplies of breathable air.

Excavation and Trenching

When there are excavation and trenching operations, there are many other considerations. Adequate drainage will prevent accumulation of water that may reduce the strength of shoring or restraints. Barriers must be installed to prevent people from falling into an excavation. There should be daily inspections of excavations and restraining systems, and emergency means for egress must be in place. Where the quality of the breathing air could be inadequate, tests will ensure that workers will not be overcome by lack of oxygen or by hazardous gases, fumes, or dusts. If people must cross over an excavation, walkways or bridges with standard guardrails are needed. Excavation and trenching were discussed in Chapter 10 in greater detail.

15-11   STORAGE OF MATERIALS

Hazards

Some of the main hazards of stored materials are items falling on someone below them, tripping or running into items protruding into traffic ways or aisles, piled materials tipping over, and physical damage to storage racks or piled materials by materials handling equipment. Not only is there a potential for injury and damage from poorly stored materials, but without good storage organization and practices, the cost of locating and keeping track of materials on hand is expensive.

Improper segregation of materials may create additional hazards from fire, explosion, and corrosion. Refer to Chapter 16.

Controls

The adage "a place for everything and everything in its place" is important for controlling hazards of materials storage. Well-planned, organized, and maintained materials storage areas are essential. There is a wealth of specialized equipment for solving materials storage problems.

Proper stacking is another control. Many kinds of materials are packaged. The packaging or the material inside can hold only so much weight before it crushes. When items are stored on damaged or crushed containers, there is a greater chance of tipping over. The higher materials are stacked, the higher the center of gravity and the less force required to tip the stack over.

Another control for some stored items is cross ties. Items such as lumber and bagged materials are made more stable by placing elements in different directions at different levels of a pile so that materials are interlocked.

Another control is stepping back materials. When materials are stacked several rows deep and several rows high, the front tier has a tendency to tip over. Cross ties may help, and stepping back the rows may help more. Using this method, the front row is not stacked as high as the row behind it, and the second row is not stacked as high as the third row.

Retaining walls can help keep bulk materials from spreading out or collapsing on someone or onto people or equipment. The retaining walls function like shoring and they must have adequate strength and bracing.

Racks keep many materials from falling. There are special racks for bar stock, piping, drums, rolls, and palletized materials. The racks must be sufficiently strong to support the loads placed on them. The Rack Manufacturers Institute[4] has structural standards and test procedures for racks.

It is not uncommon that industrial trucks run into racks. Protective barriers installed along the floor near rack columns can reduce the likelihood of physical damage to the racks. Stiffeners can be added to rack columns that are subject to physical damage by materials handling vehicles. Racks must be designed to remain standing, even when there is some physical damage to structural elements.

There must be enough aisle space between racks and items stored openly. The aisles should allow vehicles to place materials on racks or to remove them safely. Because aisle width is an important economic consideration in storage facilities, special vehicles can minimize aisle width requirements.

Aisles for walking and for materials handling vehicles should be clearly marked. Usually, they are separate aisles. Stored materials must not protrude into the marked aisle space. Some materials that are stored in elevated positions near aisles and may be unstable should be restrained by netting or other means. This is an important concern in some retail facilities where public customers are present.

Fire protection for stored materials will vary with the type and amount of materials, building design, and other factors. Chapter 16 discusses fire protection.

EXERCISES

1. A box is 8 ft high, 4 ft deep, and 6 ft wide and weighs 270 lb. A worker pushes on the box at a location midpoint along the width and 5 ft above the floor. The coefficient of friction between the box and floor is 0.53. Determine what value of the pushing force, P, will cause

   (a) tipping (rotation about the edge along the floor opposite the worker)

   (b) sliding

2. A worker picks up boxes from the floor and places them on a conveyor. This activity will occur every 30 s. The boxes, weighing 23 lb each, are located 20 in forward of the worker's ankle midpoint at the point of grasping them. The conveyor is 38 in high. There is no twisting during the lift. Determine the recommended weight limit (RWL) and the lifting index (LI) for this activity.

3. For the activity in Exercise 2, is the activity acceptable or are controls needed to prevent lifting injuries? If controls are needed, what kind would you recommend?

4. A 12,000-lb load is to be lifted with a sling. For the conditions in each of the following cases, determine what size chain or rope is needed.

   (a) Alloy steel chain, double sling, 45′ angle with vertical.

   (b) Single-legged 6 × 19 improved plow steel grade rope with fiber core, choker, mechanical splice.

   (c) Two-legged bridle sling, 6 × 19 improved plow steel grade rope with fiber core, 60° angle with vertical, hand-tucked splice.

   (d) Nylon rope, endless sling, basket hitch, 0° angle from horizontal.

   (e) Polypropylene rope, vertical hitch.

5. A sling is used to lift a 1,000-lb load. If the sling is rigged as a choker and the legs below the choke point form an angle of 60° with the vertical, what load is carried by

   (a) each leg?

   (b) the vertical portion?

6. A stationary Grove Model RT530E mobile crane (refer to Figure 15-3) will be used to lift material. The tires are fully inflated. The ground is level and firm. Determine what load can be safely lifted for each of the following conditions and identify the basis for the load limit (structural failure or tip over):

(a) Boom length: 40 ft

Radius of load: 20 ft (boom angle is 30°)

Boom position: centered over front

On fully extended outriggers

(b) Same conditions as (a) except the boom is over the side.

(c) Same conditions as (a) except the crane is on rubber (outriggers not used).

(d) Boom length: 50 ft

Radius of load: 12 ft (boom angle is 56°)

Boom position: centered over front

On fully extended outriggers

(e) Same conditions as (d), except the boom is over the side.

(f) Same conditions as (d), except the crane is on rubber (outriggers not used).

REVIEW QUESTIONS

1. Name one materials handling hazard associated with each of the following and give a control for that hazard:

(a) environments where materials handling tasks are performed

(b) maintenance of materials handling equipment

(c) loading of materials handling equipment

(d) securing loads

(e) movement of material and materials handling equipment

2. Explain the biomechanics of lifting and the load effect of

(a) stooping

(b) rapidly picking up a load

(c) distance of a load center of gravity from the body

3. How can the following contribute to injury associated with manual lifting?

(a) size of load

(b) raising a load overhead

(c) frequency of lifting

(d) asymmetrical loads

4. Explain how the recommended weight limit and lifting index are used to decide what kinds of controls are needed to minimize injury from manual lifting.

5. Describe two hazards and applicable controls related to use of

(a) jacks

(b) hand-operated materials handling vehicles

(c) powered materials handling vehicles

    (**d**) hoisting apparatus

    (**e**) conveyors

    (**f**) elevators

    (**g**) escalators

    (**h**) manlifts

6. Define ROPS and FOPS and describe what protection each provides for powered materials handling vehicles.

7. Identify at least one hazard and associated control for the following rigging components:

    (**a**) ropes

    (**b**) hooks.

8. Can steel wire ropes store dangerous amounts of energy due to elasticity during loading?

9. What is a primary source publication containing information on proper rigging of loads?

10. Describe two main hazards associated with bulk materials and applicable controls for these hazards.

11. Define angle of repose and its significance for safety of bulk materials and excavations.

12. Explain how a person can suffocate in bulk materials even though the person's head remains above the materials.

13. Name five controls to prevent accidents related to storage of materials.

NOTES

1 *Work Practices Guide for Manual Lifting,* Publication No. 81-122, National Institute for Occupational Safety and Health, Cincinnati, OH, March 1981.

2 *Scientific Support Documentation for the Revised 1991 Lifting Equation,* U.S. Department of Health and Human Services, Public Health Service, Centers for Disease Control and Prevention, National Institute for Occupational Safety and Health, Technical Contract Report, May 8, 1991, National Technical Information Service (NTIS), Springfield, VA, NTIS Document Number PB 91-226274.

3 Waters, Rhomas R., Putz-Anderson, Vern, and Gard, Arun, *Application Manual for the Revised NIOSH Lifting Equation,* U.S. Department of Health and Human Services, Public Health Service, Centers for Disease Control and Prevention, National Institute for Occupational Safety and Health, January, 1994.

4 ANSI MH 16.2, *Safety Practices for the Use of Industrial and Commercial Steel Storage Racks* (sponsored by Rack Manufacturers Institute).

BIBLIOGRAPHY

American National Standards Institute, New York:

    MH 16.2 *Safety Practices for the Use of Industrial and Commercial Steel Storage Racks* (sponsored by Rack Manufacturers Institute)

American Society of Mechanical Engineers, New York:

B56.5 Safety Standard for Guided Industrial Vehicles and Automated Functions of Manned Industrial Vehicles

B56.8 Safety Standard for Personnel and Burden Carriers

B56.11 Evaluation of Visibility from Powered Industrial Trucks

Material Handling Industry of American, Rack Manufacturers Institute, Charlotte, NC:

Specification for the Design, Testing and Utilization of Industrial Steel Storage Racks, 2002.

Specification for the Use of Industrial and Commercial Steel Storage Racks—A Manual of Safety Practices/A Code of Safety Practices, 1996.

Crane Handbook, Construction Safety Association of Ontario, Toronto, Canada, 1975.

SCHULTZ, GEORGE A., *Conveyor Safety: Safety in the Design and Operation of Material Handling Systems,* American Society of Safety Engineers, Des Plaines, IL, 2000.

MacCOLLUM, DAVID V., *Crane Hazards and Their Prevention*, American Society of Safety Engineers, Des Plaines, IL, 1999.

MacCOLLUM, D. V., "Lessons from 25 Years of ROPS," *Professional Safety*, January: 25–31 (1984).

MORGAN, CARL O., *Excavation Safety: A Guide to OSHA Compliance and Injury Prevention*, Government Institutes, Rockville, MD, 2003.

Rigging Manual, Construction Safety Association of Ontario, Toronto, Canada, 1975.

ROSSNAGEL, W. E., *Handbook of Rigging*, 4th ed., McGraw-Hill, New York, 1988.

SAE Handbook, Society of Automotive Engineers, Inc., Warrendale, PA, annual. The *Handbook* includes numerous standards for on-highway vehicles and off-highway vehicles. Many deal with safety aspects. Categories for standards are:

PASSENGER CARS, TRUCKS, BUSES, AND MOTORCYCLES

Vehicle Identification Numbers

Passenger Cars, Trucks, Buses, and Motorcycles

Transmissions

Seals

Tires

Wheels

Bumpers

Seat Belts and Restraint Systems

Passenger Car Components and Systems

Trailers

Truck and Bus

Motorcycles

OFF-HIGHWAY MACHINES AND VEHICLES

Snowmobiles

Agricultural Tractors

Construction and Industrial Equipment

MARINE EQUIPMENT

Marine Equipment

Maintenance and Repair

Specifications for the Design, Testing and Utilization of Industrial Steel Storage Racks,

Scientific Support Documentation for the Revised 1991 Lifting Equation, U.S. Department of Health and Human Services, Public Health Service, Centers for Disease Control and Prevention, National Institute for Occupational Safety and Health, Technical Contract Report, May 8, 1991, National Technical Information Service (NTIS) Springfield, VA, NTIS Document Number PB 91–226274.

SWARTZ, GEORGE, *Forklift Safety: A Practical Guide to Preventing Powered Industrial Truck Incidents and Injuries*, National Safety Council, Itasca, IL, 1999.

SWARTZ, GEORGE, *Warehouse Safety*, Government Institutes, Rockville, MD, 1999.

WATERS, RHOMAS R., Putz-Anderson, Vern, and Gard, Arun, *Application Manual for the Revised NIOSH Lifting Equation*, U.S. Department of Health and Human Services, Public Health Service, Centers for Disease Control and Prevention, National Institute for Occupational Safety and Health, January, 1994.

Work Practices Guide for Manual Lifting, Publication No. 81–122, National Institute for Occupational Safety and Health, Department of Health and Human Services, Cincinnati, OH, March, 1981.

CHAPTER 16

FIRE PROTECTION AND PREVENTION

16-1   INTRODUCTION

The Fire Problem

Each year, fire-related losses in the United States are considerable. There are approximately 1 million fires involving structures and approximately 8,000 deaths each year, and the total annual property loss exceeds $20 billion. These figures are conservative and do not include many indirect costs, such as litigation, investigation, and other costs borne by society.

Most deaths resulting from fires do not result from burns. Only approximately one fourth of fire-related deaths are the result of burns. Nearly two thirds of all fire-related deaths result from inhalation of carbon monoxide, smoke, toxic gases, and asphyxiation. Approximately one tenth of the deaths are from mechanical injuries, such as injuries from falls or falling material.

The overall death rate from fires is 2.8 per 100,000 population. The fire death rates for children younger than 5 years and adults older than 40 years exceeds the population average. Death rates increase exponentially with age for those older than 40 years. The death rate is 12.5 for those aged 75 to 85 years and 22.2 for those older than 85 years.

Studies have shown that deaths in fires are often alcohol related. In one study, evidence suggested that more than 80% of the victims had been consuming alcohol. Another study looked at age and blood alcohol level for fire victims. For those victims aged 30 to 60 years, approximately one quarter had no alcohol content in the blood and approximately two thirds had an alcohol content of more than 0.10%.

Most fatal fires involve one or two victims. However, large losses receive the greatest attention. The greatest losses have occurred in places of assembly. On December 30, 1903, 602 people died in the fire at Chicago's Iroquois Theater. On July 6, 1944, the Ringling Brothers and Barnum and Bailey Circus fire in Hartford, Connecticut, took 168 lives. There have been a number of famous fires in supper clubs, such as the fire at the Beverly Hills Supper Club in Southgate, Kentucky, on May 28, 1977, that claimed 165 lives.

Other fires and the count of victims have a place in history books. Within three weeks in 1980, two hotel fires gained national attention. On November 21, the MGM Grand fire in Las Vegas claimed 85 lives and on December 4 a fire in the Stouffer Inn near New York City killed 26 people. Both impacted building codes and sprinkler codes. Although it did not have the greatest toll for school fires, the fire at Our Lady of Angels Grade School in Chicago on December 1, 1958, had a lasting effect on codes and standards. Although it occurred nearly a century ago, the greatest human tragedy in an industrial fire occurred in New York City on March 25, 1911, at the Triangle Shirtwaist Factory. That fire took 145 lives and influenced building design standards.[1] Clearly, the attack on the World Trade Center on September 11, 2001, and the resulting fire and building collapse of the two towers holds the record for fatalities in a fire-related disaster.

Causes of Fires

In civilian fires with fatalities, the leading causes of ignition are cigarettes (35%), heating and cooking equipment (7%), matches, lighters, and candles (5%), and car crashes (4%). Leading causes of industrial fires are listed in Table 16-1.

Arson

Estimates of direct losses resulting from arson and incendiary fires in the United States range as high as $1 to $3 billion each year. The reasons for arson vary from financial problems that might be resolved through an insurance claim, to anger with an employer, school, or other person, to vandalism and other problems. Prevention of incendiary fires will not be discussed specifically in this chapter. There has been much research and study of arson cases, and most insurance companies apply specific actions that can reduce arson opportunities and potential losses. Application of fire protection designs and systems will help reduce losses for all fires.

Improving Fire Prevention and Protection

The Great Chicago Fire of 1871, many fires before it, and many since have stimulated research and action. As a result, we have the capability to prevent many fires today and to minimize the loss and damage. However, the knowledge, laws, and standards that could be used often are not applied or are ignored. Decision makers sometimes cut initial costs in hopes that fire will not strike in the future. Like other standards, fire codes are minimums, and in many cases, they can be exceeded.

TABLE 16-1 Leading Causes of Industrial Fires[a]

Percent of Fires	Causes
23	Electrical (wiring and motors, poor maintenance of equipment)
18	Smoking materials
10	Friction (bearings, jammed machines, etc.)
8	Overheated materials
7	Hot surfaces (heat from boilers, furnaces, ducts, lamps, etc.)
7	Burner flames (open flame equipment)
3–5	Combustion sparks
3–5	Spontaneous ignition
3–5	Cutting and welding
3–5	Fire from nearby facility
3–5	Arson
1–2	Sparks
1–2	Chemical action
1–2	Lightning
1–2	Molten substances

[a] Based on a Factory Mutual Engineering Corporation study of 25,000 fires over 10 years.

Outbreaks of fire were an everyday occurrence in Rome that led to the first known fire fighting organization, the Corps of Vigiles. Devastating fires in early European cities led to equipment improvements. The introduction of fire engines and pumpers in the sixteenth century began to replace bucket brigades. In 1678, Boston created the first paid fire department in the United States. Public water systems in cities emerged to provide an adequate water supply, including that needed for fire fighting. The early building codes of the seventeenth century that ensured there would be an adequate supply of ladders and buckets grew in complexity in the nineteenth and twentieth centuries. Later codes covered building design to control fire spread, water supply, and adequate means for exiting. Spurred by disasters like the Great Chicago Fire, zoning ordinances began to limit the amount of land that could be occupied by buildings to help prevent fire jumping from one property to the next. Zoning ordinances also set standards for adequate access routes for fire equipment.

Fire Codes and Standards

In the United States today, there are many fire codes. Most cities have their own, often incorporated into building codes, zoning, and other ordinances. Most states have similar codes that apply when local governments have not adopted their own standards. Most local governments adopt codes developed and maintained by standards organizations. There are several model building codes in the United States that incorporate many design requirements to minimize the chances of fire in a building, to minimize the rate at which fires spread, and to ensure exiting by occupants. Two examples are the Uniform Building Code and the code of the Building Officials Conference of America.

The National Fire Protection Association publishes and maintains a wide range of standards for many aspects of fire prevention, protection, engineering, and extinguishment, including standards for design of buildings and other facilities. This collection of standards is called the National Fire Code. Each element of this collection has an identifying number. For example, NFPA 101 is the *Life Safety Code*, which addresses safety of occupants and safe egress. There is a separate committee for each of the many codes, and each committee continually reviews and updates the code's provisions. The National Fire Code is the primary source of standards in the United States regarding matters associated with fire. Other standards include model building codes and local codes.

16-2 PHYSICS AND CHEMISTRY Of FIRE

Combustion

Fire is defined as the rapid oxidation of material during which heat and light are emitted. Combustion is an exothermic (gives off heat), self-sustaining reaction involving a solid fuel, a gaseous fuel, or both. (A fuel is any combustible material.) Most often, the reaction is oxidation of a fuel by oxygen in air. Combustion usually involves the emission of light: solid fuels usually appear to glow or smolder; gaseous fuels usually emit a visible flame.

Ignition

Ignition is the initiation of combustion. Each material has some minimum temperature, called the ignition temperature, that must be reached for ignition to occur. When some external source, such as a flame, spark, ember, or heat, causes ignition, the process is called piloted ignition. When there is no external source, the process is called autoignition, spontaneous ignition, or spontaneous combustion. The piloted ignition temperature of a fuel is lower than the autoignition temperature.

Spontaneous combustion or autoignition is often caused by heat buildup resulting from the oxidation of organic material. Bacteria and decomposition can contribute to the process in some materials. The material must reach the autoignition temperature and have enough oxygen for combustion. Among other materials, those made from agricultural products that have a high vegetable oil content, in general, are susceptible to autoignition. When left on open piles, oily rags from furniture finishing and other activities have led to many fires.

Fire Triangle and Pyramid

The components necessary for fire or combustion are (1) a fuel, (2) a source of oxygen, and (3) a source of heat or flame or some minimum temperature. As soon as combustion starts and there are sufficient amounts of the three components, combustion will continue. These three components form the fire triangle, which is illustrated in Figure 16-1(a). Technically, this model is correct for glowing combustion at the surface of solid fuels. Sources of oxygen are air, compressed air, liquid or solid oxygen, and other chemicals. There are many kinds of fuels. Paper, wood, and cloth are examples of ordinary combustibles, and there are also flammable liquid and gaseous fuels. Fuels can even be solids, like metals. Examples of heat sources are sparks, arcs, hot surfaces, and open flames.

During the process of combustion of vapors and gases in which there is a visible flame, a better model is the fire pyramid (see Figure 16-1(b)). In addition to the three components in the fire triangle, a fourth is production of hydroxyl (OH) radicals. An uninhibited reaction breaks down molecules into hydroxyl radicals, which last for an extremely short time, on the order of 1 ms.

Heat and Heat Sources

The temperature necessary to sustain combustion varies for different materials. Combustion of ordinary hydrocarbons occurs at temperatures of approximately 3,000°F, whereas some metals burn at temperatures of 5,000°F or higher. In the exothermic process of combustion, the heat given off is dependent on the fuel. Each kind of fuel has an energy or heat value per unit weight that is called heat of combustion. For example, carbon monox-

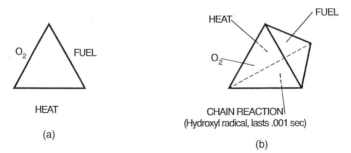

Figure 16-1. (a) The fire triangle and (b) the fire pyramid help in remembering the components involved in combustion.

ide contains 4,300 BTU/lb, paper has roughly 8,000 BTU/lb, gasoline has roughly 19,000 BTU/lb, and hydrogen has more than 51,000 BTU/lb.

The rate of heat produced during combustion is a function of the rate of combustion. Data on combustion often are available for common fuels being burned in air. When high-pressure air or gases with oxygen content more than the normal 21% (160 mmHg partial pressure) found in air are applied to combustion, the rate of heat production will increase significantly. In addition, the ignition temperature may be lowered. Limited data are available on enriched combustion.

If there is not enough oxygen available, some fuels will not burn completely. Not only will the rate of combustion slow, but partially burned fuel will become airborne. A very common product of combustion that could oxidize further is carbon monoxide. Much of the content of smoke is unburned or incompletely burned fuel. If a new source of oxygen reaches the fire, the rate of combustion can increase rapidly. For example, someone may open a window or door or a fire may break through a wall or ceiling and allow air to rush in. A flame front may actually travel very quickly along the new mixture of airborne fuel and air. In fact, the flame front may move faster than a person can run. In building fires, the sudden combustion of smoky air when a window or door is opened is called a back flash.

The rate of heat production and the location of combustion changes constantly during a fire. The total heat produced during a fire is the product of the weight of fuel consumed and the heat of combustion for the fuel. For example, if 20 lb of gasoline (a little more than 2 gal) were consumed in a fire, the total heat produced would be 20 lb × 19,000 BTU/lb = 380,000 BTU. Approximately 50 lb of paper would produce the same amount of heat, but the gasoline would burn much more quickly.

The heat from a fire is transferred to the surroundings by convection, radiation, and conduction. For most fires, heat is lost to surrounding air by convection. Because of large temperature differences between the combustion process and the air entrained into the process, tremendous upward air currents can occur. The upward movement of hot gases is driven by the thermal gradient.

Much heat is also lost through radiation. Radiation is highly dependent on geometry. One body must "see" the other for radiation heat transfer to occur. We do not feel the heat of the sun's rays when we stand in the shade because the tree prevents our body from "seeing" the sun. The surfaces surrounding a fire in a room "see" the fire and absorb much of the heat, and as the heat is absorbed by the surfaces, the room surface temperatures go up. The surfaces of a building near a fire can ignite if their temperature rises enough from the heat absorbed through radiative transfer. The rate of radiative heat transfer is a function of the fourth power of the absolute temperatures of the surfaces involved.

Methods for Controlling Combustion and Extinguishing Fires

The theory for fire found in the fire triangle and fire pyramid gives us insights into methods for extinguishing fire and stopping combustion. The fire triangle gives us three approaches: cool the fire, limit the oxygen supply, and remove the fuel. Cooling the fire reduces the temperature of the process or reduces the flame or heat source. Shutting off air supplies to a fire can stop the combustion by limiting the oxygen supply. Sometimes one can shut off the fuel by closing a valve or letting the fuel be consumed so there is none to burn. In oil fires, an explosion can sometimes separate the flame from the fuel to stop the fire. The fire pyramid yields a fourth approach: inhibiting the reaction producing hydroxyl radicals.

Products of Combustion and Their Hazards

Depending on the fuel, availability of oxygen and other factors, there can be a wide variety of products produced from combustion. Not only does combustion produce flames, heat, and smoke, but also many kinds of gases. Smoke is technically very fine solid particles suspended in air, and it may include droplets of steam. Two of the most common gases of combustion are carbon monoxide, which forms when there is not enough oxygen for complete combustion of the fuel, and carbon dioxide, which forms when combustion is complete. Other gases formed can include hydrogen sulfide, sulfur dioxide, ammonia, hydrogen cyanide, hydrogen chloride, nitrogen dioxide, acrolein, and phosgene.

Carbon monoxide is a major cause of death in fires because, first, it is a very common product of combustion and, second, relatively low concentrations can be lethal. A concentration of 1.0% or higher can cause death in 1 min or less. Carbon monoxide attaches easily to the hemoglobin of the blood's red cells, the cells that normally transport oxygen. It does not release easily from the hemoglobin, and in fact, significant levels of carbon monoxide are commonly found in fire fighters several days after a fire. As a result, at high carbon monoxide levels the oxygen supply to tissues of the body quickly deteriorates.

Carbon dioxide can contribute to fire-related deaths. At low concentrations it is not harmful. It is a normal product of combustion in cell metabolism of the body. A physiological response to increasing carbon dioxide in the blood is increased respiration rate. When one exercises, the increase in carbon dioxide signals the body that more oxygen is needed for the exercising muscles. Externally supplied carbon dioxide produces the same effect. However in a fire, increased inhalation of other combustion products creates greater danger for a person.

Lack of oxygen is another hazard in fires. A fire, particularly one in a confined space, may consume much of the oxygen available. As the normal 21% oxygen content declines, the capability of the hemoglobin of the red cells to transport oxygen to body tissues decreases. Reduced oxygen supply can impair tissue function, and when muscles are affected, motor performance goes down; when nerve cells are affected, motor performance can be impaired further and mental processes reduced. If oxygen supply is reduced enough, unconsciousness results.

Other combustion gases can have a variety of effects. The most important factors are the toxicity of each material and the level of concentration. Some materials, like hydrogen cyanide, are highly toxic and can cause death when inhaled. Acrolein, for example, is a strong respiratory irritant and can cause death in small concentrations. Ammonia is an irritant to the eyes and nose, both of which can contribute to breathing difficulties.

16-3 BEHAVIOR OF FIRE

General Movement of Hot Gases and Smoke

During a fire, the hot gases and smoke rise above the flame because of their lower density. The presence of smoke has no significant effect on the movement of the gases emanating from a fire. If the gases reach an obstruction, they move laterally until either they find an opportunity to continue a vertical movement or until they cool sufficiently to move downward.

Because of their increase in temperature, gases in a fire expand in volume. As the gases cool, they return to their original volume or near it. Combustion can create small overpressures in a closed space, and the amount of overpressure can affect movement of gases into or out of the space.

Horizontal Movement

In a closed space, the hot gases rise until they reach a ceiling, where they form a layer floating above the cooler air. There is limited direct mixing between them. The hot gases will move horizontally along the ceiling very quickly. As soon as the hot layer is confined, it increases in thickness, pushing into the cool layer below. Figure 16-2 illustrates this concept. The mass of hot gas, M, entering the upper layer is related to the size of the fire:

$$M = \frac{ph^{3/2}}{40}, \tag{16-1}$$

where

   M is in pounds per second,

   p is the perimeter of the fire area in feet, and

   h is the distance from the floor to the base of the hot layer in feet.

Equation 16-1 assumes that the fire is large enough for flames to reach the hot gas layer.

   In 1953, a fire at a General Motors plant in Livonia, Michigan, demonstrated the disastrous effects of rapid horizontal movement of hot gases. The fire destroyed the 34-acre plant, which was under one relatively flat roof. Studies after that fire resulted in methods for limiting horizontal movement of heat and smoke.

Vertical Movement

Hot, expanded gases from a fire move vertically. In tall buildings, the movement of heat and smoke can be important. Several factors affect the vertical movement: tightness of construction, external winds, the difference between internal and external temperature, and the presence of vertical openings, such as stairways, elevator shafts, or ventilation shafts. A major factor is the stack effect, named after the movement of heat and smoke up a chimney or smoke stack.

   Consider a tall narrow container. If the temperature inside the container is warmer than the temperature of air outside of it, a column of air outside will weigh more than one inside. The pressure at the bottom of the column inside the container will be less than that for the column outside the container. If the container has an opening at the bottom and at the top, the pressure outside the container will want to push air in at the bottom, and because the air inside is warmer and less dense, it will want to flow out at the top. This is called the stack effect. There are pressure differences between the two adjacent columns

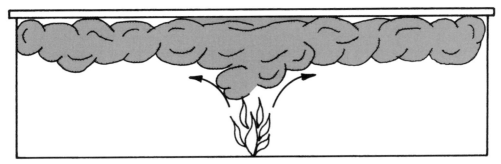

Figure 16-2.   Hot gases build up and move laterally in horizontal buildings.

of air. The pressure difference, P_d, between the bottom of the container and the top defines the draft created:

$$P_d = 7.63 \text{H} \left(\frac{1}{T_o} - \frac{1}{T_i} \right), \tag{16-2}$$

where

P_d is inches of water,

H is the vertical distance between inlet and outlet in feet,

T_o is the absolute temperature of the external air in degrees Rankine, and

T_i is the absolute temperature of the internal air in degrees Rankine.

Equation 16-2 assumes standard pressure and air density.

If $T_o < T_i$, air will flow into the lower opening and out the top; if $T_o > T_i$, the air will flow into the top opening and out the bottom; if $T_o = T_i$, there is no air flow.

When there is a difference between inside and outside temperatures, there is some point between the top and bottom openings where the internal and external pressures are equal. This is referred to as the neutral pressure plane. An overpressure from a fire inside a building will move the neutral pressure plane downward. If the temperature outside a building is less than that inside, a vent opening can move the neutral pressure plane upward.

The location of the neutral pressure plane can influence the distribution and buildup of smoke in a tall building. However, the movement of heat and smoke in tall buildings is not fully understood and is difficult to model. If air enters at the bottom near a fire, the fire can become more intense, adding smoke. External winds can change patterns of movement. On a hot summer day when the external air temperature is higher than the air-conditioned interior, the stack affect is reduced or eliminated.

Smoke Produced

The amount of smoke produced in a fire is difficult to predict. Studies have shown that the rate of smoke produced is related to the perimeter of the fire and the height of a clear zone of air above the fire (see Figure 16-3). During a fire, the amount of smoke produced diminishes as the layer of clear air becomes smaller. By reducing the size of a fire, the amount of smoke produced decreases.

16-4 FIRE HAZARDS OF MATERIALS

Flammable and Combustible Liquids

Classification Many of the common fuels, such as gasoline, diesel fuel, and heating oil, are liquid. Interest in the properties of these fuels resulted in a classification system and formulation of properties of liquid fuels.

NFPA developed a classification system for flammable liquids and combustible liquids (see Table 16-2) that uses flash point, vapor pressure, and anticipated ambient temperature conditions. The flash point of a liquid is the lowest temperature at which the vapor pressure of the liquid is just sufficient to produce a flammable mixture at the lower limit of flammability. Flash points are affected somewhat by laboratory test methods and other factors. The flash point is the lowest temperature of a liquid in an open container at which vapors evolve fast enough to support continuous combustion.

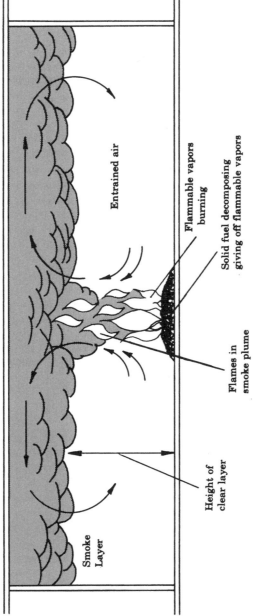

Figure 16-3. The rate of smoke production is a function of the fire perimeter and clear zone height.

289

TABLE 16-2   NFPA Classification for Flammable and Combustible Liquids

Category	Description
Flammable liquid	Flash point <100°F; vapor pressure <40 lb/in² at 100°F
Class IA	Flash point <73°F; boiling point <100°F
Class IB	Flash point <73°F; boiling point ≥100°F
Class IC	Flash point ≥73°F and <100°F
Combustible liquid	Flash point ≥100°F
Class II	Flash point ≥100°F and <140°F
Class IIIA	Flash point ≥140°F and <200°F
Class IIIB	Flash point ≥200°F

Vapor pressure is a property of a liquid in a closed container. The atmosphere above the liquid is a mixture of air and vapors of the liquid. The portion of vapor that will form in this mixture is a function of the vapor pressure of the liquid. The boiling point is the temperature at which the equilibrium vapor pressure of a liquid equals the total pressure on the surface. More simply, it is the temperature at which a liquid boils under some surrounding total pressure. Tables give vapor pressures of liquids at particular temperatures and normal boiling points (1 atm) of liquids.

In most locations, an indoor temperature could reach 100°F at some time, moderate heating is necessary to produce higher temperatures to an arbitrary limit of 140°F, and considerable heating is necessary to produce ambient temperatures higher than 140°F. These factors form the basis for the classes of flammable and combustible liquids in Table 16-2.

Vapor Volume   It is known that oxygen, most commonly in air, must be present with a fuel for combustion. If there is too much or too little fuel in the mixture of fuel and air, combustion will not occur. Normally, flammable and combustible liquids themselves will not burn—the liquid fuel must vaporize.

Vapor volume is the volume of gas formed when a liquid fuel evaporates. The vapor volume can be computed from other properties of the liquid and vapor:

$$V_v = \frac{8.33 \times \text{specific gravity of the liquid}}{0.075 \times \text{vapor density of the vapor}} \tag{16-3a}$$

$$= \frac{111 \times \text{specific gravity}}{\text{vapor density}}, \tag{16-3b}$$

where

V_v is in cubic feet per gallon,

8.33 is the weight (pounds) of 1 gal of standard water, and

0.075 is the weight (pounds) of 1 ft³ of air at standard temperature and pressure.

Some tables listing properties of flammable and combustible liquids give vapor volume directly.

Flammable Limits   The mixture of fuel and standard air necessary for combustion of fuel vapors must be within certain limits. The lower flammable limit (LFL) is the minimum concentration of vapor-to-air below which propagation of a flame will not occur in the presence of an ignition source. The upper flammable limit (UFL) is the maximum vapor-to-air mixture above which propagation will not occur. Mixtures below the lower flammable limit are said to be too lean; those above the upper flammable limit too rich. The term *lower explosive limit* is equivalent to LFL, and the term *upper explosive limit* is equivalent to UFL. Flammable limits vary somewhat with temperature and pressure. The flammable range is the mixture of fuel and air between LFL and UFL. Table 16-3 gives LFL and UFL along with other properties of some flammable and combustible liquids.

Examples It is often necessary to estimate whether evaporating fuel will form a combustible mixture. Such mixtures may result from a spill in a closed space or loss of flammable material in a process. An exact determination is not always possible, because vapors may not mix uniformly throughout the space. Vapors from a solvent tank may be heavier than air, may evaporate at the surface of the liquid, may slide over the edge of the tank, and may concentrate near the floor. In another case, vapors may be lighter than air and may concentrate near a ceiling. For some computations of this type, it is useful to apply an adjustment factor for incomplete mixing and local concentrations.

Example 16-1 Consider the case of a child who used gasoline to clean his hands in a bathroom. There was a gas water heater in the same room. While cleaning his hands, the gasoline vapor ignited and the child was severely burned. Assume the room was $6 \times 10 \times 8$ ft.

(a) Assuming full mixing (uniform vapor–air mixture throughout the room), how much gasoline had to evaporate to reach a combustible mixture?

(b) Assume there was incomplete mixing (nonuniform vapor–air concentrations in the room). Apply an adjustment factor of 5 and estimate the minimal amount of vapor required to develop a locally combustible mixture in the room.

(c) Will the vapors of gasoline concentrate at the floor or at the ceiling when evaporating in air?

The room volume is $480\,ft^3$. From Table 16-3, the LFL for gasoline is assumed to be 1.4% in air. The vapor volume of gasoline necessary for combustion is then $480 \times 0.014 = 6.72\,ft^3$. From Table 16-3, the vapor equivalent of gasoline is assumed to be $24\,ft^3/gal$. Thus, the minimum volume of liquid gasoline that must evaporate to produce the LFL is $6.72/24 = 0.28\,gal$. For unequal mixing, the estimated minimum is $0.28/5 = 0.056\,gal$, assuming a factor of 5 is a reasonable adjustment factor. This is approximately 14 tablespoons of gasoline. From Table 16-3, the vapor density of gasoline is 3 to 4. Because the vapors are much denser than air, they will concentrate quickly at the floor, where the source of ignition for a gas water heater is located and a combustible mixture will form quickly.

Flammable Gases

Flammable gases refer to those gases that exist at standard temperatures and pressures (above the normal boiling point of a gas) and burn in normal concentrations of oxygen in air. Flammable gases must combine in appropriate mixtures with air for combustion to occur and they must be at or above their ignition temperature to burn. Any substance that has a vapor pressure of more than $40\,lb/in^2$ (absolute) at 100°F in its liquid state is considered a gas. Like vapors of flammable and combustible liquids, the specific gravity of a gas can be important when analyzing fire potential. Full mixing in a space may not occur, but local concentrations may build up and reach combustible mixtures.

TABLE 16-3   **Properties of Selected Flammable Liquids**[a]

Liquid	Flash Point °F	Boiling Point °F	Ignition Temp. °F	Flammable Limits Vol. %		Specific Gravity (Water = 1)	Vapor Density (Air = 1)	Vapor Volume (ft³/gal)
				LFL	UFL			
Acetaldehyde	−38	70	347	4.0	60	0.8	1.5	58
Acetone	−4	133	869	2.5	13	0.9	2.0	44
Acrolein	−15	125	428	2.8	31	0.8	1.9	—
Allylamine	−20	128	705	2.2	22	0.9	2.0	—
Amyl acetate	60	300	680	1.1	7.5	0.9	4.5	22
Benzol (benzene)	12	176	928	1.3	7.9	0.9	2.8	37
Butadiene monoxide	<−58	151	—	—	—	0.9	2.4	—
Butyl alcohol	98	243	650	1.4	11.2	0.8	2.6	—
Butyl chloride	15	170	860	1.8	10.1	0.9	3.2	—
Carbon disulfide	−22	115	194	1.3	50	1.3	2.6	54
Cyclohexane	−4	179	473	1.3	8	0.8	2.9	30
Denatured alcohol	60	175	750	—	—	0.8	1.6	—
Dibutyl ether	77	286	382	1.5	7.6	0.8	4.5	—
Dichloroethylene-1,2	36	119	860	5.6	12.8	1.3	3.4	43
Diethylamine	−9	134	594	1.8	10.1	0.7	2.5	32
2,2-Dimethylbutane	−54	122	761	1.2	7.0	0.6	3.0	—
2,3-Dimethylpentane	<20	194	635	1.1	6.7	0.7	3.5	—
P-dioxane	54	214	356	2.0	22	1.0+	3.0	39
Divinyl ether	<−22	102	680	1.7	27	0.8	2.4	—
Ethyl acetate	24	171	800	2.0	11.5	0.9	3.0	33
Ethyl alcohol	55	173	685	3.3	19	0.8	1.6	56

Ethylamine	<0	62	725	3.5	14	0.8	1.6	50
Ethyl chloride	−58	54	966	3.8	15.4	0.9	2.2	46
Ethyl ether	−49	95	356	1.9	36	0.7	2.6	31
Gasoline	−45	100–400	536–853	1.4	7.6	0.8	3–4	24–32
Hexadiene-1,4	−6	151	—	2.0	6.1	0.7	2.8	—
Hexane	−7	156	437	1.1	7.5	0.7	3.0	25
Isopropyl alcohol	53	181	750	2.0	12.7	0.8	2.1	—
Jet fuel (JP-4)	−10–+30	—	464	1.3	8.0	—	—	—
Kerosene	100–162	304–574	410	0.7	5	<1	—	—
Methyl alcohol	52	147	867	6.07	36	0.8	1.1	80
Methylcyclohexane	25	214	482	1.2	6.7	0.8	3.4	26
Methyl ethyl ether	−35	51	374	2.0	10.1	0.7	2.1	—
Methylethylketone	16	176	759	1.4	11.4	0.8	2.5	36
Naphtha V.M. & P.	28	212–320	450	0.9	6.0	<1	—	—
Nitroethane	82	237	778	3.4	—	1.1	2.6	46
Paraldehyde	96	255	460	1.3	—	1.0–	4.5	—
Pentane	<−40	97	500	1.5	7.8	0.6	2.5	29
Petroleum ether	<0	95–140	550	1.1	59	0.6	2.5	—
Propanol	−22	120	405	2.6	17	0.8	2.0	—
Propylene oxide	−35	94	840	2.3	36	0.8	2.0	—
Toluol	90	231	896	1.2	7.1	0.9	3.1	—
Turpentine	95	300	488	0.8	—	<1	1.8	—
Vinyl ethyl ether	<−50	96	395	1.7	28	0.8	2.5	—
Xylene-o	90	292	867	1.0	7.0	0.9	3.7	27

a Most properties from NFPA 325M.

Nonflammable gases are those that do not burn in any concentration of air or oxygen. Inert gases are those that will not support combustion. Some gases, called reactive gases, will react with other materials in processes other than combustion. An example is chlorine reacting with hydrogen. Chlorine is a nonflammable (but toxic) gas and hydrogen is a flammable gas, and although their reaction will produce a flame, the reaction does not involve oxygen. Besides being combustible or reactive, gas toxicity is another hazardous property (see Chapter 24). Another hazard of gases is related to the Boyle-Charles law (see Chapter 19). Gases are classified in other ways, based on physical properties and use.

Other Materials

Besides flammable and combustible liquids and flammable gases, there are many other kinds of materials that will burn. Some of the major ones are wood, metals, and plastics.

Wood   Wood and wood products, like pulp, paper, and cardboard, will burn. There is considerable variation in the heat value of different wood species and products. During combustion, wood will char, that is, form a layer of partially burned material (similar to charcoal) that insulates material below it from the heat of combustion and slows the burning rate. In fact, wood can retain many of its structural properties in a fire for some time because of, to a large extent, the insulating effect of the char formation.

People have studied the ignition and charring of wood. Ignition temperatures vary significantly, depending on the moisture content, density, and other factors. When wood burns, moisture and other noncombustible gases are driven from it initially. As combustion progresses, the temperature increases and water vapor and carbon monoxide are produced. Until this point, wood has absorbed heat, but the wood reaches higher temperatures, flammable vapors and particulates form an exothermic reaction, and charcoal is formed. In general, the ignition temperature of wood declines as combustion moves through the preceding stages. Ignition temperatures for test blocks are on the order of 300° to 400°F. Charcoal may ignite at significantly lower temperatures.

Metals   Most metals burn under certain conditions. Small cross sections and fine particles burn more readily than thick solids. Some metals are called combustible because they are relatively easy to ignite. Combustible metals include magnesium, titanium, zinc, sodium, lithium, some radioactive metals, and others. Burning metals create special extinguishment problems. For example, dumping water on a titanium fire may add to the fire. The water may break down into oxygen and hydrogen and the oxygen combining with titanium and hydrogen becomes another fuel. Magnesium burns in a carbon dioxide atmosphere. Aluminum, iron, and steel do not burn easily because they do not react with oxygen easily. Fine metal powders may ignite easily and explode like other dusts (see Chapter 17).

Plastics   There is a wide variety of plastics, some with special additives to achieve particular properties. They vary in many ways, including fire-related properties. However, some generalizations illustrate their hazards. Plastics tend to have higher ignition temperatures than wood. Some have a rapid flame spread rate and some are easily ignited and burn rapidly. Many plastics produce dense, black smoke during combustion, often because of the additives that inhibit flammability. Like other materials, most plastics produce carbon monoxide and many also produce other toxic gases. During combustion, plastics tend to melt, which may result in drippings that spread the fire. Cellular plastics without flame retardants tend to create fast-spreading, high-intensity, dense smoke fires.

Identification of Hazards of Materials

NFPA developed a system for identifying fire hazards of materials that has been in use for some time.[2] The system presents information on labels and placards about three types of hazards, which are subdivided into five levels of severity. The information is useful for fire fighters and others and is presented in a four-quadrant diamond symbol. As illustrated in Figure 16-4, three quadrants are for the three kinds of hazards: health, flammability, and reactivity. The lower quadrant contains special information and symbols. One symbol, a W with a horizontal line through it, shows that a material has a hazardous reaction with water. Another symbol is the radioactive pinwheel symbol (see Figure 22-2). Numbers in each quadrant give the degree of the hazard. General interpretations for degree of hazard are as follows:

0. No special hazards, therefore no special measures for fire fighting.

1. Nuisance hazards are present that require some care; standard fire fighting procedures can be used.

2. Hazards are present that require certain equipment or procedures to handle these materials safely; can be fought with standard procedures.

3. Fire can be fought using methods intended for extremely hazardous situations, such as unmanned monitors or personal protective equipment that prevents all bodily contact.

4. Too dangerous to approach with standard fire-fighting equipment and procedures; withdraw and obtain expert advice on how to handle.

Degree of hazard information is included in Figure 16-4. Further data and interpretations for specific types of hazards are found in NFPA 704.

16-5   FIRE SAFETY IN BUILDINGS

Fundamentals

There are at least two important lessons learned from the Great Chicago Fire of 1871. One resulted from the inability of fire equipment to move down congested streets to a location where they could fight parts of the fire effectively and the other resulted from the rate at which the fire spread. The density of frame structures on lots was a big factor. Fed by the wind, fire leaped from building to building very quickly. Undoubtedly, radiation, convection, and wind-blown sparks all played a large part in the process. As a result, rules were written for layout of communities, streets, building sites, water supplies, and construction of buildings. Other fires added to these lessons and influenced today's standards as well.

The main objectives for fire safety in buildings are (1) getting occupants out safely, (2) minimizing property loss for structures and contents, and (3) minimizing interruption of operations. Through continued study, knowledge of fire behavior and building design makes it possible to minimize fire losses. Often the goal is to confine a fire to the site of origin, then to the building of origin. Through proper design of facilities, one can confine most fires to the floor and even the room of origin.

Site Planning and Accessibility

Fire departments and equipment should have access to all sides of a building. Access roads should be adequate even during peak traffic loads. Landscaping, external structures, and

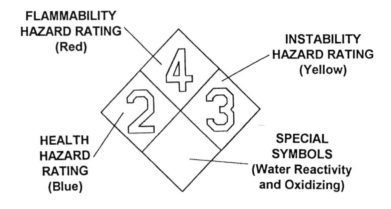

DEGREE OF HAZARD	HEALTH HAZARD RATING	FLAMMABILITY HAZARD RATING	INSTABILITY HAZARD RATING
4	Materials that, under emergency conditions, can be lethal.	Materials that will rapidly or completely vaporize at atmospheric pressure and normal ambient temperature or that are readily dispersed in air and will burn readily.	Materials that in themselves are readily capable of detonation or explosive decomposition or explosive reaction at normal temperatures and pressures.
3	Materials that, under emergency conditions, can cause serious or permanent injury.	Liquids and solids that can be ignited under almost all ambient temperature conditions. Materials in this degree produce hazardous atmospheres with air under almost all ambient temperatures or, though unaffected by ambient temperatures, are readily ignited under almost all conditions.	Materials that in themselves are capable of detonation or explosive decomposition or explosive reaction, but that require a strong initiating source or that must be heated under confinement before initiation.
2	Materials that, under emergency conditions, can cause temporary incapacitation or residual injury.	Materials that must be moderately heated or exposed to relatively high ambient temperatures before ignition can occur. Materials in this degree would not under normal conditions form hazardous atmospheres with air, but under high ambient temperatures or under moderate heating could release vapor in sufficient quantities to produce hazardous atmospheres with air.	Materials that readily undergo violent chemical change at elevated temperatures and pressures.
1	Materials that, under emergency conditions, can cause significant irritation.	Materials that must be preheated before ignition can occur. Materials in this degree require considerable preheating, under all ambient temperature conditions, before ignition and combustion can occur.	Materials that in themselves are normally stable, but that can become unstable at elevated temperatures and pressures.
0	Materials that, under emergency conditions, would offer no hazard beyond that of ordinary combustible materials.	Materials that will not burn under typical fire conditions, including intrinsically noncombustible materials such as concrete, stone, and sand.	Materials that in themselves are normally stable, even under fire conditions.

Figure 16-4.   NFPA symbol system for identification of hazards of materials. (Refer to NFPA 704.)

vehicle parking should not create barriers to access. Water supplies, hydrants, and valves should be located conveniently to support fire-fighting strategies. Adequate pressures and quantities of water must be available. If special hazards exist and other kinds of extinguishing agents are needed, they must be planned into the site in adequate amounts and at effective locations. Gas lines and supply lines for other fuels entering a facility should have shutoff valves located where it is not dangerous to operate them in a fire. An example is a vehicle driving away from a gas pump, catching a fuel hose, and starting a fire. If the shutoff is at the pump, it would be difficult to approach and shut off the fuel.

Separation of Structures

Buildings should have sufficient separation to minimize fire traveling from one building to another. Distance between buildings is one way to provide separation. Shielding by fire walls is another way. Other design features, such as parapets, minimal wall openings to adjacent buildings, fire resistant door and window materials for openings, automatically closed doors, dampers, and shutters, all will help limit fire movement between buildings. So will sprinkler systems and water curtains. Many factors affect design decisions made to meet separation requirements. Type of construction, building height, size of exposed walls, and amount of openings in building walls are a few considerations.

Building Construction

A designer has a wide choice of materials and methods of assembly for a building design. The materials and the methods of assembly can affect the ability of a building to meet fire safety objectives and also can affect insurance premium rates.

Fire Resistance Ratings   Fire resistance is a rating for building assemblies and components based on laboratory tests. Given in units of minutes or hours, the fire resistance ratings indicate how long an assembly or component will withstand a particular test fire. The objective is to confine a fire long enough and to ensure that a structure will not collapse to allow occupants to escape and fire fighting to begin containment and extinguishing tasks.

Building Construction Classification   There are many ways to classify the materials and methods of assembly used in a building. The current NFPA classification scheme considers the ability to retain structural integrity and to provide fire barriers and to recognize the fuel contribution of the structure and the enclosing walls and roof. Classifications consider other features to limit fire and smoke movement and to meet other fire safety requirements within the building. The classifications and a brief description of each are found in Table 16-4.

Structural Integrity

Many factors affect the ability of a building and the assemblies of materials in it to continue to carry loads during a fire. Designers use analytical design, data from design tables and guides, codes requirements, and tests results to reach decisions.

Materials expand when heated and contract when cooled, and different materials have different coefficients of thermal expansion. If expansion during a fire is not incorporated into a design, structural damage may result from excessive loads that members place on each other, often related to buckling and other failures not created by direct heat damage to materials or by combustion.

TABLE 16-4   NFPA Classifications of Building Construction

Classification	Description
A. Type of construction (from NFPA 220)	
Fire-resistive	Structural members including walls, partitions, columns, floors, and roofs are of noncombustible or limited-combustible materials and have fire resistance ratings that meet or exceed particular fire resistance ratings.
Noncombustible/ limited-combustible	Walls, partitions, and structural members are of noncombustible or limited-combustible materials, but do not qualify as fire-resistive construction.
Protected noncombustible/ limited-combustible	Bearing walls or bearing portions of walls, exterior or interior, are of noncombustible or limited-combustible materials and have minimum hourly fire resistance ratings and stability under fire conditions, and floors and roofs and their supports have minimum hourly fire resistance ratings.
Heavy timber	Bearing walls or bearing portions of walls are of noncombustible or limited-combustible materials and have minimum hourly fire resistance ratings and stability under fire conditions; nonbearing exterior walls are of noncombustible or limited-combustible materials; columns, beams, and girders are of heavy timber, solid or laminated; and floors and roofs are of wood without concealed spaces. In addition, components and assemblies must meet certain dimensional and other criteria.
Ordinary	Exterior bearing walls or bearing portions of exterior walls are of noncombustible or limited-combustible materials and have minimum hourly fire resistance ratings and stability under fire conditions; nonbearing exterior walls are of noncombustible or limited-combustible materials; and roofs, floors, and interior framing are wholly or partly of wood of smaller dimensions than required for heavy timber construction.
Protected ordinary	Ordinary construction may be designated as "protected" when roofs and floors and their supports have minimum hourly fire resistance ratings.
Wood frame	Exterior walls, bearing walls and partitions, and roofs and their supports are wholly or partly of wood or other combustible materials, when the construction does not qualify as heavy timber construction or ordinary construction.
Protected wood frame	Wood frame construction may be designated "protected" when roof and floors and their supports have minimum hourly fire resistance ratings.
B. Class of occupancy (primary classifications from NFPA 101)	
Assembly	An occupancy used for a gathering of 50 or more persons for deliberation, worship, entertainment, eating, drinking, amusement, awaiting transportation, or similar uses or used as a special amusement building, regardless of occupant load.
Educational	All buildings used for educational purposes through the twelfth grade by six or more persons for 4 hours per day or more or more than 12 hours per week.
Health care	Used for purposes of medical or other treatment or care of four or more persons where such occupants are mostly incapable of self-preservation because of age, physical or mental disability, or because of security measures not under the occupants' control. They include hospitals, nursing homes, and limited care facilities.

TABLE 16-4   continued

Classification	Description
Detention and correctional	Used to house four or more persons under varied degrees of restraint or security where such occupants are mostly incapable of self-preservation because of security measures not under the occupants' control.
Residential	An occupancy that provides sleeping accommodations for purposes other than health care or detention and correctional. Residential occupancies are further divided into one- and two-family dwellings, lodging and rooming houses, hotels and dormitories, apartments, residential board, and care facilities.
Mercantile	Used for the display and sale of merchandise.
Business	Used for account and record keeping or the transaction of business other than mercantile.
Industrial	An occupancy in which products are manufactured or in which processing, assembling, mixing, packaging, finishing, decorating, or repair operations are conducted.
Storage	Used primarily for the storage or sheltering of goods, merchandise, products, vehicles, or animals.
C. Hazard of contents (from NFPA 101)	
Low hazard	Those of such low combustibility that no self-propagating fire therein can occur.
Ordinary hazard	Those that are likely to burn with moderate rapidity or to give off a considerable volume of smoke.
High hazard	Those that are likely to burn with extreme rapidity or from which explosions are likely.

The strength properties of many materials are affected by temperature. For example, when heated, steel quickly loses its ability to carry a load, even its own weight, if the temperature is high enough. To slow the rate of temperature rise, steel normally is covered with materials (plaster, gypsum, and other materials) that insulate it from the high gas temperatures in a fire.

Wood is a combustible material. However, depending on moisture content, it may not collapse rapidly because of the insulating effect that char provides. Some materials, like wood and other combustibles, can be treated with substances that slow the rate of burning.

Concrete is a common structural element. Although it has some insulating properties, heat can damage it, causing loss of strength, spalling, and other effects from heat. The kinds of materials in concrete and their mix affect its ability to withstand a fire. If heat reaches reinforcing materials in concrete, significant structural capacity can be lost.

Confinement

A major objective in building design is limiting a fire to the area of origin. Confining a fire to a small area is best. Some call this strategy compartmentation. A building and each portion of it are designed to restrict horizontal and vertical movement of fire, smoke, and heat. Partitioning assemblies, doors, windows, duct runs, and other openings are designed to meet fire ratings. The confinement should function until a fire is extinguished or burns itself out. When confinement is the criterion, potential fire severity determines what fire resistance is needed. As noted earlier, time to exit and time for fire fighters to begin control and extinguishing actions are also criteria for determining fire resistance.

Fire Load

A fire is characterized by three stages: growth from a small origin, full development, and decay. In a fully developed fire, the temperature in a confined space will reach 1500° to 2300°F. Before or during full fire development, the contents of a room may burst into flames. This is called flashover.

The severity of a fire in terms of intensity and duration is a function of the quantity of combustibles available, their burning rate, and the air available for combustion. The surface area of fuel and the amount of oxygen control combustion. The greatest heat load occurs when just enough ventilation is present so the rate of combustion is controlled by the fuel surface area. At low oxygen supply rates, ventilation limits the rate of combustion. At high ventilation rates, heat is removed from the area of the fire, which reduces the heat transfer rate. Plots of time and temperature provide a basis for assessing fire severity. Fires with different temperature histories are compared with standard time-temperature profiles. A test fire is equivalent in severity to the standard when the areas under the time-temperature curves are equal. Barriers, such as walls, floors, ceilings, and doors, must be able to withstand the desired fire severity.

Studies have shown a relationship between fire severity and fire load. One can determine fire load during design or use of a building and have some idea of the severity potential. Fire load is the maximum heat released if all combustibles in a fire area burn, including heat from combustible contents and combustible interior finish, floor, and structural materials. Fire load is usually expressed as equivalent combustible weight divided by the fire area and is given in pounds per square foot. Actual fire loads are adjusted to the equivalent heat of combustion of ordinary combustibles, assumed to be 8,000 BTU/lb. Table 16-5 gives approximate heat of combustion data for some common materials.

Example 16-2 Assume that a warehouse contains 1,000 lb of epoxy stored in a 50 ft² area. What is the fire load? The heat of combustion of epoxy is approximately 14,400 BTU/lb. The fire load is 1,000 lb × 14,400 BTU/lb/8,000 BTU/lb/50 ft² = 36 lb/ft².

When ordinary combustibles, such as paper, are stored in steel containers, they will not burn completely and will not contribute the full heat value of the materials. Usually, ordinary combustibles stored in partially and completely enclosed steel containers are derated when estimating fire loads.

For fully enclosed containers, the ratio of fully enclosed combustible weight, W_E, to the total weight of all combustibles, F_T, determines what derating factor, K, is used:

W_E/F_T	K
<0.5	0.4
0.5–0.8	0.2
>0.8	0.1

For containers that have one side open (partially enclosed), the weight of ordinary combustibles, W_P, is derated by $K = 0.75$.

The total derated fire load, F_L, is the sum of combustibles in the open, derated fully enclosed combustibles, and derated partially enclosed combustibles.

TABLE 16-5 Approximate Heat of Combustion for Some Common Materials

Material	Approximate Heat of Combustion	
	(MJ/kg)	(BTU/lb)
Charcoal	33.7–34.7	14,492–14,879
Coal		
Anthracite	30.9–34.6	13,288–14,879
Bituminous	24.7–36.3	10,621–15,610
Cotton	16.5–20.4	7,096–8,773
Gasoline	46.8	20,126
Kerosene	46.4	19,954
Leather	18.2–19.8	7,527–8,515
Oil, linseed	34.2–39.4	16,857–16,943
Paper		
Brown	16.3–17.9	7,010–7,698
Magazine	12.7	5,461
Newsprint	21.5	9,246
Rubber (auto tires)	32.6	14,019
Starch	17.6	7,569
Straw	15.6	6,708
Wheat	15	6,451
Wood		
Birch	20	8,600
Douglas fir	21	9,031
Maple	19.1	8,214
Red oak	20.2	8,687
Spruce	21.9	9,375
White pine	19.2	8,257
Hardboard	19.9	8,558
Wool	20.8–26.6	8,945–11,439

Fire Spread

The objective of compartmentation is to confine a fire to the room or area of origin. Wall, floor, ceiling, and opening barriers are rated to help achieve that goal. When rated barriers are in place, fires do not spread readily by heat transfer or structural failure of the barriers. They spread primarily through horizontal and vertical openings, such as open doors and unenclosed stairs and shafts.

Many buildings have interstitial spaces between floors for electrical and communication lines, heating and cooling ducts, and steam and water lines. Fires in these hidden areas can burn for some time. When they break through a ceiling, for example, they can spread very rapidly across an open room. A flame front often will travel faster than occupants can run.

Vertical openings may exist between wall surface materials and vertical shafts may house electrical, communication, and other building services. Laundry chutes, elevator shafts, ventilation shafts, or atriums are all examples of vertical openings.

Preventing fire spread through horizontal and vertical openings in buildings may be difficult, but a number of approaches can reduce the rate of spreading. Where possible, concealed spaces should not have combustible materials. Like other spaces, they should be enclosed with rated barriers. Fire-retardant coatings may help. Vertical openings should be blocked where possible. Fire stops refer to a wide range of methods for sealing concealed spaces, including blocking between wall studs. Fire stops also include protective devices and systems for sealing conduits, pipe and other penetrations, foam-in-place sealants, and other noncombustible materials, even sand. Fire doors and shutters are useful for larger openings for escalators and conveyors. Limiting fire spread in atriums requires a combination of design features.

16-6 INDUSTRIAL AND PROCESS FIRE HAZARDS

General

In many ways, the fire hazards in industrial facilities are the same as in other facilities. One major difference is the quantity of materials, fuels, and power present in one location. Each kind of operation and process presents particular fire prevention and protection problems. This text cannot discuss in detail each operation, process, and kind of material. A few operations, principles, and procedures give examples. One should consult specific references to find out more about particular fire hazards and controls.

Venting

One problem found across many industries and processes is the use of single-story facilities. As noted earlier, heat and smoke can move rapidly under the roof of a horizontal building. There are some techniques that can help confine fires in industrial plants. Ventilation is an important method. Roof vents allow hot gases to escape and do have some effect on confinement. Compartmentation will help limit the horizontal spread of fire. In industrial plants, full partitions may interfere with the flow of materials and production activities. Where large open spaces are needed and to avoid interfering with manufacturing and material handling processes, partial partitions may hang from the ceiling. Curtain boards are vertical panels constructed of noncombustible materials that are suspended from the ceiling. When they fit tightly to the ceiling and extended as far to the floor as possible, they can prevent the lateral movement of a hot gas layer. In effect, curtain boards can partition a large plant into smaller areas to control fire spread. Figure 16-5 illustrates the use of curtain boards.

Vents should be located within each area or section created by curtain boards. From knowledge of fire behavior and the development of a hot gas layer below a ceiling, an expression for the amount of vent area, A, required for a section is

$$A = 0.14 \frac{ph^{3/2}}{d^{1/2}}, \tag{16-4}$$

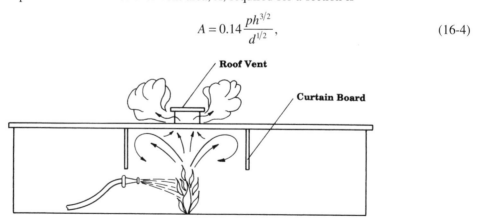

Figure 16-5. Example of curtain boards for limiting fire spread in horizontal construction.

where

A is the vent area required in square feet,

p is the perimeter of the fire area in feet,

h is the distance from the floor to the base of the hot gas layer in feet, and

d is the depth of the layer in feet.

Example 16-3   In a factory, the curtain boards hang 10 ft from the ceiling. The bottoms of the curtain boards are 14 ft from the floor. If a fire is located in a 4 ft × 5 ft area and one assumes that the hot gas layer formed along the ceiling extends half the distance to the bottom of the curtain boards, how much vent area is needed?

The perimeter of the fire is $(2)(4 + 5) = 18$ ft. Then, applying Equation 16-4,

$$A = \frac{0.14(18)(14+5)^{3/2}}{5^{1/2}} = 93.3 \text{ ft}^2.$$

Fire Walls

Another way to prevent lateral spread of fire in large buildings or between buildings is by using fire walls. Fire walls must withstand the potential for complete burnout on both sides. They require special structural construction so heat and collapse of other building components will not affect their stability. Fire walls often are visible from outside a building, because they have a parapet extending approximately 3 ft above the roof line. The parapet helps prevent fire from burning along the roof and across the fire wall. Fire walls have some limitations, and other partitioning methods may be substituted.

Welding and Cutting

Common industrial processes are welding and cutting. Unsafe welding and cutting procedures and equipment too often lead to fires. Heating certain materials with welding equipment may produce toxic gases and vapors. Welding processes also can damage eyes. Later chapters discuss these hazards. One should also reference special publications on welding and cutting equipment and operations that provide more details on hazards and controls.

The main types of welding are electric welding and oxygen-fuel welding. The most common form of electric welding is arc welding. An arc is made to pass between the metal to be welded and an electrode. There are several kinds of electrodes, some containing filler material for the joint. The electric arc creates a high, localized temperature that melts the filler material and heats the surrounding metal. There are several variations to the basic arc welding process for different metals and for improved control in automated welding.

In oxygen-fuel welding a fuel (the most common fuel is acetylene, although others are sometimes used) is mixed with oxygen in a nozzle device or torch. Both oxygen and fuel move from separate compressed gas tanks through pipes and hoses to the torch. When ignited, the rate of fuel can be adjusted to achieve very high temperatures that heat the metals being joined and the filler material to form a bond.

Metals also can be cut by using the high temperatures achieved by arc and gas welding processes. A narrow band of metal is oxidized to form a cut.

Fire hazards result from the high temperature and open flame equipment, scattering of sparks and hot metal particles, the welding equipment itself, the objects being worked on and the environments. In welding and cutting operations, sparks and hot particles fly from the operation and may fall to lower locations or into cracks and openings. The sparks and hot particles can travel as far as 30 to 50 ft and are capable of igniting flammable and combustible material within such a range. Because persons performing welding wear high-density glasses or face shields, they cannot see what is going on in lower light levels. They also may not be able to monitor such a large area. When the welding operation cannot be confined or performed in a safe location free of combustibles, a fire watcher assists the welder. The fire watcher's job is to monitor the area around the operation for lingering sparks or fire and to extinguish them.

Welding equipment needs special care to ensure that it is not accidentally operated. Oxygen must be handled carefully and maintained properly. Leaking oxygen could get into clothing and other material and cause rapid flame up if ignited. Electrical equipment must be grounded, and all equipment should comply with electric codes. Cylinder valves should be closed when not in use. Leaking fuels, such as acetylene and butane, have led to many explosions related to welding and cutting.

The objects worked on and the environment must have proper controls as well. Quite often one hears of a person who attempts to repair a container that had a fuel in it. The vapors ignite and explode, usually resulting in severe injury or death. By filling such containers with water or sand or by cleaning them and flushing out residual fuel, welding can be performed safely. The environments for welding and cutting must be free of flammable gases. In confined spaces, adequate ventilation and breathing air is needed. Welders should wear proper protective clothing, and welders and others involved in welding and cutting operations must have training in the operations, their hazards, and proper controls.

Hot Work Permit Procedures

Many organizations use a permit procedure for all hot work, except that involving normal operations or processes. Hot work is any kind of welding, cutting, burning, or activity that involves or generates sparks or open flame. It includes heated equipment that may provide an ignition source for a fire. Hot work often involves people from a maintenance department going to other departments to perform activities. The main idea in a hot work permit procedure is to ensure that supervisors of all departments involved and workers who may be involved in any way in the work participate in the decision to start work and to conduct it safely. A sample hot work permit procedure is outlined in Figure 16-6.

Storage

Storage of flammables and combustibles requires special attention. A few fundamental storage principles are covered here. References and applicable sections of the National Fire Code have much more to say about particular storage matters related to fire protection.

Indoor Storage of Flammable Liquids

Flammable liquids are used in many operations for cleaning, as fuel for equipment, and for other purposes. To prevent fires, there are many requirements for storage and dispensing of flammable liquids indoors. One general principle is to limit the quantity of flammables stored in an occupied or operational area to that needed for specific jobs or a single day. Larger quantities should be stored in special facilities separate from occupied and operational areas.

PERMIT
Cutting and Welding
with Portable Gas or Arc Equipment

Date _____   Building _____

Dept _____   Floor _____

Work to be completed _____

_____

Special precautions to be taken _____

_____

Fire watch required?   ☐ Yes   ☐ No

I have examined the location where this work is to be performed, identified
the necessary precautions, and therefore grant permission for this work as
show on the reverse side of this permit.

Permit expires _____

Signature _____
(Person responsible for authorizing welding and cutting work.)

Time and Date: Work started _____

                 Completed _____

FINAL CHECK-UP

I have inspected the work area and all surrounding areas for sparks and heat
(including floors above and below, opposite sides of walls, and enclosed
areas) at least 30 minutes after the work was completed and found no signs
of combustion.

Signed _____
(Supervisor or Fire Watcher)

Figure 16-6.   An example of a hot work permit form.

Storerooms   Special storerooms confine flammable liquids. NFPA has detailed design
standards for these storerooms. Fire codes limit the quantities and sizes of containers that
can be stored indoors. Some design features for flammable storerooms are explosion-proof
switches and electrical fixtures, ventilation, self-closing doors with fusible links, a static
electricity grounding system, special signage, raised door sill, and special floor contours.

Storage Cabinets   Storage cabinets may be located outside special storerooms. Standards and codes specify their design and construction and limit the quantities of flammables they may contain. Cabinets help protect stored liquids from fires outside them, confine
spills within them, and keep flammables organized. There are also various styles of safety
cans and vertical or horizontal dispensing drums.

Grounding and Bonding   When flammables are transferred from storage drums to
small dispensing containers or when there are transfers of large quantities of flammables,
the containers involved must be connected to each other (bonding) by an electrical con-

---

PROCEDURE SUMMARY

Before approving any cutting and welding permit, the fire safety supervisor or his appointed alternate shall inspect the work area and confirm that all necessary precautions have been taken to prevent fire in accordance with NFPA standards and company policy.

PRECAUTIONS
☐ Sprinklers in service
☐ Cutting and welding equipment is in good repair

WITHIN 35 FEET OF WORK
☐ Floors are swept clean of combustibles
☐ Combustible floors are wet down, covered with damp sand, metal or other
  shields
☐ No combustible material or flammable liquids are present
☐ Combustibles and flammable liquids are protected with covers, guards, or
  metal shields
☐ All wall and floor openings are covered
☐ Covers are suspended beneath work to collect sparks

WORK ON WALLS OR CEILINGS
☐ Construction is noncombustible and without combustible covering
☐ Combustibles are moved away from opposite side of wall(s)

WORK ON ENCLOSED EQUIPMENT
(Tanks, containers, ducts, dust collectors, etc.)
☐ Equipment is cleaned of all combustibles
☐ Containers are purged of flammable vapors

FIRE WATCH
☐ To be provided during and for 30 minutes after operation
☐ To be supplied with extinguisher and small hose
☐ Fire watcher is trained in use of equipment and in sounding fire alarm

FINAL CHECK-UP
☐ To be made 30 minutes after completion of any operation unless fire watch
  is provided

Signed _____
        (Supervisor)

Figure 16-6. *continued*

ductor or both must be connected to a grounding rod or line. (See Chapter 12 for a discussion of bonding and grounding.)

Drums Flammable liquids must be stored in closed containers and may be dispensed from drums by gravity or suction pump, not by pressurizing the drum. Valves for gravity dispensing must be approved by a recognized testing laboratory and must close automatically. Drip containers are required for catching leaked fluids.

Safety Cans Safety cans contain 5 gal or less of flammable liquids and are used for moving fluid to the point of use. They have several features for safety. They close automatically after tilting or pouring, they have a pressure relief valve to vent vapor, they have a flame arrestor in the spout, and in the event of a flame on the outside, the arrestor absorbs heat and prevents the flame from passing to the inside of the can. Safety cans have leak-resistant designs.

Plunger Cans   Plunger cans are designed to wet cleaning cloths or wipes with flammable liquid. A shallow pan rests on the end of a shaft that is spring supported on the top of the can. By placing a cloth or wipe on the pan and pressing down, a small amount of fluid is pumped into the cloth through the dispensing shaft. Excess fluid drains back into the can.

Cleaning and Dip Tanks   There are many fire protection features for cleaning and dip tanks. Tanks vary in size, capacity, and design features. They have covers that protect the fluid from ignition in a fire. Some have foot-operated lids that close when no one is stepping on the lever that opens the lid. Others are normally open and have fusible links that melt at relatively low temperatures, causing the lid to close. Some have drain boards so that dripping fluid from washed parts stay in the tank. Additional protection for large tanks includes sprinklers and automatic drains to holding tanks that operate in the event of a fire. Other fire extinguishing equipment may be needed.

Waste   Cloths and wipes contaminated with flammable liquids should be stored in self-closing containers. Waste containers should be small to limit the quantity accumulated, and they should be emptied regularly to prevent spillage outside the can. A foot lever opens a lid and the lid closes when the lever is released. Other design features prevent heat transfer to the contents.

Warehouses and Other Facilities

There are many fire protection problems in warehouses. One should reference applicable NFPA codes and other sources for more details about particular materials. Some important factors for fire protection in warehouse storage are type of commodity, ease of ignition, rate of fire spread, and rate of heat produced. Other factors are quantities of material stored, how they are stored, height of storage, and accessibility. Distance to other commodities also can be important, particularly if one commodity is a fuel and another is an oxidizer. All these characteristics help determine the fire hazards and suitable controls.

For warehouse storage, codes group commodities that have similar fire hazard and control characteristics into classes. Standards specify the height to which commodities may be piled because the height of materials affects fire growth, intensity, and control.

Because warehouses have high densities of materials, a combination of extinguishing equipment is needed. Sprinklers are essential. Because water from sprinklers may not reach fires between, under, or within units of stored materials, fire hoses and portable extinguishers may be needed. Still, some materials may need to be moved to gain access to burning units. Sprinkler design standards mandate clearances between stacked material and sprinkler heads to allow spray patterns to develop. Standards also give flow rates, sprinkler head density, temperature ratings of sprinkler heads, water flow rates of auxiliary fire hose, overall water supply, and other requirements for various commodities and classes of commodities. Sprinkler system requirements and recommendations also are affected by type of storage: storage on pallets, in boxes and bins, on racks, in bulk, or in packaged units. Requirements vary somewhat for different types of facilities, such as indoor versus outdoor, cold storage, bulk tanks, or bins.

16-7 LIFE SAFETY

As previously noted, the first priority in a fire is protection of human life. Life safety deals with providing people with (1) a reasonable degree of safety from fire in a facility and (2) an adequate opportunity to exit facilities if a fire occurs. There are many life safety codes adopted by organization and government units, but the one most often cited is NFPA 101 and NFPA 101B.[3] Sometimes codes do not explain the theory or concepts behind them or how they are applied. For NFPA 101, the *Life Safety Code Handbook* supplies such details.

Human Behavior in Fires

Under the stress of a fire situation, people do not always behave logically. The behavior of one person may affect the behavior of others, and the ability to perform correctly during a fire is confounded by incomplete information about fire conditions and routes to safety. Behavior is affected by personal conditions, such as age, mobility, ability to see and hear, and previous training. Physical conditions in a fire, like smoke, loss of power for lights, rate of fire spread, and heat buildup, can affect visibility, options for movement, and decision making. Density of people, the number of routes, capacity of routes, and distance to the exterior can affect movement and travel time. Today, computer modeling systems allow designers to model exiting behavior with some degree of precision and to evaluate some building features related to life safety.

General Principles of Life Safety

The degree of fire hazard for a building determines the risk to occupants. The degree of fire hazard is based on building contents, the rate of fire propagation for the contents, and activities performed in a building. Life safety codes recognize three classes of hazard: low, ordinary, and high. Life safety codes divide regulations by type of occupancy, for example, residential, places of assembly, hospitals, and industrial. Life safety standards vary by hazard class and type of occupancy (refer to Table 16-4).

Provisions of life safety codes address properties of interior finishes, size, number and location of exits, protection of exit routes from fire and smoke, alarm systems, emergency lighting, signage for exit routes, compartmentation, construction, horizontal and vertical openings, extinguishing systems, and other factors. Some of these provisions are summarized and explained in the following text.

Interior Finishes

Interior finishes are the materials that make up exposed interior wall, column, and ceiling surfaces of buildings. Interior floor finishes refer to the floor covering. Finish materials are tested in laboratory procedures for two fire characteristics: (1) how quickly flame spreads across the material and (2) the amount of smoke produced. Interior finishes are divided into three classes, class A, B, or C, determined by controlled laboratory tests that rate the materials. Ratings by class are as follows:

Class A: flame spread, 0–25; smoke developed, 0–450

Class B: flame spread, 26–75; smoke developed, 0–450

Class C: flame spread, 76–200; smoke developed, 0–450

Floor finishes are divided into class I and class II, based on critical radiant heat flux ratings from controlled tests. The *Life Safety Code* specifies which classes of finishes are allowed for each occupancy and surfaces for exits, access to exits, and other spaces.

Means of Egress

Many factors affect the ability to egress and the time required to do so. A means of egress is a continuous and unobstructed way of travel from any point in a building or structure to a public way (street, alley, or other similar parcel of land open to the outside air). There are three parts to the means of egress: exit access, exit and exit discharge. Exit access is a path leading to an exit. An exit is that portion of a means of egress that is separated from all other spaces of a building or structure by construction or equipment to provide a protected way of travel to the exit discharge. It may consist of doorways, stairs, ramps, corridors, or similar components that are bounded by walls, floors, and doors. An exit is bounded by one or more entrances to it and one or more doors to leave it at ground level. An exit discharge is the last segment of a means of egress between the protected exit and the land outside.

The code specifies a number of attributes for means of egress, such as capacity, number, travel distant to exits, discharge from exits, illumination, emergency lighting, and marking (such as signs and their features). Based on occupancy, it also specifies characteristics of means of egress components (doors, doorways, stairs, ramps, corridors, etc.). For example, the occupancy determines what panic hardware and fire exit hardware is acceptable. It specifies details for revolving doors, turnstiles, sliding doors, illumination (including emergency lighting), and other components.

Capacity The capacity for means of egress is determined from the occupancy load, which is not less than the number of persons determined from occupant load factors. NFPA 101 lists occupant load factors for various occupancies in square feet per person or square meters per person.

Occupant load is computed from

$$\text{Occupant load} = \frac{\text{floor area}}{\text{occupant load factor}}. \tag{16-5}$$

For example, the occupant load factor for a casino is $11\,\text{ft}^2$ per person and $100\,\text{ft}^2$ per person for an office. Thus, a $10,000\,\text{ft}^2$ casino would have an occupancy load of 909 and the same office space would have an occupancy load of 100.

Occupant load values are used to determine the capacity of components of a means of egress, depending on the occupancy. The total capacity of a stairs, for example, is expressed as

$$\text{Total stair capacity} = \frac{\text{width}}{\text{capacity of means of egress for stairs}}. \tag{16-6}$$

Consider a new health care occupancy, which must have a capacity of means of egress for stairs of $0.3\,\text{in}$ per person and $0.2\,\text{in}$ per person for doors, ramps, or horizontal exits. Thus, stairs that are the minimum 44 in wide each would have a capacity of $44/0.3 = 146$ people. If a floor has an occupancy load of 200 people, there would need to be two stairs or wider stairs to meet the occupant load. Other components of the means of egress would require similar analysis.

Number of Means of Egress   The minimum is two. For occupant loads between 500 and 1,000, three are required, and for occupant loads more than 1,000, there must be four.

Exit Access   Occupants should be able to travel to an exit without obstructions. Some occupancies have distance limits for dead-end routes to an exit, but dead ends should be avoided. Routes should not require passage through doors that can be locked and should not pass through areas of more severe fire hazard. Maximum travel distance permitted from any point to an exit varies with occupancy and whether the building has a sprinkler system. Minimum corridor widths vary with occupancy, but are generally 36 in or more. If access involves use of stairs, the stairs must meet design standards.

Width   The minimum width of any means of egress is 36 in and other specifications may apply to egress components.

Stairs   The minimum width clear of all obstructions for stairs is typically 44 in. Minimum tread depth is 11 in, and riser height can be no more than 7 in. Headroom must be at least 6 ft 8 in, and the maximum height between landings is 12 ft.

16-8   FIRE DETECTION AND ALARM SYSTEMS

Fire Protection

Fire protection refers to methods for controlling and extinguishing fires. It involves working against time. Figure 16-7 illustrates the process. Indicators of combustion are smoke, flame, and heat, which must be detected in some manner. Then warnings are needed to begin appropriate action for preservation of life and property. The actions needed are exiting or getting people to safety and fighting the fire. The fire-fighting objectives are organized to minimize the amount of property involved and to achieve extinguishment.

Fire Detection and Alarms

There are many kinds of equipment for detecting fires and giving alarms. The devices may be quite simple, applying only to certain aspects of the process. Devices may depend on human activation or may be automatic; they can be combined in sophisticated sensor, annunciator, and alarm systems. Systems generally require regular testing to be sure that components are working properly. Systems can be computer controlled. The computer may perform internal checks constantly for component failures and report which ones are not working properly. A number of NFPA codes establish standards for sensor and alarm components and systems.

There are several kinds of detectors. There are detectors for heat, smoke, flame, and gas content. Each type is suited to particular applications, depending on the type of fire

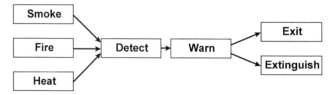

Figure 16-7.   Typical actions in response to a fire.

that can occur and the kind of—flammable or combustible materials present. Placement of detectors during installation can be critical. Some require more maintenance than others, and testing is essential to ensure their reliability.

Heat Detectors There are several kinds of heat detectors. They include fixed temperature detectors, rate-of-rise detectors, rate compensation detectors, and others.

Fixed Temperature Fixed temperature heat sensors are designed to operate when a preset temperature is reached. They are available in a variety of temperature settings. Because there is some mass in the sensor, the fixed temperature sensors take some time to respond to conditions. The air surrounding the device will reach the trigger temperature at some time before the sensor elements do. The time lag depends on the device, and because of the lag, fixed temperature sensors are not suitable for fires that develop quickly.

Many fixed temperature sensors depend on fusible elements made from metal that melt at a preset temperature. The break in an element can be coupled mechanically or electrically to other action, such as turning on sprinkler heads or an alarm system.

Another type of fixed temperature element uses two continuous wires, separated by insulation. When a predetermined temperature is reached, the insulation melts and the wires come into contact, which can activate an electrical circuit.

A third type of fixed temperature element uses bimetallic strips or disks. The two metals forming the strip each expand at different rates with temperature increases, which causes the strip to bend or the disk to snap to a different curvature. The movement is coupled to electrical circuits.

Rate of Rise Rate-of-rise detectors respond to fires that flame up quickly but do not react to slower changes in ambient temperature that would normally be expected for slow-developing fires. They typically respond to temperature changes on the order of 12°F/min. One design involves two physical phenomena. First, the air inside a tube expands and builds up pressure inside the tube. When the pressure reaches some predetermined level, a switch closes or opens. Second, the tube has a small hole in it that allows expanded air to leak at a controlled rate. The difference between these two phenomena establishes the rate of temperature rise sensed.

Rate Compensation A rate compensation detector responds to a preset temperature, but is less sensitive to thermal lag than a fixed temperature device. Thus, temperatures that increase rapidly and exceed the preset temperature will be detected sooner.

Other Other heat detectors use thermocouples and gas release from solids to sense temperature changes and to trigger switches and controls. In addition, some detectors have multiple elements that respond to both rapidly and slowly developing fires.

Smoke Detectors In many fires, smoke is present before there is a significant heat buildup. As a result, smoke detectors usually detect fire before heat sensors. Smoke detectors operate on one of two principles: ionization or photoelectricity.

Ionization Ionization detectors contain a very small quantity of radioactive material. They ionize the zone of air around the radioactive material, making the air conductive. When smoke particles enter the zone, the conductivity decreases. An electrical circuit that monitors the conductivity can detect sufficient change and trigger an alarm. Ionization detectors are most sensitive to small smoke particles found in high energy, open-flame fires.

Photoelectricity Photoelectric smoke detectors depend on a source and receiver of light. A beam is directed from the source to the receiver or sensor. Smoke particles entering the beam reduce the light arriving at the sensor and scatter the light of the beam. If the receiver is located in the beam, reduction in light is sensed when smoke is present. The reduction can trigger an alarm. If the receiver is located outside the normal beam, the smoke scatters the light, causing some light to fall on the sensor. When sufficient light arrives at the sensor, an alarm is triggered. This principle also applies to several locations. In some types of photoelectric detectors, a vacuum pump draws air from one or more locations to a central sensing device. Any one of the sources can trigger a response from the sensor.

Flame Detectors Flame detectors monitor for certain wavelengths in the field of view of the sensing device. Infrared and ultraviolet wavelengths are most common. These devices are sensitive to the glow from flames or embers. Some infrared sensors are programmed to monitor a wide band of wavelengths; others measure frequency of infrared present in its field of view, such as in a flickering flame. Broadband infrared sensors are susceptible to false readings from sunlight or other radiant sources that are not sources of fire. In such cases, shielding from alternate sources is necessary but should be used judiciously, because if a shielded flame detector cannot see the fire source, it cannot respond.

Gas Sensors These detectors sense the presence of certain gases produced by combustion in most fires. Electrical circuits coupled to sensing devices trigger alarms.

Alarms and Controls There are many kinds of alarm and control devices activated directly or indirectly from detectors or manual signalling devices. Alarms may be auditory or visual. Visual alarms may simply be a light on an annunciator panel located at the entrance to a building or at some central monitoring station. Audio alarms may be continuous or intermittent tones, variable pitch tones, or voice instructions. Alarms may activate fire and smoke control doors or hatches, may release security systems, and may affect elevator controls or heating, ventilating, and air conditioning (HVAC) equipment. Voice communication may be live instructions from a central station or recorded messages from recordings or computer chips. Each application dictates the kind of alarms and controls needed.

16-9 FIRE EXTINGUISHMENT

Fire extinguishment is the application of agents to control fire spread and ultimately to put a fire out. By far the most common agent is water. For some applications, additives enhance the capabilities of water. Other agents are carbon dioxide, halogens, dry chemicals, foams, and other special-purpose agents.

Kinds of Extinguishment

Fire is extinguished by eliminating any or all of the four components that make up the fire triangle and the fire pyramid. For most materials, combustion is stopped if heat, fuel, or oxygen is removed. In addition, if the creation of the hydroxyl radical can be stopped, combustion will not occur.

Portable Fire Extinguishers

Most fires start small and can be extinguished easily during their early stages. Portable fire extinguishers contain small quantities (for ease of carrying and handling) of an extinguishing agent suitable for suppression of small fires. Extinguishers must be located strategically for quick response, be suitable for the kind of fire encountered, and work properly. The user must know how to use them correctly. A variety of extinguishing agents are available. NFPA 10 contains standards for fire extinguishers.

Classes of Fires   There are four classes of fires. The classification scheme helps determine what agents in extinguishers are suitable for different fires.

Class A   Class A fires involve ordinary combustibles (wood, paper, cloth, rubber, plastics). Extinguishment is caused by cooling or smothering.

Class B   These fires involve flammable and combustible liquids, flammable gases, greases, and oils. Extinguishment is accomplished by inhibiting the release of combustible vapors or the development of the hydroxyl radical.

Class C   Electrical equipment fires are class C fires. Extinguishing agents for electrical fires must not conduct electricity.

Class D   Class D fires involve combustible metals. Extinguishing agents must absorb heat and not react with the metals.

Portable Extinguishers   Extinguishers may contain agents that are effective on one or more classes of fires. Labels on extinguishers identify what classes they are suitable for. Extinguisher labels also contain information about effectiveness of the extinguisher. Examples for class A and B fires are 4-A and 20-B, where 20 is better than 4. These ratings only give relative effectiveness and are not absolute ratings; the 20 is not 5 times more effective than the 4 rating. An extinguishing agent may have ratings for more than one class of fire, and each rating may differ in effectiveness.

Extinguisher Requirements   Portable extinguishers do not replace fixed extinguishing systems. Standards define the number and distribution of extinguishers. Important considerations are the class of fire anticipated, the occupancy classification, the class (light, ordinary, high) of hazard in the building, the floor area served and the travel distance from any location to an extinguisher. Extinguishers require inspection, maintenance, testing, and record keeping. Standards detail frequencies and procedures for these activities.

Water Extinguishment

Water can cool the burning surface of many materials and can stop vaporization of the material necessary to support combustion. Through a change of state, water takes up much heat, and thus, a water spray applied to the point of combustion is usually more efficient than a solid stream. Water also can extinguish some materials by smothering the fuel and preventing air from reaching it. However, because of danger from electric shock, water may not be suitable for electrical fires. Also, water may react with certain materials and create hazardous chemicals or conditions.

Additives improve the effectiveness of water as an extinguishing agent for certain fire conditions. They may affect viscosity and surface tension, may cause foaming or create other characteristics, and may prevent freezing.

Water Supplies

Engineering of water supplies for fire protection is an important task. Designs must consider the total amount available in a period of time, the rate of supply at various locations, and the distribution. NFPA codes detail water supply requirements for fire protection in communities and special facilities. Standards also cover hydrants, pumps, fire hose, nozzles, and other components of water supply systems for fire protection.

Chapter 10 discussed some hydraulics. Bernoulli's equation is an essential engineering principle in fire hydraulics. Losses in pipes, fittings, and other components are adjusted to equivalent pipe length for use in Bernoulli's equation. Distribution systems need regular testing to ensure that water supplies are available when and where needed.

Sprinkler Systems

Sprinkler systems are automatic or semiautomatic extinguishing systems for buildings and other facilities. Studies and experience have shown them to be the most effective means for controlling fires in buildings. In most fires where sprinklers are present, activating only a few sprinklers near the fire is sufficient for control. The cost of sprinkler systems is balanced by reduced insurance premiums and reduced losses if a fire does occur. NFPA codes cover sprinkler systems.

In general, sprinkler systems distribute extinguishing agents to the locations where fires occurs. Sprinkler heads operate independently and determine when they should release the extinguishing agent. Placement of sprinkler heads, type of head, appropriate agent, and proper maintenance and testing are essential to the success of a system.

Kinds of Sprinkler Systems There are many kinds of sprinkler systems. Most are water based. However, they may have other agents.

Wet-Pipe Systems A wet-pipe system contains water under pressure at all times. Any sprinkler head that is opened will allow water to pass immediately. Wet-pipe systems can be damaged by water freezing in the pipes, but antifreeze solution stored in unprotected portions of the piping can prevent this threat.

Dry-Pipe Systems A dry-pipe system contains air or nitrogen under pressure at all times. A valve separates water supplies from the dry pipes. An open head reduces the pressure in the gas-charged pipes and allows the valve to open, releasing water to the open sprinkler head and the fire. Dry-pipe systems are suitable for areas subject to freezing. Compared with wet-pipe systems, dry-pipe systems respond more slowly to a demand for water at a sprinkler head, and more heads are likely to open in a fire. Special features prevent inadvertent operation of the water valve.

Preaction Systems This is a special form of dry-pipe system. The piping may or may not be under pressure. Sensors in the protected area besides those at each sprinkler head sense a fire and open a valve to fill the pipe with water. The special sensors operate before those in the sprinkler heads. This design reduces the delay found in a dry-pipe system.

Combined Systems   These systems combine features of a dry-pipe system and a preaction system. Pipes are filled with air under pressure. Supplementary sensors open a water valve and air exhaust ports. This allows the piping to fill completely with water before the sprinklers open.

Deluge Systems   A deluge system is similar to a preaction system. All sprinklers are open at all times. Fire detecting devices activate the water valve, allowing water to emerge from all sprinkler heads.

Other Systems   Some systems are designed for limited water supplies and have one or more limited capacity pressure tanks to supply water. Related systems produce a water curtain to protect outside walls. Other variations may provide reduced protection. An example is fixed water spray protection to provide cooling for tanks exposed to fire.

Components   The main components of a sprinkler system are piping, sprinkler heads, and hangers. Piping distributes water throughout the system. Its main components are risers (major vertical pipes), crossmains, and branches. Hangers of various types support the components. The branches extend from the cross mains, and the sprinkler heads attach to branches. There are many kinds of sprinkler heads, the features of which affect how quickly they open and the spray pattern and distribution of water developed. Fusible links activate most sprinkler heads. The sprinkler code contains standards for all system components.

Sprinkler System Design

There are two methods for designing sprinkler systems. One is hydraulic design; the other works from tables, charts, and data provided in the code. The code specifies the water pressure required at each head, and achieving those pressures ensures the proper water flow rate at the heads. Selection of sprinkler heads is important to get the water where it is needed and to make sure that all areas or locations are protected.

Hydraulic Design   In hydraulic design, calculations must show that all head pressures will be achieved when water at some pressure and flow rate is applied at the inlet to the system. Hydraulic designs normally reduce system components and costs compared with the table method. Computer programs help to analyze sprinkler designs to determine whether they meet hydraulic requirements and to simulate performance. The general formula in hydraulic design of sprinklers is

$$P_i \geq P_s + P_f + 0.434h, \tag{16-7}$$

where

P_i is the pressure at the system inlet (pounds per square inch),

P_s is the pressure at the sprinkler head of interest (pounds per square inch),

P_f is the friction loss from pipe length, bends, valves, and other fittings (pounds per square inch), and

h is the vertical rise from inlet to sprinkler head (feet).

Computer programs help analyze sprinkler designs to determine whether they meet hydraulic requirements.

Table Method Tables, charts, and other data in the code define what size pipe is needed, how many branches can be on a cross main, and how many sprinkler heads can be on any branch. They also specify the maximum distance between sprinkler heads and between a sprinkler head and a wall, the maximum distance between branches, and the maximum area any one head may serve.

Criteria differ by type of hazard. For the table method, a trial-and-error approach will determine the minimum number of sprinkler heads and other elements necessary to meet the requirements.

Fire Suppression Systems

Suppression systems that do not use water or modify properties of water are useful in particular applications. One should refer to applicable standards for design and use of these systems and agents, which include carbon dioxide, halons, dry chemicals, foams, and combustible metal agents.

Carbon Dioxide Carbon dioxide is stored under pressure as a gas or liquid. It extinguishes by reducing the oxygen content of air and by cooling. It is most suitable for class B fires. It is also useful, but less effective, for class A fires. It can be toxic, it has a noisy discharge, and in enclosed spaces, it can reduce oxygen content for breathing. Carbon dioxide can be piped from storage containers to points of application. Application is by filling an entire enclosure (total flooding) or by local application to burning material.

Halons Halons are hydrocarbons in which one or more hydrogen atom is replaced by atoms from halogens, such as fluorine, chlorine, bromine, or iodine. A variety of halons have been used for fire suppression, particularly in clean rooms and for protecting electronic and computer systems. Some were discontinued because of toxicity and corrosion effects. Today, two halons are in general use: halon 1211 (bromochlorodifluoromethane) and halon 1301 (bromotrifluoromethane). Extinguishing involves interruption of the hydroxyl radical of combustion. Effectiveness is a function of many factors. Halon 1301 is best suited for total flooding applications, whereas halon 1211 is well suited for local application systems. Halons are normally delivered under pressure with nitrogen. Low concentrations seem to have little toxic effect on humans, but do contribute to environmental problems (ozone depletion). The design of halon systems must include allowable exposure times and concentrations.

Studies have determined that halon 1301 has a high ozone depletion potential and therefore is detrimental to the environment. Although the U.S. Environmental Protection Agency allows existing halon 1301 systems to stay in service, new installations are discouraged. To reduce release of halon into the atmosphere, pressure and puff tests or door fan tests replace previously used full-flooding halon acceptance tests. Some new agents are replacing halon 1301 and conventional sprinklers and carbon dioxide suppression systems or new water mist systems may be viable alternatives.

Dry Chemicals There are a number of dry chemicals that are effective in extinguishing fires. They are most effective for flammable liquids and electrical fires. Certain types are also effective for ordinary combustibles. The ingredients are not toxic. They are stable materials in fine powder form. They act primarily by smothering, cooling, and shielding fuel from the radiant heat of a flame. They are expelled by a gas, such as nitrogen, under pressure.

Foams Foams are gas-filled bubbles formed from water-based and other materials. They are primarily used on flammable or combustible liquid spills and fires. The foam forms a layer that prevents vaporization of the liquid. They are used also in applications involving class A and class B materials. Some foams are called high expansion, because they expand in volume by factors of 100 to 1,000. These foams can fill locations that are difficult to reach. The foams are applied by mixing materials in nozzles or foam makers. Fixed foam systems can be actuated automatically. Portable equipment is used for aircraft rescue and industrial fires. The general design formula for high-expansion foam systems for surface fires of flammable and combustible liquids with flash points higher than 100°F is

$$R = \left(\frac{V}{T} + R_s\right) \times C_N \times C_L,$$ (16-8)

where

R is the rate of foam discharge (cubic feet per minute),

V is the submergence volume (cubic feet) or volume of space to be protected,

T is submergence time (minutes; normal range from design standards is 2–8 min),

R_s is the rate of foam breakdown by sprinklers (cubic feet per minute),

C_N is the compensation for normal foam shrinkage (normally approximately 1.15), and

C_L is the compensation for leakage around door openings and so on. C_L varies from 1.0 to 1.2.

R_s is normally determined from test data. Where data are not available, R_s may be estimated from the total discharge (gallons per minute) from the maximum number of sprinklers expected to operate times 10.

Combustible Metal Agents For combustible metal fires, agents are available that will not react with the metals. They are generally specific to particular metals; no one agent is suitable for every kind of combustible metal. Most agents are proprietary. Some non-proprietary agents, such as talc, sand, soda ash, and others, vary in effectiveness, but may be useful in certain applications.

16-10 FIRE DEPARTMENTS

In the United States, public fire departments, whether paid or voluntary, provide most fire protection services. Fire departments play important roles in prevention of fires and enforcement of code, training and education of fire fighters and the community, handling communication of fire alarms and other emergencies, responding to fires, and reporting and administration activities. Today, fire departments are likely to respond more often to nonfire than fire emergencies. Hazardous materials and emergency medical responses have become an important part of fire department operations.

There are many opportunities for engineers to contribute to fire department effectiveness. Engineers are needed for design of water systems, site location for response units, modeling and monitoring response times and capabilities, design of fire equipment, and designing alarm and communication equipment and systems.

In-Plant Organizations

Many industries and large facilities cannot depend solely on public fire departments. Much effort is needed within private organizations to ensure that fire prevention and protection is adequate. Many organizations have fire brigades. They handle immediate responses to fire calls, conduct training of employees, and conduct simulations and fire drills. They monitor facilities for fire hazards, check exit routes to make sure they are clear, and may handle inspection and testing of extinguishers, detectors, and other equipment.

Mutual Aid Agreements

Mutual aid agreements are statements that one fire protection organization will assist another and vice versa when major responses are needed. This reduces the personnel and equipment required by any one organization for severe situations. Not only do public fire departments establish such agreements, but many companies set them up with local departments and other nearby companies.

EXERCISES

1. Toluol (molecular weight = 92.13) will be used as a solvent in an operation. Compute the volume of vapor (cubic feet) produced by the evaporation of 1 gal of liquid.

2. Using a dilution factor of 5 to account for nonuniform mixing, what ventilation rate (cubic feet per hour) will be required to keep toluol from reaching the LFL, if 3 gal are evaporated each hour?

3. Eight gallons of turpentine are lost through evaporation because of a spill in an enclosed factory room. The room is 50 ft long, 120 ft wide, and 20 ft high. Using a dilution factor of 3 for unequal mixing, determine if there is a danger of fire if all of the turpentine evaporates.

4. A drying oven has a volume of 50,000 ft^3. A production line runs through it. Parts suspended from the line are dipped in a degreasing tank and drained before entering the oven. The oven is vented with exhaust ventilation and air is replaced by clean air at a rate of 3,000 ft^3/min. When the line and oven are running, 3 gal/hr of solvent are evaporated from the parts. At the beginning of the shift, there is no vapor present in the oven. A design distribution constant $K = 4$ is used to allow for incomplete mixing of vapor in the oven. The solvent has a vapor equivalent of 23 ft^3/gal. The LFL and UFL for the solvent are 1.4 and 8.3, respectively.

   (a) Assume that at startup there is no delay in the solvent soaked parts filling the oven. How long will it take after startup before the concentration of solvent in the oven reaches 100 ppm?

   (b) What will the concentration of solvent vapor be in the oven after 1.5 hr?

   (c) If the line starts but the exhaust ventilation system is not turned on, how many gallons of solvent will evaporate before a flammable mixture is reached? Assume uniform distribution of vapors.

   (d) If the line stops but the exhaust ventilation system keeps running, how long will it take to reduce a 200 ppm concentration to 50 ppm? Assume the evaporation stops when the line stops.

5. The area formed by combustibles in a manufacturing facility is 35×40 ft. There are curtain boards extending 12 ft from the ceiling that end 15 ft above the floor. What area of roof venting is needed between curtain boards spaced 175 ft apart?

6. An office room measures 40×60 ft. It contains 2,000 lb of paper stored openly, 1,500 lb of paper stored in file cabinets, and 1,200 lb of paper manuals stored in open book shelves. Assume the paper has a heat content of 8,000 BTU/lb. What is the total fire load in the room?

7. A three-story motel is on the drawing board. The accompanying illustration shows guest room layouts for each floor. Consider only the residential section. Assume the lobby is unoccupied. Determine the following from the current *Life Safety Code* (NFPA 101®):

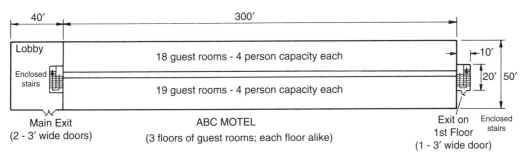

Diagram for Exercise 7.

(a) What is the occupant load for determining the amount of fire exits?

(b) What exit capacity is required for each floor of the motel?

(c) Is the travel distance to exits from any room exceeded?

(d) Assume the travel distance for rooms in the center of a floor is too long. What redesign alternatives would make the travel distance satisfactory.

(e) What fire-resistance rating must the walls enclosing the stairways have?

(f) What fire-resistance rating must the walls between guest rooms have?

(g) The designers are considering carpet for the corridors. The manufacturer certifies that the carpet selected has a flame spread rating of 81. Can it be installed in the corridors?

(h) Could the same carpet be used in the guest rooms?

(i) Do the exit stairs and door widths satisfy the code?

8. A sprinkler head has a discharge rate of 22.5 gal/min and a coefficient $K = 5.65$. What pressure is required at the sprinkler head?

9. For the wet-pipe sprinkler line shown in the accompanying diagram, what pressure is required at the inlet if the discharge rate must be 22.5 gal/min and the sprinkler head has a coefficient $K = 5.65$? Assume that fittings connecting two different sized pipes are the size of the smaller pipe. Assume that pipe diameters, d, equal the nominal diameters.

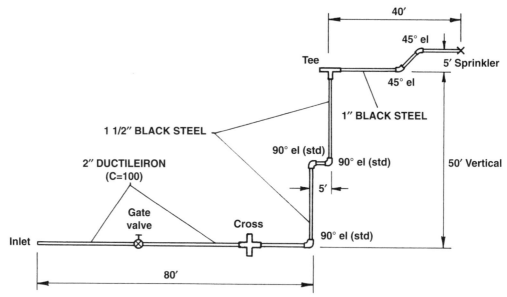

Diagram for Exercise 9.

10. For the facility layout shown in the accompanying sketch, complete the sprinkler system layout using NFPA 13. The cross mains are already located, placed on one side of the building, and the branch lines should extend across the room from the cross main.

(a) Locate branch lines and sprinkler heads on branch lines. Dimension locations along branches and between branches and walls.

(b) How many sprinklers are required for each of the three zones to meet all design criteria?

(c) What size copper tubing is required for the branch lines in the extra hazard zone?

11. Identify the occupancy classification for

(a) a retail store

(b) a grade school

12. What type of construction is

(a) a frame structure that uses standard wood studs?

(b) a frame structure that uses steel studs?

(c) a structure with 8-in concrete block walls and partitions?

REVIEW QUESTIONS

1. Describe total fire losses for the United States each year in cost and deaths.

2. What is the cause of most deaths from fires?

3. For what age groups is death from fire most prevalent?

4. Describe the relationship between alcohol use and fires.

5. What is the leading cause of civilian fires?

6. What are the two leading causes of industrial fires?

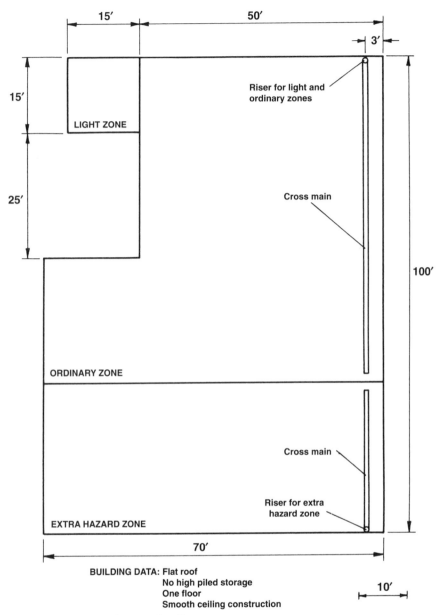

Diagram for Exercise 10.

7. To what extent are arson and incendiary fires a significant element in the fire losses?

8. What organization is the main source for fire codes and standards in the United States?

9. Define
  (a) combustion
  (b) ignition
  (c) spontaneous combustion

     (d) the fire triangle

     (e) the fire pyramid

10. Name four methods for controlling combustion and for extinguishing fires.

11. Name four products of combustion and the danger of each.

12. Describe the movement of hot gases from a fire in a horizontal building and a vertical building.

13. Define or characterize each of the following for a flammable or combustible liquid:

     (a) flash point

     (b) fire point

     (c) vapor pressure

     (d) vapor volume

     (e) lower flammable limit

     (f) upper flammable limit

     (g) flammable range

14. Define

     (a) flammable gas

     (b) char

15. Describe the NFPA symbol and method for identifying hazardous materials and their properties.

16. What are the objectives for fire safety in buildings?

17. Identify at least three requirements for site planning that are important for fire safety.

18. What are four characteristics of buildings that are important in designing for fire safety?

19. Explain the following:

     (a) compartmentation

     (b) fire load

     (c) fire resistance rating

     (d) flame spread rating

20. How can the buildup and lateral movement of heat from a fire in a single story factory be minimized?

21. What are the fire hazards of welding and cutting and how can they be controlled?

22. Identify at least four ways to minimize fire hazards of flammable liquids stored indoors.

23. What is life safety?

24. What aspects of building design do life safety features address?

25. Define

     (a) means of egress

     (b) exit

     (c) exit access

     (d) exit discharge

(e) exit capacity

(f) occupant load

25. Name at least four kinds of detectors for fire protection and describe the function of each.

26. Describe the materials included in each for the four classes of fires.

27. Identify and describe five kinds of sprinkler systems.

28. What are the two approaches for designing sprinkler systems?

29. How does each of the following accomplish fire extinguishment?

(a) water

(b) carbon dioxide

(c) halons

(d) dry chemicals

(e) foams

(f) agents for combustible metals

(g) water mists

NOTES

1 Stein L., *The Triangle Fire*, Carroll & Graf Publishers, Inc., New York, 1962. Von Drehle, David, *Triangle—The Fire that Changed America*, Grove Press, New York, 2003.

2 NFPA 704, *Identification of the Hazards of Mate-*

rials, National Fire Protection Association, Quincy, MA.

3 NFPA 101, *Life Safety Code* and NFPA 101B, *Means of Egress*, National Fire Protection Association, Quincy, MA.

BIBLIOGRAPHY

BENEDETTI, R. P., ed., *Flammable and Combustible Liquids Code Handbook*, 3rd ed., National Fire Protection Association, Quincy, MA, 1987.

BOUCHARD, J. K., ed., *Automatic Sprinkler Systems Handbook*, National Fire Protection Association, Quincy, MA, 1988.

BRYAN, J. L., *Automatic Sprinkler and Standpipe Systems*, 3rd ed., National Fire Protection Association, Quincy, MA, 1997.

BUKOWSKI, R. W., O'LAUGHLIN, R. J., and ZIMMERMAN, C. E., eds., *Fire Alarm Signaling Systems Handbook*, National Fire Protection Association, Quincy, MA, 1987.

CANTER, D., ed., *Fires and Human Behavior*, Wiley, New York, 1978.

COTE, A., and BUGBEE, P., *Principles of Fire Protection*, National Fire Protection Association, Quincy, MA, 1987.

DRYSDALE, D., *An Introduction to Fire Dynamics*, Wiley, New York, 1985.

DUBAY, CHRISTIAN, *Automatic Sprinkler Systems Handbook*, National Fire Protection Association, Quincy, MA, 2002.

ERVEN, L. W., *Techniques of Fire Hydraulics*, Glencoe Publishing Co., Mission Hills, CA, 1972.

Fire Protection Handbook, 19th ed., National Fire Protection Association, Quincy, MA, 2003.

FRIEDMAN, R., *Principles of Fire Protection Chemistry*, 2nd ed., National Fire Protection Association, Quincy, MA, 1989.

GAGNON, ROBERT M., and KIRBY, RONALD H., *A Designer's Guide to Fire Alarm Systems*, National Fire Protection Association, Quincy, MA, 2003.

GRANT, C. E., and PAGNI, P. J., *Fire Safety Science*, Hemisphere Publishing Corp., New York, 1986.

HICKEY, H. E., *Hydraulics for Fire Protection*, National Fire Protection Association, Quincy, MA, 1980.

Industrial Fire Hazards Handbook, 3rd ed., National Fire Protection Association, Quincy, MA, 1990.

COTE, RON, and HARRINGTON, GREGORY E., eds., *Life Safety Code Handbook*, National Fire Protection Association, Quincy, MA, 2003.

LEMOFF, T. C., ed., *National Fuel Gas Code Handbook*, National Fire Protection Association, Quincy, MA, 1988.

Operation of Fire Protection Systems, National Fire Protection Association, Quincy, MA, 2003.

PATTON. A. J., and RUSSELL, J. C., *Fire Litigation Sourcebook*, Garland Publishing, New York, 1986.

PLANER, R. G., *Fire Loss Control*, Marcel Dekker, New York, 1979.

PURINGTON, R. G., *Fire-Fighting Hydraulics*, McGraw-Hill, New York, 1974.

SCHULTZ, N., *Fire and Flammability Handbook*, Van Nostrand Reinhold, New York, 1985.

The National Fire Code, National Fire Protection Association, Quincy, MA, updated regularly. Now also includes NFPA 5000, *Building Construction and Safety Code*.

The SFPE Handbook of Fire Protection Engineering, 3rd ed., National Fire Protection Association, Quincy, MA, 2002.

VON DREHELE, DAVID, *Triangle—The Fire That Changed America*, Grove Press, New York, 2003.

WALLS, W. L., ed., *Liquified Petroleum Gases Handbook*, National Fire Protection Association, Quincy, MA, 1988.

CHAPTER 17

EXPLOSIONS AND EXPLOSIVES

17-1 EXPLOSIONS

General Characteristics

The term *explosion* is difficult to define precisely. In general, it refers to a group of phenomena in which there is a sudden expansion or a bursting effect. One definition is a rapid increase of pressure or an excessively high pressure in a confined space followed by its sudden release resulting from rupture of the container. In some explosions, there are visible flames and a flash of light. In others, one observes matter flying in many directions. Explosions produce a noticeable sound described as a "crack" or "boom," depending on the location of the person relative to the source and other conditions.

Controlled explosions can be very useful. For example, explosions regularly occur in internal combustion engines. Accidental or uncontrolled explosions can have disastrous effects. Spectacular explosions of various types gain widespread attention. One of the most notable was the space shuttle Challenger explosion on live television in 1986. Others are grain elevator explosions and explosions of railroad tank cars and storage vessels containing liquid petroleum gas (LPG) and other materials. Although the visible effects appear similar, the phenomena are not exactly alike.

Kinds of Explosions

There are various schemes for classifying explosions. For the relatively common explosions produced by exothermic chemical reactions (sometimes called combustion explosions), there are two main types: deflagration and detonation.

A deflagration is an exothermic reaction that expands rapidly from the burning gases to the unreacted material by conduction, convection, and radiation. The combustion zone progresses through the material at a rate that is less than the velocity of sound. A deflagration may not always produce sufficiently rapid increases in pressure to produce an explosion.

A detonation is an exothermic reaction characterized by the presence of a shock wave in the material that establishes and maintains the reaction. A detonation usually results in sufficiently rapid increases in pressure to produce an explosion. The reaction zone expands at a rate greater than the speed of sound in the unreacted material.

A more detailed classification scheme considers a variety of phenomena. Classifications differ somewhat among sources.

Condensed Phase Detonations Condensed phase materials are high explosives and propellants. Detonation of such materials has occurred during manufacturing, transportation, storage, and use. Certain compounds decompose almost instantly in a violent reaction. Most produce hot gases. Rapid decomposition can occur with acetylene, hydrogen, and certain metallic azides. For example, in 1947 in Texas City, Texas, *S. S. Grandcamp*, packed with well over 1,400 tons of ammonium nitrate fertilizer, accidentally ignited, exploded, and killed 400 people outright. People more than one mile away were injured and all houses within 0.9 miles were destroyed. Explosives are discussed further in the ensuing text.

Combustion Explosion of a Gaseous or Liquid Fuel in an Enclosure   These explosions can be divided into two groups based on the length-to-diameter ratio (L/D) of the container. Many buildings, ship holds, and boilers fall into this group. For containers with an L/D ≈ 1, there is a relatively slow rise in pressure. The overpressure causes the container to rupture.

In containers with large L/D ratios, such as pipes, certain buildings, or tanker ships, the dynamics of the flame front and the turbulence of the gases in front of it are important. As a flame propagates, it creates turbulence in front of it, and this turbulence improves mixing and expands the flame area and speed of travel. Pressures can increase very rapidly by as much as 20 times, achieving pressures of 15 to 20 atm. Damage often is greatest at points distant from the source of ignition. If the rapidly accelerating flame front and pressure wave are reflected from a surface, pressures may double again. This type of explosion can occur in compressed-air lines, where fuel from compressor lubricants and enriched oxygen usually are present.

Combustion Explosions of Dusts in an Enclosure   Dust explosions most often occur in containers where the dust is distributed in the atmosphere. They also can occur when some activity suddenly creates a cloud of airborne dust and a source of ignition is present. Similar to gas and vapor explosions, dust explosions show effects related to the L/D ratio of the container.

Boiling Liquid-Expanding Vapor Explosions   When a container holding a liquid at a temperature well above its boiling point ruptures, the liquid will evaporate or boil rapidly into a vapor state. The sudden expansion, a physical phenomenon, can throw parts of the container considerable distances. This phenomenon includes sudden releases of pressurized steam and may be accompanied by fire if the material is combustible, thereby producing thermal as well as physical effects. The heat of the fire can increase the rate of pressure rise and the pressure achieved. Boiling liquid-expanding vapor explosions (BLEVEs) involving LPG produce spectacular fireballs.

Explosions of Pressure Vessels Containing Nonreactive Materials   This type of explosion is closely related to a BLEVE. It is typical for a pressure vessel with a weak structure that will fail at quite low pressures (less than 2 lb/in^2 [gauge]). Materials are not thrown as violently as in a BLEVE. The pressure vessel itself may have severe damage from the rapid expansion of steam or from a combustion explosion within the vessel.

Unconfined Vapor Cloud Explosions   Unconfined vapor clouds are open-air concentrations of fuels in vapor form. The cloud can dissipate to a harmless condition (from a flammability viewpoint) in which the concentration is too low to burn or it can be ignited as it is released from a container, thereby burning off the fuel at a somewhat controlled rate, with little overpressure resulting. However, if the cloud is ignited and the flame accelerates rapidly enough, a dangerous blast wave can result.

Deflagrations of Mists   If fuels are dispersed in air in the form of a fog or mist and concentrations fall within flammable limits, ignition can produce violent deflagrations.

Chemical Reactor Runaway   Chemical reactions may create too much pressure for the container they are in. Inadequate cooling, insufficient stirring, too much catalyst, and other factors may cause the reaction to go out of control. In a sealed container, pressure increases may result from the reaction itself and from temperature increases following the Boyle-Charles law (see Chapter 19).

Nuclear Reactor Runaway or Nuclear Detonations   A nuclear detonation occurs when the structure of nuclei is rapidly rearranged by either fission or fusion. In an air detonation of a nuclear device, energy is converted into a blast wave, thermal radiation, and nuclear radiation. In the event of a nuclear reactor runaway, the rate of decomposition is much less than that of a weapon. The likelihood of a runaway reactor is very low because control systems included in the reactor prevent such an occurrence. However, the Chernobyl reactor explosion on April 26, 1986, was a runaway event for one type of reactor design. The explosion was caused by heat from the reaction, not from a nuclear explosion, such as that in a bomb.

17-2   EXPLOSION HAZARDS

There are three main causes of damage and injury from explosions. The most common cause of damage or injury is a blast wave or pressure wave that radiates from the explosion. Another cause of damage or injury is thermal radiation from combustion. A third source of damage and injury stems from projectiles (flying metal, glass, wood, etc.) created by the explosion. A blast wave impacting on a secondary object may produce additional projectiles.

Blast Wave Effects

A blast wave emanating from an explosion is called a free-field blast wave until it reaches an object and interacts with it. Sources of explosions that have very high energy and power densities produce ideal blast waves that have predictable properties. The blast wave decays with distance from the source. For high explosive materials, the distance for a blast wave is related to the cube root of the charge weight. In some cases, one can estimate the forces involved in an explosion from the fragment distribution pattern.

The interaction of a blast wave with an object is complex. The reflections can produce local pressures that are much higher than that of the blast wave itself. The object struck may bend or break. The damage produced is related to pressure, impulse (duration), and drag force. Table 17-1 lists typical damage resulting from various overpressures.

Thermal Effects

A combustion explosion produces a fireball. Radiation damage from the fireball is related to the size of the ball and its duration. Most fireballs reach a temperature on the order of 2,400°F. The radiant energy dissipates in relation to the distance squared, but accurate prediction of thermal damage is quite difficult.

TABLE 17-1 Blast Wave Effects Resulting from Overpressure

Approximate Overpressure (lb/in²)	Effect
0.5–1.0	Shatter glass
1.0	Knock a person down
2–3	Shatter 8–12 in block or concrete wall
5	Snap utility pole
7	Overturn loaded railroad car
11	Threshold for lung damage
15	50% of eardrums are ruptured

Scatter of Fragments

Fragment damage is also difficult to predict. The scatter depends on the size of the explosion and failure modes for materials. It is known that glass will fragment quite easily and scatter, whereas tougher, more ductile materials may not scatter as much. However, tougher materials may be thrown farther because they do not fragment. The orientation of a container (building, tank, etc.) can affect the likely points for failure. An internal pressure wave is likely to be greater along a long axis or orientation. Damage will be greatest at the ends of the axis, and fragments are likely to move in the direction of the internal pressure wave. The presence of venting and pressure relief devices can affect the degree of fragmentation and the scatter pattern. In some cases, the release of gases in an explosion can cause a container, such as a cylinder or tank, to act like a missile, particularly when there is a failure at one end of the container.

17-3 DUST EXPLOSIONS

Description

Dust explosions result when fine particles are dispersed in air and ignited. A flame front spreads rapidly through the contaminated air and pressure and temperature increase. Virtually all organic dusts, some inorganic dusts, and certain metallic dusts are combustible in air and can explode. Inert dusts, like limestone, are sometimes used as extinguishing agents.

Dust must be airborne to burn readily. Dusts can become airborne from some operation, or dust that has settled on equipment in ducts or on parts of structures can become airborne if disturbed. Dust explosions frequently occur as a sequence of explosions. A first explosion may cause accumulated dust to become airborne and result in a second explosion. A fire may start in a pile of dust on a hot piece of equipment, and a fire hose or extinguisher may throw the dust into the air, producing an explosion.

The severity of dust explosions is a function of the material. Compared with combustible dusts, oxidizing dusts can accelerate the combustion process. If oxidizing agents are mixed with combustible dust, the resulting explosion will be even more severe. Some materials will burn more readily than others, producing higher rates of pressure and temperature buildup.

Other properties of dusts that affect the likelihood of ignition and the severity of combustion include panicle size, concentration, oxygen presence, presence of impurities, moisture content, and air turbulence. Small particles tend to be easier to ignite, and fine dusts tend to produce higher rates of pressure rise during combustion. Similar to flammable gases and vapors, certain concentrations of dusts are combustible. Particle size affects the concentration required for combustion, and if a concentration is too low, combustion is not likely. The presence of inert material in a dust tends to reduce its ability to burn. As the moisture content of dust increases, the ignition temperature of the dust increases. The rate of combustion increases with partial pressure of oxygen. Therefore, the presence of inert gas can help reduce the likelihood of dust explosions. Combustion will be faster and an explosion more violent when a combustible dust and air are mixed turbulently.

Indices of Dust Explosion Hazard

Explosion properties of dusts are available in tables, such as those found in the *National Fire Code* and the *Fire Protection Handbook*. The Bureau of Mines developed some indices to represent the relative hazard of various dusts. There are three indices: ignition sensitivity, explosion severity, and explosibility index. The latter is the product of the first two. Ignition sensitivity and explosion severity are derived from laboratory tests in comparison with a standard Pittsburgh coal dust. The ignition sensitivity is a function of ignition temperature, minimum energy of ignition, and minimum concentration. Explosion severity is a function of maximum explosion pressure and maximum rate of pressure rise. The rating scale in Table 17-2 classifies the relative hazard of a dust. Table 17-3 lists the explosion properties of a few dusts.

17-4 CONTROLS FOR EXPLOSIONS

The appropriate controls for preventing an explosion or reducing the damage vary with each application. In the following text, controls are summarized. Training, procedures, and enforcement are also important. Today, one can also use specialized software to analyze the impact of explosion or the risk of damage to surrounding buildings and structure resulting from explosions of various types.

Limit Quantities of Materials

If the amount of materials that can explode is minimized in any location, the likelihood and degree of damage will be small should an explosion occur. The likelihood of an explosion also will be reduced in many cases, because there is less opportunity for ignition. Many examples could be given. OSHA and NFPA have rules about the amount of flammables and combustibles stored in work areas. Large amounts should be stored in remote areas. As detailed later in this chapter, standard reference tables prescribe limits on the amounts of explosive materials in one place. In locations where dust can accumulate, regular cleaning can reduce dangers of explosions. The design of such locations also can

TABLE 17-2   Relative Explosion Hazard of Dusts

Type of Explosion	Ignition Sensitivity	Explosion Severity	Explosibility Index
Weak	<0.2	<0.5	<0.1
Moderate	0.2–1.0	0.5–1.0	0.1–1.0
Strong	1.0–5.0	1.0–2.0	1.0–10
Severe	>5.0	>2.0	>10

TABLE 17-3 Explosion Properties of Selected Dusts

Dust Type	Explosibility Index	Ignition Sensitivity	Cloud Explosion Severity	Ignition Temperature (°C)	Minimum Cloud Ignition Energy (J)	Minimum Explosion Concentration (oz/ft^2)
Aluminum fines	>10	1.4	7.7	760	0.05	0.045
Benzoic acid	>10	5.4	2.1	620	0.02	0.03
Cellulose	2.8	1.0	2.8	480	0.08	0.055
Cellulose acetate	>10	8.0	1.6	420	0.015	0.04
Charcoal (hardwood mixture)	1.3	1.4	0.9	530	0.02	0.14
Cinnamon	5.8	2.5	2.3	440	0.03	0.06
Coal (Pittsburgh)	1.0	1.0	1.0	610	0.06	0.055
Corn	6.9	2.3	3.0	400	0.04	0.055
Grain dust	9.2	2.8	3.3	430	0.03	0.055
Magnesium (milled)	>10	3.0	7.4	560	0.04	0.03
Nylon polymer	>10	6.7	1.8	500	0.02	0.03
Peanut hull	4.0	2.0	2.0	460	0.05	0.045
Polystyrene molding compound	>10	6.0	2.0	560	0.04	0.015
Polyurethane foam, not fire retardant	>10	6.6	1.5	510	0.02	0.03
Rubber, crude, hard	7.4	4.6	1.6	350	0.05	0.025
Rubber, synthetic, hard, contains 33% sulfur	>10	7.0	1.5	320	0.03	0.03
Salicylanilide	5.8	4.1	1.4	610	0.02	0.04
Sugar, powdered	9.6	4.0	2.4	400	0.03	0.05
Sulfur	>10	20.2	1.2	190	0.015	0.035
Titanium	>10	5.4	2.0	510	0.025	0.045
Vitamin C (ascorbic acid)	2.3	1.0	2.2	460	0.06	0.07
Wheat flour	4.1	1.5	2.7	440	0.06	0.05
Wood flour (white pine)	9.9	3.1	3.2	470	0.04	0.035

help by minimizing places for dust to accumulate. Compressed air lines with traps and accumulators will minimize buildup of flammable oils in lines, and the lines should be cleaned or purged regularly to minimize fire and explosions within the pipes.

Prevent Combustible Concentrations

Processes and operations where flammable mixtures of gases, vapors, and dusts with air could occur should be designed and managed to prevent combustible and explosive mixtures with air. This principle includes reducing oxidizers that add to the explosive energy of a combustible material. For this reason, fuels and oxidizers are stored separately from each other. Monitoring equipment can help detect explosive concentrations and provide warnings or actuate ventilation or other equipment to reduce the hazards.

Eliminate Sources of Ignition

If there is a potential for an explosive atmosphere, sources of heat, flame, and spark must be kept away or carefully controlled. Some companies do not allow workers to carry any kind of smoking material, matches, or lighters on company premises. A hot work permit system is needed. Moving belts and other sources of static electricity must be grounded, and electrical equipment (switches, fixtures, motors, etc.) must be sealed to prevent a dusty atmosphere from reaching sources of sparks. Such equipment must meet electrical code requirements for explosive atmospheres. By itself, this means of prevention is not dependable. In many operations, such as those in grain elevators and factories, it is very difficult to eliminate every source of ignition.

Reduce Oxygen

Some environments that could produce explosive mixtures can be protected by keeping oxygen concentrations low. An inert gas, such as nitrogen, may be used to replace air. However, people cannot enter such environments without an independent source of breathing air. Oxygen concentration above standard atmospheric concentrations and compressed oxygen mixtures that elevate the partial pressure of oxygen and elevate combustion behavior should be avoided to reduce the potential for accelerated rates of burning.

Provide Overpressure Relief

In containers, including tanks and buildings, where there is a potential for explosive mixtures or releases, pressure relief devices should be in place. Venting is a passive means for minimizing damage from an explosion. In some cases, such as pressurized vessels and boilers, venting can prevent an explosion. Pressure relief valves, fusible plugs, and other devices are discussed further in Chapter 19. NFPA 68[1] provides guidance for explosion venting. In buildings, vents to relieve explosive pressures can be built into walls and roofs, and sashes, window panels, and doors also can be designed to relieve explosive pressures. The quantity and design for explosion relief depends on the rate of pressure rise expected, the L/D ratio of the container, the type of explosive material, and other factors.

Install Extinguishing and Suppression Systems

If a fire occurs that could lead to an explosion, extinguishing equipment may put out the fire and prevent an explosion. However, some kinds of extinguishing equipment for certain situations may add to the potential for explosion. For example, equipment that distributes dust into the air in a fire can lead to an explosion. There are some special suppression systems for explosions that are an active means for minimizing damage. They have to act very quickly to detect an explosion and minimize any effects. NFPA 69[2] gives details on explosion prevention systems, including suppression systems.

Use Distance and Barriers

One means to reduce the severity of an explosion is to separate quantities of materials by distance and barriers. A blast wave deteriorates with distance. A barrier that can withstand a blast wave will reduce the energy of the wave and will reduce the impact on objects shielded by the barrier.

Provide Remote Controls

If there are conditions where combustion can lead to explosion, there should be remote controls for valves and equipment. During normal operation, manual controls on equipment may not be hazardous, but if there is a danger of explosion, remote controls will avoid the need for operators to enter dangerous areas and operate valves usually.

17-5 EXPLOSIVES

An explosive is a chemical compound or mixture of substances used or intended for the purpose of creating a rapid self-propagating reaction and explosion. Energy is released in the form of heat and pressure. Explosive materials include explosives, blasting agents, slurries, and detonators.

Depending on the type of substance, there are various ways to initiate a reaction within an explosive. Ignition may be by fire, friction, concussion, percussion, or detonation.

Explosives are classified by the Department of Transportation (DOT) into classes.

> Class A. These explosives possess a detonating hazard. They are divided into many types, including black powder, low explosives, high explosives, dynamite, nitroglycerine, picric acid, lead azide, blasting caps, detonating primers, shape charges, and ammunition.

> Class B. These are explosives that, in general, function by rapid combustion rather than detonation. Included are fireworks, flash powders, pyrotechnic signal devices, some liquid or solid propellant explosives, some smokeless powders, and certain ammunition.

> Class C. These are certain types of fireworks or manufactured articles that contain class A or class B explosives or both as components in restricted quantities and certain types of fireworks.

Many organizations establish standards and regulations for explosives. Besides the DOT, the Bureau of Alcohol, Tobacco, and Firearms (ATF), OSHA, MSHA, and state and local governments have regulations on explosives. Standards are set by the Institute of Makers of Explosives, NFPA, and other organizations. A longstanding reference on explosives and their use is the *Blasters' Handbook*.

Hazards

Explosives are dangerous materials. They create damage from heat and pressure waves. The primary hazard is release of their energy in the wrong place and at the wrong time. The quantity and type of material activated determines the degree of damage for a particular place and time. Release in the presence of people can cause serious injury or death. The injury may be from the heat or pressure wave or from materials thrown by the pressure wave. Properly used, explosives can serve useful functions. They are essential for mining excavation, demolition, and similar activities.

Controls

Minimizing explosive hazards includes applying controls during manufacture, distribution, and use. Controls include training and licensing of handlers, users, and distributors.

Manifesting explosive materials from factory to final use minimizes their getting into the hands of unqualified users or being improperly used. Many states require that those involved in explosives be licensed after adequate training and examination.

Another control is proper storage. Blasting caps and detonating devices are not stored with explosives. Standards limit the quantities in magazines or in daily use and specify the distance certain quantities must be located from transportation routes and buildings. They also specify the distances between stored quantities in magazines. Most quantity-distance tables are based on the American Table of Distances for Storage of Explosives (see Table 17-4) developed by the Institute of Makers of Explosives.

Magazines, storage facilities for explosives, must meet design specifications. Design criteria include fire resistance, impedance to firearms, physical security, location, and other factors. Magazines must meet standards that vary with the quantity of material stored. Distances between magazines or between a magazine and a building, highway, or railway may be lower when there is a proper natural or artificial barricade, such as a mound of earth or wall of timber that will prevent material from an exploding magazine from affecting the adjacent structures, highway, or railway.

Implosions

An implosion is a sudden, inward collapse of a building or closed container when the external pressure is greater than that inside the container sufficient to produce structural failure in some element of the structure. An implosion can also be the inward collapse of a building or structure during controlled demolition using explosive charges that create failure in structural elements (columns, beams, etc.).

If an implosion is planned, humans should be barricaded and removed from any area in which debris could create a hazard. For structures containing hazardous materials, demolition procedure may require removing the hazardous materials before the implosion to reduce risks to surrounding areas.

TABLE 17-4 American Table of Distances for Storage of Explosives (Abbreviated)[a]

Quantity of Explosives (lb) Over-Less Than	Minimum Distance between Magazines (ft)	Distance from Inhabited Building (ft)	Distance from Public Highway (ft)	Distance from Passenger Railways and High-Traffic Highways[b] (ft)
2–5	12	140	60	102
10–20	20	220	90	162
40–50	28	300	120	230
100–125	36	400	160	300
200–250	46	510	210	378
500–600	62	680	270	506
1,000–1,200	78	850	330	636
2,000–2,500	98	1,090	380	816
5,000–6,000	130	1,460	470	1,092
10,000–20,000	164	1,750	540	1,374
20,000–25,000	210	2,000	630	1,752
50,000–55,000	280	2,000	880	2,000
95,000–100,000	370	2,000	1,090	2,000

[a] Distances shown are for unbarricaded magazines. They may be reduced by $1/2$ for quantities under 10,000 lb, if properly barricaded. For large quantities, allowable reductions in distance vary.

[b] More than 3,000 vehicles per day.

If a container could experience implosion because of a process failure, changes in the process design, or in the structural elements may be needed to reduce or eliminate risks from release of process materials or damage to surrounding structures or containers.

EXERCISES

1. An industrial plant is 750 ft wide (along the south and north) and 350 ft deep. It is located on a rectangular lot that is 1,750 ft wide (along the south and north) and 1,500 ft deep. The lot is bordered on the south by a highway, on the west by a rail line (passenger service), on the east by a residential area, and the north by an uninhabited flood plain of a stream. The plant sets back 150 ft north of the highway to provide a parking lot and is offset 250 ft from the west lot line for a shipping and receiving area. Fifteen pounds of explosives are to be used daily in the plant and a 60-day supply is to be maintained at the site. Prepare a layout of the site and using the American Table of Distances, show which areas on the site can be used for

    (a) a daily-use magazine

    (b) a 60-day supply

2. Cinnamon dust has an ignition sensitivity of 2.5 and an explosion severity of 2.3. What is its explosibility index?

REVIEW QUESTIONS

1. Briefly describe each class of explosion in the following list. Note any unique characteristics for each class.

    (a) combustion explosion

    (b) deflagration

    (c) detonation

    (d) condensed phase detonation

    (e) combustion explosion of a gaseous or liquid fuel in an enclosure

    (f) combustion explosions for dust in an enclosure

    (g) BLEVE

    (h) explosions of pressure vessels containing nonreactive materials

    (i) unconfined vapor cloud explosions

    (j) deflagrations of mists

    (k) chemical reactor runaway

    (l) nuclear reactor runaway or nuclear detonations

2. What are the three main causes of damage and injury from explosions?

3. Describe factors that contribute to the occurrence and severity of dust explosions.

4. What three indices are used to characterize the explosion hazard of dusts?

5. What controls help prevent explosions?

6. What controls reduce damage resulting from explosions?

7. Describe characteristics of each class of DOT explosives.

(a) Class A

(b) Class B

(c) Class C

8. What are the hazards of explosives?

9. What controls reduce the hazards of explosives?

NOTES

1 NFPA 68, Venting of Deflagrations.

2 NFPA 69, *Explosion Prevention Systems.*

BIBLIOGRAPHY

BARTKNECHT, W., *Explosions—Course, Prevention, Protection*, H. Burg and T. Almond, transl., Springer, New York, 1981.

Blasters' Handbook, E. I. DuPont de Nemours & Co., Wilmington, DE, current edition.

CONKLING, J. A., *Chemistry of Pyrotechnics*, Marcel Dekker, New York, 1985.

Fire Protection Handbook, 19th ed., National Fire Protection Association, Quincy, MA, 1986.

HARRIS, R. J., *The Investigation and Control of Gas Explosions in Buildings and Heating Plants*, 3rd ed., Methuen, New York, 1983.

NAGY, J., and VERAKIS, H. C., *Development and Control of Dust Explosions*, Marcel Dekker, New York, 1981.

National Fire Protection Association Codes, National Fire Protection Association, Boston, MA:

NFPA 68 Venting of Deflagrations

NFPA 69 Explosion Prevention Systems

NFPA 77 Static Electricity

NFPA 495 Explosive Materials Code

Natural Gas Safety Handbook for Utility Workers and Contractors, National Safety Council, Itasca, IL, 2003.

STREHLOW, R. A., "Accidental Explosions," *American Scientist*, 68:420–428 (1980).

CHAPTER *18*

HEAT AND COLD

18-1 INTRODUCTION

Humans tolerate a limited range of thermal environments. At one extreme, there is excessive cold and low temperatures; at the other, there is excessive heat and high temperature. Only a narrow region in the middle is thermally comfortable. The farther thermal conditions deviate from the region of comfort, the greater the likelihood of injury and the faster injury will occur.

Humans, like other warm-blooded animals, have internal thermal regulation systems. The rate at which metabolic heat is produced in the body must be balanced by the rate at which heat is lost to the environment. If heat is lost too quickly, one becomes cold; if heat is lost too slowly or is added to the body, one becomes hot. The body has limited means for adjusting the rate at which heat is lost. The rate of cooling is increased by sweating and more blood flowing near the skin. To prevent heat loss, peripheral blood flow is reduced and shivering occurs.

Heat and cold injuries and illnesses are related to the ability of the body to transfer heat to or from the environment or objects the body contacts. The thermal environment can create heat-exchange problems for the entire body (heat stress or cold stress conditions) or for local areas of the body. Thermal injuries and illnesses resulting from excessive heat and high temperatures are more frequent than those associated with cold conditions.

18-2 HEAT TRANSFER

Heat exchange between the body and the environment primarily involves convection, radiation, and evaporation. Conduction is another method of heat transfer, but it is of little significance in air environments. However, in an underwater environment, it is the dominant method of heat transfer. Conduction is also important when the body contacts an object of extreme temperature.

For whole-body heat exchange, metabolic heat, M, must be balanced with the environment through convection, C, radiation, R, and evaporation, E. Heat exchange between the body and the environment can be expressed in simplified form as

$$M \pm C \pm R - E = O. \tag{18-1}$$

Some heat is gained or lost through storage in the body mass, as evidenced by increases or decreases in body temperature. A precise heat balance equation also would include expressions for change in body mass and the resultant heat gain or loss. Body mass increases through intake of food and drink and decreases through excretion and urination. A precise balance would include an expression for peripheral blood flow and an expression for evaporative loss through respiration. It also would recognize that the body is seldom in a steady-state condition. A person's activity and the resulting heat produced by metabolism changes frequently. Changes in blood flow and sweating occur often. The heat balance equation would have to be adjusted for different individuals. The effect of clothing, acting as insulation, would need to be included. Nevertheless, Equation 18-1 provides the basis for understanding the heat transfer concept. It allows us to make quantitative estimates of what occurs in a thermal environment. It allows one to compute with reasonable accuracy whether an environment is likely to cause illness or injury.

Metabolism

The rate at which heat is produced in the body is determined by the activity being performed. Cells in the body burn oxygen and nutrients in performing their functions, and heat is produced in the chemical process of combustion. In general, the body is inefficient at converting fuel energy to work. As a result, most metabolic energy is converted to heat. A minimum amount of cellular activity is required just to maintain life. Cells produce more heat with increased activity, and the total amount of heat produced by the body is determined by the activity of the body. During sleep, the body of an average person burns approximately 70 to 75 kcal/hr and converts it to heat, whereas during very heavy exercise or work, 720 kcal/hr or more of heat may be produced. Typical values of oxygen consumption for various activities are listed in Table 18-1.

Convection

Convection is the transfer of heat by movement of air over the surface of a body. One empirically derived equation for convective heat transfer between a human body and the environment in warm-to-hot environments is

$$C = 1.0 \ V^{0.6}(T_a - T_s), \tag{18-2}$$

where

C is convection heat transfer (kilocalories per hour),

V is air speed (meters per minute),

T_a is air temperature, dry bulb (degrees Celsius), and

T_s is skin surface temperature (degrees Celsius).

This is an approximation and, although this equation may not be accurate for all thermal conditions, it helps us understand the convective component in Equation 18-1. Equation 18-2 assumes a surface area for the human body of $1.8 \, m^2$, which is an average value, and it needs to be corrected for actual body surface area and the effect of clothing when more precise calculations are necessary.

From Equation 18-1, it should be noted that convection, C, can add or remove heat from the body. The direction of heat transfer and the appropriate sign in Equation 18-1 can be established by noting the temperature difference between air and the body surface (see Equation 18-2). If the skin temperature is higher than the surrounding air, heat will be removed from the body. If the air temperature is higher then skin temperature, heat will be added to the body, adding to the burden of metabolic heat that must be removed through radiation or evaporation to maintain a constant body temperature.

TABLE 18-1   Metabolic Costs (Oxygen Consumption) for Selected Activities

Activity	Cost (kcal/hr)
General	
Light work	Up to 200
Moderate work	200–350
Heavy work	350–500
Resting	
Sleeping	70–75
Sitting quietly	80–100
Standing, relaxed	110
Walking, running	
Walking on the level	
3.2 km/hr	190
4.0 km/hr	230
4.8 km/hr	265
5.6 km/hr	300
6.4 km/hr	350
Running, 11.3 km/hr	810
Work	
Desk work	115
Driving a car	
Light traffic	80
Heavy traffic	190
Sheet metal work	200
Carpentry	230
Truck and automobile repair	250
Welding	180
Shoveling	410
Sweeping floors	235
Cleaning windows	225
Sawing wood by hand	480
Recreation	
Volleyball	210
Tennis	425
Swimming	400–550
Dancing, moderately	150
Basketball	515

Adapted from Conzolazio, C. F., Johnson, R. E., and Pecora, L. J., *Physiological Measurements of Metabolic Functions in Man*, McGraw-Hill, New York, 1963.

Air speed affects the rate of heat transfer by means of convection. Whether heat is being added or removed, a fourfold increase in air speed will approximately double the rate of heat transfer. Note that having a fan blow air over the body when the air temperature is higher than skin temperature actually adds heat to the body by convection. As we will see, increased air speed also can affect the rate of evaporative loss, which may produce a net heat loss when air temperature is higher than skin temperature.

The temperature of the skin is not constant with time and is not uniform over the entire body surface. Vasoconstriction and vasodilation influence cutaneous blood flow and skin temperature. Air temperature, too, affects the skin temperature. Because Equation

18-2 requires a single estimate of skin temperature, a value of 35°C is used for heat stress conditions.

Radiation

The body exchanges heat with its surroundings through radiation. Thermal radiation is highest for wavelengths in the infrared region. Radiation involves the geometric relationship between the surfaces of two bodies. The surface of one body must be able to "see" the other surface for radiant energy to be exchanged. For infrared radiation, air is essentially transparent.

Radiative heat transfer is a function of the fourth power of the absolute temperature of the surfaces involved. An approximation used for heat transfer from the human body in warm-to-hot environments is

$$R = 11.3(T_w - T_s), \tag{18-3}$$

where

R is radiation (kilocalories per hour),

T_w is mean radiant temperature of the solid surroundings (degrees Celsius), and

T_a is skin surface temperature (degrees Celsius).

Although Equation 18-3 is accurate in a limited range of thermal conditions, like Equation 18-2, it helps us to understand the radiative heat transfer component of Equation 18-1.

Like convection, radiation can add heat to the body or remove it. If the mean radiant temperature of the surrounding surfaces (usually walls) is higher than the skin temperature, heat will be added. If the skin temperature is greater than the mean radiant temperature of the surrounding surfaces, heat will be removed. By carefully noting the direction of heat transfer in Equation 18-3, the appropriate sign for R can be inserted in Equation 18-1. In Equation 18-3, a single estimate of skin temperature is required. A value of 35°C is commonly used for heat stress conditions.

Evaporation

Humans have the capability to sweat as a means for cooling the body. Sweat glands in the skin secrete sweat, which is primarily water containing some dissolved salts. Sweat gland activity is controlled by the hypothalamus in the brain. The level of physical activity and other factors influence the number of sweat glands active at one time. Sweat increases as the thermal regulation system in the body requires increased cooling to remove heat. As the water evaporates from the skin, cooling results from the change of state from liquid to vapor. There is considerable variation in sweat rates among individuals.

The maximum amount of cooling that can be achieved through sweating is a function of air speed and the ability of the surrounding air to accept additional moisture. An estimate of maximum cooling capacity for warm-to-hot environments is

$$E_{max} = 2.0V^{0.6}(PW_s - PW_a), \tag{18-4}$$

where

E_{max} is the maximum evaporative heat loss (kilocalories per hour),

V is the air speed (meters per minute),

PW_s is the vapor pressure of water at skin temperature (millimeters of mercury), and

PW_a is the vapor pressure of water at air temperature (millimeters of mercury).

Equation 18-4 helps us understand evaporative heat loss, even if it may not be accurate for all thermal conditions. In hot, humid conditions, the vapor pressures in air and at the skin surface are nearly the same. Cooling through evaporation of sweat then is limited by the environment. In hot, dry environments, the difference in vapor pressures is large and evaporation is rapid. In hot, dry conditions where evaporation is rapid, the actual cooling of the body may be limited by the maximum rate at which sweat is produced.

Vapor pressures can be determined through the use of a psychrometric chart. Dry-bulb air temperature and wet-bulb air temperature or relative humidity must be known. A psychrometric chart is provided in Figure 18-1.

Clothing and Insulation

Clothing slows down the rate of heat transfer between the body and the environment. In most cases, clothing is used as insulation to slow the loss of heat from the body. Various fabrics have different insulation value. Aluminized reflective clothing may help to reduce radiant heat gain from intense radiant sources, such as a fire or open flame operation. Fabrics that inhibit moisture loss may eliminate evaporation as a means of heat loss when the air inside the fabric becomes saturated with moisture. Equipment, such as vortex coolers and water-cooled underwear, may be required to remove heat from the "microclimate" inside some types of clothing assemblies.

Heavy or restrictive clothing may add to thermal problems by increasing the metabolic work required to move the clothing. An activity that is considered light work may become heavy work when special or heavy clothing is worn. Restrictive clothing may make wearers less agile, and loose clothing is more likely to become caught in equipment

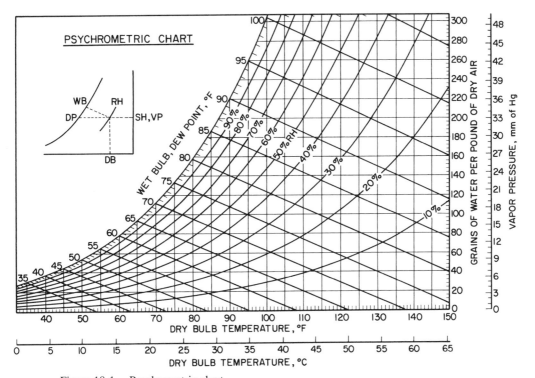

Figure 18-1. Psychrometric chart.

and machines. Special clothing may be used to insulate a person from local heat transfer problems, such as burns from hot objects and hot particles or frost injury from very cold objects.

The insulation value of a clothing assembly is given in units called clo. Clo values for clothing assemblies are determined by measuring the heat transfer from a heated manikin in controlled laboratory conditions. Table 18-2 provides clo values for some clothing assemblies.

For cold environments, the amount of clothing required can be estimated from

$$I_{req} = \frac{13.3(T_s - T_a)}{M},\tag{18-5}$$

where

I_{req} is the clothing insulation required (clo),

T_s is the skin surface temperature (degrees Celsius),

T_a is the air temperature (degrees Celsius), and

M is the metabolic rate (kilocalories per hour).

Equation 18-5 assumes a body surface area, SA, of $1.8 \, m^2$ as an average value. Adjustments in the body surface area may be required for accuracy. Body surface area is computed from the empirical equation of DuBois:

TABLE 18-2   Insulation Value of Different Clothing Assemblies

Clothing Assembly	I_{cl} (clo)
Nude	0
Shorts	0.1
Typical tropical clothing assembly	
Shorts, open-neck shirt with short sleeves, light socks, and sandals	0.3–0.4
Light summer clothing	
Long light-weight trousers, open-neck shirt with short sleeves	0.5
Light working assembly	
Athletic shorts, woolen socks, cotton work shirt (open-neck), and work trousers, shirttail out	0.6
U.S. Army fatigues, men's	
Light-weight underwear, cotton shirt and trousers, cushion sole socks, and combat boots	0.7
Typical business suit	1.0
Typical business suit plus cotton coat	1.5
Light outdoor sportswear	
Cotton shirt, trousers, t-shirt, shorts, socks, shoes, and single-ply poplin (cotton blend) jacket	0.9
Heavy traditional european business suit	
Cotton underwear with long legs and sleeves, shirt, woolen socks, shoes, suit including trousers, jacket, and vest	1.5
Heavy wool pile assembly	
Polar weather suit	3–4

From Fanger, P. O., *Thermal Comfort*, Danish Technical Press, Copenhagen, 1970.

**TABLE 18-3   Approximate Insulation Value of Air Film at
Different Air Speeds**

I_f (clo)	Air Speed (km/hr)
0.7	Calm
0.5	1.8
0.3	6.0
0.1	16.1
Negligible	65.0

$$SA = 71.84 \times 10^{-4}(W^{0.425})(H^{0.725}), \qquad (18\text{-}6)$$

where

SA is the body surface area (square meters),

W is the weight of the person (kilograms), and

H is the height of the person (centimeters).

Skin temperature for comfort in the cold is approximately 33°C. This value can be used as an average skin temperature in Equation 18-5, although exposed skin would have a lower temperature.

The available insulation is made up of the insulation value of the clothing assembly plus a film of still air at the outer surface of the clothing,

$$I_a = I_{cl} + I_f, \qquad (18\text{-}7)$$

where

I_a is the available insulation (clo),

I_{cl} is the insulation value of a clothing assembly (clo), and

I_f is the insulation value of the film of still air at the clothing surface (clo).

The film of still air and its insulation value diminishes rapidly as air speed increases. Table 18-3 lists some typical insulation values for the film of air at different air speeds.

For a known environment and activity, the insulation that must be provided by a clothing assembly can be determined by combining Equations 18-5 and 18-7 and solving for I_{cl}:

$$I_{cl} = I_{req} - I_f. \qquad (18\text{-}8)$$

18-3   HEAT

Hazards

The effects of high temperatures and hot environments on humans can be grouped into two categories: heat illnesses and burns. Heat illnesses mainly are caused by excessive exposures to hot environments. Burns result from contact with hot materials or surfaces or from excessive irradiance of the skin by heat-producing wavelengths of radiant energy.

Heat Illnesses

Excessive exposure to hot environments can result in behavioral changes, an elevated core temperature of the body (hyperthermia), failure of the temperature regulatory mechanism, circulatory failure, depletion of water and body salts, and inflammation of sweat glands. These and other symptoms characterize the heat illnesses described in the following text.

Heat Stroke or Sunstroke   This illness is primarily a failure of the thermal regulatory mechanism in the body and is manifested by the termination of sweating. The skin becomes hot and dry, the body temperature rises, and mental confusion, loss of consciousness, convulsions, or coma may result. Immediate and rapid cooling is required. Delays in treatment can be fatal.

Sunstroke is heatstroke resulting from excessive exposure to the sun.

Heat Hyperpyrexia   This illness is a milder form of heatstroke in which there is partial rather than complete failure of sweating. Some sweating may continue to occur. Other manifestations of heatstroke are also less severe.

Heat Syncope   Standing individuals who are not acclimatized to hot environments may faint because the redistribution of blood to peripheral tissue reduces the blood flow to the brain. The condition is corrected by removal to a cooler location and having the victim lie down.

Heat Exhaustion or Heat Prostration   Excessive loss of water through sweating and inadequate circulation may result in fatigue, nausea, headache, or giddiness. This illness is characterized by cold, clammy skin. Urine is concentrated and low in volume. The primary cause is inadequate water intake during exertion in hot environments. Victims can be treated by removing them to a cooler location, having them rest, and replacing body fluids.

Heat Cramps   This illness is characterized by painful muscle cramps during or after exertion in hot environments. The cramps are caused by excessive loss of body salts during sweating. Treatment involves replacing depleted body salts. Commercial fluids, sometimes called sports drinks, containing chemicals lost through sweating are useful.

Heat Rash   This disorder, often called prickly heat, is characterized by small, blister-like eruptions on the skin that have a prickly sensation during heat exposure. Sweat glands become plugged, sweat is retained, and inflammation results. The skin must be kept dry and heat exposure avoided during treatment.

Heat Fatigue   Some individuals, particularly those who are not acclimatized, may exhibit reduced performance of sensorimotor, mental, and vigilance tasks when exposed to heat. Other behavioral changes (reduced work effort, reducing clothing, or seeking cooler conditions) also may result as discomfort and physiological strain from hot environments are noted. In many cases, acclimatization and training may reduce behavioral effects to some extent.

Indices of Heat Stress

The idea of thermal comfort is based on subjective evaluations of thermal environments by room occupants. Individuals rate conditions as cold, cool, neutral, warm, or hot. An early thermal comfort scale was the Effective Temperature Scale. More recently, comfort information is based on Operative Temperature. The comfort zone for a given set of humidity, air speed, metabolic rate, and clothing insulation conditions is the range of operative temperatures that provide acceptable thermal environmental conditions or is the combination of air temperature and mean radiant temperature that people find thermally acceptable. ASHRAE publishes procedures for calculating operative temperature.[1]

Heat stress is concerned with hot environments outside the comfort zone, primarily those that result in physiological (blood flow, internal body temperature, sweating) changes. These changes are indicators of heat strain, a term referring to the physiological changes brought on by heat stress conditions of the physical environment.

A number of heat stress indices have been developed to predict whether exposures to hot environments will result in excessive heat strain. One index is the heat stress index (HSI). Another is the wet-bulb globe temperature index (WBGT). The HSI is probably a more accurate predictor of heat strain. However, instrumentation and calculations required for a WBGT assessment are much simpler. Because of its simplicity for field use, the WBGT is used more often. Both the HSI and WBGT decision tables apply to 8-hr exposures.

For weather purposes, the National Weather Service established the Heat Index (HI). It takes into account both heat and humidity to give an indication of how hot it feels to people when the air temperature is more than 57 degrees. The HI is computed from the following:

$$\begin{aligned} HI = {}&-42.379 + 2.04901523 \times T_a + 10.14333127 \times RH - 0.22475541 \times T_a \times RH \\ &- 0.00683783 \times T_a^2 - 0.05481717 \times RH^2 + 0.00122874 \times T_a^2 \times RH \\ &+ 0.00085282 \times T_a \times RH^2 - 0.00000199 \times (T_a \times RH)^2, \end{aligned} \tag{18-9}$$

where T_a is the air temperature (degrees Farenheit) and RH is the relative humidity (percent).

Heat Stress Index The HSI compares the amount of sweat that must be evaporated to balance the heat loss equation (Equation 18-1) for a given set of environmental conditions with the maximum amount of sweat that can actually be evaporated for those conditions:

$$HSI = E_{req} \times \frac{100}{E_{max}}, \tag{18-10}$$

where

HSI is a dimensionless index number,

E_{req} is the evaporative heat loss required (kilocalories per hour), and

E_{max} is the maximum evaporative heat loss (kilocalories per hour).

E_{req} can be computed by combining Equation 18-1 with Equations 18-2 and 18-3; E_{max} is computed from Equation 18-4. In Equations 18-2 and 18-3, the skin surface temperature can be assumed to be 35°C.

The implications for various values of HSI are provided in Table 18-4. The decision of whether an environment is safe or whether corrective actions are needed can be based on an assessment of the environment in question using HSI and Table 18-4. Environmental conditions that result in values for HSI greater than those explained in Table 18-4 may exist, and they indicate that conditions are excessive for everyone. Values less than those in Table 18-4 indicate cold conditions to which HSI does not apply.

Wet-Bulb Globe Temperature Index The WBGT index was developed as a simple-to-use method for determining if military troops were likely to suffer from heat illnesses in hot environments. Only two or three measurements are needed: wet-bulb (static) tem-

TABLE 18-4 Heat Stress Index Implications[a]

HSI	Implications of 8-hr Exposures
−20	Mild cold strain.
−10	
0	No thermal strain.
+10	Mild to moderate heat strain. Small decrements in motor, mental, and vigilance tasks.
20	Little effect for performance of heavy physical work.
30	
40	Severe heat strain. A threat to health unless workers are physically fit. Unacclimatized
50	workers need a break-in period. Medical selection of workers desirable to screen out
60	those with existing physiological impairments.
70	Very severe heat strain. Only a small percentage of the population can be expected to
80	qualify for this work. Special protective measures needed to prevent heat illnesses from
90	occurring.
100	The maximum strain tolerated daily by fit, acclimatized young men.

[a] Adapted from Belding, H. S., and Hatch, T. F., "Index for Evaluating Heat Stress in Terms of Resulting Physiological Strains," *J. Amer. Soc. Heating and Ventilating Eng.*, 27:129ff (1950).

perature (WB), dry-bulb temperature (DB), and globe temperature (GT). WBGT values are computed from one of two equations, depending on the presence of a solar load:

$$\text{WBGT} = 0.7\text{WB} + 0.2\text{GT} + 0.1\text{DB} \text{ (with a solar load)} \tag{18-11}$$

and

$$\text{WBGT} = 0.7\text{WB} + 0.3\text{GT} \text{ (with no solar load).} \tag{18-12}$$

If a person is exposed to a sequence of differing thermal environments during an 8-hour period, an average WBGT value can be computed from

$$\text{WBGT}_{\text{avg}} = \frac{(\text{WBGT}_1 \times t_1) + (\text{WBGT}_2 \times t_2) + \cdots + (\text{WBGT}_n \times t_n)}{t_1 + t_2 + \cdots + t_n}. \tag{18-13}$$

Permissible heat exposure threshold limit values (TLVs) have been established by the American Conference of Governmental Industrial Hygienists (ACGIH)[2] and are presented in Table 18-5. Refer to the recommendations for additional details. Knowledge of the metabolic activity (see Table 18-1) and the WBGT allows a person to decide from the TLVs if adjustments in work are needed. The TLVs assume a person is wearing light summer clothing (long-sleeved shirt and pants with reasonable air flow), has adequate water and salt intake, and has a deep-body temperature at or below 38°C. The thermal conditions in the rest area are assumed to be roughly the same as the work area.

If a worker is wearing clothing that is heavier than a light summer assembly, the computed WBGT is adjusted upward depending on the clothing assembly. For woven cloth overalls, 3.5° are added to WBGT and 5° are added for double-cloth overalls. The use of Table 18-5 and these adjustments are not applicable to encapsulating suites or garments that are impermeable or highly resistant to water vapor or air movement through the fabric.

Full acclimatization results when someone has experience continued physical activity under heat-stress conditions similar to those anticipated for a task. Typically, full acclimatization requires up to 3 weeks of exposure. Table 18-5 differentiates between acclimatized and unacclimatized workers.

TABLE 18-5 Screening Criteria for Heat Stress Exposures[a,b] (°C, WBGT)

	Acclimatized				Unacclimatized			
Work Demands	Light	Moderate	Heavy	Very Heavy	Light	Moderate	Heavy	Very Heavy
100% work	29.5	27.5	26		27.5	25	22.5	
75% work, 25% rest	30.5	28.5	27.5		29	26.5	24.5	
50% work, 50% rest	31.5	29.5	28.5	27.5	30	28	26.5	25
25% work, 75% rest	32.5	31	30	29.5	31	29	28	26.5

[a] From *Threshold Limit Values for Chemical Substances and Physical Agents & Biological Exposure Indices*, American Conference of Governmental Industrial Hygienists, Cincinnati, OH, 2004.
[b] Refer to the publication for additional details related to implementation.

Burns

The body is capable of removing heat from the skin or other body tissue at a rate that is mainly related to blood flow through the tissue. If heat is added to local tissue through contact with a hot object (conduction), from exposure to sources of high radiant energy (radiation), or from exposure to hot air (convection) at a rate that is faster than the removal rate, the tissue must store the energy. The tissue temperature will rise, resulting in discomfort, pain, or tissue damage. If the heat transfer rate is high, the elevation of local skin temperature can occur in seconds or less.

Human skin has a reflectance ranging from 5% up to 70% for wavelengths in the region from $0.2–2.0 \times 10^{-6}$ m. This reflectance depends to some extent on skin color. Above and below these wavelengths, the skin acts essentially as a black body and absorbs all radiant energy. Radiation in the microwave and infrared range causes heating of tissue.

Burn Classifications Burn classifications describe the depth of tissue that has been damaged. The outer layer of the skin is called the epidermis and the inner layer, containing hair follicles and sweat glands, is called the dermis. Below the skin, in the subcutaneous region, is a network of blood vessels serving the skin.

In the past, burns have been classified as first, second, or third degree, based mainly on visual characteristics of the wound. More recently, burns are classified as partial-thickness or full-thickness burns. These two classifications schemes are detailed in Table 18-6.

Other classifications of burns account for the portion of the body surface area injured, the part of the body injured, age of the injured person, and other factors. Beside thermal burns, other kinds of burns can occur. Electricity flowing over or through the body causes a variety of injuries, including thermal damage. Contact with chemicals may produce damage to the skin and other tissues that are classified as chemical burns. Electrical and chemical injuries are discussed elsewhere. Contact with very cold, cryogenic liquids also can cause tissue damage similar to that resulting from high temperatures.

Heat and Temperature The normal internal temperature of the body is approximately 37.5°C, whereas the skin temperature is approximately 35°C. Elevation of tissue temperatures will result in damage or destruction, depending on the length of time the tissue remains at an elevated temperature. When the total heat transfer is such that the core temperature of the body is elevated for a sufficient length of time, death can result, even though

TABLE 18-6   Burn Depth Classifications

Degree of Burn	Wound Thickness	Characteristics
First	Superficial	Erythema (reddening), pain; partial healing occurs in 5–10 days
Second	Deep	Blisters, pain; partial healing occurs in 2 weeks to 1 month
Third	Full	Skin destroyed, subcutaneous and possibly deeper tissue destroyed, lack of pain; extended period of healing

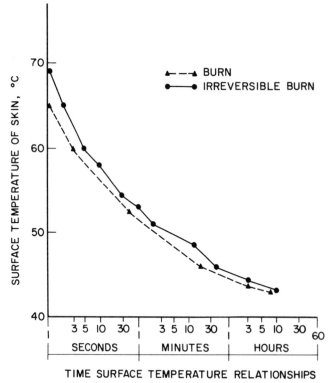

Figure 18-2.   Relationships between skin surface temperature and time of exposure required for burns. (Used with permission from Artz, C. P., Moncrief, J. A., and Pruitt, B. A., *Burns*, W. B. Saunders Co., Philadelphia, PA, 1979.)

the surface temperature does not produce pain. An example is immersion in a bath for an extended period at a temperature of 43°C. People have died of such an exposure in hot tubs. However, local tissue damage will occur when the skin surface temperature is 44°C or more. Pain occurs when the temperature reaches 45°C, and a skin temperature of 46°C is reported to be intolerably painful. The relationship between skin surface temperature and exposure time that will cause burns is shown in Figure 18-2. Figure 18-3 presents the relationship between irradiance of the skin and exposure time in terms of pain.

Controls

Heat stress and thermal injuries can be reduced or eliminated by controlling the source, modifying the environment, adjusting the work or activity, providing protective equipment, and meeting physiological and medical needs of workers.

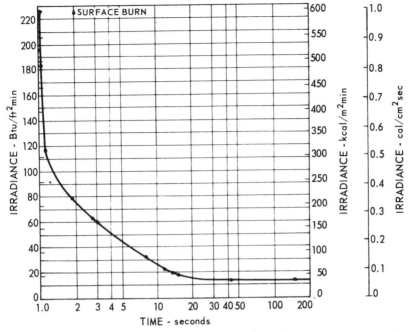

Figure 18-3. Relationship between irradiance and time of exposure for strong pain. (From Webb, P., *Bioastronautics Data Book*, NASA SP3006, National Aeronautics and Space Administration, Washington, DC, 1964.)

Controlling the Source Keeping heat sources and hot surfaces away from occupied areas will reduce the chance of heat buildup or contact. For example, placing an oven in a separate area can reduce the heat added to a work area. Insulation applied to a hot surface or object will prevent air from becoming hot and people from coming into contact with the hot surfaces. Not only is the hazard reduced, but in many cases, insulation conserves energy.

In some operations, materials and processes have temperatures that are higher than necessary. Reducing temperatures may reduce or eliminate thermal hazards.

Modifying the Environment The physical parameters involved in heat stress are air temperature, air velocity, mean radiant temperature, and vapor pressure. When these parameters cannot be modified by adjustments at the source, ventilation may reduce heat stress. Ventilation, the process of supplying air from some location other than the stressful environment, may be limited as a solution by the heat capacity and temperature of the ventilation air. An assessment of the supply air and thermal conditions must be made to estimate the effectiveness of ventilation. When ventilation is not effective, a final control is cooling of air with air conditioning equipment.

For environments with intense radiant sources, shielding may reduce the radiation reaching a person, because radiant energy travels in a straight line and air is not heated by it directly. In other cases, the distance between a person and the radiant source can be lengthened to reduce radiation levels, because radiation intensity diminishes with the square of the distance.

Barrier guards will prevent people from coming into contact with hot surfaces or becoming close to operations where hot material is found.

Adjusting Activities People can modify activities to reduce heat stress. Metabolic heat generation can be reduced by making work easier and by providing power tools and equipment to reduce the work effort required.

Another way to adjust the work is to limit the time of exposure to hot environments and to provide adequate periods of rest, both of which reduce the heat buildup in the body. The TLVs in Table 18-5 recommend this approach. Workers may be rotated through different jobs so that only a portion of their work is in stressful environments. That way productivity is not lost.

Providing Protective Equipment For hot environments, water-cooled clothing (helmets, underwear, and full uniforms) can be used. Air-cooled clothing, which relies on air lines and vortex devices for cooling, is usually less expensive and may provide sufficient cooling capacity in some environments. Pressurized air suitable for cooling is not necessarily safe for breathing.

In environments with intense radiation, reflective clothing may be helpful, and protective eye wear that reflects or filters harmful wavelengths may be needed. A wide variety of aluminized fabrics are available. Two types of reflective suits used for fires are called proximity suits and entry suits. Use of proximity suits is limited to approaching the heat source. Entry suits have an insulation layer inside to prevent contact burns and to reduce the rate of heat buildup. Without internal cooling, most full garments would create excessive conditions inside the suit in a short time.

Protective clothes may solve some heat problems and may add to others. In selecting protective clothing, one should remember that protective clothing does not remove the hazard; its effectiveness depends on the wearer's cooperation for effectiveness and may increase the metabolic rate for the activity. The metabolic work required to do physical work while wearing heavy clothing can be significantly higher than for the same activity without the clothing. Heat and moisture buildup inside a garment can create a hazardous local thermal environment that is not present outside the garment.

Insulated materials can be used to prevent burns that result from contact with hot objects or splashes of hot material or from heat transfer by conduction. Gloves, pads, and other kinds of insulated clothing items are available in fabrics with a variety of thermal characteristics.

Physiological and Medical Controls Heat strain can be reduced to some extent by ensuring that people replace lost body salts and water. Medical examinations may help identify those who are greater risks for heat stress because of age, physical condition, or existing health problems. High-risk people can be kept out of heat-stress environments. A program of acclimatization can also be used to improve to some degree the capacity of individuals to perform in hot environments. In extreme environments, continuous medical monitoring may be required to ensure that deep body temperatures do not exceed a limit of 38°C (as recommended by ACGIH and others) and that other physiological responses (heart rate, sweat rate) are not excessive.

18-4 COLD

The body has a very limited capacity to adjust for cold environmental conditions. Adjustments in blood circulation may conserve body heat, and shivering slightly increases metabolic heat production.

Hazards

Cold can produce local tissue damage and can reduce the core temperature of the body (hypothermia). Cold-related illnesses and diseases are described in the following text.

Trenchfoot When a person is exposed for several days to temperatures sensed as cold (but well above freezing), when the skin is moist, and when there is inactivity, trenchfoot is likely to occur. The prolonged exposure causes vasoconstriction in the feet and legs, which initially produces a pale appearance and numbness. After the initial paleness, swelling and pain occur. The numbness may continue for several weeks after the feet are warmed.

Chilblains An itching and painful reddening of the skin is caused by congestion of the capillaries when tissues are exposed to the cold. Symptoms are particularly found in exposed areas, such as the fingers, ears, or toes.

Cold Urticaria (Hives) In some people, coldness in tissue causes histamine to be released in the tissue, producing itchy, red blotches in the skin. In some cases, the exposure results in swelling of tissue and other histamine reactions (vomiting, rapid heart rate, swelling of breathing passages).

Frostbite When the temperature of tissue reaches or goes below the freezing point, frostbite results and tissue is damaged. The degree of damage depends on the depth of tissue frozen. Mild damage may result if only superficial layers are frozen. Skin is usually white or grayish yellow in appearance. Pain may be present, but often a victim is unaware of the frostbite.

Hypothermia A general lowering of the body temperature results in a variety of symptoms. As the core temperature is lowered from a normal temperature of approximately 37.5°C, shivering appears initially. Numbness, disorientation and confusion, amnesia, and impairment of judgment occur as core temperatures are lowered. When temperatures reach 26° to 30°C, unconsciousness, cardiac arrhythmia, muscular rigidity, ventricular fibrillation, respiratory arrest, and death are likely to occur. However, people have survived when the body temperature has reached lower values.

Indices of Cold Stress

There are few indices of cold stress. Some time ago, the National Safety Council published recommended exposure limits for cold environments (see Table 18-7). Earlier in this chapter, the effect of wind speed on the rate of convective cooling was noted. In cold environments, heat loss by convection is the most significant means of cooling. The effect of wind action on cooling has resulted in an indicator of cold stress called windchill, recently updated to Wind Chill Temperature Index. OSHA[3] uses a conceptual equation to help visualize danger from cold exposure called the Cold Stress Equation. The concept is: low temperature + wind speed + wetness = injuries and illnesses. The concept groups combinations of air temperature or water temperature (from wet or damp clothing) and wind speed into zones of danger: little danger, danger, and extreme danger. The combinations appear in Table 18-8.

TABLE 18-7 Exposure Limits for Cold Temperatures[a]

Temperature Range (°C)	Minimum Daily Exposure
−1 to −18	No exposure time limit if the person is properly clothed.
−18 to −35	Total cold-room work time: 4 hr; alternate 1 hr in and 1 hr out of the chamber.
−35 to −57	Two periods of 30 min each, at least 4 hr apart. Total cold-room work time allowed: 1 hr (*Note:* Some difference exists among individuals. One report recommends 15-min periods not more than four periods per work shift; another limits periods to 1 hr out of every 4, with a low chill factor, i.e., no wind; a third says that continuous operation for 3 hr at −54°C has been experienced without ill effect.)
−57 to −73	Maximum permissible cold-room work time: 5 min over an 8-hr working day. For these extreme temperatures, the wearing of a completely enclosed headgear, equipped with a breathing tube running under the clothing and down the leg to preheat the air, is recommended.

[a] Adapted from Alpaugh, E. L., *Fundamentals of Industrial Hygiene*, National Safety Council, Chicago, IL, 1971.

TABLE 18-8 OSHA Cold Stress Equation Zones (approximate)

Temperature, °F (°C)	Wind Speed (mph)			
	0–10	10–20	20–30	30–40
30 (−1.1)	LD	LD	LD	LD
20 (−6.2)	LD	LD	LD	LD
10 (−12.2)	LD	LD	D	D
0 (−17.8)	LD	D	D	D
−10 (−23.3)	LD	D	D	D
−20 (−28.9)	D	D	D	ED
−30 (−34.4)	D	D	ED	ED
−40 (−40)	D	ED	ED	ED
−50 (−45.6)	D	ED	ED	ED

LD = little danger (freezing to exposed flesh within 1 hr); D = danger (freezing to exposed flesh within 1 minute); ED = extreme danger (freezing to exposed flesh within 30 seconds).

Windchill

This index was derived from measurements of the rate of cooling of a container of water. The cooling power of wind at some temperature is compared with the equivalent cooling power of relatively still air at another temperature. The mathematical equation developed from the study (its validity has been questioned) is

$$\text{windchill} = [(100V)^{1/2} - V + 10.45] \times (33 - T_a), \tag{18-14}$$

where

> windchill is that part of the total cooling that is primarily the result of the wind action in cold environments (kilocalories per square meter per hour),
>
> V is the wind speed (meters per second),
>
> 10.45 is the arbitrary constant,
>
> 33 is the skin temperature (degrees Celsius), and
>
> T_a = air temperature (degrees Celsius).

	Actual Thermometer Reading - °F									
	50	40	30	20	10	0	-10	-20	-30	-40
Wind speed - mph	Equivalent Temperature - °F									
calm	50	40	30	20	10	0	-10	-20	-30	-40
5	48	37	27	16	6	-5	-15	-26	-36	-47
10	40	28	16	4	-9	-21	-33	-46	-58	-70
15	36	22	9	-5	-18	-36	-45	-58	-72	-85
20	32	18	4	-10	-25	-39	-53	-67	-82	-96
25	30	16	0	-15	-29	-44	-59	-74	-88	-104
30	28	13	-2	-18	-33	-48	-63	-79	-94	-109
35	27	11	-4	-20	-35	-49	-67	-82	-98	-113
40	26	10	-6	-21	-37	-53	-69	-85	-100	-116
Over 40 mph	Little Danger			Increasing Danger			Great Danger			
(little added effect)	(for properly clothed person)			(Danger from freezing of exposed flesh)						

Figure 18-4. Windchill chart.

A windchill chart showing equivalent temperatures for different wind speeds is provided in Figure 18-4. Calm means that the wind velocity is very low, but not zero.

In 2001, the National Weather Service published a new Wind Chill Temperature or Wind Chill Temperature (WCT) Index equation that is now used as a better estimate for warning people about the danger of cold air temperatures combined with wind. The new formula, based on exposure of the human face, is

$$\text{NWS Wind Chill Temperature Index} = 13.12 + (0.6215 \times T_a) - (11.37 \times V_{10}^{0.16}) + (0.3965 \times T_a \times V_{10}^{0.16}), \qquad (18\text{-}15)$$

where

WCT is degrees Celsius,

T_a is the air temperature (degrees Celsius), and

V_{10} is the wind speed at 10 meters (standard anemometer height), in kilometers per hour).

Compared with windchill temperatures, the WCT Index results in a slightly higher temperature for most wind speeds and is a more reliable indicator of potential danger.

Controls

Controls for preventing injury from cold environments include changes to the environment, adjustments in activities, and protective clothing.

Modify the Environment   For environments that can be controlled, it may be feasible to warm the air temperature or provide radiant heat sources. In some situations, reductions in wind speed may be possible by providing windscreens or enclosures.

Adjust Activities   The easiest way to protect someone from the cold is to minimize the duration of exposure. Frequent breaks to warm up may be needed and rests should be in warm environments.

Provide Protective Clothing   Adequate amounts of clothing can provide the insulation necessary to retain body heat (previously discussed in Section 18-2). The outer layer of clothing should provide a windscreen, but also allow moisture to escape. Sweating because of too much insulation or condensation of sweat in clothing should be avoided. Inner fabrics should absorb moisture so that the skin is not wet. Attention should be given to protection of body extremities (hands, head, and feet). Because the hands have a relatively large surface area and a large portion of the blood flow through the body goes to the head, both can have a high rate of heat loss. If practical problems of power supply, connections, and controls can be resolved, electrically heated clothing can be used.

When contacting, handling, or using solids and liquids at subfreezing temperatures, protective clothing should be worn to prevent freezing of local tissue or freezing of moist skin to frozen objects. Fabrics, clothing, and protective equipment should be selected to provide protection appropriate to the hazards involved.

18-5   MEASUREMENT

Before the degree of hazard for a thermal environment can be determined, measurements of conditions must be made. A variety of instruments are needed to assess the thermal environment.

Air Temperature

The simplest device for measuring air temperature (dry bulb) is a thermometer. Battery-powered, portable instruments with analog or digital readouts can also be used. Sensors are often thermocouple or thermistor devices.

Humidity

Relative humidity is difficult to measure directly. Although accurate humidity instruments are available, humidity levels often are determined through the use of dry-bulb and wet-bulb temperature readings and a psychrometric chart.

Wet-bulb temperatures are determined by placing a clean, wetted sock over the bulb of a thermometer and rapidly moving the bulb through the air (or blowing air over the bulb) to provide cooling by evaporation of the moisture from the sock. The temperature reading will be lower than a dry-bulb reading. Special kits are available for holding two thermometers (one wet-bulb and one dry-bulb) on a sling and whirling them around. This is a sling psychrometer. Other devices are stationary and air motion is provided by a fan that blows air over the wetted bulb. When making wet-bulb readings for WBGT, however, the wet-bulb thermometer must be held stationary.

Mean Radiant Temperature

The mean radiant temperature of surrounding surfaces can be determined by measuring the temperature and area of each surface and averaging them. This is a rather cumbersome and often impractical approach. An alternative is to use a black globe thermometer. The temperature inside a 15-cm (6-in) diameter hollow copper sphere, painted matte black on the outside, provides a good estimate of the mean radiant temperature. After a 25- to 30-min period in which the globe has been at a fixed location in an environment, a steady-state condition is reached inside the globe. The rate at which radiant energy is absorbed is balanced by convective heat loss from the sphere. The temperature inside the sphere can be measured by the sensor of any dry-bulb instrument located at the approximate center of the sphere.

Air Speed

Many different types of anemometers are available for measuring air speed. Desirable features for assessing thermal conditions are portability, durability, and ruggedness. Also important are having a nondirectional sensor, sensitivity at low speeds, and fast response time. Instruments with a heated sensing device (heated thermocouple, heated thermistor, hot wire) provide these features.

Body Temperature

For most hot and cold environments there is no need to monitor the body temperature. In extremely hot or cold environments and where people are confined or isolated from assistance, core temperature monitoring may be necessary to ensure that upper or lower limits are not exceeded and that personnel are removed from stressful conditions before severe physiological damage occurs. Activities may be restricted by direct leads between the person and the instrument, a difficulty that can be solved through the use of a telemetry system.

Another problem is tolerance of core temperature sensors by the wearer. Because oral temperature is not very accurate and reliable when measurement of core temperature is critical, rectal probes often are used. The least objectionable sensor is a tympanic sensor, which uses a thermistor bead or small thermocouple that is placed in the ear canal against or very near the eardrum and is held in place by a custom-molded ear plug. Tympanic temperature closely follows core temperature. Digital thermomemters also can add convenience for temperature readings.

EXERCISES

1. Estimate how long a person can be exposed to each of the following conditions before burns will result or strong pain occurs:

   (a) direct skin contact with a surface at 60°C

   (b) thermal radiation of 300 kcal/m²/min on the skin

2. With air movement at 1.46 km/hr and a relative humidity of 50%, what combination of dry-bulb and wet-bulb temperatures will be comfortable for most people?

3. Compute the heat stress index for the following conditions and determine if any precautions are warranted:

   (a) $M = 350$ kcal/hr, $V = 14$ km/hr, $T_a = 31$°C, $T_w = T_a$, relative humidity at the skin $= 100\%$, and relative humidity of the air $= 80\%$.

   (b) $M = 200$ kcal/hr, $V = 5$ km/hr, $T_a = 21$°C, $T_w = T_a$, relative humidity at the skin $= 70\%$, and relative humidity of the air $= 40\%$.

4. Compute the WBGT index for the following and recommend any controls that might be appropriate:

   (a) WB $= 23$°C, GT $= 35$°C, DB $= 31$°C, continuous heavy workload, solar load.

   (b) WB $= 15$°C, GT $= 40$°C, continuous light workload, no solar load.

5. Compute the time weighted average WBGT if a worker doing continuous light work is subjected to the following conditions as a sequence during an 8-hr workday:

   Condition 1: WBGT = 35°C, duration = 100 min

   Condition 2: WBGT = 24°C, duration = 20 min

   Condition 3: WBGT = 30°C, duration = 60 min

   Condition 4: WBGT = 40°C, duration = 40 min

6. Compute the windchill cooling rate for the following conditions and find an equivalent temperature for calm air (1 km/hr). What danger exists, if any, for each windy condition?

   (a) $V = 25$ km/hr, $T_a = 0°C$

   (b) $V = 10$ km/hr, $T_a = -10°C$

7. An employee is required to work in the cold. Determine the insulation required and estimate the insulation that must be provided by the clothing assembly for each set of conditions.

   (a) Hard work, $T_a = -15°C$, calm air

   (b) Moderate workload, $T_a = -25°C$, V = 6 km/hr

8. For the conditions in Exercise 7, estimate the adjustment in insulation required if the employee had the following height and weight:

   (a) height = 152 cm, weight = 39 kg

   (b) height = 191 cm, weight = 115 kg

9. Power line workers experience the following conditions while repairing electrical distribution lines following a storm: −10°F air temperature, 18 mph winds. Determine the degree of danger for these conditions.

10. For the conditions in Exercise 9, what is the WCT Index?

REVIEW QUESTIONS

1. Describe the fundamentals of heat exchange between the human body and the environment for

   (a) convection

   (b) conduction

   (c) radiation

   (d) evaporation

2. Identify the physical parameters involved in heat transfer from the human body.

3. What mechanisms can the body use for increasing heat loss? Discuss the limitations of each.

4. Discuss the advantages and disadvantages of clothing in terms of the impacts on thermal balance.

5. What effect does air motion have on heat exchange? How effective is air as an insulator?

6. Under what conditions will turning on a fan add to the heat of metabolism?

7. What is mean radiant temperature?

8. What injuries and illnesses can result from exposure to heat? Describe their symptoms.

9. What instrumentation is required for determining the heat stress index and the WBGT index? How are they different?

10. Describe the classifications schemes for burns and typical characteristics for each class.

11. What controls can be used to eliminate or reduce hazards from heat?

12. What is the maximum deep body temperature that should not be exceeded when working in hot environments?

13. What injuries and illnesses can result from exposure to cold? Describe their symptoms.

14. What is windchill? How is it different from Wind Chill Temperature Index?

15. What controls can be used to minimize or eliminate hazards from cold?

16. What instrumentation is required to measure

    (a) dry-bulb air temperature?

    (b) wet-bulb air temperature?

    (c) mean radiant temperature?

    (d) air speed?

    (e) body temperature?

17. Define

    (a) hyperthermia

    (b) hypothermia

18. What are the symptoms of the following disorders?

    (a) heat stroke

    (b) heat exhaustion

    (c) heat syncope

    (d) heat cramps

    (e) heat rash

    (f) trenchfoot

    (g) chilblains

    (h) cold urticaria

    (i) frostbite

NOTES

1 ANSI/ASHRAE Standard 55, *Thermal Environmental Conditions for Human Occupancy*, American Society of Heating, Refrigerating and Air-Conditioning Engineers, Inc., Atlanta, GA, 2004; and *ASHRAE Handbook—Fundamentals*, Chapter 8, American Society of Heating, Refrigerating and Air-Conditioning Engineers, Inc., Atlanta, GA, updated periodically.

2 *Threshold Limit Values for Chemical Substances and Physical Agents & Biological Exposure Indices*, American Conference of Governmental Industrial Hygienists, Cincinnati, OH, updated annually.

3 OSHA 3156, *The Cold Stress Equation*, U.S. Department of Labor, Occupational Safety and Health Administration, 1998.

BIBLIOGRAPHY

ARTZ, C. P., MONCRIEF, J. A., AND PRUITT, B. A., *Burns*, W. B. Saunders Company, Philadelphia, PA, 1979.

ASHRAE Handbook of Fundamentals, American Society of Heating, Refrigeration and Air Conditioning Engineers, New York, current edition.

ANSI/ASHRAE Standard 55-2004. *Thermal Environmental Conditions for Human Occupancy*, American Society of Heating, Refrigeration and Air Conditioning Engineers, Atlanta, GA, 2004.

BELDING, H. S., AND HATCH, T. F., "Index for Evaluating Heat Stress in Terms of Resulting Physiological Strains," *J. Amer. Soc. Heating and Ventilating Eng.*, 27:129ff (1955).

BURTON, A. C., AND EDHOLM, O. B., *Man in a Cold Environment*, Arnold, London, 1955.

FANGER, P. O., *Thermal Comfort*, Danish Technical Press, Copenhagen, 1970.

HARDY, J. D., ed., *Temperature, Its Measurement and Control in Science and Industry*, vol. 3, Reinhold, New York, 1963.

HARDY, J. D., GAGGE, A. P., AND STOLWIJK, J. A. J., *Physiological and Behavioral Temperature Regulation*, Charles C. Thomas, Springfield, IL, 1970.

PARSONS, KEN, *Human Thermal Environments: The Effect of Hot, Moderate and Cold Environments on Human Health, Comfort and Performance*, CRC Press, Boca Raton, FL, 2003.

Protecting Workers in Cold Environments, Fact Sheet No. OSHA 98–55, U.S. Department of Labor, Occupational Safety and Health Administration, December, 1998.

The Industrial Environment-Its Evaluation & Control, National Institute for Occupational Safety and Health, Cincinnati, OH, 1973.

WEBB, P., ed., *Bioastronautics Data Book*, NASA SP-3006, National Aeronautics and Space Administration, Washington, DC, 1964.

CHAPTER 19

PRESSURE

19-1 INTRODUCTION

We live in an environment in which the air in the atmosphere creates pressure around us. We experience changes in pressure when we fly in airplanes or climb mountains. Some people experience changes in pressure when they scuba dive or work in tunnels or caissons. The human body can function within a particular range of pressures, and it is limited in the rate of change it can tolerate. In explosions, there are rapid changes in pressure that can injure people and destroy buildings and other property. There are many products and processes involving elevated pressures where sudden or uncontrolled releases, or both, can cause injury and damage.

19-2 LOW-PRESSURE ENVIRONMENTS

Physics

Low-pressure environments are those that have a pressure less than sea level. (At sea level, standard atmospheric pressure is 760 mmHg or approximately 14.7 lb/in^2). The air around us is composed of oxygen, nitrogen, and small amounts of other gases. Our environment contains approximately 21% oxygen. As we go up in altitude from sea level, the pressure around us decreases (see Table 19.1). However, the gas mixture stays the same. Dalton's law of partial pressure states that the partial pressure, P, of any gas in a mixture (x), is equal to the total pressure, P_{tot}, times the percent of the gas in the mixture:

$$P_x = P_{tot} \times \%\text{gas} x. \tag{19-1}$$

For example at sea level, the partial pressure of oxygen P_{O_2} is

$$P_{O_2} = 760 \times 21\% = 160\,\text{mmHg}.$$

At altitudes above sea level, the total pressure is lower than that at sea level. The partial pressure of oxygen is also lower. Table 19-1 lists the pressure at altitudes and depths underwater.

Physiology

The partial pressure of oxygen affects the ability of the blood to transport oxygen throughout the body. The red cells in the blood perform the transport function. They contain hemoglobin, which forms a loose bond with oxygen. When we inhale, the red cells pick up oxygen from the lungs and release it to cells in the body as the blood circulates. During

Safety and Health for Engineers, Second Edition, by Roger L. Brauer
Copyright © 2006 John Wiley & Sons, Inc.

TABLE 19-1   Pressures at Altitude and Underwater[a]

Altitude or Depth Below Sea Level (ft)	Total Pressure (mmHg)	Total Pressure (lb/in²)	Partial Pressure of Oxygen for Standard Gas Mixture in Air (mmHg)
45,000	111	2.15	23.3
40,000	144	2.79	30.2
30,000	223	4.3	46.7
20,000	349	6.8	73.1
15,000	424	8.2	88.8
10,000	523	10.1	110
5,000	632	12.2	132
Sea level	760	14.7	159
100 underwater	3,040	58.8	637
200 underwater	5,320	103	1,115
300 underwater	8,360	162	1,751

[a] Table data assume fresh water and 33 ft of depth is 1 atm.

circulation, the hemoglobin also picks up carbon dioxide, a waste product of cellular metabolism, from the cells. The red cells return to the lungs, release carbon dioxide, and bond with oxygen once again. The carbon dioxide is exhaled.

The percent of the red cells actively bonding with and transporting oxygen is normally approximately 97% at sea level. However with increasing altitude, there is a reduction in the portion of red cells that are effectively transporting oxygen. This is the oxygen dissociation curve shown in Figure 19-1.

The body has some ability to improve oxygen transport at increased altitude by increasing breathing rate and heart rate. These forms of compensation are limited. The body also will produce higher concentrations of red cells in the blood, but this adjustment takes nearly 1 month to occur fully. People who live at high altitudes have a higher density of red cells in the blood than those who live at sea level. Athletes who must perform at high altitude often will train in such an environment for some time to allow the red cell adjustment to occur.

Hazards

Hypoxia is a lack of metabolic oxygen. A reduction in oxygen transport affects cell metabolism. One can express the oxygen deficiency in terms of altitude, saturation of red cells, or partial pressure. The effects are a result of the degree of hypoxia.

One of the first effects exhibited is loss of night vision. There are two kinds of receptor cells in the retina of the eye. One type (cones) senses color; the other type (rods) senses black and white. Rods are most sensitive at low light levels. As one goes up in altitude, the ability to see in low light levels is reduced. This effect begins to appear at approximately 6,000 ft.

As one progresses to higher altitudes, other effects are impaired memory, judgment and coordination, drowsiness, euphoria, syncope (unconsciousness), and death.

As one goes up in altitude, the middle ear vents through the Eustachian tube and pressure in the middle ear that is greater than the surroundings is reduced. When one moves rapidly from altitude to sea level, the surrounding pressure is higher than that in body cavities and venting of the middle ear is more difficult because the opening of the Eustachian

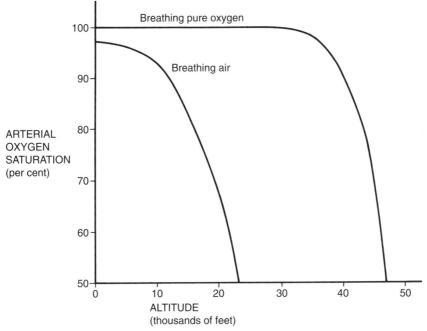

Figure 19-1.   The effects of oxygen deficiency caused by low-pressure atmospheres are delayed by increasing the oxygen content of a breathing gas mixture.

tube in the pharynx may seal shut. The increase in atmospheric pressure can create pain in sinuses blocked by mucus.

Controls

The primary objective is to maintain the partial pressure of oxygen near that at sea level. For reduced-pressure environments, one can accomplish this in two ways: increase the total pressure or increase the portion of oxygen in the breathing air.

There is a preferred range of partial pressures of oxygen. The lower limit is equivalent at standard sea level conditions to approximately 16% to 17% oxygen. The body's reaction to high levels of oxygen in air when it is inhaled for extended periods is called oxygen toxicity. At 60% oxygen mixtures at sea level, often the upper limit of recommended oxygen concentrations, people start to cough. At higher concentrations of oxygen, tolerance is reduced further.

Commercial jet aircraft fly at 25,000 to 40,000 ft. At these altitudes, adjustments in breathing air are essential. In most commercial aircraft, the total pressure is increased above that outside the aircraft and there is an emergency oxygen system in the event that cabin pressure is lost.

American manned spacecraft balance the demands for a suitable breathing atmosphere against the structural weight necessary to sustain the pressure difference between inside and outside the spacecraft. A cabin pressure of approximately 260 mmHg with 60% oxygen achieves a desired partial pressure similar to sea level. Higher total pressures require stronger and heavier structural elements for spacecraft flying in pressures that are near zero. Breathing 100% oxygen can extend the oxygen dissociation curve to new limits (see Figure 19-1). However, oxygen toxicity symptoms will appear with extended exposure.

One should also note that increasing the portion of oxygen in an environment elevates the rate of combustion should a fire start.

19-3 HIGH-PRESSURE ENVIRONMENTS

Physics

Pressures above those at sea level are encountered in underwater diving and in certain kinds of construction work. In tunnelling and caisson work, water may seep into the underground work area. So the work area is sometimes sealed and pressurized to minimize the water seepage and to avoid its interference with the work. Workers enter the pressurized area through air locks.

Hyperbaric chambers in which medical activities or other work are performed under elevated atmospheric pressure are other examples of high-pressure environments.

Dalton's law of partial pressure applies to high-pressure environments as well. Pressures under water increase at the rate of 1 atm for every 32 to 33 ft of depth, depending on the density of water. Salt water is slightly more dense than fresh water. Thus, a worker in a pressurized tunnelling project that is 33 ft below the surface is working at a total pressure of approximately 2 atm. Note that the pressure at the surface is already approximately 1 atm and pressure resulting from depth adds to it.

Another important phenomenon for high pressure environments involves Henry's law. This law says that gases will dissolve in fluids that are under pressure. When pressure surrounding the fluid is reduced, the gas escapes, typically forming small bubbles. We observe this when we open carbonated soft drinks that are under pressure.

Hazards

The allowable range of oxygen partial pressures also applies to high-pressure environments. If surface air is pumped under pressure to a diver. The portion of oxygen in the air must be reduced if the diver is more than 65 ft (3 atm) below the surface. Otherwise, the equivalent of the 60% oxygen limit at sea level is exceeded. At significant depths, controlling oxygen content of breathing gases becomes difficult. Limits become very tight. Enriched oxygen mixtures (partial pressures for oxygen above that at sea level) also increase dangers from fire.

Another problem is that of inert gases used in breathing mixtures. At high pressures, nitrogen induces narcotic effects (euphoria, drowsiness, and muscular weakness), called nitrogen narcosis. In deep diving, helium, another inert gas, often replaces nitrogen in breathing gases. Helium does not produce narcotic effects until much higher pressures are reached. However, helium does create communication and heat transfer problems. Speech sounds are at much higher frequencies in a helium atmosphere and convective heat loss increases compared with nitrogen. Contaminant gases in breathing mixtures may add other dangers.

As pressure increases, gas going into solution in the body according to Henry's law is not a problem. However, when someone returns from high-pressure to normal environmental pressure, there are dangers. The family of decompression disorders has various names: decompression sickness, dysbarism, the bends, or caisson's disease. Bubbles or embolisms may form in tissues or in the blood. Bubble formation in tissue frequently occurs in body joints and can be very painful. Bubbles can interrupt blood flow in the heart or lungs if they are large enough.

Controls

Breathing gases supplied to people in high-pressure environments must be well controlled for proper oxygen partial pressures. In addition, contaminant gases and particles from compressor equipment and lubricants may be dangerous and must be removed from breathing mixtures. Carbon monoxide, for example, can be particularly dangerous, because it bonds easily with hemoglobin and prevents red cells from providing adequate oxygen transport.

To prevent nitrogen narcosis, substitute other inert gases when pressures exceed that necessary for nitrogen narcosis onset. Also consider the adverse effects of alternate inert gases.

To prevent decompression sickness, one can either limit the time of high-pressure exposure or control the rate of decompression. Diving tables give information about allowable exposure time at various depths or pressures and the procedures and durations for decompression. Diving tables are based on diving experience. The U.S. Navy diving tables are the most authoritative. (Most other diving tables are derived from Navy tables.) OSHA[1] diving tables give total decompression time for exposures to certain pressures and durations. They also detail how many stages of decompression are needed, the length and pressure of each stage, and the total decompression time after exposures to particular pressures and durations.

There are other controls that are important for work in high-pressure areas and for diving. Divers must be medically qualified and have regular medical examinations. Diving procedures must be planned carefully and followed during predive, dive, and post-dive phases. Equipment must be tested and examined regularly. One must plan for emergencies and practice emergency procedures.

19-4   PRESSURIZED CONTAINERS

There are many kinds of pressurized containers in processes and products, ranging from aerosol cans to inflated tires, water heaters, tanks, compressed gas cylinders, and cookers. In explosions, even buildings and pipes can be pressurized containers. Pipes, pipelines, and process equipment are pressurized containers when they are pressurized to test for leaks and other faults.

Hazards

A container is no stronger than its weakest member, such as a joint, cover, seal, wall, or relief device. Corrosion or physical damage from handling may create a weak point. Containers that have pressurized contents may reach pressures too high for the container, any of its parts, or both. Consequently, the container may leak slowly or suddenly.

The pressure in a container may exceed its limits from exposures or failure of a relief device to function. For example, exposure to sun or some other heat source may increase the pressure in a container. The Boyle-Charles law helps us predict the pressure change:

$$\frac{P_1 V_1}{T_1} = \frac{P_2 V_2}{T_2},$$

$$(19\text{-}2)$$

where

P is the pressure (absolute),

V is the volume, and

T is the temperature (absolute),

and subscripts define the initial and final conditions.

Example 19-1 Suppose a gas cylinder is pressurized to 2,200 lb/in^2 (gauge) at 70°F. After exposure to direct sun, the temperature increases to 180°F. What is the pressure after exposure?

Because the volume is constant, the final pressure is

$$P_2 = \frac{P_1 T_2}{T_1} = \frac{(2200+15)(180+460)}{(70+460)}$$
$$= 2,675 \, \text{lb/in}^2 \, (\text{absolute}) = 2,660 \, \text{lb/in}^2 \, (\text{gauge}).$$

A leak in an unrestrained, pressurized container may create sufficient force from the released gas to put the container in motion. A broken valve fitting on a pressurized gas cylinder can send the container in motion like a missile, the momentum of which can send the cylinder through a wall. Similar events occur in overheated and overpressured water heaters.

Leaking material may have dangers inherent to its temperature or toxic or reactive properties. Contact with hot water or high-pressure steam will cause burns. A material may inflict injury if it is a caustic or acid. Inhalation of toxic materials or contact with them may cause harm, depending on the material.

Compressed air in tires, air lines, and equipment may pose fire hazards because of the presence of enriched oxygen. This is particularly true if fuel and sources of heat or sparks are also present.

A sudden release of materials under pressure may produce a shockwave capable of knocking someone down or causing other damage, as noted in Chapter 17. The shockwave may cause materials in its path to strike someone and cause injury and the flying materials may damage other property.

Certain kinds of truck tires and other large tires pose a danger from sudden release of inflating air. This is particularly true for multipiece rims that are inadequately fastened together. If a person is in the path of the rim or other parts that may fly during the release, a serious injury can result.

Controls

Application of Haddon's energy theory (Chapter 9) helps identify controls for pressure release hazards. First, if there is a need for pressure in an application, the amount of pressure should be limited. There are various devices that prevent overpressures. Avoid pressure buildup in containers. For example, do not expose pressurized containers to direct sun or other sources of heat. If pressures are released, controlling the location and direction can prevent injuries to nearby people. The release should be controlled so that hazardous temperatures or materials do not come into contact with people. Releases of steam or other material should be routed to an area where there is little danger. Barriers may help in some cases. For example, tire cages and other tire restraining equipment can prevent some injuries from accidents while inflating truck and other large tires.

Avoid pressures that are not needed and reduce them for certain activities. Equipment should be deenergized and depressurized before working on it. Observe tag out and lock out procedures when servicing pressurized equipment. Train workers about the dangers of pressurized equipment. They must learn how to protect themselves from these dangers.

Normally, aircraft tires are inflated with nitrogen to prevent fire that may otherwise occur if they were inflated with air and heat severely from skids during landing.

Overpressure Devices

There are a number of overpressure devices, each with particular applications. Some devices are suitable for gas and some for steam. Others are suitable for liquids and some for gases, vapors, or liquids. Some require more maintenance and testing than others to ensure that they operate correctly. Some are subject to corrosion, scale buildup, and other problems that could render them inoperative or make them operate poorly. Valves are suitable for relatively clean materials because they must reseat after relieving pressure. Valves minimize loss of material. Rupture discs are better for relieving large volumes and corrosive, dirty, and viscous fluids. Pressure relief devices and components like pipes that direct releases to a safe location must be sized for the flow and type of materials that will be released. The American Society of Mechanical Engineers (ASME) Boiler and Pressure Vessel Code gives specifications for installation and certification testing of overpressure devices.

Safety Valves A safety valve is actuated when the upstream pressure exceeds some predetermined value. The valve rapidly opens fully or pops open to relieve the pressure. Safety valves are used for gas, steam, and vapor.

Relief Valves When the upstream pressure exceeds some predetermined level, a relief valve opens in proportion to the amount of overpressure and then closes when the pressure has returned to an acceptable level. Relief valves are used primarily for liquids.

Safety Relief Valves These valves are activated by upstream pressures that exceed some value. The valves are suitable as safety valves or relief valves. They are used in liquid or in gas, steam, or vapor applications.

Frangible Discs Frangible discs or rupture discs are relatively flat metal pieces. Each disc is designed to burst at a particular pressure. They are mounted between two flanges along a vent pipe and they range in size from less than $^1/_2$ in in diameter to approximately 4 ft. Rupture pressures range from a few ounces to very high pressures. Other considerations that may be included in design are the number of pressure cycles and selecting material because of potential corrosion, temperatures in which they are placed, and other factors. Rupture discs generally release large quantities of process material when they fail and perform their function, whereas valves release small quantities. Failure of a disc may produce significant downtime for replacement of the disc, replacement of lost materials, and restart of a process. Some rupture discs are actuated by a quantity of explosive material connected to a sensing device and a detonator control. Frangible discs are used for liquids, gases, steam, or vapor.

Fusible Plugs A fusible plug is a plug made of a metal that will melt at a selected temperature and will relieve a container of pressure. Fusible plugs are used in boilers, compressed gas cylinders, and other pressure vessels. They allow overpressures from fire or other causes of overheating to be vented. A plug may consist entirely of the alloy that will melt or have a high-temperature metal core surrounded by the low-temperature alloy.

Discharge Lines or channels approaching or leaving a pressure relief device must be sized to provide adequate flow of materials. Discharged materials must flow to some location where there is no danger to people because high temperatures and high flow rates may cause injury. High volumes must be discharged to adequately sized holding areas or containers.

Vacuum Failures Pressure reduction can damage containers. For example, tanks can collapse as they are drained. Even if venting is provided, the vent line must provide adequate flow rates to prevent a vacuum from occurring.

Freeze Plugs Another type of pressure relief device is a freeze plug. Water and many water-based liquids expand near the freezing point. A freeze plug will allow the liquid to expand and drain, thereby reducing the likelihood of damage and controlling the location of failure. It may be possible to reduce the risk of damage even further by using antifreeze solutions together with freeze plugs.

Temperature Limit Devices In accordance with Boyles-Charles Law, pressure will increase and decrease in closed containers with temperature changes in contents. Temperature limit sensors and control systems often are used in connection with processes and containers where pressure limits create dangers.

19-5   HIGH-PRESSURE FLUIDS

Gases and liquids under pressure are very common. A few examples are hydraulic lines, compressed air for many purposes, paint sprayers, grease guns, hydraulic and pneumatic tools, spray applicators for agricultural chemicals, water hoses used in fire fighting and landscaping, and fuel injection devices in engines. Gases and liquids under pressure and the lines and hoses used to distribute them have dangers. Pressure testing of pipelines and process equipment can produce explosive releases of gases.

Hazard

Pressurized gases and fluids can cause injury. Major hazards are air and gas injuries, injection injuries, and whipping of lines.

Air and Gas Injuries Getting pressurized gas or air into the viscera can cause injury or can rupture tissues. For example, some people use compressed air to clean parts. If the nozzle of a compressed air line is placed in the mouth with air flowing, the air can inflate and injure tissues.

Injection Injuries Injection injuries occur when a fine stream of gas or fluid enters the body. Fine streams of air, gas, or liquid can penetrate the skin. Medical inoculations with injection guns apply this principle. Fine, high-pressure streams of water (sometimes mixed with abrasive particles) are used to cut stone in excavation projects. A fine stream may look safe because of its size and seem harmless because of the familiarity with water or some other common liquid. However, it can cause serious injection injuries. High-pressure streams of fluids can easily make incisions in skin and other tissue. Injuries from 650 to more than $7,000 \, \text{lb/in}^2$ streams are noted in medical literature. Such pressures can be created from equipment operating at much lower pressures. Fluid injected through even a tiny hole in the skin can migrate throughout several layers of tissue and is extremely difficult to remove. If the fluid is toxic or contaminated with infectious microorganisms, amputation often is the only solution. Radical treatments for injection injuries are related to the delay in onset of treatment. Injection injuries typically involve the fingers and hands, but have included the arms, face, and other parts of the body. A gas injected under the skin may create embolisms in the blood stream that can interrupt lung or heart functions if they migrate to these organs.

Whipping of Lines Fluid moving through a nozzle creates reactive forces on the nozzle. If the forces are large enough, they can cause the nozzle and hose to move or whip. If the hose or nozzle strikes someone, it can cause serious injury, and if it strikes something, it can cause damage.

Controls

One control for hazards of compressed air lines is reducing the pressure to a low level. OSHA requires compressed air for cleaning purposes to be $30 \, \text{lb/in}^2$ or lower. There are pressure-reducing nozzles that drop line pressure to $30 \, \text{lb/in}^2$ or less. Setting pressure regulators on general use air lines to $30 \, \text{lb/in}^2$ or less minimizes danger to users.

Controls for reducing the hazards of compressed gas and fluid lines include distance, guarding, and the use of solid lines. Increasing the distance of hydraulic lines from people or body parts reduces the force of released fluid by the time it contacts a person. Leaking hydraulic hoses have caused injection injuries. For example, a line leading to a control box may be within a few inches of an operator's hands. Keeping lines away from hands or fingers can reduce the chance of injury.

Solid lines do not leak as readily as hoses, which develop leaks from vibration, pressure, bending cycles and aging. Where lines must come close to people, solid lines and tight, well-maintained fittings reduce the chances of a leak.

Hose or lines that come near people should have the extra protection of guarding. A shield of metal or other materials that a fluid stream cannot cut easily gives this protection.

People who work around compressed air lines, hydraulic systems and other pressurized fluid and gas equipment should learn about the hazards. They should learn not to place fingers or hands against a fluid stream, and they should learn not to place the stream near anyone else. Protective gloves and clothing may help reduce injection injuries.

When pipelines and process equipment are pressure tested, workers should be clear of potential rupture points along the line. Instruments for reading pressures should not require workers to remain close to pressurized lines. If a compressor and its operators are removed from other crews preparing pipelines for testing, radio communication between crews is essential. Otherwise, one crew may not be aware of what the other is doing. The preparation crew may not be aware of pressure in a line. Department of Transportation regulations and other references give detailed procedures for pressure testing of pipelines transporting hazardous liquids.[2]

EXERCISES

1. A cylinder is filled with nitrogen gas at 70°F to the maximum allowed pressure (2,640 lb/in² [gauge]). If it were left in the sun and the contents reached a temperature of 210°F, what would be the resulting pressure in the cylinder?

2. A caisson worker is required to work under water at a depth of 60 ft digging a tunnel under a river. Assuming a 33-ft column of water is equivalent to 1 atm, determine the following:

   (a) the pressure for the worker in millimeters of mercury

   (b) the partial pressure of oxygen if air is pumped from the surface (assume standard sea level conditions) to the worker

3. In a diving operation, if the upper limit for oxygen at sea level is 40% (where oxygen toxicity starts to appear) and the lower limit is 16% (hypoxia occurs), what is the allowable range for oxygen in the breathing gas mixture (percent of total) when a diver is submerged to 600 ft below sea level?

REVIEW QUESTIONS

1. What is Dalton's law of partial pressure?

2. What effect does low atmospheric pressure and low oxygen partial pressure have on the body?

3. What are the hazards of high altitude?

4. What controls can eliminate or reduce the dangers of high altitude?

5. What is Henry's law?

6. What is the significance of Henry's Law for people who work in high-pressure atmospheres?

7. What are the hazards of high-pressure atmospheres?

8. What controls can reduce the dangers of these hazards?

9. What are alternate terms for dysbarism?

10. Identify three hazards associated with high pressure containers.

11. What controls can eliminate or reduce these hazards?

12. Explain the operating principle for the following overpressure devices:

    (a) safety valve

    (b) relief valve

    (c) safety relief valve

    (d) frangible disc

    (e) fusible plug

    (f) freeze plug

13. When overpressure is released, identify a hazard and control for the release.

14. By what principle of physics are temperature and pressure interrelated?

15. Identify a hazard associated with high pressure

(**a**) hydraulic lines

(**b**) air lines

16. What is a suitable control for each hazard in question 15?

NOTES

1 29 CFR 1926.804.　　　　　　　　　　　**2** 49 CFR 195.300.

BIBLIOGRAPHY

ASME Boiler and Pressure Vessel Code, American Society of Mechanical Engineers, New York, regularly updated.

BREGLIA, R. J., "Toxicology of High-Pressure Injection or Grease Gun Injuries," National Lubricating Grease Institute (NLGI) *Spokesman*, March: 424–427, (1984).

Guide to Safe Handling of Compressed Gases, Matheson Tri-Gas, Secaucus, NJ, 1983.

Handbook of Compressed Gases, 2nd ed., Compressed Gas Association, Van Nostrand Reinhold, New York, 1981.

HILLS, B. K., *Decompression Sickness*, Wiley, New York, 1977.

Medical Problems of Man at High Terrestrial Elevations, Technical Bulletin MED 288, Department of the Army, Washington, DC, October 15, 1975.

MEGYESY, E. F., ed., *Pressure Vessel Handbook*, 6th ed., Pressure Vessel Handbook Publishing, Inc., Tulsa, OK, 1983.

CHAPTER 20

VISUAL ENVIRONMENT

This chapter addresses the visual environment, which includes lighting, color, and signage. Many aspects of the visual environment contribute to accidents, whereas other characteristics help prevent them. Engineers need to understand the visual environment and how it relates to accidents and their consequences.

20-1 ILLUMINATION

Lighting or lack of lighting can contribute to accidents. People need to see what they are doing and where they are going. Some aspects of lighting are distracting or interfere with tasks. One study suggests that approximately one fourth of all accidents involve poor lighting. Another study found that falls are much more frequent in the evening and at night, which suggests a relationship to lighting conditions.

In 1973, all three engines in a jumbo jet failed because O-ring seals were missing. The failure nearly caused the pilot to ditch the plane and its 172 passengers in the Atlantic Ocean. During the investigation of this incident, mechanics for the airline testified that it was too dark in the service area to see if the seals were in place.

A patron entered a restaurant from the bright sunlight for a noon lunch. The restaurant was dimly lit to create an aesthetic effect. The patron fell down a step that was only a few feet from the entrance, claiming he was unable to see it.

Illumination and Lighting

The human eye detects light in wavelengths within the range of 380 to 750 nm (see Figure 21-1). Ultraviolet light lies below and infrared light lies above the visual range.

Figure 20-1 illustrates a light source and what happens to light distributed from it. The intensity of a source is measured in Candela or candlepower. The output of a light source, such as a lamp, is expressed in units of lumens (lm), which originally was defined as the light from a standard candle falling on a 1 ft^2 area 1 ft from the candle.

Light leaving a source goes in many directions and may not scatter uniformly in all directions. As light travels farther from its source, its energy diminishes. Like other point source radiant energy sources, light intensity decreases in relation to the square of the distance from the source. The light at one unit of distance from a source will be reduced to one quarter the amount at two units from the source. Flux is the light travelling through some unit of area. Luminous flux has units of lumens (lm).

Eventually, light from a source arrives at a surface. Illumination is light falling on a surface. Illumination has units of footcandles (fc) or lux (lx). If 100 lm arrive at a 1-ft^2

Safety and Health for Engineers, Second Edition, by Roger L. Brauer
Copyright © 2006 John Wiley & Sons, Inc.

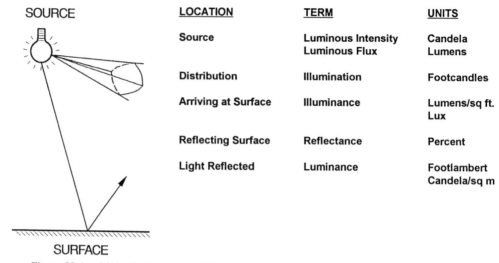

SURFACE

Figure 20-1. Light distribution and illumination terminology.

surface, the illumination is 100 fc. If 100 lm arrive at a 1-m² surface, the illumination is 9.29 lx (1 fc = 10.764 lx).

The light arriving at a surface is absorbed, transmitted, or reflected. The reflectance of a surface is the portion of arriving light that is reflected. Reflectance, a property of the surface, is reported in percent. Luminance is the light emitted or reflected from a surface per unit area. Luminance has units of foot lamberts.

Light sources are daylight and artificial light, normally from electric lighting devices. Typically, a lighting fixture has a lamp and a luminaire. Luminaires help distribute the light in some desired pattern. There are several types of artificial light sources. Major types are incandescent, fluorescent, and high-intensity discharge (mercury and sodium vapor and halogen). Each type produces a different spectrum of wavelengths.

Lighting is classified as general or supplementary. General lighting provides lighting to a large area. It may be direct, which allows light to move from the source directly to surfaces, or indirect, which provides light by reflecting it from surfaces such as walls or ceilings. Task lighting is a common form of supplementary lighting. Located near a particular task, task lighting supplements general lighting to provide the amount and quality of light required.

Emergency lighting is another class of lighting important for safety. Emergency lighting typically is backup lighting that provides light when normal light sources fail. Emergency lighting has an alternate power supply, typically batteries. Interiors of buildings and exit routes have emergency lighting as required by fire codes and other standards.

Hazards

The major hazards associated with lighting involve illumination levels, changes in illumination levels, qualitative aspects of lighting, and flicker of some light sources.

Illumination Levels   If there is too little light, one cannot see well. Insufficient light leads to errors and sometimes to accidents. An error may be placing a hand into a dangerous location or machine, not seeing a step and falling, or failing to detect a problem or a faulty part. An error may result from not seeing a situation that is dangerous and not being able to react quickly enough.

Extremely bright light sources can injure the receptor cells in the eyes. For example, even a brief look at the sun can damage the eye. Similarly, looking at the light from arc welding can cause eye injury very quickly.

Even when bright lights are not sufficient to cause injury, they can create afterimages that obscure portions of the visual field until the receptor cells have had a chance to recover. The afterimage from a flash of a camera or similar bright light is a common experience. The afterimage may lead to errors in vision and to accidents.

A bright light in the visual field may interfere with the ability to see or detect an object. The result may be visual errors and accidents.

Changes in Illumination Level   Changes in illumination level interfere with the ability of the eyes to adjust quickly enough to permit seeing without error. Examples of changing light levels are the transition from bright outdoor light to dark interiors or from a bright area of a building to a dark one. Another example is looking at a brightly lit task, then turning the eye to focus on a location that is darker.

There are two ways the eyes adjust to changing light levels. In one form of adjustment, the iris increases and decreases in diameter to adjust the size of the pupil. This occurs quite quickly to changes in illumination. The pupil size also changes in response to startle and interest: pupils enlarge when a person is interested in someone or something or when someone is surprised.

The eyes also adjust photochemically to changes in light level. Photochemical changes provide sensitivity changes as great as 100,000 times or more. The concentration of rhodopsin in receptor cells of the retina is related to the sensitivity of the cells to light. Increases in light arriving at the retina reduce light sensitivity by reducing the concentration of rhodopsin. This is called light adaptation. One achieves full light adaptation in approximately 3 min. After being exposed to bright light and then the level of light droping to low levels for a period, the retina must adjust to the dark. This is called dark adaptation. Changes in rhodopsin concentration in dark adaptation may take 30 min to several hours.

Qualitative Aspects of Lighting   Some qualitative aspects of lighting are glare and luminous contrast. Glare is the presence of a bright light in the visual field. *Direct glare* occurs when the light in the visual field is a light source. An example of direct glare is the headlights of an oncoming car at night. *Reflected glare* occurs when a bright light reflects from a surface. An example of reflected glare is the reflection of a light on a glossy page, which obscures the print. Glare can lead to errors in perception and detection that result in accidents and may produce afterimages or delay visibility resulting from adaptation.

Luminous contrast refers to the changing light levels of an environment. For example, one may look at work on a desk that has a certain illumination. Shifting the eyes to a wall presents a much darker or lighter level of illumination. When there is too much difference between the two surfaces, the eyes have difficulty adapting, which may lead to visual errors.

Flicker and Strobe Effects   Some light sources are not constant and rapidly flicker on and off. The gas in a fluorescent tube actually turns on and off 120 times per second under 60-Hz electrical power. Most people do not notice this because the eyes tend to fuse images that flash at frequencies higher than 15 to 18 times per second. A flickering light source may be distracting or disturbing. Some people seem to be sensitive to fluorescent light flicker.

When an object oscillates or rotates and a bright light flashes on and off at the same frequency as the motion, the object appears to stand still. One can measure the speed of rotating machine parts with a strobe light that has an adjustable frequency for its flashes. A person who does not realize that the light source is flashing on and off may perceive that an object is not in motion. For example, placing a hand into the moving equipment that appears to be stationary may cause injury.

Other Lighting Effects   There are other effects of light that can be distracting or can lead to errors in vision and visual perception. Some workplaces use sodium vapor lights to reduce energy costs. Sodium vapor lights produce light that is primarily in the yellow region. The light makes normal color appear different. Color coded information may not be perceived correctly.

Recent studies determined that a small portion of the population suffers from seasonal affective disorder. This condition affects people during the winter months in northern climates when there is limited daylight. Depression may result and can be severe and can severely affect performance. A daily dose of high-intensity, full-spectrum (nearly matching the spectrum of sunlight) fluorescent light for 2 to 4 hr eliminates the depression in 90% of the cases.

When there is a bright light source in the visual field, one tends to turn the head and eyes toward it. This phenomenon is called phototropism. The presence of such lights can be distracting.

Shadows may cause errors in perception of an object, but they also may be beneficial. Shadows from side lighting of difficult to see objects may make these easier to see.

Characteristics of People   Individual variations among people add to potential problems associated with lighting. For example, in low light levels, the rod cells in the retina are the receptors. The rod cells are not sensitive to color and the cone cells, which sense color, are not effective in low light. Studies indicate that night vision, which refers to the function of rod cells in low light levels, diminishes with age. Compared with someone who is 20 years old, a 45-year-old person needs four times the light to achieve the same level of perception, and by age 60 years, the light levels required are double those required at age 45 years.

Many people aged 55 and older have some degree of cataracts. Cataracts are an opaqueness that develops in the lens and its encapsulating tissue. They also may occur on the cornea. Cataracts can occur in children or at young ages. Cataracts filter the light entering the eye and reduce the amount reaching the retina. They also cause light to scatter. If cataracts are severe enough, the scattering can produce multiple images, reduced acuity and color perception, and cause other vision problems. In extreme cases, cataracts cause blindness.

There are many visual disorders that require correction. Some result from aging. Without correction, visual errors increase.

Controls

Illumination Levels   Illumination standards determine the amount of light suitable for various tasks. They are based primarily on the type of task. Current standards include adjustment factors for age, speed or accuracy, and reflectance of the task background. Table 20-l is a summary of recommended illumination levels for interior lighting. Table 20-2

TABLE 20-1 Illuminance Categories and Illuminance Values for Generic Types of Activities in Interiors

Type of Activity	Illumination Category	Ranges of Illuminances	
		(lx)	(fc)
General lighting throughout space			
Public spaces with dark surroundings	A	20–30–40	2–3–4
Simple orientation for short, temporary visits	B	50–75–100	5–7.5–10
Working spaces where visual tasks are only occasionally performed	C	100–150–200	10–15–20
Illuminance on task			
Performance of visual tasks of high contrast or large size	D	200–300–500	20–30–50
Performance of visual tasks of medium contrast or small size	E	500–750–1,000	50–75–100
Performance of visual tasks of low contrast or very small size	F	1,000–1,500–2,000	100–150–200
Illuminance on task, obtained by a combination of general and local (supplementary lighting)			
Performance of visual tasks of low contrast and very small size over a prolonged period	G	2,000–3,000–5,000	200–300–500
Performance of very prolonged and exacting visual work	H	5,000–7,500–10,000	500–750–1,000
Performance of very special visual tasks of extremely low contrast and small size	I	10,000–15,000–20,000	1,000–1,500–2,000

TABLE 20-2 Weighting Factors To Be Considered in Selecting Specific Illuminance within Ranges of Values of Each Category

	Weighting Factor		
For illuminance categories A through C			
Room and occupant characteristics	−1	0	+1
Occupant ages (yrs)	Less than 40	40–55	Over 55
Room surface reflectances[a]	>70%	30–70%	<30%
For illuminance categories D through I			
Task and worker characteristics	−1	0	+1
Worker ages (yrs)	Less than 40	40–55	Over 55
Speed and/or accuracy[b]	Not important	Important	Critical
Reflectance of task background[c]	>70%	10–70%	<30%

[a] Average weighted surface reflectances, including wall, floor, and ceiling reflectances, if they encompass a large portion of the task area or visual surround.

[b] In determining whether speed and/or accuracy is not important, important, or critical, the following questions need to be answered: What are the time limitations? How important is it to perform the task rapidly? Will errors produce an unsafe condition or product? Will errors reduce productivity and be costly?

[c] The task background is that portion of the task on which the meaningful visual display is exhibited.

gives data for determining adjustment factors used in selecting levels within tasks, and Table 20-3 provides the data for deciding what value to select for a particular category. The *IESNA Lighting Handbook* is a long-standing, recognized authority on lighting standards and is the source for the recommendations above.

TABLE 20-3   Decision Data for Selecting Illumination Values within Categories

Categories	Sum of Weighting Factors	Value to Use
A to C	−2 or −1	Low
	0	Medium
	+1 or +2	High
D to I	−3 or −2	Low
	−1, 0, or +1	Medium
	+2 or +3	High

TABLE 20-4   Minimum Illuminance Levels for Safety[a]

Hazards requiring visual detection	Slight		High	
Normal[b] activity level	Low	High	Low	High
Illuminance levels				
Lux	5.4	11	22	54
Footcandles	0.5	1	2	5

[a] Minimum illuminance for safety of people. These are absolute minimums at any time and at any location on any plane where safety is related to seeing conditions. Refer to *IESNA Handbook* for recommendations on particular types of facilities and areas.
[b] Special conditions may require different illuminance levels. In some cases, higher levels may be required as, for example, where security is a factor. In some other cases, greatly reduced levels, including total darkness, may be necessary, specifically in situations involving manufacturing, handling, use or processing of light-sensitive materials (such as photographic products). In these situations, alternate methods of ensuring safe operations must be relied on.

Example 20-1   Assume that a visual task involves small objects and medium contrast. It is also known that some of the people who will perform the task are older than 55 years of age. Speed for the task is critical. The reflectance of the task background is 50 percent. What is the preferred illumination level?

From Table 20-1, the task falls within category E. There are three illumination levels possible. To decide which of the three is preferred, determine the adjustment factor. From Table 20-2, age has a +1 factor, speed a +1 factor, and reflectance a value of zero. The total is +2. According to Table 20-3 for category E, when a weighting factor is +2 or +3, the high value from Table 20-1 is used. Therefore, the preferred illumination level is 1,000 lx.

The American National Standards Institute publishes standards of the Illuminating Engineering Society of North America (IESNA). NFPA 101 identifies minimum lighting requirements for various occupancies and exiting activities. In some cases, the minimum is 1 fc, which is a very low level and may not be sufficient for transition zones where one experiences a sudden change in illumination. This is particularly true if the transition involves dark adaptation. The life safety and exit codes also have standards for emergency lighting. IESNA has a list of illuminance levels regarded as *absolute minimums for safety alone*; see Table 20-4. Note that in all illumination designs, one must start with illumination levels higher than desired because the luminaires and lamps normally become dirty and reduce actual illumination levels with time. Cleaning will restore original lighting levels.

Quality of Lighting The quality of lighting may affect error rates and cause accidents. For many applications, control of illumination levels alone is not sufficient. Designs must control illumination quality as well.

One should analyze workstations to avoid direct or reflected glare. This is particularly true where tasks are critical or continue for extended periods. For example, a surface of a machine that has a dangerous action may become polished. A reflected light source may obscure the dangerous location, which could lead to placing hands erroneously in the danger zone at the wrong time. Screens on computer monitors illustrate the continuous problem. Reflections of light sources on the screen can cause eye strain, changes in head position that lead to fatigued neck and shoulder muscles, and related effects. This family of cumulative disorder is sometimes called VDT (visual display terminal) syndrome. Recent monitor designs significantly reduce such problems.

One also should analyze workstations for brightness ratio or luminance ratio. This refers to the difference in lighting level between the task and the surround. One may be brighter than the other. The goal is to minimize the difference in illumination levels between a task and its surround and keep the difference within certain ratios. This reduces the degree of adaptation required in a visual task as eyes focus on the task, then the surround, and back to the task. There are often practical limitations that may not permit achievement of this goal. Table 20-5 lists recommended luminance or brightness ratios for industrial settings.

Guarding can reduce dangers of oscillating and rotating equipment. This is particularly true when there is gas-vapor lighting that may produce strobe effects.

Example 20-2 An industrial task is to be illuminated to a level of 100 lx. What is the minimum illumination required for an adjacent surface? For a remote surface?

From Table 20-5, the recommended maximum luminance ratio between a task and an adjacent darker surround is $3:1$. Therefore, the minimum illumination required for the adjacent area is $100 \times 1/3 = 33.3$ lx. From Table 20-5, the ratio between a task and remote darker surfaces is $10:1$. The minimum illumination recommended for the remote surfaces is $100 \times 1/10 = 10$ lx.

The reflectance of surfaces should be controlled to prevent reflective glare from interfering with visual tasks. Table 20-6 lists recommended reflectance values for industrial, interior environments.

TABLE 20-5 Recommended Maximum Luminance Ratios[a]

Surfaces	Ratio by Environmental Class[b]		
	A	B	C
Between tasks and adjacent darker surroundings	3:1	3:1	5:1
Between tasks and adjacent lighter surroundings	1:3	1:3	1:5
Between tasks and more remote darker surfaces	10:1	20:1	—
Between tasks and more remote lighter surfaces	1:10	—	—

[a] From ANSI/IESNA RP7, *Practices for Industrial Lighting.*
[b] A = interior areas where reflectance of surfaces can be controlled. B = areas where reflectances in immediate work area can be controlled, but control of remote surround is limited. C = areas (indoor and outdoor) where it is impractical or difficult to control reflectances.

TABLE 20-6   Recommended Reflectance Values[a]

Surface	Reflectance (%)
Ceiling	80–90
Walls	40–60
Desk and bench tops, machines, and equipment	25–45
Floors	Not less than 20

[a] From ANSI/IESNA RP7, *Practices for Industrial Lighting.*

20-2   COLOR

Color and Safety

Color is useful in safety for marking hazards and coding information. For example, color can mark edges of steps and other changes in walking surfaces, and it can code classes of information, including safety information and signs.

One limitation for color coding is that a small portion of the population is color blind. Color blindness occurs primarily among males. Approximately 6% of adult males have a marked reduction in color sensitivity, and less than 1% of adult males are totally color blind. Color blindness can lead to errors. For example, color-blind electricians have attached the green ground wire in equipment to a hot lead and energized the equipment. A color-blind driver depended on the position coding of red and green to interpret traffic signals correctly. At one location, the red light was at the bottom and led to a motor vehicle accident.

Another limitation of color coding is that one must remember the meaning of the colors. However, if color coding is applied consistently over time, the general meaning of the colors becomes well known in a culture. We call such well-known conventions population stereotypes. For example in the United States, standard traffic signals have led to a general knowledge among the population. Most people understand that red means stop or danger, yellow means caution, and green means go. The nuclear power plant accident at Three Mile Island, Pennsylvania, taught us some lessons about the importance of applying color coding consistent with learned population stereotypes.[1] The color coding of signal lights on the Three Mile Island control room panels did not always follow the red-green convention. Chapter 33 discusses this accident further.

Color Standards

There are several standards that include color codes for safety information and information that could help one recognize hazards. American National Standards Institute (ANSI) standards in the Z535 series define color codes, signs, tags, and symbols for safety applications. Table 20-7 lists standard color markings for hazards. DOT has color standards for highway signage and markings. OSHA regulations include color coding for respiratory protective equipment, signage, and accident prevention tags. ANSI A13.1 defines standard markings (including colors) for identification of piping systems. The Army-Navy Aeronautical Specification AN-C-56 is a standard for three colors recognizable by most color-blind people. Other organizations have developed recommendations for the application of color in other safety communication applications.

TABLE 20-7 Standard Color Codes for Marking Hazards[a]

Color	Use
Red	Fire equipment, danger, emergency stops
Yellow	Marking hazards
Green	First aid, safety equipment
Black/white	Traffic marking, housekeeping
Orange	Dangerous parts of machines or equipment
Blue	Informational signs
Reddish purple	Radiation hazards

[a] Excerpt from ANSI Z535.1, *Safety Color Codes.*

The National Institute for Standards and Technology (NIST) studied 59 safety colors under seven different types of light sources.[2] The study resulted in recommended changes to certain colors in ANSI Z535 to make them easier to differentiate and recognize.

20-3 SIGNAGE

Signage and Safety

Signs take many forms. They may be large to provide safety information for a large area or they may be in the form of tags for lockout and tagout procedures. They mark hazards at particular locations on machines and equipment and in buildings and they provide warnings on products. They need to incorporate standard color codes where applicable. For many use environments, signs should be multilingual or include symbols that are not language dependent. Because not everyone understands or recalls the meaning of many symbols, symbols are more effective if coupled with text information.

Signage Standards

There are number of standards for marking hazards and communicating safety information through signage. ANSI publishes several standards on signage. Regulations of the Department of Transportation, OSHA,[3] Consumer Products Safety Commission (CPSC), and other federal agencies also contain signage standards important for safety. There is a United Nations system for labeling hazardous materials. The International Standards Organisation (ISO) publishes standards for signs and symbols. Many consensus standards contain specifications for signage. Examples are the National Electric Code and the National Fire Code. The Association for the Advancement of Medical Instrumentation (AAMI) has established standards for symbols and signs on medical equipment.[4]

Signs must meet many characteristics to be readable (see Chapter 33), particularly when used as warnings and instructions (see Chapter 7).

EXERCISES

1. What illumination level is recommended for public areas with dark surroundings, where all age groups will use it, accuracy of information is important, and reflectance of the task background is 35%?

2. What illumination level is recommended for an industrial task involving very small items for an entire workday, where items are of low contrast? Several workers are older than 55 years and speed is critical. The task background has a reflectance of 65%.

3. An assembly task is to have 40 fc of light. If more remote surfaces are darker than the task, what is the least amount of light you would recommend for the surrounding dark surfaces?

4. You are to design the entrance to a restaurant. The interior decor is to be dimly lit to create an intimate dining mood. The decor is to be in place even at lunch time, when full sunlight may be present outdoors. Part of the dining area is to be sunken relative to the entry level and other parts are to be elevated from it. Determine what lighting levels and lighting placement you would recommend for different areas of the restaurant. Justify your decisions, allowing for visual adaptation of patrons. Some patrons may be older than 65 years, have some form of cataracts, or have macular degeneration. What other design features would reduce hazards?

REVIEW QUESTIONS

1. What units of measure are used for the following forms of light?
   (a) light output from a source
   (b) luminous flux
   (c) light arriving at a surface
   (d) light emitted or reflected from a surface
   (e) reflectance of a surface

2. Characterize the following:
   (a) general lighting
   (b) supplemental lighting
   (c) direct lighting
   (d) indirect lighting

3. For each of the following, identify at least one hazard and control:
   (a) illumination level
   (b) change in illumination level
   (c) qualitative aspects of lighting
   (d) flicker of a lighting source

4. How long does full light adaptation take?

5. How long does full dark adaptation take?

6. Which receptor cells in the eyes are sensitive to color?

7. Which receptor cells in the eyes are most sensitive to low light?

8. How do cataracts affect vision?

9. How is color used for safety?

10. Name two limitations for using color coding.

11. Name one limitation of text signage.

12. Name one limitation of symbols.

NOTES

1 Sheridan, T. B., "Human Error in Nuclear Power Plants," *Technology Review*, February: 23–33 (1980).

2 NSBIR-86-3493, *Safety Color Appearance Under Selected Light Sources*, National Bureau of Standards, Washington, DC, 1987.

3 29CFR1910.144 Safety Color Code for Marking Physical Hazards; 29CFR1910.145 Specification for Accident Prevention Signs and Tags.

4 ANSI/AAMI/IEC TIR 60878, *Medical Equipment Symbols and Safety Signs*, Association for the Advancement of Medical Instrumentation, Arlington, VA.

BIBLIOGRAPHY

ANSI Standards, American National Standards Institute, New York:
A13.1 Scheme for the Identification of Piping Systems
IESNA RP-1 Recommended Practice on Office Lighting
IESNA RP-3 Recommended Practice on Lighting for Educational Facilities; High School and College
IESNA RP-7 Practice for Industrial Lighting
IESNA RP-8 Practice for Roadway Lighting
IESNA RP-11 Design Criteria for Lighting Interior of Living Spaces
IESNA RP-16 Nomenclature and Definitions for Illuminating Engineering
IESNA RP-22 Recommended Practice for Tunnel Lighting
IESNA RP-27.1 Recommended Practice for Photobiological Safety for Lamps and Lamp Systems—General Requirements
IESNA RP-27.2 Recommended Practice for Photobiological Safety for Lamps and Lamp Systems—Measurement Techniques

IESNA RP-27.3 Recommended Practice for Photobiological Safety for Lamps—Risk Group Classification and Labeling
IESNA RP 28 Recommended Practice on Lighting and the Visual Environment for Senior Living
IESNA RP-29 Recommended Practice on Lighting for Hospitals and Health Care Facilities
IESNA RP-30 Recommended Practice on Museum and Art Gallery Lighting
Z535.1 Safety Color Code
Z535.2 Environmental and Facility Safety Signs
Z535.3 Criteria for Safety Symbols
Z535.4 Product Safety Signs and Labels
Z535.5 Accident Prevention Tags
BOYCE, P. R., *Human Factors in Lighting*, 2nd ed., Macmillan, New York, 2000.
IESNA Lighting Handbook, 9th ed., Illuminating Engineering Society of North America, New York, 2000.

CHAPTER *21*

NONIONIZING RADIATION

21-1 THE ELECTROMAGNETIC SPECTRUM

Electromagnetic radiation is arranged in a spectrum of wavelengths; see Figure 21-1. Wavelengths range from long (3×10^8 m) and very low frequency (1 Hz) to short (3×10^{-15} m) and very high frequency (10^{23} Hz). The spectrum is divided into several bands: gamma rays, x-rays, ultraviolet, visible, infrared, microwave, television, and radio waves, induction heating, and power waves.

There are many sources of radiant energy. Some, like the sun and fires, are natural sources. Others, like microwaves, radio transmission, atomic reactors, lamps, and lasers, are manufactured sources. Some sources have energy levels high enough to ionize atoms or molecules or break the bond of molecular elements. Approximately 10 to 12 eV or more are needed to break these bonds. Because photon energies of electromagnetic radiation are proportional to radiation frequency and inversely proportional to wavelength, wavelengths in the lower portion of the spectrum generally are below this minimum energy level.

Some effects of electromagnetic waves that are dependent on frequency include visibility, penetration, and heating of materials, including human tissue. Some properties apply across the spectrum; for example, energy from a radiation source diminishes as a function of the distance squared (refer to Chapter 22).

Although energy levels for nonionizing radiation do not affect molecular structure, nonionizing radiation can affect biological tissue by changing energy levels in tissue molecules, often producing heat. Heat most easily affects certain tissue, like that of the eye, because the eye has little blood circulation and little ability to remove heat through blood movement. Tissue absorbs some wavelengths, but for other wavelengths, tissue is essentially transparent. The depth of penetration for absorbed wavelengths varies as well. Figure 21-2 illustrates the general absorption properties of the eye for electromagnetic radiation.

Because of the increase in radiation in products and equipment, Congress enacted legislation to control radiation emissions.[1] Today, several government standards (OSHA, Consumer Products Safety Commission [CPSC], and the Food and Drug Administration [FDA]) and consensus standards apply to equipment, exposures, and measurement.

21-2 MICROWAVES

Microwaves have wavelengths from approximately 1 mm to 10 m and frequencies from approximately 30 MHz to 300 GHz. Microwaves are used in communications, navigation, medical diathermy, microwave ovens, in drying equipment, and other applications. A variety of devices produce microwaves.

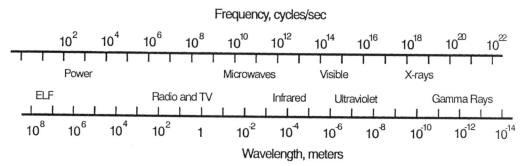

Figure 21-1. Electromagnetic spectrum.

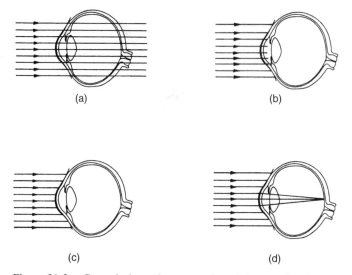

Figure 21-2. General absorption properties of the eye for electromagnetic radiation. (a) X-ray and gamma ray. (b) Ultraviolet A. (c) Ultraviolet Band C and infrared Band C. (d) Visible and infrared A.

Hazards

Microwaves less than 3 cm are absorbed in the outer skin, 3 to 10 cm wavelengths penetrate from 1 mm to 1 cm into the skin, and 25 to 200 cm wavelengths penetrate to deeper tissue and organs. Human tissue is essentially transparent to microwaves with wavelengths of more than 200 cm.

Absorbed microwave radiation is primarily converted to heat. Tissue and body temperatures rise, depending on exposure and location exposed. Temperature increases in deeper tissue may cause damage before one senses heating. Absorption by other materials also will produce heating, possibly raising temperatures high enough to be a hazard.

Because people may come into contact with microwave equipment and to avoid electric shock and burn hazards, grounding is essential, particularly if power densities are near acceptable limits. Objects near power sources, such as fences and vehicles, also must be grounded because they may couple to radio frequencies.

The greatest danger from microwaves is to the eyes, where microwaves seem to have a cumulative effect on the lens of the eyes, ultimately producing cataracts. The onset of cataracts depends on frequency, power density, duration of exposure, and intervals between exposures. There is some evidence that microwaves also affect the central nervous system in various ways. They also affect the performance of some types of cardiac pacemakers.

For a time, a beneficial use of microwaves on humans was diathermy. Diathermy is a therapeutic means of generating heat in body tissues. Localized application of levels up to $100\,\text{mW/cm}^2$ helped promote healing in joints, tendons, muscle, and other tissue. Diathermy devices operate in shortwave radio frequencies and ultrasonic frequencies in addition to microwave frequencies. If not applied carefully for limited times, the treatment can burn skin and injure deeper tissues. In addition, diathermy can cause severe injury or even death when used on patients with implanted leads or other objects or devices.

Controls

One control to prevent microwave injury is limiting exposure. This is accomplished by limiting the intensity of microwaves (the frequency or wavelength one is exposed to) or limiting the duration of exposure. Distance from a source and shielding also can control intensity of exposure. Table 21-1 lists attenuation of various shielding materials. Before implementing controls, one must determine what potential exposures exist through analysis, measurements, or both. These are compared with exposure standards to determine if controls are needed.

Other controls are signs to warn about radiation hazards or dangers (see ANSI C95 for details), special protective clothing that uses metallized fabric to provide some shielding, and protective eyewear that has a closely woven screen for shielding. Workers should handle equipment near microwave sources with insulated gloves to minimize shock and burn hazards.

High-power microwave equipment must be grounded to reduce electrical hazards. Access to hazardous microwave locations can be protected by interlocks on doors and regular inspections, and testing ensures that the interlocks are working. During servicing of microwave equipment, lockout and tagout procedures are important.

Standards

Microwave exposure standards appear in several publications. One source is OSHA standards,[2] which limit power density to $10\,\text{mW/cm}^2$ for exposure periods of $0.1\,\text{hr}$ or more. For exposures shorter than $0.1\,\text{hr}$, energy density is limited to $1\,\text{mW-hr/cm}^2$, with limited excursions higher than $10\,\text{mW/cm}^2$.

TABLE 21-1 Microwave Shielding Attenuation Factors

	Frequency (GHz)			
Material	1–3	3–5	5–7	7–10
60×60 mesh screening	0.01	0.003	0.006	0.01
32×32 mesh screening	0.016	0.006	0.006	0.016
16×16 window screening	0.016	0.01	0.01	0.006
$^1/_4$-in mesh (hardware cloth)	0.016	0.032	0.06	0.1
Window glass	0.63	0.63	0.50	0.45
$^3/_4$-in pine sheathing	0.63	0.63	0.63	0.45
8-in concrete block	0.01	0.006	0.006	0.001

TABLE 21-2 Electromagnetic Field[a] Radio Frequency and Microwave Threshold Limit Values[b,c]

Frequency	Power Density (mW/cm^2)	Electric Field Strength, E (V/m)	Magnetic Field Strength, H (A/m)	Averaging Time E^2, H^2, or S (min)
30 kHz–100 kHz	614	163	6	
100 kHz–3 MHz	614	16.3/f	6	
3 MHz–30 MHz	1,842/f	16.3/f	6	
30 MHz–100 MHz		61.4	16.3/f	6
100 MHz–300 MHz	1	61.4	0.163	6
300 MHz–3 GHz	f/300			6
3 GHz–15 GHz	10			6
15 GHz–300 GHz	10			$616,000/f^{1.2}$

[a] The exposure values in terms of electric and magnetic field strengths are obtained by spatially averaging over an area equivalent to the vertical cross section of the human body (projected area).
[b] f = frequency in MHz.
[c] *2004 Threshold Limit Values for Chemical Substances and Physical Agents and Biological Exposure Indices*, American Conference of Governmental Industrial Hygienists, Cincinnati, OH. (Note: This publication is updated annually and standards may change. Refer to the current edition for additional details.)

The American Conference of Governmental Industrial Hygienists (ACGIH) publishes exposure limits based on power density, electric field strength, and magnetic field strength for particular frequency bandwidths.[3] This standard includes radio and microwave frequencies. Table 21-2 lists the ACGIH threshold limit values (TLVs), which should be used as guides in the evaluation and control of exposure to radio frequency and microwave radiation and should not be regarded as a fine line between safe and dangerous levels. Radio frequency radiation exposures should be kept as low as reasonably possible.

The FDA sets standards for microwave ovens (21 CFR 1030.10). Included in the standard is a limit for the amount of microwaves that can leak from an oven throughout its lifetime. The limit is $5 mW/cm^2$ at approximately 2 in from the oven surface.

Measurement

One can measure microwaves with thermal or electrical detector instruments. Thermal detectors assess temperature rise in a material, whereas electrical detectors convert microwaves into direct current. Instruments are usually factory calibrated to particular microwave frequencies and intensities.

21-3 ULTRAVIOLET RADIATION

Ultraviolet (UV) radiation of significance to safety falls in the region between 200 and 400 nm. In this region, photon energy levels range from 4.4 to 3.1 eV. Air, water, and window glass are transparent for wavelengths in the 300- to 400-nm range, and ordinary window glass and skin absorb wavelengths from 200 to 320 nm. The ozone layer absorbs much of the UV energy from the sun in the wavelengths less than 290 nm, and although UV wavelengths extend below 200 nm, these wavelengths are poorly transmitted or fully absorbed in air.

One classification scheme for UV radiation has three spectral bands:

1. UV-A or near UV includes wavelengths from 315 to 400 nm
2. UV-B extends from 280 to 315 nm
3. UV-C extends from 200 to 280 nm

Wavelengths from 200 to 315 nm are of greatest concern in safety and health. This band is called the actinic UV region.

The sun is the most notable source of UV radiation. Other sources include heliarc welding, mercury and xenon discharge lamps, certain lasers, and full-spectrum fluorescent lamps.

Hazards

One hazard of UV exposure is skin burns or erythema (reddening). With extended exposure, blistering can occur. Skin does not absorb all UV wavelengths the same, and the erythemal effects vary across the absorbed spectrum. Wavelengths less than 280 nm are absorbed in the most outer layer of the skin (stratum corneum of the epidermis), wavelengths between 280 and 320 nm are absorbed in deeper layers of the skin (dermis), and those between 320 and 380 nm are absorbed in the outer layer of the skin (epidermis). Thus, exposures to wavelengths at or near 300 nm are most likely to produce erythema. Erythema has a latent period of 2 hr or more, depending on the degree of UV exposure. Some people expose themselves to UV to get a "tan." Such exposures cause the pigment in the skin (melanin granules) to migrate toward the surface (tanning). The likelihood of erythema is reduced with increased natural or conditioned (tanned) levels of pigmentation.

Other hazards of extended UV exposure are skin cancer and skin aging (photoaging). Skin cancer is more prevalent among people who spend considerable time in sunlight and it occurs more readily among those who have little skin pigment. People with a history of severe UV burns, particularly if severe burns occur in childhood (young skin), have a higher incidence of skin cancer, as do people with occupations requiring extensive work in sunlight and those who live in sunny regions.

Another hazard of UV exposure is keratitis—inflammation of the cornea of the eye. As noted in Figure 21-2(b), UV in the 280- to 400-nm range and some UV in the 200- to 280-nm range is absorbed by the cornea. The threshold for injury is greatest at or near 280 nm. Initially, one has a sensation of sand in the eyes. In addition, the eyes may water and the eyelids may swell shut.

Controls

The primary controls for UV exposures are limiting exposure, particularly to most harmful wavelengths, and using absorbing materials. In some cases, it may be possible to limit the UV energy at the source.

Length of Exposure Limiting exposures to UV prevents erythema or keratitis. The duration depends on the intensity of UV irradiation. Limiting extended exposures seems to reduce the likelihood of skin cancer. However, there are other factors involved in skin cancer risks.

Selective Wavelengths All UV wavelengths can produce harm. However, the effects that are most likely to appear differ according to wavelength. Controlling exposure to the most harmful wavelengths can help reduce risks of skin burns and cancer. For example, informed selection of UV lamps, which differ in wavelengths emitted, will lessen risks. Similarly, protective coating for the skin, such as suntan lotions and creams, may filter certain wavelengths. They have a sun protective factor rating that rates filtering of UV rays in general. The ratings of products range from zero to more than 40. Some filtering ingredients are more effective than others, because each ingredient protects for different wavelengths.

Absorption of Ultraviolet Wavelengths   It is quite easy to shield the skin and eyes from UV exposure. Shielding limits the energy level reaching the eyes or skin. Because UV radiation is readily absorbed by many materials, one should select those that eliminate or minimize harmful UV exposures. Tanning lotions have a sunscreen index and sunglasses have a UV absorption index, both of which indicate degrees of shielding. Workers, such as welders and other who may be exposed to high levels of UV, must wear appropriate eye and skin protection (see Chapter 28 on personal protective clothing). High-intensity mercury vapor discharge lamps have an outer protective glass envelope that absorbs UV wavelengths less than 320 nm. Protective eyewear must accompany sunlamp use.

Standards

There are several standards on UV exposures. The American Conference of Government Industrial Hygienists (ACGIH) maintains exposure standards based on wavelength, exposure time, and irradiance.[4] Table 21-3 lists the ACGIH TLVs. The FDA regulates sunlamp and UV lamps (21 CFR 1040.20) and high-intensity mercury vapor discharge lamps (21 CFR 1040.30).

Measurement

There is a variety of UV detection devices. They convert radiation arriving at a sensing device or medium to some form of display. Conversion methods include electrical (photovoltaic cells and phototubes), thermal (thermopile), and chemical (photographic plates). For most devices, selective filters determine what wavelengths arrive at the sensor.

21-4 INFRARED RADIATION

Infrared radiation has wavelengths from 700 nm to 1 mm and is characterized by smaller bands. IR-A (near infrared) is the spectral region from 701 to 1400 nm. IR-B includes wavelengths from 1.4 to 100 μm. The IR-C spectrum is 0.1 to 1 mm. Most radiative heat transfer involves the infrared region. Sources of infrared typically are sources of radiative heat, including fires and open flames, stoves, electrical heating elements, certain lasers, and many other sources.

Hazards

Near infrared radiation (700–1400 nm) passes through the lens of the eye to the retina or is refracted from other tissues (see Figure 21-2(d)). High energy levels can cause a variety of eye disorders, among which is scotoma. Scotoma is loss of vision in a portion of the visual field resulting from damage to the retina where radiation is absorbed. Other disorders range from simple reddening from low-level exposures to swelling of the eye, hemorrhaging, and lesions. Extended exposures to infrared radiation can cause cataracts.

TABLE 21-3 Sample Exposure Limits for Ultraviolet Radiation on Unprotected Skin and Eyes[a]

Wavelengths (nm)	TLV (J/m²)	Relative Spectral Effectiveness S_λ
180	2,500	0.012
190	1,600	0.019
200	1,000	0.030
210	400	0.075
220	250	0.120
230	160	0.190
240	100	0.300
250	70	0.430
260	46	0.650
270	30	1.000
280	34	0.880
290	47	0.640
300	100	0.300
310	2,000	0.015
320	2.9×10^4	0.010
330	7.3×10^4	0.00041
340	1.1×10^5	0.00028
350	1.5×10^5	0.00020
360	2.3×10^5	0.00013
370	3.2×10^5	0.000093
380	4.7×10^5	0.000064
390	6.8×10^5	0.000044
400	1.0×10^6	0.000030

Note: for t_{max} = maximum exposure time in seconds = $0.003/E_{eff}$,
where $E_{eff} = \Sigma E_\lambda S_\lambda \Delta_\lambda$ and E_{eff} = effective irradiance relative to a monochromatic source at 270 nm in watts per centimeter squared.
E_λ = spectral irradiance in watts per centimeter squared per nanometer,
S_λ = relative spectral effectiveness (unitless), and
Δ_λ = band width in nanometers.
[a] *2004 Threshold Limit Values for Chemical Substances and Physical Agents and Biological Exposure Indices*, American Conference of Governmental Industrial Hygienists, Cincinnati, OH. (Note: This publication is updated annually and standards may change. Refer to the current version for additional details.)

Common examples are glassblower's and bottlemaker's cataracts that result from looking into fire and heat sources. Iron workers who peer into furnaces extensively have a high incidence of cataracts.

High levels of infrared heat also can cause ignition of materials and fire. Refer to chapter 16.

Controls

To limit the danger from infrared radiation, limit the duration of exposure and the intensity of exposure. Because the danger is mainly to the retina of the eye, looking into infrared sources should be avoided. The intensity of exposure is most easily reduced by shielding. Eyewear that absorbs and reduces the amount of infrared reaching the eye should be worn. Lenses in glasses, goggles, or faceshields must have the correct shade to reduce harmful levels (see Chapter 28).

21-5 HIGH-INTENSITY VISIBLE LIGHT

Visible light occurs in the region from 380 (violet) through 750 nm (red). It passes through the cornea and lens and is focused on the retina (see Figure 21-2(d)). There are many sources of visible light, including natural light from the sun and artificial light sources. Most objects we see are reflected light.

Hazards

If energy levels for visible light are too high, they can cause injury to the eyes. Injury may involve the retina and other parts of the eye. Sources of high-intensity visible light are welding, carbon arc lamps, and some lasers.

Controls

Controls include enclosing the source, limiting source intensity, and shielding or filtering the source from the eyes. For example, portable panels screen welding operations so that the bright source does not reach people. Welders should wear protective eyewear with the applicable filter density for the type of operation (see Chapter 28). Carbon arc lamps are enclosed except for the aperture to direct the light where it is needed. Lasers with dangerous levels of visible light also have enclosures.

21-6 LASERS

The word *laser* is an acronym for light amplification by stimulated emission of radiation. A laser is a source of intense, coherent, and directional optical radiation. Lasers are usually composed of an energy source, a resonant cavity, and an active lasing medium. A laser system is an assembly of electrical, mechanical, and optical components that includes a laser.

Lasers and laser systems are becoming quite common in today's high-technology society. There are lasers for welding and machining and there are lasers for accurate measurement of distance and alignment of equipment. Lasers are used in mining and stress analysis and in holographic devices at grocery checkouts, and they have many medical applications. There are laser pointers for lectures and presentation. The military uses lasers in weapon systems to measure distance to targets or in guidance.

Hazards

Some lasers are dangerous, whereas others are not. The hazard depends on intensity and wavelength of light beams, duration of exposure, means of exposure, and the part of the body exposed.

Materials exposed to laser radiation can burn if energy levels are high enough and heat builds up sufficiently to start combustion. Lasers can cut or remove materials. Controlled local heating with laser beams in certain liquids causes the material to change state. Some machines apply this principle to create prototype plastic parts. In medical applications, lasers remove unwanted tissue or cut tissue.

The effects of lasers on people are essentially the same as for visible, ultraviolet, and infrared radiation. Lasers fall in these bands of wavelengths. Like other nonionizing radiation, the greatest danger is to the eye. Dependent on wavelength and the location where energy is absorbed, damage may be to the cornea, lens, retina, or other parts of the eye (refer to Figure 21-2). Dangers are greatest when a beam is viewed directly and may be reduced when a beam is reflected. As illustrated in Figure 21-3, properties of the reflecting surface may keep the beam essentially intact (specular reflection) or scatter the light (diffuse reflection). A polished, flat surface (like a mirror or window glass) reflects a beam with minimal absorption and little diffusion and poses the greatest reflected danger.

Damage ranging from erythema (reddening) to blistering and charring also may be inflicted on the skin. Compared with the region from 315 nm to 1 mm, effects to the skin are somewhat reduced in the 200 to 315 nm region.

A classification system for laser hazards divides lasers into five categories. Table 21-4 details this classification scheme.

Controls

The controls required depend on the class of laser. Where protection is heeded, controls may include enclosure of the laser source, control of potentially reflective surfaces, inter-

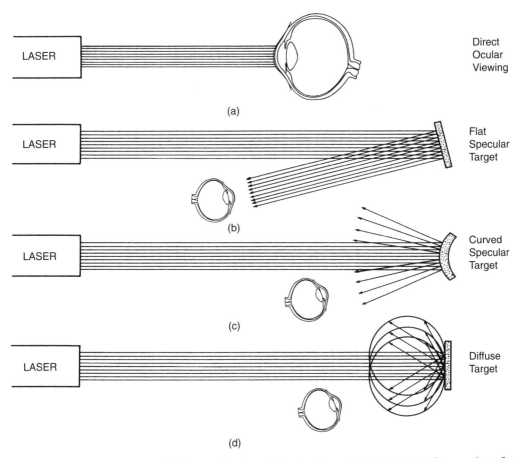

Figure 21-3.   Laser radiation may be viewed directly (a) or reflected (b–d); (b) flat specular reflection; (c) curved specular reflection; (d) diffuse reflection.

TABLE 21-4 Laser Device Hazard Classification Definitions[a]

Category	Definition
Class 1	Any laser device that cannot emit laser radiation levels in (exempt) excess of the AEL for the maximum possible duration inherent to the design of the laser or laser system. The exemption from hazard controls strictly applies to emitted laser radiation hazards and not to other potential hazards.
Class 2	Visible (400–700 nm) CW laser device that can emit a power exceeding the AEL far Class 1 for the maximum possible duration inherent to the design of the laser or laser system but not exceeding 1 mW.
	Visible (400–700 nm) repetitively pulsed laser devices that can emit a power exceeding the appropriate AEL for Class 1 for the maximum possible duration inherent to the design of the laser device but not exceeding the AEL for a 0.25-s exposure.
Class 2a	A visible (400–700 nm) laser or laser system that is not intended for intrabeam viewing and does not exceed the exposure limit for 1,000 s of viewing time.
Class 3a	Lasers or laser systems that have (1) an accessible output power or energy between one and five times the lowest appropriate AEL for Class 2 for visible wavelengths and between one and five times the AEL for Class 1 far all other wavelengths or (2) do not exceed the appropriate exposure levels as measured over the limiting aperture (2.5 mW/cm^2) for visible CW lasers.
Class 3b	(1) *Infrared (1.4 μm–1 mm) and ultraviolet (200–400 nm) laser devices.* Emit a radiant power in excess of the AEL Class 1 for the maximum possible duration inherent to the design of the laser device. Cannot emit an average radiant power of 0.5 W or greater for T_{max} greater than 0.25 s or a radiant exposure of 10 J/cm^2 within an exposure time of 0.25 s or less.
	(2) *Visible (400–700 nm) CW or repetitive pulsed laser devices.* Produce a radiant power in excess of the AEL Class 1 for a 0.25-s exposure (1 mW for a CW laser). Cannot emit an average radiant power of 0.5 or greater for T_{max} greater than 0.25 s.
	(3) *Visible and near-infrared (400–1,400 nm) pulsed laser devices.* Emit a radiant energy in excess of the AEL Class 1. Cannot emit a radiant exposure that exceeds that required to produce a hazardous diffuse reflection (refer to standards for precise properties).
	(4) *Near-infrared (700–1,400 nm) CW laser devices or repetitively pulsed laser devices.* Emit power in excess of the AEL for Class 1 for the maximum duration inherent in the design of the laser device. Cannot emit an average power of 0.5 W or greater for periods in excess of 0.25 s.
Class 4	(1) *Ultraviolet (200–400 nm) and infrared (1.4 μm–1 mm) laser devices.* Emit an average power of 0.5 W or greater for periods greater than 0.25 s, or a radiant exposure of 10 J/cm^2 within an exposure duration of 0.25 s or less.
	(2) *Visible (400–700 nm) and near-infrared (700–1,400 nm) laser devices.* Emit an average power of 0.5 W or greater for periods greater than 0.25 s, or a radiant exposure in excess of that required to produce a hazardous diffuse reflection (refer to standards for precise properties).

Note: AEL (accessible emission limit) is the maximum accessible emission level within a particular class. The Class 1 AEL is that radiant power or energy of a laser under consideration such that no applicable exposure limit for exposure of the eye for a specified exposure duration can be exceeded under any possible viewing conditions with or without optical instruments, whether or not the beam is focused.

[a] From Technical Bulletin TB MED 524, Department of the Army, Washington, DC, June 1985.

locks on doors to locations where lasers are used, fail-safe pulsing controls to prevent accidental actuation, remote firing room and controls, use of baffles to limit locations of beams, and wearing of suitable protective eyewear and clothing. The FDA has standards for the classification and safety design features of lasers (21 CFR 1040.10 and 1040.11).

Other controls include warnings, training of users, and medical surveillance of users. Where lasers present dangers, warning signs and alarm systems are needed. Dangerous laser devices must have warning labels. Operators must learn how to operate laser systems correctly and safely. The training should include procedures to avoid pointing them where other people may pass or controlling access to the location where they are used. People who do not operate laser systems but work with teams using them must learn proper procedures. They must learn the hazards the systems present. If protective eyewear or clothing is required, users must wear them.

Medical surveillance includes eye and vision examinations and proper selection of protective eyewear and clothing.

The equipment itself and the operation for which it is used may present other hazards that must be controlled. Electrical hazards must be controlled with grounding, compliance with electrical codes, and lockouts for maintenance and servicing. Where lasers vaporize material in machining, often ventilation is required. If lasers cause materials to fly, such as in cutting and machining operations, guarding is needed to contain the flying particles. Ionizing radiation from some lasers requires proper shielding. Cryogenic systems to cool lasers must be properly designed, operated, and maintained to prevent injury from the extremely cold coolants.

Standards

Tables 21-5 through 21-7 give U.S. Army exposure standards for various wavelengths based on potential damage to the eyes and skin. ANSI publishes laser exposure standards, as does the ACGIH. There are some differences among standards.

Measurement

The measurement of laser radiation is expensive and difficult to make. Manufacturers normally certify the source energy and wavelengths emitted from lasers, and these data normally are adequate to recognize hazards and establish suitable controls for laser systems. Some apply a factor of safety in selecting and implementing controls.

21-7   OTHER NONIONIZING RADIATION

There are devices and systems that use or emit nonionizing radiation for which there are no consensus standards for safety. In some cases, little is known about the dangers to people, particularly over extended periods of exposure at levels that are less than immediate injury levels. Some devices are in wide use and have drawn public concern. An example is the widespread use of monitors or visual display terminals (VDTs) in office automation. Many manufacturers have reduced the emission levels in newer monitors and the change to flat panel displays reduces emissions even further.

Recent studies evaluated extended exposures to nonionizing radiation. Some report evidence that extended exposures to microwaves may be related to development of cancerous tumor in rats. Statistical studies suggest increased rates of leukemia among power station operators, aluminum workers, power and telephone linemen, and other workers chronically exposed to electric and magnetic fields. As yet, results of such studies are not conclusive regarding harmful effects to humans.

The current microwave standard is based mainly on thermal effects in tissue. Other biological effects are under investigation. Not all effects are harmful; some are beneficial. Incorporation of new information into exposure standards generally is a slow process.

TABLE 21-5 Exposure Limits[a] for Direct Ocular Exposures (Intrabeam Viewing) from a Laser Beam[b]

Spectral Region	Wavelength (nm)	Range of Exposure Times[c] (s)	Exposure Limit	Defining Aperture (mm)
UV-C	200–302	10^{-9} to 3×10^4	$3\,\text{mJ/cm}^2$	1^d
UV-B	303	10^{-9} to 3×10^4	$4\,\text{mJ/cm}^2$	1^d
	304	10^{-9} to 3×10^4	$6\,\text{mJ/cm}^2$	1^d
	305	10^{-9} to 3×10^4	$10\,\text{mJ/cm}^2$	1^d
	306	10^{-9} to 3×10^4	$16\,\text{mJ/cm}^2$	1^d
	307	10^{-9} to 3×10^4	$25\,\text{mJ/cm}^2$	1^d
	308	10^{-9} to 3×10^4	$40\,\text{mJ/cm}^2$	1^d
	309	10^{-9} to 3×10^4	$63\,\text{mJ/cm}^2$	1^d
	310	10^{-9} to 3×10^4	$100\,\text{mJ/cm}^2$	1^d
	311	10^{-9} to 3×10^4	$160\,\text{mJ/cm}^2$	1^d
	312	10^{-9} to 3×10^4	$250\,\text{mJ/cm}^2$	1^d
	313	10^{-9} to 3×10^4	$400\,\text{mJ/cm}^2$	1^d
	314	10^{-9} to 3×10^4	$630\,\text{mJ/cm}^2$	1^d
UV-A	315–400[e]	10^{-9} to 10	$0.56t^{1/4}\,\text{J/cm}^2$	1^d
	315–400	10 to 10^3	$1.0\,\text{J/cm}^2$	1
	315–400	10^3 to 3×10^4	$1.0\,\text{mW/cm}^2$	1
Light	400–700	10^{-9} to 1.8×10^{-5}	$5 \times 10^{-7}\,\text{J/cm}^2$	7
	400–700	1.8×10^{-5} to 10	$1.8t^{3/4}\,\text{mJ/cm}^2$	7
	400–550	10 to 10^4	$10\,\text{mJ/cm}^2$	7
	550–700	10 to T_1	$1.8t^{3/4}\,\text{mJ/cm}^2$	7
	550–700	T_1 to 10^4	$10\,C_B\,\text{mJ/cm}^2$	7
	400–700	10^4 to 3×10^4	$C_B\,\mu\text{W/cm}^2$	7
IR-A	701–1,049	10^{-9} to 1.8×10^{-5}	$5C_A C_p \times 10^{-7}\,\text{J/cm}^2$	7
	701–1,049	1.8×10^{-5} to 10^3	$1.8C_A t^{3/4}\,\text{mJ/cm}^2$	7
	1,050–1,400	10^{-9} to 5×10^{-5}	$5C_p \times 10^{-6}\,\text{J/cm}^2$	7
	1,050–1,400	5×10^{-5} to 10^3	$9t^{3/4}\,\text{mJ/cm}^2$	7
	701–1,400	10^3 to 3×10^4	$320C_A\,\mu\text{W/cm}^2$	7
IR-B & C	1.4–$10^3\,\mu\text{m}$	10^{-9} to 10^{-7}	$10^{-2}\,\text{J/cm}^2$	$1, 11^f$
	1.4–$10^3\,\mu\text{m}$	10^{-7} to 10	$0.56t^{1/4}\,\text{J/cm}^2$	$1, 11^f$
	1.4–$10^3\,\mu\text{m}$	>10	$0.1\,\text{W/cm}^2$	$1, 11^f$

[a] Exposure limits are for maximum permissible exposure to laser radiation under conditions to which nearly all persons may be exposed without adverse effects. The values should be used as guides in the control of exposures and should not be regarded as fine lines between safe and dangerous levels.

$C_A = 10^{[0.002(\lambda - 700)]}$ for $\lambda = 700$–1049 nm
  $= 1$ for $\lambda = 400$–700 nm
  $= 5$ for $\lambda = 1,050$–1,400 nm
$C_B = 1$ for $\lambda = 400$–550 nm
  $= 10^{[0.015(\lambda - 550)]}$ for $\lambda = 550$–700 nm
$T_1 = 10 \times 10^{[0.02(\lambda - 550)]}$ for $\lambda = 550$–700 nm
$C_p = 1/(F)^{1/2}$ for PRF (pulse repetition frequency) ≤ 100 Hz is determined from charts for PRFs from > 100 Hz $\leq 1,000$ Hz
  $= 0.06$ for PRFs $> 1,000$ Hz

These values of C_p only apply for $t \leq 10\,\mu\text{s}$. Determining C_p is more complex for $t > 10\,\mu\text{s}$. Refer to standards.
[b] From Technical Bulletin TB MED 524, Department of the Army, Washington, DC, June 1985.
[c] The exposure limit at 1,540 (erbium) for a single pulse ($<1\,\mu\text{s}$) is 1 J/cm².
[d] Or $0.56t^{1/4}$ J/cm².
[e] Or not to exceed 1 J/cm² over 24 hr.
[f] 1 mm for 1,400 to 105 nm; 11 mm for 105 to 106 nm.

TABLE 21-6 Exposure Limits for Viewing a Diffuse Reflection of a Laser Beam or an Extended Source Laser[a]

Spectral Region	Wavelength (nm)	Exposure Times[b] (S)	Exposure Limit[b]
Light	400–700	10^{-9} to 10	$10t^{1/3}$ J/cm² sr⁻¹
	400–550	10 to 10^4	21 J/cm² sr⁻¹
	550–700	10 to T1	$3.83t^{3/4}$ J/cm² sr⁻¹
	550–700	T1 to 10^4	$21C_B$ J/cm² sr⁻¹
	400–700	10^4 to 3×10^4	$2.1C_B$ mW/cm² sr⁻¹
Near	700–1,400	10^{-9} to 10	$10C_A t^{1/3}$ J/cm² sr⁻¹
Infrared	700–1,400	10 to 10^3	$3.83C_A t^{3/4}$ J/cm² sr⁻¹
	700–1,400	10^3 to 3×10^4	$0.64C_A$ W/cm² sr⁻¹

[a] From Technical Bulletin TB MED 524, Department of the Army, Washington, DC, June 1985.
[b] For C_A, C_B, and T_1, see Table 21-5.

TABLE 21-7 Protection Standards for Skin Exposure to a Laser Beam[a]

Spectral Region	Wavelength (nm)	Exposure Time[b] t(s)	Exposure Limit[b]
UV	200–400	10^{-9} to 3×10^4	Same as Table 21-5
Light and infrared A	400–1,400	10^{-9} to 10^{-7}	$2 C_A \times 10$–2 J/cm²
	400–1,400	10^{-7} to 10	$1.1 C_A t^{1/4}$ J/cm²
	400–1,400	10 to 3×10^4	$0.2 C_A$ W/cm²
Infrared B and C	1.4 μm–1 mm	10^{-9} to 3×10^4	Same as Table 21-5

[a] From Technical Bulletin TB MED 524, Department of the Army, Washington, DC, June 1985.
[b] The limiting aperture for all exposure limits is 1 mm for wavelengths less than 0.1 mm. The limiting aperture for wavelengths more than 0.1 mm is 11 mm. Whole-body exposure should be limited to 10 mW/cm². The above limits refer to a laser beam having a cross-sectional area less than 100 cm².
[c] For C_A, see Table 21-5.

The rapid expansion in use of cell phones and their compact designs that place them against the head have raised some long-term health concerns related to high use levels. This concern has not been fully resolved, but the level of exposure for most low-frequency users does not seem to be significant. One can obtain radiation data for phone models.

Visual Display Terminals and Computer Monitors

When office automation exploded in the 1980s, there was controversy of many kinds related to computer equipment. Some believed that radiation emitted by computer monitors created risks to fetuses. Much of the concern arose from clusters of female office workers who miscarried or bore children with birth defects. One 5-year study found no effect on miscarriages for women who work on VDTs for fewer than 20 h per week and a 5% greater incidence of miscarriage for those who work at VDTs more than 20 hr per week. The controversy seems to have disappeared, because studies showed no strong link between VDT use and miscarriages and birth defects. In addition, manufacturers reduced the emission levels.

The attention on hazards of VDT workstations has had a number of beneficial effects. Researchers identified a number of problems with VDT workstations and environments that contribute to cumulative trauma, visual, and other disorders. Conventional furniture did not provide a suitable fit between worker and work stations for computers and VDTs. The quality of the visual image on VDT screens, glare, lighting, and other factors were identified as contributing to VDT workstation problems. Newer display technologies and designs and higher screen resolutions have reduced the visual problems considerably.

Low-Frequency Electric and Magnetic Fields

For a period in the 1980s, there was significant attention on low-frequency electric and magnetic fields. Power transmission frequencies (50–60 Hz) and below are called extremely low frequency (ELF) and most often occur around extra high-voltage (EHV) transmission lines. They also occur in special radio transmission frequencies that penetrate underwater for submarines. There are also many nontransmission sources for power frequency fields, such as wall wiring, appliances, and lighting fixtures.

Public concern over power transmission lines initially involved aesthetics of towers, property right-of-way issues, and nuisance effects, which include interference with radio and television reception, audible noise, and induced shocks that can occur when a person standing beneath an EHV line touches a large ungrounded metal object such as a truck or farm vehicle. Potential health effects were first noted in the early 1970s in scientific and medical literature. Under certain circumstances, the membranes of cells can be sensitive to even fairly weak externally imposed low-frequency electromagnetic fields, and biochemical responses can be triggered. A government-sponsored review of the ELF health effects literature found the existing information complex and inconclusive. However, effects are clearly demonstrated at the cellular level, and epidemiological evidence is beginning to provide a basis for concern about risks from chronic exposure. Some states have established limits on field strength in right of ways.[5]

EXERCISES

1. A worker is exposed to microwave radiation. Properties of the exposure are as follows: type of source, continuous wave; power density at worker location, 15 mW/cm^2; duration of exposure, 3 min. Is the exposure allowable under microwave radiation exposure standards?

2. A microwave oven has a 6-GHz source. If the energy incident on the inside of the door is 200 mW/cm^2 and the door has a 32×32 mesh screen, what is the energy level passing through this screen?

3. A worker with unprotected eyes and skin is exposed to UV radiation. The properties of the exposure are as follows: spectrum, 290 nm; duration, 6 h; irradiance, 8 mJ/cm^2. Does the exposure exceed recommended standards?

4. It is known that a work environment has a broadband ultraviolet source, uniformly distributed over the range 215 to 285 nm and having a uniform irradiance of $7 \text{ mW/cm}^2 \text{nm}^{-1}$ over the range.

   (a) Compute the effective irradiance (E_{eff}), using bandwidths of 10 nm.

   (b) Determine the allowable exposure time.

5. A worker receives a direct ocular exposure to a laser. Determine if the TLV is exceeded for each of the following conditions:

Exposure wavelength	Time	Irradiance
(a) 307 nm	10 s	28 mJ/cm^2
(b) 410 nm	15 s	10 mJ/cm^2
(c) 800 nm	1.5 s	10 mJ/cm^2
(d) 560 nm	200 s	15 mJ/cm^2.

6. A worker receives skin exposure to a laser beam. Determine whether the TLV is exceeded for the following conditions: wavelength, 1,200 nm; exposure time, 1×10^{-8} s; irradiance, 0.5 J/cm^2.

7. Conduct a review of the literature to identify the current state of knowledge related to cellular phone use and identify any exposure standards that have emerged.

REVIEW QUESTIONS

1. Which forms of nonionizing radiation are

    (a) absorbed by the cornea of the eye?

    (b) pass through eye tissue?

    (c) absorbed by the lens of the eye?

    (d) absorbed by the retina of the eye?

2. Compared with other body tissues, why is the eye highly susceptible to damage from nonionizing radiation?

3. What are five effects of microwaves on humans?

4. What controls are used to limit exposures to microwaves?

5. What are the dangers of UV radiation?

6. What are the dangers from exposure to infrared radiation?

7. What controls can reduce or eliminate UV and infrared dangers?

8. What are the dangers of high-intensity visible light?

9. What controls limit these dangers?

10. What are the dangers associated with lasers?

11. What parts of the body are most susceptible to laser injury?

12. What controls reduce the hazards of lasers?

13. What are possible dangers associated with VDTs?

NOTES

1 Public Law 90-602, Radiation Control for Health and Safety Act of 1960.

2 29 CFR 1910.97.

3 *Threshold Limit Values for Chemical Substances and Physical Agents & Biological Exposure Indices*, American Conference of Government Industrial Hygienists (ACGIH), Cincinnati, OH, annual update.

4 *Threshold Limit Values for Chemical Substances and Physical Agents & Biological Exposure Indices*, American Conference of Governmental Industrial Hygienists, Cincinnati, OH, annual update. (Note: This publication is updated annually and standards in it may change.)

5 Nair, I., Morgan, M., and Florig, H. K., *Biological Effects of Power Frequency Electric and Magnetic Fields*, Office of Technology Assessment, Congress of the United States, May, 1989.

BIBLIOGRAPHY

American National Standards Institute, New York:

C95.1 Safety Levels with Respect to Human Exposure to Radio Frequency Electromagnetic Fields, 3 kHz to 300 gHz.

C95.2 Standard for Radio Frequency Energy and Current Flow Symbol

C95.3 Measurement of Potentially Hazardous Electromagnetic Fields—RF and Microwave

C95.4 Recommended Practices for Determining Safe Distances from Radio Frequency Transmitting Antennas When Using Electric Blasting Caps During Explosive Operations

Z136.1 Safe Use of Lasers

Z136.2 Safe Use of Optical Fiber Communication Systems Utilizing Laser Diode and LED Sources

Z136.3 Safe Use of Lasers in Health Care Facilities

Z136.5 Safe Use of Lasers in Educational Institutions

Z136.6 Safe Use of Lasers in an Outdoor Environment

Control of Hazards to Health from Laser Radiation, Technical Bulletin TB MED 279, Department of the Army, Washington, DC, May 30, 1975.

GILCHREST, B. A., *Skin and Aging Processes*, CRC Press, Boca Raton, FL, 1984.

MICHAELSON, S. M., and LIN, J. C., *Biological Effects and Health Implications of Radiofrequency Radiation*, Plenum, New York, 1987.

ORN, MICHAEL K., *Handbook of Engineering Control Methods for Occupational Radiation Protection*, Prentice Hall, Englewood Cliffs, NJ, 1992.

PEARCE, B., ed., *Health Hazards of VDTs*, Wiley, New York, 1984.

POLK, C., and POSTOW, E., *Handbook of Biological Effects of Electromagnetic Fields*, CRC Press, Boca Raton, FL, 1986.

SLINEY, D., and WOLBARSHT, M., *Safety with Lasers and Other Optical Sources*, Plenum, New York, 1980.

STENECK, N. H., COOK, H. J., VANDER, A. J., and KANE, G. L., "The Origins of U.S. Safety Standards for Microwave Radiation," *Science*, 208:1230–1237 (1980).

WAXLER, M., and HITCHENS, V. M., *Optical Radiation and Visual Health*, CRC Press, Boca Raton, FL, 1986.

WEBER, M. J., ed., *Handbook of Laser Science and Technology*, Vols. 1–5, CRC Press, Boca Raton, FL, 1982–1987.

WINBURN, D. C., *Practical Laser Safety*, Marcel Dekker, New York, 1985.

CHAPTER 22

IONIZING RADIATION

22-1   INTRODUCTION

On March 28, 1979, an accident sequence took place at the Three Mile Island Unit 2 nuclear power facility. Widespread media attention, investigations, and a presidential commission (the Kemeny Commission) focused the attention of the nation on this event for a long time. The commission report cited many contributing factors, including operator error, lack of training, complexity of the failure, lack of information to operators, and design deficiencies in the control room. In addition, there was a bureaucratic snarl that withheld information about a similar malfunction at another plant. Among the materials released were 15 Ci of iodine (^{131}I).

On April 26, 1986, the nuclear power plant at Chernobyl, in the northern Ukraine region of the Soviet Union, exploded. Several people were killed and more than 100,000 people were evacuated. More radioactive iodine and cesium was released from this accident than from all nuclear bombs and atmospheric testing since 1945. Airborne radioactive iodine and cesium contaminated large portions of Europe. The explosion produced much international public discussion about resulting dangers from low-level exposures, cleanup costs, and responsibilities across national borders. By 2000, funding from many parts of the world started a clean up of the remaining plant.

During October and November of 1988, national attention focused through the news media on the safety of Department of Energy nuclear plants used to produce weapons for national defense. Reports addressed the dangers to workers (past and present) and surrounding communities at Fernald, Ohio, Hanford, Washington, and other locations. Government officials acknowledged the release of 230 tons of radioactive materials into the air and water at the Fernald plant and the release of 530,000 Ci of ^{131}I at Hanford between 1944 and 1957. One of the key issues in this discussion was the need to replace aging facilities and the multibillion dollar cost to the nation. With the end of the Cold War, projects began to clean up these plants.

Ionizing radiation draws a great deal of emotional response from the public. There are many factors contributing to this response, including fears related to the devastation of the atomic bombs dropped in World War II and fears of cancer and other health effects. There are economic issues, concern for terrorism, liability limits, long-term dangers of some isotopes, lack of understanding about ionizing radiation, and many other aspects that contribute to the opinions and beliefs expressed in the public debate. The complex issues confound the safety aspects of ionizing radiation. Public attention and discussion continues on such issues as low- and high-level waste disposal and decommissioning nuclear facilities after their useful life is over.

Safety and Health for Engineers, Second Edition, by Roger L. Brauer
Copyright © 2006 John Wiley & Sons, Inc.

The purpose of this chapter is to explore the hazards of and controls for ionizing radiation, not to address the many issues this topic involves. However, readers should perform an evaluation of this complex subject and, as best they can, separate the emotional elements involved in written material before forming their own opinions.

22-2 PHYSICS OF RADIATION

Ionizing radiation is any electromagnetic or particulate radiation capable of producing ions when it interacts with atoms and molecules.

Types of Radiation

The main types of radiation are x-rays, gamma rays, alpha particles, beta particles, neutrons, and other high-energy particles.

Alpha particles have the same structure as the nuclei of helium atoms: two protons and two neutrons. Relative to other forms of ionizing radiation, alpha particles are large. They have little penetrating power and are stopped by a piece of paper or the outer layer of skin.

A beta particle is an electron (negative charge) or positron (positive charge) separated from the nucleus of an atom. Smaller than alpha particles, beta particles travel at higher speeds and have enough penetrating power to pass through nearly $1/4$ in of tissue.

Gamma radiation is not a particle; it is energy waves in the electromagnetic spectrum. During radioactive decay, certain materials emit gamma rays from the nucleus of decaying atoms. Gamma rays have much greater penetrating power than beta particles and are best stopped by dense materials.

X-rays are similar to gamma rays. They are energy waves generated from outside the nucleus of an atom during decay or by impact from external electrons. X-rays have penetrating powers similar to gamma rays.

Neutrons are particles that have high mass and no charge. They are not produced spontaneously; they are produced by nuclear reactions. Neutron energies are classified as slow, medium, and fast. Because neutrons readily penetrate matter, high-density materials containing high levels of hydrogen atoms are necessary to stop them.

Sources of Radiation

There are two main sources of ionizing radiation: natural and artificial. Natural radiation sources include cosmic and gamma radiation found in certain soils. Building products made from these radioactive soils also give off natural radiation. These sources create external exposures. Natural radiation also is present in some ingested food and water and some inhaled air. These are internal exposures. Estimates indicate that people in the United States receive approximately 125 mrem of natural radiation per year: 100 mrem externally and 25 mrem internally. Natural exposures vary considerably by geographic location.

Artificial sources that expose the general population are certain consumer products, medical sources, occupational sources, and general environment sources. Products with ionizing radiation include television sets, computer monitors, some smoke detectors, luminous dials on clocks, and luminous signs. Medical sources are gamma rays, beta rays, and x-rays for diagnostic and treatment procedures. Occupational sources involve jobs related to manufacturing of products with radiation sources, medical services and research, mining, production of radioisotopes, nuclear fuel and weapons, transportation, and waste handling. There are both exposures during normal job functions and exposures during accidental releases.

Units of Measure

There are several kinds of measures related to ionizing radiation: measures for physical characteristics, for biological effects, and for converting physical measures to units of biological effects. There are also measures that express the amount of ionizing radiation absorbed by an exposed person.

Physical   The energy levels of ionizing radiation are measured in megaelectron volts (MeV). Units of measure of ionized air are roentgens (R), and a curie (Ci) is the amount of radioactive material that has a disintegration rate of 3.7×10^{10} atoms/s.

   Radioactivity decays with time. The time required for a material to lose half of its activity is its half-life. Half-life values are found in tables of physical properties of radioactive materials.

Radiation Dose   The amount of radiation absorbed by the body per unit of tissue mass is the rad (radiation absorbed dose). The dose corresponding to the absorption of 100 erg/g of tissue is 1 rad. A millirad (mrad) is one thousandth of a rad. Absorption may occur over the whole body or in local tissue.

Biological Effects   Radiation produces biological effects on the human body. Different types of radiation produce effects at different rates. Rems (roentgen equivalent man) are units of biological effect. The effect produced by a dose of 1 R of x-rays is 1 rem. Different types of radiation are absorbed at different rates. As a result, different doses are needed for different materials to produce similar biological effects. The relative biological effect (RBE) is the measure of biological effect (rem) divided by the dose (rad). Table 22-1 lists doses necessary for equivalent biological effects.

22-3   HAZARDS

Exposure to radiation may produce a variety of effects in humans. Damage is a function of the type of radiation, dose, the tissue and organs exposed, and age. In general, radiation affects rapidly developing cells most, which makes radiation therapy useful for rapidly growing cancer cells. It also makes radiation more dangerous for infants and children who have many rapidly developing cells.

   Some organs concentrate materials. For example, iodine concentrates in the thyroid gland. Physicians have treated hyperthyroidism with radioactive iodine, which, when con-

TABLE 22-1   Dose Equivalents for 1 rem of Radiation

Dose (rad)	Type of Radiation
1	1 R of gamma rays or x-rays
1	x-rays, gamma rays, or beta particle radiation
0.1	Neutrons and high-energy protons
0.05	Particles heavier than protons and with sufficient energy to reach the lens of the eye

centrated by the thyroid, destroys cells. Another example is concentration of strontium 90 (^{90}Sr) by mammary glands. Cows that consume food contaminated with ^{90}Sr produce highly contaminated milk. ^{90}Sr has a fairly long half-life and becomes a more persistent contaminant than other radioactive materials, such as ^{131}I, which has a short half-life. High-dose rates are usually acute exposures. Their effects include ulceration of skin and intestinal tissue and reductions in white cell production. Symptoms of acute radiation sickness are weakness, sleepiness, and eventually stupor, tremors, convulsions, and death. Symptoms may include nausea, vomiting and diarrhea, loss of hair, and bleeding. Death may occur after 1 or 2 days or may be delayed up to several weeks.

Low-dose rates may be acute or chronic. Effects of low-dose rates often are delayed. There may be damage to cell structure and function and cells may die. Low levels may produce reddening of skin or damage to internal organs. Ionizing radiation may produce genetic effects, leukemia and other cancers, cataracts, and reduction in life span. Statistical studies of populations form the basis for knowledge of low-level effects.

For whole-body radiation, the most critical organs and tissues are the lens of the eye, the blood-forming organs (red bone marrow), and the gonads. Internal radiation sources may affect several vital organs.

Linearity Hypothesis

High-dose rates, on the order of 500 to 600 rems, are fatal to approximately half of an exposed population. At the other extreme, people receive low levels of radiation from natural and other sources throughout their lifetime. Statistical studies have compared effects of low-level exposures to the general population, but it is difficult to estimate the actual effect, particularly when the main concern is cancers that appear as much as 30 to 40 years after exposure. Consequently, the picture at the low end is unclear. For convenience in estimating effects, the linearity hypothesis extrapolates the effects at high levels to effects at low levels. The effects lessen at lower levels. The hypothesis suggests that the likelihood of dying of cancer increases with exposure. Experts do not agree on the use of the linearity theory and the effects from low-level exposures.

Studies of effects became more controversial in 1981, when there was a recalculation of data from the explosions at Hiroshima and Nagasaki. The recalculations suggested that cancers resulted from lower-energy gamma rays, whereas previous studies had suggested that high-energy level neutrons and protons were the primary forms of radiation. These data suggested that low-energy forms of radiation were more dangerous than previously thought.[1]

22-4 EXPOSURE STANDARDS

In general, standards for ionizing radiation assume that effects are cumulative. Therefore, exposure limits include a period of exposure. The limit can be exceeded by a single, acute exposure or by repeated exposures over an extended time. Several organizations have exposure standards. A few will be noted in succeeding text.

Table 22-2 gives a summary of exposure standards from selected federal agencies. Some agencies, such as the Tennessee Valley Authority (TVA) and the Department of the Navy, have adopted standards more stringent than some found in Table 22-2. There are additional standards for emissions from electronic products,[2] emissions from nuclear power plants,[3] and transportation, marking, and packaging of radioactive materials.[4]

TABLE 22-2   Exposure Limits for Ionizing Radiation

Exposure	Limit	Source
Whole body; head and trunk; active blood-forming organs, 1.25 rem per calendar quarter lens of eyes; gonads	1.25 rem per calendar quarter	OSHA[a], NRC[b]
Hands and forearms; feet and ankles	8.75 rem per calendar quarter	NRC[b]
	18.75 rem per calendar quarter	OSHA[a]
Skin of whole body	0.5 rem per calendar quarter	NRC[b]
	7.5 rem per calendar quarter	OSHA[a]
Individuals younger than 18 yr of age	1/10 of above limits	OSHA[a], NRC[b]
Drinking water: Radium (^{226}Ra and ^{228}Ra) and gross alpha particle	5 pCi/l	EPA[c]
Drinking water: gross alpha particle activity (excluding radon and uranium)	15 pCi/l	EPA[c]
Drinking water: beta particles and photons from artificial sources to total body or any internal organ	4 mrem/yr	EPA[c]
Air quality, exposure of the public		
Whole body	25 mrem/yr	EPA[d]
Critical organ	75 mrem/yr	EPA[e]
Average person	500 mrem/yr, 2 mrem/hr	NRC[f]
	100 mrem/7 days	
Nuclear industry worker	5(N–18) rem/yr, 3 rem/quarter	NRC[g]

[a] 29 CFR 1920.1096.
[b] 10 CFR 20.101: 1/10 for persons younger than 18 yr of age, 10 CFR 20.104.
[c] 40 CFR 141.15.
[d] 40 CFR 141.16; based on 21 intakes per day for most radioactive materials.
[e] 40 CFR 61.102.
[f] 10 CFR 20.105.
[g] 10 CFR 20.101, where N is age in years at last birthday.

22-5   CONTROLS

There are several types of engineering controls to limit dangerous exposures to radiation: limiting radiation emissions at the source, limiting time of exposure, extending the distance from a source, and shielding. Other controls also help prevent dangerous exposures.

Limit Radiation Source

The best way to prevent radiation exposure is to limit the amount of radiation from a source. Limiting the quantity of ionizing material achieves this goal.

Time

Another method for minimizing risk to people is to limit the duration of exposure. One can prevent access to locations where radiation sources exist, which prevents unnecessary exposures. Procedures also can limit the duration of exposure. Because exposures are considered cumulative and because radiation cannot be sensed, one must use measurement and dosimetry. Measurement assesses quantity, whereas dosimetry incorporates the duration of exposure.

Radioactive materials decay with time. Each material has a radioactive half-life. A half-life is the time required for a material to decay to half of its original energy level. Because radioactive materials decay, the danger of a material decreases with time. This is particularly true for materials having relatively short half-lives.

Distance

Generally, airborne particulates and gases that are contaminated are diluted with increasing distance. Particulates that are large enough will settle out of the air. Thus, distance will reduce the likelihood of exposure to radioactive materials released from an operation.

Radiation levels decrease with the square of distance from the source, the inverse square law. As shown in Figure 22-1, a person at one unit of distance from a source has some level of radiation exposure. At double the distance, the amount of radiation will be one fourth that of the first location. The level of radiation, I_r, is

$$I_r = \frac{1}{n_d^2},\qquad(22\text{-}1)$$

where n_d is the number of distance units relative to some reference location. Distance is an appropriate solution for certain kinds of exposures. For example, distance is helpful in reducing external exposures but of little value for internal exposures.

Example 22-1   A dental technician takes x-rays of patients' teeth. The control cord for the x-ray machine allows the technician to stand 5 ft away. If the control cord is lengthened and allows the technician to stand 15 ft away, what reduction in radiation level would result?

Applying Equation 22-1 and assuming the exposure level for the initial position to be unity, the radiation at the new location would be $I_r = 1/(3)^2 = 1/9$ of that for the first position.

Shielding

Reducing radiation levels with shielding is another form of protection. Shielding effectiveness, the ability of a specific material to attenuate radiation, varies with different forms of radiation. For example, to some extent, air attenuates low-energy beta waves, but it has little effect for other forms of radiation. Also, hydrogen is an effective attenuation medium for low-energy level neutrons.

Attenuation properties of materials are measured in half-value thicknesses. One half-value thickness will cut the energy level of radiation arriving at the material to one half as it leaves the material after passing through it. The amount of radiation absorbed, R_a, and the amount transmitted, R_t, are given by

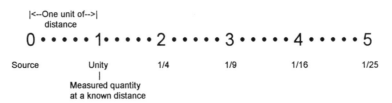

Figure 22-1.   Radiation decreases with the square of the distance from a source.

$$R_a = R_0\left[1 - \left(\frac{1}{2}\right)_n\right]$$ (22-2a)

and

$$R_t = R_0\left(\frac{1}{2}\right)_n,$$ (22-2b)

where

   n is the number of half-value thicknesses and

   R_0 is the amount of arriving radiation.

Published tables give attenuation properties of materials. Table 22-3 contains selected attenuation properties.

Example 22-2 One foot of water shields neutrons at 60 MeV. By what percent is the energy level of the neutrons reduced? What radiation level passes beyond the water shield?
    From Table 22-3, the half-value thickness of water for neutrons is 9.25 cm. The number of half-value thicknesses is (1 ft × 30.48 cm/ft/9.25 = 3.295 half-value thicknesses. The attenuation of the shield is $1 - (1/2)^{3.295} = 1 - 0.102$ or 89.8% reduction. The radiation level leaving the shield is 60(0.102) = 6.11 MeV.

Barriers

Barriers can protect many radiation sources. For example, walls or fences around operations involving radiation sources will keep people out who should not be there. Liners under holding ponds of contaminated waste will prevent leaching into streams or groundwater.

Other Safeguards

There are many other methods that help prevent or reduce exposures to radiation sources. Warnings, a variety of procedures, security systems, training of personnel, and analyzing systems for potential failure modes are some methods. American National Standards Institute (ANSI) standards and government regulations provide detailed guidance for many of these safeguards.

TABLE 22-3 Radiation Absorption Properties of Selected Materials

Material	Radiation Type	Half-Value Thickness	Attenuation Coefficient
Concrete	^{137}Cs	1.9 in	
	^{60}Co	2.6 in	
	^{198}Au	1.6 in	
	^{192}Ir	1.7 in	
	^{226}Ra	2.7 in	
Lead	^{137}Cs	0.65 cm	
	^{60}Co	1.20 cm	
	^{198}Au	0.33 cm	
	^{192}Ir	0.60 cm	
	^{226}Ra	1.66 cm	
Steel	^{137}Cs	0.64 in	
	^{60}Co	0.82 in	
	^{192}Ir	0.50 in	
	^{226}Ra	0.88 in	
Water	Neutrons	9.25 cm	0.602/cm

Warnings Warnings should mark locations and equipment where there are ionizing radiation sources. Figure 22-2 shows the standard ionizing radiation symbol. Packaging and labeling of radioactive material requires this symbol as well. Visual warnings, such as flashing lights and audio signals, help people in an area recognize when there is a danger associated with radiation operations or material is present that could be dangerous.

Evacuation Should there be a significant release of radioactive material that endangers people in a facility or the public located outside a facility, it is important to lessen their chance of exposure. Nuclear Regulatory Commission (NRC) and other regulations require evacuation plans for accidental releases.

Security Should radioactive material get into the wrong hands, there is a danger of exposure to numerous unknowing people. This has happened with medical materials. It has happened with reprocessed irradiated metal, which was distributed as furniture components. The metal was discovered when it triggered radiation alarms during a delivery to a site that monitored all departing and entering vehicles and people.

Procedures play a major role in controlling entry or exit of radioactive material from sites. Typical procedures include security, physical monitoring and manifest systems. Security procedures and physical security systems also can prevent unauthorized persons from entering a facility or entering dangerous locations.

Dosimetry Because one cannot sense radioactive material, it is essential to monitor exposures of people who work with and around such material. Various measurement methods are described in Section 22-6.

Training People who work with and around radioactive material need training to understand the hazards of ionizing radiation. They need to understand how to protect themselves

Figure 22-2. Standard radiation symbol.

and what procedures to follow. They need to develop skill in performing activities correctly and to know the value of protective clothing.

The public and people responsible for the public need training to deal with emergencies. If there are evacuation plans, police and other emergency organizations need to know what to do and need to have skill in completing assignments.

Complex systems involving radioactive materials often have a simulator, which allows trainers to create a variety of situations for teaching operators and others how to react, make judgments, and act correctly.

System Design and Analysis   In operations and facilities involving radioactive material, there is seldom the luxury of waiting for things to go wrong. Analytical techniques, such as those described in Chapter 36, are useful for anticipating what might go wrong. A notable study of nuclear power plant failures[5] applied such methods. Although failures in such complex systems are difficult to predict and foresee, such methods are essential in ensuring safety.[6]

22-6   MEASUREMENT

Instruments

Instruments for measuring ionized radiation typically include a sensing device and a readout device. Some are useful for field measurement, whereas other combinations come in small packages useful for dosimetry. Sensors are most critical, because different types of sensors are appropriate for different types of radiation. Sensors include Geiger-Mueller tubes (used in Geiger counters), ionization chambers, luminescent detectors, scintillation detectors, and photographic emulsions.

A Geiger-Mueller tube is a gas-filled chamber used to measure alpha, beta, and gamma radiation. Radiation entering the tube ionizes the gas and creates small currents that the instrument measures.

Ionization chambers measure beta rays, gamma rays, and x-rays using a charge placed on an electrode in a tube. Radiation ionizes the air surrounding the electrode and allows charge to leak away. The amount of charge lost is related to the amount of radiation arriving at the chamber. Pocket ionization detectors are common dosimetry devices.

Luminescent detectors measure exposures to neutrons. Arriving radiation changes the energy content of solids in the detectors. The energy change causes them to emit light. The amount of light is proportional to the energy change.

In scintillation detectors, incoming radiation strikes a thin layer of crystals or a solution of organic materials that produce light. Light output is proportional to the radiation absorbed. These devices measure alpha, beta, gamma, and slow neutron sources.

Radiation-sensitive photographic film detects gamma rays and x-rays. The radiation affects the emulsion similar to the way light does. The developed films are compared with standards to establish the radiation exposure. Film badges are common dosimetry devices that use this method.

Surveys

OSHA and other agencies require surveying of areas that will be occupied to determine if they comply with regulations and standards for ionizing radiation. A survey may include background readings for radiation with follow-up readings at various times. It can also include analysis of operations and methods for each area of concern.

Dosimetry

One must monitor the exposure for each worker involved with ionizing radiation. Visitors' exposures are also important. OSHA and other organizations require such records. Typically, individuals wear film badges, pocket ionization detectors, or other instruments while in areas where exposures are possible. Exposures are recorded and the records of exposure are retained for each person.

22-7 RADON

Radon is a colorless, odorless, and radioactive gas that is formed from the complex decay of uranium. It is found naturally in many locations in soil, rock, and water and is often found in mining and ore tailings. The decay products of radon, called radon daughters, emit alpha, beta, and gamma radiation. Most have very short half-lives. In the late 1980s, radon gained much national attention, and programs to map the presence of radon in homes and buildings spread across the country. The concern resulted from recognition that radon gas can accumulate in houses and other buildings and the variety of decay products can be inhaled over extended periods. Concentrations may build within buildings to levels that are considered dangerous and could contribute to increased lung cancer rates.

Current standards recommend radon levels of 4 pCi/l or less. Methods to prevent buildup of radon concentrations include ventilating crawl spaces, foundation drain tile, and other locations. Assessment of construction sites for radon and control methods for radon are now an important element of facility design and real estate transactions.

22-8 IRRADIATED FOOD

A potentially useful application of radiation is in food preservation. Procedures for irradiating food to prevent spoilage and extend shelf life have been recognized for some time. The Food and Drug Administration has standards for the production, processing, handling, and packaging of food.[7] However, the public has been very reluctant to accept irradiated food products.

EXERCISES

1. A gamma radiation source, radium (^{226}Ra), is used in a hospital laboratory. If shielding is considered as a means of control, how many inches or centimeters are needed to reduce the radiation level to 1% of what a worker would be exposed to without shielding? Assume the shielding material is

   (a) concrete

   (b) steel

   (c) lead

2. An alternative to Exercise 1 is to extend the distance between the radiation source and the worker. The initial design placed the worker 2 ft from the source. If a revised design placed the worker 10 ft from the source, what percent reduction in radiation exposure would there be for the second position relative to the first?

3. Another worker must reach into the instrument in Exercise 1 each day to make a calibration check. In so doing, the worker is exposed to some radiation. If only his hands and forearms are exposed and there are 60 working days per calendar quarter, how much radiation is the worker allowed to receive per work day to reach just the allowable limit in a quarter?

4. Visit a facility that uses radioactive material (power plant, laboratory, medical facility, etc.). Discuss with the staff what precautions they take in preparation for using a fissile source, during its use, and in disposing of used materials. Find out what contingency plans they have in place to deal with a release to the atmosphere or a loss of material by theft or other means.

REVIEW QUESTIONS

1. Name five kinds of radiation and characterize each.
2. Name two natural sources of radiation.
3. Name three products that produce or are sources of radiation.
4. Explain the following units of measure for radiation:
    (a) MeV
    (b) half-life
    (c) rad
    (d) roentgen
    (e) rem
    (f) RBE
5. What factors determine whether a radiation exposure is safe or harmful?
6. What effects may result from high dose rates of radiation?
7. What effects may result from low-dose rates of radiation?
8. What organs are most susceptible to damage from whole-body radiation exposure?
9. What is the linearity hypothesis?
10. What is a half-value thickness and for what means of control is it used?
11. List eight controls and safeguards that can reduce the exposure to or injury from radiation exposure.
12. Explain how each of the following works in measuring radiation:
    (a) Geiger-Mueller tube
    (b) ionization chamber
    (c) luminescent detectors
    (d) radiation sensitive photographic film
13. What is radon? Where is it found? How is it controlled?

NOTES

1 Marshall, E., "New A-Bomb Studies after Radiation Estimates," *Science*, 212:48–51 (1981).

2 21 CFR 1020.

3 40 CFR 190, 10 CFR 20.105, and 10 CFR 20.106.

4 49 CFR 172.

5 *Reactor Safety Study: An Assessment of Accident Risks in U.S. Commercial Nuclear Power Plants,* WASH-1400, U.S. Nuclear Regulatory Commission, U.S. Government Printing Office, Washington, DC, 1975. Available from National Technical Information Service, Springfield, VA.

6 H. W. Lewis, "The Safety of Fission Reactors," *Scientific American*, 242:3:53–65 (1980).

7 21 CFR 179.

BIBLIOGRAPHY

American National Standards Institute, New York: There are numerous standards on safety in construction and operation of nuclear facilities, radiation protection, radiation instrumentation, and radioactive equipment. Readers should refer to a complete listing of ANSI standards and standards of the Nuclear Regulatory Commission (NRC), 10 CFR.

ATTIX, F. H., *Introduction to Radiological Physics and Radiation Dosimetry*, Wiley, New York, 1986.

BRODSKY, A., ed., *Handbook of Radiation Measurement and Protection*, CRC Press, Boca Raton, FL, 1980.

GUSEV, IGOR, GUSKOVA, ANGELINA, and METTLER, FRED A., eds, *Medical Management of Radiation Accidents*, 2nd ed., CRC Press, Boca Raton, FL, 2001.

HOPKE, P. K., ed., *Radon and Its Decay Product: Occurrence, Properties, and Health Effects*, American Chemical Society, Washington, DC, 1987.

KLEMENT, A. W., Jr., *Handbook of Environmental Radiation*, CRC Press, Boca Raton, FL, 1982.

LEWIS, E. E., *Nuclear Power Reactor Safety*, Wiley, New York, 1977.

MILLER, K. L., and WEIDNER, W. A., *Handbook of Management of Radiation Protection Programs*, CRC Press, Boca Raton, FL, 1986.

ORN, MICHAEL K., *Handbook of Engineering Control Methods for Occupational Radiation Protection*, Prentice Hall, Englewood Cliffs, NJ, 1992.

PROFIO, A. E., *Radiation Shielding and Dosimetry*, Wiley, New York, 1979.

SHAPIRO, J., *Radiation Protection,* Harvard University Press, Cambridge, MA, 1972.

STEWART, D. C., *Data for Radioactive Waste Management and Nuclear Applications*, Wiley, New York, 1985.

TOTH, L. M., MALINAUSKAS, K. P., EIDAM, G. R., and BURTON, H. M., eds., *The Three Mile Island Accident: Diagnosis and Prognosis,* American Chemical Society, Washington, DC, 1986.

WHICKER, F. W., *Radioecology: Nuclear Energy and the Environment*, 2 vols., CRC Press, Boca Raton, FL, 1982.

CHAPTER 23

NOISE AND VIBRATION

23-1 SOUND

Hearing is an essential method by which humans receive information. Of all the human senses, only vision has a higher rate of information transfer. Through hearing, we receive critical information, such as the cry of a child or the wail of a siren, and we derive information from speech and enjoy the pleasure of music.

Sound is the propagation, transmission, and reception of waves in some medium, most commonly air. Noise refers to unwanted signals.

23-2 PHYSICS

Terms, Definitions, and Units of Measure

Sound waves may have a single frequency (pure tone) or may be a combination of frequencies (a spectrum). White noise is a spectrum with a uniform distribution of energy over the bandwidth. Most sounds, however, have a complex assembly of frequencies. Frequency analysis or spectral analysis instruments can sort complex sounds into a distribution of frequencies. The range of frequencies of greatest interest is that which humans can hear: approximately 16 Hz to 20,000 Hz.

Noise may be distributed over a wide range of frequencies (broad band) or over a narrow range of frequencies (narrow band). Short-duration noise pulses, which can occur once or be repetitive, are impulse noise.

The period for one cycle, T, is the inverse of frequency, f:

$$T = \frac{1}{f}. \tag{23-1}$$

The distance travelled in one period is the wavelength, λ. The speed that sound travels, c, in air is given by

$$c = \lambda f. \tag{23-2}$$

The approximate speed of sound in air at 70°F is 1,130 ft/s. It varies with temperature because temperature affects air density. Another expression for the speed of sound in air is

$$c = 49(T_a + 460)^{1/2}, \tag{23-3}$$

where T_a is the temperature of air in degrees Fahrenheit. The speed of sound in water is 4,700 ft/s, and in many metals it is more than 16,000 ft/s.

Safety and Health for Engineers, Second Edition, by Roger L. Brauer

Amplitude is the intensity of sound at a source or at some distance from the source. Units of measure for amplitude are given as force per unit area, commonly expressed as dynes per square centimeter (dyn/cm^2) and Newtons per square meter (N/m^2).

Sound power describes the energy radiated from a sound source, and it is expressed in units of watts (W). Sound power level, L_w, in decibels is the power relative to a value of 10^{-12} W.

Sound pressure refers to the pressure changes arriving at some location. The sound pressure level, L_p, in decibels is the pressure level relative to a value of 2×10^{-5} N/m^2. Sound intensity at some location is the average rate at which sound energy moves through a unit area normal to the direction of propagation.

Sound intensity is expressed in joules per square meter per second (J/m^2/s). Sound intensity level, L_I, in decibels is the sound intensity relative to 10^{-12} W/m^2.

Decibels

There is a very wide range of sound pressures encountered in our world. The range extends at least six orders of magnitude. For example, the threshold of hearing is 0.00002 N/m^2 and the threshold for auditory pain is 20 N/m^2. For convenience, a logarithmic scale describes the range of sound measures. Decibels (dB) are dimensionless ratios of a measured value of sound relative to some reference value. For sound pressure level, L_p, decibels are derived from

$$L_p = 20 \log_{10} \frac{p}{p_o}, \tag{23-4}$$

where

p is the measured root mean square (rms) sound pressure and

p_o is the reference sound pressure (0.00002 N/m^2).

Similar expressions represent the decibel units of sound intensity, L_I,

$$L_I = 10 \log_{10} \frac{I}{I_o} \tag{23-5}$$

and sound power level, L_w,

$$L_w = 10 \log_{10} \frac{W}{W_o}. \tag{23-6}$$

Because of logarithmic scaling, sound pressure levels in decibels from two or more sources cannot be added arithmetically to determine the resulting sound pressure level. The data in Table 23-1 helps determine the resulting sound pressure level from two sources. It gives the relationship between the difference in decibels for two sources and the amount to add to the higher of the two sources. If there are more than two sources, one pair of values at a time must be added sequentially.

Frequency

Octaves divide the entire audible range into smaller bandwidths. The highest frequency in an octave has twice the frequency of the lowest frequency, and the frequency at the center of each octave designates the bandwidth of interest. The center frequency is the square root of the product of the lowest and highest frequency. The center frequency of an octave

TABLE 23-1 Factors for Combining Two Sound Sources

Procedures

A. Find the difference in decibels between two sound sources L_1 and L_2 (column 1).

B. Add the number in column 2 corresponding to the difference in column 1 to the highest of the two sources to find the resulting combined sound level L_s.

(1) Difference between two sound sources (dB)	(2) Amount to add to greater sound source (dB)
0.0–0.1	3.0
0.2–0.3	2.9
0.4–0.5	2.8
0.6–0.7	2.7
0.8–0.9	2.6
1.0–1.2	2.5
1.3–1.4	2.4
1.5–1.6	2.3
1.7–1.9	2.2
2.0–2.1	2.1
2.2–2.4	2.0
2.5–2.7	1.9
2.8–3.0	1.8
3.1–3.3	1.7
3.4–3.6	1.6
3.7–4.0	1.5
4.1–4.3	1.4
4.4–4.7	1.3
4.8–5.1	1.2
5.2–5.6	1.1
5.7–6.1	1.0
6.2–6.6	0.9
6.7–7.2	0.8
7.3–7.9	0.7
8.0–8.6	0.6
8.7–9.6	0.5
9.7–10.7	0.4
10.8–12.2	0.3
12.3–14.5	0.2
14.6–19.3	0.1
19.4 or more	0.0

adjacent to one of interest is either twice or half the center frequency of the octave of interest. Center frequencies of some standard octaves are 125, 250, 500, and 1,000.

Sound Distribution

Sound pressure decreases inversely with distance from a source. In a free field, sound moves from a source in a spherical pattern. However, most sources and environments are not free fields, so sound may project unevenly. It may strike objects, panels, walls, and ceilings and be reflected; it may be trapped and absorbed in holes in materials and lose energy; it may reflect and create standing waves; it may travel through panels and objects;

it can travel around objects and through openings. Sound may cause the mass of a panel to move and transmit pressure waves on the opposite side of the panel.

It is very difficult to predict the sound incident at some location in an actual field. It is easier to determine the actual sound pressure level by measurement. Absorption can be predicted. It is measured in sabins. One sabin is the absorption of $1\,\text{ft}^2$ of a perfectly absorptive surface. The noise reduction, *NR*, possible from absorption is

$$NR = \log_{10} \frac{\text{absorption before treatment}}{\text{absorption after treatment}}. \tag{23-7}$$

Sound transmission loss, *TL*, is the reduction in sound energy across a barrier

$$TL = \log_{10} \frac{\text{acoustic energy transmitted}}{\text{incident acoustic energy}}. \tag{23-8}$$

23-3   HEARING

Anatomy and Physiology of the Ear

The ear is a fantastic instrument and it is quite complex. A basic knowledge of its anatomy and physiology is essential to understanding the hazards of noise and in applying controls.

As illustrated in Figure 23-1, there are three main segments of the ear: the external, middle, and inner ear. The external ear is the part we see. The main portion is the pinna, which is a cartilage covered by skin that helps focus sound toward the ear canal

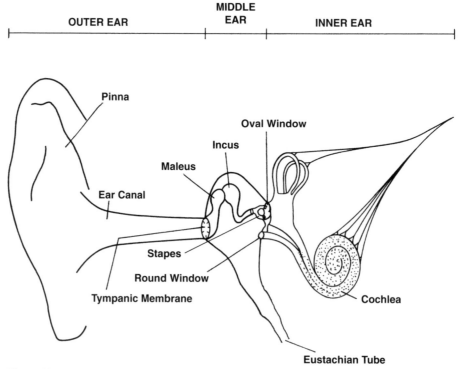

Figure 23-1.   Major anatomical features of the ear.

or external auditory meatus. The external ear amplifies the arriving sound by as much as 23 dB. At the inner end of the ear canal is the tympanic membrane or eardrum. Arriving sound waves transfer sound energy into motion of the eardrum.

The middle ear is an air-filled chamber that vents to the throat through the Eustachian tube. It contains three small bones (ossicles) and two suspensory muscles for them. The three bones are the maleus (hammer), the incus (anvil), and the stapes (stirrup). Sound energy travels across the three bones from the tympanic membrane of the outer ear to the oval window of the inner ear. The action of these three bones produces an amplification of approximately 2.5 dB. The suspensory muscles are the tensor tympani and the stapedius. When there is a loud sound, these muscles contract and provide some attenuation of energy transmitted across the three bones.

The main element of the inner ear is the cochlea. It is a fluid-filled coil that is divided into two chambers by thin tissue. Sound energy travels from the oval window or base of the cochlea to the distant end or apex of the coil. The waves continue back through the second chamber to the round window, which deflects to adjust for the pressure from the waves. The round window and oval window both separate the inner ear from the middle ear.

Figure 23-2 shows a cross section of the cochlea. There is actually a third chamber extending along the length of the coil that contains the sound receptor mechanisms. The membrane separating this chamber from the first is Reissner's membrane, and the membrane separating this chamber from the second is the basilar membrane. Located along the basilar membrane is the organ of Corti. It contains rows of hair cells and a layer of tissue, called the tectoral membrane, that rests on the ends of the receptor hairs. The rows of receptors are classified into two groups: inner hair cells and outer hair cells.

When sound travels along the chambers of the cochlea, the membranes enclosing this third chamber deflect. The deflection of the tectoral membrane against the inner hair cells is converted into nerve impulses by the receptor hair cells. In general, the hair cells at the base of the cochlea sense high frequencies, whereas those near the apex sense low frequencies. Recent theories suggest that the inner hair cells are the receptors, whereas the outer hair cells serve an amplification function. Noise damage to the outer cells is more prevalent than to the inner hair cells.

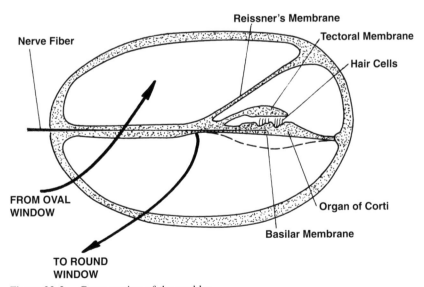

Figure 23-2.   Cross section of the cochlea.

Audition and Audiology

The primary means for hearing is sound travelling through the outer and middle ear to the inner ear. A small amount of hearing occurs through bone conduction to the inner ear.

Physical sound arriving at the ear is not converted directly to perceived sound. The ear is not equally sensitive to tones of equal sound pressure. As a result, there is a scale called loudness. Loudness is the perceived amplitude of sound. Studies have produced equal loudness curves for pure tones. In general, the ear is most sensitive at frequencies toward the middle of the auditory range.

Audiology is the assessment of hearing ability. Typically, a subject sits in a room that has a very low sound level and is isolated from external sound sources. The subject wears earphones connected to sound equipment called an audiometer. The operator sends a sequence of pure tones from the equipment to the subject and increases the amplitude of the tone until the subject responds, using hand gestures or a response switch, that the tone is audible. Indicators on the equipment show the tone and level. The procedure continues through a range of standard octave band center frequencies, and the responses of the subject are noted and compared with a standard. A plot of results, called an audiogram, notes the reduction in hearing threshold level in negative decibels relative to the reference standards.

23-4 HAZARDS OF NOISE

There are many effects of noise on people. However, any effects are detrimental. Although the best-known effect is noise-induced hearing loss, there are many other effects that contribute to accidents and health problems.

Types of Hearing Loss

There are several kinds of hearing loss. Some are the result of disease, such as infections of the middle ear and scar tissue formation on the eardrum; others are the result of trauma, such as perforation of the eardrum or separation of the ossicles of the middle ear. Conductive losses involve the outer and middle ear. Sensory losses involve the organ of Corti and sensory nerves. Hearing loss that can occur in the brain involves difficulties in interpreting sound. There are also psychological hearing impairments that have no physiological basis.

Noise-Induced Hearing Loss

Exposure to noise can produce hearing loss. Such losses are a function of duration of exposure and sound intensity. High frequencies are more damaging than low frequencies, and continual noise is more damaging than intermittent noise. There are also individual differences among people. An exposure may not produce the same losses in two people.

Temporary Threshold Shift   One type of noise-induced hearing loss is a temporary threshold shift (TTS). The hearing threshold increases after exposure to excessive noise because sound damages the organ of Corti and its receptor hair cells. Relatively short exposures to loud noise can produce TTS. One example is listening to a highly amplified music group for an hour or two. This disability is temporary and hearing sensitivity returns to pre-exposure levels after a recovery period.

Permanent Threshold Shift   Continued exposures to noise that produces TTS will cause a permanent decrease in hearing sensitivity or permanent threshold shift. The sensitivity does not return after a recovery period and the sensory damage to the organ of Corti becomes permanent. There are many examples of hearing loss among riveters, such as World War II aviators and tractor-driving farmers exposed to engine noise.

Acoustic Trauma   Loud noise from explosions or similar sources of pressure waves may rupture the eardrum or damage the structure of the middle or inner ear. Such conductive damage is called acoustic trauma. In some cases the damage is temporary. The injured tissue may heal, restoring hearing to full or near-full sensitivity.

Other Effects of Noise

Communication Effects   Noise can interfere with audio warnings intended to prevent accidents. For example, a person may not hear the horn of a car or train in a noisy environment. Noise may interfere with cries for help and may prevent rescue attempts.

Noise also can interfere with speech communication. One worker may not understand or may misunderstand what another says. For example, two people operating an injection machine had to use verbal communication in a noisy workplace to manage the machine operation. There was only one set of controls. Worker A said "Wait a second," but worker B, the one with the controls, thought A said to go ahead. Worker A then reached into the danger zone to remove a part that did not fall from the die and worker B tripped the die and closed it on workers A's hand.

One index for assessing the ability to communicate verbally is the Preferred-Octave Speech Interference Level (PSIL). PSIL is the numerical average sound level (dB) measured for each of three octave bands (500, 1,000, and 2,000 Hz). A study of the relationship between PSIL and speech intelligibility in different ambient noise conditions helps estimate potential communication problems (see Figure 23-3).

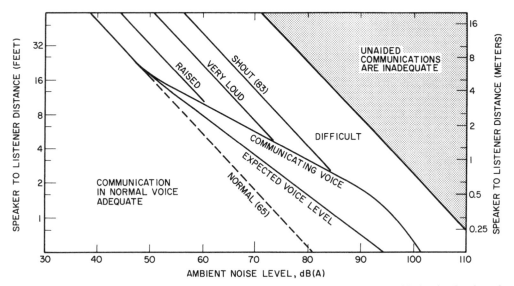

Figure 23-3.   Permissible distance between a speaker and listeners for specified voice levels and ambient noise levels. The levels in parentheses refer to voice levels measured 1 m from the mouth. (From MIL-STD-1472D.)

Reduced Learning Studies show that noise in classrooms and homes may hinder language skill development in children. Not only is attention interrupted, but speech and particular speech sounds are not communicated well.

Startle Response A sudden noise will cause a startle response in people. The reaction is involuntary and it is characterized by vasoconstriction, an increase in blood pressure, dilation of pupils, and muscle contraction. Every time a loud noise occurs, these reactions occur, but within a few minutes after a loud noise, conditions return to normal. The startle reaction may create errors in critical situations.

Physiological Reactions People do not become accustomed to noise. Studies of people who work in noisy environments suggest they have higher incidence of circulatory problems, heart disease, ulcers, and medical problems in general. Blood pressure is higher in children exposed to aircraft noise in schools. Noise levels during pregnancy have been linked to low infant birth weights. Many of these effects occur at noise levels below those related to hearing loss. Exposures to very high noise levels can produce vertigo and nausea.

Sleep Disruption Noise can prevent someone from falling asleep, and it can rouse one from sleep to lighter levels of sleep or to wakefulness. Disruption of sleep may affect performance during daily activities.

Social Effects Noise affects social behavior. Interference with communication can raise frustration levels and can affect relationships. Noise can distract attention away from critical activities and can lead to headache, irritability with others (including parents/children and workers/supervisors), anger, and fatigue and can, on occasion, lead to extreme behavior. Noise may be annoying. Annoyance is subjective and, in part, is a function of noise content. What may be meaningful or pleasant to one person may be annoying to others because of intensity or content. Annoyance is related to time and place. People are more irritated by noise when relaxing or trying to sleep.

23-5 STANDARDS

Origin of Exposure Standards

Although there was prior knowledge that extended exposures to noise caused hearing loss, standards limiting exposure were not suggested until the 1950s. The first standard was issued by the Air Force in 1956. Subsequent studies and standards committees wrestled with the problem for more than another decade before noise exposures were included in federal legislation. The 1969 Walsh-Healey Public Contracts Act incorporated the initial threshold limit values suggested by the American Conference of Governmental Industrial Hygienists (ACGIH) that same year. The initial OSHA standard was identical to that in the Walsh-Healey Act, and the same standard appeared in the 1971 Federal Coal Mine Health and Safety Act. In 1981, OSHA made a change in its standard, requiring employers to implement hearing conservation programs for employees in certain noise environments.

OSHA Standards

Table 23-2 lists the OSHA permissible exposures, which consider noise levels that produce hearing loss through exposures of an 8-h workday. When exposures vary during the

TABLE 23-2 OSHA Permissible Noise Exposures[a]

Duration of Exposure (hr)	Sound Level (DBA slow response)
8	90
6	92
4	95
3	97
2	100
1.5	102
1	105
0.5	110
0.25 or less	115
Impulse or impact noise	140 peak level

[a] 29 CFR 1910.95.

workday, a computed time-weighted average noise (TWAN) helps determine if the exposures in Table 23-2 are exceeded:

$$\text{TWAN} = \frac{C_1}{T_1} + \frac{C_2}{T_2} + \cdots + \frac{C_n}{T_n}, \tag{23-9}$$

where

C_x is the total time of exposure at a specified noise level and

T_x is the total time of exposure permitted at the specified noise level.

If the value of TWAN is greater than 1.0, the exposure exceeds the OSHA limits.

Example 23-1 A worker is exposed to the following:

Noise Level (dBA[1])	Duration (hr)
110	0.25
100	0.5
90	1.5

Is the OSHA limit exceeded?

From the preceding exposure data, allowable exposure times in Table 23-2, and Equation 23-9,

$$\text{TWAN} = \frac{0.25}{0.5} + \frac{0.5}{2} + \frac{1.5}{8} = 0.938.$$

Because 0.938 is less than 1.0, the exposure is allowable.

OSHA considers continuous noise as that involving noise level maxima that occur at intervals of 1 s or less.

Other Standards

The threshold limit values of the ACGIH have slightly more restrictive exposure limits at all sound levels than do OSHA standards. In addition, ACGIH limits the number of impact noises per day based on peak sound level.

23-6 CONTROLS

There are many controls for noise. Engineering controls can be grouped into four classes:

1. prevent noise sources before they are introduced into a work or other environment
2. if noise sources exist, replace equipment, processes, or materials with quieter ones
3. if noise sources cannot be replaced, modify them to reduce noise emitted
4. modify sound distributed in the environment

Another way to classify noise controls is

1. treat the source
2. treat the path
3. treat the receiver

Treating the source includes the first three options in the first classification scheme; treating the path is the same as the fourth option. Controlling noise at the receiver usually involves an enclosure for people or personal protection equipment, both of which limit sound reaching people from outside sources.

Many times a combination of controls is best. Some solutions have physical limits; others have economic limits.

Avoid Noise Sources

One way to control noise problems is to analyze processes and equipment during planning and design prevent noise problems from occurring. For example, set noise specification limits for purchased equipment and make sure equipment orders include such specifications. Watch for high-velocity flow of gases and fluids, high-speed equipment, and high pressure processes, which often are sources of noise. Establish source noise specifications and analyze processes and systems to reduce noise generation. Often, it is cheaper to design facilities and buy equipment with noise control than to try to control noise sources later.

Look for noisy locations in a floor plan or plant layout and for ways to enclose potentially noisy activities, processes, and equipment to prevent noise from travelling to less noisy areas. Group noise sources to lessen controls costs and try to separate people from noise sources by distance and barriers. Look for and avoid routes or channels by which sound can travel from one location to another.

To control noise in a room, one must analyze sound travel. Noise will travel away from a source, and the energy level in a sound field will decay with the square of the distance from a source. However, most rooms have reflective surfaces, such as floors, walls, and ceilings, and placement of noise sources in a corner in conjunction with a highly reflective floor and walls will concentrate noise energy into an emitted area. Conversely, placement of noise sources away from reflective surfaces gives noise a chance to dissipate before reaching reflective walls, ceilings, and floors.

One can design rooms to absorb sound. A room constant, R, is a measure of the ability of a room to absorb sound. A high room constant will have a high level of sound absorption. A room constant can be computed as

$$R = \frac{\alpha_m S_t}{1 - \alpha_m},$$ (23-10)

where

α_m is the mean sound absorption coefficient for all room surfaces (dimensionless),

S_t is the total area of room boundary surfaces (square feet),

and

$$\alpha_m = \frac{S_1\alpha_1 + S_2\alpha_2 + \cdots + S_n\alpha_n}{S_1 + S_2 + \cdots + S_n}, \tag{23-11}$$

where subscripts refer to particular surfaces that make up the total surface and their respective absorption coefficients.

During the design phase, one can specify and select sound-absorbing surface finishes for ceilings, walls, and floors. This will help reduce sound to acceptable levels. However, there is a limit to the benefit from absorbing surfaces.

One also can analyze the effect of distance of people from sound sources. Keeping people away from sources can lessen potential violations of exposures standards and adverse effects.

Some materials, processes, and equipment have components that transmit sound or have natural frequencies that can amplify sounds. Metal panels, pipes, and tubes often vibrate and ring. The flow of high-pressure and high-velocity liquids and gases often produce higher noise levels than those with low pressures and velocities. One should look for these potential sources of problems and seek to reduce them during design of processes, buildings, and equipment.

Replace Noise Sources

If noise sources are in place, one can seek to substitute alternate equipment, materials, processes, and activities that will reduce noise. Many of the actions suggested for design and planning apply here. The main difference is usually cost, because retrofits and modifications usually are more expensive than initial built-in control features.

Modify Noise Sources

There are many options for modifying a sound source. Analysis of a sound source will determine which options are feasible, practical, and economical. Because many control options are frequency dependent, one must conduct a noise survey of an environment and the sources within it to help define the problem and potential solutions. Normally, the survey data will give records of total sound level, usually in A scale decibels, and sound levels for each octave band.

Control Direction of Source   One option is to direct the sound from a source away from locations where people will be. For example, one can direct an opening in a machine (vent, duct, material feed opening, etc.) away from operators or extend a vent or duct through a roof or wall.

Reduce Flow Rates   High-speed and high-pressure gases and liquids passing through pipes and ducts often create high sound levels. This is particularly true for flow-through valves, bends, and other transitions where turbulence is high. Reducing pressures and speeds can reduce sound sources.

One example is the use of air to eject partially filled boxes of material passing in a production line. A small stream of high-pressure air can be used to remove the low-weight boxes. Alternatively, a low-velocity stream of low-pressure air acting over a larger area can create the same ejection force at lower sound levels and lower frequencies. One can avoid transitions in flow lines and turbulence that produce noise. One can also place transitions in locations away from people.

Reduce Driving Forces The higher the force producing oscillations and vibrations, the greater the amplitude of sound generated. Reducing shaft speed and balancing rotating equipment reduce forces in rotating shafts that are slightly eccentric. Separating vibrating sources from sheet metal panels and structures through flexible couplings and flexible connections reduces sound sources. Examples are placing motors and vibrating equipment on independent footings and connecting air handlers from ductwork with fabric connections. There are many kinds of isolating pads that prevent transfer of vibrations from machines to the structures they rest on or are attached to. One should obtain manufacturer's data and engineering information to help determine if commercial isolators will be satisfactory for the frequencies and forces involved with particular applications.

Control Vibrating Surfaces Panels, walls, and other elements of equipment and buildings that vibrate perpendicular to their surface create large pressure waves. One can control these sources by reducing the amplitude of vibrating members or reducing their area. Adding damping material, bracing, or stiffness can reduce motion. One can also add mass. If the source frequency and the natural frequency of a driven member coincide, consider changing one of them.

A few examples illustrate some of these options. It is standard practice in the sheet metal industry to place an X crease on large duct panels to stiffen the panel and reduce the amplitude possible for vibration.

To prevent transfer of sound across a standard stud-drywall wall panel, mount resilient metal strips on the studs and attach the drywall panels to the strips, rather than directly to the studs. This dampens vibration of the panels and limits the ability of a panel on one side of the wall to excite the panel on the other side.

For some equipment with large surfaces or panels, it is common to spray the surface with an agent that helps to reduce panel vibration. In addition, a second panel is attached over the sprayed agent to form a sandwich panel. This stiffens the panels and reduces noise produced. Tumblers filled with shot that are used to finish stamped metal parts have lower sound levels when covered with foam spray and lined with resilient rubber material.

One can increase stiffness by using heavier gauge material. The heavier material requires greater forces to produce displacement.

The folding vinyl room partitions that also have effective sound reduction typically have lead panels built into their core. The lead panels add mass to the folding elements of the partition and reduce the amplitude of panel motion. This is one feature that helps reduce sound transmission across the panels.

Dividing large panels into smaller elements reduces their sound production because the smaller panels reduce the range of frequencies transmitted. One also can reduce the area by perforating a panel, which may not be practical for some panels that function as enclosing material.

Modify Sound Paths

Enclosures The ideal enclosure is a full one that has no openings and fully encloses a sound source. However, we seldom achieve full enclosures because there are openings for feeding materials or for other purposes, and there may be cracks around temporary openings or doors. Considerable amounts of sound can leak through small openings and cracks.

Inside the enclosure, there is reverberant buildup. By lining the inside of an enclosure with sound-absorbing materials, reverberant buildup is minimized. An average absorption coefficient inside the enclosure of at least 0.7 makes buildup insignificant.

An expression for the sound pressure level, L_p, inside an enclosure is

$$L_p = L_w + 10 \log_{10}\left(\frac{Q}{4r^2} + \frac{4}{R}\right) + 10.5 \text{ dB}, \tag{23-12}$$

where

L_w is the sound power level of the source relative to a reference value of 10^{-12} W,

Q is the directivity factor,

r is the distance from the acoustic center of the source (feet), and

R is the room constant (square feet).

Equation 23-12 is helpful for developing the most effective and economical enclosure. Equation 23-11 is applied to a number of possible conditions. The differences in results between conditions determines the transmission loss required and noise reduction achieved.

First, consider initial, unenclosed conditions for some predetermined distance from the sound source. For the unenclosed sound source, assume the difference between L_p and L_w is D_u. Then consider the enclosed sound source and compute a new difference between L_p and L_w. This value indicates the enclosure causes an increase in L_p because of reverberant buildup. The difference between the enclosed condition and the unenclosed condition, $D_e - D_u$, is the transmission loss required because of the enclosure, TL_e.

The actual decibel level for the unenclosed condition may be above some standard to be achieved. Let us call this difference TL_s. It is common to adjust TL_s upward by approximately 5 dB to ensure that the final solution is effective. The total transmission loss, TL_t, that the enclosure must achieve is $TL_s + TL_e$. One can then select and analyze enclosure materials and sound-absorbing lining materials to determine the design that is most cost effective. Repeat this procedure for each octave band center frequency.

Controlling sound transmission through the enclosure materials is another important consideration. One can select materials that effectively limit sound transmission. Table 23-3 lists some sample materials and sound transmission properties, assuming well-sealed enclosures.

TABLE 23-3   Sound Transmission Loss in Decibels of Various Materials

	Octave Band Center Frequency (Hz)					
Material	125	250	500	1,000	2,000	4,000
Door, heavy wooden, special hardware, rubber gasket all around edge, $2\frac{1}{2}$ in thick	30	30	24	26	37	36
Door, solid oak, ordinary hung, $1\frac{3}{4}$ in thick	12	15	20	22	16	—
Glass, $\frac{1}{4}$ in	27	31	33	34	34	42
Sheet steel, $\frac{1}{4}$ in	23	38	41	46	43	48
Sheet lead, $\frac{1}{16}$ in	32	33	32	32	32	—
Gypsum wall board, $\frac{1}{2}$ in	18	22	26	29	27	26
Wall $\frac{1}{2}$-in gypsum wallboard on both sides of 2×4 studs 16-in O.C. (on center)	20	27	37	43	48	43
Concrete, reinforced, 4 in thick	37	36	45	52	60	67
Concrete block, 4 in hollow, dense aggregate, no surface treatment	30	39	43	47	54	50
Concrete block, 8 in hollow, dense aggregate, no surface treatment	40	53	54	58	58	50

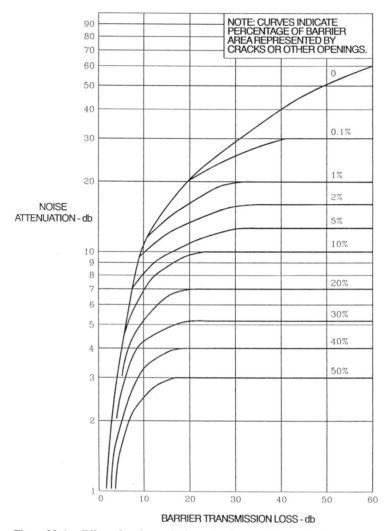

NOISE ATTENUATION - db

BARRIER TRANSMISSION LOSS - db

NOTE: CURVES INDICATE PERCENTAGE OF BARRIER AREA REPRESENTED BY CRACKS OR OTHER OPENINGS.

Figure 23-4. Effect of enclosure leaks on barrier sound transmission loss (TL). Enter the chart with the barrier transmission loss for a fully enclosed noise source. Move vertically to intersect with the curve representing the percentage of the barrier area presented by cracks or openings. From the intersection, move left to read the noise attenuation actually achieved.

The sound transmitted through an incomplete enclosure is largely a function of the unenclosed portion of the full enclosure, whether from cracks or other openings. Estimates of sound reduction may be made from curves shown in Figure 23-4. Enter the chart at the bottom using data on the maximum transmission loss for a full enclosure. Move vertically to intersect with a curve representing the opening areas (percent of full enclosure), and then read the actual reduction along the left of the chart.

Room Absorption   One can place sound-absorbing materials on the surfaces of a room or surfaces of objects within it to reduce sound levels, thereby replacing hard, smooth, reflecting surfaces with porous, sound energy-absorbing materials. There is a practical limit to this approach. To a great extent, the effectiveness of additional treatment depends on the lack of sound-absorbing surfaces before treatment. The noise reduction is computed as

$$NR = 10\log\frac{A_2}{A_1},\qquad(23\text{-}13)$$

where

A_2 is the total absorption units (sabins) in a room after treatment and

A_1 is the total absorption units (sabins) in a room before treatment.

Total absorption units, A_x, are the sum of products of various surface areas involved times the absorption coefficients, α_x, at specific frequencies for the respective surfaces.

The absorption properties of room finish materials vary with frequency. Manufacturers of sound-absorbing materials usually can provide data on absorption coefficients by bandwidth center frequency. Table 23-4 gives sound absorption coefficients for selected building materials. The sound absorption for a clothed person in a room varies with frequency and is approximately 3 to 5 sabins.

Barriers or Shields   Another approach for reducing sound transmission along a path is to insert a barrier, often some type of panel, along the path. The purpose of a barrier is to deflect sound waves that would otherwise move between a source and a receiver. A barrier extends from the floor to some height short of a ceiling. As shown in Figure 23-5, the sound absorbed is a function of the geometry of the panel relative to the direct path between noise source and receiver and the wavelength. The noise reduction is estimated from the data in Figure 23-6. Barriers are not very effective for low frequencies. They should be as high as practical and close to the source for maximum effectiveness. Barriers should have good transmission loss properties so that sound does not pass through them.

Absorption Along a Transmission Path   In many buildings, there are paths between one location and another through which sound can travel. Typical paths are heating and ventilation ducts. Ducts lined with sound-absorbing material can reduce the amount of sound they transmit. The noise reduction, NR, is a function of the absorption properties

TABLE 23-4   Sound Adsorption Coefficients for Selected Building Materials

Material	Octave Band Center Frequency (Hz)					
	125	250	500	1,000	2,000	4,000
Brick, unglazed	0.03	0.03	0.03	0.04	0.05	0.07
Carpet, $1/8$-in pile	0.05	0.05	0.10	0.20	0.30	0.40
Carpet, $1/4$-in pile	0.05	0.10	0.15	0.30	0.50	0.55
Floor, concrete	0.01	0.01	0.01	0.02	0.02	0.02
Floor, wood	0.15	0.11	0.10	0.07	0.06	0.07
Glass fiber, 1 in thick (3 lb/ft^3, mounted with impervious backing)	0.14	0.55	0.67	0.97	0.90	0.85
Glass fiber, 3 in thick (3 lb/ft^3, mounted with impervious backing)	0.43	0.91	0.99	0.98	0.95	0.93
Gypsum board, $1/2$ in nailed to 2 × 4 studs, 16 in O.C. (on center), painted	0.10	0.08	0.05	0.03	0.03	0.03
Paneling, hardwood, $1/4$ in on wood framing	0.58	0.22	0.07	0.04	0.03	0.07
Fabric, light velour, 10 oz/yd^2, hung straight in contact with wall	0.03	0.04	0.11	0.17	0.24	0.35
Fabric, heavy velour, 18 oz/yd^2, draped to half area	0.14	0.35	0.55	0.72	0.70	0.65

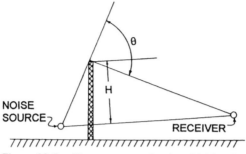

Figure 23-5.   The geometry of noise shields. (By permission from *The Noise Manual*, 5th ed., American Industrial Hygiene Association, Fairfax, VA, 2000.)

NOISE REDUCTION OF SHIELDS

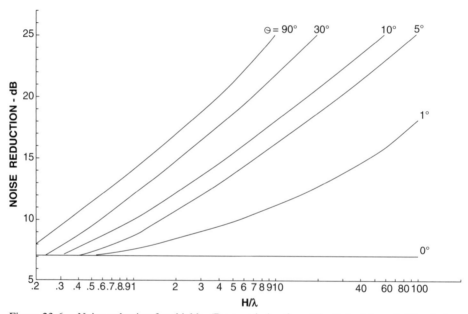

Figure 23-6.   Noise reduction for shields. (By permission from *The Noise Manual*, 5th ed., American Industrial Hygiene Association, Fairfax, VA, 2000.)

of the lining material, the length of duct, and the cross section of the duct measured in dB reduction per foot of duct length:

$$NR = \frac{12.6 \, P\alpha^{1.4}}{A}, \tag{23-14}$$

where

P is the perimeter of the duct in inches,

α is the absorption coefficient of the lining material at a particular frequency, and

A is the cross-sectional area of the duct in square inches.

Mufflers   Mufflers reduce the sound along a path and reduce the sound released from flow of fluid. There are several ways to design them. The preferred approach depends on physical dimensions, design goals, and other factors.

One type of muffler is a straight-through type of dissipative muffler (see Figure 23-7(a)) that creates very little pressure drop. Sound is absorbed by placing material around the outside of the pipe and perforations in the pipe wall. The sound-absorbing material may have an outer shell. This type of design is most effective for pipes with less than a 6-in diameter.

A second type of muffler, better suited to pipes with diameters larger than 6 in, is a center-body type of dissipative muffler. Channels distribute the gas flowing through the muffler, and perforations in the channel walls allow surrounding absorptive material to reduce sound energy. The channels have center and outer absorbing material that creates more area for sound absorption.

A third type of muffler is a reactive type (see Figure 23-7(b)) that uses baffles and resonance phenomena to reduce noise levels. Resonance mufflers are suited to narrow band frequencies because of design limitations.

Treat the Receiver

A temporary means for controlling noise at the receiver is the use of earplugs and earmuffs. A permanent means is a partial or full enclosure. Enclosed workspaces are much like enclosed sound sources. The enclosing walls must minimize sound transmission: doors or windows must have means to seal cracks around their edges, and the intersections of ceilings and walls or walls and floors must have sealed joints. Placing openings in enclosures on sides away from sound sources lessens the chance of sound penetrating them or the cracks along their edges.

The noise reduction of an enclosure depends on the transmission loss of the enclosure and the room constant (see Equation 23-9). It can be computed as

$$NR = TL - 10 \log_{10}\left(\frac{1}{4} + \frac{S_w}{R_2}\right),$$
(23-15)

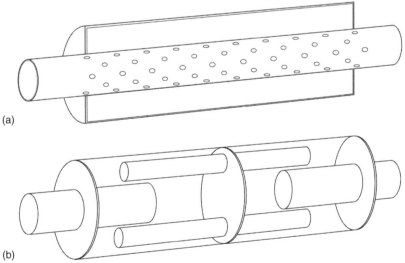

(a)

(b)

Figure 23-7.   Two types of muffler designs. (a) A straight-through dissipative design. (b) A reactive type with baffle and tubes.

where

S_w is the exposed surface area of the enclosure (square feet) and

R is the room constant (square feet).

Desired noise levels in decibels are compared with measured noise levels to determine the reduction, *NR*, required at each band-center frequency. Often an adjustment factor of approximately 5 dB is added to each difference to ensure that higher noise levels do not render the enclosure ineffective. The transmission loss, *TL*, is the sum of the required reduction and the value resulting from the expression involving the room constant in Equation 23-15.

23-7 MEASUREMENT

Measures

The human ear does not sense each frequency in the audible range with the same sensitivity. As a result, actual sound pressure levels are adjusted to represent typical ear sensitivities. The most common weighting scale is the A scale. Other weighting scales are B, C, and D. Table 23-5 lists frequency weighting adjustments to convert physical sound pressure levels (decibels) to A scale values (A scale decibels [dBA]).

Dosimetry

Many workers receive noise exposures that vary almost constantly during a workday. In 1981, OSHA required hearing conservation programs for workers exposed to TWAN levels of more than 85 dBA. Noise dose, *D*, measures varying exposures over a period of time and is equivalent to an exposure of 90 dBA for an 8-hr period. OSHA exposure standards in Table 23-2 indicate that this is the maximum sound level permitted for an 8-hr period. Higher levels exceed the allowable dose. Noise dose in percent is

$$D = 100 \frac{C}{T}, \tag{23-16}$$

TABLE 23-5 Frequency Weighting Adjustments for Converting Sound Pressure levels to A-Scale Readings

Octave Band Center Frequency (Hz)	Correction (dB to dBA)
31.5	−39
63	−26
125	−16
250	−9
500	−3
1,000	0
2,000	+1
4,000	+1
8,000	−1
16,000	−6.6

where

C is the total length of the workday in hours and

T is the reference duration in hours for a measured A-weighted sound level, L, in decibels.

T is derived from an expansion of the OSHA exposure limits (Table 23-2) and is computed as

$$T = \frac{8}{2^{(L-90)/5}}.$$ (23-17)

When noise exposure for a work shift consists of two or more periods of noise at different levels, the total noise dose over the workday is given by

$$D = 100 \left(\frac{C_1}{T_1} + \frac{C_2}{T_2} + \cdots + \frac{C_n}{T_n} \right).$$ (23-18)

The 8-hr TWAN sound level, in decibels, can be computed from noise dose:

$$\text{TWAN} = 16.61 \log_{10} \left(\frac{D}{100} \right) + 90.$$ (23-19)

Example 23-2   A worker is exposed to a 107 dBA constant noise source during a 7-hr shift. What is the noise dose?

From Equation 23-17, the reference duration is

$$T = \frac{8}{2^{(107-90)/5}} = \frac{8}{2^{3.4}} = 0.758 \text{ hr.}$$

From Equation 23-16, the dose is

$$D = 100 \frac{C}{T} = 100 \left(\frac{7}{0.758} \right) = 923\%.$$

Instruments

Sound-Level Meter   The most common sound-measuring instrument is the sound-level meter, which measures the sound level for each octave or for a broad band, normally approximately 20 to 20,000 Hz. On many of these meters, the user may select broad band or the octave of interest. The instrument measures sound arriving through a microphone. The signal is amplified and may be filtered before the sound level in decibels appears on a digital display or a moving indicator meter. Depending on the instrument features, the user may have a choice of filters or weighting values. There are standard weighting values for three filters: A, B, or C. The three weighting scales attempt to mimic the response of human hearing to low-, medium-, and high-intensity sounds, respectively. Users also may have a choice of fast or slow response to control swings in the display. Meters must be calibrated against standard sound sources. Meter suppliers usually have such a calibrating device, which is placed over the microphone on the meter and generates controlled sound sources.

Impulse Meter   Sound level meters do not respond quickly enough for measurement of impulse sounds. Impulse meters provide this quick response, giving readouts in decibels for peak levels of transient sound.

Frequency Analyzers A frequency analyzer measures the distribution of sound across one or more bandwidths. Frequency analyzers work on much the same principle as sound-level meters. Typically, frequency analyzers give distribution readings by octave, although some can subdivide octaves into even finer distributions. Distributions are very helpful in assessing noise problems and pinpointing corrective action for high-amplitude frequencies.

Dosimeters People wear noise dosimeters, many of which are pocket sized, throughout their work day. The dosimeter records the sound level continuously and integrates sound level over time relative to an 8-hr day. By reading the integrated value (noise dose) at the end of the day, it is easy to determine if someone has exceeded the allowable noise dose.

Surveys

The only way to determine whether noise exceeds an exposure standard is to make measurements in a survey. There are two kinds of surveys. The intent of one is to establish whether an individual has an excessive exposure. Noise dosimeters are well suited for this kind of survey. The other type of survey has the purpose of finding out if conditions could create excessive exposures. In this kind of survey, measurements taken at various locations help determine noise sources or potential noise exposures at locations where people may be.

23-8 VIBRATION

Background

Vibrations of the human body or parts of the human body can be annoying, can affect performance, or can cause trauma. There are many conditions that produce local or general vibration of the body. Some effects are known, but because experiments may cause injuries, the data available in the literature are not complete. However, the data do give us a picture of hazards and suggest controls that can be applied to prevent vibration-induced health and safety problems.

Vibration forces applied to the whole body or to body parts cause motion. The motion is described in terms of amplitude, frequency, and force (usually relative to gravity). The body is a complex system of masses restrained by tissues with varying properties, including damping of motion and resistance similar to springs. Trauma to tissues results when external vibrating forces accelerate the body or some part so that amplitudes and restraining capabilities by tissues are exceeded. Each body part has a natural or resonant frequency. A natural or resonant frequency is the frequency at which an object will vibrate if deflected and left to vibrate on its own. If the exciting frequency is a natural or resonant frequency of some body part, the resulting motion can produce severe damage. Amplitudes less than necessary for trauma can impede performance or be uncomfortable.

Hazards

Studies on whole-body vibration suggest that truck drivers, tractor drivers, and heavy equipment operators have a greater incidence of back troubles than do people in other kinds of jobs. Factors other than vibration may be involved. Drivers of such vehicles experience low-frequency vibration in the 4- to 8-Hz range while seated. Spinal compression, length of exposure, and other factors may contribute to damage.

Consider the eye, for example. The eyeball has a natural or resonant frequency of 60 to 90 Hz. Vibrations in this frequency range may interfere with vision tasks and may damage the retina or nerves connected to the eye. Large amplitudes may damage the supporting muscles and ligaments.

An operator sits with arms extended to operate controls. If the operator's body is moved vertically, the muscles of the arm may not be strong enough to hold the hands steady, resulting in incorrect operation of the controls. Vibration of the chest may create breathing difficulties, and vibration of the hands can produce Raynaud's syndrome (see Chapter 13).

Table 23-6 lists approximate natural or resonant frequencies for various body parts. Figure 23-8 gives human response regions for vibration. These regions are not precise because effects are often a function of vibration direction. The standard convention for acceleration of the human body is as follows:

Direction of Body Motion	Symbol
Forward	$+G_x$
Backward	$-G_x$
To right	$+G_y$
To left	$-G_y$
Headward	$+G_z$
Footward	$-G_z$

Figure 23-9 gives the International Organization for Standardization (ISO) standard for safety for whole-body vibration in a vertical direction ($\pm G_z$). The curves are lower for fatigue-decreased deficiency and comfort. Note the lowest area for each curve is at the whole-body natural frequency. This standard, although helpful, is not without criticism. The American National Standards Institute also publishes guidance on whole-body vibrations.[2]

Chapter 13 included a discussion of vibrations on the hand and a resulting cumulative trauma disorder called vibration-induced white finger or Raynaud's syndrome.

TABLE 23-6 Approximate Natural or Resonant Frequencies of Human Body Elements

Element	Natural or Resonant Frequency (Hz)
Whole body, seated vertical	4–6
Whole body, prone	3–4
Whole body, standing vertical	5–12
Body seated on cushion	2–3
Head relative to body	20–30
Skull	300–400
Eyeball	60–90
Lower jaw relative to head	100–200
Shoulder and head, transverse	2–3
Hand	30–40

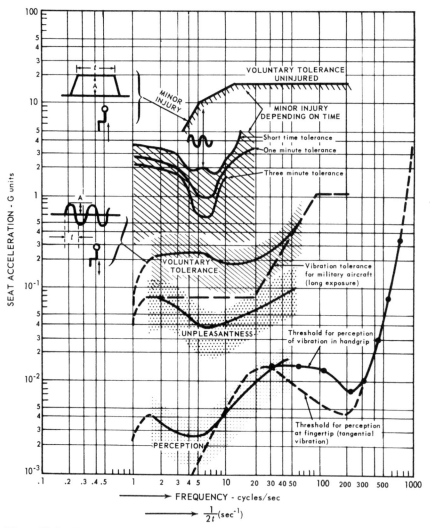

Figure 23-8. Human response regions for vibration. (From Webb, P., *Bioastronautics Data Book*, NASA SP3006, Washington, DC, 1964.)

Controls

One control for vibration problems is to prevent vibration from being applied to the body or its parts. Another control is to reduce amplitudes to levels below those that can produce injury or interfere with performance. One must consider carefully what body part may be involved. A third control is to model the vibrating system including the human body and then apply damping, springs, or both that reduce the vibration.

Measurement

Accelerometers measure vibration and acceleration. One should refer to a good reference on vibration instrumentation for detailed information on vibration measurement. Instrumentation includes sensors, amplifier, and recording equipment.

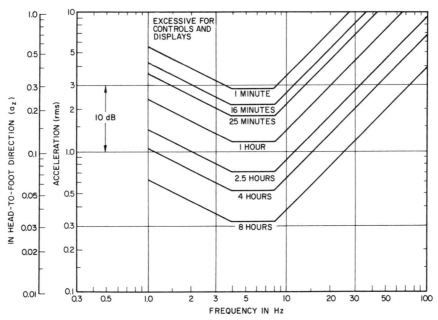

Figure 23-9.   ISO standard for whole-body vibration in a vertical direction. The curves give fatigue-decreased proficiency boundaries for vertical (head-to-toe) vibration. For reduced comfort, subtract 10 dB; for safe exposures, add 6 dB. (From MIL-STD-1472D.)

EXERCISES

1. Noise assessments were made for several operations in a plant. Some of the data appear in the accompanying table. Measurements were made at operator locations. For which operations are corrections mandatory according to OSHA exposure rules?

Operator Location	Duration of Daily Exposure (hr)	Full-Scale Reading (dBA)
A	3	100
B	1	97
C	8	88
D	5	96
E	1/2	110

2. A worker moves around among three operations each day. The duration and sound level for each are as follows: operation 1, 3 hr, 95 dBA; operation 2, 4 hr, 90 dBA; operation 3, 1 hr, 97 dBA.

   (a) What is the time-weighted average exposure value?

   (b) Based on OSHA standards, is any protective action required?

3. For the exposure in Exercise 2, determine the noise dose, D.

4. A worker's noise dosimeter reads 65 at the end of an 8-hr day. What is the equivalent time-weighted average?

5. A machine produces a noise level $Lp = 100\,dBA$ for a worker located 5 ft from it ($r = 5$), where $R = 500\,ft^2$ for the room and $Q = 4$. The machine will be installed in a new plant in a different arrangement than in the current plant. The worker distance from the machine, r, will be 15 ft, the room will have $R = 2{,}000\,ft^2$, and $Q = 2$. What will the worker's noise exposure, L_p, be in A scale decibels?

6. A room measures 50 ft wide, 70 ft long, and 12 ft high. It has a 90-dBA sound level from some source at a frequency of 1,000 Hz. The surfaces in the room and their absorption coefficients are:

Surface	Absorption Coefficient at 1,000 Hz
Ceiling	0.02
Floor	0.02
Walls	0.01
Covered pipe (180 ft^2)	0.50
Machinery (200 ft^2)	0.02
8 people	4 sabins/person absorption

To reduce sound levels, the room is modified by covering the ceiling with acoustical tile. The tile has an absorption coefficient of 0.80 at 1,000 Hz.

(a) What noise reduction, NR, in decibels, is achieved?

(b) What is the resulting sound level in the room after modification?

7. A 25-ft long duct has a cross section of 6 ft × 4 ft. At one end, a fan produces 100 dBA at all frequencies from 125 to 4,000 Hz. If the duct is lined with 1-in thick glass fiber, compute the reduction in sound coming out of the duct for the following frequencies:

(a) 125 Hz

(b) 500 Hz

(c) 2000 Hz

8. An enclosure is made of 1/2-in thick plywood. Inside the enclosure is a noisy machine. Determine what reduction in noise will actually occur at 1,000 Hz if

(a) there is a 20% opening in the enclosure

(b) there is a 5% opening in the enclosure

9. Noise in an automated production area exceeds OSHA limits. One solution is to provide a full enclosure room in the production area to reduce the duration of potential exposure to noise. The workers can spend the greater part of the day performing tasks within the enclosure, observing operations in general through windows, and can wear hearing protection when they enter the automated production area to take samples or make adjustments. The sound levels in the production area are as follows:

Octave Band Center Frequency (Hz)	Production Area Sound Level (dBA)
125	88
250	90
500	91
1,000	92
2,000	93
4,000	89

Compare the resulting noise levels for two alternate wall materials:

(a) 4-in dense aggregate concrete block

(b) $^1/_2$-in gypsum board on 2 × 4 studs, 16-in O.C. (on center)

Plot the original and resulting sound levels by octave band.

10. A highway construction crew worker drives a machine that breaks up concrete pavement. If the machine vibrates the driver at 4 Hz and produces a seat acceleration of 0.5 m/s^2, is the operation safe or must one provide some damping for the seat to protect the operator?

REVIEW QUESTIONS

1. Define noise.

2. What frequencies do people hear?

3. Name the units of measure used for the following:

    (a) sound amplitude

    (b) sound power

    (c) sound pressure

    (d) sound intensity

    (e) sound pressure level

    (f) sound frequency

    (g) sound absorption

4. What is the significance of the A weighting scale?

5. Give the main anatomical parts of the following:

    (a) outer ear

    (b) middle ear

    (c) inner ear

    (d) cochlea

6. What is an audiogram?

7. Identify five effects of noise on people.

8. Explain the following:

    (a) bone conduction

    (b) temporary threshold shift

    (c) permanent threshold shift

    (d) acoustic trauma

    (e) startle response

9. What average sound pressure level in dBA is the maximum allowed under OSHA regulations during an 8-hr period?

10. At what sound level does OSHA require a hearing-conservation program?

11. What are the main types of controls for noise?

12. Give an example for controlling noise.

    (a) at its source

    (b) along its path

    (c) at the receiver

13. Explain noise dose.

14. What three parameters describe vibration induced motion?

15. List the hazards of vibration for

    (a) the whole body

    (b) the arms

    (c) the head

    (d) the hands

    (e) the chest

16. What controls can reduce vibration dangers?

NOTES

1 Refer to Section 23-7 for a discussion of the A noise weighting scale.

2 ANSI S3.18, *Guide for the Evaluation of Human Exposure to Whole-Body Vibration* and ANSI S3.29, *Guide to the Evaluation of Human Exposure to Vibration in Buildings*, American National Standards Institute, New York.

BIBLIOGRAPHY

American National Standards Institute, New York:

S1.8 Reference Quantities for Acoustical Levels

S1.9 Instruments for the Measurement of Sound Intensity

S3.1 Maximum Permissible Ambient Noise Levels for Audiometric Test Rooms

S3.14 Rating Noise with Respect to Speech Interference

S3.18 Guide for the Evaluation of Human Exposure to Whole-Body Vibration (also ISO 2631)

S3.29 Guide to the Evaluation of Human Exposure to Vibration in Buildings

S12.1 Guidelines for the Preparation of Standard Procedures to Determine the Noise Emission from Sources

S12.10 Measurement of Airborne Noise Emitted by Information Technology and Telecommunications Equipment

S12.12 Engineering Method for the Determination of Sound Power Levels of Noise Sources Using Sound Intensity

S12.14 Methods for the Field Measurement of the Sound Output of Audible Public Warning Devices Installed at Fixed Locations Outdoors

S12.15 Measurement of Sound Emitted from Portable Electric Power Tools, Stationary and Fixed Electric Power Tools, and Gardening Appliances

S12.16 Guidelines for the Specification of Noise of New Machinery

S12.17 Impulse Sound Propagation for Environmental Noise Assessment

BERGER, ELLIOTT H., ROYSTER, LARRY H., ROYSTER, JULIA D., DRISCSOLL, DENNIS P., and LAYNE, MARTHA, *The Noise Manual*, 5th ed., American Industrial Hygiene Association, Fairfax, VA, 2003.

BRAMMER, A. J., and TAYLOR, W., *Vibration Effects on the Hand and Arm in Industry*, Wiley, New York, 1989.

FELDMAN. A. S., and GRIMES, C. I., *Hearing Conservation in Industry*, Williams & Wilkins, Baltimore, MD, 1985.

GASAWAY, D. C., *Hearing Conservation: A Practical Manual and Guide*, Prentice-Hall, Englewood Cliffs, NJ, 1985.

Hearing Conservation in the Workplace, National Safety Council, Itasca, IL, 1992.

KNOWLES, EMORY E III, ed.,*Noise Control*, 3rd ed., American Society of Safety Engineers, Des Plaines, IL, 2003.

OLISHIFSKI, J. B., and HARFORD, E. R., eds., *Industrial Noise and Hearing Conservation*, National Safety Council, Chicago, IL, 1975.

STATLOFF, R. T., and STATLOFF, J., *Occupational Hearing Loss,* Dekker, New York, 1987.

WEBB, P., ed., *Bioastronautics Data Book*, NASA SP-3006, National Aeronautics and Space Administration, Washington, DC, 1964.

CHAPTER 24

CHEMICALS

On December 3, 1988, in Bhopal, India, during the predawn hours, methyl isocyanate gas leaked from a chemical plant and settled over a 15-square mile area populated by more than 200,000 people. This "devil's day," as termed by the local people, left more than 2,500 dead and approximately 20,000 injured. The injuries often included damage to eyes and mild to severe pulmonary disorders.

In Michigan in 1973, an estimated 1,000 lb of polybrominated biphenyl (PBB), a flame-retardant chemical, were accidentally mixed into livestock feed. The feed was distributed to farms throughout the state and found its way into millions of animals. As a result, the state ordered the destruction of more than 30,000 cattle, 1,400 sheep, 5,900 swine, and 1.5 million chickens. Researchers found PBB in 97% of tissue samples taken from human adults across the state.

Chemicals are the basic elements of life and the world around us. Chemical products fill our homes. Materials made from chemicals are elements of our cars, clothing, furniture, tools, and other things we come in contact with daily. Most chemicals are not dangerous. Many chemicals and compounds are beneficial to humans; others are dangerous and may cause harm.

Today, however, public concern over chemicals is at an all-time high. There are more books written about chemicals and the dangers of chemicals for people than about any other subject covered in this book. Is the concern valid? Are people unnecessarily afraid of chemicals? Certainly events like those cited raise public concern. There are many other events involving chemicals: Love Canal, New York, Times Beach, Missouri, and Institute, West Virginia, to list a few.

The purpose of this chapter is to explore the hazards and controls for dangerous chemicals, not to pass judgment on these public issues. Readers should form their own opinions.

There are more than 3,000,000 registered compounds. About 60,000 compounds have significant economic value and are in the marketplace, and an estimated 700 to 1,000 new compounds enter the marketplace each year. There are published exposure standards for approximately 500 chemical compounds and the National Institute for Occupational Safety and Health (NIOSH) has compiled a list of 5,000 chemicals that have some inherent hazard. When considering the hazards of chemicals, several factors are important:

1. compounds that have known hazards may not be hazardous at low concentrations
2. compounds that are not normally dangerous may become so for certain uses
3. some compounds become dangerous when combined with other compounds

Like other hazards, there is much to learn about some of these chemicals, their use, and disposal.

This chapter cannot cover all the compounds important for safety and health and all the conditions in which they may be found or used. This discussion focuses on an understanding of chemical hazards.

24-1   CHEMICAL REGULATIONS AND STANDARDS

Indoors

Concern about indoor air quality continues to grow. Several federal agencies regulate chemicals in indoor air. OSHA regulates the workplace, the Environmental Protection Agency (EPA) regulates some public places, and the Department of Transportation (DOT) regulates public transportation, such as aircraft, trucking, and railways. Many local governments have additional laws and regulations, and some individual organizations have standards for indoor air quality. For example, in 1987 and 1988, many local government organizations and companies banned smoking in public places and workplaces or restricted it to particular areas. In the late 1990s, people became concerned with molds and their effects on people (see Chapter 26), so that a number of state governments have begun regulating mold inspectors.

For many years, the American Conference of Governmental Industrial Hygienists (ACGIH) has published standards for exposures to hazardous materials in the workplace. The American Industrial Hygiene Association (AIHA) also publishes guidelines for exposures to certain materials.

For a long time, little attention was given to indoor air quality other than for workplaces. However, some organizations have begun to address the problem. In the 1930s and even earlier, the American Society of Heating, Refrigeration and Air Conditioning Engineers (ASHRAE) recommended ventilation rates for indoor space, and their recommendations were incorporated into many building codes. Energy conservation in buildings, reduced ventilation rates, and tighter construction that limits air infiltration produced a renewed interest in indoor air. ASHRAE, the American Institute of Architects, and other organizations held conferences on indoor air quality. The American Society for Testing and Materials (ASTM) has an indoor air subcommittee. Indoor air quality will remain an active topic for some time and will result in new standards.

The American Institute of Chemical Engineers (AIChE) and its Center for Chemical Process Safety (CCPS) provide a variety of publications on the identification and control of hazards in chemical process plants and facilities.

Outdoors

For outdoor environments, there are many laws and regulations governing air and water quality and the handling and disposal of hazardous material. In addition, there are laws governing the cleanup of sites containing hazardous materials. Some of the environmental laws have expired before Congress resolved additional provisions under renewal acts.

Air   Congress passed the Clean Air Act in 1970 and renewed it in 1977 and 1991. Its purposes are to

1. protect and enhance the quality of the nation's air
2. initiate and accelerate research to achieve prevention and control of air pollution
3. provide technical and financial assistance to state and local governments for air pollution prevention and control programs

4. encourage and assist the development and operation of regional air pollution prevention and control programs

Water   Congress passed several laws affecting water quality, including the Clean Water Act of 1972, which set limits on toxic discharges of industry, and created a large public works program to build sewage treatment plants. The Safe Drinking Water Act of 1974 (and modified in 1992) sets standards for the taste, color, and appearance of public drinking water and maximum limits for certain chemicals and bacteria it contains. The Federal Water Pollution Control Act of 1977 (and modified in 1992) extended previous regulation of water quality. The Marine Protection, Research and Sanctuary Act of 1972 banned ocean dumping of radioactive, biological, and chemical warfare wastes. It established a permit system for ocean dumping of other wastes, such as sewage sludge and dredged materials.

Control of Hazardous Materials   The Federal Insecticide, Fungicide and Rodenticide Act of 1972 (and modified in 1994) required all pesticides to be registered and required manufacturers to submit detailed information about pesticide ingredients and safe use. The Resource Conservation and Recovery Act of 1976 (often referred to as RCRA and modified in 1992) authorized regulation of the generation, treatment, storage, and disposal of hazardous wastes. It established a manifest system for tracking hazardous materials from creation to disposal or use and it eliminated open landfills. The Toxic Substances Control Act (often called TSCA) was passed in 1976 and was modified in 1992. It required testing of new chemicals for safety before they reach the marketplace. The Hazardous Materials Transportation Act (discussed in Chapter 14) controls the packaging, labeling, and transportation of hazardous materials.

Cleanup   In 1980, Congress passed the Comprehensive Environmental Response, Compensation and Liability Act (sometimes called CERCLA, modified in 1992), which addressed the cleanup of existing hazardous waste sites. This Superfund bill required industry to contribute to the cost of cleaning up sites containing hazardous materials. An EPA survey of states identified approximately 35,000 hazardous waste sites. In 1986, Congress passed the Superfund Amendments and Reauthorization Act (called SARA), which extended the Superfund and hazardous waste site cleanup.

Right-to-Know   Public demand for more information about the dangers of particular chemicals led to federal and state right-to-know laws. Under the Emergency Planning and Community Right-to-Know Act of 1986 (also known as SARA Title III and amended in 1992), employers must inform employees within and citizens outside a plant about the dangers of chemicals at the site. Material safety data sheets (MSDSs) standardize the information and give details about chemicals or compounds and applicable protection from harm when exposed to them (see Figure 24-1). SARA Title III also covers labeling and storage requirements and allows states to pass right-to-know laws that require emergency planning for communities around a plant.

Companies analyze process plant safety and develop emergency response plans that they coordinate with local police, fire departments and other organizations in a neighboring community. The analysis and emergency response plans cover both exposures within the plants and in the surrounding community.

Products

There are also regulations covering chemicals in certain products. For example, food additives and coloring agents fall under the control of the Food and Drug Administration

Material Safety Data Sheet

May be used to comply with OSHA's Hazard Communication Standard, 29 CFR 1910 1200. Standard must be consulted for specific requirements.

U.S. Department of Labor

Occupational Safety and Health Administration (Non-Mandatory Form)
Form Approved
OMB No. 1218-0072

IDENTITY *(as Used on Label and List)*	Note: Blank spaces are not permitted. If any item is not applicable or no information is available, the space must be marked to indicate that.

Section I

Manufacturer's name	Emergency Telephone Number
Address *(Number, Street, City, State and ZIP Code)*	Telephone Number for Information
	Date Prepared
	Signature of Preparer *(optional)*

Section II—Hazardous Ingredients/Identity Information

Hazardous Components (Specific Chemical Identity, Common Name(s))	OSHA PEL	ACGIH TLV	Other Limits Recommended	% (optional)

Section III—Physical/Chemical Characteristics

Boiling Point		Specific Gravity (H$_2$0 = 1)	
Vapor Pressure (mm Hg)		Melting Point	
Vapor Density (AIR = 1)		Evaporation Rate (Butyl Acetate = 1)	
Solubility in Water			
Appearance and Odor			

Section IV—Fire and Explosion Hazard Data

Flash Point (Method Used)		Flammable Limits	LEL	UEL
Extinguishing Media				
Special Fire Fighting Procedures				
Unusual Fire and Explosion Hazards				

(Reproduce locally)   OSHA 174 Sept. 1985

Figure 24-1.   OSHA format for material safety data sheets.

(FDA), which also controls certain chemicals in cosmetics. The Consumer Products Safety Commission (CPSC) regulates the chemicals in a variety of consumer products.

One action of the CPSC that garnered much public attention involved the recall and banning of children's sleepwear that contained a fire-retardant material known as TRIS. After pressure to prevent the horrors of sleepwear (particularly dacron) burning and melting into the skin of children, Congress required a fire retardant for the sleepwear. TRIS was the material most often used. Not only was it an effective fire retardant, but it also retained properties in the clothing that made the sleepwear soft and saleable. However, it was suspected of being a cancer-causing agent. Before researchers could complete health studies to establish if TRIS was a carcinogen, Congress, facing public pressure to protect children's health, pressured the CPSC to act. The Commission tried to balance eco-

Section V—Reactivity Data				
Stability	Unstable		Conditions to Avoid	
	Stable			
Incompatibility *(Materials to Avoid)*				
Hazardous Decomposition or Byproducts				
Hazardous Polymerization	May Occur		Conditions to Avoid	
	Will Not Occur			

Section VI—Health Hazard Data			
Route(s) of Entry	Inhalation?	Skin?	Ingestion?
Health Hazards *(Acute and Chronic)*			
Carcinogenicity	NTP?	IARC Monographs?	OSHA Regulated?
Signs and Symptoms of Exposure			
Medical Conditions Generally Aggravated by Exposure			
Emergency and First Aid Procedures			

Section VII—Precautions for Safe Handling and Use

Steps to Be Taken in Case Material Is Released or Spilled

Waste Disposal Method

Precautions to Be Taken in Handling and Storing

Other Precautions

Section VII—Control Measures			
Respiratory Protection *(Specify Type)*			
Ventilation	Local Exhaust		Special
	Mechanical *(General)*		Other
Protective Gloves		Eye Protection	
Other Protective Clothing or Equipment			
Work/Hygienic Practices			

Figure 24-1.   *continued*

nomic, health, and political factors. TRIS retained the feel necessary to induce customers to buy sleepwear, whereas other additives gave fabric a coarse feel. With each washing, the TRIS content of sleepwear decreased about 50% so that a few washings would remove any potentially dangerous levels from already purchased clothing. The sleepwear industry was a collection of small firms dedicated to producing this clothing. The CPSC ultimately decided to recall TRIS-treated sleepwear. The action seriously harmed the industry, while protecting children against an undetermined hazard. Sleepwear with flame-retardant protection as required by law was no longer available, so as an alternative, many parents purchased non–fire-retardant insulated underwear. The problem was back to where it started.

Processes

Chemical engineers and other specialists work on the safety of equipment, systems, and processes for the manufacture of chemicals, petroleum, and other products. The processes may use heat, pressure, chemical reactions, and other methods to achieve the end product after several stages in the process. Many processes are based on continuous and controlled flow of materials, rather than batch approaches. Part of the design responsibility is to reduce or eliminate risks in the processes and to include sensors, warning systems, automated or manual system adjustments, or shutdown when processes go outside the acceptable range of operating parameters.

OSHA has established a performance standard[1] for evaluating the hazards and risks of such processes and defining controls for the hazards. A number of states have enacted similar process safety standards and regulations. The Center for Process Safety and the American Institute of Chemical Engineers have publications and references that help achieve safe processes. Those involved in process safety often apply techniques and methods associated with system safety (covered in Chapter 36).

Workplaces

There are several standards for exposures to chemicals on the job, for levels of contaminants in drinking water and in air, for use of chemicals in consumer products (cosmetics, food, etc.), and for agricultural purposes. One should refer to appropriate federal agencies for current exposure standards for chemicals. This discussion focuses on standards for workers.

For many years, the ACGIH has published a booklet[2] listing the recommended exposure limits for workers. It is intended as a guide to help limit harmful chemical exposures for workers; it does not distinguish safe from harmful environments. The threshold limit values (TLVs) include exposures to airborne particulates and gases or vapors. An ACGIH committee updates the TLV Guide annually. Changes result from careful review of research literature and monitoring of reported experience. The ACGIH also publishes documentation for recommendations found in the TLV Guide.

Early OSHA standards listed permissible exposure limits (PEL) for workers. Both ACGIH and OSHA standards generally are based on 8-hr time-weighted averages (TWA). The TWAs allow for reasonable exposure excursions during a workday. For particular chemicals, ACGIH lists ceiling limits: these concentrations are never to be exceeded. The TLVs have a "skin" notation for substances that are easily absorbed through the skin. For many years, ACGIH's TLV Guide has included a second set of time-weighted average exposure limits based on a 15-min exposure period. These are short-term exposure limits (STEL).

In January 1989, OSHA completed initial rule-making on a new table of exposure standards for more than 400 substances.[3] Appendix A contains a sample of the current OSHA exposure standards. The final rule-making report includes justification for the exposure limits for substances. The main difference between ACGIH and OSHA exposure standards is a legal consideration: ACGIH standards are recommended practice for industrial hygienists practicing in industry, whereas OSHA standards are enforceable as government regulations. Since initial publication, OSHA has been working on updates to its exposure standards, including a more global update. However, the process of changing them has been very slow.

Mixtures   Some people are exposed to more than one substance during an 8-hr period, either as independent substances or mixtures. In some cases, the combination of chemicals involved in the exposures may have independent effects. In other cases, the chemicals may react with each other and have a synergistic effect. When exposures involve two or more hazardous substances that act on the same organ, one should address combined effects. When there is little or no information about combined effects, the additive effect should be considered. One can evaluate additive effects by using

$$X = \frac{C_1}{T_1} + \frac{C_2}{T_2} + \cdots + \frac{C_n}{T_n}, \tag{24-1}$$

where

C_x is the atmospheric concentration of a substance and

T_x is the threshold limit value.

If $X < 1$, the mixture does not exceed the TLV; if $X > 1$, the mixture exceeds the TLV.

Example 24-1 During an 8-hr workday, a worker is exposed to the following mixture of substances: acetic acid, 4 ppm; stoddard solvent, 150 ppm; ethyl ether, 230 ppm. From the OSHA permissible exposure limits (Appendix A), the allowable exposures for these substances are 10, 500, and 400 ppm, respectively.

Applying Equation 24-1,

$$X = \frac{4}{10} + \frac{150}{500} + \frac{230}{400} = 1.275.$$

Because $X > 1$, the mixture exceeds the threshold limit value.

Biological Exposure Indices Biological monitoring consists of an assessment of overall exposure to chemicals that are present in a workplace through measurement of appropriate determinants in biological specimens collected from workers. Biological determinants can be the chemical itself or its metabolite(s) or a characteristic reversible biochemical change induced by the chemical. Appropriate measurements are made on exhaled air, urine, blood, or other biological specimens. Biological exposure indices (BEIs) serve as reference values. They represent the levels of determinants that are most likely to be observed in specimens collected from a healthy worker who has been exposed to chemicals to the same extent as a worker with inhalation exposure to the TLV-TWA. BEIs published by ACGIH are reference values intended as guidelines for the evaluation of potential health hazards in the practice of industrial hygiene. BEIs apply to 8-hr exposures, 5 days a week. Individual differences may account for occasional measurements above BEIs, but if a sample exceeds a BEI consistently, there is cause for investigation of the workplace. BEIs should be used as a backup check on exposures, not the primary means for determining if a hazard exists.

24-2 HAZARDS

Chapter 9 introduced hazard recognition and control. The discussion notes that one must know at least three items of information about an agent to determine if it is hazardous:

1. what the agent is and what form it is in
2. the concentration
3. the duration and form of exposure

Having this information available is particularly important when seeking to understand hazards and controls for chemicals.

The main hazards for chemicals are (1) health effects, (2) fires and explosions, and (3) reactivity with other materials. This chapter mainly discusses health effects, and Chapters 16 and 17 discussed fire and explosion hazards. Reactivity refers to the relative stability of a material: in a fire, a material may become unstable from the heat or products of combustion; some materials may be sensitive to mechanical disturbance from pressure or physical impact; and extinguishing agents, such as water, can react with certain materials. A reaction may involve releases of large amounts of energy and the release may be sudden, as in an explosion.

Classification

There are several ways to classify hazardous materials. Chapter 14 discussed the DOT classification and labeling of hazardous materials, including chemicals. In the United States, a commonly used classification scheme for hazardous chemicals is that of NFPA 704.[4] A diamond-shaped symbol, divided into quadrants, conveys health, flammability, and reactivity information (see Chapter 16).

Identification

There are several ways to identify chemicals. Each chemical or compound has a chemical name and chemical formula. Examples are ozone (O_3) and methyl isocyanate (CH_3NCO). The Chemical Abstract Service has a registration system for assigning unique identifying numbers in the format xxx-xx-x: ozone and methyl isocyanate are 10028-15-6 and 624-83-9, respectively. NIOSH maintains the Registry of Toxic Effects of Chemical Substances (RTECS), which assigns each chemical a unique number: RS8225000 and NQ9450000 denote ozone and methyl isocyanate. As noted in Chapter 14, DOT has an identifying number for hazardous materials. Numbering systems for procurement purposes, such as the federal National Stock Number, also identify chemicals.

Type of Airborne Contaminants

There are two main forms of airborne contaminants: particulates and gases or vapors. Particulates include dusts, fumes, smoke, aerosols, and mists that are classified additionally by size and chemical makeup. Shape also can be important. Some particulates are fibrous, having long, thin shapes, whereas others may be more spherical and have a fairly uniform cross section. Figure 25-5 provides size characteristics of some airborne contaminants.

Dusts Dusts are airborne solids, typically ranging in size from 0.1 to 25 μm. Dusts larger than 5 μm settle out in relatively still air because of the force of gravity. Many dusts are created by processes that break materials into small sizes, such as grinding and mixing.

Fumes Fumes are fine solids less than 1 μm in size that are often formed by condensation of vapors. For example, heating of lead vaporizes some lead material that quickly condenses to small, solid particles.

Smoke Smoke is carbon or soot particles, generally less than 0.1 μm in size, that results from incomplete combustion of carbonaceous material.

Aerosols Aerosols are airborne solid or liquid particulates dispersed in air.

Mists   Mists are fine liquid droplets suspended in or falling through air. Mist is generated by condensation from the gaseous to the liquid state or by breaking up of liquid into fine particles through atomizing, splashing, or foaming.

Gases   A state of matter separate from solids and liquids.

Vapors   The gaseous phase of a substance that is liquid at normal temperature and pressure.

Health Effects

Health effects for different chemicals vary considerably. The likelihood and degree of damage depend on type and form of substance, the type and rate of exposure, and what happens to the substance in the body. Most hazardous materials affect particular organs of the body. For example, some damage tissue, such as skin and eyes, on contact, some affect respiration, some damage nerves and other elements of the central nervous system, and some affect oxygen transport of the blood or other blood functions. Often general symptoms, such as headache and nausea, are confused with symptoms of other diseases. Chemical exposures may not be recognized immediately as the cause.

Some effects of chemicals appear as behavioral changes. For example, lead exposure may lead to forgetfulness, hallucinations, and lethargy. Again, behavioral symptoms resulting from chemical damage often are confused with other causes.

Latency Period   Some chemicals have immediate effects. An example is strong acid or caustic contacting tissue and destroying it. These are sometimes called chemical burns, because they exhibit properties similar to thermal damage of tissue. Other chemicals may not manifest their effects for some time. The delay between exposure and observable effect is a latency period. Some carcinogens have a latency period as long as 20 to 40 yr.

Acute Versus Chronic Exposures   Exposures can be acute or chronic. For some chemicals, disease or effects do not appear until after repeated exposures; in other cases, a single exposure may be sufficient to induce effects. Some materials will not cause significant permanent damage with a single exposure; some may. Others may not cause permanent damage at all.

Local Versus Systemic Effects   Local effects occur when substances cause injury to skin, eyes, or respiratory tract after one or more exposures. Systemic effects occur when substances enter the body and produce damage to organs or biological functions. The effects may be behavioral or physical. Examples of damage include kidney dysfunctions or failures, clotting of blood, damage to liver tissue, and ulcerations in the digestive tract.

Asphyxiants   Asphyxiant materials do not have direct effects on the body or its organs, but they do displace oxygen in a breathing atmosphere. The reduced oxygen content affects the partial pressure of oxygen and inhibits oxygen transport in the blood (see Chapter 19). Some asphyxiants may interfere with oxygen transport or breathing in other ways.

Nuisance Dusts   Some materials are simply a nuisance. They may cause irritation, coughing, or similar symptoms, but have no long-term effects. Certain dusts are classified as nuisance dusts.

Individual Differences Not everyone exhibits the same effects or degree of effects from a chemical exposure. Two people may have the same exposure, but only one may have a reaction. Some people are allergic to certain materials in their environment. The allergic reaction may be inherited or may develop during life. Some people become sensitized to a substance. Initially, they do not react to an exposure, but then suddenly they show reactions. Removal of the substance may stop the reaction. However, a single, subsequent exposure after sensitization usually produces the reaction.

Pneumoconiosis Pneumoconiosis is a disease of the lungs resulting from the inhalation of various kinds of dusts and other particles. The term means dusty lung. The disease has several names depending on the material one is exposed to. Table 24-1 lists several forms of pneumoconiosis.

There are other forms of lung disorders related to exposures to hazardous materials. Fibrosis is the formation of scar tissue, which forms when the body attempts to engulf foreign material that lodges in the lung. Bronchitis is the overproduction of mucus, which often results in coughing. Asthma is the constriction of the bronchial tubes caused by a histamine reaction to some toxin that produces swelling. Hives on the skin is a similar reaction.

Carcinogens A substance that produces cancer in animals or humans under certain quantified exposures is a carcinogen. Some define a carcinogen as any agent that increases tumor induction in humans or animals. Even the induction of benign tumors may be enough to characterize a substance as a carcinogen. Irreversibility and a long latency period after the initial exposure to a carcinogenic agent characterize carcinogenesis. There are specific tests to determine when a material is to be classified a carcinogen or a suspected carcinogen.

Mutagens A mutagen is any substance that causes changes in the genetic structure in a current generation of animals or humans such that it can cause cancer or some mutation in a later generation. Mutations may not show up until several generations later. Mutagens cause inheritable changes in the chromosomes, changes that may not be observable deformities. Radiation, for example, has been associated with sterility.

Teratogens A teratogen is any substance that causes malformations or serious deviations from the normal in a human or animal fetus. Congenital malformations or abnor-

TABLE 24-1 Types of Pneumonconiosis

Disease	Material Inhaled
Asbestosis	Asbestos fibers
Silicosis	Free silica (SIO_2) dust from mining, sandblasting, quarrying, and in ash from volcanic eruptions
Berylliosis	Beryllium particles
Byssinosis	Cotton dust
Metal fume fever	Particulates of zinc, magnesium, copper, and their oxides (other metals have also been known to produce metal fume fever)
Siderosis	Iron oxide (often from welding and mining)
Kaolinosis	Kaolin (china clay) dust from grinding and handling
Mica pneumoconiosis	Mica dust from grinding
Bauxite pneumoconiosis (Shaver's disease)	Aluminum oxide fumes from smelting bauxite

malities in offspring resulting from exposure of a mother or fetus to some agent is teratogenesis. Typically, there is no exposure effect on the mother. Teratogens interfere with normal embryonic development.

24-3 TOXICOLOGY

Toxicity is the capacity of a material to produce injury or harm after it reaches a site in or on the body where harm can result. In contrast, a health hazard is the possibility that exposure to a toxic material will cause harm under ordinary use circumstances or when a specific quantity is used under particular conditions. Toxicology is the science that deals with the nature and effects of poisons. Tables 24-2 and 24-3 summarize some toxicity rating schemes.

Sources of Data

There are many publications that list toxic properties of chemical substances. One of the primary references is *Dangerous Properties of Industrial Materials*, edited by N. I. Sax. This publication lists toxicity and other data for more than 18,000 substances. Another primary reference is *Patty's Industrial Hygiene and Toxicology*. This long-standing, multivolume publication includes discussions of evaluation and control methods as well as toxicity data. The OSHA Hazard Communication Standard[5] requires that chemical manufacturers and importers provide customers with material safety data sheets (MSDSs). These data sheets are good sources of information about the toxicity and other physical, chemical, and hazardous properties of chemicals.

There are also some computer accessible data banks of toxicity data for substances. The most noteworthy ones are the National Library of Medicine's Hazardous Substances Data Bank and the NIOSH Registry of Toxic Effects of Chemicals and the NIOSH Pocket Guide to Chemical Hazards.[6] Also there are computer data banks and CD-ROM collections of MSDSs, many available over the Internet.

Routes of Entry

There are three main routes of entry for hazardous substances into the body: inhalation, ingestion, and absorption through the skin. An occasional route of entry is injection. The most common route of entry in work environments is by inhalation. As soon as materials enter the body and reach the blood, they can be distributed throughout the body. Some materials have an affinity for certain kinds of tissue, and consequently, concentrate in particular organs.

Inhalation During respiration, airborne gases and particulates are carried into the upper respiratory system and lungs. The body may absorb the materials into the bloodstream or may encapsulate the materials in the lung tissue. The inhaled materials or portions of them may be exhaled as well.

Figure 24-2 illustrates the structure of the respiratory system. Air passes through the pharynx and trachea to the bronchi, bronchioles, and ultimately to the terminal air sacks or alveoli, where inhaled gases may enter the blood stream.

Gases and vapors disperse with oxygen and nitrogen in normal air. Consequently, hazardous gases will travel with normal air deep into the lungs. Depending on solubility and other properties, a gas may go into solution in the blood or may attach to red cells or elements of the blood. The blood transports the material to other tissues in the body for

TABLE 24-2 Toxicity Rating System[a]

Rating	Description
U	**Unknown.** Insufficient data are available to enable a valid assessment of toxic hazard to be made.
0	**None, no toxicity.** This rating applies to chemicals that (a) produce no toxic effects under any conditions of normal usage or (b) require overwhelming doses to produce any toxic effects in humans.
1	**Low, slight toxicity.** The rating is characterized under four types of exposure: (a) *Acute local.* Chemicals that on a single exposure lasting seconds, minutes, or hours cause only slight effects on the skin or mucous membranes or eyes, regardless of the extent of exposure. (b) *Acute systemic.* Chemicals that can enter the body by inhalation, ingestion, or dermal contact and produce only slight toxic effects, regardless of the duration of exposure or after the ingestion of a single dose, regardless of the amount absorbed or the extent of the exposure. (c) *Chronic local.* Chemicals that on repeated or continuous exposure covering days, months, or years cause only slight and reversible damage to the skin or mucous membranes. The extent of the exposure can be great or small. (d) *Chronic systemic.* Chemicals that on repeated or continuous exposure covering days, months, or years cause slight and usually reversible toxic effects on the skin, mucous membranes, or eyes. The exposure can be by ingestion, inhalation, or skin contact and may be great or small. Slightly toxic chemicals produce changes readily reversible once the exposure ceases with or without medical intervention.
2	**Moderate toxicity.** Chemicals may cause reversible or irreversible changes in the human body not necessarily sever enough to cause serious physical impairment or threten life. Ratings are characterized under four types of exposure: (a) *Acute local.* Chemicals that on a single exposure lasting seconds, minutes or hours produce moderate toxicity to the skin, mucous membranes, or eyes. The effects can be the result of an intense exposure for seconds or a moderate exposure for hours. (b) *Acute systemic.* Chemicals that after being absorbed by inhalation, ingestion, or skin contact produce moderate toxicity after a single exposure lasting seconds, minutes, or hours or after the ingestion of a single dose. (c) *Chronic local.* Chemicals that on continuous or repeated exposure over days, months, or years cause moderate toxicity to the skin, mucous membranes, or eyes. (d) *Chronic systemic.* Chemicals that on absorption by ingestion, inhalation, or skin contact cause moderate toxicity after continuous or repeated exposures over days, months, or years.
3	**High, severe toxicity.** Ratings are characterized under four types of exposure: (a) *Acute local.* Chemicals that on a single exposure covering seconds or minutes can cause injury to the skin, mucous membranes, or eyes of sufficient severity to threaten life, cause permanent physical impairment, or cause disfigurement. (b) *Acute systemic.* Chemicals that after a single exposure by inhalation, ingestion, or skin contact cause injury of sufficient severity to threaten life. The exposure may last seconds, minutes, or hours or may be a single ingestion. (c) *Chronic local.* Chemicals that on continuous or repeated exposures covering days, months, or years can cause injury to the skin, mucous membranes, or eyes of sufficient severity to threaten life or produce permanent impairment, disfigurement, or irreversible change. (d) *Chronic systemic.* Chemicals that on continuous or repeated exposures by inhalation, ingestion, or dermal contact to small amounts for days, months, or years can produce death or serious physical impairment.

[a] From Sax, N. I., *Dangerous Properties of Industrial Materials*, 7th ed., Van Nostrand Reinhold, New York, 1988.

TABLE 24-3 Degree of Toxicity Ratings

Toxicity Rating	Probable Lethal Dose for a 70-kg Human	Experimental LD_{50}: Dose per kg of Body Weight
Dangerously toxic	A taste	<1.0 mg
Seriously toxic	A teaspoonful	1–50 mg
Highly toxic	An ounce	50–500 mg
Moderately toxic	A pint	0.5–5 g
Slightly toxic	A quart	5–15 g
Extremely low toxicity	More than a quart	>15 g

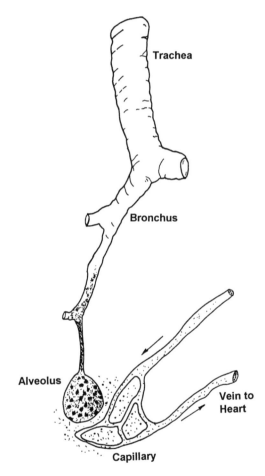

Figure 24-2.   General structure of the human respiratory system.

which there may be more affinity. For example, carbon monoxide forms a bond with the hemoglobin of the red cells more readily than oxygen. In addition, because it does not release from the hemoglobin easily, the red cells are unavailable to bond with oxygen, and therefore become ineffective in transporting oxygen throughout the body.

   Although gases and vapors move freely with inhaled air, particulates may not reach the alveoli. Because of their mass or shape, larger particulates (generally those larger than 5–7 μm in diameter) are not able to make sharp turns with air movement. They impinge on the mucus membrane of the upper respiratory tract and the cilia, which line the mucus membrane, move the foreign matter upward along the respiratory tract in a cleaning process. The foreign material is expectorated or swallowed. Smaller particulates (less than 5–7 μm in diameter) do reach the alveoli. Because particulates smaller than 0.5 μm tend to remain airborne, they are likely to pass from the lungs as one exhales. Those between approximately 0.5 and 5 to 7 μm have a good chance of lodging in the alveoli.

Ingestion   Ingestion involves eating or drinking materials. Ingested materials are absorbed into the blood after traveling to the intestinal tract. Although people ingest little toxic material directly from work environments, often food is contaminated by hands or from contaminated eating areas. Some toxic material may be ingested from material originally inhaled and moved through ciliary action to the throat, where it is swallowed.

Absorption through the Skin   Some materials enter the bloodstream through the skin. The skin may be abraded or have lesions that foster absorption, although many materials are absorbed directly through intact skin. Elevated skin temperature or moisture on the skin may enhance cutaneous absorption. Some skin areas, such as the back of the hand and follicle-rich areas, exhibit higher absorption rates than areas like the palm of the hand and the forearms.

Injection   Materials may be injected purposely or accidentally. Accidental injection is more important for occupational settings. Refer to the discussion of injection injuries in Chapter 19.

Direct Contact with Tissue

Beside the three main routes of entry, some materials may be harmful in direct contact with external tissue.

Dermatitis   There are many forms of skin disorders, many of which result from exposures to chemicals. Dermatitis is a general term for skin disorders. Some forms of dermatitis exhibit reddening; others involve cracking, sores, acne, and other disorders. Some forms of dermatitis, called contact dermatitis, result from direct contact with a substance that may be in gaseous, liquid, or solid form and may be a direct irritant or an allergen. Chloracne results from chlorinated naphthalenes and polyphenyls acting on sebaceous glands. The acne-like appearance can appear on skin of the hands and arms, as well as other areas of the body.

Eye Irritation   Substances may contact eyes or tissues around the eyes and cause irritation or more severe injury. Conjunctivitis is an inflammation of the conjunctiva, the delicate membrane that lines the eyelids and covers the exposed surface of the eyeball.

24-4 TOXICOLOGICAL DATA

Methods for Assessing Toxicity

Toxicity data for substances come from many sources, most often from controlled studies. Because all chemicals can be toxic, the studies manage not only the amount of chemical involved, but also the conditions of exposure. Nearly all substances fail to exhibit effects at very low exposure levels, but at some level, effects begin to appear. This suggests a threshold level for effects, above which increased concentrations will produce more severe effects. Also, as exposure levels increase, the effects are likely to occur in a larger portion of the population. One cannot assume that the relationship between exposure and effects is linear.

Human Experimentation One way to collect data on toxicity of materials is through experimentation with humans. In general, society does not condone human experimentation, but on occasion, particularly when there is a strong national concern for some disease or illness, society may accept human testing of substances for medical treatment of terminally ill patients who approve. When a pharmaceutical substance appears to have high benefit and relatively little risk as a result of many other tests, the FDA may approve testing in humans. Because there is virtually no opportunity to perform general testing of substances for toxicity on humans, toxicity data must come from other than human tests.

Human Experience Sometimes accidental exposures provide opportunities to compile data on the toxicity of a substance. The exposures may be acute or chronic. In accidental exposures, there is no control over the exposure, so the difficulty is knowing what the exposure level actually was. Researchers must estimate exposure levels and conditions from limited information using epidemiological procedures. (Epidemiology is the study of disease in human populations.) After a pattern of disease appears that could be related to some exposure, further testing may result. For example, workers exposed to certain pesticides exhibited similar disease patterns that stimulated further testing, which ultimately lowered exposure standards and even led to a ban on certain materials.

Animal Studies Most toxicity data come from controlled and replicated animal studies. The physiology of animal species varies from humans, and as one moves farther away from humans in the zoological chain, the meaningfulness of test results is less valid for generalization to humans. However, standard tests and methods for rating toxicity help estimate human effects. Controlled breeding and raising of test species also helps improve reliability and comparability of results.

A problem in toxicity testing involves time. Some substances produce disease in a portion of a population after a long latency period or chronic exposures. To replicate slow exposures or wait for latency periods would be very expensive. As a result, many toxicity test procedures involve high-dose rates. Measures for controlled doses to experimental animal populations are quantities per unit body weight, per unit of skin area, or per unit of volume of inspired air. Toxicity theories assume that the product of the concentration and the time of exposure has a linear or near linear relationship with the effects. This assumption has been the source of much debate.

Studies report toxicity data for animal populations in various units. One unit is lethal concentration (LC), which applies to airborne concentrations. Another unit of measure is lethal dose (LD), which involves ingestion, injection, or other means of applying a substance. Effects are often reported as the portion of the exposed population that dies as a result of the controlled exposure. A subscript indicates the portion that died. For example, LD_{50} means that 50% of a test population died of a particular dose; LC_{10} means that 10% died of an inhaled concentration. The designation "toxic concentration—lethal" is the lowest published lethal concentration. Similarly, "toxic dose—lethal" is the lowest published lethal dose. Other notations in toxicity tables indicate effects on skin, blood, nervous system, muscles, or other tissue and organs. Notations also indicate whether a substance produces irritation or mutagenic, carcinogenic, teratogenic, or other effects.

Microorganism Testing Because animal testing is expensive and time consuming, researchers have developed some short-term tests for chemical toxicity, many of which involve bioassays of microorganisms. Studies monitor the growth patterns of particular bacteria and contrast bacteria exposed to a chemical compared with unexposed samples.

These tests may help screen substances rapidly for particular characteristics. In general, short-term tests have limitations. They do not include variables that animal studies do in the mathematical models for generalizing test results to humans. For example, one of the notable short-term tests for carcinogens is the Ames test. The procedure uses Salmonella and the results involve dose, but not time or duration of exposure. In addition, the Ames test has been criticized for its lack of reliability. It demonstrates high correlations between certain known carcinogens and human experience, whereas correlations between test results for other carcinogens relative to human experience are low.

24-5 CONTROLS

There are three classes of controls for protecting people from exposure to hazardous materials in the workplace:

1. engineering controls
2. work practices and administrative controls
3. personal protective equipment

One should consider these controls in the order presented, as discussed in Chapter 9.

Engineering Controls

Engineering controls include process changes and substitution of nonhazardous substances, enclosing a source, isolation of a worker, and the use of ventilation.

Substitution By replacing a hazardous material with a nonhazardous one, the danger of exposure to a hazardous material can be eliminated. Even a less hazardous material may form a desirable control by substitution. If the proposed alternate material does not perform as well as the original material, substitution may not be a desirable solution.

Isolation Isolation uses a barrier between a source of contaminants and workers. Often, enclosing the process equipment that generates a contaminant provides the barrier. The enclosure usually involves an exhaust ventilation system as well. Sometimes there is a port with built-in gloves that a worker's hands can slip into to allow the worker located on one side of the enclosure to perform some function inside the enclosure. This setup is called a glove box. In other cases, workers are located in an enclosure with a controlled, uncontaminated atmosphere from which they can see an operation and control it remotely. The separation prevents potential exposures to contaminants. Placing workers in an enclosure may not be desirable if the workers must enter the contaminated process area frequently. They may fail to protect themselves as they move from the protected area to the unprotected or to decontaminate them selves when returning from the unprotected area. Another form of isolation is separating hazardous operations from nonhazardous ones.

Ventilation Ventilation is useful for airborne contaminants. There are two types of ventilation for controlling hazardous substances: one is general ventilation or dilution ventilation, and the second is local exhaust ventilation. Chapter 25 gives details on ventilation.

In general ventilation, one moves fresh outside air into the general work space to dilute or displace contaminants. The goal is to keep concentrations at or below allowable levels. There are several limitations for general ventilation:

1. It normally requires movement of large volumes of air.
2. The outdoor air may already be contaminated and not fresh or clean.

3. Heating, cooling, or dehumidifying outdoor air is costly.

4. The general movement may not dilute substances at all locations in a space to keep air quality within limits everywhere.

Local exhaust ventilation requires much smaller volumes of air than general ventilation. Local exhaust systems capture contaminants at or near their source, before hazardous substances reach the breathing zone of people, and moves the air to locations where people will not be present. To avoid violation of air pollution standards or exposure of more people, exhausted air may need treatment to remove and collect contaminants.

Work Practices and Administrative Controls

Work practices and administrative controls include housekeeping procedures, materials handling or transfer procedures, leak detection programs, training, modifying the work, and personal hygiene. Administrative controls may involve several controls organized into programs.

Housekeeping Housekeeping activities include removal of dust accumulation and rapid cleanup of spills. Regular removal of hazardous dust from floors, walls, or other surfaces is important. Otherwise, the material can become airborne and can pose greater danger. Removal should avoid dispersing a material into the air. Therefore, vacuum cleaning equipment that properly traps the contaminants should be used. Sweeping, compressed air, or blowers should not be used to remove dust from surfaces because they will cause material to become airborne. In some operations, spraying water on materials may eliminate sources of dust.

Materials Handling or Transfer Procedures Loading and unloading of materials can cause materials to become airborne, thereby exposing workers to hazardous levels of the materials. For example, transferring liquid into an empty tank truck or car will displace residual or evolving vapors from the container. Pouring or dumping dusts or dusty materials into open containers or on piles may generate a great amount of airborne dust. These transfer operations may require closed transfer systems or exhaust systems to prevent exposures of workers or others.

Another form of control involving materials transfer is use of containers to collect overfill spills or leaking material between transfers. For example, one could place a drip pan under a fill spout for liquid material.

Leak Detection Programs Leak detection may involve visual inspections and automatic sensor devices, as well as scheduled inspections of valves, piping, and other potential sources for leaks. Quick repair will minimize potential exposures. Sensor systems for particular materials can determine if a hazardous material is airborne and can trigger visual or audio alarms. The signal should initate corrective action or the system should automatically shut down a process.

Training Proper training of workers and supervisors is necessary to supplement other controls. OSHA standards and some state right-to-know laws require training workers to understand the hazards associated with materials they may encounter on their jobs and how to protect themselves from those hazards.

Modifying the Work One can reduce the exposure of a person by limiting the duration of exposure. Keeping the exposure time during a day or longer period low by having more workers share an activity will ensure that any one worker remains below the exposure limits. Exposing additional workers may not be an appropriate procedure.

Personal Hygiene   Personal hygiene involves cleaning skin that becomes contaminated during normal work activity or as the result of a spill or accident. One should evaluate carefully which soaps, waterless hand cleaners, washing facilities, and emergency showers provide the proper cleaning agents to remove particular contaminants. If there is a possibility for contamination and injury of eyes, there should be emergency eyewash fountains. Workers may need places to change clothes to prevent carrying hazardous materials home or outside a controlled area. People should not eat or drink in areas where there are toxic materials.

Personal Protective Equipment

Use of personal protective equipment may be necessary when adequate engineering controls, work practices, or administrative reforms cannot be achieved. Maintenance workers and those involved in spill cleanup activities usually need personal protective equipment. Other controls may be too expensive or difficult to achieve for such operations. Depending on the hazards associated with certain substances, personal protective equipment may involve eye and face protection, protective clothing, protective creams and lotions, respiratory protection, or other equipment. Chapter 28 discusses personal protective equipment in more detail.

24-6   MEASUREMENT

To determine what chemicals are present in the air, water, or solids, one must collect samples and analyze them. Analysis will also provide data on quantity or concentration of the contaminants. Gases and vapors require different collection and analysis procedures than do particulates. Some methods help determine if a worker's exposure exceeds the 8-hr TWA; others find out if a contaminant is present and how much its concentration varies over time. The kind of assessment needed will vary. There are laboratory instruments intended for highly accurate analysis. There are field instruments for quick assessment of conditions. There are instruments attached to workers (personal samplers or dosimeters) to monitor exposures. Also, there are automatic sensor, analysis, and alarm systems intended to monitor releases and to warn people to evacuate or to correct faulty equipment or conditions. Many of the instruments must be calibrated regularly to ensure accurate assessment. There are a wide range of collection procedures and devices. This discussion is a limited review of collection and analysis equipment.

OSHA and NIOSH have published sampling and analysis procedures appropriate for particular substances.[7] ASTM has standards on instrumentation, sampling, and analytical procedures. ACGIH publishes a book, *Air Sampling Instruments*, and there are many other treatises on the subject. With technological changes in instrument design and features, there are also changes in their use and the methods associated with making assessments.

Sampling for Gases and Vapors

There are short- and long-term sampling devices for gases and vapors. Some short-term samples are grab samples. Short-term samples involve drawing air into an evacuated, contaminant-free container (plastic, glass, or stainless steel) or pumping a sample into an evacuated container or through an impregnated filter paper or gas detection tube. Most often one uses a small hand pump or squeeze bulb.

There are many kinds of glass detection tubes that contain material designed to change color when they react to particular contaminants. This method is called colorimetry. This form of sampling and analysis device gives quick indications of the presence and concentration of substances. The glass tubes are sealed until used. The user breaks the ends off to allow air to pass through the tube and its indicator material. A scale on the tube or the hand pump holder gives a quantitative reading. One must know what substances to test for before selecting the proper indicator tube.

Continuous or long-term samples require more elaborate equipment. Sample times are typically 15 to 30 min. Some systems continually sample many ports or sensors to detect if a leak or release of hazardous material occurs. Portable, powered air-sampling pumps draw air through a variety of collecting devices. They are calibrated so that one knows the volume of air drawn through the collection device. Some collectors contain an absorbent material (charcoal or silica gel) that accumulates the substance. After collection, samples are analyzed in a variety of ways to assess the type and concentration of contaminants.

There are a variety of portable devices that give direct readings of certain gases or vapors. There are oxygen meters, combustible gas meters that indicate the presence of particular flammable gases, and portable infrared analyzers and gas chromatographs for assessing many different materials.

Chemical Analysis

There are many ways to identify a type of contaminant and to assess concentration. The proper method depends on the form of the material, type of material, and collection method used. Colorimetric methods cause a reactive material to produce a color in proportion to the quantity of a substance that is present. Ion exchange methods separate elements by passing a sample through calibrated columns of eluting agents. Gravimetric methods weigh the presence of a substance collected or the product of a reaction. Volumetric methods involve reaction with definite volumes of standard solutions or reagents. Gas chromatography measures the length of time materials take to pass through a column of detector material. X-ray diffraction produces a unique spectrum when a beam of x-rays impinges on a crystalline sample. X-ray fluorescence is a similar technique. Spectroscopy involves affecting a sample with carbon arc, infrared radiation, electron beams, or high-temperature flame. Each spectroscopy process produces a spectrum that gives a unique signature for particular materials.

Sampling for Particulates

Particulates may be liquids or solids, and most often air passes through a collection device to obtain a sample of particulates. Collection may involve filtration, impactors, impingers, or other devices. The method used depends on the information desired and other factors, such as particle mechanics in an airstream.

There are a variety of special filtration devices for air sampling. Impactors involve a sudden change in direction for an airstream. The airborne particulates impact on a surface because they cannot change direction because of their mass. There are single or multi-staged impactors. Impingers move air through a small opening, causing materials to impinge on a plate (or the bottom of a collector tube) immersed in a liquid. The particles become trapped and can be analyzed various ways.

Some other devices are centrifugal separators, electrostatic precipitators, particle sizers, and counters. Centrifugal separators cause air to spin. Because of their mass, the particulates move to the outside of the rotating air column, where they are collected or removed from the column. Electrostatic precipitators place a charge on particulates in an airstream passing through the device. The charged particles move to collection surfaces with an opposite charge and lodge there. Some analysis requires knowing the number of particles per volume of air in a gas stream or the distribution of particles by size. There are several kinds of instruments for obtaining particle size, the number of particulates, or the distribution by size. Some of these instruments can obtain data as air passes through the device.

24-7   CONFINED SPACES

Confined spaces are enclosures having limited means of access and egress. They can be storage tanks, tank cars, pressure vessels, boilers, bins, silos, and similar enclosures that have access through a manhole or door. Open pits, vaults, and vessels with limited ventilation, as well as underground utility tunnels, storm sewers, pipelines, septic tanks, and similar containers are also confined spaces. Some partial enclosures, such as railroad boxcars, can be confined spaces.

A long time ago, miners sent a canary into a mine. If the canary returned, it was assumed that the atmosphere was satisfactory for workers; if the canary did not return, it was assumed the atmosphere was not suitable for workers. Too often, workers are sent into confined spaces and become the canaries that do not return. Today, confined spaces require special procedures to ensure that they are safe before people enter or work in them.

Hazards

There are three main hazards of confined spaces. The first is oxygen deficiency. Oxygen-deficient atmospheres are those with less than 18% oxygen. This reduced concentration can result from other gases mixed in air that reduce the normal 21% of oxygen in the atmosphere. Another way that oxygen deficiencies arise is the presence of heavier-than-air gases that settle in a closed container, especially where there is no opening at the bottom. Heavier-than-air gases also create problems when there is little ventilation and mixing of gases within a confined space. Some confined spaces have oxygen-depleting bacteria, have oxidation processes (such as rusting) that consume available oxygen, or have combustion.

The second hazard of confined spaces is flammable and combustible gases, vapors, or dusts. Sources of heat or spark may ignite these materials, and a fire or an explosion may result. The gases may result from residual fuels, methane produced by decay (anaerobic bacteria), or heavier-than-air vapors that flow into the space.

The third hazard is toxicity. The toxic materials may have direct effects, such as pulmonary paralysis from hydrogen sulfide. They can be asphyxiants, such as carbon monoxide, that interrupt oxygen transport or they can be irritants at very low concentrations and lethal at higher levels.

Another hazard of confined spaces is a pressurized atmosphere that can produce injury when opened. Confined spaces may contain moving parts that can cause injury if external controls are not locked out and tagged out. There also may be hazards from electrical equipment or dangers from ducts, pipes, or drains that connect to sources of hazardous materials.

Controls

Before entry, one must evaluate a confined space for hazards. It should be depressurized, connections to potentially hazardous materials must be isolated and sealed, energy sources must be locked out and tagged out, stored energy must be released or controlled to prevent inadvertent release, and the atmosphere must be tested for oxygen content, toxic materials, and flammable gases and vapors. Atmospheric testing must not be limited to locations near the entry because often there are widely varying conditions resulting from layering of gases.

Ventilation systems for confined spaces can achieve several purposes. They must provide adequate breathable air supplies unless workers wear self-contained breathing equipment. Often confined spaces are vented with elevated oxygen levels to produce breathable mixtures. The ventilation also must reduce flammable hazards to 10% or less of lower flammable limits. The intake for the ventilation system must not take in exhausted contaminants.

If heated processes, such as open flames, welding, and cutting, are used in a confined space, precautions for fire protection and removal of smoke and fumes are necessary. If workers use solvents for cleaning, the hazards of the materials must be assessed and suitable precautions must be taken. Electrical equipment should use low voltage.

Activities in a confined space may create noise, ionizing radiation, or heat or cold hazards. Precautions for these must be in place.

Workers involved inside confined spaces or in support of confined space activities must be trained in the hazards associated with the operations. They must learn what procedures to follow before entry, during occupancy, and in emergencies; they must know rescue procedures; and they must know what protective or rescue equipment is required and how to use it.

There should be at least two workers involved in confined space work, one of whom should be an observer. The observer is the prime rescue person and cannot enter the space without a replacement observer. With proper planning, rescue can be completed without a second person entering the space. Use of lifelines and harnesses or belts is critical.

Before beginning work in a confined space, workers should receive entry permits. A confined space permit system requires the identification of the space as hazardous and includes an evaluation of the potential hazards. Supervisors and safety specialists must each agree and attest that all preentry evaluations and procedures are complete and that the space is ready for safe entry. Work should be thoroughly planned, workers trained, and emergency procedures and equipment in place.

Standards

OSHA has established a standard for confined spaces which require a permit for entry.[8] The standards establishes procedures for determining whether a confined space poses any hazard for which the employer must issue a permit after providing the necessary protection to any workers who may enter it. The American National Standards Institute has a standard on working in tanks and other confined spaces.[9]

EXERCISES

1. A worker is exposed to naphthalene vapor on the job. Based on OSHA permissible exposure limits, what is the allowable concentration for

    (a) an 8-hr day?

    (b) a 15-min period?

    Is there any danger of absorbing the material through the skin?

2. A worker is exposed to silica dust (crystalline quartz). What is the allowable exposure (based on OSHA standards) if the substance is respirable?

3. During an 8-hr day, workers in one plant are exposed to a mixture of three chemical vapors considered to have additive effects: nitrobenzene, 0.3 ppm; ethyl acetate, 100 ppm; 2-butanone, 75 ppm. Is the OSHA exposure standard for the mixture exceeded?

4. A city has an infestation of mosquitos. There is a public fear of meningitis, and demands for the city to take some remedial action grow. A city worker sprays neighborhoods with malathion mixed with kerosene. What dangers, if any, are there for the worker? What are the exposure limits, if any apply?

5. A worker in a pharmaceutical plant helps produce iodine. What OSHA standard applies to this worker for an 8-hr day? Is the standard a time-weighted average (TWA)?

6. Locate an organization that has activities involving confined space entry. After reviewing the OSHA confined space entry standard, develop a written procedure for

   (a) a permit required confined space

   (b) a confined space for which a permit is not required

7. A worker is exposed to methyl formate during an 8-hr shift with the following exposures: 2 hr at 150 ppm, 2 hr at 75 ppm, and 4 hr at 50 ppm. Based on OSHA permissible exposure limits and associated computational procedures, what is the 8-hr time-weighted average limit? Is the exposure for the shift acceptable?

REVIEW QUESTIONS

1. Approximately how many chemical compounds are there? How many have economic value? How many have some inherent hazard? How many have published exposure standards?

2. What federal agency regulates the use of chemicals related to the following:

   (a) outdoor air quality

   (b) water quality

   (c) public transportation

   (d) the workplace

   (e) indoor air quality

   (f) consumer products

3. What do the following acts address or what significant elements do they cover?

   (a) Clean Air Act

   (b) Clean Water Act

   (c) Safe Drinking Water Act

   (d) Federal Water Pollution Control Act

   (e) Marine Protection, Research and Sanctuary Act

   (f) Federal Insecticide, Fungicide and Rodenticide Act

   (g) Resource Conservation and Recovery Act

   (h) Toxic Substances Control Act

   (i) Hazardous Materials Transportation Act

    (j) Comprehensive Environmental Response, Compensation and Liability Act

    (k) Superfund Amendments and Reauthorization Act

    (l) Emergency Planning and Community Right-to-Know Act

4. What are the three main types of hazards for chemicals?

5. What are the two major forms of airborne contaminants?

6. Define the following:

    (a) dusts

    (b) fumes

    (c) smoke

    (d) aerosols

    (e) mists

    (f) gases

    (g) vapors

7. What is a latency period for a health effect?

8. Compare acute and chronic exposures.

9. What is the difference between local and systemic effects from chemicals?

10. What is an asphyxiant?

11. What is a nuisance dust?

12. What are some individual differences among people who are exposed to chemicals?

13. What are the following?

    (a) pneumoconiosis

    (b) carcinogen

    (e) mutagen

    (d) teratogen

    (p) dermatitis

    (f) chloracne

    (g) conjunctivitis

14. What are the three main routes of entry into the body for chemicals?

15. Name and briefly explain four methods for assessing toxic properties of chemicals.

16. Explain the following:

    (a) LD_{50}

    (b) LC_{10}

    (c) TCL

    (d) TDL

17. Define the following related to chemical exposure standards:

    (a) PEL

    (b) STEL

    (c) TWA

    (d) TLV

    (e) "skin" notation

    (f) ceiling value

    (g) BEI

18. Name and briefly explain three engineering controls for chemicals.

19. Name and briefly explain six work practices and administrative methods for controlling exposures to chemicals.

20. Why is personal protective equipment a last choice method for controlling exposures to chemicals?

21. Briefly explain the following:

    (a) grab sample

    (b) colorimetry

    (c) long-term sampling for gases and vapors

    (d) impinger

    (e) centrifugal separator

    (f) electrostatic precipitator

NOTES

1 29 CFR 1910.119, Process Safety Management of Highly Hazardous Chemicals.

2 *Threshold Limit Values for Chemical Substances and Physical Agents and Biological Exposure Indices*, American Conference of Governmental Industrial Hygienists, Cincinnati, OH, annual.

3 *Federal Register*, Volume 54, No. 12, Book 2, pp. 2329–2984, January 19, 1989.

4 NFPA 704, *Identification of the Hazards of Materials*, National Fire Protection Association, Quincy, MA.

5 29 CFR 1910.1200.

6 *NIOSH Pocket Guide to Chemical Hazards and Other Databases*, U.S. Department of Health and Human Services, Centers for Disease Control and Prevention, National Institute for Occupational Safety and Health, DHHS (NIOSH) Publication No. 2002-440, June 2002.

7 See *Federal Register*, Vol. 54, No. 12, Book 2, pp. 2960–2983, January 19, 1989, for recommended procedures. Also see Schlecht, P. C., and O'Connor, P. F., *NIOSH Manual of Analytical Methods*, 4th ed., DHHS (NIOSH) Publication 94-113, and supplements available from the Superintendent of Documents, Washington, DC.

8 29 CFR 1910.146, Permit-Required Confined Spaces.

9 ANSI Z117.1, *Safety Requirements for Confined Spaces*.

BIBLIOGRAPHY

American National Standards Institute, New York:
    Z129.1 Precautionary Labeling of Hazardous Industrial Chemicals.

ADAMS, R. M., *Occupational Skin Disease*, Grune and Stratton, New York, 1982.

Air Sampling Instruments for Evaluation of Atmospheric Contaminants, 6th ed., American Conference of Governmental Industrial Hygienists, Cincinnati, OH, 1983.

ALDERSON, M., *Occupational Cancer*, Butterworths, London, 1986.

BARRETT, J. C., *Mechanisms of Environmental Carcinogenesis*, CRC Press, Boca Raton, FL, 1997.

BLACKMAN, WILLIAM C., JR., *Basic Hazardous Waste Management*, 2nd ed., CRC Lewis Publishers, Boca Raton, FL, 1996.

BRETHERICK, L., ed., *Hazards in the Chemical Laboratory*, 4th ed., American Chemical Society, Washington, DC, 1986.

CLAYSON, D. B., DREWSKI, D., and MUNRO, I., eds., *Toxicological Risk Assessment*, 2 vols., CRC Press, Boca Raton, FL, 1985.

CROWL, DANIEL A., and LOUVAR, JOSEPH F., *Chemical Process Safety—Fundamentals with Applications*, 2nd ed., Prentice Hall, Upper Saddle River, NJ, 2002.

DEISLER, P. F., JR., ed., *Reducing the Carcinogenic Risks in Industry*, Dekker, New York, 1984.

DE RENSO, D. J., *Solvents Safety Handbook*, Noyes Publications, Park Ridge, NJ, 1986.

DINARDI, SALVATORE, R., *The Occupational Environment: Its Evaluation, Control, and Management*, 2nd ed., American Industrial Hygiene Association, Fairfax, VA, 2003.

Direct Reading Colorimetric Indicator Tubes Manual, American Industrial Hygiene Association, Akron, OH, 1976.

DUTKA, B. J., and BITTON, G., *Toxicity Testing Using Microorganisms*, 2 vols., CRC Press, Boca Raton, FL, 1986.

Engineering Solutions to Indoor Air Problems, Proceedings of the ASHRAE Conference (IAQ88), April 11–13, 1988, American Society of Heating, Refrigerating and Air Conditioning Engineers, Atlanta, GA, 1989.

FISHER, G. L., and GALLO, M. A., eds., *Asbestos Toxicity*, Dekker, New York, 1987.

GAMMAGE, R. B., and KAYE, S. V., eds., *Indoor Air and Human Health*, Lewis Publishers, Chelsea, MI, 1985.

GILBERT, STEVEN G., *A Small Dose of Toxicology: The Health Effects of Common Chemicals*, CRC Press, Boca Raton, FL, 2004.

GODISH, THAD, *Air Quality*, 4th ed., CRC Press, Boca Raton, FL, 2004.

GODISH, T., *Indoor Air Pollution Control*, Lewis Publishers, Chelsea, MI, 1989.

Guidelines for Process Safety Fundamentals in General Plant Operations, Center for Chemical Process Safety of the American Institute of Chemical Engineers, New York, 1995.

HODGSON, E., MAILMAN, R. B., and CHAMBERS, J. F., eds., *Dictionary of Toxicology*, Van Nostrand Reinhold, New York, 1988.

Indoor Air Quality Conference Proceedings, Consumer Federation of America, Washington, DC, January 1985.

Indoor Pollution: The Architect's Response, American Institute of Architects Symposium, November 9–10, 1984, San Francisco, AIA Item Number M700, American Institute of Architects, Washington, DC, 1984.

KAMRIN, M. A., *Toxicology: A Primer on Toxicology Principles and Applications*, Lewis Publishers, Chelsea, MI, 1988.

KEIL, CHARLES B., ed., *Mathematical Models for Estimating Occupational Exposure to Chemicals*, American Industrial Hygiene Association, Fairfax, VA, 2000.

KIRSCH-VOLDERS, M., ed., *Mutagenicity, Carcinogenicity and Teratogenicity of Industrial Pollutants*, Plenum, New York, 1984.

LAUWERYS, ROBERT R., and HOET, PERRINE, *Industrial Chemical Exposure: Guidelines for Biological Monitoring*, 3rd ed., CRC Press, Boca Raton, FL, 2001.

LEWIS, RICHARD J., SR., ed., *Sax's Dangerous Properties of Industrial Materials*, 10th ed., John Wiley & Sons, New York, 2000.

LODGE, J. P., JR., *Methods of Air Sampling and Analysis*, 3rd ed., Lewis Publishers, Chelsea, MI, 1989.

LOWRY, G. G., and LOWRY, R. C., *Lowrys' Handbook of Right-to-Know and Emergency Planning—SARA Title III*, Lewis Publishers, Chelsea, MI, 1988.

MANAHAN, S. E., *Toxicological Chemistry*, Lewis Publishers, Chelsea, MI, 1988.

MULHAUSEN, JOHN R., and DAMIANO, JOSEPH, *A Strategy for Assessing and Managing Occupational Exposures*, 2nd ed., American Industrial Hygiene Association, Fairfax, VA, 1998.

NRIAGU, J. O., *Lead and Lead Poisoning in Antiquity*, Wiley, New York, 1983.

Odor Thresholds for Chemicals with Established Occupational Health Standards, American Industrial Hygiene Association, Fairfax, VA, 1989.

PARKES, W. R., *Occupational Lung Disorders*, 2nd ed., Butterworths, London, 1981.

Patty's Industrial Hygiene and Toxicology, 5th ed., 13 volumes, John Wiley & Sons, New York.

Vol. 1, Harris, Robert, ed., *Recognition and Evaluation of Chemical Agents*, 2000.

Vol. 2, Harris Robert, ed., *Recognition and Evaluation of Physical Agents, Biohazards, Engineering Control and Personal Protection*, 2000.

Vol. 3, Harris, Robert, ed., *Legal, Regulatory, and Management Issues*, 2000.

Vol. 4, Harris, Robert, ed., *Specialized Topics and Allied Professions*, 2000.

PIPITONE, DAVID A., ed., *Safe Storage of Laboratory Chemicals*, Wiley-Interscience, New York, 1984.

PLUMMER, R., STOBBE, T. J., MOGENSEN, J. E., and JERAM, L. K., *Minimizing Employee Exposure to Toxic Chemical Releases*, Noyes Publications, Park Ridge, NJ, 1987.

POHANISH, RICHARD P., and GREENE, STANLEY A., *Wiley Guide to Chemical Incompatibilities*, 2nd ed., John Wiley & Sons, New York, 2003.

RAPPE, C., CHOUDHARY, G., and KEITH, L. H., *Chlorinated Dioxins and Dibenzofurans in Perspective*, Lewis Publishers, Chelsea, MI, 1986.

REEVES, A. L., ed., *Toxicology*, Vol. 1, Wiley, New York, 1981.

Reproductive Health Hazards in the Workplace, Office of Technology Assessment Task Force, Lippincott, Philadelphia, PA, 1987.

RICHARDSON, M., ed., *Toxic Hazard Assessment of Chemicals*, American Chemical Society, Washington, DC, 1986.

SALEM, H., *Inhalation Toxicology*, Dekker, New York, 1986.

SCOTT, R. M., *Chemical Hazards in the Workplace*, Lewis Publishers, Chelsea, MI, 1989.

SHELDON, L., et al., *Biological Monitoring Techniques for Human Exposure to Industrial Chemicals*, Noyes Publications, Park Ridge, NJ, 1986.

SHERMAN, J. D., *Chemical Exposure and Disease*, Van Nostrand Reinhold, New York, 1988.

SITTIG, M., *Handbook of Toxic and Hazardous Chemicals and Carcinogens*, 2nd ed., Noyes Publications, Park Ridge, NJ, 1985.

SPERLING, F., ed., *Toxicology*, Vol. 2, Wiley, New York, 1981.

STACEY, NEILL H., and WINDER, CHRIS, eds., *Occupational Toxicology*, 2nd ed., CRC Press, Boca Raton, FL, 2004.

Standards, Regulations and Other Technical Criteria Related to Indoor Air Quality, National Institute of Building Sciences, Washington, DC, February, 1986.

TURIEL, I., *Indoor Air Quality and Human Health*, Stanford University Press, Stanford, CA, 1985.

WADDEN, R. A., and SCHEFF, P. A., *Indoor Air Pollution*, Wiley-Interscience, New York, 1982.

WALSH, P. J., DUDNEY, C. S., and COPENHAVER, E. D., *Indoor Air Quality*, CRC Press, Boca Raton, FL, 1983.

WEILL, H., and TURNER-WARWICK, M., *Occupational Lung Diseases*, Dekker, New York, 1991.

WEISS, G., ed., *Hazardous Chemicals Data Book*, 2nd ed., Noyes Publications, Park Ridge, NJ, 1986.

WILLIAMS, P. L., and BURSON, J. L., *Industrial Toxicology*, Van Nostrand Reinhold, New York, 1985.

CHAPTER 25

VENTILATION

25-1   TYPES OF VENTILATION

There are several types of ventilation and each has different uses. Several previous chapters discussed the need for ventilation. Chapter 16 identified the need for ventilation to keep flammable gases and vapors below the lower flammable limit (LFL). Chapter 18 discussed the role of air movement in reducing heat stress. Chapter 24 identified a requirement to keep toxic contaminants at or below certain concentrations and noted the importance of ventilation in making confined spaces safe for entry.

Ventilation can reduce odors in a room and can dilute cigarette smoke. Older air quality standards found in building codes have their origin with studies of acceptable levels of body odor and cigarette smoke.[1] Ventilation can control microorganisms, dusts, and other particulates in hospitals and clean rooms. Some clean rooms use laminar flow to prevent particulates from getting distributed in the room.

There is also a need to use ventilation to limit carbon dioxide buildup in a closed, occupied space. One inspires oxygen and expires carbon dioxide as a product of cellular combustion. A person expires approximately $0.7\,ft^3$ of CO_2 per hour. If the standard CO_2 content of air is 0.03% and the upper limit is 0.6%, then the amount of ventilation air required is $4\,ft^3/min$ per person. This amount is very small. Infiltration of outdoor air into a building often provides enough air to meet this requirement, but tightly sealed buildings may not.

Thermal Control

One will recall from Equations 18-2 and 18-4 that air velocity is one of the key physical parameters contributing to control of heat stress. Air velocity strongly influences convective and evaporative cooling. When it is warm indoors and cool outdoors, we may open a window in a building to let air move through. Not only is there a temperature difference, but there is air movement. We turn on a fan or set a fan in a window to increase air velocity. Most often we use thermal comfort ventilation to provide cooling. However, if the conditions are right, ventilation can be used to warm a space and its occupants.

General or Dilution

General ventilation and dilution ventilation are the same. They refer to the process of using clean air (often outside air) to reduce the level or concentration of contaminants in a building or space.

Dilution ventilation reduces the concentration of flammable or combustible gases and vapors below the LFL. When used for this purpose, a factor of safety is applied to the

---

Safety and Health for Engineers, Second Edition, by Roger L. Brauer
Copyright © 2006 John Wiley & Sons, Inc.

air requirement, because concentrations may not be uniformly distributed. The rate of contaminant generation may vary. The gases or vapors may be lighter or heavier than air and tend to concentrate locally. Factors of safety range from 3 to 10. High values apply to those contaminants with high toxic or flammable hazards, whereas low values apply to those contaminants with low toxic or flammable hazards.

There are practical limits to dilution ventilation. They include cost, effectiveness, and risk. If a contaminant is generated at a high rate, a very large amount of air is required to keep the contaminant at or below the LFL or below some allowable toxic concentration. It is expensive to move large quantities of air, and it is often more expensive to heat, cool, or remove moisture from entering air.

General ventilation does not always reach local sites in a space, particularly near contaminant sources, where concentrations may exceed safe levels for fire or health.

General ventilation may move contaminated air to locations that are not otherwise contaminated and it could expose more people by moving toxic mixtures into breathing zones of people. Contaminants may evaporate at varying rates because of varying process temperatures or particular use activities. General ventilation works best where contaminant generation is uniform and the rate of generation is low.

If contaminants are highly toxic or very flammable, then dilution ventilation is not a good choice for contaminant control. Failure to keep concentrations below limits can have serious consequences. A system failure could produce very dangerous conditions. Local concentrations could exceed limits and lead to injury, fire, or explosion. Someone could interfere with planned circulation and air flow patterns and produce a dangerous condition. For example, someone could stack supplies near a work station, restrict air movement and dilution, and cause local conditions to exceed limits.

Principles   There are several principles for dilution ventilation arrangements. Contaminated air should move away from occupants and fresh air should pass by occupied areas first as it moves toward a source of contaminants. Supply air should be as widely dispersed as possible to be sure that all areas of a space receive fresh air and to reduce the possibility of local concentration buildup.

Units of measure for dilution ventilation are units of air flow, such as cubic feet per minute. Air changes per hour are not appropriate, because they reflect room volume only. Dilution ventilation requirements should be based on the amount of air needed to control some contaminant level or rate of contaminant generation. When computing the ventilation required, one normally adjusts data to standard temperature (0°C) and pressure (460 mmHg).

Steady-State Concentrations   One can compute the dilution ventilation air, V_r, required to keep some evaporating solvent below a prescribed limit [lower flammable limit (LFL) or threshold limit value (TLV)] from

$$V_r = \frac{403 \times \text{specific gravity} \times 10^6 \times E \times K}{\text{molecular weight} \times L}, \tag{25-1}$$

where

      V_r is in ft³/min,

      E is the constant evaporation rate in pints/hr,

      K is a safety factor (normally from 3 to 10),

      L is the limiting concentration (LFL or TLV) in parts per million, and the specific gravity and molecular weight are properties of the solvent. If the evaporation rate is in pounds per minute, the coefficient in Equation 25-1 is 387 instead of 403.

Example 25-1 Suppose workers apply adhesive in a certain operation. The adhesive has a solvent base, methyl ethyl ketone (MEK). If the process allows 6 p of MEK to evaporate per hour, how much dilution air is required to prevent combustion? The molecular weight of MEK is 72. Referring to Table 16-3, the specific gravity of MEK is 0.8. The LFL is 1.8, and because this is fairly low and therefore quite flammable, a high value (8) for a factor of safety, K, is selected. From Equation 25-1, we find that the required dilution air is

$$V^r = \frac{403(0.8)(10^6)(6)(8)}{72(1.8 \times 10^4)}$$

$$= 11{,}940 \text{ ft}^3/\text{min}.$$

Concentration Buildup Suppose a contaminant is evaporating at some rate in a space that initially contains clean air. Also suppose there is a fixed rate of dilution air for the space. The concentration of the contaminant in the space may increase over time. One can determine what the concentration is after a period of time from

$$\ln \frac{G - \left(\dfrac{Q}{K}\right)C}{G} = \frac{\left(\dfrac{Q}{K}\right)t}{V} \tag{25-2a}$$

or

$$\frac{G - \left(\dfrac{Q}{K}\right)C}{G} = e^{\frac{-(Q/K)t}{V}}, \tag{25-2b}$$

where

C is the concentration of gas or vapor at time t,

G is the rate of generation of the contaminant,

Q is the rate of ventilation,

K is the factor of safety to allow for incomplete mixing, and

V is the volume of the room or space.

Example 25-2 A process of degreasing metal furniture starts into operation. Initially, the air in the 20,000-ft³ dryer is free of solvent vapors. The ventilation rate in the dryer is 2,500 ft³/min, and there is a factor of safety $K = 4$. The degreasing solvent evaporates at a rate 0.8 ft³/min. How long will it take for the concentration to reach 500 ppm?

One can rearrange and apply Equation 25-2a to determine the time from no solvent to a concentration of 500 ppm:

$$t = -\frac{VK}{Q}\left\{\ln\left[G - \frac{\left(\dfrac{Q}{K}\right)C}{G}\right]\right\}$$

$$= -\frac{20{,}000(4)}{2{,}500} \times \left\{\ln\left[\left(0.8 - \frac{\left(\dfrac{2{,}500}{4}\right)(500 \times 10^{-6})}{0.8}\right)\right]\right\}$$

$$= 15.85 \text{ min}.$$

Example 25-3 For the process in Example 25-2, what is the concentration after 1 hr? Rearranging Equation 25-2b, the concentration is

$$C = \frac{G - Ge^{\frac{-(Q/K)t}{V}}}{\dfrac{Q}{K}}$$

$$= \frac{0.8 - 0.8e^{\left[\frac{-\left(\frac{2,500}{4}\right)(60)}{20,000}\right]}}{\dfrac{2,500}{4}}$$

$$= 1,084\,\text{ppm}.$$

Purging If the space is contaminated at some concentration and ventilation is started, the concentration will decrease. The rate of purging is

$$\ln\left(\frac{C_2}{C_1}\right) = \frac{-Q(t_2 - t_1)}{VK}, \tag{25-3}$$

where

C_1 is the initial concentration at the time ventilation starts, t_1, and

C_2 is the final concentration at time t_2.

Example 25-4 For the dryer in Example 25-2, assume that the degreasing line stops. No more furniture enters the dryer. The generation of contaminants stops. At that time, t_1, the concentration in the dryer is 50 ppm. If the ventilation continues, how long will it take to reduce the concentration in the dryer to 10 ppm?

Applying Equation 25-3,

$$(t_2 - t_1) = -\frac{VK}{Q} \times \ln\left(\frac{C_1}{C_2}\right)$$

$$= -\frac{20,000(4)}{2,500} \times \ln\left(\frac{50}{10}\right)$$

$$= 51.5\,\text{min}.$$

Local Exhaust

The main purpose of local exhaust ventilation is capturing contaminants at their source before they contaminate a room or work station. The captured contaminants are moved through ducts to another location. It may be necessary to remove the contaminants before dumping the air outdoors or before recirculating the air. Figure 25-1 illustrates the key components in a local exhaust ventilation system. Later sections of this chapter give more details about local exhaust ventilation systems.

One advantage of local exhaust ventilation is complete or nearly complete capture of contaminants. The capture is independent of the rate of contaminant generation, toxicity, flammability, or the type of contaminant. Another advantage of local exhaust ventilation is the relatively low volume of air required compared with dilution ventilation.

Disadvantages of local exhaust ventilation systems are complexity of design, system cost, and difficulty in modifying or moving them. Failures of both dilution ventilation and local exhaust ventilation systems could create dangerous conditions. Because local exhaust

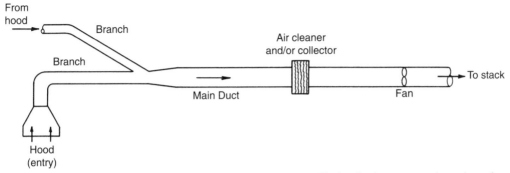

Figure 25-1.   Key components of a local exhaust system. If a local exhaust system is used to exhaust toxic contaminants, an alarm may be important to alert people when there is an interruption of air flow.

systems are well suited for toxic contaminants, an alarm may be important to alert people when there is an interruption of air flow or a dangerous contaminant level.

25-2   PRINCIPLES OF VENTILATION

Air Flow

Air moves when there is a pressure difference between two locations and the air moves from the high pressure location to the low pressure one. The quantity of air flow Q is

$$Q = VA, \tag{25-4}$$

where

Q is cubic feet per minute,

V is the velocity in feet per minute, and

A is the cross-sectional area through which air flows in square feet.

This is a restatement of Equation 10-6 and principles related to it.

For a pipe or duct of constant cross-sectional area, the velocity of air moving in it is constant over its entire length. If the cross-sectional area changes over the length of a pipe, the velocity is different at each different area of the pipe. The amount of air flowing is constant over the length of a pipe, regardless of cross-sectional changes.

The pressure creating air movement is the total pressure, TP. Total pressure has two components: static pressure, SP, and velocity pressure, VP. For air flow, all three are normally measured in inches of water. The three pressures are related:

$$TP = SP + VP. \tag{25-5}$$

Figure 25-2 illustrates the measurement of the three pressures.

Static pressure is the potential pressure exerted in all directions by a fluid at rest. In a duct, static pressure tends to expand or collapse a pipe, depending on whether static pressure is positive or negative. Static pressure is measured normal to the direction of air flow.

Velocity pressure is the kinetic pressure that causes a fluid to flow at some velocity. It is always positive and acts in the direction of air flow. It exists only when air is in motion.

The velocity pressure is related to air velocity as follows:

$$V = 4{,}005(VP)^{\frac{1}{2}}, \tag{25-6}$$

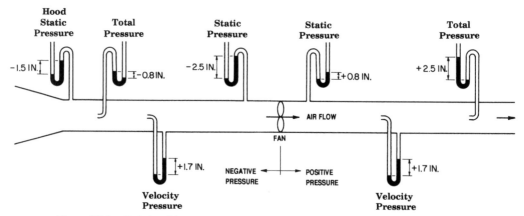

Figure 25-2. Relationships among total pressure, static pressure, and velocity pressure.

where

> V is in feet per minute and
>
> VP is in inches of water.

Losses in Systems

When air moves, there are pressure losses because some of the energy involved in potential flow is converted to heat. There are several types of losses, each resulting from different phenomena.

One type of loss is friction loss. As air moves through a pipe, the surfaces create some friction. The rougher the surfaces, the greater the friction loss; the higher the velocity, the greater the friction loss. Friction loss in a pipe or duct varies directly with pipe length, inversely with pipe diameter, and directly with the square of the velocity. Friction losses are often given per unit of pipe length.

Another kind of loss is dynamic loss. Turbulence results when there is a bend in a pipe or the cross section changes. Dynamic losses increase with increasing abruptness of the bend or change. Dynamic losses are expressed in units of equivalent pipe length. A bend or transition has the same loss as that resulting from friction loss over some length of pipe or duct of the same size. Sometimes, dynamic losses are reported as a fraction of the velocity pressure. For example, a loss for an elbow might be $0.13VP$.

Dynamic losses also result from acceleration of air at rest. Most often this occurs at the entrance into an exhaust system. Turbulence at the entry to the system adds to dynamic losses. The coefficient of entry, C_e, is a measure of the efficiency at the entry of a hood or pipe. The efficiency indicates how well static pressure is converted to velocity pressure.

Not only are there friction losses along pipe walls and dynamic losses at the system entry and at each bend or transition, but there are also losses at filters or other air cleaning devices that are part of the system. In summary, Bernoulli's equation (Equation 10-8) applies to air flow in ducts. A related form for the equation is the sum of the static pressure and velocity pressure of a point upstream in a ventilation system is equal to the sum of the static pressure, the velocity pressure and friction losses (FL) and dynamic losses (DL) at a point downstream in the system:

$$SP_1 + VP_1 = SP_2 + VP_2 + FL + DL. \tag{25-7}$$

One must expend energy to create a pressure difference between the ends of the system. A fan normally creates the pressure difference for an exhaust system by creating a static pressure great enough to overcome the resistance of the system.

Flow of Jets

A jet of air, blown from a small pipe into a space or room with a large volume of still air, can penetrate a limited distance into the large space. If the velocity of the air jet is V as it leaves a pipe of diameter d, the velocity at a distance $30d$ from the face of the pipe is approximately $0.1V$ (see Figure 25-3).

Flow at Pipe Entry

If the flow is reversed so that air enters a pipe of diameter d with a velocity of V at the pipe face, the velocity of air drops off rapidly as a function of distance upstream from the pipe face. For a plain pipe, the velocity at a distance of $1d$ from the pipe face is less than $0.075V$ (see Figure 25-4). Placing a flange around the pipe entrance extends the velocity profile only a small amount.

Make Up Air

Regardless of the purpose for ventilation, the air used in ventilation must be supplied from somewhere. Make-up air replaces air removed by a ventilation system. If the volume of make-up air is less than the volume of exhausted air for a given space, there will be a negative pressure in the space. Conversely, if the volume of make-up air is greater than the volume of exhausted air in a space, there will be a slight positive pressure. In general, it is desirable to have a negative pressure in a contaminated space or in a space where there is a source of contaminants. The negative pressure will draw air from adjacent spaces through cracks in windows, doors, ducts, or pipes and the contaminants will stay in the contaminated space. If there are contaminants in adjacent spaces, tight seals between spaces will prevent transfer of the contaminants between spaces. The presence of a positive pressure in a contaminated space will spread contaminated air to adjacent spaces.

Cleaning Air

When air is removed from a space through dilution or local exhaust, the contaminants are moved elsewhere, usually outdoors. Outdoor air quality standards limit the dumping of contaminants. As a result, it is common to clean the contaminants from the exhausted air. Some contaminants collected by local exhaust systems may have economic value. There

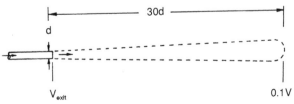

Figure 25-3.   A jet of air can penetrate deeply into a space. Air exiting a pipe has approximately 10% of the exit velocity at a distance of $30d$ from the exit.

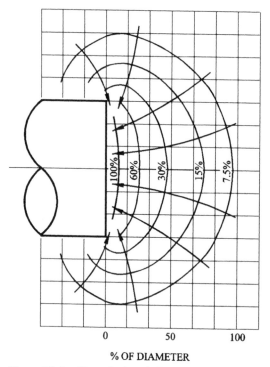

% OF DIAMETER

Figure 25-4. The velocity of air entering a pipe or hood decreases rapidly with distance from the pipe entrance.

are several types of equipment for removing airborne particulates and gases or vapors from exhausted air (see Section 25-6).

Recirculating Air

Because it is expensive to heat, cool, or remove moisture from ventilation air, it may be more economical to clean exhaust air and recirculate it. In deciding on recirculation, one must consider the possible effects of recirculation on occupants. If the contaminated air has potential health consequences, recirculation is not normally recommended, even if the cleaning process adequately removes contaminants. The more dangerous a contaminant, the more care one should take in deciding on recirculation. Failures in the system and inadequate maintenance can lead to hazardous conditions, and the more dangerous a contaminant, the greater the protection needed to ensure that system failures will not circulate hazardous air into occupied areas.

If air is recirculated and there is a health hazard from a potential failure in the cleaning process, several design factors must be met.

1. Contaminants in recirculated air that have a health hazard should not exceed recommended concentrations. One can estimate the permissible concentration, C_r, of a contaminant in air exiting a cleaning device prior to mixing with air in a workspace from

$$C_r = 0.5(\text{TLV} - C_o)\left(\frac{Q_T}{Q_R}\right)\left(\frac{1}{K}\right), \qquad (25\text{-}8)$$

where

C_o is the concentration of contaminant in a worker's breathing zone when local exhaust is discharged outdoors,

Q_T is the total ventilation flow through the affected space (cubic feet per minute),

Q_R is the recirculated air flow (cubic feet per minute),

K is a factor of safety related to incomplete mixing (range is 3 to 10), and

TLV is the threshold limit value of the concentration.

For recirculation of nuisance contaminants, the applicable coefficient in Equation 25-7 is 0.9 instead of 0.5.

2. There must be a primary and secondary cleaning system in series, each with equal efficiency. An alternative to a secondary cleaning system is a fail-safe monitoring system that must monitor the level of contaminant in the cleaned air being recirculated.

3. There must be a warning system that indicates problems in the cleaning systems. A problem may be inefficiency or failure of the secondary system or excessive levels of contaminants exiting the system.

4. If the warning system indicates a problem, either the recirculated air must be diverted immediately to the outdoors or the contaminant generating process must be completely shut down.

5. Periodic testing of recirculated air is necessary to ensure that the system is working properly.

6. Warning signs must tell occupants of the potential danger from a failure of the recirculation system. It must explain the meaning of a warning signal and the actions required for protection. AGCIH[2] gives additional factors to consider in designing recirculation systems.

Location of Exhaust Vents and Inlets

Local exhaust systems often have exhaust vents located on roofs of buildings where there are also inlets for air conditioning and air recirculation systems. When inlet and exit vents are close to each other and when wind conditions are just right, exhausted contaminants may travel directly to inlets and return to the building interior. There should be adequate separation of exhaust vents from any type of air inlet to ensure that contaminants do not reenter the building.

Protecting the Breathing Zone

Protecting the breathing zone of occupants is a basic concept for design of any type of ventilation system. Air flow patterns must move contaminated air away from a breathing zone, not near or through a breathing zone. If air is moved through a breathing zone, it should be clean air.

25-3 CAPTURING PARTICULATES AND GASES

Flow Requirements for Capture

The main idea of local exhaust ventilation is capturing contaminants at their source. One can accomplish this most easily by enclosing the source as much as possible. Enclosures increase the efficiency of capture, reduce operating cost, and require less air flow to capture contaminants. If the source of contamination cannot be enclosed, the entry (hood) to the local exhaust system should be as close to the source as possible. The farther a hood is from the source of contaminants, the less efficient it will be. Air volume required to accomplish capture increases with distance between a contaminant source and the face of a hood. The shape of a hood also can affect the likelihood of capture. The profile of air movement at the entry extends farther upstream from the hood for certain types of hoods.

The force on contaminants created by moving air must overcome other forces acting on contaminants. For gases, vapors, and particulates, the forces include thermal air currents, room air currents, and motion created by a process or operator. Diffusion, buoyancy, and gravity also apply to gases and vapors. Buoyancy acts on gases lighter than air, and gravity acts on gases heavier than air. Both buoyancy and gravity produce small forces on gases and vapors compared with other sources. Depending on their size and mass, particulates may act differently than gases and vapors. Very small particulates have little mass, act much like gases, and remain entrained in air, whereas large particulates settle out of the air because of gravity. Forces from processes that place particulates in motion, such as grinding, may be difficult to overcome with air movement. Capture is more effective if the process motion is in the same direction as air movement for capture.

The velocity of air required to capture contaminants varies with contaminants and processes. Table 25-1 lists a range of capture velocities. Capture velocity is the air velocity at any point in front of a hood or at a hood opening necessary to overcome opposing air currents and to capture contaminant air at that point by causing it to flow into the hood.

Types of Hoods and Hood Properties

Hoods can be plain openings of round or square pipe, can have flanges, can have very narrow slots, and can enclose a process or form a canopy over a tank or process. For most, the air velocity at some distance in front of the hood is a function of the shape and air

TABLE 25-1   Range of Capture Velocities[a]

Condition of Dispersion	Examples	Capture Velocity of Contaminant (ft/min)
Released with practically no velocity into quiet air	Evaporation from tanks; degreasing, etc.	50–100
Released at low velocity into moderately still air	Spray booths; intermittent container filling; low speed conveyor transfers; welding; plating; pickling	100–200
Active generation into zone of rapid air motion	Spray painting in shallow booths; barrel filling; conveyor loading; crushers	200–500
Released at high initial velocity into zone	Grinding; abrasive blasting	500–2,000

In each category above, a range of capture velocity is shown. The proper choice of values depends on several factors:

Lower End of Range	Upper End of Range
1. Room air currents minimal or favorable to capture	1. Disturbing room air current
2. Contaminants of low toxicity or of nuisance value only	2. Contaminants of high toxicity
3. Intermittent, low production	3. High production, heavy use
4. Large hood, large air mass in motion	4. Small hood, local control only

[a] *Industrial Ventilation*, 25th ed., American Conference of Governmental Industrial Hygienists, Cincinnati, OH, 2004.

TABLE 25-2 Properties of Various Hood Types[a,b]

Type	Aspect Ratio (W/L)	Air Volume	C_e
Plain opening: square, rectangular, and round	≥0.2	$Q = V(10x^2 + A)$	0.72
Flanged opening: square, rectangular, and round	≥0.2	$Q = 0.75V(10X^2 + A)$	0.82
Slot (rectangular)	<0.2	$Q = 3.7LVX$	
Flanged slot (rectangular)	<0.2	$Q = 2.8LVX$	
Booth	To suit work	$Q = VA = VWL$	
Canopy	To suit work	$Q = 1.4PDV$	

[a] From *Industrial Ventilation*, 25th ed., American Conference of Governmental Industrial Hygienists, Cincinnati, OH, 2004.
[b] W = width of rectangular opening, ft; L = length of rectangular opening, ft; C_e = entry coefficient; X = distance in front of hood face, ft; A = cross-sectional area, ft^2; P = perimeter of work or tank, ft; D = height above work or tank to canopy face, ft.

flow. Table 25-2 lists properties of various hood types. Slots (small aspect ratio openings) help to distribute uniformly the velocity of air in front of a larger hood face. The face velocity is the air velocity at the hood opening.

Example 25-5 A flanged opening rectangular hood ($W = 4$ in and $L = 16$ in) is placed 6 in from a contaminant source. The contaminant is released at low velocity into moderately still air. (a) What volume of air is required to capture the contaminant? (b) If the opening dimensions were $W = 2$ in and $L = 32$ in, what volume of air would be required?

From Table 25-1, the upper recommended capture velocity is 200 ft/min. For (a), the rectangular hood has an aspect ratio more than 0.2 (4/16 = 0.25). From Table 25-2, the applicable flow equation for this hood is $Q = 0.75 V(10X^2 + A)$. The area of the opening is $(4 \times 16)/144 = 0.444$ ft^2. The flow required is $200\{(6/12)^2 + 0.444\} = 104$ ft^3/min.

For the alternate hood (b), the aspect ratio is less than 0.2 (2/32 = 0.06). The flow equation from Table 25-2 is $Q = 2.8LVX$. Although the face area is the same as the previous case, the volume of required air is $Q = 2.8(32/12)(200)(6/12) = 747$ ft^3/min.

As air moves into a hood, the area of moving air decreases and velocity increases. Changing static pressure (in velocity pressure) causes a loss at the entry. The coefficient of entry, C_e, represents that loss. For standard air, the static pressure at the hood throat is

$$Q = 4005AC_e(SP_h)^{1/2}, \tag{25-9}$$

where

  SP_h is the hood static pressure in inches of water,

  V is in feet per minute,

  C_e is dimensionless,

  A is the area of the hood opening in square feet, and

  Q is the air flow rate in cubic feet per minute.

25-4 FLOW IN PIPES AND DUCTS

Flow Requirements in Pipes

After being captured, a contaminant moves from the hood through ducts to a point of discharge. Particulates may settle out of the flowing air and collect in the ducts, causing plug-

TABLE 25-3   Design Velocities for Moving Contaminants in Ducts

Velocity Nature of Contaminant	Examples	Design (ft/min)
Vapors, gases, and smoke	All vapors, gases, and smokes	Any velocity (1,000–2,000 is common)
Fumes	Zinc and aluminum oxide fumes	1,400–2,000
Very fine, light dust	Cotton lint, wood flour, litho powder	2,000–2,500
Dry dusts and powders	Fine rubber dust, Bakelite modeling powder dust, jute lint, cotton dust, shavings (light), soap dust, leather shavings	2,500–3,500
Average industrial dust	Sawdust (heavy and wet), grinding dust, buffing lint (dry), wool jute dust (shaker waste), coffee beans, shoe dust, granite dust, silica flour, general material building, brick cutting, clay dust, foundry (general), limestone dust, packaging and weighing asbestos dust in textile industries	3,500–4,000
Heavy dusts	Metal turnings, foundry tumbling barrels and shakeout, sand blast dust, wood blocks, hog waste, brass turnings, cast iron boring dust, lead dust	4,000–4,500
Heavy or moist dusts	Lead dust with small chips, moist cement dust, asbestos chunks from transite pipe cutting machines, buffing lint (sticky), quick-lime dust	4,500 and up

ging. The duct velocity is the air velocity in the duct, and it must be high enough to prevent settling and plugging. Table 25-3 lists design velocities for ducts. Duct velocities that are too high can cause denting, damage as particles impinge on duct walls, and possible leaks. Some airborne materials are sticky or have electrostatic properties that cause clinging to duct surfaces. Velocity will not overcome these problems.

Designing Ducts and Pipes

A designer must size ducts, select bends and elbows, combine several ducts into one system, select fans and motors, and move air through cleaning devices and on out through exhaust stacks. The design process requires detailed analysis through each element of the system. Different operations, types of contaminant, and degrees of hazard all impact design decisions. This text does not cover the design process in detail. In general, a designer must ensure that design velocities are maintained along the system. A design should have sufficient pressure at each location to ensure that there is proper air movement in each branch and at each hood.

25-5   FANS

Fans, blowers, or ejectors provide air movement in local exhaust systems. Fans are most common. Normally, they are located downstream of air cleaning devices. Ejectors are pneumatic conveyors that prevent contaminants from flowing through an air-moving device.

Types of Fans

There are many kinds of fans. Some are combined with stacks or other elements of a duct system and sold as a package. Some fans include motors; others require separate selection of motors. Some fans have enclosed motors to prevent ignition of flammable dusts, gases, and vapors (see Chapters 12, 16, and 17). Table 25-4 lists types of fans and some key features of each.

Fan Selection and Fan Laws

Manufacturers rate fans for flow and static pressure produced. A designer must match system flow and pressure requirements to fan rating curves. The resistance of an exhaust system may vary during its operation. For example, the cleaning device may plug up and significantly increase losses. A design must include anticipated changes in operating characteristics. Changing the speed of a motor may compensate for the additional losses and maintain the velocity and flow required.

Flow rate, pressure produced, and horsepower vary with fan speed. However, fan speed has a different relationship with each parameter. These relationships are fan laws (refer to Equations 25-10 through 25-12). Flow rate varies directly with fan speed, whereas total pressure (TP) and fan static pressure (FSP) vary with the square of fan speed. Air

TABLE 25-4   Types of Fans and Key Features[a]

Type of Fan	Key Features
Axial Flow	
Propellar fan	Moves large quantities of air; low static pressure; used for relatively clean air and no duct resistance; common for general ventilation
Tubeaxial (duct) fan	Fabricated in a round duct; used for condensable fumes, pigments, and other materials that collect on blades; larger diameters at slow speeds are better for abrasives and accumulating material
Vane axial fan	Develops higher pressures than other axial flow fans; more economical in horsepower and space
Centrifugal	
Forward-curved blade	Squirrel cage wheel; leading edges curve toward the direction of rotation; develops low to moderate static pressure; not recommended for dusts or fumes that adhere to blades (causes imbalance and is difficult to clean)
Straight or radial blade	Paddle wheel; most commonly used fan in exhaust systems; used for materials that clog a fan wheel; medium tip speed; medium noise factor; used for heavy dust load
Backward blade	Blades are inclined in opposite direction from fan rotation; high tip speed, high fan efficiency; blade shape is conducive to buildup of material; not suited to condensable fumes or vapors.
Special	
Airfoil-backward curved	Airfoils vary with manufacturer; quiet; high efficiency; blade usually has low vibration
In-line flow centrifugal	Backward-curved blades; special housing to fit ducts
Power exhausters, power roof ventilators	Packaged unit with fan, stack and weather; may be axial flow or centrifugal; various discharge patterns available
Combination fan and dust collector	Wide variety available; proper application important

[a] Derived from *Industrial Ventilation*, 25th ed., American Conference of Governmental Industrial Hygienists, Cincinnati, OH, 2004.

horsepower (AHP) or brake horsepower (BHP) varies with the cube of fan speed. Selecting components and applying fan laws to achieve an efficient and economical to operate system is a complex process, particularly when the properties of the system vary with time.

$$FSP = SP_{out} - SP_{in} - VP_{in},$$ (25-10)

where

FPS is fan static pressure (inches of water),

SP is static pressure (inches of water), and

VP is vapor pressure (inches of water).

$$AHP = \frac{5.2Q(TP)}{33,000} = \frac{Q(TP)}{6,350},$$ (25-11)

where

Q is flow rate (cubic feet per minute) and

TP is total pressure (inches of water).

$$BHP = \frac{5.2Q(TP)}{33,000(E)} = \frac{AHP}{E},$$ (25-12)

where E is mechanical efficiency (dimensionless).

25-6 AIR CLEANING DEVICES

Very often a local exhaust ventilation system must include a capability to remove contaminants before air is dumped to the outdoors. Cleaning is essential if air is to be recirculated. There are several types of air cleaning devices. Figure 25-5 illustrates properties of aerosols and related cleaning equipment.

Types of Air-Cleaning Devices

The main types of air-cleaning equipment are mechanical separators, filtration devices, wet collectors, electrostatic precipitators, gas adsorbers, and combustion incinerators.

Efficiency ratings for air-cleaning devices can be misleading. If efficiency is based on mass, collecting only a few large particles can achieve high efficiency even if nearly all small particles pass through. Efficiency data is most accurately represented by a curve showing the portion of each size of particle actually captured. For gases passing through an adsorption bed, efficiency varies with the concentration of the gas. Low concentrations have lower efficiencies per pass than do high concentrations, and efficiency drops as the adsorption bed loads up with contaminants.

Mechanical Separators One type of mechanical separator is a gravity chamber. Air moves through an enclosure and, because its cross-sectional area is large, the air velocity is very slow. Gravity acts on the suspended particles as they pass through the chamber and pulls them to the bottom of the enclosure, where they stay until removed. Particles smaller than $40\,\mu$m in diameter pass through a gravity chamber and are not collected. Gravity chambers are low-cost collection devices for large particles.

Impingement separators are another type of mechanical separator where dust-laden air passes through a network of baffles. Because the air changes direction quickly, partic-

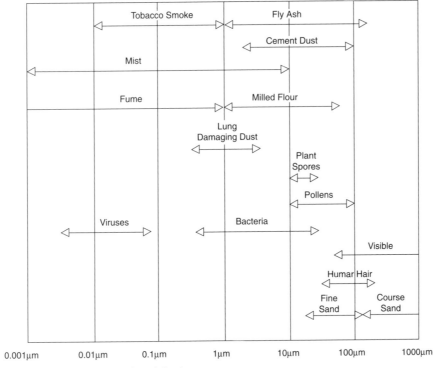

Figure 25-5.   Examples of particle sizes.

ulates have more momentum and cannot make the quick turns. Consequently, they impinge on the baffles, and the baffles direct them to one side of the flow. The clean air separated from the particulates passes out the less contaminated side of the baffles. Overall efficiency depends on particle size, gas velocity, and particle density. Particles smaller than $20\,\mu m$ are not collected. An advantage for impingement separators is low cost.

Cyclone collectors or separators are a common mechanical collector. Contaminated gas enters tangentially into a circular chamber. The rotating gas causes particulates to move to the outside of the rotating column. The particulates fall to the bottom and exit through a port. The partially cleaned air escapes through a vent at the top and center of the cylinder. The rotation can generate forces on the particles many times the force of gravity, and efficiency increases as the radius of a cyclone separator decreases. These separators have relatively low cost. Particles smaller than $5\,\mu m$ are not collected and impinging materials will erode the cylinder walls.

Filtration Devices   There are several forms of filtration devices. Mat filters are very porous and have low efficiency. Some filters, like those made of glass fibers, are disposable; others are washable. Ultrafiltration filters, such as high-efficiency particulate air filters (HEPA), remove a wide range of particles, but they require considerable maintenance and have high-pressure drops across them.

The most common filtration devices are fabric filters. There are many kinds of fabrics. The type of contaminant and the temperature of the air are but two factors that affect selection. Some filters are in the form of tubes or stockings; others have an envelope or pleated form. Air moves through the fabric bags and dust collects inside them. The more material that collects, the greater the efficiency, the smaller the particles collected, and the higher the pressure drop across the filter. Many of the large fabric filters are self-cleaning. For some, cleaning is accomplished by agitation or motion that shakes off the collected material and cleans the filters; for others, reverse air flow knocks material loose. Air from an exhaust system must be diverted to an alternate collector during the self-cleaning cycle of a collector.

Wet Collectors   The idea of wet collectors is to put contaminants in contact with a liquid, usually water. After being trapped in the liquid, the contaminants may accumulate in it. For some collectors, the contaminants and liquid may pass through cleaning elements in the wet collector system. Examples of wet collectors are spray chambers, wet centrifugal collectors, wet filters, orifice collectors, venturi collectors, and packed towers. Some wet collectors, like packed towers, also may contain adsorption material for collecting contaminant gases and vapors. Wet collectors have advantages, such as constant pressure drop, capability to handle high temperatures and humidities, compact design, and moderate cost. Some wet collectors remove 90% of 1-μm particles. Water used in wet collectors may need treatment before disposal.

Electrostatic Precipitators   Air containing solid or liquid particles passes through a bank of discharge electrodes that place a high negative charge on the particles. Collecting electrodes or plates with the opposite charge attract the charged particles. Some precipitators have more than one stage. Electrostatic precipitators have high efficiencies, even for small particulates, and they have very little pressure drop, but they are expensive to operate compared with other devices.

Gas Collection   There are both absorbing and adsorbing gas collectors. Gas passing through a liquid may react with or dissolve in the liquid. This is absorption. Some materials, like activated carbon and alumina, adsorb certain gases and vapors at the surface of the material. The adsorbing medium may hold up to half its weight in captured gases and vapors. The medium can be reactivated by heating it to drive off the captured gases and vapors. If the gases and vapors have economic value, collecting or condensing them may be desirable. The efficiency of gas collection varies somewhat with concentration of gas or vapor in the air. Another means for removing gases and vapors from air is cooling and condensation. The incoming air is cooled to form condensation and the resulting liquids are removed.

Combustion Incineration   Combustion incinerators use oxidation to convert gases and vapors into less harmful material. However, not all gases and vapors end up in a harmless form. Combustion may involve direct flame or catalytic combustion. For some gases and vapors, efficiencies may reach 98%.

Selection of Air-Cleaning Devices

There are many factors to consider in selecting air-cleaning devices. Volume of air flow, concentration of contaminants, kind of contaminant and contaminant properties, temperature, pressure drop, contaminant hazards, and other things are important. National, state, and local pollution control laws affect the choice of collection device as well.

25-7   VENTILATION MEASUREMENT

There are many instruments for assessing air flow and distribution patterns of moving air.

Smoke tubes are hand-help pumps that disperse smoke or powder visible to the eye. One can watch the movement of the smoke to see what movement patterns exist in a space or near an exhaust hood. Smoke-producing candles also are useful.

Anemometers measure air velocity. Some anemometers have a small impeller moved by the air. The impeller drives a gauge that displays velocity. Another type of anemometer has a vane that moving air deflects. An indicator connected to the vane gives air velocity. Most anemometers are very directional. They must be in line with the direction of air flow for proper readings.

Heated wire, thermocouple, or thermistor anemometers measure air velocity, temperature, or static pressure. Some brands are not directional, others are directional.

A pitot tube is a tube inserted into an airstream to measure total, velocity, or static pressure. For correct readings, the mouth of the tube must be pointed against the airstream. The position varies with the pressure of interest. The tube involves a water manometer, water gage, or other readout device. Figure 25-2 illustrates a basic pitot tube.

25-8   STANDARDS

OSHA Standards

OSHA has several standards requiring ventilation. They involve ventilation for abrasive blasting, electrostatic spraying, grinding, polishing, buffing, spray finishing, spraying operations, powder coating, textiles, asbestos, and other activities. There are ventilation requirements for bulk oxygen systems, bulk plants, confined spaces, dip tanks, laundries, open surface tanks, processing buildings, sawmills, exhaust duct systems, storage rooms, and open surface and other tanks. 29 CFR 1910 contains these regulations. In 29 CFR 1926, there are ventilation regulations for tunnels and shafts, compressed air, preservative coatings, temporary heating devices, welding, and cutting.

ACGIH

A long-standing publication of the ACGIH is *Industrial Ventilation*. It is the primary reference for design of general and local exhaust ventilation systems.

Others

The American National Standards Institute, the National Fire Protection Association, and other organizations have standards on ventilation. The Mine Safety and Health Administration details requirements for ventilation of mines in 30 CFR 75.300. One can also refer to the ASHRAE Handbook[3] for standards, procedures, and design information relating to ventilation systems and associated components.

EXERCISES

1. A flanged slot hood exhausts particulates in an operation. The flow volume for the exhaust system is 10,000 ft^3/min. The hood is 6 ft wide × 1 ft long.

   (a) What is the velocity at a distance 2 ft in front of the hood face?

   (b) Neglecting entry losses, what is the face velocity for the hood?

   (c) If the exhaust duct has a 2.5 ft diameter, what is the velocity in the duct?

2. A flexible, round, plain-opening pipe removes welding fumes. The pipe is 8 in in diameter and is positioned so the pipe face is 8 in from the welding. Assume the capture velocity produced is 200 ft/min at the point of welding.

   (a) What is the flow rate in the pipe?

   (b) Estimate the velocity at the face of the pipe?

3. A canopy hood hangs over an automated welding operation. The welding table and hood are both 4 ft wide and 3 ft long. The hood is 3 ft above the welding operation and the capture velocity is 200 ft/min. What flow rate is required in the hood?

4. For the following particle sizes, which type(s) of air cleaners are likely to be effective and economical?

   (a) 50 μm

   (b) 15 μm

   (c) 1 μm

5. A solvent leaks into a production room of 125,000 ft^3. Personnel are evacuated and an exhaust fan is set up in an external doorway that is sealed with a plastic sheet with an opening just large enough for the fan. A window is opened across the room to provide make-up air. If the fan moves air at 4,000 ft^3/min and the solvent concentration is 125 ppm when the fan is turned on, how long will it take to reduce the concentration to 20 ppm? Assume a factor of safety of 3 and that no additional amount of solvent has leaked into the room and evaporated after the fan is started.

6. Investigate how different brands of home, room-type air-cleaning devices work. Compare the brands in terms of efficiencies based on the amount of air cleaned per hour and based on the size of particle collected by the device.

REVIEW QUESTIONS

1. Name three purposes for ventilation.

2. What are the two major types of ventilation for controlling airborne contaminants?

3. Name two principles that dilution ventilation systems should meet.

4. What are the main components in a local exhaust system?

5. Define the following:

   (a) static pressure

   (b) velocity pressure

   (c) total pressure

   (d) friction loss

   (e) dynamic losses

   (f) make up air

   (g) recirculated air

   (h) capture velocity

   (i) face velocity

   (j) duct velocity

6. Compare the velocity of air from a jet blowing into a space and the movement of air being drawn into an exhaust entry.

7. What factors should be considered in deciding if air should be recirculated?

8. What factors should be included in design of an air recirculation system when the air contains contaminants with health hazards?

9. What are the fan laws?

10. Identify four types of air cleaning devices. Describe how each works. Identify an advantage or disadvantage for each.

11. What air-cleaning devices are suitable for gases and vapors?

12. Name four devices for measuring air flow or air distribution.

NOTES

1 Brauer, R. L., and Kuehner, R. L., "The Variability of Ventilation Codes," in *Symposium on Odors and Odorants: The Engineering View*, Chicago, January 27–30, 1969, American Society of Heating, Refrigerating and Air Conditioning Engineers, Atlanta, GA, 1969.

2 *Industrial Ventilation—A Manual of Recommended Practice*, 25th ed., American Conference of Govern-mental Industrial Hygienists, Cincinnati, OH, 2004.

3 *ASHRAE Handbook*, American Society of Heating, Refrigerating and Air-Conditioning Engineers, Inc., Atlanta, GA. Volumes on fundamentals, equipment, and systems and applications are updated regularly.

BIBLIOGRAPHY

American National Standards Institute, New York:

AIHA Z9.2 Fundamentals Governing the Design and Operation of Local Exhaust Systems

AIHA Z9.4 Ventilation and Safe Practices for Fixed Location Enclosures for Exhaust Systems for Abrasive Blasting Operations.

AIHA Z9.5 Laboratory Ventilation

AIHA Z9.6 Exhaust Systems for Grinding, Polishing and Buffing

AIHA Z9.7 Recirculation of Air from Industrial Process Exhaust Systems

ASHRAE 62 Ventilation for Acceptable Indoor Air Quality

ASHRAE 62.2 Ventilation and Acceptable Indoor Air Quality in Low-Rise Residential Buildings

UL 441 Gas Vents

UL 680 Emergency Vault Ventilators and Vault-Ventilating Ports

UL 705 Power Ventilators

National Fire Protection Association, Quincy, MA:

NFPA 91 Exhaust Systems for Air Conveying of Gases, etc.

NFPA 96 Ventilation Control and Fire Protection of Commercial Cooking Operations

NFPA 211 Chimneys, Fireplaces, Vents, and Solid Fuel Burning Appliances

CONSTANCE, J. D., *Controlling In-Plant Airborne Contaminants*, Dekker, New York, 1983.

HARRIS, MICHAEL K., BOOHER, LINDSAY, and CARTER, STEPHANIE, *Field Guidelines for Temporary Ventilation of Confined Spaces With and Emphasis on Hotwork*, American Industrial Hygiene Association, Fairfax, VA, 1996.

HAYASHI, T., HOWELL, R. H., SHIBATA, M., and TSUJI, K., *Industrial Ventilation and Air Conditioning*, CRC Press, Boca Raton, FL, 1985.

Industrial Ventilation—A Manual of Recommended Practice, 25th ed., American Conference of Governmental Industrial Hygienists, Cincinnati, OH, 2004.

MCDERMOTT, H. J., *Handbook of Ventilation for Contaminant Control*, 2nd ed., Butterworths, London, 1985.

MODY, V., AND JAKHETE, R., *Dust Control Handbook*, Noyes Publications, Park Ridge, NJ, 1988.

CHAPTER 26

BIOHAZARDS

26-1 INTRODUCTION

In July, 1976, at the height of the nation's bicentennial celebration, members of the American Legion descended on Philadelphia for their national convention; 182 conventioneers staying at one hotel became ill and 29 died. The outbreak, which displayed symptoms similar to pneumonia, led researchers to discover a new pathogenic bacterium: *Legionella pneumophila*. This was the first pathogen identified in more than 25 years. Pursuant investigations found that other outbreaks of Legionnaire's disease had occurred earlier. The bacterium thrives in cooling tower sumps and condenser valves of air conditioning equipment and is also found in water supplies and hot water tanks of buildings.

In April, 1985, *Salmonella* contamination of 1-gal cartons of 2% milk infected more than 6,000 people in five states. Many lawsuits resulted, and the Chicago area dairy that processed the milk closed its doors. The cause may have involved the design and operation of the processing equipment. Raw, contaminated milk somehow got mixed with pasteurized milk during the processing.

In June, 1985, cheese contaminated with *Listeria monocytogenes* produced flu-like symptoms in at least 87 people in the Los Angeles area. The bacterium affected mostly women and children. Some reports linked the deaths of 28 people to the contaminated cheese.

In approximately 1987, American public attention turned to acquired immune deficiency syndrome (AIDS). The virus known as HTLV-III/LAV causes this disease. It attacks the immune system by affecting a type of white blood cell called T lymphocytes and prevents antibody development, which leaves the victims open to infections and diseases that would not otherwise be a threat. People may not be aware that they are carriers of the virus. Nonsexual contact that generally occurs among workers and clients or consumers in the workplace does not pose a major risk for transmission of the AIDS virus. However, workers engaged in emergency treatment of people and those who may come into contact with human blood and other body fluids are at higher risk than the normal population. Although much has been learned about this disease and its treatment, it remains a worldwide health threat.

In June, 1989, investigative reporters discovered a new source of potential contamination. When trucking companies that hauled food from the Midwest to the East Coast traveled in the other direction, they hauled garbage from eastern cities to Midwest landfills. There were few regulations covering these activities and minimizing food contamination.

In the late 1990s, the public became sensitive to exposure to molds that may grow in homes, often inside wall structures, and feared toxic effects of certain types. As a result,

Safety and Health for Engineers, Second Edition, by Roger L. Brauer
Copyright © 2006 John Wiley & Sons, Inc.

states have passed home inspection requirements during real estate transactions and other situations, often engaging inspectors of limited expertise.

During 2001, there were a handful of letters delivered to government officials or companies containing anthrax spores. Two or three people died, but postal systems and company mail departments were taxed to change procedures to protect the public and employees from potential exposures. Fortunately, the number of actual cases of contaminated mail were very limited, but the breadth of effects was very large.

These events illustrate the dangers of biohazards. Improvements in sewage systems, sanitation, and sanitary engineering practices contributed to major reductions of dangers from many diseases. Food inspection and handling practices reduced dangers even farther. However, dangers from biohazards remain. There are other current examples, such as bacteria in hot tubs and spas, parasitic infection from sushi (raw fish), and similar popular habits that can lead to illnesses from biohazards. Whether one is a consumer or worker, there are many activities that have biohazards.

There are specialty areas of practice for biological hazards. Groups that deal with dangers, exposures, and toxicity of biohazards typically have a background in biology, especially microbiology, environmental health, or industrial hygiene.

26-2 AGENTS AND SOURCES

Biohazards are biological hazards from plants, animals, or their products that may be infectious, toxic, or allergenic. Agents are bacteria, viruses, fungi, rickettsia, or parasites.

Bacteria

Bacteria are simple, one-celled organisms. They are not visible to the eye and they multiply by simple division. Not all bacteria are harmful; many are useful. Bacteria are characterized by their shape: cylindrical or rod shaped (bacilli), shaped like a string of beads (cocci), and spiral or corkscrew shaped (spirilla).

Viruses

A virus is an organism that depends on a host cell for development and reproduction. Viruses are parasitic and are so small that they are not visible with an optical microscope. Viruses are transmitted in many ways, including contact with infected people, from animals and insects, from contact with equipment and diseased specimen, and other means.

Fungi

There are many species of fungi. They are parasitic and grow in a living host or on dead plant or animal matter. Fungi may be microscopic in size or large (mushrooms are an example).

Rickettsia

Rickettsia are microorganisms that are rod shaped and smaller than bacteria. They depend on a host for development and reproduction and they must live within a host cell. Fleas, ticks, and lice transmit them, although they are sometimes airborne.

Parasites

A variety of protozoa, helminths, and arthropods are parasites. They are different from other organisms that are parasitic because they live in or on other plants or animals. Some well-known parasites are tapeworms, liver flukes, and hookworms.

Bloodborne Pathogens

Bloodborne pathogens are microorganisms that are present in human blood and can cause disease in humans. Examples are the hepatitis B virus and the human immunodeficiency virus. OSHA established a standard addressing safety and health of workers subject to potential bloodborne pathogen exposures.[1]

Sources of Biohazards

Diseases transmitted from animals to humans, zoonoses, are a major source of biohazards. People who work with animals, animal products, or animal waste have a greater risk of infection from biohazards. Another source of biohazards is work in hospitals, other medical facilities, or medical-related research laboratories. It may include any situation in which someone can contact the blood or body fluids of someone else, such as in an accident, first aid situation, or work with patients or hotel guests, cleaning of waste receptacles or washrooms, and so forth. Exposure can be from equipment and materials contaminated with blood, such as needle sticks, handling of linens during laundry, or waste. Table 26-1 lists examples of biohazards, occupations in which they may occur, and other relevant information. The bibliography at the end of this chapter gives more complete listings.

26-3 HAZARDS

The main danger from biohazards is infection, and there are different symptoms for different infections. There is considerable knowledge about infections from some biohazards. However, it must be assumed that a biohazard exists for work with biological agents or related materials for which disease is not known or not understood. Biohazard agents may enter the body or skin by ingestion, skin contact, puncture wounds, or inhalation of aerosols. Infections from some biohazards can be cured or at least treated. For some infections, there is no treatment or cure.

Classification

Federal agencies developed a classification scheme for biohazards. There are five categories. The first four describe increasing hazard levels. Each of the four classes requires increasing levels of controls. The fifth class includes animal pathogens excluded or restricted from import to the United States. Table 26-2 lists the five categories of biohazards.

26-4 GENERAL CONTROLS

One can prevent many occupational infections with training, procedures, and special equipment and facilities. This is particularly true for laboratories involved with infectious

TABLE 26-1   Examples of Biohazards[a]

Biohazard	Sources and Comments
Bacterial	
Anthrax	Direct contact with infected animals, hides, and wool. Risk is higher for veterinarians, farmers, butchers, leather and wool workers, carpet workers.
Brucellosis	Ingestion, inhalation, contact with infected animals, cuts, and scratches. Those who work with cattle and hogs, meat packing workers.
Salmonellosis	Oral; Food service workers and patrons, meat and poultry workers.
Tetanus	Entrance through breaks in the skin from penetrating or crushing trauma. Persons handing jute or contact with manure.
Tuberculosis	Inhalation, contact with lesions. Silica workers, people exposed to heat and organic dusts, medical personnel, animal caretakers.
Fungal	
Dermatophytoses	Contact; People involved with farm and domestic animals and handlers of hides.
Histoplasmosis	Inhalation and ingestion, Roof demolition workers and workers in barns and chicken houses.
Parasitic	
Creeping eruption	Penetration of skin by infected larvae. People involved with digging in soil (ditch diggers, utility workers, laborers, masons, gardeners, plumbers).
Hookworm	Penetration of skin (particularly bare feet) by larvae. Barefoot farmers and ditch diggers, sewer workers, tunnel workers, recreation (children and adults).
Schistosomiasis	Contact with infected water. Farmers and others who stand and work in flooded areas.
Swimmer's itch	Penetration of wetted skin by snails. Workers in and around fresh water, divers, dock workers, lifeguards, and recreational swimmers.
Rickettsial	
Ornithosis	Inhalation or contact with infected bird droppings. Zoo workers, taxidermists, poultry farmers and processors, pet workers and owners, people in locations where there is an accumulation of bird droppings.
Q fever	Inhalation of contaminated dusts and contact with infected animals (cattle, pigs, or sheep) or contaminated substances. Farmers, veterinarians, slaughter house workers, and hide and wool workers.
Rocky Mountain spotted fever	Tick bites, skin contact with tick tissue or feces. People who work outdoors, such as in lumbering, construction, forestry, ranching.
Viral	
Cat-scratch disease	Break in skin, usually from animal scratch. People who work or play with cats and dogs.
Hepatitis (viral)	Fecal-oral transmission. Health workers, particularly pediatrics, oral surgeons.
Milker's nodules	Breaks in skin. Dairy farmers and workers.
Rabies	Bite of infected animal (dogs, cats, bats, pigs, rats, etc.) People who work and play with animals or come near animals (letter carriers, delivery workers, etc.).

[a] Adapted from Feiner, B., "Occupational Biohazards," in *Dangerous Properties of Industrial Materials*, 6th ed., Sax, N. I., ed., Van Nostrand Reinhold, New York, 1984, and other sources.

TABLE 26-2   Classification of Biohazards[a]

Class	Definition
1	Agents of *no hazard or minimal hazard* under ordinary handling conditions that can be handled safely without special apparatus or equipment, using techniques generally acceptable for nonpathogenic materials.
2	Agents of *ordinary potential hazard* that may produce disease of varying degrees of severity through accidental inoculation, injection, or other means of cutaneous penetration, but which can usually be adequately and safely contained by ordinary laboratory techniques.
3	Agents involving *special hazard* that require special conditions for containment or require a USDA permit for importation and are not in a higher class.
4	Agents that are *extremely hazardous to personnel or can cause serious epidemic disease* and require the most stringent conditions for containment. Included are certain types of class 3 agents imported into the United States.
5	Foreign animal pathogens that are excluded from the United States by law or whose entry is restricted by the USDA.

[a] Adapted from *Classification of Etiologic Agents on the Basis of Hazard*, Centers for Disease Control, Public Health Service, U.S. Department of Health and Human Services, Atlanta, GA, 1974.

agents. The U.S. Department of Health and Human Services, Public Health Service, developed guidelines for preventing laboratory infections.[2] The principles they developed are helpful in assessing risk and preventing infections in other activities.

The overriding principle for preventing laboratory infections is containment, the purpose of which is to reduce exposure of laboratory personnel and other persons to potentially hazardous agents. Containment includes preventing escape of potentially hazardous agents outside the laboratory to such persons as workers laundering laboratory clothing, visitors, and family members of laboratory workers. The three elements of containment include laboratory practice and techniques, safety equipment, and facility design. Primary containment addresses protection of personnel and the immediate laboratory environment from exposure, which is achieved with proper techniques and safety equipment. Vaccines may provide additional protection for personnel. Secondary containment refers to protection of environments outside the immediate laboratory, which requires both proper procedures and facility design.

Warnings that laboratories and containers have biohazards are important. The biohazard symbol (see Figure 26-1) is an essential part of a biohazard warning. Several organizations have biohazard warning requirements and standards.[3]

Laboratory Practice and Technique   Laboratory workers must learn what hazards exist for particular agents in a laboratory and must receive training in proper handling and operations of laboratory materials. Workers must periodically update their knowledge and skills to ensure high retention levels. There should be an operations and biosafety manual for each laboratory that identifies hazards that may be encountered and what protection is needed for these hazards. Persons knowledgeable in hazards, safety procedures, and laboratory techniques should direct the work. Knowledgeable people should complete a risk assessment before starting any work to identify dangers and to implement appropriate protection.

Biosafety related to laboratory work extends into criminal laboratories and the investigator and laboratory personnel handling and analyzing criminal evidence. Specialists in

Fiugre 26-1. Biohazard symbol.

this field are becoming more concerned with safety in their work with biological materials from crime scenes.

Safety Equipment   Enclosed containers, biological safety cabinets, and personal protection equipment are the main kinds of safety equipment. Safety containers prevent release of unsafe substances during normal activities and operations. An example is a safety centrifuge cup. Biological safety cabinets are partial or full enclosures where the air flow is designed to retain agents within the cabinets. There are three classes of cabinets.

A Class I biological safety cabinet has an open front. Air moves inward across the face at 75 ft/min and exhaust air is filtered through high-efficiency particulate air (HEPA) filters.

A Class II biological safety cabinet also has an open front with a 75 ft/min face velocity. It has vertical laminar flow air movement and the air is both HEPA filtered and recirculated within the cabinet and HEPA filtered and exhausted. The filtered recirculation prevents contamination of agents by air drawn into the cabinet. Class II cabinets must meet the National Sanitation Foundation standard.[4]

Class III biological safety cabinets are totally enclosed. Workers complete activities in the cabinet via rubber gloves built into the cabinet walls. To prevent contamination of the contents, supply air enters the cabinet through HEPA filters. The cabinet operates under at least 0.5 in of water-negative pressure. Exhaust air moves through two stages of HEPA filters. Typically, a Class III cabinet has its own exhaust fan, which is independent of any other ventilation systems. Other equipment, such as refrigerators, dunk tanks, and centrifuges, is a part of the cabinet or is contained in the work area within the cabinet.

One can achieve Class III standards another way. Workers wear one-piece, positive-pressure, full-body protective suits that have a life support system and work inside Class

I or II cabinets. In this case, the work area must have an airlock with airtight doors. Workers must pass through a chemical shower to decontaminate the suit before leaving the work area. Exhaust from the suit must pass through a two-stage HEPA filter.

Facility Design   Facilities play an important role in containment. Designs protect both those working in a facility and those outside the laboratory and the surrounding community. There are three classes of facility design, each providing a different level of safety. The three facility classes are basic, containment, and maximum containment laboratories. Design features are based on four levels of biosafety for infectious agents and for work with vertebrate animals. Tables 26-3 and 26-4 summarize the four biosafety levels. Designs include easily cleaned surfaces, special features for furniture, cleaning facilities for workers, and other features.[5]

Basic laboratories are intended for work with agents not associated with disease in healthy adults and work in which standard laboratory practices provide adequate protection. Separate basic laboratories from public and office areas. Containment equipment is not normally required.

A major feature of a containment laboratory is a controlled access zone. Containment laboratories have specialized ventilation systems and may be separate buildings or controlled access modules within a building.

Maximum containment laboratories support work with agents that are extremely hazardous or may cause epidemics. Often these laboratories are separate buildings. A main design feature is highly effective barriers, which may include sealed openings, airlocks or liquid disinfectant barriers, clothing-change and shower rooms, double-door autoclave, biowaste treatment system, separate ventilation system, and treatment system to decontaminate exhaust air.

TABLE 26-3   Summary of Recommend Biosafety Levels for Infectious Agents[a]

Biosafety Level	Practices and Techniques	Safety Equipment	Facilities
1	Standard microbiological practices	None: Primary containment provided by adherence to standard laboratory practices during open bench operations	Basic
2	Level 1 practices plus: laboratory coats; decontamination of all infectious wastes; limited access; protective gloves and biohazard warning signs as indicated	Partial containment equipment (class I or II biological safety cabinets) used to conduct mechanical and manipulative procedures that have high aerosol potential that may increase the risk of exposure to personnel	Basic
3	Level 2 practices plus: special laboratory clothing; controlled access	Partial containment equipment used for all manipulations of infectious material	Containment
4	Level 3 practices plus: entrance through change room where street clothing is removed and laboratory clothing is put on; shower on exit; all wastes are decontaminated on exit from the facility	Maximum containment equipment (class III biological safety cabinet or partial containment equipment in combination with full-body, air-supplied, positive-pressure personnel suit) used for all procedures and activities	Maximum Containment

[a] From *Biosafety in Microbiological and Biomedical Laboratories*, HHS Publication (CDC) 84-8395, Centers for Disease Control, Public Health Service, U.S. Department of Health and Human Services, Atlanta, GA, March, 1984.

TABLE 26-4 Summary of Recommended Biosafety Levels for Activities in Which Experimentally or Naturally Infected Vertebrate Animals Are Used[a]

Biosafety Level	Practices and Techniques	Safety Equipment	Facilities
1	Standard animal care and management practices	None	Basic
2	Laboratory coats; decontamination of all wastes and of animal cages before washing; limited access; protective gloves and hazard warning signs as indicated	Partial containment equipment and/or personal protective devices used for activities and manipulations of agents or infected animals that produce aerosols	Basic
3	Level 2 practices plus: special laboratory clothing; controlled access	Partial containment equipment and/or personal protective devices used for all activities and manipulations of agents or infected animals	Containment
4	Level 3 practices plus: entrance through clothes change room where street clothing is put; shower on exit; all wastes are decontaminated before removal from the facility	Maximum containment equipment (class III biological safety cabinet or partial containment equipment in combination with full-body, air-supplied positive, pressure personnel suit) used for all procedures and activities	Maximum containment

[a] From *Biosafety in Microbiological and Biomedical Laboratories*, HHS Publication (CDC) 84-8395, Centers for Disease Control, Public Health Service, U.S. Department of Health and Human Services, Atlanta, GA, March, 1984.

Robotics Another control to reduce the dangers of contact with biohazards is the use of robotics for analysis and processing of biological samples. A robot placed in an enclosure can perform many functions, thereby reducing human handling and potential contacts. Automatic or manual controls operated from outside the enclosure direct the actions of the robot. There are companies that produce robots for processing biohazards.

26-5 SICK BUILDING SYNDROME AND INDOOR AIR QUALITY

Sick building syndrome is a term from the 1980s. It stems from a number of incidents where many occupants of an entire building or a certain portion of a building exhibited a rash of physical complaints, including headaches, muscle pains, chest tightness, nausea, fever, cough, allergic asthma, allergic rhinitis, pneumonitis, and pneumonia. Often the symptoms diminished over weekends. Some individuals became sensitized or exhibited allergic reactions to conditions. ASHRAE defines the term *sick building* as a building in which a significant number (more than 20%) of building occupants report illnesses perceived as building related.[6]

Researchers began looking into these problems, and there appear to be a number of causes, including contaminants entering air conditioning units from birds. Investigations found development of microbial slimes in air-handling units. Condensate pans, humidifiers, and tanks of coolant from machining can form places for organisms to grow. Hot water supplies that are not hot enough can incubate bacteria. Beside biological contaminants, there are often chemical contaminants from smoking, off-gassing of new furniture and carpet, asbestos particles from insulation, and other applications and release of polychlorinated biphenols from exploding or burning electrical transformers. Poor lighting and acoustics also have been implicated as contributors to sick building syndrome, as has tight building construction, where there is little infiltration of outdoor air. Other factors may be the release of formaldehyde from certain foam insulations and the release of contaminants from heating, cooking, and power machinery. Emissions from powered vehicles can contribute. Researchers continue in their efforts to identify the complexities of this problem and its corrections.

Controls vary with the problem. Chlorination and other chemical treatments for water-based problems in heating, ventilating, and air conditioning (HVAC) equipment have not always been effective. Segmented smoking areas have not always removed contaminants from the general office environment. When contaminants build up as a result of an inadequate supply of outdoor air, attention should be directed to the HVAC system. Specific causes, when located, should be removed at the source.

ASHRAE's ventilation standard[7] may be helpful. The Environmental Protection Agency (EPA) offers a number of publications on indoor air quality and radon. Table 26-5 lists some actions for preventing and controlling sick building syndrome.

TABLE 26-5   Some Actions for Preventing and Controlling Sick Building Syndrome[a]

Contamination clean up
  Remove harmful chemical sources when they exceed recognized limits
  Remove dirty air filters in HVAC systems
  Empty all condensate drainage trays
  Use hot water to clean microbial growth from condenser coils, tubing, etc.
  Swab down suspected ductwork with antimicrobial solution
  Remove materials in locations found infested with microbial growth where cleaning is not possible
  Clean carpeting and furniture that has microbial growth
  Make sure drains in HVAC equipment are working

Preventive maintenance
  Keep hot water supply temperatures higher than 120°F
  Provide drains for air handling packages to prevent stagnant water
  Limit relative humidity to less than 70%
  Abandon air washers that use recirculating water systems
  Use steam from fresh water for humidifiers, not recirculated water
  Abandon spray coil systems
  Keep coils, pans, drainage systems, and duct work clean
  Check air filters regularly and replace
  Prevent stagnant water

Preventive design
  Locate intake vents where they receive fresh air and not contaminated air
  Use only steam humidifiers, not recirculating ones
  Use prefilters to clean air upstream of high efficiency filters
  Design HVAC systems to handle varying resistance to air flow in buildings
  Locate HVAC system components where it is easy to inspect and service them

[a] Adapted from Bishop, V. L., Custer, D. E., and Vogel, R. H., "The Sick Building Syndrome: What It Is and How to Prevent It," *National Safety and Health News*, Vol. 132, No. 6:31–38 (1985).

26-6 GENETIC ENGINEERING

In the 1950s, researchers discovered the double helix of deoxyribonucleic acid (DNA), the building block of life. Since then, research has mapped the complete DNA molecule and many companies work with modified DNA molecules. Genetic engineering, cloning, and gene splicing are becoming common and in some cases commercialized. In 1980, the U.S. Supreme Court decided that a live, laboratory-made microorganism is patentable. This decision let stand a lower court decision for the first patent on genetically engineered materials. The patent recognized a General Electric product—an oil-eating organism.

Genetic engineering, gene splicing, and cloning led to public fears about modifying DNA. Modified organisms, plant, and animal species produced in error or for destructive purposes made some people fear incidents similar to the black plague of the Middle Ages and the worldwide influenza epidemic of 1918 that killed 20 million people. The National Institutes of Heath (NIH) developed guidelines for recombinant DNA research[8] and a review committee assesses potential hazards of proposed research. Genetic engineering has produced biological growth of insulin, interferon, and bacteria capable of digesting 2,4,5-T, the key chemical component in Agent Orange. Other useful products have also emerged from biotechnology research.

A major control to prevent dangerous releases of new organisms is careful review by government agencies, such as the EPA. There are few methods for evaluating the safety of genetically engineered products. Risk analysis techniques are useful.

26-7 OTHER BIOHAZARDS AND CONTROLS

There are many other ways biohazards can threaten people in daily living and in special environments. Biohazards in the food chain threaten many people. Biohazards in hospitals require careful control to prevent transfer in infections and disease.

Food

Biohazards enter the food chain at many points. For example, a natural bacterial growth in corn stored in grain bins and elevators can produce a toxic substance called aflatoxin. Farmers, grain elevators, and grain companies must monitor corn for this toxin. Some seasons have a greater problem than others. The U.S. Department of Agriculture has numerous standards for and conducts inspections on grain and other food materials. These standards and other controls ensure that plant and animal foods pose little risk of biohazards for consumers.[9] The Department of Commerce has standards for fish and seafood products.[10] The Food and Drug Administration sets standards on food for human consumption[11] and food for animals intended for human consumption.[12]

Restaurants and Food Establishments

State and local governments have public health inspectors who regularly inspect restaurants and related food establishments. Regulations require workers to handle, refrigerate, and process food properly for customers. They also prevent spread of biological agents by insects and rodents through sanitary practices. Establishments that do not meet inspection standards must resolve deficiencies or face temporary or permanent closure. A major problem in food preparation and service at restaurants and hotels is ensuring that employees wash their hands and adhere to other basic sanitation practices.

Hospitals and Other Health Care Facilities

Federal regulations for construction of medical facilities[13] include features that help minimize hazards of biological agents. State regulations and certification programs for medical facilities also help establish controls to minimize dangers of biological agents. The American Hospital Association and other healthcare organizations have their own guidelines to assist operators of facilities to minimize biohazards and any resulting infections. Refer also to Chapter 30.

Swimming Pools, Spas, Saunas, Therapy Pools, and Tanning Booths

A variety of public facilities create opportunities for transfer of biohazards; proper design, operation, cleaning, and maintenance are important in controlling biohazards. Each recreational facility may have other safety and health hazards requiring additional controls.

Water in swimming pools becomes contaminated by microorganisms from swimmers' skin, mucus, feces, and urine and from dirt, plant material, and other sources. The water must be cleaned and disinfected. Recirculation systems pump water through filters to remove hair, lint, and other large particulates. Filters are backwashed to clean them and the collected materials are flushed down sewer lines. Chlorine and other chemicals are used to disinfect the water. The rate of filtration and disinfecting is based on swimmer load and tests of water samples. There are state and local codes for operation of public swimming pools, and there are many sources of criteria for design of large pools.

Spas, therapy pools, hot tubs, and similar water containers that have multiple uses require treatments similar to swimming pools. Simply draining the containers, flushing out solids, and scrubbing them regularly with disinfecting detergents will prevent transfer of biohazards.

Other recreational facilities that many people use require care in preventing transfer of biohazards. Cleaning, disinfecting, and other means can be effective.

Plumbing, Sanitary Sewer Systems, and Water Supplies

The implementation of sanitary sewer systems removed many kinds of biohazards. State and local governments have plumbing codes, codes for sanitary sewer systems, and codes for water supply systems. In many locations, plumbing and sanitary lines must be inspected when they are initially installed or modified. Strict code enforcement prevents potential disease transfer. Careful separation of potable water from untreated water supplies and sewer lines prevents any cross-contamination or back flow to the potable water.

Ventilation and dehumidification in closed spaces can help prevent the growth of molds. Treatment of standing water in air conditioning equipment also can minimize the opportunity for molds to grow.

EXERCISES

1. Visit a water treatment facility, biological laboratory, or other facility to find out how biohazards are controlled.

2. Find out how hospital biosafety hazards are controlled.

3. Research the events surrounding a major biohazard incident.

4. Research the scope of biohazard problems in the world, particularly developing countries that have poor water supplies, limited sanitary sewers and treatment plants, and practices that lead to biohazard infections and illnesses.

5. Visit a food processing plant and find what measures are in place to control biohazards.

6. Investigate state and local regulations governing food handling and service in restaurants and eating establishments. Determine how public health and sanitary inspection ratings are completed and scored. Obtain records of inspection scores for a group of local restaurants for the last 6 to 12 months.

7. Investigate various types of molds and identify the hazards each presents. Review methods for removal or treatment and the safety and effectiveness of each.

REVIEW QUESTIONS

1. What are the five kinds of biohazard agents?

2. What are zoonoses?

3. What is the main danger from biohazards?

4. What are the four classes of biohazards in order of hazard severity?

5. What are the three major types of controls for biohazards in laboratory work?

6. Characterize class I, class II, and class III biological safety cabinets.

7. Define sick building syndrome. What symptoms do people typically exhibit with sick building syndrome? What agents are commonly involved?

8. What are potential hazards for genetic engineering? What controls are there?

NOTES

1 29 CFR 1910.1030, Boodborne Pathogens.

2 *Biosafety in Microbiological and Biomedical Laboratories*, HHS Publication (CDC) 84-8395, Public Health Service, Centers for Disease Control, U.S. Department of Health and Human Services, Atlanta, GA, March, 1984.

3 Examples are 29 CFR 1910.145, 49 CFR 172.444, 49 CFR 173.388, and ANSI 35.2, *Specifications for Accident Prevention Tags*.

4 *Class II (Laminar Flow) Biosafety Cabinetry*, NSF/ANSI 49-04a, National Sanitation Foundation International, Ann Arbor, MI, 2004.

5 See details for each class of facility in *Biosafety in Microbiological and Biomedical Laboratories*, HHS Publication (CDC) 84-8395, Public Health Service, Centers for Disease Control, U.S. Department of Health and Human Services, Atlanta, GA, March, 1984.

6 *Indoor Air Quality—Position Paper*, American Society of Heating, Refrigerating and Air-

Conditioning Engineers, Atlanta, GA, approved by the ASHRAE Board of Directors, August 11, 1987.

7 ANSI/ASHRAE Standard 62.1-2004, *Ventilation for Acceptable Indoor Air Quality*; and ANSI/ASHRAE Standard 62.2-2004, *Ventilation and Acceptable Indoor Air Quality in Low-Rise Residential Buildings*, American Society of Heating, Refrigerating and Air-Conditioning Engineers, Atlanta, GA.

8 "Recombinant DNA Research. Proposed Revised Guidelines," National Institutes of Health, U.S. Department of Health, Education and Welfare, *Federal Register*, Vol. 42, No. 187 (1977).

9 7 CFR and 9 CFR contain regulations on food safety.

10 50 CFR 260.

11 21 CFR 100.

12 21 CFR 500.

13 42 CFR 124.

BIBLIOGRAPHY

Biohazards Reference Manual, American Industrial Hygiene Association, Akron, OH, 1985.

Biosafety in Microbiological and Biomedical Laboratories, HHS Publication (CDC) 84-8395, Centers for Disease Control, Public Health Service, U.S. Department of Health and Human Services, Atlanta, GA, March, 1984.

Boss, Martha J., and Day, Dennis W., eds., *Biological risk Engineering Handbook: Infection Control and Decontamination*, CRC Press, Boca Raton, FL, 2003.

Burroughs, H. E., and Hansen, Shirley J., *Managing Indoor Air Quality*, 3rd ed., CRC Press, Boca Raton, FL, 2004.

Chakrabarty, A. M., *Genetic Engineering: Benefits and Biohazards*, CRC Press, Boca Raton, FL, 1978.

"Coordinated Framework for Regulation of Biotechnology: Announcement of Policy and Notice for Public Comment," *Federal Register*, Vol. 51:23,302–23,393 (1986).

Feachem, R. G., Bradley, D. J., Garelick, H., and Mara, D. D., *Sanitation and Disease*, Wiley, New York, 1983.

A Guide to the Work-Relatedness of Disease (revised), Publication 79-116, National Institute for Occupational Safety and Health, Washington, DC, 1979 (NTIS PJ-298-561).

Biosafety Reference Manual, 2nd ed., American Industrial Hygiene Association, Fairfax, VA, 1995.

Dillon, H. Kenneth, Heinsohn, Patricia A., and Miller, J. David, eds., *Field Guide for the Determination of Biological Contaminants in Environmental Samples*, American Industrial Hygiene Association, Fairfax, VA, 1996.

Environmental Mold: State of the Science, State of the Art, American Industrial Hygiene Association, Fairfax, VA, 2003.

Flannigan, Brian, Samsom, Robert A., and Miller, J. David, *Microorganisms in Home and Indoor Work Environments: Diversity, Health Impacts, Investigation and Control*, CRC Press, Boca Raton, FL, 2002.

"Guidelines for Research Involving Recombinant DNA Molecules," *Federal Register*, Vol. 47:38,048–38,068 (1982).

Heinsohn, Robert Jennings, and Cimbala, John M., *Indoor Air Quality Engineering: Environmental Health and Control of Indoor Pollutants*, CRC Press, Boca Raton, FL, 2003.

Hunter, D., *The Diseases of Occupations*, English Universities Press, London, 1975.

Indoor Air Quality Research: New Directions, American Industrial Hygiene Association, Fairfax, VA, 2003.

Infection Control in the Hospital, 4th ed., American Hospital Publications, Inc. (subsidiary of the American Hospital Association), Chicago, IL, 1979.

Koren, Herman, and Bisesi, Michael S., *Handbook of Environmental Health*, 4th ed., 2 vols., CRC Press, Boca Raton, FL, 2002.

Mims, C. A., *The Pathogenesis of Infectious Disease*, 3rd ed., Academic, New York, 1987.

Occupational Diseases: A Guide to Their Recognition (revised), Publication 77–110, National Institute for Occupational Safety and Health, Washington, DC, 1975 (NTIS PB-83-129-528).

Occupational Disease: The Silent Enemy, Publication 75-110, National Institute for Occupational Safety and Health, Washington, DC, 1975 (NTIS PB-83-179-812).

"Proposal for a Coordinated Framework for Regulation of Biotechnology: Notice," *Federal Register*, Vol. 49:50,856–50,907 (1984).

Rafferty, Patrick J., ed., *The Industrial Hygienist's Guide to Indoor Air Quality Investigations*, American Industrial Hygiene Association, Fairfax, VA, 1998

Reinhardt, P. A., and Gordon, J. G., *Infectious and Medical Waste Management*, Lewis Publishers, Chelsea, MI, 1990.

Report of Microbial Growth Task Force, American Industrial Hygiene Association, Fairfax, VA, 2001.

Salvato, Joseph A., Nemerow, Nelson L., and Agaedy, Franklin J., *Environmental Engineering*, John Wiley & Sons, Hoboken, NJ, 2003.

Smith, P. W., ed., *Infection Control in Long-Term Care Facilities*, Wiley, New York, 1984.

Wormser, G. P., Stahl, R., and Bottone, E. J., *AIDS—Acquired Immune Deficiency Syndrome and Other Manifestations of HIV Infection*, Noyes Publications, Park Ridge, NJ, 1997.

Zhang, Yuanhui, *Indoor Air Quality Engineering*, CRC Press, Boca Raton, FL, 2004.

CHAPTER **27**

HAZARDOUS WASTE

27-1   INTRODUCTION

Love Canal

On August 2, 1978, state officials ordered the emergency evacuation of 240 families from the Love Canal neighborhood of Niagara Falls, New York. After heated confrontations among the local residents and city, state, and federal officials, some action finally was taken. With Love Canal, the eyes of the nation opened to the problems of hazardous waste in the United States.

Love Canal began in 1880 as the dream of William T. Love. He began construction on a dream city surrounding an electrical generating plant that was to use diverted water from the Niagara River to produce direct current power. A canal from the river would provide water to the plant. Economics and the invention of alternating current that could travel long distances ended the dream during its early development. Between 1942 and 1952, the abandoned canal became a dumping ground for waste from local chemical plants. In 1953, Hooker Chemical Company, the owner of the site, deeded the land to the local school district for $1.00. A school playground and housing were constructed on and adjacent to the site. The new owners did not know what the site contained.

After Love Canal received national attention, a study by New York State[1] reported that nearly 22,000 tons of chemical waste were dumped in Love Canal (see Table 27-1). The study found that 152 of 215 waste disposal sites in Niagara and Erie Counties of New York were known to have or were suspected of containing hazardous waste. Some contained even more waste than Love Canal. At one site, entire tank cars of waste were buried.

Times Beach

The Centers for Disease Control began investigating Times Beach, Missouri, in 1971. On December 23, 1982, they issued a warning to residents of Times Beach that their town was unsafe because of dioxin (2,3,7,8-tetrachlorodibenzoparadioxin, or TCDD) contamination. Sax[2] lists TCDD as a "very, very toxic material." Just 2 weeks before, the town's residents had faced a record flood of the Merimac River. After the 1982 warning and evacuation of residents, the federal government removed contaminated soil from the community at a multimillion dollar cost.

Other Sites

Although the above two episodes gained national attention, there are many other stories related to hazardous waste. In the "Valley of the Drums" near Shephardsville, Kentucky, between 17,000 and 100,000 drums were abandoned illegally. On April 22, 1980, a major

Safety and Health for Engineers, Second Edition, by Roger L. Brauer
Copyright © 2006 John Wiley & Sons, Inc.

TABLE 27-1   Waste Content of Love Canal

Type of Waste	Estimated Quantity (tons)
Miscellaneous acid chlorides	400
Thionyl chloride	500
Miscellaneous chlorinations	1,000
Mercaptans	2,400
Trichlorophenol	200
Benzoyl chloride	800
Metal chlorides	400
Liquid disulfides	700
Benzyl hexachloride	6,900
Chlorobezenes	2,000
Benzyl chlorides	2,400
Sulfides	2,100
Miscellaneous	2,000
Total	21,800

fire on the waterfront of Elizabeth, New Jersey, consumed thousands of containers of hazardous materials and gave off noxious smoke and gases. In several locations, midnight dumping operations left roads and communities with hazardous waste problems. Radioactive mine tailings form the foundations of houses and yards in some uranium mining communities in Colorado and Wyoming. In the summer of 1988, medical waste washed ashore at several public beaches along the east coast of the United States. The waste was improperly dumped at sea.

In 1990, the Department of Energy estimated that the immediate costs to clean up contaminated nuclear production facilities and their sites would be $30 billion. Cumulatively, these stories have alerted the public to the problem. New laws continue to emerge. Even disposable baby diapers, a product for the modern family, have been banned in some locations unless they are made of biodegradable materials. Even biodegradable plastics do not degrade easily in landfills, where they are isolated from air. Engineers face a growing challenge to create products from materials that create fewer waste problems, as well as a need to find more effective methods for managing waste problems, particularly hazardous waste. The public in the "throwaway" society faces a challenge to reduce the volume of waste through recycling, changed packaging, and other means.

27-2   HAZARDOUS WASTE

The Hazardous Waste Problem

Hazardous wastes comprise a small portion of the total waste generated in the United States. The U.S. Environmental Protection Agency (EPA) estimated that 80 billion pounds of hazardous waste are produced each year; others believe the amount is higher. The EPA estimated that for a time, only 10% of these wastes were properly disposed of through on-site disposal methods and secured landfills. The remainder were placed in unlined lagoons and ponds, in nonsecure landfills, or were disposed of in other ways. Alternate disposal methods include ocean dumping, placing waste in sewer systems, dumping on roads, deep-well injection, or burning in ordinary incinerators. An EPA survey in 1980 produced a list of more than 32,000 known waste sites containing hazardous waste. These results led to tighter disposal regulations, a Superfund cleanup program for existing hazardous waste sites, and increased liability for owners of contaminated land. More recent legislation has reduced the dispoal options and tightened standards that have controlled management of hazardous waste.

There are two categories of wastes. Producer wastes are those generated by industry. They are usually concentrated and found in particular locations. Consumer wastes are those disposed of by the ultimate user of products. They are usually low in concentration, but widely dispersed.

On June 13, 1989, the EPA released a list of 595 waterways (rivers, creeks, streams, oceans, and lakes) across the United States that are polluted above acceptable levels by 126 chemicals considered harmful to the environment. More than 17,000 other waterways are contaminated at lower levels or by other substances. The EPA created a cleanup plan by 1992. In a number of areas, cleanup plans and implementation actions have led to reduced hazardous waste problems.

Hazardous Waste Definition

Under the Resource Conservation and Recovery Act (RCRA) of 1976, the EPA has a complicated definition for a hazardous waste. Figure 27-1 is a summary flow chart of the definition. In short, a hazardous waste is any material that meets the RCRA definition of a solid waste and is not excluded as a hazardous waste.

Referring to Figure 27-1, materials are one of three categories:

1. garbage, refuse, or sludge
2. solid, liquid, semisolid, or containing gaseous material
3. other

These categories help determine if a material is an RCRA solid waste. Materials that are a RCRA solid waste are evaluated further to see if they are RCRA hazardous wastes; materials that are not RCRA solid wastes are not RCRA hazardous wastes. Category 1 materials are RCRA solid wastes, and some materials in category 2 are RCRA solid wastes. Category 3 materials (other) are excluded. They include domestic sewage, Clean Water Act point source discharge, irrigation return flow, source, special nuclear or by-product materials regulated under the Atomic Energy Act of 1954 and amendments, or in situ mining waste.

RCRA solid wastes appearing in an EPA list of hazardous materials[3] or containing a waste found in the list may be RCRA hazardous wastes. They are evaluated further to see if they are excluded from the list by other parts of the regulation. In some cases, one can also petition to have them removed.

Materials that are not in the list must be checked for other characteristics to determine if they are hazardous wastes. Hazardous characteristics are ignitability, corrosivity, reactivity, and toxicity. To be an RCRA hazardous waste, materials with any of these characteristics must also (1) cause, or significantly contribute to, an increase in mortality or an increase in serious irreversible, or incapacitating reversible, illness or (2) pose a substantial present or potential hazard to human health or the environment when improperly treated, stored, transported, disposed of, or otherwise managed. Furthermore, the materials must be reasonably measured or detected.

If a material is an RCRA hazardous waste, there are regulations governing it. The regulations vary for different conditions, including the following:

1. Whether it is generated by a small-quantity generator.
2. Whether it is intended to be reused, recycled, or reclaimed.
3. Whether it is sludge or is or contains material in the EPA hazardous waste list.
4. Whether the handler is (a) generator, (b) transporter, or (c) owner or operator of a treatment, storage, or disposal facility.

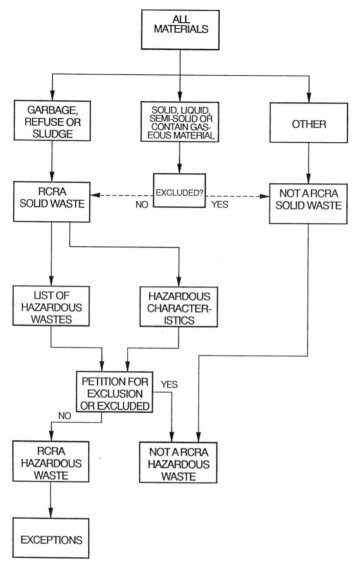

Figure 27-1. Logic tree for EPA definition of a hazardous waste.

Beside the EPA definitions, there are many ways to classify hazardous waste. One classification for hazardous waste relates to how long a material is hazardous. Some materials, like heavy metals, are toxic for extended periods. Most often the hazard is an intrinsic property of the material. Some materials lose their hazardous property over time at a moderate rate, whereas other materials, such as acids or bases, lose their hazardous property rather quickly in contact with other materials.

Hazards

The major hazards of waste are those already identified for other substances in Chapter 24: ignitability (flammability), reactivity, corrosivity, and toxicity. In addition, damage to the environment is important. Hazardous materials may contaminate groundwater and water supplies, cause closing of wells, destroy natural habitats, contaminate soil, kill fish or livestock, incapacitate sanitary sewer treatment facilities, damage crops, or contribute to air pollution. Hazardous waste may also have biohazards.

Problems

A number of practices add to the general waste problem. The hazardous materials portion of the waste has gained public attention and brought about new laws and regulations accompanied by new technologies. However, even though significant efforts have been underway to clean up hazardous waste sites, many waste sites resulting from poor practices of the past will be with us for some time.

A serious problem in the past was hazardous waste management. Not that long ago, people disposed of hazardous waste by passing it along. One simply called a waste hauler and paid to have it removed. There was little concern on the part of waste generators, whether consumer or producer, about what happened to waste after it was hauled away. Engineers, among others, received little training about what to do with materials that were not needed or left over from manufacturing processes. Laws and regulations have changed things, but improper disposal has not disappeared totally.

Another problem, still with us, is that we are a "throwaway" society. We design products for disposal after we are through using them or after they fail. Even products which become obsolete, such as personal computers, cannot be upgraded or repaired economically and become waste. A large portion of the waste generated in this country could be used again or recycled for additional uses, but not always economically. The throwaway products and packaging add to the waste problem.

Another problem of the past was selection of waste sites. We located dumps on the least desirable land, land of little value for agriculture, business, or residences. Unfortunately, the undesirable lands were often wetlands or land with porous, sandy soil, which allowed materials dumped in these sites to migrate into surface and underground water supplies. Recovery of these leaching materials is extremely difficult.

Beside selection, waste site design created additional problems because there was very little design. One dumped waste or piled it up and sometimes covered it. Little attention was given to preventing dump contents from leaching into the soil below or to nearby areas.

Another problem is liability for past dumping. It was not uncommon in the past that companies transferred waste to waste management and disposal companies. Later, the waste company went out of business. Under current law, the parties involved, including the generator, have a financial liability for the cleanup of sites contaminated by their waste. The cost for past practices, even if well managed under previous standards, can be very high. Selling a plant to another company does not always relieve the first owner of financial responsibility if contamination is discovered after the sale. Although the purchase price for the new owner may be very low, failing to investigate the property for hazardous waste may drive the real price high when cleanup costs are figured in.

27-3   CONTROLS FOR HAZARDOUS MATERIALS

There are many choices involved in reducing the dangers from hazardous waste, and the appropriate choices will vary with particular materials. Methods include eliminating the hazardous material, reducing quantities generated, restricting the area contaminated by containing the waste, storing the waste, separating wastes by degree of hazard, pretreatment, treatment, detoxification, and disposal. Other methods include reusing, recycling, or reclaiming materials and managing distribution.

There are many factors that affect the choices available. One factor is laws and regulations. Federal and state law and regulations specify acceptable methods for different hazardous materials. Some methods are not technically feasible or practical for particular materials. Economics is very important, so it is necessary to keep the waste management and treatment costs as low as possible. Social and political factors play an important role. Some communities will not accept locating landfills, incinerators, recycling, or other hazardous materials processes in their area. "Not in my backyard" is a common public theme. There are many court cases where existing and proposed waste control facilities were challenged by individuals and communities. Even at the national level, the process of selecting a disposal site for hazardous waste and radioactive waste can drag on and on.

Cost is another critical factor. The cost to dispose of hazardous waste has increased dramatically. Some companies pay several hundred dollars or more per 55-gal drum for hazardous waste removal and disposal. Even consumers are charged for disposal of certain products, such as tires, batteries, and other auto parts, when their repair shops replace them with new products. Rising costs are the result of many factors. There are few new companies entering the hazardous waste business and there are fewer and fewer approved waste sites. Haul distances are long for many parts of the country, particularly the East and Northeast. Hazardous waste companies or would-be companies have difficulty obtaining liability insurance, and the insurance is expensive and usually does not exceed 1 million. Because the generator of waste has a responsibility for it, even when it is transferred to someone else, the processes for managing waste and waste records are expensive.

Eliminate Hazardous Wastes

One way to minimize the dangers of hazardous wastes is to prevent their generation, which can be accomplished by substituting less hazardous material in a process, using materials that do not produce hazardous waste, or changing a process to reduce the amount of hazardous material used or to prevent generating hazardous waste.

Reduce Quantities

Often complete elimination of hazardous waste is not possible. However, the economics of hazardous waste management and disposal provide an incentive to produce as little as possible. Several approaches are possible: some are preventive; some apply to waste after it is produced.

Analysis and Plans An important approach is analysis of waste streams and formulation of a hazardous waste reduction plan. A National Academy of Sciences report suggests three phases in a waste minimization program. Each phase must deal with hazards, risk, and economics of options, and the involvement of management and workers is important for each phase. Follow-up for each phase must evaluate the effectiveness of actions taken or to be taken.

The first phase is reviewing and improving operating practices. It is important to document the processes by which hazardous materials are created. Improved practices may result from employee training, management changes, hazardous material inventory control from procurement through disposal, and keeping different waste streams separate. A process change may be as simple as buying the minimum quantity needed instead of large quantities that have a low purchase costs and high disposal costs. Spills also can be avoided through improved materials handling practices, and leaks can be prevented through preventive maintenance programs.

The second phase addresses processes in greater depth through a feasibility study. The options for reducing hazardous waste may involve changes in processes and equipment, quality control, and instrumentation. Options must evaluate cost, process reliability, on-site and off-site safety, potential waste reduction, regulatory compliance, and other factors.

The final phase is the implementation phase. The selected changes in processes are completed through redesign and modification or new construction and replacement. Implementation may require participation of process equipment vendors. This phase requires evaluating vendors carefully to ascertain what experience they have in waste minimization and how well their products will work when installed.

Compaction   Compaction is an approach to reducing solid waste volume after it is generated. However, some types of waste do not lend themselves to compaction.

Reuse, Recycle, Reclaim

One way to minimize hazardous waste is to reuse, recycle, or reclaim materials. For example, there are many processes that require solvents. In some cases the solvent may evaporate, but because it cannot be exhausted to the atmosphere, it must be recovered through condensation, adsorption, or other means. It can be reused. Some processes may contaminate a solvent, but a user may be able to reprocess it or, perhaps, a supplier will reprocess it and provide a credit on new purchases. Another process may modify the solvent and its properties, and the waste may have economic value to someone else. The generator may sell it for use by another company. There are a number of waste exchanges among industrial groups and companies that help waste producers find potential users.

Containment

Another important option for controlling hazardous waste generation is containment. Processes and operations should be designed to minimize the amount of air, water, and other materials that become contaminated by hazardous material. The less material contaminated, the less there is to manage. For example, if one allows waste materials to mix with storm water, the task of removing contaminants increases substantially because a large volume of water must be treated to remove the contaminants.

Special equipment and devices may help contain contaminants. For example, sensors may be placed in sewer lines to detect the presence of hazardous waste. If hazardous waste is present, the lines divert to holding tanks and prevent spills into sewer systems. Automatically or manually activated valves also can prevent hazardous materials from entering sewer lines.

Storage

There is not always a place to put hazardous waste. After it is generated, it must be held for subsequent transport and processing, which may be in batches. Removal from the site may require waiting until there is a full truck load or a reasonable quantity for transport.

Liquid storage may involve tanks (underground or above ground) or open lagoons. Solid waste may be waste piles or other forms of surface impoundment. Wastes should be compatible. For example, strong acids should not be mixed with strong bases. Different wastes or types of wastes should be in separate containers when they are incompatible, and containers must be in good condition and not leak during the storage time.

Separation

Because materials contaminated with hazardous wastes are considered hazardous waste also, it is a good strategy to separate hazardous materials from general waste. Separating hazardous wastes into compatible groups is important, as is separating them into practical groups. For example, a supplier may give credit for return of a material for reprocessing. Therefore, mixing other materials with it may reduce the amount of credit and may make reprocessing more difficult or costly.

Pretreatment

Pretreatment is any process that makes final treatment more economical, feasible, or effective. The final treatment may not affect or be suitable for certain hazardous materials. Pretreatment may include any form of treatment, may involve several processes in sequence, often involves separation, and may reduce the variety, volume, or concentration of contaminates. One example of pretreatment is removal of oil from storm water runoff in parking lots. Holding tanks and skimmers, sludge pits, or other means are used to help extract the oil from the runoff water.

Detoxification

One form of pretreatment or treatment is detoxification. In detoxification, hazardous properties, particularly those that are toxic, are removed from a material. There is growing interest in this capability. Biologists are finding microorganisms that perform this function. For example, researchers have identified a fungus that will digest dioxin, DDT, benzopyrene, and polychlorinated biphenyls.

Treatment

There are many kinds of treatment for wastes, which may involve physical, chemical, or biological processes. Treatments may include disposal methods. Certain methods are applicable for only certain wastes.

Biological treatments typically involve organic wastes. Microorganisms of various types, including bacteria, break down substances into alternate, more desirable forms.

Chemical treatment may include adjusting pH with acids or bases. It may involve extraction of oils, heavy metals, ion exchange, oxidation, and other techniques.

Physical treatments primarily isolate or concentrate particular materials, reduce their volume, solidify, detoxify, or perform some combination of these functions. For example, evaporation can reduce volume, as can compaction. Adsorption may remove certain materials. In solidification, liquids or slurries are mixed with stabilizing or binding agents to prevent leaching of materials when they are buried in landfills. Cementation processes appear useful for binding high concentrations of inorganics and heavy metals. Thermoplastic processes mix paraffin, bitumen, or asphalt with dried sludge wastes at elevated temperatures. Polymerization and encapsulation are other forms of solidification processes.

Disposal

There are three primary methods for waste disposal: burial in landfills, deep-well injection, and incineration. Laws and regulations often limit the methods available.

Burial   There are two kinds of landfills: conventional and secure. Municipalities use conventional landfills for general waste because they are relatively cheap to operate. They also have a high potential for leaching of materials to the surface for runoff, lateral leaching into adjacent ground, and leaching into deeper soil. Leaching into deeper soils can contaminate aquifers and other underground sources of drinking water. Today, landfills must meet strict design standards. The standards help prevent leaching. Operating standards require dumped materials to be covered daily, which prevents air pollution and scattering of materials by wind and animals.

Secure landfills for hazardous waste must meet stringent standards of the EPA. A secure landfill must have an impervious clay base or an artificial liner. These features are intended to contain any leachate. A drainage system around the landfill collects groundwater and prevents it from entering the landfill. The drainage system and groundwater near and under the landfill are monitored for leachate.

A problem for many communities is having sites or access to landfills within a convenient distance. Establishing a new site or finding a suitable location for a site is not always easy. The public near a site usually resists having a site in their neighborhood. Some locations have haul contracts with sites several states away from their own state.

Deep-Well Injection   Today, EPA regulations generally ban deep-well injection. In deep-well injection, liquid wastes are pumped through pipes deep into the earth into porous rock formations or natural underground domes. The depth is normally below any useful underground water sources. If suitable sites are available, deep-well injection can avoid transportation costs and processing costs. However, some deep-well sites have not been without problems. Layers of rock and soil above the disposal sites have become contaminated. There are questions about the reliability of geological sites to seal the waste adequately from movement and the potential failure of piping leading to the deposit site. Potential earthquakes, failures in overpressured sites, and other problems are not fully resolved for many sites.

Incineration   Except for heavy metals and a few other forms of hazardous waste, incineration can be a safe method of disposal. Incineration processes are tailored to the kind of waste involved and they can produce heat and steam for other processes. Today, incinerators must have scrubbers downstream of the combustion process to ensure that hazardous materials do not escape. Researchers continue to study combustion methods and methods for recovering any dangerous gases or particulates remaining from combustion. A major problem for incinerators is public acceptance. Few are being built and communities that are potential sites for hazardous waste incineration plants frequently do not want such a facility in their neighborhood.

Manage Distribution

Another important control for hazardous waste is careful management. The Resource Conservation and Recovery Act of 1980 requires that anyone who generates, stores, transports, or disposes of hazardous waste must obtain an identifying number. The waste generated, stored, transported, or disposed of must follow a manifest system. The manifest system allows tracking of the type and amount of hazardous material generated until it is disposed of properly and the person who transfers hazardous waste to a disposal company receives information on where and how waste was disposed. Tracking of hazardous material purchased, received, stored, and used is essential in ensuring that hazardous waste is properly identified and disposed.

27-4 LAWS AND REGULATIONS

It is impossible to list all the laws and regulations that affect waste management and disposal. They change constantly at federal, state, and local levels. Some foreign laws and regulations are even more stringent than those in the United States. This section reviews selected federal laws and regulations that apply to hazardous waste.

Resource Conservation and Recovery Act

The Resource Conservation and Recovery Act of 1976 (RCRA) was the first major federal legislation on hazardous waste. The act does not cover radioactive waste, which falls under the Atomic Energy Act of 1954. Later, the Uranium Mill Tailings Radiation Control Act of 1978 gave the EPA responsibility for cleanup of radioactive materials from inactive uranium processing sites. The RCRA does not include water pollutants regulated by the Clean Water Act of 1972. Also excluded by the RCRA were boiler fuel incinerators. Many industrial or commercial boilers recover heat energy from materials defined as hazardous wastes.

The main portion of the law dealing with hazardous waste contains eight sections.

- The first required the EPA to develop criteria for identification of hazardous waste materials. It also required the EPA to develop a list of substances meeting these criteria. That list began to define hazardous waste.
- The second addressed record keeping, labeling, packing, and transporting of hazardous wastes. It created the requirement for the manifesting procedures.
- The third section required the EPA to develop regulations for transporters to ensure that hazardous waste does not endanger human health or the environment.
- The fourth required development of regulations for hazardous waste disposal facilities.
- The fifth required the EPA to issue permits to all operators of hazardous waste disposal facilities. These permits specify the type and amount of waste to be received and processes allowed for a particular site. They also state methods of disposal for particular wastes.
- The sixth allowed states to establish their own programs if they meet or exceed EPA programs.
- The seventh required inspection of hazardous waste sites and facilities.
- The eighth established civil and criminal penalties for violators of hazardous waste regulations.

1980 EPA Regulations   In 1980, the EPA completed final rules on compliance with the RCRA. Earlier in this chapter, some provisions were discussed. Most of the regulations deal with the estimated 15,000 large-quantity generators (LQGs) of hazardous waste. LQGs are those producing 1,000 kg per month or more of waste.

1984 Hazardous and Solid Waste Amendments   The original RCRA excluded from strict EPA regulations those small businesses generating less than 2,200 lb (1,000 kg) of hazardous waste monthly. In 1984, Congress changed the law to exclude small businesses producing less than 220 lb (100 kg) of hazardous waste per month. In 1986, the EPA issued regulations affecting approximately 175,000 small quantity generators (SQGs) that produce between 100 and 1,000 kg per month. These generators produce approximately 800,000 metric tons of hazardous waste per year, which is less than 0.5% of all hazardous waste produced in the United States.

Regulation of Underground Storage Tanks   Revisions to the RCRA established controls for underground storage tanks. Many that were placed in the ground for storage of gasoline, petroleum products, and other substances developed holes from corrosion and other factors and leaked their contents into the ground. The material migrated and contaminated ground water. This portion of the RCRA caused a great deal of removal of unsued storage tanks and ones with designs that would not contain leaking materials. In addition, the standards established requirements for the design and use of new storage tanks that would prevent, monitor for, and remedy leaks.

Marine Protection, Research and Sanctuary Act of 1972

This act bans dumping of radioactive, biological, and chemical warfare wastes in the ocean. It requires permits for dumping of sewage sludge and dredged materials.

Medical Waste Tracking Act of 1988

This law established a 2-year demonstration program for tracking medical waste from generator to disposal. Although only a few states are included, others may volunteer. The law requires the EPA to establish regulations for medical waste generators, transporters, and treatment, storage, and disposal facilities. The agency is to monitor the program and report on results, costs, and benefits.

CERCLA (Superfund)

The problems of hazardous waste extend back many years. What to do with waste already disposed of also became a national issue. The result was development of a Superfund. Contributions by both government and private industry created a source to cover the cost of cleaning up the worst of the many hazardous waste sites of the past. Industry contributed 75% of the initial $1.6 billion fund. The Comprehensive Environmental Response, Compensation and Liability Act (CERCLA), passed in late 1980, took action to clean up some of the hazardous waste sites. The act also addressed liability. It stated that those causing or contributing to a release or threatened release from an inactive hazardous waste site shall have strict, joint, and several liability for cleanup, containment, and emergency response activities at the site. Liable parties included generators and transporters of the waste and owners and operators of the disposal site.

The CERCLA legislation expired in 1985, but was reauthorized in 1986 under the Superfund Amendments and Reauthorization Act (SARA).

SARA Title III

Another problem addressed by legislation was informing citizens who might be exposed to hazardous materials. This idea is called community right-to-know. The Emergency Planning and Community Right- to-Know Act of 1986 (also known as SARA Title III) established the requirement to inform communities. The act also authorized states to pass laws requiring emergency planning for potential hazardous material release events. Under these state laws, companies must prepare emergency response plans and make them available to authorities in communities around a plant. Coordination with local medical services, fire departments, police departments, and other officials is essential for them to provide adequate time for people to protect themselves against the hazards of a release.

OSHA Regulations for Hazardous Waste Site Activities

As people began to clean up hazardous waste sites, workers faced dangers of exposure to unknown contaminants. Planning and management of cleanup activities must include the protection of workers. In December, 1986, OSHA created interim final rules for hazardous waste site and emergency responses. It protects public and private sector employees involved in handling hazardous waste materials. The final rule went into effect in early 1989.[4] The standard, sometimes referred to by the acronym HAZWOPER, includes site analysis and control, training, medical surveillance, air monitoring, protective equipment, informational programs, decontamination procedures, and emergency response plans.

California Proposition 65

In November, 1986, California voters overwhelmingly approved Proposition 65, which enacted the Safe Drinking Water and Toxic Enforcement Act. It requires that no business may expose people to chemicals that can cause cancer, birth defects, or reproductive harm without giving a clear and reasonable warning. One provision prohibits businesses from knowingly discharging these chemicals into the drinking water. However, the key provision of this law is that an employer has the burden of proof to prove that his products and emissions are safe; if not, the employer must give public warnings. This is in contrast to normal tort procedure where the burden of the proof on harm rests with the employee or consumer.

Nuclear Waste Policy Act of 1982

Late in 1982, Congress passed the Nuclear Waste Policy Act (NWPA). It addresses the national problem of what to do with high-level nuclear waste, which is mainly spent nuclear fuel from nuclear power plants. It also includes other forms of highly radioactive waste. Low-level nuclear waste includes contaminated clothing, waste from medical treatment, contaminated water, and other forms of waste with limited radioactivity.

NWPA provisions involved protecting public health and safety and the environment, acceptance of waste for disposal not later than January 31, 1998, creating a repository of permanent disposal of spent fuel and high level waste, safe transportation of waste to the repository, interim storage of spent fuel for utilities, public participation in the nuclear waste disposal solution, and costs recovered from waste generators. Establishing national waste sites for nuclear waste has yet to be resolved.

The Department of Energy (DOE) is the agency responsible for most provisions of the act. DOE recommended three sites to the president, who in turn, in 1991, was to recommend one site to the Congress. If Congress approved the recommended site, DOE was to build the national repository on this site. A second site was to be selected also, but a repository was not to be constructed on it. In addition, DOE was to develop the transportation system for moving the waste to the site. By 2005, the repository has not been completed and most nuclear fuel waste is stored at nuclear power plants where it is generated.

The act created the Nuclear Waste Fund to finance the waste disposal program. Since 1983, the federal government has collected from nuclear utilities fees that are determined by the amount of electricity generated.

EXERCISES

1. Investigate the design, special features, and design limitations for each of the following:

   (a) hazardous waste landfill

   (b) hazardous waste incinerator

   (c) high-level nuclear waste repository

   (d) high-level nuclear waste transport containers

   (e) deep-well injection facilities

2. Find out what plants or facilities in your community have emergency response plans. Obtain a copy of them and review them.

3. Find out if there are any hazardous waste sites in your area. Determine if any are on the Superfund National Priority List for cleanup or have been cleaned up.

4. Visit a hazardous waste generator, transporter, or disposal site. Find out how they manage manifest requirements and what procedures are involved.

5. Contact a local manufacturing plant to find out how the company manages hazardous waste.

6. Contact a hazardous waste hauling company and obtain their current waste hauling and disposal fees.

7. Check with a university in your area to find out how hazardous waste is managed across the entire campus.

REVIEW QUESTIONS

1. Describe the hazardous waste problem in the United States.

2. What is a hazardous waste?

3. What are the major dangers of hazardous waste?

4. What problems contributed to the hazardous waste issues of today?

5. Describe the controls for hazardous waste. Characterize each. Which controls are most preferred?

6. Describe the four-step process to reduce quantities of hazardous waste.

7. Briefly describe the role of each of the following in protecting the public against hazardous waste:

(a) RCRA

(b) Marine Protection, Research and Sanctuary Act of 1972

(c) Medical Waste Tracking Act of 1988

(d) CERCLA

(e) SARA

(f) SARA Title III

(g) NWPA

8. What does HAZWOPER refer to?

NOTES

1 "Draft Report on Hazardous Waste Disposal in Erie and Niagara Counties, New York," State of New York, Department of Environmental Conservation, Interagency Task Force on Hazardous Wastes, March 1979.

2 Lewis, Sr., Richard J., ed., *Sax's Dangerous Properties of Industrial Materials*, 3 volumes, 10th ed., John Wiley & Sons, New York, 2000.

3 40 CFR 261, subpart D.

4 *Federal Register*, Vol. 54, No. 42: 9294–9336 (1989). 29 CFR 1910.120, Hazardous Waste Operations and Emergency Response.

BIBLIOGRAPHY

Accident Prevention Manual: Environmental Management, 2nd ed., National Safety Council, Itasca, IL, 2000.

A Guide to the U.S. Department of Energy's Low-Level Radioactive Waste, National Safety Council, Itasca, IL, 2002.

ANDLEMAN, J. B., and UNDERHILL, D. W., *Health Effects from Hazardous Waste Sites*, Lewis Publishers, Chelsea, MI, 1987.

BENNETT, G. F., et al., *Handbook of Hazardous Materials Spills*, McGraw-Hill, New York, 1982.

BLACKMAN, WILLIAM C., Jr., *Basic Hazardous Waste Management*, 2nd ed., CRC Lewis Publishers, Boca Raton, FL, 1996.

BLOCK, MARILYN R., and MARASH, I. ROBERT, *Integrating ISO 14001 into a Quality Management System*, 2nd ed., American Society for Quality, Milwaukee, WI, 1999.

BROWN, M., *Laying Waste: The Poisoning of America by Toxic Chemicals*, Pantheon Books, New York, 1979.

CACCAVALE, SLAVATORE, *A Basic Guide to RCRA— Understanding Solid and Hazardous Waste Management*, American Society of Safety Engineers, Des Plaines, IL, 1998.

CARSON, RACHEL, *Silent Spring*, Houghton-Mifflin, Boston, 1962.

DAWSON, G. W., and MERCER, B. W., *Hazardous Waste Management*, Wiley-Interscience, New York, 1986.

DOMINUEZ, G. S., and BARTLETT, K. G., *Hazardous Waste Management*, CRC Press, Boca Raton, FL, 1986.

Environmental Statues, Government Institutes, Inc., Rockville, MD, updated regularly.

The EPA Manual for Waste Minimization Opportunity Assessments, U.S. Environmental Protection Agency, Washington, DC, April, 1988.

EPSTEIN, SAMUEL S., *The Politics of Cancer*, Sierra Club Books, San Francisco, CA, 1978.

EPSTEIN, S. S., BROWN, L. O., and POPE, C., *Hazardous Waste in America*, Sierra Club Books, San Francisco, CA, 1982.

ESPOSITO, M. P., et al., *Decontamination Techniques for Buildings, Structures and Equipment*, Noyes Publications, Park Ridge, NJ, 1987.

FAWCETT, H. H., *Hazardous and Toxic Materials: Safe Handling and Disposal*, Wiley-Interscience, New York, 1984.

GRIFFIN, R. D., *Principles of Hazardous Materials Management*, Lewis Publishers, Chelsea, MI, 1988.

HIAKI, S., and BROSCIOUS, J. A., *Underground Tank Leak Detection Methods*, Noyes Publications, Park Ridge, NJ, 1987.

HIGGINS, T. E., *Hazardous Waste Minimization Handbook*, Lewis Publishers, Chelsea, MI, 1989.

Infectious Waste: The Complete Resource Guide, Bureau of National Affairs, Washington, DC, 1988.

KAYS, W. B., *Construction of Linings for Reservoirs, Tanks and Pollution Control Facilities*, 2nd ed., Wiley, New York, 1986.

LEVINE, S. P., and MARTIN, W. F., *Protecting Personnel at Hazardous Waste Sites*, Butterworths, London, 1984.

LINDGEN, G. F., *Managing Industrial Hazardous Waste—A Practical Handbook*, Lewis Publishers, Chelsea, MI, 1989.

MAJUMDAR, S. K., and MILLER, E. W., eds., *Hazardous and Toxic Waste: Technology, Management and Health Effects*, Pennsylvania Academy of Science, Easton, PA, 1984.

MARTIN, E. J., and JOHNSON, J. H., Jr., *Hazardous Waste Management Engineering*, Van Nostrand Reinhold, New York, 1996.

Occupational Safety and Health Guidance Manual for Hazardous Waste Site Activities, DHHS (NIOSH) Publication 85-115, National Institute for Occupational Safety and Health, U.S. Department of Health and Human Services, Cincinnati, OH, October, 1985.

Reducing Hazardous Waste Generation, National Academy of Sciences, National Academy Press, Washington, DC, 1985.

SHIFER, R. W., and McTIGUE, W. R., Jr., *Handbook of Hazardous Waste Management for Small Quantity Generators*, Lewis Publishers, Chelsea, MI, 1988.

Storage and Treatment of Hazardous Wastes in Tank Systems, U.S. EPA, Noyes Publications, Park Ridge, NJ, 1987.

Toxic Substance Storage Tank Containment, Noyes Publications, Park Ridge, NJ, 1985.

WAGNER, K., et al., *Remediation Technology for Waste Disposal Sites*, 2nd ed., Noyes Publications, Park Ridge, NJ, 1986.

WAGNER, K., WETZEL, R., BRYSON, H., FURMAN, C., WICKLINE, A., and HODGE, V., *Drum Handling Manual for Hazardous Waste Sites*, Noyes Publications, Park Ridge, NJ, 1997.

CHAPTER *28*

PERSONAL PROTECTIVE EQUIPMENT

In discussing controls for hazards in many of the previous chapters, one alternative often noted was personal protective equipment. There are many activities where people wear protective equipment as a primary control or as a backup protection if other controls fail. In some cases, personal protective equipment is available to workers if conditions change such that hazards suddenly exist. There are many manufacturers of personal protective equipment. The personal protective equipment industry has approximately $10 billion or more in sales each year in the United States.

28-1   GENERAL PRINCIPLES

Priorities

As noted in Chapter 9, personal protective equipment falls at the bottom of the list of priorities for controlling hazards. It is in the same priority class with procedures. In fact, its use is a procedure. Personal protective equipment is low in the priority list because it does not remove hazards. In general, personal protective equipment creates a barrier between the hazard and the wearer; the hazards, however, remain. Personal protective equipment also is low in the priority list because it requires user behavior to be sure that it is in place when needed; it is not automatically in place.

Problems

Personal protective equipment is essential protection for many hazards, although it may not be sufficient. There are several things that may prevent it from being adequate. One problem is effectiveness. One must have the right equipment for the hazard. Another problem is fit. Poor fit may result in inadequate protection. A third problem is use. Users must wear the personal protective equipment, even though the hazard it protects against is not present at all times. A fourth problem is maintenance. Some equipment has a limited life, has replaceable parts, or requires regular cleaning and testing.

Effectiveness   For personal protective equipment to be effective, one must know the hazard. For example, some respirators only protect against gases, some protect only against particular gases, and others protect against particulates. Some glove coatings may dissolve in the presence of certain solvents, but be effective for others. Some hazards are life threatening; others are not. The proper equipment for the hazard or hazards at hand must be selected carefully.

Safety and Health for Engineers, Second Edition, by Roger L. Brauer
Copyright © 2006 John Wiley & Sons, Inc.

Fit For personal protective equipment to be effective, it must fit the user. People come in various sizes and shapes. Respirators that do not fit the face or are placed over beards will leak and will not achieve the protection desired. Improperly adjusted hard hat suspensions can lead to injury if something strikes the hard hat. An acid suit and gloves that do not meet properly at the wrist will expose skin to possible contact with spilled acid. Poorly fitted equipment may create discomfort and discourage use.

Women have traditionally had difficulty getting personal protective equipment that fits well. With the increasing number of women in hazardous jobs, the availability of sizes for small women has improved.

Use Issuing employees personal protective equipment does not mean that it will be worn. Work rules and enforcement of them is one important means for gaining user compliance. Feedback on use increases use. Studies have shown that giving wearers results of hearing tests after use of equipment compared with nonuse contributed significantly to wearing of hearing protection. Worker acceptance also is an important factor in use. Workers need to participate in selection. Style and other choices as well as the chance to try different products and check their comfort will improve acceptance and use. Use of company logos or worker team logos can contribute to self-image and user acceptance.

Maintenance Proper maintenance is essential to ensure personal protective equipment effectiveness. Some respirators require testing and periodic replacement of valves, seals, filter elements, straps, canisters, and other components. Some personal protective equipment requires regular cleaning. Some personal protective equipment is designed for a single use and disposal and may be dangerous to reuse.

Regulations and Standards

Various government organizations require the use of personal protective equipment. Agencies include OSHA, The Mine Safety and Health Administration (MSHA), and others. There are National Highway Traffic Safety Administration (NHTSA) standards for seatbelts in vehicles (see Table 14-1). OSHA, for example, places the burden for proper personal protective equipment on the employer. Even if workers must buy their own equipment, the employer is responsible for it being proper for the hazard, being in good condition, working properly, and achieving a good fit.

There are several organizations that have recommendations and consensus standards on personal protective equipment design and performance: The National Institute for Occupational Safety and Health (NIOSH) has numerous criteria and performance publications and reports; American Society for Testing and Materials (ASTM) has standards for testing personal protective equipment; American National Standards Institute (ANSI) has several standards for design, performance, and use of personal protective equipment; Society of Automotive Engineers (SAE) has standards for seat belts in various vehicles.

Sources of Help

There are many publications that help buyers of personal protection equipment locate manufacturers and suppliers of these products. A major annual publication devoted to safety equipment is the Grey House Publishing Safety and Security Directory, which provides information on regulations and selection procedures. It also contains lists of suppliers and advertising from many of them. Another major publication for locating supplies and equipment of all kinds is the Thomas Register. In addition, many safety and industrial hygiene magazines have an annual issue devoted to suppliers of personal protective equipment.

There are some computer programs that help check materials for certain types of protective equipment. Software users identify the environment, activity, or material and the software recommends the proper protection equipment and, if applicable, what materials they should have. There are data banks that recommend protective equipment for particular hazards. Some manufacturers, insurance companies, and other organizations publish charts and guides for selection of personal protective equipment.

Requirements

Personal protective equipment should meet several requirements: the equipment should not create additional hazards to the user; the materials of construction should hold up under reasonable use; they should withstand conditions for which they are intended; in some cases, they should be cleanable and they should be comfortable. Where possible, style and appearance are important for user acceptance.

User Involvement

User involvement in selection of personal protective equipment is important for acceptance and use. Users need to understand the hazards that create the need for the equipment and the dangers faced if the equipment is not worn. They need to learn the importance of the equipment in protecting themselves and to understand the rules related to enforcement of use. Users need training in proper use of equipment and proper care and maintenance. They need to test equipment for fit and comfort.

Ensuring Performance

Personal Protective Equipment Program   It is essential for an organization to establish written policies and procedures for selection, management, use, and maintenance of personal protective equipment. The documentation should cover management of all aspects of the personal protective equipment program.

Inspection   User acceptance and use of personal protective equipment are essential for equipment performance. Other factors, such as regular inspection for condition and function, help ensure performance, too. There should be a management plan and process to track inspection and condition of equipment.

Maintenance, Repair, and Cleaning   There needs to be a maintenance and repair program. Personal protective equipment that is not in good condition or is not properly adjusted for fit needs corrective action. Filters in respirators need regular replacement, protective eyewear needs regular cleaning, some items need disinfecting, and some need decontamination.

Replacement   Equipment that is not in good condition and not repairable must be removed from use. Disposable equipment intended for one time use should not be reused, and equipment that is not up to standards should be replaced.

Testing   Inspection may not be enough to determine performance of some personal protective equipment; some equipment must be tested. For example, there are devices available to test respirator fit. Lifelines must meet periodic tests to be sure they meet load-carrying capacity.

Certification The buyer or user of personal protective equipment cannot always tell if purchased items meet published standards. There are certification programs for some equipment. NIOSH certifies the performance of some respirators. The Safety Equipment Institute in Arlington, Virginia, is a not-for-profit organization that tests and certifies a broad range of industrial safety products, including personal protective equipment. Its certification is voluntary for manufacturers of products. OSHA has established procedures for accreditation of testing laboratories.[1]

28-2 HEAD PROTECTION

Hazards

One danger to the head is falling or flying objects. Falling or flying objects also can strike the neck and shoulder area. In tight spaces, one can bump the head against something and cause an injury. Bureau of Labor Statistics (BLS) data show that only 16% of workers receiving head injuries were wearing hard hats. There are also dangers of hair becoming caught in machines or hair being set on fire. Sanitation rules may require preventing hair from falling into food, and clean room work requires that hair and skin particles not contaminate the work.

Types of Head Protection

Helmets There are a wide variety of helmets or hard hats. Figure 28-1 has one example of a helmet. Helmets vary in materials of construction and features. Some can accommodate other protective equipment for eyes or hearing. ANSI Z89.1 classifies helmets as having a continuous brim all the way around (type I or hard hat) or having only a visor brim in the front (type II or hard cap). Different types and thicknesses of material vary in ability to prevent penetration of an object through the outer shell. Some materials for helmets have high dielectric properties and protect users who work around electrical lines and equipment; some helmet materials are conductive. Helmet materials are also rated for weight, flammability, and water absorption properties. ANSI standards divide helmets into class A, B, C, and D based on certain combinations of properties.

Headbands and suspension webs inside a helmet should have a $1^{1}/_{4}$-in clearance between the helmet shell and the suspension. The suspension system will distribute the forces from a blow to the helmet over a large area of the skull and help absorb the energy of a blow, thereby preventing injury. For cold weather or sanitation reasons, many manufacturers offer liners and ear covers.

There are special helmets for firefighters, riot police, motorcyclists, athletics, and recreation activities. Each has special features and may have other standards for their design. For example, firefighters' helmets have extended brims at the rear to protect the neck from falling debris.

Hoods Hoods protect the head, face, and neck from heat, flame, sparks, molten metal, liquids, dusts, and chemicals. The type of hazard dictates the degree of protection and the kind of materials that are appropriate. Hoods may include hard hat protection and other features. Figure 28-1 illustrates one type of hood with an air supply line.

Figure 28-1. Examples of a helmet with visor (top) and a hood (bottom). (Photos provided by and reprinted with the permission of Mine Safety Appliances.)

Bump Caps Bump caps, which are lighter in weight than helmets, protect users from bumping their head on objects, not from falling objects. Bump caps are not substitutes for helmets. Many models accommodate eye protection, cold weather liners, and other protective equipment.

Soft Caps Soft caps protect users from sparks, open flames, heat, dust, and molten metal splashes. They are made with fire-resistant fabrics and materials.

Hair Nets and Caps Paper, fabric, or net caps or covers prevent hair from falling into food or assemblies that must not be contaminated. They also prevent hair from becoming caught in machinery. Beards may need similar protection.

28-3   EYE AND FACE PROTECTION

Hazards

Flying objects and particles, airborne dusts, splashing liquids, excessive light, and radiation may injure the eyes. People must have protection from these dangers. The same dangers may injure facial or neck tissues. The dangers are compounded when materials are hot or can react chemically with human tissue. Selection and use are complicated in some cases when the user already has corrective eye wear. Impact injuries to eyes can occur in many off-the-job activities and in athletics. Protection is important in all activities where hazards to eyes are significant.

Types of Eye Protection

Table 28-1 lists various eye and face protectors. The table forms a guide for selecting protectors for particular operations and hazards.

Spectacles To prevent frontal impact injuries, frames and lenses of eye glasses should meet performance standards of ANSI Z87.1. Lenses must meet specific criteria to be called safety lenses, and industrial standards are more stringent than those for safety lenses for home and recreational use. Some plastic and hardened glass lenses meet these specifications. Different materials withstand pitting, heat, and chemicals better than others. Shading in lenses can reduce glare.

Spectacles with Side Shields Where there is a danger of falling or flying particles entering the eye from the side, side shield protection is needed. Depending on the size and type of particle, side shields may be solid material, perforated (for ventilation), or wire mesh. Turning the head during grinding operations, for example, creates a danger of particles reaching the eye from the side. Spectacles suitable for welding and cutting must also have adequate radiation filtering in the lenses (see Table 28-2).

Goggles Goggles protect the eye from flying particles, splashes, molten metal, heat, and glare. There are many types, suitable for particular applications (see Table 28-1). Some activities need the protection of close-fitting eye cups; for others, different forms of goggles are suitable. Ventilation helps prevent fogging of goggles, but ventilation openings should be suitable for the hazards present because particles can pass through large holes. Goggles also protect prescription glasses that do not provide adequate protection. Goggles used for welding and cutting also must have adequate ultraviolet and optical

TABLE 28-1   Eye and Face Protection Selection Guide (29 CFR 1910, Subpart I, Appendix B)

Source	Assessment of Hazard	Protection
IMPACT: Chipping, grinding, machining, masonry work, woodworking, sawing, drilling, chiseling, powered fastening, riveting, etc.	Flying fragments, objects, large chips, particles, sand, dirt, etc.	Spectacles with side protection, goggles, face shields. See notes 1, 3, 5, 6, & 10. For severe exposure, use face shield.
HEAT: Furnace operations, pouring, casting, hot dipping, and welding	Hot sparks	Face shields, goggles, spectacles with side protection. For severe exposure use face shield. See notes 1, 2, & 3.
	Splash from molten metals	Face shields worn over goggles. See notes 1, 2, & 3.
	High temperature exposure	
	High temperature exposure	Screen face shields, reflective face shields. See notes 1, 2 & 3.
CHEMICALS: Acid and chemicals handling, degreasing, plating	Splash	Goggles, eyecup, and cover types. For severe exposure, use face shield. See notes 3 & 11.
	Irritating mists	Special-purpose goggles.
DUST: Woodworking, buffing, general dusty conditions.	Nuisance dust	Goggles, eyecup, and cover types. See note 8.
LIGHT and/or **RADIATION**		
Welding: Electric Arc	Optical radiation	Welding helmets or welding shields. Typical shades: 10–14. See notes 9 & 12.
Welding: Gas	Optical radiation	Welding goggles or welding face shield. Typical shades: gas welding, 4–8; cutting, 3–6; brazing, 3–4. See note 9.
Cutting, torch brazing, torch soldering	Optical radiation	Spectacles or welding face shield. Typical shades: 1.5–3. See notes 3 & 9.
Glare	Poor vision	Spectacles with shaded or special-purpose lenses, as suitable. See notes 9 & 10.

Notes:

1. Care should be taken to recognize the possibility of multiple and simultaneous exposure to a variety of hazards. Adequate protection against the highest level of each of the hazards should be provided. Protective devices do not provide unlimited protection.

2. Operations involving heat may also involve light radiation. As required by the standard, protection from both hazards must be provided.

3. Face shields should be worn only over primary eye protection (spectacles or goggles).

4. As required by the standard, filter lenses must meet the requirements for shade designations in 29CFR 1910.133(a)(5). Tinted and shaded lenses are not filter lenses unless they are marked or identified as such.

5. As required by the standard, persons whose vision requires the use of prescription lenses must wear either protective devices fitted with prescription lenses or protective devices designed to be worn over regular prescription eyewear.

6. Wearers of contact lenses also must wear appropriate eye and face protection devices in a hazardous environment. It should be recognized that dusty and/or chemical environments may represent an additional hazard to contact lens wearers.

7. Caution should be exercised in the use of metal frame protective devices in electrical hazard areas.

8. Atmospheric conditions and the restricted ventilation of the protector can cause lenses to fog. Frequent cleansing may be necessary.

9. Welding helmets or face shields should be used only over primary eye protection (spectacles or goggles).

10. Non–side-shield spectacles are available for frontal protection only, but are not acceptable eye protection for the sources and operations listed for "impact."

11. Ventilation should be adequate, but well protected from splash entry. Eye and face protection should be designed and used so that it provides both adequate ventilation and protects the wearer from splash entry

12. Protection from light radiation is directly related to filter lens density. See note 4. Select the darkest shade that allows task performance.

TABLE 28-2   Filter Lens Shade Numbers for Protection Against Radiant Energy[a]

Welding Operation	Shade No.[b]
Shielded metal-arc welding	
$^1/_{16}$, $^3/_{32}$, $^1/_8$, or $^5/_{32}$ inch electrodes	10
Gas-shielded arc welding (nonferrous)	
$^1/_{16}$, $^3/_{32}$, $^1/_8$, or $^5/_{32}$ inch electrodes	11
Gas-shielded arc welding (ferrous)	
$^1/_{16}$, $^3/_{32}$, $^1/_8$, or $^5/_{32}$ inch electrodes	12
Shielded metal-arc welding	
$^3/_{32}$, $^7/_{32}$, or $^1/_4$ inch electrodes	12
$^5/_{16}$ or $^3/_8$ inch electrodes	14
Atomic hydrogen welding	10–14
Carbon arc welding	14
Soldering	2
Torch brazing	3 or 4
Light cutting, up to 1 inch	1 or 4
Medium cutting, up to 6 inches	4 or 5
Heavy cutting, 6 inches and more	5 or 6
Gas welding (light) up to $^1/_8$ inch	4 or 5
Gas welding (medium) $^1/_8$ to $^1/_2$ inch	5 or 6
Gas welding (heavy) $^1/_2$ inch and more	6 or 8

[a] 29 CFR 1926.252.
[b] Note: In gas welding and cutting where the torch produces a high yellow light, it is desirable to use a filter or lens that absorbs the yellow or sodium line in the visible light of the operation.

TABLE 28-3   Selecting Laser Safety Glass[a] **(OSHA Table E-3**[b]**)**

Intensity, CW Maximum Power Density (W/cm²)	Attenuation Optical Density (O.D.)	Attenuation Factor
10^{-2}	5	10^5
10^{-1}	6	10^6
1.0	7	10^7
10	8	10^8

[a] Note: This table lists the maximum power or energy density for which adequate protection is afforded by glasses of optical densities 5 through 8. Output levels falling between entries in the table require the higher optical density.
[b] See 29 CFR 1926.102(b)(2) for additional requirements for protection from lasers.

radiation protection (see Table 28-2). Goggles for use around lasers must have adequate protection for the energy and wavelength of the laser beam (see Table 28-3).

Face Shields   Pouring liquids, working with molten metal, and other activities require protection of the face. A face shield has a large, transparent panel that extends over the front and sides of the face. A face shield alone is never adequate protection for the eyes. It is used in addition to eye protection.

Welding Helmets   Welding helmets protect the face against ultraviolet radiation, sparks, and molten metal during electric arc welding. The helmet may have a window with radiation protection for the eyes (see Table 28-2).

Laser Safety Glasses   Employees potentially exposed to laser beams need eye protection. One form of protection is laser safety goggles with adequate filtering capacity for the wavelength and intensity of the laser beams.

28-4   HEARING PROTECTION

Hazards

As noted in Chapter 23, exposure to excessive noise will produce temporary or permanent hearing loss.

Types of Hearing Protection

There are two kinds of hearing protection devices (see Figure 28-2): muffs, which fit over the ears to keep sound from entering the ears, and plugs of various types that are inserted into the ear canal. In either case, a good seal between the device and the head or ear canal is important. Plain cotton is not acceptable for hearing protection. The Environmental Protection Agency has a noise reduction rating (NRR) and labeling standards for hearing protectors.[2] These standards provide a reliable way to rate hearing protectors. The ratings are for continuous noise and may not represent the noise reduction from impulse noise. An NRR of 10 would have the effect of reducing the noise in an environment at 90 dBA to an equivalent of 80 dBA if the protectors fit properly. For frequencies less than 500 Hz, an adjustment in rating is necessary. Manufacturers can provide performance data by octave for their products.

Muffs   Muffs are best for severe noise environments and are more effective for high frequencies than for low ones. They have a cushion that fits against the head. There are various patented designs for attenuation. Muffs can be attached to helmets, and some manufacturers offer muffs with receivers and communication equipment built into them. They are also available with high dielectric material for work around high-voltage sources.

Figure 28-2. Example of ear muffs. (Photo provided by and reprinted with the permission of EAR Brand of Aearo Company, Indianapolis, IN.)

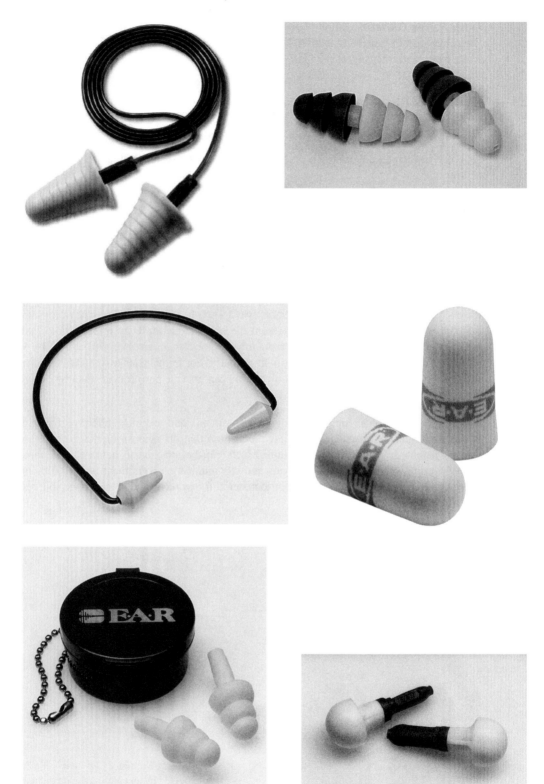

Figure 28-2. *continued* Examples of ear muffs. (Photos provided by and reprinted with the permission of EAR Brand of Aearo Company, Indianapolis, IN.)

Plugs   There are many types of ear plugs. Some are reusable; others are disposable. There are custom molded ones that are molded in the ear canal of a user. Others have properties or features that adjust to the user. Plugs are more effective for high frequencies than low ones. Users will find some types of plugs more comfortable than others.

Combination   For the most severe noise environments, a combination of both muffs and plugs will effect the greatest sound attenuation and protection.

28-5   RESPIRATORY PROTECTION

Hazards

The physiological function of respiration is essential to life. Lack of oxygen can produce syncope (fainting) or death. Excessive concentrations of certain gases and particulates can interfere with breathing. In addition, certain gases and particulates create health hazards if inhaled. An atmosphere that is *immediately dangerous to life or health* (IDLH) poses an immediate threat to life, irreversible adverse health effects, or impairment of a person's ability to escape from a dangerous atmosphere.

Types of Protection

There are two approaches to respiratory protection. One approach is to ensure that breathing air is of good quality. This is accomplished by supplying air that has the right mixture of oxygen and inert gases and is free of contaminants. An alternate approach is to clean the air before it is inhaled into the lungs. If there are particulates present, it may be possible to filter them from inhaled air. If there are undesired gases, it may be possible to adsorb them or chemically capture them before they are inhaled. The method selected depends on the degree of danger and potential failure of the equipment.

There are three classes of respirators: self-contained sources of breathable air; supplied-air respirators, which provide breathable air from other sources through a supply line; and air-purifying respirators, which remove particulates and gases from inhaled air. Examples of respiratory protective equipment are shown in Figure 28-3.

SCBA   Self-contained breathing apparatuses (SCBA) are devices that provide clean, breathable air to a user. Users usually carry SCBA equipment on a backpack, and the packs have limited supplies and use times. A closed-circuit SCBA device recovers oxygen from expired breath to extend the time of use; open-circuit devices do not recover expired air.

Hose Mask   This is one form of supplied-air respirator. An air line provides breathable air to a user from an outdoor source. A hose mask with a blower makes breathing easier during inhalation, because the blower aids the air flow in the line. The length of hose limits the distance a user can move from the supply. The wearer must be able to escape unharmed without the air of a respirator if the air line fails. This also limits the distance from a source. If air is compressed or pumped to a user, contaminants (such as compressor lubricating oil) must be removed from the breathable air.

Figure 28-3.   Examples of respiratory protective equipment.

Figure 28-3. *continued*

Air-Line Respirator   This is another type of supplied-air respirator. Breathable air is supplied directly to a respirator through a hose. Breathing regulators match the wearer's need.

Air-Supplied Suits and Hoods   This type of supplied-air respirator directs air to the breathing zone of the user. It may supply a hood or a full body suit. The air may also provide cooling air to a full body suit. The source for the air supply is independent of the ambient atmosphere.

Air-Purifying Respirators   These devices have filters, cartridges, or canisters that remove particulates and gases. They can have a full face piece, which covers the mouth, nose, and eyes, a half-mask device, which covers the nose and mouth, or a mouthpiece respirator, which covers the mouth only and requires users to wear a nose clip to prevent inhalation through the nose.

Canisters or Chemical Cartridge   This is one type of air-purifying respirator. Breathing air moves through a container that removes gases or vapors. Because no single process removes all types of hazardous gases, canisters work on particular gas contaminants. Table 28-4 lists the color coding of canisters. Canisters must also have labels that state what

TABLE 28-4 Color Codes for Gas Mask Canisters (OSHA Table I-1[a])

Atmospheric Contaminants to Be Protected Against	Colors Assigned[b]
Acid gases	White
Hydrocyanic acid gas	White with $^1/_2$-in green stripe completely around the canister near the bottom
Chorine gas	White with $^1/_2$-in yellow stripe completely around the canister near the bottom
Organic vapors	Black
Ammonia gas	Green
Acid gases and ammonia gas	Green with $^1/_2$-in white stripe completely around the canister near the bottom
Carbon monoxide	Blue
Acid gases and organic vapors	Yellow
Hydrocyanic acid gas and chloropicrin vapor	Yellow with $^1/_2$-in blue stripe completely around the canister near the bottom
Acid gases, organic vapors, and ammonia gases	Brown
Radioactive materials, except tritium and noble gases	Purple (magenta)
Particulate (dusts, fumes, mists, fogs, or smoke) in any combination with any of the above gases or vapors	Canister color for contaminant, as designated above, with $^1/_2$-in gray stripe completely around the canister near the top
All of the above atmospheric contaminants	Red with $^1/_2$-in gray strip completely around the canister near the top

[a] 29 CFR 1910.134(g).
[b] Gray shall not be assigned as the main color for a canister designed to remove acids or vapors. Note: Orange shall be used as a complete body or stripe color to represent gases not included in this table. The user will need to refer to the canister label to determine the degree of protection the canister will afford.

gases and gas concentrations they handle. Because an analysis of the contaminants in an environment is not always available, using canister respirators requires careful application. The kinds and concentration of contaminants may change over time and make a canister respirator ineffective.

Filter Respirators   Respirators remove particulates through mechanical filters. There are filters for particular types and sizes of particulates. Filter respirators do not protect against gases and vapors. Some filter respirators have replaceable filters and some are disposable.

Other Respirators   People may work in areas that are free of contaminants. However, a leak in a system may produce dangerous breathing atmospheres. In such situations, escape respirators issued to workers provide protection for the applicable danger for a very short time. Escape respirators are not intended for general use.

28-6   HAND, FINGER, AND ARM PROTECTION

Hazards

There are many hazards for hands, fingers, and arms. One hazard is hot or cold material and objects, in which case thermal insulation is needed to protect tissue. Some operations may create fire or flame hazards for the protective clothing, so the clothing must minimize the danger of catching on fire. Another hazard is sharp objects and equipment, such as when handling sheared metal or metal objects with burrs or cutting meat with sharp knives and tools. The protective clothing must be tough and resistant to cuts and tears. Chemicals that can damage tissue or be absorbed through skin are other hazards. Protective material that prevents chemical penetration of materials is essential. Other hazards are damage to tissue from solvents or even water that dries oils from skin and causes cracking; bumping into objects that are sharp or pointed; and radiation burns, such as in welding. Slipping and loss of grip may require that protective clothing have slip-resistant properties. However in some operations, touch and feel are as important as the need for worker protection, and the protective clothing must minimize the loss of touch. Another hazard is electric shock. Sometimes protective clothing must provide electrical insulation. Hands may reach into biological or radiation boxes through protective gloves, and the material must provide the necessary protection.

Types of Protection

Gloves and Mittens   Gloves and mittens cover the hands and fingers. The hazards present and the materials that create the hazards must be analyzed, and then the proper material for gloves or mittens can be selected. Most manufacturers of gloves and mittens produce their products in a wide range of materials. Many publish guides and give direct help in making selections.

Some types of gloves protect the fingers; others are fingerless and protect the hand. Some extend protection to the wrist and lower arm. There are gloves resistant to solvents, water, acids, caustics, salts, fats and greases, detergents, cuts, and abrasions. Some gloves are lined with cotton or other material for comfort and to reduce sweating; some are impregnated with lead to shield x-rays and other forms of radiation. Leather gloves stop sparks and molten metals found in welding and foundry operations. Some gloves are woven from steel or other metal for meat cutting operations.

Pads   At home, people use hot pads to remove hot pans from a cooking oven. Workers in bakeries, foundries, and other hot processes may find pads more convenient than gloves.

Finger Guards and Cots   Finger guards or cots cover an entire finger or a portion of a finger. They are suitable in operations where finger protection is needed, but full gloves are not desired.

Sleeves   When protection must extend to the arms, workers may need special sleeves. Sleeves may cover the wrist and forearm only, may extend to the elbow or to the shoulder, or may have gloves and mittens attached. Materials for sleeves may be the same as those for gloves and mittens.

Creams and Lotions   Lotions and creams provide some protection to hands and fingers from water, solvents, fats, irritants, and other substances. Lotions and creams can replace skin oils, soften chapped skin, and kill germs. Where severe exposures occur, protective gloves are more effective than creams and lotions.

28-7   FOOT AND LEG PROTECTION

Hazards

A major hazard for the foot is falling objects. In addition, there are hazards of contact with chemicals, slipping, and stepping on protruding nails, hot materials, or wet materials. There are also cutting hazards of tools, such as axes and chain saws, and dangers from cold. Where these hazards are present, footwear and clothing for the legs can prevent many injuries. BLS analysis shows that in up to 75% of foot injuries, workers were not wearing safety shoes.

Types of Protection

Safety Shoes   Standard safety shoes (see Figure 28-4a) have steel toes that meet crushing tests found in ANSI Z41.1. The standard calls for safety toes to withstand a 75-lb impact load and a 2,500-lb compression load. Safety shoes are available in almost any style of work or dress shoe. Safety-toe shoes are available in the form of rubber and plastic boots. Steel-toe shoes can sometimes create a hazard. Exposure to heat may store enough heat in the steel toe plate to cause radiation burns to the toes.

Metatarsal or Instep Guards   For work involving heavy objects, one can attach metal guards that extend further over the foot rather than just over the toe. These guards are metatarsal or instep guards (see Figure 28-4b).

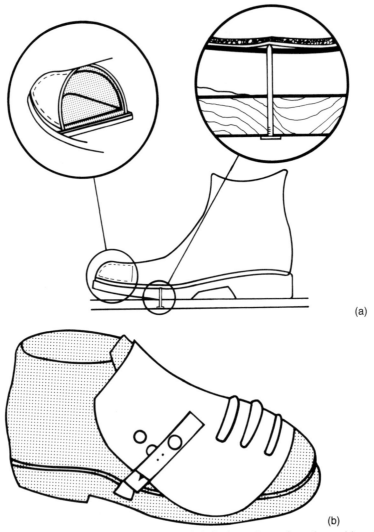

(a)

(b)

Figure 28-4. Foot protection: (a) exploded view of a safety shoe with safety toe and instep protector; (b) metatarsal guard.

Steel Insoles   For those who work in construction or other locations where there is a danger of stepping on sharp objects that can penetrate the sole of a shoe, metal insoles that are usually built into the shoe structure provide protection.

Rubber or Plastic Boots   For work with wet and muddy processes and exposure to chemicals, there are rubber and plastic boots. They may be ankle high or extend over the entire thigh. In some environments, waist-high boots are used.

Insulation   For those who work in the cold, electrically heated insoles are available. For those who walk on hot surfaces or face the danger of splattered molten metal from welding or foundry work, shoes that have various forms of insulation or insulated soles are available. To provide for quick removal, some foundry shoes have quick release closures. The tops of foundry boots fit snugly around the lower leg to prevent any hot material from dropping into the shoe.

Shin Guards   For operations that require protection of shins from falling or moving objects, there are padded shin guards. The guards are made from metal, plastic, or other materials.

Leggings and Leg Protection   For loggers and others who work with chainsaws, piked poles, or axes, protective panels of metal and ballistic nylon protect the sides and top of the foot and the front and sides of the lower leg.

Conductive Soles and Nonsparking Shoes   People who work in hospital operating rooms where there is a danger of fire or explosion from flammable gases can wear conductive shoes. When used with a conductive floor, any charge buildup on a person is removed. Spark-resistant shoes have no metal parts in them other than a steel safety toe. A shoe nail in an ordinary shoe may generate a spark that could be dangerous in an explosive atmosphere.

Nonconductive Shoes   People who work with high-voltage electrical equipment can obtain shoes with electrical insulation. They prevent electrical shock and flow of current through the shoes.

Slip-Resistant Soles   There are many surfaces where workers may slip and fall. Friction is created by the combined properties of two interfacing materials. Proper shoe soles and heels play an important part in preventing slips. (See Chapter 11 for additional information on slip resistance.)

28-8 BODY PROTECTION

Hazards

Hazards that require the use of personal protective equipment for the body include exposure to hazardous materials or biohazards. Substances may have many forms, such as liquid, dust, mist, or other forms. There are also dangers of work in confined spaces where atmospheres may not support life, dangers from fire or high heat sources that require protection, and dangers of sparks, molten metal, or hot or dangerous liquids that require special protective clothing. In many cases, there is a need to protect the body and wear other personal protection equipment at the same time. Full protective clothing creates a minienvironment for a worker that is sealed from the surround. This closed environment requires supplying breathing air and removing heat and moisture from the suit. Many kinds of personal protective equipment and clothing for the body are made from disposable materials. For clothing contaminated by hazardous materials and biohazards, it may be more economical to dispose of the clothing than to decontaminate it properly for reuse. Work in explosive environments requires static-free fabrics for protective clothing.

Types of Protection

Coats and Smocks   Coats and smocks extend to the knee or below. They create a barrier for spills of various substances on personal clothing. Depending on the fabric used, the coat or smock may be suitable for splatters of water, acids, oils, solvents, or other materials. The coats and smocks may be collected and cleaned for reuse or disposed of after use.

Coveralls   Coveralls extend over the arms, body, and legs and some have hoods and boots. They may be useful in clean rooms or to provide greater protection than coats or smocks. Coveralls are usually one-piece garments, although in some applications, users may prefer a two-piece garment (pants and shirt or jacket).

Aprons   Aprons cover the front of a person from the upper chest to below the knees. Aprons may protect against splatters of hazardous substances, molten metals, oils, greases, or other materials. The type of hazard determines the kind of fabric and coating appropriate for the application. Leaded aprons help prevent radiation from reaching the body of the wearer.

Full Suits   When substances create a danger to life or may cause immediate or latent health problems, full body suits are useful. The full suit provides a barrier between the danger and the user.

   Because of the gravity of the hazard, the integrity of the barrier is important. Tears or holes may allow contaminants inside the suit. The user of a full suit needs breathing air and, depending on activity level and duration in a suit, cooling and moisture controls also may be essential. Special suits are available for protection in radioactive environments.

Fire Entry and Proximity Suits   People may need to approach or enter burning locations for rescue or critical tasks. They need insulation from the heat. There are two kinds of suits for fire: a proximity suit, which allows one to approach a fire, and an entry suit, which permits walking within the fire itself. The entry suit has heavier insulation. Both suits have a limited use time, and users need their own air supply. Fabrics normally are coated material to reflect radiant energy and to slow the rate of heat transfer from the fire. The insulation and suit fabric are noncombustible or are made from fire-resistant materials.

Cooling   As noted in Chapter 18, the heat produced from metabolism must be removed or the body temperature will rise. The metabolic cost of an activity may double when wearing protective clothing. The environment within a closed suit will limit heat transfer from the body. If a suit must be worn for more than 15 or 20 min, cooling can be important. Cooling can be provided by pumping breathable air into the suit and distributing it before it is exhausted to the surround or returned to a remote location. There are self-contained refrigeration or air conditioning units available for full body suits. There is special underwear with small tubes that allow coolant to flow to legs, arms, and body. There are also vortex coolers in which compressed air passes through a device that causes the air to spin in a vortex. The temperature of the air along the outer surface of the vortex drops as a result of rotation, and bleeding it off can provide a significant amount of cooling capacity. There are also air- and water-cooled helmets. A very large portion of the body's blood flow goes to the head and brain. Studies have shown that cooling the head alone can remove a large amount of heat from the body.

Rainwear   Another kind of personal protective clothing is rainwear. There are many people who must work in wet outdoor conditions, and rainwear keeps users dry. Wet clothing increases body cooling. For cold weather, rainwear also may need insulation.

High-Visibility Clothing   Vehicles may strike people working on road construction, parking cars, and managing vehicle traffic. There are luminescent orange vests, arm bands, and jackets that help make them visible.

Joggers, emergency workers, and others who could be struck at night by vehicles can wear reflective stripes on clothing and shoes. Visibility is important so drivers have time to react and avoid hitting someone.

Personal Flotation Devices   For activities around or on the water where there is a danger of drowning, people need personal flotation devices. There are several types of personal flotation devices, including jackets and vests, which are worn in activities near water. The U.S. Coast Guard has standards for personal flotation devices.[3]

Puncture-Resistant and Cut-Resistant Clothing   Police, security personnel, bomb squads, and others need protection from ballistic objects. Body armor clothing provides this protection. Workers using chainsaws need protection from chains cutting through clothing. For these hazards, there are fabrics that resist severe cutting and shearing. Workers in meat cutting operations need protection from powered saws and other cutting equipment. Clothes made from woven metal fabric give them some protection.

28-9   FALL PROTECTION

Hazards

Chapter 11 discussed the dangers of falls. There are a variety of fall protection devices available (Figure 28-5 shows one type of harness). Most interrupt falls in progress. Because they must interrupt a body in motion, the devices must withstand certain loads prescribed by standards.

Types of Fall Protection Equipment

There are several components that form a fall protection system, and together they prevent or limit falls. The person who could fall wears a safety harness. Safety belts are not often used. Some ANSI standards limit the deceleration force for body belts to 10 times the force of gravity. For full-body harnesses, the limit has been 35 times the force of gravity. A lanyard or lifeline attaches to the belt's D ring and must be securely anchored. Lifelines are anchored independently of any scaffolding a person is on. Safety belts or harnesses and lifelines also are used for rescue of people from confined spaces, grain bins, or similar

Figure 28-5. Example of a fall protection harness. (Photos copyrighted by and reprinted with the permission of Klein Tools, Inc.)

locations. With a lifeline, a rescuer can retrieve an incapacitated worker without endangering himself or others by entering the same location.

Safety Belt There are several classes of safety belts worn around the waist. Safety belts are useful for keeping someone from falling. When used for protection of falls, a belt can cause injury to the wearer during the stop or potentially slip free of the belt. Window washing safety belts have special fittings that attach to the window units.

Safety Harness There are several types of safety harnesses. One style has a belt at the waist and a harness over the chest and shoulders. Another style has additional loops that support the upper legs. A third style has a sling support that forms a seat, but it is not intended as a fall arresting harness. Compared with safety belts (not often used or allowed in some applications), harnesses distribute the forces over a greater portion of the body, and therefore are less likely to cause injury in an arrested fall.

Lanyard A lanyard is a short, flexible rope or strap that connects a safety harness to an acceptable anchor point or a grabbing device on a lifeline. Lanyards must be $1/_2$-in nylon rope or equivalent, have a static load capacity of 5,400 lb,[4] and not have a fall distance of more than 6 ft. Some lanyards are designed to absorb energy in arresting a fall, thus reducing the impact load on a person.

Hardware All hardware must be free from sharp edges and must withstand 4,000-lb static loads.[5] The most common hardware are D rings on safety belts and harnesses and snap hooks on lanyards.

Grabbing Device A grabbing device connects a body belt or lanyard to a lifeline. Some grabbing devices move freely along a lifeline when there is no load, but when there is a sudden load or movement, they lock onto the line.

Lifeline A lifeline is a rope that extends from an appropriate anchor point to a body harness or lanyard. The anchor point and lifeline must be capable of a static load of 5,400 lb.[6]

Fall Arrestor   There are several patented fall-arresting devices. These devices are incorporated into a lanyard and create a controlled deceleration force for the person being stopped. They are used for longer free falls and where lanyards do not have elastic properties.

Climbing Safety Systems   Workers climbing fixed ladders or poles need protection from falling. There are patented systems that attach permanently or temporarily to ladders on towers, bridges, antennas, or other equipment. A safety belt or harness connects to the climbing safety device. If a person falls, the device locks and stops the fall.

Safety Nets   Safety nets may not be classified as personal protective equipment because they are not worn by workers. However, for workers involved in construction or bridge work, safety nets may be an important form of fall protection equipment. They do not replace lifelines and related fall protection equipment.

28-10   ELECTRICAL WORKER PROTECTION

People who work around electrical equipment face the danger of current flowing through them. Their personal protective equipment needs electrical insulating properties. Personal protective equipment must have nonconductive properties and must prevent current from flowing through it.

28-11   EMERGENCY SHOWERS AND EYE WASH FOUNTAINS

Hazard

Time is critical in responding to a spill of certain chemicals on the body or in the eyes. Flushing the affected area quickly with a lot of water will dilute or remove the dangerous substance and minimize injury.

Types of Equipment

There are three devices that must be readily available in areas where people work with substances that are dangerous to eyes. These devices are emergency showers (Figure 28-6a), eyewash fountains (Figure 28-6b), and eye- and facewash fountains. OSHA requires these devices for work areas, but does not specify exact locations. ANSI Z358.1 gives detailed requirements for design and location. Periodic testing ensures that the equipment meets standards. One should review location and signage periodically because operations change or signage may become defaced or damaged or be removed.

Emergency Showers   An emergency shower is a unit that floods a person's entire body when an actuating valve is tripped. Most emergency showers are permanently installed to a water line to ensure an unlimited source of water. There are also portable units that have limited water supplies, but do provide at least 15 min of flooding. An emergency shower should be located not more than 10 s travel time away from potential users and should be clearly identified by signs and markings. In a spill emergency, there is not enough time to look for the shower. In some locations, showers need protection from freezing, and if conditions would make the water very cold, it should be heated enough to make it comfortable for use.

 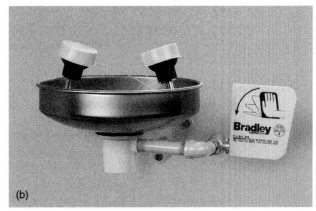

Figure 28-6. Examples of (a) a safety shower and (b) safety eyewash fountain. (Photos provided by and reprinted with permission of the Bradley Corporation, Menomonee Falls, WI.)

Emergency Eyewash Fountains When users bend over them, emergency eyewash fountains deliver two streams of water in an upward direction directly into each eye. An emergency shower cannot flush the eyes as well because the flow is downward. There are both fixed and portable eyewash units available. Fixed units have unlimited water supplies, whereas portable ones deliver a required flow rate of 1.5 l/min for 15 min. The fixture has room for holding both eyelids open to the streams of water and the control valve does not require a hand to hold it open. Again, fountains must be convenient to locations where there are dangerous materials, and there should be clear signs and markings because vision is probably impaired when foreign material enters the eyes and an emergency fountain is needed.

Emergency Eye- and Facewash Fountain Emergency eye- and facewash equipment floods the face and eyes. It has a higher flow rate than an eyewash fountain, but other design and location features are similar.

EXERCISES

1. A person wearing fall protection falls, and a fall-arresting device and lifeline interrupt the fall. The person weighs 185 lb and the arresting force is 1,263 lb. What G load does the person experience? Is it acceptable for a body harness?

2. Obtain literature, including performance and cost, on ear protection devices. Assume that disposable plugs are issued every day, reusable plugs are issued every 2 weeks, and muffs are issued once every 6 months. If a noise reduction of 15 dBA is required, find out which form of ear protection is adequate and which is most economical.

3. Select a job or operation, each having one of the following hazards:
   (a) fall from a roof with a 5 : 12 pitch
   (b) asbestos removal from community schools
   (c) dishwashing operation in a major restaurant
   (d) work in an operation where acid may splatter
   (e) removal of steel parts by a cutting torch from inside an old, cylindrical container that is 12 ft high and 6 ft in diameter and has a 30-in access hole in the top

   Recommend personal protective equipment for these activities.

REVIEW QUESTIONS

1. Where does personal protective equipment fall in the priorities for controls for hazards?

2. Does personal protective equipment remove hazards?

3. What are three important problems in the use of personal protective equipment?

4. Where does one find standards for selection, use, maintenance, and testing of personal protective equipment?

5. Where can one obtain help in selection and application of personal protective equipment?

6. What can be done to ensure performance of personal protective equipment?

7. Briefly describe the functions and characteristics for each of the following:

   (a) helmet

   (b) bump cap

   (c) hood

   (d) hair nets and caps

   (e) spectacles

   (f) side shields

   (g) goggles

   (h) face shield

   (i) welding helmet

   (j) laser safety glasses

   (k) ear muffs

   (l) earplugs

   (m) self-contained breathing apparatus

   (n) supplied-air respirators

   (o) air-purifying respirators

   (p) canisters

   (q) sleeve

   (r) finger cots

   (s) safety shoes

   (t) steel insoles

   (u) leggings

   (v) nonconductive shoes

   (w) aprons

   (x) full body suits

   (y) fire entry and proximity suits

   (z) high visibility clothing

   (aa) puncture-resistant and cut-resistant clothing

   (bb) safety belt

   (cc) safety harness

(**dd**) lanyard

(**ee**) grabbing device

(**ff**) fall arrestor

(**gg**) lifeline

(**hh**) climbing safety system

NOTES

1 29 CFR 1907.

2 40 CFR 211.

3 46 CFR 75.

4 29 CFR 1926.104.

5 29 CFR 1926.104.

6 29 CFR 1926.104.

BIBLIOGRAPHY

American National Standards Institute, New York:
ISEA 101 Limited-Use and Disposable Coveralls—Size and Labeling Requirements
ISEA 105 Selection Criteria for Hand Protection
ISEA 107 High Visibility Safety Apparel
ISEA 110 Air-Purifying Respiratory Protective Escape Devices
IWCA I14.1 Window Cleaning Safety
S12.6 Methods for Measuring the Real-Ear Attenuation of Hearing Protectors
Z41 Personal Protection—Protective Footwear
Z87.1 Practice for Occupational and Educational Eye and Face Protection
Z88.7 Color-Coding of Air-Purifying Respirator Canisters, Cartridges and Filters
Z88.10 Respirator Fit Testing Methods
Z89.1 Industrial Head Protection
Z359.1 Safety Requirements for Personal Fall Arrest Systems, Subsystems, and Components
National Fire Protection Association, Quincy, MA:
NFPA 1404 Fire Department Self-Contained Breathing Apparatus Program
NFPA 1851 Standard on Selection, Care, and Maintenance of Structural Fire Fighting Protective Ensemble Elements
NFPA 1852 Standard on Selection, Care, and Maintenance of Open-Circuit SCBA
NFPA 1971 Protective Clothing for Structural Fire Fighting
NFPA 1976 Protective Clothing for Proximity fire Fighting
NFPA 1977 Protective Clothing and Equipment for Wildland Fire Fighting
NFPA 1981 Open-Circuit Self-Contained Breathing Apparatus for Fire Fighters
NFPA 1898 Standard on Breathing air Quality for Fire and Emergency Services Respiratory Protection
NFPA 1991 Vapor Protective Suits for Hazardous Chemical Emergencies
NFPA 1992 Liquid Splash Protective Suits for Hazardous Chemical Emergencies
NFPA 1994 Standard on Protective Ensembles for Chemical or Biological Terrorism Incidents
NFPA 1999 Protective Clothing for Emergency Medical Operations
ANNA, DANIEL H., *Chemical Protective Clothing*, 2nd ed., American Industrial Hygiene Association, Fairfax, VA, 2003.
BOMAN, ANDERS, ESTLANDER, TUULA, WAHLBERG, JANE, and MAIBACH, HOWARD I., eds., *Protective Gloves for Occupational Use*, 2nd ed., CRC Press, Boca Raton, FL, 2004.
Grey House Publishing Safety and Security Directory, Millerton, NY, annual.
ELLIS, J. NIGEL., *Introduction to Fall Protection*, 3rd ed., American Society of Safety Engineers, Des Plaines, IL, 2001.
FORSBERG, KRISTER, and MANSDORF, S. Z., *A Quick Selection Guide to Chemical Protective Clothing*, 4th ed., American Industrial Hygiene Association, Fairfax, VA, 2003.
Personal Protective Equipment for Hazardous Materials Incidents: A Selection Guide, Report DHHS 84-114 (NTIS No. PB85-222230), National Institute for Occupational Safety and Health, Department of Health and Human Services, Washington, DC, 1985.
RAJHANS, G., and BLACKWELL, D. S., *Practical Guide to Repirator Usage in Industry*, Butterworths, London, 1985.
Respiratory Protection: A Manual and Guideline, 3rd ed., American Industrial Hygiene Association, Farfax, VA, 2001.
SCHWOPE, A. D., COSTAS, P. P., JACKSON, J. D., and WEITZMAN, D. J., *Guidelines for the Selection of Chemical Protective Clothing*, 3rd ed., American Conference of Governmental Industrial Hygienists, Cincinnati, OH, 1987.
STULL, JEFFREY O., *PPE Made Easy: A Comprehensive Checklist Approach to Selecting and Using Personal Protective Equipment*, Government Institutes, Rockville, MD, 1998.

CHAPTER 29

EMERGENCIES

29-1 INTRODUCTION

Disasters and emergencies fill the pages of history. The burning of Rome, the Great Chicago Fire, the Johnstown flood, the Three Mile Island incident, the Bhopal tragedy, World Trade Center events of September 11, 2001, the tsunami of December 26, 2004, and many other events have generated lessons learned, many of which led to changes in standards, codes, and laws.

One problem remains for every disaster and emergency situation: what to do when it occurs. People must decide what to do and how to act. Their actions should help minimize the danger and losses. Quick and proper action can prevent unnecessary losses. Proper design can reduce losses as well.

An emergency is any event that (1) happens suddenly, (2) disrupts the routine of an organization or community and affects its ability to function normally, and (3) requires immediate action. A disaster is an emergency that results in multiple injuries or deaths, produces major property damage, or both.

No one is isolated from emergency situations. Emergencies can occur anywhere and can affect anyone. One can prevent many emergencies, but not all.

Different emergencies have different lead times. For some there is reasonable time to act and prevent at least some losses; for others, there is little or no time to act before the emergency is present.

29-2 TYPES OF EMERGENCIES

There are several types of emergencies. Some result from the forces of nature, some involve fire and explosion, and others may involve system failures. Some emergencies entail traffic or transportation problems and some result from the behavior of people. On occasion, there are also police and military actions.

Natural Emergencies

Natural emergencies include floods, hurricanes, tornadoes, wind storms, snow, sleet, earthquakes, mud slides, avalanches, volcanic eruptions, and even dust or sand storms. In some locations, an insect infestation may create an emergency.

Safety and Health for Engineers, Second Edition, by Roger L. Brauer
Copyright © 2006 John Wiley & Sons, Inc.

Fire and Explosion

Chapters 16 and 17 discussed fires and explosions. A fire in a hotel, theater, or other high-occupancy facility may injure many people. Explosions can damage buildings far from the explosion site and glass and other flying materials can inflict injury. Fires involving hazardous materials have caused the evacuation of entire communities. Fires in compressed air lines can lead to explosions.

System Failures

There are many kinds of system failures that can create emergencies. For example, interruption of operations may create hazardous conditions: boiler overheating can cause dangerous conditions; failure of temperature-limit controls can lead to runaway processes; failure of pressure-limit controls can lead to rupture of pipes, gaskets, vessels, and other equipment; sudden releases of steam, gas, fuel, or hazardous chemicals can create dangers for plant personnel and surrounding communities. System failures may lead to fire and explosion. In 1988, a fire in a telephone switching center near Chicago interrupted businesses in a wide area, affecting some for more than 1 month. Applying water to an industrial magnesium fire intensified the fire.

Traffic Problems

Transportation accidents can interfere with traffic movement. Overturned tank trucks, multiple car accidents, and derailed railroad cars may block traffic for extended periods of time. If there are hazardous materials in a mishap, clearing, or evacuating the area may be important. Spilled materials may require proper treatment to prevent further disaster. For example, gasoline running into a storm sewer could lead to an explosion.

Behavior of People

The behavior of people can lead to emergencies. Some behaviors intend to cause harm; others do not, but have the same result. Strong feelings may lead to riots and mob behavior. Strikes and work stoppages sometimes lead to problems. Crowds rushing to sales when goods are in short supply and crowds fighting for tickets to public events or pressing to enter auditoriums and stadiums have led to disasters. In April, 1989, people at a soccer stadium in England continued to fill a standing room area near the playing field. In a few minutes, more than 100 fans were killed because the people at the front were pressed against a crowd-restraining fence and crushed so they were unable to breath and died. There are episodes where disgruntled workers or terrorists commit sabotage that leads to emergencies.

Military Action

Military action in time of war creates emergency situations. Even when the military is used for police actions, there is often an emergency situation and there are dangers to the general population. Keeping people clear of the dangerous areas is important.

29-3   PRIORITIES IN EMERGENCIES

There are well-established priorities for emergencies, the first of which is safety of people. The people may be employees, customers, visitors, or the public. Evacuation of people who could be injured and care for those injured have the highest priority. Actions to prevent involvement of additional people are also important. Isolating an area from inadvertent involvement or keeping the curious away can avoid further injury. If a snow storm or hurricane is imminent, people should be moved from dangerous areas. In emergencies, evacuation routes from buildings, sites, or communities must stay clear.

The second priority is protection of property. This may involve turning off power, fuel, or supplies to prevent further damage. Processes may be shut down manually or automatically to render them safe or to minimize loss of materials and products. Controlling and extinguishing fires will keep losses down. The proper actions depend on the kind of emergency and the kind of facility, process, or location. For example, if there is danger of a flood, creating dikes may prevent flooding of property.

The third priority is cleanup and salvage. Spilled hazardous materials must be removed to make an area safe. Fires sometimes leave building walls standing without support, which could collapse on passers by. Damaged equipment may be restorable with proper treatment. In removing debris or rubble, careful action will prevent further damage or injury. Managing cleanup and salvage is an important task. Communities struck by tornadoes often receive generous but misplaced help from volunteers. As a result, people waste resources and destroy salvageable items. Cleaning up damage to power, communications, fuel, and processes requires trained and qualified people.

The fourth priority is restoring operations and returning things to normal. For companies or businesses, there are losses in income and production until operations begin again. After an emergency, the condition and safety of equipment must be checked and items must be repaired. Startup procedures require extra care.

29-4   PREVENTING LOSSES IN EMERGENCIES

The main objective in dealing with emergencies is to be prepared to take proper actions. The actions may involve company, community, state, medical, and other organizations and participants. Preparedness for emergencies involves analysis to identify potential emergency situations, planning to detail the actions and participants, design to remedy physical deficiencies, training to ensure proper implementation, and having prepositioned contracts or agreements for specialized equipment or personnel.

Regulations and Programs

Some federal regulations require preparation for emergencies. They may be supplemented by state and local requirements. However, many emergencies are not covered by regulations.

Chemical Releases   Title III of the Superfund Amendments and Reauthorization Act of 1986 (SARA) required chemical plants to develop emergency response plans. Title III, called the Emergency Planning and Community Right-To-Know Act of 1986, created state Emergency Response Commissions and emergency planning districts. Facilities subject to SARA Title III participate in a local emergency planning process. Each facility also submits material safety data sheets to local or state elements, or both. Each district completes and maintains an emergency plan.

In response to SARA Title III, the Environmental Protection Agency (EPA) developed the Chemical Emergency Preparedness Program (CEPP).[1] The Synthetic Organic Chemical Manufacturers Association (SOCMA) also developed the Community Awareness and Emergency Response (CAER) Program.[2] These programs help organizations prepare for emergencies.

Radiological Accidents   There are regulations involving several federal agencies (the Federal Emergency Management Agency [FEMA], the Nuclear Regulatory Commission [NRC], the EPA) and state and local governments regarding preparedness for radiation releases and accidents.[3] Much of the emergency response responsibility of the federal government falls under the Department of Homeland Security.

Analysis

It is not a simple task to identify what natural and human-made conditions may lead to emergencies. One must brainstorm, work through operations, evaluate historical evidence from past events, involve people with experience, seek help of experts, and apply other techniques to list events that lead to emergencies. The situations may involve on-site activities, off-site activities, and events that one cannot control. The situations may involve activities and equipment of others. Examples are a railroad that passes near a plant, delivery companies that enter a plant site, children or visitors on a plant site, activities on adjacent properties, drivers losing control on a nearby road, or a plane crashing into a plant. One must look at interruption of power and system failures that could create an emergency. One must consider security and terrorist threats, even hacker efforts and sabotage of critical computer and communication systems.

It is not enough to identify the scenarios that spawn emergencies. The situations that can result also should be considered: whether is there danger to life, danger to property, or both and how severe the situation could become. The time of day or the presence of other conditions that could make things worse or less dangerous also should be taken into account.

There may not be enough time or resources to plan for every potential emergency, but the more situations one is prepared for, the better. A scheme to rank situations allows one to start with those situations that could have the greatest impact on life and property or have the greatest likelihood of occurrence.

Computer tools may help visualize potential problems in an emergency. For example, there are programs to aid in analyzing the flow of people exiting a fire. There are programs to estimate the dispersion of chemical vapors and gases and to aid in locating routes of travel and alternate routes when some are blocked. These and similar tools may help determine if emergencies create other problems. There are programs that show the historical distribution of lightning and the paths of tornadoes and hurricanes on record.

Planning

A key to emergency preparedness is planning. Plans can be strategic or tactical, general or specific. Planning should include actions, participants, authority and agreements, communication, data and information resources, supplies and equipment, locations for actions teams, and training procedures. Table 38-1 provides an example outline of an emergency response plan.

Actions   The first component of an emergency plan is what actions should be taken. The actions can be general or very specific. For example, an evacuation of a plant can begin with an automatic or manual alert. Department and section heads could proceed to lead their people to a safe location based on preestablished maps and routes. The plan could assign responsibility for checking all areas to ensure that no one lingers.

Besides evacuation, the plan may call for actions on the part of particular people to trip fire doors or shut off power or fuel sources or to shut down computers. The plan may call for individuals to notify local fire departments, police, or nearby facilities. Actions should include site security and control.

Participants   For each action, there should be people to perform them. The person or group responsible for an action must be assigned to the action. Examples are the plant fire brigade, first aid team, an emergency response committee, or all department heads.

Participants may be contractors who are used only for an emergency. For example, at the time of a flood, there is a great demand for earth-moving equipment. If the only source is a contractor, arrangements for services must be in place in advance. A call while water levels are rising may be too late because others may have already arranged for the contractor's services.

Participants also may include arrangements with local fire and police departments through mutual aid agreements or other instruments.

Authority   Participants must have authority when an emergency occurs. If they have authority for assigned actions during normal conditions, there is no need for additional authority. If responsibilities change from normal conditions, the authorities must be established clearly for emergency conditions. For example, because someone else has been delegated, a section head may lose responsibility for his or her staff during an evacuation. After an emergency occurs, it may be too late to contact the necessary specialists. That authority must be in place, as should the authority for mutual aid units to enter a site.

Communication   Communication is one of the most critical components in an emergency because information flow is essential. The status of conditions must be current because decisions must be made quickly and accurately. The more information available and the more accurate it is, the better the decisions are likely to be. There should be a command center to lead emergency operations. Information for public release should emanate from this center. Communication systems should be functional during emergencies or separate systems may be needed.

Data and Information Resources   During an emergency, there is a need for key information. Not only is the status of the current situation essential, but so is data about the site, utilities, evacuation routes, road conditions, materials and equipment involved, people injured, location of resources, and other elements. Decision makers need access to maps, charts, tables, data bases, phone directories, and other information sources.

For chemical releases, there are computer programs that will track a chemical cloud. The value of these programs is limited by the accuracy of real-time data and the accuracy of computer modeling. The actual events may not follow the projections made by the model. At least these tools are helpful in developing plans because they can generate various scenarios that may be similar to real releases. Other computer tools are helpful in managing communication and data during emergencies.

Supplies and Equipment Emergency supplies and equipment must be available when an emergency occurs. It is usually too late to obtain the correct items after an emergency exists. Emergency plans need to identify what supplies and equipment are needed, in what quantities, and at what locations. If supplies and equipment are not on site, contracts must be in place for immediate delivery on demand. Planning should consider potential problems with deliveries because if delivery delays are possible, planners should consider propositioning and storing needed items where they are easier to obtain and near the points where they will be needed.

Supplies could include medical and first aid supplies, neutralizing agents for spills, sand or salt for sleet or snow conditions, and sand for sandbags. Equipment could include rescue equipment, fire extinguishing equipment, traffic control devices, and barricades. Powered or manual materials-handling equipment, like cranes, wreckers, end loaders, and shovels, also may be needed.

Supplies and equipment may include food, drink, and cooking equipment, tables and chairs, and cots for workers to rest or to temporarily house stranded people.

Locations and Facilities There should be locations for emergency teams to prepare for work or from which communications teams operate. Designed locations in existing areas may serve nicely and easily may be converted to alternate uses. In some cases, there should be mobile facilities. Sanitary facilities should be available. All emergency facilities should be safe from dangers. There may be a designated and equipped command and control location.

Medical Services Any emergency planning must consider the potential for injuries to people. Medical staff, first aid staff and evacuation teams, rescue equipment, and vehicles should be part of any emergency plan. Hospital emergency rooms, operating rooms, or decontamination facilities may be necessary. Travel routes and alternate routes must be considered. There should be alternate sites to support special types of injuries, like burns or radiation injuries, and there should be backup sites to support overloads at primary support sites.

Training Emergency plans need to include training requirements for participants. Planners should identify clearly the knowledge, skills, and abilities for each "player." There should be training records for each participant or team and there should be training scenarios that one can test. Emergency 911 call centers need information to alert appropriate authorities when callers report certain kinds of emergencies.

Design

After analysis identifies potential emergencies and plans detail what needs to be in place, it may be necessary to change some facilities and equipment to meet the identified needs. Building evacuation routes and exits may need changes or special features may need to be installed to ensure that proper emergency components are in place. For example, regular phone lines may go out when there is loss of power. Even cell phones may not operate under certain kinds of emergencies. A backup power supply and supporting fuel supplies should be in place. Emergency lighting may be needed or there may be a need for backup power for computer system or processing equipment to be sure that operations are reduced to safe levels.

Access routes for emergency equipment may be inadequate and need change. The capacity of pedestrian routes or shelters may have to be updated to meet the occupancy level of expanded facilities and operations.

Training and Execution

A key to making any emergency plan work is training. To attempt execution of a plan without training will probably result in failure, at least in part. Each plan may be different, and individuals and teams need training for their role in each plan. There should be opportunities to test equipment and to assess supplies that can deteriorate and become unusable if stored for too long.

Training programs should make participants in emergency responses knowledgeable about the hazards. For example, local fire departments may not be familiar with fire equipment on site or they may not have training in dealing with the unique hazards within a plant.

29-5   RESOURCES

There are a variety of resources available in planning and dealing with emergencies. This section will note a few. One resource is governmental bodies. At the federal level, there is FEMA. There are similar organizations at state and local levels that can help. Local police, fire, and medical services may provide help.

There are private organizations that provide contracted services for planning and management of emergencies. There are information resources on hazardous substances, such as the Chemical Transportation Emergency Center (CHEMTREC).[4] It offers a public service hotline for fire fighters, law enforcement, and other emergency responders and for manufacturers and shippers to obtain information and assistance for emergency incidents involving chemicals and hazardous materials. Callers can obtain information on nearly three million material safety data sheets. The American Institute of Chemical Engineers in New York has a Center for Chemical Process Safety that prepares guidance documents.

There are a number of industry and professional membership organizations that provide publications or other assistance with emergency planning. A few examples are the Amercian Chemistry Council and the Center for Chemical Process Safety of the American Institute of Chemical Engineers.

29-6   A DISASTER DILEMMA

Technology solves some problems, but creates others. Trust in technology may lead to emergencies and disasters that may not have otherwise occurred. This is not something new. An engineer sold the town elders of Dixon, Illinois, a solution to their 600-ft river span: an iron bridge. On May 3, 1873, during the dedication of the new bridge, 200 people fell into the river and at least 38 died when the bridge collapsed.

The invention of the automobile was hailed as a solution to the traffic congestion and pollution of horse-drawn vehicles. At the turn of the century in New York City, horses deposited 2.5 million pounds of manure and 60,000 gallons of urine per day on the streets. Now the automobile has led to pollution problems, gridlock, and nearly 50,000 deaths per year in the United States.

During the twentieth century, the United States spent billions of dollars to control the erratic behavior of waterways. It appeared that levees could control rivers and coastal areas. However, hurricanes and abnormal water loads still produce wide destruction. Some of the damage results from people building on reclaimed land with an expectation that the area is safe.

The 1984 chemical plant tragedies in Mexico City and Bhopal, India, would probably have been much less destructive had there been restrictions on dense populations living adjacent to the plants. Both locations had squatters' shacks in close proximity to the industrial complexes.

The past events should not dampen the use of technology to solve the problems of the human race. Perhaps with greater foresight and a wider view of potential consequences, we would need fewer emergencies and disasters to teach us how to make technology safer. Planning for emergencies should also lead to preventive actions so that emergencies are less likely to occur.

EXERCISES

1. Locate the local public emergency response organization. Find out how it is organized and operated. Determine what kinds of emergencies it has plans for.

2. Obtain a copy of an emergency response plan from a company. Evaluate its contents.

3. Select a natural and a human-made emergency for your area. Develop an emergency plan for each.

4. Select an emergency or disaster, such as a hurricane that struck an east cost or Gulf of Mexico city. Identify what damage may have resulted from a trust in technology and what regulations were modified after the disaster to prevent similar damage in the future.

REVIEW QUESTIONS

1. What is an emergency?

2. What is a disaster?

3. Where and to whom can emergencies happen?

4. Name five kinds of emergencies and give an example of each.

5. List the four priorities of an emergency in order from highest to lowest importance.

6. What kinds of emergencies do federal regulations cover?

7. List four actions necessary to prepare for emergencies.

8. List seven factors that must be considered in planning for emergencies.

NOTES

1 Chemical Emergency Preparedness Program, U.S. Environmental Protection Agency, Washington, DC.

2 Community Awareness and Emergency Response (CAER) Case Studies, Synthetic Organic Chemical Manufacturers Association, Washington, DC.

3 44 CFR 315 and *Federal Register* (1984) 49, No. 191:38596.

4 CHEMTREC, Arlington, VA.

BIBLIOGRAPHY

BURKE, ROBERT, *Hazardous Materials Chemistry for Emergency Responders*, 2nd ed., CRC Press, Boca Raton, FL, 2003.

CHARLES, M. T., CHOON, J., and KIM, K., *Crisis Management: A Casebook*, Charles C. Thomas, Springfield, IL, 1988.

Great Fires of America, Country Beautiful Corporation, Waukesha, WI, 1973.

HEALY, R. J., *Emergency and Disaster Planning*, Wiley, New York, 1969.

HOFFMAN, J. M., and MASER, D. C., eds., *Chemical Process Hazard Review*, American Chemical Society, Washington, DC, 1985.

KELLY, R. B., *Industrial Emergency Preparedness*, Van Nostrand Reinhold, New York, 1989.

KENNETT, F., *The Greatest Disasters of the 20th Century*, Castle Books, distributed by Book Sales Inc., Secaucus, NJ, 1975.

KLETZ, T., *What Went Wrong: Case Histories of Process Plant Disasters*, 2nd ed., Gulf Publishing Company, Houston, TX, 1989.

KRIVAN, S. P., "Avoiding Catastrophic Loss: Technical Safety Audit and Process Safety Review," *Professional Safety*, 31, No. 2:21–26 (1986).

LAUGHLIN, JERRY, and TREBISACCI, DAVE, eds., *Hazardous Materials Response Handbook*, 4th ed., National Fire Protection Association, Quincy, MA, 2002.

LEVY, MATTHYS, and SALVADORI, MARIO, *Why Buildings Fall Down—How Structures Fail*, W. W. Norton & Company, New York, 1992.

LOWRY, G., and LOWRY, R., *Handbook of Right-to-Know and Emergency Planning*, Lewis Publishers, Chelsea, MI, 1988.

McCULLOUGH, D., *The Johnstown Flood*, Simon & Schuster, Inc., New York, 1968.

MUSSELMAN, V. C., *Emergency Planning and Right-to-Know: An Implementor's Guide to SARA Title III*, Van Nostrand Reinhold, New York, 1989.

NAEIM, F., *The Seismic Design Handbook*, Van Nostrand Reinhold, New York, 1989.

PERRY, R. W., and LINDELL, M. K., *Handbook of Disaster Response Planning*, Hemisphere Publishing Corporation, New York, 1988.

REED, R. C., *Train Wrecks*, Superior Publishing Company, Seattle, WA, 1968.

STEIN, L., *The Triangle Fire*, Carroll & Graf Publishers, Inc., New York, 1962.

TERRIEN, E. J., *Hazardous Materials and Natural Disaster Emergencies, Incident Action Guidebook*, Technomic Publishing Co., Inc., Lancaster, PA, 1984.

The AIHA 2003 Emergency Response Planning Guidelines and Workplace Environmental Exposure Level Guides Handbook, American Industrial Hygiene Association, Fairfax, VA, 2003.

VULPITTA, RICHARD T., *On-Site Emergency Response Planning Guide*, National Safety Council, Itasca, IL, 2003.

ZEIMET, DENIS E., and BALLARD, DAVID N., *Hazardous Materials Behavior and Emergency Response Operations*, American Society of Safety Engineers, Des Plaines, IL, 2000.

CHAPTER 30

FACILITY PLANNING AND DESIGN

The best time to incorporate safety into a facility is during the planning and design of a new facility or the modernization of an existing facility. Too often, designers consider safety as a design afterthought. Some designers do not find safety important until after construction is underway or after a facility is completed. Incorporating safety during design makes economic sense because it is much cheaper to make changes during design than to negotiate change orders with a contractor or to modify a facility after completion.

30-1 FACILITY DEVELOPMENT PROCESS AND SAFETY

Figure 30-1 represents one way to describe the facility procurement process. Because conditions and requirements change, the process of keeping facilities supportive of an organization and its operations is cyclical and continual. Organizations obtain facilities through new construction, leasing, and modification of existing facilities.

Safety, health, and environmental factors must be considered throughout the process design and development process. Safety cannot be assigned to one step or saved until after a project is completed.

Recognizing the Need

Typically, the process begins with recognition that existing facilities are inadequate. The need for a new or upgraded facility may result from inadequate safety features. The likelihood or consequences of a potential loss or the likelihood of a loss occurring may justify corrective measures in economic or other terms. Changes in regulations and standards may require facility modifications.

Planning and Budgeting

As soon as a facility need has been determined, preliminary planning sufficient to develop and justify a budget request must be carried out. This may involve a needs or feasibility study, which may include analyzing needs and evaluating alternatives for obtaining the needed facility. The planning process also should include a risk analysis. The major systems in a facility must be analyzed because safety affects these systems. For example, there may be a need for special fire protection features (sprinklers, sensors, and alarm devices). Other requirements may include special ventilation, lighting, emergency power, or shielding equipment. There may be a need for barrier walls or physical distance to separate hazards. All of these safety essentials add to the budget for the project.

Safety and Health for Engineers, Second Edition, by Roger L. Brauer
Copyright © 2006 John Wiley & Sons, Inc.

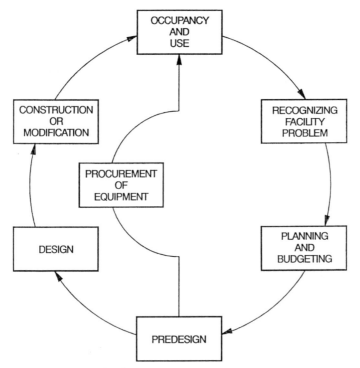

Figure 30-1. The facility procurement process.

Predesign

As soon as a course of action is selected, there are further details to work on, including preparation of detailed requirements and technical information for design. For example, one needs to compute the water supply requirements, to identify which spaces or operations need emergency lighting or power, and to complete a preliminary hazard analysis for each operation or space. From the analysis, there are corrective features to be incorporated into the design of the facility and its systems.

Design

After compiling detailed requirements, project engineers and architects prepare one or more preliminary design concepts. Then, they pursue one approved design concept in detail. Designers try to meet as many requirements as possible. Not all can be met and there are tradeoffs. However, there are some requirements that are not negotiable and must be met. Analyzing and prioritizing safety requirements will help designers deal with important safety features.

    The design process must include hazard analysis, and where possible, designers must select corrective measures and integrate them into the design. However, the selected controls for some hazards may not be a part of a facility design. For example, the controls may be procedures or personal protective equipment. The design must deal with these controls through design documentation, warnings, instruction manuals for the facility, and similar means.

Selection and Procurement of Equipment

Closely associated with the facility design is selection and procurement of equipment that will be used in the facility. For process facilities, the planning, budgeting, requirements analysis, and design of equipment are an integral part of the facility design. Designers identify hazards associated with the equipment and specify what controls to procure with purchased or fabricated equipment for the facility project. Decisions about hazards and controls cannot be left to purchasing agents. Safety requirements and responsibility for determining compliance with safety requirements in facilities belongs to designers.

For facilities that have uses other than manufacturing and production, analysis of activities that will occur in a facility and equipment that users may bring in should be part of the design process. What hazards equipment users may bring into a facility and what to incorporate in the facility design also should be considered. It may be important to identify what hazards do not have controls in the facility and what controls user equipment must have.

Construction or Modification

After design and funding, the facility project moves to construction. The project may be new construction or modification of an existing facility. Contractors also install new or modified equipment. Part of the construction or modification is quality assurance. Designers should implement steps to ensure that a design is built correctly. This may require testing, evaluation, and formal acceptance procedures. The designer must be sure that contractors do not make changes that introduce dangers and that safeguards are correctly constructed and installed.

Occupancy and Use

After completing a facility project, occupants move in and the facility is put to use. It now should meet its functions safely, although after moving in, there will be some fine adjustment, which may include safety features. Users may change operations and activities, thereby reducing the effectiveness of some safety features or requiring new ones. Continual monitoring of hazards and controls after moving in will prevent introduction of hazards from changes in operations and activities or minimizing them.

30-2 TOOLS TO HELP SAFETY IN DESIGN

There are a variety of tools and procedures that help designers reach solutions that satisfy as many requirements as possible. Some tools, such as the energy theory (see Chapter 3) and the four Ms and goal accomplishments models (see Chapter 9), aid in hazard recognition.

Later chapters (Chapters 35–38) describe procedures and methods for identifying hazards and selecting and ranking controls. These apply knowledge about hazards and controls covered in other portions of the book.

There are many references that deal with the control of hazards in facilities, and this book covers many in other chapters. There are laws, regulations, and consensus standards for many kinds of hazards. There are references containing design guidance and criteria that capture the experience of others. Many of these references deal with particular kinds of facilities. Table 30-1 is a checklist derived from a survey of process company experi-

TABLE 30-1   Checklist of Hazard Factors for Facility Design and Operation[a]

Plant site problems
Unusual exposure to natural emergencies
Inadequate water supply and other utilities
Exposure to hazards from nearby plants
Unreliability of public fire and emergency
  protection
Traffic difficulties limit emergency equipment
  access
Air and water pollution problems
Inadequate waste disposal facilities
Climate problems requiring moving hazardous
  processes indoors
Poor drainage

Inadequate plant layout and spacing
Congested process and storage areas
Lack of isolation for very hazardous
  operations
Exposure of high value and difficult-to-replace
  equipment
Lack of adequate emergency exit facilities
Insufficient space for maintenance or
  emergency operations
Source of ignition too close to hazardous
  materials
Critical plant areas exposed to hazards
Inadequate hazard classification of plant areas
Lack of isolation of critical plant areas from
  community

Structure not in conformity with use requirements
Disregard for code requirements
Lack of fire-resistive structural support
Failure to provide blast walls or cubicals to
  isolate extra hazard operations
Inadequate explosion venting and ventilation of
  buildings
Insufficient exit facilities
Electrical equipment does not meet code
Unprotected critical wiring
Inadequate hazard anticipation (i.e.,
  explosion)

Inadequate material evaluation
Fire, health, and stability of materials not
  evaluated
Inadequate controls for quantities of materials
Inadequate evaluation of processing
  environment for hazards of materials
Lack of information on dust explosion
  properties of materials
Inadequate evaluation of health hazards of
  materials
Incomplete inventory of hazardous materials in
  the plant

Lack of long-term exposure information
Improper packaging and labeling of
  chemicals

Chemical process problems
Lack of information on process temperature
  and pressure variations
Hazardous by-products or side reactions
Inadequate evaluation of process reactions
Lack of identification of processes subject to
  explosive reactions
Inadequate evaluation of environments
Overlooking conditions for extreme process
  conditions
Lack of evaluation of vapor cloud hazard

Material movement problems
Inadequate control of chemicals during night
  operations
Inadequate control for hazardous dusts
Piping problems
Improper identification of hazardous material
  during transportation
Loading and unloading problems inadequate
  control of heat transfer
Flammable gases and vapor problems in
  pneumatic conveyors
Waste disposal and air pollution problems
Vapor cloud problems

Operational failures
Lack of detailed descriptions and procedures
  for operating all sections of the plant
Poor training program
Lack of training on health hazards
Lack of supervision
Inadequate start-up and shutdown procedures
Poor inspection and housekeeping
Inadequate permit system
Lack of emergency control plans
Inadequate drills
Lack of medical and biological surveillance

Equipment failures
Hazards built into equipment
Corrosion or erosion failures
Metal fatigue
Defective fabrication
Inadequate controls
Process exceeded design limitation
Poor maintenance program
Inadequate repair and replacement program
Lack of "fail-safe" instrumentation
Poor check on construction criteria or material
  specification

TABLE 30-1   continued

Equipment not capable of toxic or hazardous materials	Poor check on boiler and machinery risks
Ineffective loss prevention program	Lack of preemployment physical examinations and periodic checkups
Inadequate support of top management	Lack of training on health hazards and use of personal protective equipment
Lack of assigned responsibility	
Poor accident and hazard prevention program	Lack of conformance with government regulations
Insufficient fire protection manpower, equipment, and organization	Insufficient in-plant monitoring of physical and chemical hazards
Ineffective explosion prevention and control program	Individual hazard approach instead of "total loss control" concept
Lack of emergency planning	

[a] Derived from *Hazard Survey of the Chemical and Allied Industries*, American Insurance Association, New York, 1979. The items resulted from a questionnaire survey of failures of many plants and operations.

ences. The survey compiled data on things that went wrong with process facilities. Organizations, like Underwriters Laboratory[1] and Factory Mutual System,[2] offer both publications and experience that can help identify safety problems with facility designs.

Another important tool for incorporating safety in facility design involves the makeup of the design team. Safety and health specialists should be a part of the design team.

Ideally, there should be a safety specialist on the design team who should be knowledgeable in safety, health, the environment, and fire protection. The practice of having safety specialists involved in design is growing. For example, many petroleum and chemical companies have safety engineering specialists involved in the planning and design of processing plants and facilities. The Department of Defense requires that system safety be applied to major facility projects. The Army Corps of Engineers created a computer data bank of lessons learned about facility safety that allows people working with them to search in various ways for facility safety problems that others have experienced.

If safety specialists are not part of a design team, at least there should be a thorough safety review of a design by a specialist. The review should cover safety features built into the facility and safety during use of the facility. The review should consider operation, maintenance, repair, and servicing of the facility, its subsystems, and equipment. The design must cover routine use of a facility and activities of those who are its caretakers. It must cover special activities important for emergencies and rescue.

A design can be broken into several components, and one way to organize the components is by scale. One can evaluate the *site* and *siting*; one can evaluate the *building* or *facility*; one can check the facility *interior* and *occupancy*. Sometimes evaluation of *work stations* is important. One must look at particular *equipment*. One must consider the *operations*, *processes*, or *activities* for each of these components. Table 30-2 provides a checklist covering some considerations when planning a site and a building.

A review of some facility factors about safety is given in the following section. This review gives an example of an approach by facility scale. Particular operations, processes, and activities will include additional considerations. The review reflects some information presented in other chapters.

TABLE 30-2 Incomplete List of Safety Considerations in Facility Planning

Site

Drainage	Prevention of spills, leaks, or activities that may contaminate storm water or to control runoff
	Flood control and protection
Utilities	Prevention of damage to utilities (power lines, pipelines, communication lines)
	Remote shutoff of utilities
	Barricades to protect people from utilities
	Water supply for operations and fire protection
	Storage of fuels on site
Traffic	Traffic load on or adjacent to site (pedestrian, vehicular, railroad, public, employees, delivery, etc.)
	Need to separate kinds of traffic
	Types and quantities of materials entering or stored on site
	Access for emergency vehicles and equipment
Hostile conditions	Wind loads and conditions
	Snow loads
	Earthquake zones
	Lightning protection
	Protection from sun, water, weather, or other hazards
	Physical security
Site storage	Storage of fuels, water, hazardous materials on site (type and quantity)
	Separation requirements
Adjacent properties and community	Barricades and fences to keep unauthorized people from dangerous areas or materials
	Protection from hazards of adjacent or nearby properties or transportation routes
	Dangers to adjacent properties from site or operations
	Emergency response plan
	Community right-to-know requirements
	Emissions and controls
Fire and emergency	Access for emergency vehicles and equipment
	Fire load
	Fire response management
	Fire suppression and extinguishment

Buildings

Walkways, stairs, and access	Adequate lighting at entries, transition points, stairs, landings, etc.
	Guard rails for elevated surfaces
	Handrails for stairs and certain types of pedestrians
	Handrail dimensions for firm grip
	Floor finishes to minimize slipperiness, prevent corrosion, etc.
	Drainage or raised flooring for wet or oily areas or where foreign material is slippery
	Spill isolation to prevent distribution of contaminants
	Minimize changes in surfaces (elevations, irregularities, slipperiness)
	Aisle and area markings for pedestrian areas, vehicles, materials, handling, or hazardous operations
	Door sizes for vehicles and materials
	Safe access for building and equipment maintenance
Hazard zones	Enclosure, isolation, separation, or other controls for hazardous sources (noise, gaseous or particulate contaminants, etc.)
	Exhaust ventilation, hoods, or cabinets for hazardous contaminants
	Barricades for hazardous operations
	Sensors and monitoring equipment for hazardous materials

TABLE 30-2   continued

	Warning signs that meet various standards and markings to identify hazardous areas
	Communication systems
Fire protection	Fire loads
	Smoke and heat detectors
	Alarm systems and evacuation management systems
	Sprinklers and other fire suppression equipment
	Fire ratings for materials, finishes, and partitions (walls, ceilings, floors, doors)
	Exit signage
	Roof vents and curtain boards
	Explosion venting
	Fire control equipment and systems for flammable operations
	Utility shut off systems for emergencies
Sanitation and first aid	Sufficient washrooms
	Clean areas for food and eating, including isolation from sources of contamination
	Lockers, change rooms, and showers for control of contaminants
	Emergency showers
	Emergency eyewash fountains
General	Enclosure of dangerous equipment and building systems
	Maintainability to minimize repair, cleaning, and servicing hazards
	Storage areas for hazardous materials
	Proper separation of noncompatible materials
	System, equipment, and operations shutoffs in nonhazardous areas

30-3   SITE CONSIDERATIONS

Location

There are many location factors that impact plant design and potential hazards.

Climate and Natural Conditions   In planning a facility, there is a need to know about wind, particularly if there are any potential releases from the facility. Drainage is important and there should not be an accumulation of water from processes, leaks, or storms. Hazardous materials should not contaminate runoff water, and stored materials should not create pollution. It is important to know if there is a flood potential or dangers from flight paths and other forms of transportation. Soil characteristics, water table, and ground water are important factors in case of spills.

Neighborhood and Population   The population living or working in the vicinity should be known. It is necessary to know if there are periods when there are high densities of people from traffic or special activities, if there are playgrounds or schools nearby, and if there are hazards from nearby plants and operations.

Size of Site

If there are hazards such as noise, explosions, or heat, the distance to the periphery of the site and between parts of the plant or facility must be great enough for protection.

Access and Circulation

The site should have adequate access for facility occupants, emergency, equipment and delivery vehicles. If possible, circulation routes for pedestrian traffic should not have to cross vehicular traffic routes. Routes for emergency access should not be blocked by normal or peak traffic. Materials handling, delivery, and rail traffic should be isolated from employee and visitor parking and walkways.

Layout

The site arrangement should isolate hazardous materials and operations from nonhazardous activities. The operations should have smooth flow. As few people as possible should have exposure to any hazard. Quantities of materials, particularly those with high energy content, should be separate from activities that could trigger fire.

Utilities

Utilities should be located to avoid creating hazards. For example, gas lines should not be near rail lines or public or employee areas. Shutoff locations should be separated from areas where the utilities may be hazardous, and shutoff controls should be accessible to emergency crews even during incidents. Lines should be protected from vehicle damage. Utilities, such as water, should be sufficient for emergency needs from supply mains to use points in the facility. Consideration should be given to the effects of power outages or shutdown of ventilation systems, lighting, process equipment, and computers. There may be a need for uninterrupted power supply, alternate power, or generator sets and battery systems, such as those for emergency lighting.

Storage

Sufficient storage capacity for quantities of materials should be available. Hazardous materials should be isolated from other materials and areas of the site. Storage equipment and layouts should minimize traffic problems for materials handling equipment.

Security

The need for security of the plant site and elements on it should be considered. For example, the facility may have control points for anyone entering the site, for access to plant areas but not administrative areas, or for areas where there are special hazards. Hazardous areas may require fencing and other physical security equipment to monitor unauthorized access.

Many of the aforementioned factors apply to the design of a building on a site. Additional considerations are listed in the following text. They are certainly not all the safety factors that are important in designing a building.

30-4   BUILDINGS AND FACILITIES

Layout

Designers should separate processes that have noise, heat, or cold, require ventilation, or have other environmental hazards from areas that do not have these hazards. Partitions most be provide to isolate fire hazards.

Access and Circulation

Review whether emergency equipment and personnel have easy access to all locations. Arrange circulation to prevent traffic congestion and conflicts between vehicular and pedestrian traffic. There should be sufficient exit routes and exit units. Doors should also have enough width and height for vehicles and materials handling. Circulation areas should be marked or delineated clearly.

Materials of Construction

Select materials to meet fire protection requirements or withstand corrosive or reactive materials that may be used and minimize dangers from operations.

Flooring

Flooring should be selected for many factors, especially safety. Ensure that floors will carry the anticipated loads and analyze the structure before authorizing new uses for a facility designed for other uses. Surfaces should be slip resistant and there should be no sudden changes in slipperiness coefficients. Small changes in elevation (one or two steps) should be avoided, but if necessary, clearly visible. Open stairs and other openings should be protected. Avoid patterned finishes on stairs and sudden changes in view on stairs that create visual distractions.

Ventilation

Check operations for generation of heat, gases, vapors, or airborne contaminants and place them in locations to minimize the portions of the building affected. Capture contaminants at the source to minimize the volume of air that must be treated. Consider the effects of spills and leaks and how they will be managed.

Lighting

Besides adequate lighting for routine activities, watch for transition zones near entries. Make sure these zones allow for eyes to adjust, particularly if there are stairs or other walking hazards. Determine if interior spaces have adequate emergency lighting to allow for safe exiting.

Storage

Analyze the types and quantities of materials that may be present. Plan storage locations for each type of item. Separate incompatible materials, such as oxidizers and fuels. Provide adequate storage equipment and racks to keep materials organized. Storage areas should be clearly marked.

Communication

Consider routine and emergency communications. If there are potential emergencies, provide communication systems that can reach occupants to inform them of dangers and corrective actions to take. Voice systems may need equipment dedicated for emergencies to ensure that emergency communication does not compete with regular equipment use.

Video systems may be needed for security and computer systems for monitoring equipment for safe operations.

Fire Protection

Analyze the fire loads and fire controls. Determine if compartmentation is adequate. Analyze water supplies for adequacy at all locations where extinguishing water is needed. Consider the value of sprinkler systems and the type best suited to each location.

30-5 WORK STATION CONSIDERATIONS

Work station design must address details of particular tasks. There should be room to sit and stand. Seating should be comfortable and adjustable. Handling of materials should be minimized. Furniture and layouts should avoid the need to twist, turn, bend, or stoop. Ergonomic considerations (Chapter 33) are important in work station designs. Facility designers should specify safety features for furnishing. Requirements should not be left to procurement people.

30-6 EQUIPMENT AND PROCESS CONSIDERATIONS

There are many safety considerations for equipment. Proper controls, guarding, noise characteristics, electrical grounding, and other factors are important. One must apply techniques to identify hazards and risk and options for eliminating or reducing risks. Chapters 35 and 36 review some of these approaches.

For process equipment, designers need to specify what safety features are needed and what tests will determine if requirements are met. For process equipment, there should be fail-safe features. Fire protection, overpressure, excess heat, runaway reactions, dust control, exhaust ventilation, dangers of flammable liquids, leaks, sensing devices to report status, and many other safety features are important. Designers need to consider access for setup, maintenance, and cleaning. Access by stairs, fixed ladders, or platforms should be part of equipment where applicable. Chapter 36 covers some of the approaches for complying with the OSHA Process Safety Standard. References from the Center for Chemical Process Safety provide considerable details for identifying and controlling hazards process plants.

EXERCISES

1. Obtain the drawings and specifications for a public assembly, commercial, or retail building. If documents are not available, visit an existing building. Evaluate the design for safety of employees and public users and their activities. Prepare a report of findings indicating to the designer what is wrong or can be improved for safety and what corrections you recommend.

2. Meet with designers from an architect-engineering firm. Ask them how they incorporate safety into their facility planning and design process.

3. Review a process safety audit report. Summarize the procedures used and the resulting recommendations. Find out how the results were actually implemented.

REVIEW QUESTIONS

1. When is the best time to put safety features into a facility?

2. Describe major steps in the facility development process and safety considerations for each.

3. What are some resources for information about safety in facility design?

4. Identify three safety considerations for each of the following:

(a) site

(b) building

(c) work station

(d) equipment

NOTES

1 Northbrook, IL.

2 Norwood, MA.

BIBLIOGRAPHY

BAASEL, W. D., *Preliminary Chemical Engineering Plant Design*, 2nd ed., Van Nostrand Reinhold, New York, 1989.

CHARNEY, W., *Complete Guide to Hospital Safety*, Lewis Publishers, Chelsea, MI, 1990.

CRALLEY, L. V., and CRALLEY, L. J., *Industrial Hygiene Aspects of Plant Operations*, vol. 1, *Process Flows*, 1982; vol. 2, *Unit Operations and Product Fabrication*, 1984; vol. 3, *Engineering Considerations in Equipment Selection, Layout and Building Design*, 1985; Macmillan, New York.

CROWL, DANIEL A., and LOUVAR, JOSEPH F., *Chemical Process Safety—Fundamentals with Applications*, 2nd ed., Prentice Hall PTR, Upper Saddle River, NJ, 2002.

Guidelines for Process Safety Fundamentals in General Plant Operations, Center for Process Safety of the American Institute of Chemical Engineers, New York, 1995.

Hazard Survey of the Chemical and Allied Industries, American Insurance Association, New York, 1979.

HOPF, P. S., *Designer's Guide to OSHA*, 2nd ed., McGraw-Hill, 1987.

KLEIN, B. R., ed., *Health Care Facilities Handbook*, 2nd ed., National Fire Protection Association, Boston, MA, 1988.

LEES, F. P., *Loss Prevention in the Process Industries*, 2 vols., Butterworths, Boston, MA, 1980.

MACKIE, J. B., and KUHLMAN, R. L., *Safety and Health in Purchasing/Procurement/Materials Management*, International Loss Control Institute, Loganville, GA, 1981.

MECKLENBURGH, J. C., ed., *Process Plant Layout*, Wiley, New York, 1985.

PIPITONE, D. A., *Safe Storage of Laboratory Chemicals*, Wiley, New York, 1984.

STANLEY, P. E., *Handbook of Hospital Safety*, CRC Press, Boca Raton, FL, 1981.

STEERE, N. V., *Handbook of Laboratory Safety*, 2nd ed., CRC Press, Boca Raton, FL, 1971.

STONER, D. L., SMATHERS, J. B., HYMAN, W. A., DUNCAN, D. D., and CLAPP, D. E., *Engineering a Safe Hospital Environment*, Wiley, New York, 1982.

WELLS, G. L., *Safety in Process Plant Design*, Wiley, New York, 1980.

PART IV

THE HUMAN ELEMENT

THIS SECTION of the book deals with contributions of people in achieving safety.

There are two causes for accidents: unsafe conditions and unsafe acts. Engineers deal mainly with unsafe conditions. The major role of engineers is prevention through hazard recognition and controls in design of equipment, environments, vehicles, and facilities.

Preventing unsafe acts often is viewed as largely a people problem. Rarely, however, is either an unsafe act or an unsafe condition the lone cause of an accident; the causation and correction approaches normally involve an interaction between the two. To prevent accidents by preventing unsafe acts, one must prevent behaviors that lead to accidents or mitigate the effects of unsafe acts in the causal chain. To deal with unsafe acts and their roles in accidents requires an understanding of human behavior.

Engineers can contribute to safe behaviors through design because designs can eliminate the need for unsafe behaviors. Designers need to understand human behavior and human capabilities and limitations. By making designs fit people, rather than changing people to fit designs, engineers can reduce the role of unsafe acts in the accident equation.

This section looks at procedures and training to be sure that people follow safe procedures. There are methods for identifying what behaviors are correct and safe for various jobs. This section also considers some characteristics of people and the importance of design features that can help minimize behavioral impact on accidents.

CHAPTER 31

HUMAN BEHAVIOR AND PERFORMANCE IN SAFETY

Case 1

The local newspaper reported a head-on vehicle crash that went virtually unnoticed. Robert W., the driver of an automobile, suddenly swerved into the path of an oncoming semi-trailer rig. The automobile was demolished and Robert W. was pronounced dead at the scene. The truck driver received facial injuries and two broken ribs.

Why did the accident happen? Did Robert W. suddenly become ill and lose control? Was there any icy patch on the pavement? Both were possibilities. However, only family and close friends had another theory for the accident: Robert W. had terminal cancer and had been depressed. He may have committed suicide.

Case 2

Maria B. operated a molding machine that turned out plastic parts. The machine normally operated in automatic mode. Dies came together horizontally, molten plastic was injected between the dies, and when the dies opened, the formed part fell to a parts box below. Somehow, Maria put her hand between the dies, the dies closed and crushed her hand.

Why did the accident happen? Did someone accidentally trip the machine? Why did Maria put her hand in the machine anyway? Why was the guard not in place? Would it have prevented the accident? The accident seemed unnecessary. It appeared that Maria had made a stupid mistake. Did she not read the warning sign that said, "Never reach into the machine while it is turned on"?

Maria was new on the job. In fact, it was her first day. She had been taught how to do the job just that morning and had been on her own for only 45 min. Maria did not speak English very well and could not read English. She was thrilled to have gotten the job so she could help out her poor family. Compared with previous work, this job paid well, and she wanted to do a good job for her new boss. She did not want to make a mistake.

The guard had been removed to set up a new process and the interlock, had been wired down to allow for testing by the tooling department during the morning shift. The idea was to make sure the machine operated correctly during the night shift using manual mode and then the next day to have the maintenance crew (who only worked days) replace the guard, reactivate the interlock, and return it to automatic mode.

When the machine started to malfunction, the dies opened but the part did not fall down like it was supposed to. Maria had learned nothing about that. She was trying to do her best. She was afraid her boss would be angry about bad parts. He was anxious when he showed her how to run the machine and was obviously under stress. As she reached into the machine to get the part loose, she accidentally bumped the foot control that tripped the machine.

To compound all the other factors, the control was the wrong one. It was designed for automatic mode. Manual operations for this machine required a hand control for protection, but the previous owners had set it up differently for automatic mode only and the hand controls were no longer on the machine. Maria's company was having trouble meeting production demands, had a large order to fill, and had just added a night shift. They purchased the machine at a bargain price because it was used and had the tooling department refurbish it and rush it into operation.

Case 3

The accident report seemed simple enough. Gary did not see the step, fell, and broke his arm. Gary came to work that morning, entered the building, and started down the hall to his office. Approximately 10 ft from the door, there was a step down into the new part of the building. Somehow he missed the step and fell forward. He put out his hand to stop the fall and broke his arm. That seemed straightforward. He made a dumb mistake; he was not paying attention.

You see, the day started terribly. Gary had not been getting along well with his wife and they were coming close to divorce proceedings. The night before, they had had another argument. In the morning, one of the children was sick and could not go to school. Gary had to find a baby sitter. A neighbor who usually helped out was gone for a couple of days, so Gary called his sister, who agreed to help out. However, she lived 30 min away. He rushed the child to his sister's and then drove to work.

That morning, there was an important meeting and Gary was scheduled to report on the performance of his production team. Already late, Gary arrived to find a delivery truck blocking his parking place, which forced him to park way in the back part of the lot.

As he ran from his car, the morning sun was virtually in line with the door of the building. He had trouble looking up as he approached the building and the sun glared into his eyes. After coming inside the entrance from the parking lot, it was somewhat dark. Gary had a lot on his mind thinking about the report, the sick child, and his strained relationship. Then he fell.

31-1 HUMAN BEHAVIOR

Why do people behave as they do? What makes them tick? Some people never seem to have a problem, whereas others seem to be accidents waiting to happen; some get along with everyone, whereas others are impossible. Human behavior is very complex and it is not fully predictable. Often, behaviors contribute to accidents. Behavior is affected by many things (see Table 31-1), including physiological condition, biochemistry, health, relationships with others, personal desires and goals, and so forth. This section looks at human behavior and important characteristics of it.

Theories of Behavior

There are many theories of behavior. Some are descriptive theories that allow characterization or classification of a person after observing that person's behavior. Some theories

TABLE 31-1   Performance-Shaping Factors[a]

Situational characteristics
  Temperature, humidity, air quality
  Noise and vibration
  Degree of general cleanliness
  Manning parameters
  Work hours/work breaks
  Availability/adequacy of supplies
  Actions by supervisors
  Actions by co-workers and peers
  Actions by union representatives
  Rewards, recognition, benefits
  Organization structure (e.g., authority,
    responsibility, communication channels)

Task and equipment characteristics
  Perceptual requirements
  Anticipatory requirements
  Motor requirements (speed, strength, precision)
  Interpretation and decision making
  Complexity (information load)
  Long- and short-term memory
  Frequency and repetitiveness
  Continuity (discrete versus continuous)
  Feedback (knowledge of results)
  Task criticality
  Narrowness of task
  Team structure
  Human-machine interface factors (design of
    equipment, job aids, tools, fixtures)

Job instructions
  Procedures required
  Verbal or written communications
  Cautions and warnings
  Work methods
  Shop practices

Psychological stresses
  Task speed
  Task load
  High jeopardy risk
  Threats (of failure, loss of job)
  Monotonous, degrading, or meaningless work
  Long, uneventful vigilance periods
  Conflicts of motives about job performance
  Reinforcement absent or negative
  Sensory deprivation
  Distractions (noise, glare, movement, flicker,
    color)
  Inconsistent cuing

Physiological stresses
  Fatigue
  Pain or discomfort
  Hunger or thirst
  Temperature extremes
  G-force extremes
  Atmospheric pressure extremes
  Oxygen deficiency
  Vibration
  Movement constriction
  Lack of physical exercise

Individual factors
  Previous training and experience
  State of current practice or skill
  Personality and intelligence variables
  Motivation and attitudes
  Knowledge of required performance standards
  Physical condition
  Influence of family and other persons or groups
  Group identification

[a] Derived from *Hazard Survey of the Chemical and Allied Industries*, American Insurance Association, New York, 1979. The items resulted from a questionnaire survey of failures of many plants and operations.

are predictive: they attempt to predict what a person will do given information about their past, their surroundings, or internal attributes. The information is obtained by introspective, subjective, or objective means.

Early theorists believed that behavior had biological origins. Theories related observable behavior to such things as instincts, habits, and conditioned reflexes from repeated stimuli. Later, other theorists looked to underlying elements within an individual that were not accessible through introspection by the individual. Others looked to many factors that together cause behavior: inherited traits and characteristics and environmental factors that lead to a person's behavior. The inherited traits may be both physiological and psychological. The environmental factors may be an accumulation of experiences and particular situations or conditions surrounding one at any moment.

Motivation

Motivation is that part of psychology that deals with getting someone to perform desired behaviors or actions. Motivation involves content and process theories. Content looks at the characteristics of an individual or his/her environment that stimulate performance or action and at what variables influence desired actions. Process looks at the linkages between content and specific actions and addresses the question of how to tap needs and outcomes to achieve desired actions. Although no theory of motivation is fully supported by research studies, some provide a framework for working with people toward desired actions and performance. A few theories are summarized in this section.

Maslow   Maslow developed a hierarchy of needs that has been quite popular. His theory is a content theory that looks within an individual for variables that effect desired performance. His hierarchy consisted of five classes of needs. He thought that needs at the base of the hierarchy had to be satisfied first, before higher ones were very meaningful. Higher ones became more important as lower ones were satisfied. His five classes of needs in ascending order are:

1. Physiological needs, such as hunger and thirst
2. Safety needs (primary body needs)
3. Social needs, such as friendship and affiliation
4. Esteem, including self-esteem and the esteem of others
5. Self-actualization, such as reaching one's potential

Research suggests that basic needs do not diminish as they are satisfied.

Herzberg   Herzberg's theory is a content theory that looks at work outcomes rather than needs. He proposed two types of outcomes that affect behavior: intrinsic factors and extrinsic factors. Intrinsic factors involve the work itself and recognition of one's work. Extrinsic factors include rewards associated with the work, such as pay, relations with co-workers and superiors, and working conditions. Whereas Herzberg believed that only attainment of intrinsic factors can sustain motivation toward organization goals, research suggests that both are important and there are significant differences among people in their preference for outcomes.

Vroom   Vroom addressed the motivation process. In his expectancy theory, there are three concepts. The first is the attractiveness of outcomes (valence of outcome). The theory does not concern itself with which outcomes. The second concept is the belief a person has about the link between an action and the outcome (instrumentality perceptions). For example, one may feel that achieving some performance deserves a raise. The third concept is a person's belief about the effort required in an activity and the likelihood of successful completion of the activity (expectancy perceptions). In summary, Vroom's expectancy theory states that when a person's expectancy perceptions for an activity are high and instrumentality perceptions linking the activity to attractive outcomes are high, the person will be highly motivated to engage in the activity. Related studies suggest that desired behavior is most often achieved if rewards are given every time the behavior is achieved, rather than occasionally. It is also important to state clearly the linkage between behavior and reward.

The implication of Vroom's theory and related work is that people can be motivated to perform when there are clearly defined linkages between behaviors, and rewards, the linkages are implemented consistently, and rewards are given regularly when a desired behavior is achieved.

Judgment

One definition for judgment is deciding or discriminating. It is the operations of the mind in which one compares information, evaluates values and formulates a decision, or reaches a conclusion. The decision or conclusion may be expressed verbally or may result in an action. Formulating judgments or reaching decisions can be deliberate or can extend over time. One may rely on information available from memory or drawn from careful compilation from various sources.

People differ in their ability to make judgments. Quick judgments and decision may be critical. In making quick decisions, one relies heavily on previous knowledge and experience available from memory. The action taken as a result of a judgment is more likely to be a desirable action when there is a rich background of knowledge, experience, and compiled information.

Emotion

People are not robots. People have feelings and emotions. Emotions may be experienced internally or exhibited through actions. Behavioral literature describes many kinds of emotions, including joy, fear, anger, grief, guilt, pride, love, hate, pity, and anxiety. Emotions may be generated by situations at home or at work and they may be associated with other people, with activities, or with conditions. Control of emotions and acceptable emotional expressions as well as control of the situations that generate them are important. Communications and management of interpersonal relations are means by which emotion-generated situations can be reduced. Emotions can be disruptive or facilitating, depending on the situation.

Attitudes, Opinions, and Beliefs

Attitudes, opinions, and beliefs are much the same thing: judgments or sentiments that the mind forms about something or someone. One also may hold attitudes about groups of people, social institutions, or issues. Attitudes may be positive or negative and are usually enduring. Attitudes an individual has can be inferred from their actions in certain situations and from verbal statements. Formal assessment of attitudes involves the use of carefully developed survey instruments. An attitude survey has many statements about situations or actions with which respondents agree or disagree. Results provide a picture of individual or group attitudes about situations covered in the survey.

Attitudes may be related to behavior. For example, one may have attitudes about another person, such as a supervisor. However, attitudes are not always a predictor of behavior. A person may know the effects of an action are bad, but continue to do it. Some call this cognitive dissonance. An example is a person who knows that smoking can lead to heart and lung disease, but continues to smoke.

Individual Differences

People are not alike. They differ in size, shape, strength, reaction time, physical condition, health, and physiological performance. They differ in ability to perform actions; in knowledge, skills, and abilities; in the ability to form judgments and make decisions; in attitudes and beliefs; in emotion; and in social and economic ways.

The differences are not fixed; they are variable. Individuals change over time. Some differences take care of themselves. The heart and respiratory rates are elevated when people exercise, but after resting, they return to normal. In other cases, people change through various means. Performance is changed through training, knowledge is changed through education, and some physical conditions are changed through medication. The important points are that people differ and individuals differ over time.

31-2 HUMAN BEHAVIOR AND SAFETY

Safe Behavior

Chapter 4 discussed the idea there are two causes for accidents: unsafe conditions and unsafe acts. Most of this book deals with unsafe conditions, their recognition, and control. A significant part of the accident formula is unsafe acts. Why do people perform unsafe acts? How does one prevent unsafe acts from occurring? These are behavioral issues. Understanding human behavior gives clues to managing behavior. The three Es of safety (see Chapter 3) suggest ideas to prevent unsafe acts. Education, enforcement, and engineering all have a role. Enthusiasm, a fourth E, has a role, too. Other concepts apply too.

Education Most behaviors are learned. Learning may be informal. Studies suggest that by age 6 years, people have acquired half their knowledge and skills. Children learn to walk and talk by trial and error. They obtain a great deal of reinforcement from those around them. Higher concepts and abstract learning usually occur in school. Education and training provide the knowledge and skills people require to act safely.

To avoid accidents and injuries, one must first recognize dangers in a situation. Not everyone brings the same knowledge and experience to a situation; not everyone will recognize or perceive a danger that may be inherently present or may develop. For example, some workers may not recognize that a guard should be in place on a machine because they may not have experience with equipment and may not recognize a danger or know what protection is appropriate. In another case, a dangerous situation develops rapidly. A child runs into the street after a ball. A driver may recognize the danger developing after seeing the ball roll into the street and the child near the curb. In another situation, the driver may not see a danger when a child is merely playing near a curb.

After recognizing a danger, individuals must act to protect themselves against an accident and possible harm. One must know what actions are correct and safe and one must complete the action required. Knowing the appropriate action and performing it correctly requires training and practice.

Enforcement Enforcement involves formalized rules and procedures and following them. Compliance can depend on self-discipline but more often, enforcement involves someone else auditing the actions of others. With enforcement, there may be some consequence for not acting properly. For example, one company gives drivers responsibility for their vehicles. Failure to drive properly results in loss of that job. Unsatisfactory performance of tasks can lead to other management-imposed outcomes both positive and negative.

Engineering In many cases, engineers can design to prevent certain behaviors from occurring. They also can design so that certain behaviors are not likely to cause the performer harm. This role for engineers will be discussed further in subsequent sections.

Communication

Communication is an important part of education. People cannot perform correctly if they are not told what dangers to look for, what procedures to follow, or how to act safely. Communication may involve training classes, supervisor instructions and comments, training videos and computer programs, published procedures and rules, and warnings and instructions. They may even involve simulations. People cannot be expected to decide and act on their own if they do not have the knowledge, skills, and experience to recognize a dangerous condition and to know what actions are appropriate as it develops or when it occurs.

Feedback

Knowledge of results—feedback—is an essential ingredient in learning. Correct behavior must be reinforced, and performance is greatly enhanced by knowledge of results. If someone does something correctly, they need to know; if they do it incorrectly, they also need to know. Safe behavior requires feedback on performance. Feedback on wearing of personal protective equipment—a safe behavior—is important in gaining user cooperation.

Several methods are used to provide feedback. Feedback may be verbal comments from someone else or reports of measured results of actions. For example, a report may contain the number of parts produced, the number of errors, or the accident rate. Feedback may be awards or rewards.

If individual performance is important, feedback should be given to individuals; if group performance is important, it should be directed to the group. In some cases, both individual and group feedback are needed.

Immediate feedback is generally better than delayed feedback. Actions can be divided into short increments or small elements, and feedback on these small components is usually better than feedback on large components. Feedback should be precise. If there is a particular task or component of a larger action, feedback should reflect correct or incorrect performance of the individual components. Reinforcement should be as often as practically possible.

Job Safety Analysis

Job safety analysis (JSA) is one technique to help identify what behaviors in an operation are safe and correct. It is a form of task analysis that is sometimes called job hazard analysis. In the analysis, one breaks down an operation into activities of workers. The analyst identifies the hazards associated with each activity in the operation, and for each activity, the analyst describes how to perform the job correctly and safely (see Figure 31-1). People have used a variety of forms for completing a JSA analysis. The hazard analysis and recommended practices can become part of a user manual, operations manual, or training program.

A JSA can be completed concurrently with other forms of task analysis common to industrial engineering practice. Such process analyses look at work flow, motion economy, time for each job element, eye movement, and hand and foot movement. A JSA should consider abnormal activities and conditions, not just normal, routine operations. It is often under the unusual situations (when things go wrong) that accidents and injuries occur. Even activities like cleaning and maintenance are nonroutine. People often make the wrong

WORKSHEET FOR JOB SAFETY ANALYSIS	Job		JSA by
	Supervisor	Date of Analysis	Reviewed by
	Department	Work Group	Approved by

Brief description of job, its beginning and end, and desired results

Required or recommended personal protective equipment for all tasks (Special requirements are noted with each task.)

SEQUENCE OF TASKS OR STEPS	POTENTIAL HAZARDS AND INCIDENTS	SAFE JOB PROCEDURES

Figure 31-1. Example of a job safety analysis worksheet.

decisions or take the wrong course of action in adverse situations. As discussed earlier, hazards during nonroutine, abnormal operations need to be protected by design.

Risk-Taking Behavior

For many activities, it is obvious that performing an action will not produce an accident or injury every time. A person will take a chance. For example, one does not get in an accident every time one rides a car. Therefore, one may reason that wearing a seat belt is not always necessary. Similar reasoning suggests that one can operate a machine without a guard in place, because one does not always become injured. People who are risk takers, are involved in accidents more frequently, and have higher absentee rates from work than those who are not risk takers.

Risk-taking behavior is greater under some circumstances than others. For example, most people will take greater risks when they have a choice, but are reluctant to take risks when it is required. Individuals are less likely to take risks when they are anxious and are more likely to take risks when they understand what is going on. For example, many people are afraid to undergo surgery. However, the more the individual knows about a surgical procedure, the less reluctance there is to undergo the surgery. Individuals are not willing to take risks when the status quo has strong value. People are reluctant to change, because there is often a fear of the unknown and they are satisfied with the way things are.

Risk taking is affected by people's perception of the risk. Table 31-2 lists risk perception factors that affect acceptance of risk situations. There are many things that affect

TABLE 31-2   Risk Perception Factors Affect Risk Acceptance[a]

More Acceptable	Less Acceptable
Voluntary	Involuntary
Natural	Human-made
Controllable	Not controllable
Delayed effect	Immediate effect
You and yours	Me and mine
Essential	Nonessential
Off-the-job	On-the-job
Misuse hazard	Proper use hazard
Affects few	Affects many
Effects reversible	Effects irreversible
Sensory perception	Unable to sense
Relates to self-worth	Does not relate to self-worth
Greater benefits	Lesser benefits
Experience	No experience
Understood	Not understood
Higher cost to fix	Lower cost to fix
Low consequence	High consequence
High probability	Low probability
No alternatives	Alternatives

[a] Buys, J. R., "Risk Perception, Evaluation, and Projection," Informal Report EGG-SHS-5975, Idaho National Engineering Laboratory, Idaho Operations Office, U.S. Department of Energy, Idaho Falls, ID, August, 1982.

people's estimates of risk. They may think that risk is greater than it really is or less than it really is. Table 31-3 lists several factors that affect the value people place on risk.

There are differences between group risk-taking behavior and individual risk-taking behavior. For example, fear tends to trigger group behavior. When fear is aroused, people choose to be together with others. Conversely, when people are anxious, they choose to be alone. Examples of groups are family, friends, work groups, command groups, or groups structured around coping with common or shared stresses or threats. Groups have informal structures, whereas organizations have formal structures.

Biorhythms

After the emergence of various behavioral theories, attempts were made to link the theoretical components to the likelihood of safe behavior or accidents. One concept that has drawn widespread attention is the use of biorhythms to predict the likelihood of accidents or other undesirable events caused or influenced by behavior or condition. Although some early studies appear to show that biorhythms affect accidents, more recent studies have not been able to show such affects.[1]

Biorhythms are not to be confused with biological rhythms. The theory of biorhythms stems from the early nineteenth century and it has so many followers that hand-held biorhythm calculators are easily purchased. The concept suggests three precise, fixed rhythms that originate at birth and affect events in an individual's life. These three rhythms have 23-, 28-, and 33-day periods or cycles, respectively. The congruence of the periods is said to affect events, moods, and actions of a person.[2]

TABLE 31-3   Risk Perception Factors Affecting Risk Estimation[a]

Underestimate	Overestimate
Known	Unknown
Understood	Not understood
Common	Uncommon
Mundane	Dramatic
Little media coverage	Much media coverage
Noncontroversial	Controversial
Me in control	You in control
Voluntary	Nonvoluntary
Fun risks	Work risks
Few injuries/events	Many injuries/events
Sensory perception	Unable to sense
Benign experience	Hurtful experience
"Scientific"	Not "scientific"

[a] Buys, J. R., "Risk Perception, Evaluation, and Projection," Informal Report EGG-SHS-5975, Idaho National Engineering Laboratory, Idaho Operations Office, U.S. Department of Energy, Idaho Falls, ID, August, 1982.

Alcohol and Drugs

Alcohol and drugs do contribute to accidents and injuries. There is a strong relationship between motor vehicle deaths and blood alcohol levels of drivers and between fire deaths and blood alcohol.

Employers face problems of employees drinking on the job or coming to work with alcohol in their blood. Results of one study indicate that employees who abuse alcohol are absent 16 times more often than those who do not, receive three times more sick leave, have four times the accidents, and are five times more likely to receive worker's compensation. Both street drugs and prescription drugs can increase the likelihood of accidents. When any drug reduces physical or mental performance, the chances of error, poor judgment, and accidents increase. Many companies have programs to assist employees with alcohol and drug problems that affect their work and the loss and claims rates. Some employers use drug and alcohol screening programs during hiring or employment. For some jobs that can affect the safety of others, laws may require drug and alcohol monitoring.

31-3   DESIGNING FOR HUMAN BEHAVIOR

There are many ways to remove or reduce hazards through design. Sometimes engineers forget to consider user capabilities and limitations, user behavior, and the use environment. Understanding people and their behavior is an important element of design. For example, running a pipe along a floor surface creates a tripping hazard. It does not make any difference that the activity near the pipe is a production activity or a maintenance and repair activity. The probability for an accident may be lower for certain activities, because walking near the pipe is less frequent. However, the hazard still exists. Any person walking or working near the pipe must avoid falling over it, that is, special actions to step over it are required.

Design problems may be even more subtle. A change in surface friction properties may create a slipping hazard. During initial steps on a surface, a walker gains a feel for the resistance underfoot. When there is a sudden change in resistance to a slipperier surface and the walker is not aware of the change, the gait must be adjusted suddenly from the first to the second condition. A failure to adjust can lead to feet slipping out from under the person and a fall. Similarly, a sudden change to a high friction surface will require adjustments. Failure to adjust may lead to a fall forward because the second surface prevents any movement between the shoe and the surface.

Designing for human behavior must anticipate foreseeable activities, and defining what is foreseeable requires a knowledge of what people do in various circumstances. It is not enough, for example, to safeguard machines for normal operations or production use. The designer must protect workers involved in cleaning, setup, and maintenance. In many cases, the designer can reduce hazards by incorporating features that are less dependent on people protecting themselves.

Designing for people must anticipate a range of ages and capabilities. Will the users be normal adults? Will the users have disabilities? What might the disabilities be? Could the users also be children? Will users be large or small? The field of human factors engineering or ergonomics addresses many of the capabilities and limitations of people and how to design with them in mind. Chapter 33 discusses ergonomics.

Dealing with these design problems requires analysis to identify the potential behaviors and errors in behaviors that can lead to accidents and injuries. Techniques to identify these behaviors include JSA or some derivative of it and testing of designs with users that adequately represent the population of potential users. Other methods may also be useful.

31-4 SAFETY AND COMMON SENSE

Some people have the notion that safety is nothing more than a lot of common sense. There are many problems with this approach when one considers human behavior.

What is common sense? The dictionary says it is sound, ordinary sense or good judgment. Common means a characteristic shared by a group at large, or belonging or pertaining to the community at large. Sense means sound perception or reasoning or correct judgment, or the ability to perceive or discern. It infers sensibility or a quick reaction to actions of objects or others.

The ability to perceive and recognize hazards is important to safety. To take corrective action, people need an ability to recognize the danger in a rapidly developing situation. People need skill in making good judgments or decisions about corrective actions to be safe.

One problem with common sense as a basic premise for safety is that human capabilities for achieving safe behavior are not universal. Individuals vary in their training, experience, knowledge, skill, and ability to recognize hazards, to perceive dangerous situations in a timely manner, to make sound judgments, and to take the correct protective action without error.

Most people would agree that children do not have common sense. When does one obtain common sense? How can you tell if someone has it?

Leaving safety to common sense suggests that somehow safety in a complex society will result if people are left to their own devices. Accidents are caused. Safety is achieved by thorough analysis, good design, and solid development of knowledge and skills through training and management. It is not an innate characteristic common to society. The desire to be safe is common; the actions required to achieve it are not.

31-5   TECHNOLOGICAL ILLITERACY

Another behavioral problem facing highly technical societies is a divergence between products and environments that depend on technology to make them work and to make them safe and the limited knowledge of the users and occupants that depend on that technology. Technology tends to raise the knowledge and skill required to use or control it. As a result, fewer and fewer people have the skills and knowledge to keep things safe; more and more people lack a good understanding of technologies and their potential dangers.

The average person who is cleaning something does not understand the chemistry of soap, solvents, acids, and alkaline materials. It is not uncommon to find people mixing different cleaning agents to gain more cleaning power, and too often the result is an accident and injury.

Many people do not understand mass, moments, inertia, potential and kinetic energy, or conservation of momentum. They wonder why they were unable to stop a falling object. They fail to recognize the danger of walking under a suspended load. They fail to see the danger of bicycling on the same pavement with 3,000-lb vehicles moving only inches away at speeds three or four times faster than they are. Lobby groups push lawmakers to make trucks 20% to 30% larger, and then the same lawmakers create rules that make passenger vehicles smaller.

Two national surveys that explored the technical knowledge and skills of the American population found that technological illiteracy is significant and somewhat related to literacy in general.

Fifty percent of American industrial workers have math skills at or below the eighth-grade level. Twenty-eight percent of people cannot make correct change when given a cash register receipt. Thirty-three percent of the respondents to one survey did not know how a telephone works, even though the device is much older than they are. Nineteen percent believed they had little understanding of radiation. Forty-one percent believed that rocket launchings and other space activities have caused changes in our weather. Twenty percent believed it was not wise to plan ahead, because many things turn out to be a matter of good or bad luck anyway.

Problems are compounded by the fact that literacy, including technological literacy, is lowest among the poor and uneducated. Studies of who is most affected by disaster indicate that the poor and poorly educated suffer the greatest losses. Perhaps there is a relationship between technological literacy and loss rates.

Given this context, designers have a challenge. They must make technology and technology-based products, systems, and environments safe, even for many who have little or no understanding of them. Recognition of hazards, making judgments, and taking corrective actions cannot be left to untrained users. Moving technology to third-world countries places even greater portions of a population at risk if safety is not built in or workers and communities are not trained.

31-6   JOB AND OTHER STRESSES

Another aspect of safety related to human behavior is job stress. Physical disorders that stem from behavioral problems, such as anxiety, fear, and other forms of psychological stress, are called psychosomatic disorders. The psychological condition manifests itself in physical disorders of various kinds.

Job stress is becoming more important in safety. It seems to increase with the increase in the number of high-paced, demanding jobs. In addition, more and more claims for job stress receive workers' compensation. Job stress is common in management positions. However, it occurs in many other kinds of positions as well.

Various conditions or situations in our lives cause our body to react. A scare causes the heart rate to increase, blood pressure to rise, and adrenalin to be secreted. Even sweating may start. Job situations may produce similar effects in the body. Certain tasks may be difficult to perform and produce similar reactions. There may be deadlines to meet, difficult social situations, difficult to handle co-workers, or presentations to make, all of which may produce similar responses. Continued stresses and chronically complex and difficult work situations may lead to more extreme health problems. Researchers identified major events in life that are related to subsequent illnesses and assigned weightings to these events. Table 31-4 contains a test for evaluating the likelihood of becoming ill from major stress events in life.

Stress may be positive or negative. The body's reactions to stress help us concentrate and perform. Some people do better under pressure. When there is no opportunity to relax or escape from the stress conditions, stress can be negative and can lead to reduced performance and health problems. The term *burnout* is closely associated with prolonged job stress. A number of factors contribute to job stress: not enough time to complete a job, lack of clear direction and goals, lack of clear instruction, absence of recognition or reward, lack of opportunity to participate, responsibility without authority, prejudice and bigotry, poor interaction with others because of differing goals and values, unpleasant or dangerous work conditions, lack of control over job performance, and job insecurity.

There are several techniques people can use to help reduce and manage stress. First, they must recognize the situations and conditions that lead to stress and they need to sense the body reactions that are symptoms of job stress. Applying one or more relaxation techniques can help reduce stress. A deep breath or simple exercises followed by relaxation can help. Another technique is getting away from certain difficult situations. Modifications in lifestyle, such as exercise, also may help.

31-7 MANAGEMENT PROCESSES

Management methods can affect how people perform or an organization performs. Management methods also affect the culture of an organization, referring to how people in the organization approach work planning, organization, and execution. The reader should refer to the wealth of literature on effective management methods. This section touches on a few concepts and approaches and how they may influence performance, including safety performance.

Total Quality Management and Six Sigma

More traditional organizational structures found in U.S. companies and organizations was a top-down approach. The organizational structure had multiple layers between the top leader of the organization and the lowest-level worker. Before the age of computers and the Internet, the layers of management provided a path for directing what was to be done down through the organization and communication up through the structure the status of what was accomplished. Many organizations continue to use such an approach.

In the 1990s, most U.S. organizations eliminated many of the layers in the top-to-bottom structure. In part, the change resulted from the use of personal computers and the Internet. Higher levels did not need many of the intermediate layers to analyze and process

TABLE 31-4 Life Stress Scale[a]

Event	Value	Score
Death of spouse	100	—
Divorce	73	—
Marital separation from mate	65	—
Detention in jail or other institution	63	—
Death of a close family member	63	—
Major personal injury or illness	53	—
Marriage	50	—
Being fired at work	47	—
Marital reconciliation	45	—
Retirement from work	45	—
Major change in the health or behavior of a family member	44	—
Pregnancy	40	—
Sexual difficulties	39	—
Gaining a new family member	39	—
Major business readjustment	39	—
Major change in financial state	38	—
Death of a close friend	37	—
Changing to a different line of work	36	—
Major change in the number of arguments with spouse	35	—
Taking on a mortgage more than $10,000	31	—
Foreclosure on a mortgage or loan	30	—
Major change in responsibilities at work	29	—
Son or daughter leaving home	29	—
In-law troubles	29	—
Outstanding personal achievement	28	—
Spouse beginning or ceasing work outside the home	26	—
Beginning or ceasing formal schooling	26	—
Major change in living conditions	25	—
Revision of personal habits	24	—
Troubles with the boss	23	—
Major change in working hours or conditions	20	—
Change in residence	20	—
Changing to a new school	20	—
Major change in usual type and/or amount of recreation	19	—
Major change in church activities	19	—
Major change in social activities	18	—
Taking on a mortgage or loan less the $10,000	17	—
Major change in sleeping habits	16	—
Major change in number of family get-togethers	15	—
Vacation	13	—
Christmas	12	—
Minor violation of the law	11	—
	Total	—

[a] Holmes, T. H., and Rahe, R. H., "The Social Readjustment Rating Scale," *J Pychosomatic Res*, 2:213–218 (1967).

Note:

Instructions

1. In the table above, check those events that occurred to you in the last 12 months.
2. For each item checked, insert the value for that item in the score column.
3. Sum all entered scores.
4. Compare your score to the following:

	Total Score	Probability of Becoming Ill
	<150	37%
	150–300	51%
	>300	80%

information into a form for the next higher level. Someone at any level could access information from a common source, and software applications provided the same analysis and reports to virtually anyone on the computer network. The change in technology impacted how organizations worked and reduced overhead costs.

The changes in management methods recognized that together, a group of people is smarter than any one person alone. The changes required a collaborative environment.

Another change was the emphasis on quality to be competitive in the marketplace. Many organizations rediscovered the work of Juran and Deming on quality and methods for achieving it. A national award was established called the Malcolm Baldridge Award that recognized the quality improvements of companies.

One of the basic elements of total quality management (referred to as TQM or quality improvement) involves listening to the customer to identify opportunities to improve. Another basic element was recognizing that organizational performance was tied to management processes given to workers and realizing that workers can perform not better than the process allows. Making individual performance the point of change could not be effective. In contrast, the approach sought participation of everyone in a work group. The group would review the processes it used. The group would identifying the opportunities for change that would contribute to meeting customer expectations and needs.

With this approach, the metrics for assessing performance changed. Some reduce the metrics to two basic ones: process time and errors. Process time refers to how long it takes to complete some function. Process time varies with specific settings.

Delivery time of mail is an example. One may write a letter, put it in an envelope with postage, and drop it in an out box on the desk. The addressee may see it several days later. Various steps in the process can affect the total delivery time. One may be affected by how frequently someone picks up mail from the out box. Another step may be how long it sits in the mail room before transfer to the post office. Then there are steps in the postal service handling and delivery. Suppose the daily pickup from the outbox occurs at 9 AM each morning at the same time that incoming mail is delivered. If the letter was completed at 10 AM on a Friday and the office is closed all weekend, the letter will not be picked up until Monday and will not move internally in the organization for nearly 3 days. Walking to the mail room and dropping the letter in the postal pickup box after completing the letter will reduce the delivery time potentially by more than 2 days. It would be easy for people to find opportunities to reduce the cycle time and increase the satisfaction of the recipient customer.

Errors, too, can have many definitions, depending on what is meaningful to a work process. In total quality management, the goal is to reduce errors. Errors produce "scrap" or end products or results that are not useable or do not meet standards acceptable to customers. Errors cause rework that elevates cost. By focusing on the management processes, identifying opportunities for improvement and incorporating quality enhancing steps into the processes, the error rates are reduced. The goal is for all participants to become involved in identifying and implementing changes and achieving results. The greater the focus on continuous improvement and the more frequently the process is reviewed, the faster the organization achieves new levels of quality.

Six sigma refers to six standard deviations out on the normal distribution and achieving one error in one million opportunities. Six sigma has emerged as the name for programs devoted to learning quality improvement techniques and implementing them.

Total quality management and six sigma also apply to safety management. Errors can refer to incidents, accidents, and damage-causing events. Incidents, accidents, and damage-causing events add to process times. A key to reducing incidents, accidents, and damage-causing events is incorporating hazard recognition, evaluation, and control into efforts to improve quality, of which safety is a component. To be successful, it is essential to train those in a work group on the essentials of hazard recognition, evaluation, and control relevant to their processes.

Use of a total quality management approach requires a change in how work is planned, organized, and accomplished. Everyone in the work group can contribute to improving results, including safety. Effectiveness results from broad participation.

Shared Leadership

Another management strategy that reflects the change from a top-down management approach is shared leadership. In a top-down setting, the individual at the top of an organization or an organization unit is the leader. The leader plans, organizes, and directs. Because the work to be accomplished is so complex, no one person has sufficient knowledge to be able to master all of the details of a process. The idea of shared leadership invites others in a group to contribute ideas for improvement by the entire group and to take responsibility for leading the rest of the group in certain aspects of the work. As a result, the roles of people in the group vary, depending on who is leading a particular activity. In one activity, a person my have a leadership responsibility and in the next be a participant. Leadership is not limited to only one person.

31-8 BEHAVIOR-BASED SAFETY

With these changes in management methods and the success that they achieved, safety management methods derived from or related to the general methods emerged that can be summarized as behavior-based safety methods. Much has been written about this aspect of safety management. Just as with the general shift in management methods, some organizations are more successful than others in implementing behavior-based safety. As a result, the work group and organization experiences a cultural change and safer work records.

Different authors have defined behavior-based safety somewhat differently. In general, behavior-based safety techniques focus on work processes. In analyzing work processes, the workgroup identifies behaviors that are critical to safe process performance. They measure how well the group completes safe behaviors. Measurement typically requires observation. Analysis of performance provides feedback to the participants. Participants also identify and resolve other process elements that impact the ability to perform safely as part of the continuous improvement process.

To be effective, those in the workgroup need training on hazard recognition, evaluation, and control as well as learning how behaviors that are part of the process can contribute to the safety of the work. The participants may need to change their approach to how safety is handled in the process. It requires a shift from a top-down management style. It requires broad participation and collaboration among members of a work group. It requires shared leadership within a work team. Instead of placing blame for wrong behavior, it works to change the process and to ensure that those engaged in the process understand the role that their behavior plays in the success of the process. It works in concert with other safety methods, which all contribute to continuous improvement of the processes.

EXERCISES

1. Contact a local hospital or clinic. Find out what employee assistance programs they provide for employers. These may include programs to deal with personal problems off the job, alcohol and drugs, stress management, and other subjects.

2. Obtain several accident reports or liability lawsuit reviews. Analyze the findings for unsafe acts and conditions. Evaluate the cases to determine if designs could have incorporated features that would have eliminated the unsafe acts or removed the harm from such acts.

3. Work with an organizational unit to apply quality methods to its processes and identify how to incorporate safety into the processes.

REVIEW QUESTIONS

1. To what factors did early behavioral theorists link behavior?

2. Briefly explain the following motivation theories:

   (a) Maslow

   (b) Herzberg

   (c) Vroom

3. Briefly characterize the following:

   (a) judgment

   (b) emotion

   (c) attitudes, opinions, and beliefs

   (d) individual differences

4. Name three things that can help prevent unsafe behaviors.

5. Why is communication important for safe behavior?

6. Why is feedback important for safe behavior?

7. How should feedback on performance be provided?

8. Explain job safety analysis.

9. Identify three factors that affect:

   (a) perception of risk

   (b) estimates of risk

10. What are biorhythms? Are they predictors of accidents?

11. How are alcohol and drugs related to accidents and unsafe acts?

12. What role can engineers play in preventing unsafe acts?

13. What is common sense? Is it the basis for safe behavior? Explain.

14. What is technological illiteracy? How is it related to safe behavior?

15. What is job stress? How is it controlled?

16. Explain how management methods have changed and how safety methods are an integral part of those changes.

17. What is behavior-based safety and how is it tied to process improvement?

NOTES

1 Wolcott, J., McKeeken, R., Burgin, R., and Yanowitch, R., "Correlation of General Aviation Accidents with Biorhythm Theory," *Human Factors*, 19:283–284 (1977).

2 Thommen, G., *Is This Your Day?*, Crown Publishers, Inc., New York, 1964.

BIBLIOGRAPHY

ADAMS, J. D., *Understanding and Managing Stress*, University Associates, San Diego, CA, 1980.

BERK, JOSEPH, and BERK, SUSAN, *Total Quality Management*, Sterling Publishing Company, Inc., New York, 1993.

BITTEL, L. R., and RAMSEY, J. E., eds., *Handbook for Professional Managers*, McGraw-Hill, New York, 1985.

BRUE, GREG, *Six Sigma for Managers*, McGraw-Hill, New York, 2002.

CAPEZIO, PETER, and MOREHOUSE, DEBRA, *Taking the Mystery Out of TQM*, 2nd ed., Career Press, Franklin Lakes, NJ, 1995.

CHOWDHURY, SUBIR, *The Power of Six Sigma: An Inspiring Tale of How Six Sigma Is Transforming the Way We Work*, Financial Times Prentice Hall, New York, 2001.

COVEY, STEPHEN R., *The Seven Habits of Highly Effective People*, Fireside Books, New York, 1989.

GELLER, E. SCOTT, *The Participation Factor—How to Increase Involvement in Occupational Safety*, Prentice Hall, New York, 2001.

GELLER, E. SCOTT, *The Psychology of Safety*, Chilton Book Company, Radnor, PA, 1996.

GELLER, E. SCOTT, *The Psychology of Safety Handbook*, Lewis Publishers, Boca Raton, FL, 2001.

GELLER, E. SCOTT, *Working Safe: How to Help People Actively Care for Health and Safety*, 2nd ed., Lewis Publishers, Boca Raton, FL, 2001.

HANNAFORD, E. S., *Supervisor's Guide to Human Relations*, 2nd ed., National Safety Council, Chicago, IL, 1987.

HARRY, MIKEL, and SCHROEDER, RICHARD, *Six Sigma: The Breakthrough Management Strategy Revolutionizing the World's Top Corporations*, Doubleday, New York, 2000.

IVANCEVICH, J. M., and MATTESON, M. T., *Stress and Work*, Scott, Foresman, Glenview, IL, 1980. Janis, I., *Stress, Attitudes, and Decision*, Praeger, New York, 1982.

JABLONSKI, JOSEPH R., *Implementing TQM: Competing in the Nineties Through Total Quality Management*, revised, 2nd ed., Technical management Consortium, Inc., Albuquerque, NM, 1994.

JURAN, J. M., *Juran on Quality by Design*, The Free Press, New York, 1992.

JURAN, J. M., and GODFREY, A. BLANTON, *Juran's Quality Handbook*, 5th ed., McGraw-Hill, New York, 1999.

KOHN, J. P., *Behavioral Engineering Through Safety Training: The B.E.S.T. Approach*, Charles C. Thomas, Springfield, IL, 1988.

KRAUSE, THOMAS R., *The Behavior-Based Safety Process*, 2nd ed., Van Nostrand Reinhold, New York, 1997.

KRAUSE, THOMAS R., *Employee-Driven Systems for Safe Behavior*, Van Nostrand Reinhold, New York, 1995.

KRAUSE, THOMAS R., general ed., *Current Issues in Behavior-Based Safety*, Behavioral Science Technology, Inc., Ojai, CA, 1999.

LANDY, F. J., and TRUMBO, D. A., *Psychology of Work Behavior*, rev. ed., The Dorsey Press, Homewood, IL, 1980.

PETERS, TOM, *The Pursuit of WOW!* Vintage Books, Random House, New York, 1994.

PETERSON, DAN, *Safety Supervision*, 2nd ed., American Society of Safety Engineers, Des Plaines, IL, 1999.

RAY, W., WILSON, P.E., and HARSIN, PAUL, *Process Mastering: How to Establish and Document the Best Known Way to Do a Job*, Quality Resources, NY, 1998.

SPATH, JOHN P., *Building a Better Safety and Health Committee*, American Society of Safety Engineers, Des Plaines, IL, 1998.

Supervisors' Safety Manual, 9th ed., National Safety Council, Itasca, IL, 1997.

WALTON, MARY, *The Deming Management Method*, Perigree Books, New York, 1986.

WEINSTEIN, MICHAEL B., *Total Quality Safety Management and Auditing*, Lewis Publishers, Boca Raton, FL, 1997.

CHAPTER 32

PROCEDURES, RULES, AND TRAINING

There are several techniques for managing, controlling, and changing human behavior. The goal is to help people make their actions safe, and the key approach is training. Proper actions in an organization begin with policy, policy is implemented with procedures, and procedures are implemented through instructions, warnings, and training. Safe behavior can be reinforced through verbal and printed reminders and rewards.

Although procedures and training are very important in safety, they do not remove hazardous conditions, but they may help people recognize hazards so they can be removed or controlled.

32-1 POLICIES AND PROCEDURES

Policy

Policies are statements of goals, objectives, and operational principles that govern an organization. They are created or approved at the highest level by the president, chief executive officer, or board of directors and provide general guidelines that govern activities. Normally, policies are written and publicized in an organization. A policy manual is the collection of all policies formally adopted by an organization. Policies have the effect of convincing top management to subscribe to, endorse, and support major operational provisions of an organization. Policy statements tell all subordinates in an organization what is important, what is to be achieved, and how to act. Typical contents include a statement of policy, implementation instructions, variations or exceptions, explanations for critical or complex situations, and forms for implementation or reporting.

Along with other policies, organizations need safety policies. Typically, the safety policy states that safety for employees, designers, visitors, customers, and the community is of highest importance to the company. It may contain goals to be achieved, and it assigns responsibility for safety to each organizational unit or individual affected by it. It will delegate authority, such as establishing who can sign permits or who makes safety decisions in an emergency. It may identify particular procedures that must be followed, such as general company rules or those in a company safety handbook, and it may reference particular standards to be followed. It should identify how accountability will be measured, may identify staff organizations that provide technical assistance to others in achieving safety, and may establish various safety committees and delegate certain authority to them. Policies need periodic review and before issuing policies, they may need legal review.

Safety Rules

Many companies or organizations have a booklet of general safety rules. Sometimes there is a document containing general rules that employees must follow. Included is a section on general safety rules during company activities. These rules apply to everyone.

Because there are many complexities in safety, many companies have additional rule books covering particular safety problems. This may be a company safety handbook or similar document. The handbook may give details about such topics as fire protection and fire response procedures, and it may list, cite, or define standards that designers, maintenance staff, or supervisors must follow. There may be separate manuals for design and maintenance personnel and particular operations or departments. The handbook may collect all safety manuals and operating procedures into one document system.

Procedures

Procedures are detailed implementation instructions for policies. Procedures give step-by-step information about what to do in particular situations. They identify who should do what and when actions are to be performed. Procedures may cover general situations or particular ones. When procedures apply to routine activities, they may be called standard operating procedures. Nonroutine activities, such as those for emergencies and maintenance, may require special procedures that are different from normal procedures. There are also special procedures that require detailed explanation and emphasis. They may be documented separately and referenced in routine and nonroutine procedures.

Keeping Procedures Current   If there is a change in a process, operation, or equipment used, associated procedures must be updated. Obsolete procedures may cause a participant or user to perform the wrong actions. When changes are made, all participants must be informed or retained. It may be important to implement the change at one time for everyone, so that two different procedures are not in place. It is most important to keep procedures that can have serious consequences up to date.

Aviation gives us an example of procedures that are not current. An airplane had four engines. If there was a failure in an engine, the pilot had to verify which engine failed. There were two models for the aircraft. Planes released during the first part of the production were called model A; the later ones were designated model B. Model A required that all engines be shut down and restarted one at a time to isolate the problem, whereas model B required that engines be shut down one at a time and restarted immediately. Pilots flying a model B aircraft during practice normally learned procedures for and flew model A. During a simulated engine failure introduced by the instructor, the aircraft crashed when trainees applied the wrong procedure.

Standardization   When there is more than one operation or item of equipment and people can work or operate on any of them, it may be critical to standardize layouts and procedures. If errors or system failures can cause serious losses, injuries, or death, then standardization is crucial. Standardization involves many things. It means that sequences of operations, controls, displays, procedures, and other elements are kept identical across systems and that operations are consistent with learned concepts and principles. Chapter 33 includes a discussion of population stereotypes and compatibility, elements important in standardization.

After the accident at the Three Mile Island nuclear power plant, investigators found several kinds of incompatibilities between control panel displays and patterns of behavior people had learned. A fundamental concept learned across most societies is the meaning of red and green: red means stop or danger; green means go or that things are satisfactory. One arrangement of controls violated this pattern[1] and led to wrong interpretation and action.

Standardization in aircraft is considered very important, and the industry devotes many people to identifying inconsistencies between airplanes of the same model and even between different models. A pilot or mechanic may apply the wrong procedure to an aircraft, resulting in disastrous consequences.

Levels of Procedures   It is wise to organize some procedures into different levels. Levels may be based on difficulty or complexity of tasks, frequency of tasks, or level of authority that must be involved. For example, daily startup of equipment may need a pre-energizing inspection and preliminary tests before the activity moves into full operation. Workers can be trained to perform such normal safety checks. However, if there is a failure or the equipment shuts down from an unsafe condition, a second-level procedure may require a supervisor to evaluate and concur in the problem and correction before the equipment is returned to full operation. A major failure or shutdown may require that specialists be brought in to diagnose the problem and to ensure that everything is safe. The second- and third-level procedures may require special tests and testing equipment, rather than observation alone.

Signature   An important part of many safety procedures is requiring a signature after they are completed. The signature may attest to individuals having read and understood the standard operating procedure or may attest to a properly trained person having evaluated a situation and made measurements with instruments. The signature also may attest to a person knowing that a procedure was properly and fully completed and that a situation is safe. For complex situations and situations where failure could lead to serious injury, illness, or death, such procedures may require independent evaluations by more than one person, in which case each person would then sign the form. Examples of such procedures are confined space entry, hot work permits, and lookout/tag out procedures.

Job Safety Analysis and Other Analysis

Developing procedures begins with an analysis of the tasks or activities of an operation. Various forms of task analysis may be used for engineering the process and equipment and for estimating cost and production. Included in the task analysis should be identification of hazards or unsafe acts that could occur with each task. For each hazard, the analysis should indicate what controls should be in place or applied, and for each unsafe act that could be performed, the analysis should identify how the task should be performed safety. The analysis also may identify any hazards that remain. Figure 31-1 illustrates one format for job safety analysis or job hazard analysis.

In analyzing an operation, it is not enough to consider the tasks involved in things going right or the operation working smoothly. One needs to forecast what can go wrong. The analysis needs to consider hazards and activities that might occur in abnormal conditions. Too often, injuries occur when people take the wrong action in an unusual event. Chapters 15, 36, and 37 discuss other methods that may be helpful.

Results of analysis are used to make decisions or to develop procedures. Procedures should explain normal or general operations first and then discuss the exceptions, abnormal operations, or special conditions.

Special Procedures and Permit Systems

Special safety procedures include those that explain what to do in emergency, abnormal, or very dangerous situations. They need to identify any changes in authority and responsibility from normal procedures. For example, during a fire, the fire chief or ranking firefighter takes charge of the situation, making the decisions and issuing instructions that others must obey.

Special procedures must explain what changes there are from normal procedures and detail what not to do. They should anticipate errors in judgment and action and explain what behavior is correct as well as explain why incorrect actions should not be followed and the consequences of following them.

A common form of special procedure is a permit system. A permit system recognizes that some conditions may be unsafe for particular activities. The activities and the environment in which they will occur must be evaluated and known to be safe before the activities are allowed to take place.

Two very common permit procedures are a hot work permit and a confined space entry permit. The goal is to make sure that everything is safe before a permit allows the activity to proceed. Both the person normally responsible for the work area and the supervisor of someone entering the area to perform work must sign the permit. Both attest to the conditions being safe for the activity. In some cases, a specialist may have to evaluate the area and sign the permit as well.

Because conditions may change, permits cover limited time periods and expire at the end of the approved period. Then the conditions must be evaluated again, and a new permit issued if everything is safe. Often the permits are for a particular shift or workday only.

It is easy to defeat a permit system. Someone can sign the permit without personally inspecting the area or completing the evaluation, which defeats the purpose of the permit procedure. An effective permit procedure requires that each party signing the permit fully perform the actions necessary to form a judgment that an activity can proceed safely.

Other kinds of special procedures are those that are unique to certain activities. An example is hand signals. There are special signals used by ground crews to guide pilots in moving aircraft, particularly when they are moving to or from parking at a gate or ramp. There are special hand signals used to guide crane operators in moving a load, and there are special hand signals for guiding excavation equipment operators.

Lockout and tag out procedures are special procedures for working on equipment that is normally energized with electricity, steam, mechanical, or other forms of power.

32-2   WARNINGS AND INSTRUCTIONS

Warnings and instructions are important elements of procedures. Warnings identify what dangers exist in equipment or operations and the dangers in normal and abnormal routine and nonroutine operations. Instructions explain how each person involved in an operation is to act and act safely and how people are to protect themselves from the dangers in normal and abnormal, routine and nonroutine operations. Instructions need to have a step-by-step format, need to be imperative, and need to state actions to be followed. Too often, people write instructions in a descriptive form, explaining how a process works.

Chapter 7 discussed warnings and instruction for products. Table 7-1 described 15 characteristics a warning must have. The use of warnings and instructions and principles for preparing them apply to operations as well as to products. Without warnings, users or participants may not recognize dangers inherent in an operation. Without instructions, users or participants must make up their own procedures.

Alarms and Signals   Alarms and signals are one form of warnings and instructions and they are integral parts of procedures. Sensors (sometimes human) must detect a problem and tell people about it. In particular, they must tell people where there is a danger. In one sense, signals and alarms are instructions to act; they are coded instructions, rather than direct, verbal instructions. People must learn what the alarms and signals mean and what actions they call for. Some signals, like hand signals for crane operations, are not automatic.

Symbols   Symbols are important elements of warnings. They help convey the message quickly and may help those who cannot read the language of the text. A combination of symbol(s) and text is better than either alone. Text is more precise. Some symbols have high recognition and understanding, and within a sample of people, most in the group will recall and understand the meaning of the symbol. However, for some symbols, the recognition and understanding is quite low. Not all symbols have obvious meaning to all viewers, which is why associated text is important. Seiden[2] provides a comprehensive list of references on symbols in warnings.

32-3   TRAINING

From childhood on, people gain knowledge, skill, and understanding through training. Training can take on many forms and may involve a variety of media. Training is essential for learning how to formulate safe decisions and take safe actions. Through training, people learn to minimize errors that lead to accidents and injuries.

Principles of Learning

Planning and development of training begin with an understanding of how people learn and what contributes to learning. The following list summarizes some principles of learning.

1. *Stimulate multiple senses.* We receive most information through vision. Hearing processes a lot of information, but cannot handle information at the same rate as visual input. Incorporating visual materials into training helps the learning process.

2. *Identify the need for training.* The trainee will understand what is being learned better if objectives and strategy for training are presented clearly.

3. *Organize the content logically.* It is better to conduct training in small modules rather than large ones. What constitutes logical order depends on the material being taught. One form of order is proper sequence, where early modules establish the background for later modules. Another form of order is level of difficulty, where easy material progresses to that which is more difficult.

4. *Teach principles with procedures.* People will understand procedures better and retain them longer if the principle or objective for the procedures is presented first.

5. *Teach the whole process first, then detailed parts.* Trainees should learn the whole procedure first. They need to see what each step leads to. Then they can go through the details of the process.

6. *Make sure trainees have time to practice, but keep practice periods short.* When trainees are learning skills and the criterion for success is meeting some performance standard, trainees need time to practice. Short practice periods with breaks are more effective than long practice sessions.

7. *Ensure participation when performance is the goal.* When training occurs in group arrangements, some trainees hold back from participating. An instructor must watch for this and find ways to involve everyone.

8. *Give trainees knowledge of results.* Trainees need to know how they are doing. It is better to evaluate trainees in small increments and give them results of evaluations, rather than delay evaluation and results.

9. *Reward correct performance.* There are many forms of feedback. Positive is generally better than negative. Praise and verbal comments can be used when trainees do things correctly. Accurate and immediate feedback is better than delayed and general feedback.

10. *Keep trainees interested and challenged.* Instructors can use various techniques to increase participation and interest in subject materials. Ask questions and stimulate discussion, and when there are skills involving several people, role playing exercises help maintain interest.

11. *Simulation should duplicate actual conditions.* When procedures and settings are simulated, they should accurately represent real situations as much as possible. Unrealistic simulation can lead to incorrect behavior in real contexts.

12. *Unique or unusual material is retained longest.* Use of examples and real situations helps people visualize what is taught. Dramatic and exotic style may be entertaining, but care must be given to make sure such activity is meaningful.

13. *Provide relearning to sustain knowledge and skill.* The idea of a learning curve tells us that the more skilled a person becomes, the slower the rate of improvement. After training, the knowledge or skill achieved by the end of training decays with time. Creating opportunities to relearn, update, or evaluate skills and knowledge will help keep performance at desired levels.

14. *Fit training to individual needs.* The knowledge or skill of each trainee can be assessed through pretests, interviews, and other evaluations. When there is too great a range in knowledge and skill in the same training session, few trainees are well served. With self-paced instruction and criterion-based training, individuals can achieve the desired level of knowledge or skill at their own pace. Slow learners or those with elementary skills are not intimidated by others who are advanced. Computer-based instruction and training systems allow for customized instruction and repeating of sessions to match the needs of individuals.

Training Needs

Training programs begin with assessment of need, of which there are two aspects. Is there a need to train people? What level of knowledge or skill do people already have? Review records of employees to see what knowledge, skill and experience they already have. Observe people on the job to determine if their actions are correct. Use interviews, questionnaires, or performance tests to determine if proper knowledge and skill exist.

Contingency Training in Safety

Too often, people learn how to do a job or operate some equipment by being taught only the procedures for normal operations or conditions. They never receive instructions about what to do when things are not normal or when an activity, like maintenance and cleaning, are not part of normal production operations. When things go wrong, people are left to make their own decisions about what is wrong and what to do. Errors in critical methods can be disastrous.

Training programs must teach about contingencies—anything out of the ordinary. Typically, maintenance, repair, and cleaning are not activities performed during production. A contingency occurs when machines start to produce faulty parts or when feed and ejection elements do not work right. A contingency occurs when something breaks, when a process overheats or pressure becomes too high, or when equipment does not work correctly. Contingencies are events and conditions that are not ordinary or routine.

Contingencies must be included in procedures and training programs. People have difficulty recognizing the symptoms of things going wrong and often fail to recognize what is happening and why it occurs. After they recognize that something is wrong and what it is, they need to know what actions are appropriate and safe, and they need the authority to act with safety for themselves and others. Too often, they place the importance of doing a good job above the importance of safety. Too often, production is paramount to safety.

Who to Train

Everyone needs training for safety. Within a company there are new employees and experienced employees, supervisors and managers, and special committees and teams. Product users and the general public need training in safety.

New Employees   Because accident rates decrease with time on the job and are highest at the start, it is very important to train new workers. For new employees, there are many things to learn. An orientation to the company and to a new job is a good way to get started. However, orientation sessions deal with general matters. New workers should understand how important safety is in the company or department and be aware of major procedures in case of an accident.

Details for performing a job correctly should be covered in a separate session from general orientation. Before a new worker begins work, thorough training in doing a job correctly and safely and in contingencies that create dangers should be completed. During the early phases of employment, new workers should have a lot of reinforcement on details of doing the job correctly. There should be review sessions and evaluations to see if the worker applies the training to the job. Extra help should be given until the worker can do the job well.

OSHA requires that employers provide many kinds of safety training. In fact, there are nearly 150 references to specific training requirements in the OSHA regulations, and OSHA requires that employees comply with safety rules and standards. This can be achieved only through training. Many federal and state agencies impose safety training requirements for employees on the employer.

Experienced Employees   Experienced employees need training when procedures change. For some aspects of safety, periodic reinforcement is not only necessary, but it is required by law and regulation. For example, many workers must be trained every 6 months or annually about hazardous materials found on their jobs. A worker who changes job assignments should have additional training on hazards and controls for the new tasks.

Supervisors and Managers Supervisors and managers represent company and employer responsibilities. Not only do they need to understand the hazards and controls for their workers, but they need to know what training they must administer. They need to understand the regulatory and legal responsibilities for safety that the employer or company bears and they need to know what responsibilities they have under normal and special procedures. If supervisors and managers have contract relationships with other companies and their employees, the supervisors and managers must learn how to deal with safety matters through contract chains of command.

Special Committees and Teams Many companies have in-house fire brigades, emergency response teams, and other groups that have special responsibilities. Individuals on these committees or teams perform other jobs and leave them when the need arises. Individuals on these teams need special training, which may include special drills and tests. The drills may involve teamwork among groups from within and outside the plant.

Contractors Contractors working on company premises need training in safety as much as company employees, although it is not uncommon for contractor supervisors and workers to be left on their own. A contract should include clauses that cover training requirements for safety.

Product Users Users need training in the safe use of products. Usually, a manufacturer is limited to manuals and instructions that go with the product. The safety information must be clear and understandable for all potential users, which may necessitate instructions in several languages. Warnings may be interspersed throughout the instructional materials.

For large, complex products, a manufacturer may want to provide additional training materials to help users operate and maintain the product correctly and safely. Field staff who perform customer service should be able to provide safety training for the products, although it is not uncommon to find field staff giving wrong advice that leads to accidents and injuries.

When there are changes in products or problems with a product that may cause injury, a manufacturer should try to reach users to provide new information. Many companies have registration cards for new users that can be mailed in or computer-based product registrations. Manufacturers of automobiles must keep records on purchasers that help implement product recalls and inspection notices.

Public Some operations and products can affect people outside the plant. Some companies provide training for the public about hazards of systems and equipment and how to prevent injury. For example, electrical utilities often provide public notices about dangers of excavating into buried lines and the dangers of substations and overhead lines. Several railroads offer safety programs for schools. Local police and fire departments offer public training and informational materials about traffic and fire safety. Any operations that could create dangers for a community should have training programs to inform local authorities and the public about warning systems, hazards, and means of protection.

Engineers Engineers and other professionals need training in safety because safety affects their sphere of responsibility. A National Council of Examiners for Engineering and Surveying survey found that virtually every engineering discipline and every type of engineering job function had safety responsibilities. In 1987, the Accreditation Board for Engineering and Technology added a requirement that safety be included throughout the engineering curriculum for all engineering disciplines.

Training Techniques

There are many training methods. The method to use depends on the content of and ability to deliver training. For example, there are limited ways to reach the public or product users. Sometimes only a warning sign or user manual are available; sometimes advertisements or radio and television talk shows can be used. By capturing the desired audience of trainees, the methods can be expanded. Standard audiovisual presentations may work well. Stand-up lectures, role playing, case studies, problem solving, special training facilities, computer-assisted instruction, and other techniques are useful. Which one to use depends on whether one is trying to convey general knowledge or to develop skills to a measurable level of performance. Selection also is limited by cost. If a person has a job that plays an important role in personal safety and the safety of others, significant time, money, and facilities may be needed.

A common, low-cost way to reinforce training with workers is to conduct periodic sessions with small work groups. The supervisor conducts the session with employees. Some call these sessions tailgate or toolbox safety meetings. Various techniques may be used. Typically, the group addresses one or two topics, uses a discussion format, and reviews hazards, controls, and safe procedures. A key to making these sessions effective is selecting topics that deal with particular hazards for the group, not just general ones.

Training Aids

For safety, there are many training aids and materials. Several companies produce very good training materials in a wide variety of formats and media. Information about them are found in safety and health periodicals and journals and in safety product directories. NIOSH has many safety training materials available for purchase.

Evaluation

Safety training is worthwhile only when it is transferred into practice. The only way to determine if implementation is accomplished is by making some measurements. One can test trainees during and at the end of training sessions. If performance is essential, there should be set standards that a trainee must achieve. Evaluations should be included to determine if each student has met the standards. There may be formal tests that trainees take or informal review and discussion with trainee groups. One can also apply statistical techniques to observed behavior recorded by trained observers or assessment of accident rates, loss rates, and other data.

Some employers have begun to use certification programs that include tests of knowledge on safety. An example of a nationally accredited certification program in safety for managers and first-line supervisors is the Safety Trained Supervisor certification.[3] The program offers tests for construction, general industry, and the petro-chemical industry.

Management of Training

The requirements for teaching safety within a company can be quite complex. The task of keeping track of who had what training is a major job. To do a good job, employers and their managers and supervisors may need a computer database for tracking safety training required and completed. The system can list who has had what training and when and who must be trained to meet regulatory or management requirements in upcoming months. The tracking system also may need to keep track of performance in particular skills and when a performance evaluation was last made.

32-4 PROMOTING SAFETY

There is a need to reinforce proper attitudes and actions. There are many approaches for promoting safety. This section addresses only a few.

Posters, Flyers and Newsletters

One method for reminding people of safety concepts and safe actions is through posters, flyers, and newsletters. Such publications keep important safety messages in front of people. In view of the competition for people's attention, are these publications effective?

One study[4] looked at a poster about sling safety for hoisting equipment. The poster gave a detailed instruction: "Hook that sling!" After a poster was on display 6 weeks, there was a 13% increase in hooked slings. Some workers could remember that the poster was up but few could remember the content.

Another study[5] looked at style of posters. A humorous poster and one with a medium threat were compared. Researchers questioned workers to find out if they recalled either of the two posters and recognized them. The results were as follows:

Poster Type	Recall (%)	Recognition (%)
Humorous	18	15
Medium threat	11	10
Both	18	49
Neither	52	27

Workers preferred the serious poster over the humorous one. The reviewer concludes that recall and preference for style of posters have little relationship with their impact on safe behavior.

Opinions about safety poster content and use vary. One safety poster researcher uses the following criteria: be specific, be accurate, be positive, design for safe behavior, site prominently, and stick to a simple message. Another suggests: have a clear strategy, target the content, attract attention, have a positive and attainable slogan or message, arouse interest, seek to have something that is remembered, and call for specific action.[6] The National Safety Council has developed recommendations for posters, bulletin boards, and safety displays.[7]

Awards and Rewards

Many people believe that recognition of individual worker's safe behavior is the best way to achieve desired actions. The behavior theories support the need to reward correct behavior. There are many ways to recognize people for what they do correctly. The methods are limited by one's imagination. In general, immediate reward or recognition is best; delayed recognition is less effective. Feedback each time a behavior is correct is also desired, but often is not possible. Verbal recognition by a supervisor is one of the best and least expensive kinds of feedback.

There are many kinds of award and reward programs. Several companies sell awards and materials for conducting award programs. There are awards for individual behavior and for group achievements. Group programs introduce peer pressure and peer support toward correct actions. Particular programs may be effective for a limited time, but then people lose interest. Changing programs can help maintain interest. Some kinds of awards incorporate company or work group logos and names. These help develop pride and commitment to the identified group and a positive attitude and image for the group.

EXERCISES

1. Work with an employee group to conduct a job safety analysis. From the results, develop a training program for each employee.

2. Review the training program for an employee group. Identify ways to improve it.

3. Develop a computer database management application for tracking training requirements and training completed for each member of an employee group.

REVIEW QUESTIONS

1. Explain the purpose for policies. For procedures, explain how they are related.

2. What are general safety rules?

3. Give one reason why each of the following are important for procedures:

    (a) keeping current

    (b) standardization

    (c) levels of procedures

    (d) signature

4. How does one start to develop a procedure?

5. What is a special procedure?

6. When someone signs a hot work permit or confined entry permit, what does the signature mean?

7. Explain briefly how each of the following are procedures:

    (a) warnings

    (b) instructions

    (c) alarms

    (d) hand signals

    (e) symbols

8. List and briefly explain five principles of learning.

9. What is contingency training and why is it important for safety?

10. Identify why training is important for each of the following:

    (a) new employees

    (b) experienced employees

    (c) supervisors and managers

    (d) special committees and teams

    (e) contractors

    (f) product users

    (g) public

    (h) engineers

  11. In what way are posters, flyers, newsletters, awards and, rewards an element of safety training?

NOTES

1 Sheridan, T. B., "Human Error in Nuclear Power Plants," *Technology Review*, February: 23–33 (1980).

2 Seiden, R. M., *Product Safety Engineering for Managers*, Prentice-Hall, Englewood Cliffs, NJ, 1984, pp. 222–223.

3 The Safety Trained Supervisor Certification is operated by the Council on Certification of Health, Environmental and Safety Technologists (CCHEST), Savoy, IL (www.cchest.org).

4 Laner, S., and Sell, R. G., "An Experiment on the Effect of Specially Designed Safety Poster," *Occupational Psychology*, 34:153–169 (1960).

5 Reported by Sell, R. G., "What Does Safety Propaganda Do for Safety? A Review," *Applied Ergonomics*, 8:203–214 (1977).

6 "What Makes an Effective Safety Poster?" *National Safety and Health News*, 134:32–34 (1986).

7 *Posters, Bulletin Boards, and Safety Displays*, Data Sheet I-616-Revision 86, National Safety Council, Chicago, IL, 1986.

BIBLIOGRAPHY

American National Standards Institute, New York:

ANSI/ASAE S351, Hand Signals for Use in Agriculture

ANSI/NFPA 72A, Installation, Maintenance and Use of Local Protective Signaling Systems

ANSI/NFPA 72C, Remote Station Protective Signaling Systems

ANSI/NFPA 72D, Proprietary Protective Signaling Systems

ANSI/NFPA 72F, Installation, Maintenance, and Use of Emergency Voice/Alarm Communication Systems

ANSI/NFPA 72G, Installation, Maintenance, and Use of Appliances for Protective Signaling Systems

ANSI/NFPA 72H, Testing Procedures for Signaling Systems

ANSI/NFPA 74, Household Fire Warning Equipment

ANSI/NFPA 171, Visual Alerting Symbols for General Public Fire Safety

ANSI/NFPA 172, Fire-Protection Symbols for Architectural and Engineering Drawings

ANSI/NFPA 174, Fire Protection Symbols for Risk Analysis Diagrams

ANSI/NFPA 178, Symbols for Fire Fighting Operations

ANSI/SAE J115, Safety Signs

ANSI/UL 38, Manually Actuated Signaling Boxes for Use with Fire-Protective Signaling Systems

ANSI/UL 864, Control Units for Fire-Protection Signaling Systems

ANSI/UL 904, Vehicle Alarm Systems and Units

ANSI/UL 969, Marking and Labeling Systems

ANSI/UL 985, Household Fire Warning System Units

ANSI/UL 1069, Hospital Signaling and Nurse-Call Equipment

ANSI/UL 1638, Visual Signaling Appliances

ANSI Z35.1, Specifications for Accident Prevention Signs

ANSI Z35.2, Specifications for Accident Prevention Tags

ANSI Z35.4, Informational Signs Complementary to ANSI Z35.1

ANSI Z53.1, Safety Color Code for Marking Physical Hazards

ANSI Z53.1 Set A, Swatches for Safety Colors Mentioned in ANSI Z53.1

ANSI Z53.1 Set B, Swatches for Highway Colors Mentioned in ANSI Z53.1

ANSI Z129.1, Precautionary Labeling of Hazardous Industrial Chemicals

Anderson, C. R., *OSHA and Accident Control Through Training*, Industrial Press, New York, 1975.

Cantonwine, Sheila, *Safety Training That Delivers—How to Design and Present Better Technical Training*, American Society of Safety Engineers, Des Plaines, IL, 1999.

The Grey House Safety and Security Directory, Grey House Publishing, Millerton, NY, updated annually.

Kelley, Stephen M., *Locktou/Tagout: A Practical Approach*, American Society of Safety Engineers, Des Plaines, IL, 2001.

McManus, Neil, *Safety and Health in Confined Spaces*, Lewis Publisher, 1998.

NIOSH Publications Catalog, 6th ed., Publication No. 84–118, National Institute of Occupational Safety and Health, Department of Health and Human Services, Cincinnati, OH, August, 1984.

Rekus, John F., *Complete Confined Spaces Handbook*, Lewis Publishers, Boca Raton, FL, 1994.

ReVelle, J. B., *Safety Training Methods*, Wiley-Interscience, New York, 1980.

Slote, L., "Occupational Safety and Health Training Programs," *Handbook of Occupational Safety and Health*, Part VI, Wiley, New York, 1987.

Society of Automotive Engineers, Warrendale, PA:

SAE J107, Operator Controls and Displays on Motorcycles

SAE J1048, Indicator and Telltale Symbols for Motor Vehicle Controls

CHAPTER 33

ERGONOMICS

4 AM, March 28, 1979

During the first minute of a loss of coolant accident at Three Mile Island, 500 or more lights go on and off; during the second minute, more than 800 lights are illuminated. Operators incorrectly diagnose the problem: for more than 2 hr, they do not recognize that they have too little cooling water; they think they have too much. Instead, the reactor was boiling dry. Investigations after the accident reveal many problems of questionable design in control rooms. For example, two digital controllers are side by side and look exactly alike. The left one concentrates borated water and the right one dilutes it. The operator has to remember that the decimal point is one digit before the end digit on the left controller and one digit after the last digit on the right controller. Another example is two auxiliary feedwater meters. One labeled A is on the left; one labeled B is on the right. There are two related switches, labeled A and B. However, switch A is on the right and B is on the left.[1]

May, 1979, to May, 1980

Four women working with visual display terminals (VDTs) at the *Toronto Star* all delivered fetuses with different birth defects. Four co-workers who did not work with VDTs gave birth to normal, healthy babies during the same period. This observation of a cluster of miscarriages raised the continuing question: are their dangers working at VDTs? The answers have not satisfied everyone. Whereas one of every six pregnancies ends in miscarriage, to date, there seems to be no direct relationship to VDTs, even though workers still discover miscarriage clusters. The use of computers in office and other work continues to grow tremendously. The question of miscarriages among office workers has focused a lot of attention on VDT workstation problems and has produced design and ergonomic solutions for office workstations.

June, 1988

OSHA proposed a $3.1 million penalty against a meat packing company that was charged with willfully ignoring a serious health hazard that had injured hundreds of employees. The health hazard covered cumulative trauma disorders, including carpal tunnel syndrome. Because OSHA had no particular standards on ergonomics, the agency later agreed on a smaller fine and required the company to conduct research into ergonomic aspects of their workplace.

January, 2004

In an unusual action, the U.S. Congress passed a law rescinding the ergonomics standard established by OSHA. After completing extensive hearings and several attempts at an ergonomics standard, OSHA had approved a standard to deal with repeated motion and musculoskeletal disorders. Many companies believed that the standard imposed liability and excessive government intervention and lobbied Congress for action.

33-1   INTRODUCTION

Definition

Ergonomics has become a major element of safety. Approximately one third of all workers' compensation claims involve repetitive motion disorders or cumulative trauma disorders. Some estimate that such claims will climb to one half before long. Ergonomics is much broader than safety and health. It addresses job performance and the ability of people to perform tasks. It extends to preferences and choice, which are important in marketing.

Ergonomics comes from the Greek words *ergon*, meaning work, and *nomos*, meaning law. Ergonomics means the laws of work. It may be defined also as the relationships between people and a variety of things: equipment, environments, facilities, vehicles, printed materials, and other informational media. Ergonomics relates human capabilities and limitations to the design of products, systems, and environments. In the field of ergonomics, there are three major kinds of relationships, which are somewhat inter-related: performance, safety and health, and satisfaction. Performance attempts to extend the abilities of a person by improving output and reducing errors. The safety and health relationship attempts to minimize accidents and injuries resulting from human limitations. Satisfaction involves designing things that people judge as comfortable, desirable, convenient, and pleasing.

Ergonomics is applied through design. Consequently, designers must understand human behavior, physiology, kinesiology, biomechanics, and other fields that study the characteristics of humans. Designers apply human characteristics in creating workplaces, furniture, vehicles, buildings and equipment, and informational products for human use. Designs must reflect normal operation, foreseeable misuses, and maintenance and repair. Ergonomics data contribute to all of these.

Ergonomics, human factors engineering, and human engineering are virtually the same, although some people may argue over subtle differences. Ergonomics in building design sometimes is called architectural psychology or environmental design.

History

To a great extent, ergonomics developed out of aviation and other military matters in World War II. Critical tasks in flying an aircraft required effective presentation of information and error-free operation of controls. Today, ergonomics has become a fairly common term and its application continues to grow. Circa 1980, the application of ergonomics to the office environment paralleled the growth of microcomputers, and circa 1985, ergonomics started to gain more attention in occupational safety because claims for cumulative trauma disorders grew rapidly.

Applications in This Book

Many previous chapters discussed hazards, injuries, and controls that involve ergonomics. For example, Chapter 7 discussed warnings and other chapters contributed information about signs. Chapter 11 discussed design of floors, stairs, ladders, and handholds; Chapter 13 included a review of cumulative trauma disorders and design of tools and machines that incorporate ergonomic features; Chapter 15 covered manual materials handling and prevention of lifting injuries; Chapters 18 through 20 covered human aspects of thermal conditions, pressure, and lighting. Not only is health a concern in hot and cold environments, so is performance. Altitude and high pressure environments impact human performance and impose dangers. People cannot function if lighting is inadequate. Both quantity and quality of light affect performance, and reduced performance and errors may lead to accidents. Chapter 23 covered noise and vibration, both of which impact performance and safety, and Chapter 28 reviewed personal protective equipment, where ergonomics is an essential element in proper and comfortable fit.

This chapter discusses major areas of the field of ergonomics and considers safety and health applications of these areas, which include anthropometry, displays and controls, work physiology, and biomechanics. The bibliography at the end of the chapter provides additional ergonomics references and data that cannot be reproduced here.

Some General Principles

A few principles of ergonomics apply to a wide variety of applications. Some general principles are introduced here. Other, more specific ones follow in later sections.

People versus Machines People and machines have different capabilities; neither is best at all functions. Table 33-1 lists some functional differences between people and machines.

Design the Job to Fit the Person People are limited in what they can do. Failure to recognize capabilities and limitations may cause people to make errors or create hazards. Errors can be errors of commission, doing something that should not have been done, or

TABLE 33-1   A Comparison of Functional Capabilities of People and Machines[a]

People Are Better at	Machines Are Better at
Detecting signals in high noise environments	Responding with minimum lag time (machines have microsecond lags, whereas people have lags of 200 ms or more)
Recognizing objects over varied conditions of perception	Precise, repetitive operations
Handling unexpected occurrences	Storing and recalling large amounts of data
Ability to reason inductively	Monitoring kinds of functions
Ability to profit from experiences	Deductive reasoning ability
Originality	Sensitivity to stimuli (the range of human sensitivity is limited)
Flexibility of reprogramming	Exerting force and power
Ability to perform when overloaded and to adjust to compensate for the overload	

[a] From Meister, D., and Rabideau, G. F., *Human Factors Evaluation in System Development*, Wiley, New York, 1965.

errors of omission, failing to do something that should be done. Asking people to adjust to a job, equipment, or environment may be asking them to exceed their capabilities. In fact, the conditions may be harmful.

Work Smart, Not Hard Productivity is not increased by only speeding up the activities and methods in place. It is improved also by finding new ways to accomplish something. Errors and accidents are not reduced by only doing a better job. Changing the job by applying knowledge of accident and error causes will reduce them.

All People Are Not Alike There are variations among people. Some differences, like size, build, and weight, are easily observed. Other differences are discerned through physiological and behavioral measures. Reaction time, strength, coordination, responses to environmental conditions, beliefs, and other attributes of people require measures. The variability among people requires that designers provide adjustment. It may require that managers treat people differently. There is no one solution that works for everyone.

33-2 ANTHROPOMETRY

Description

Anthropometry is the science of measuring the human body. There are two classes of anthropometric data: static and functional or dynamic. Static measurements include standing height, sitting height, length, breadth, and depth of body segments and postures. Functional data describes such things as reach, range of motion, and forces generated by hands and feet in different directions. Tables in Appendix B include basic anthropometric data.

Anthropometric data are normally reported for fifth, fiftieth, and ninety-fifth percentiles in males and females. For a given dimension, 95% of the population is larger than the fifth percentile value, 95% of the population is smaller than the ninety-fifth percentile dimension, and 50% of the population is larger or smaller than the fiftieth percentile dimension.

People who are tall and are at the ninety-fifth percentile in height do not necessarily have arms or legs that are at the ninety-fifth percentile. There is a moderate correlation among dimensions of body members.

Many of the static data are measured on nude or lightly clothed persons. When applying this data to design, there is often a need to make some allowance for clothing. Some of the main data are now roughly a generation old and may not fully reflect today's population. The source for data in the bibliography is primarily from young adults in the United States. Diffrient et al.[2] provide data for a wide range of populations, but generally the data available for children, the elderly, and the disabled are limited. Data also may not reflect the true dimensions for populations in other countries.

There are special instruments for making anthropometric measurements: calipers and anthropometers are used to measure many static dimensions, and goniometers measure joint angles. There are standard landmarks of the body used in defining particular measurements and movements.

Application

The primary use of anthropometric data is for fit and reach, but there are other uses, too. People come in a variety of sizes and shapes. A few principles apply to the use of anthropometric data in design, although each principle may not work for every situation.

1. *Design so things are adjustable for different users.* One size does not fit all. Office furniture manufacturers now provide many adjustable features in chairs, work surface heights, and positioning of keyboards and monitors, and adjustments are starting to appear in seating and workstation equipment in factories and shops. Barber chairs have had adjustment in height, tilt, and rotation for years to make hair cutting easier. For pallet loading, there are now adjustable height pallet platforms that allow the user to adjust the pallet height as the pallet is filled or emptied. People may need to read or observe an object. One example of this is rear view mirrors on automobiles that are adjustable for different drivers and sitting postures. Displays on equipment may need to be adjustable so a process or machine operator can view them easily. Many computer monitors now come with adjustable bases.

2. *Design for the ninety-fifth percentile male to fit and the fifth percentile female to reach.* Not everything can be designed for adjustment. Designing a doorway, seating in airplanes, or headroom in an automobile allows only one possible solution. The goal is to allow most people to fit within the dimensions. If a very large person will fit, a small person will fit as well. If a person must reach a control or a part, the distance from the person to the object should not be longer than a short person's reach; a tall person can reach that far, as well. Note that in using this principle, at least 1 of every 20 people is not accommodated by data for those within the ninety-fifth percentile. As a result, some designs may need to extend the limits for the extreme population.

3. *Know the population you are designing for.* If data on a particular population are available, a design can be fitted to them. Anthropometric data from reference tables are useful when there is no other source of information for estimating dimensions and movements of people. However, reference tables may have been derived from a population significantly different than the population for which something is being designed.

33-3   INFORMATION AND DISPLAYS

The People–Machine Interface

Displays and controls are elements in an interface between people and machines. Figure 33-1 illustrates the interface and elements involved for people and for machines. Machines perform some functions and people control the operations of most machines. The operator must know what the machine is doing to know how to adjust the controls. *Displays* provide information about the machine to the operator, and the operator must sense the information provided by the display. The operator *processes* the information, *decides* what action to take, *acts* on some control, and affects the machine *operation*. If there is a failure in the person–machine loop, it may affect the performance of the system. Failures on the person side are usually called errors.

Sensory Reception of Information   For a display to be useful, information must be in a form that can be sensed. If it is outside human sensory capabilities, the display is useless. For example, humans cannot see wavelengths in the infrared or ultraviolet region, and few can hear ferquencies near or higher than 20,000 Hz.

By far, vision is the richest sensory mode because it has a high rate of reception. People can discriminate among many shades, shapes, and textures. The second best sense is audition, but it is well below vision in capacity. The rate of information flow is quite

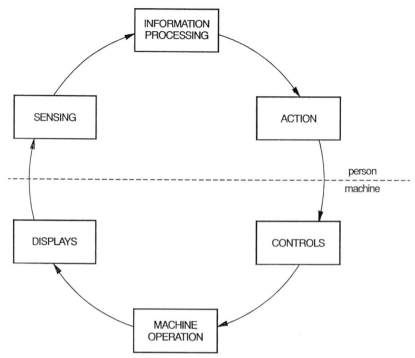

Figure 33-1. Example of a job safety analysis worksheet. Schematic of the interface between people and machines.

limited, but people can learn to discriminate among many different sounds. The tactile sense is less capable than audition. Humans can discriminate accurately among limited shapes and textures through touch.

Processing of Information   Information received through the senses needs to be interpreted and processed: letters of the alphabet or symbols must be converted into some meaning; what the sound of an alarm means must be recalled. The processing of information in the brain involves understanding the meaning of the stimuli, long-term and short-term memory, problem solving, making judgments, and deciding. Not only would the sound of 500 alarms in a nuclear power plant control room create sensory overload, it would produce processing overload. When there are multiple elements of information coming in, one must prioritize them and select which is the most important.

There are many potential sources of error as people process information. One may have difficulty selecting and understanding competing information or one may have to integrate information from several sources to recognize a pattern for an event that is starting to occur. One must also remember what information means. An example is what to do in an emergency when there is an audio alarm. When the same alarm signals fire and tornado, the wrong action could be fatal. The correct action for fire is to get out, whereas the correct action for a tornado is to stay inside and take cover. Another processing error is difficulty solving a problem, such as diagnosing what is wrong with a machine. One may not have the experience (previous information) to make correct decisions or judgements that are safe.

Types of Displays

Displays are classified first by the sensory mode of their information. Visual displays are most common, followed by auditory displays, and occasionally, displays for other senses. No matter what sensory mode is selected for a display, the conditions must be suitable for information transfer. For visual displays, there must be sufficient quantity and quality of light for the displayed information to be seen. Auditory signals must be loud enough to be heard and, preferably, should not compete with other sounds.

Visual Displays

Visual displays are classified as quantitative and qualitative. Quantitative displays present numerical values; qualitative displays present conditions. There are also status displays, signals, lights, and representational displays. Several types of displays may be integrated into a complex display.

For quantitative displays, three common styles of indicators are direct reading or digital display, moving pointer–fixed scale, and fixed pointer–moving scale. Figure 33-2 gives examples of these displays and Table 33-2 summarizes the usefulness of each type of indicator.

Status indicators give qualitative information about the status of a system or component. The red-yellow-green traffic lights are a common status indicator that tells which lanes of traffic are moving and which are stopped.

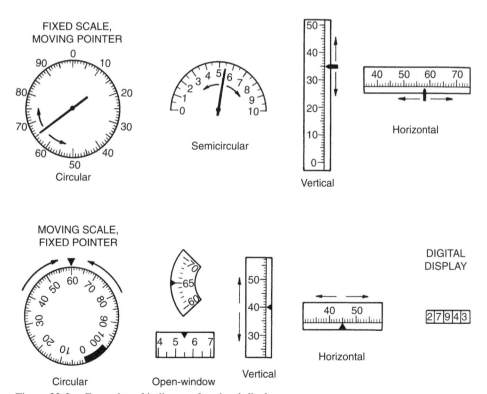

Figure 33-2. Examples of indicators for visual displays.

TABLE 33-2   A Comparison of Different Indicator Styles[a]

Used for	Counter or Digital Display	Moving Pointer–Fixed Scale	Fixed Pointer–Moving Scale
Quantitative	GOOD. Requires minimum reading time with minimum reading error	FAIR	
Qualitative	POOR. Position changes not easily detected	GOOD. Location of pointer and change in position easily detected	POOR. Difficult to judge direction and magnitude of pointer deviation
Setting	GOOD. Most accurate method of monitoring numerical settings, but relation between pointer motion and motion of setting knob is less direct	GOOD. Has simple and direct relation between pointer motion and motion of setting knob, and pointer-position change aids monitoring	FAIR. Has somewhat ambiguous relation between pointer motion and motion of setting knob
Tracking	POOR. Not readily monitored, and has ambiguous relationship to manual-control motion	GOOD. Pointer position is readily monitored and controlled, provides simple relationship to manual-control motion, and provides some information about rate	FAIR. Not readily monitored and has somewhat ambiguous relationship to manual-control motion
Orientation	POOR	GOOD. Generally moving pointer should represent vehicle or moving component of system	GOOD. Generally moving scale should represent outside world or other stable frame of reference
General	FAIR. Most economical in use of space and illuminated area; scale length limited only by number of counter drums; difficult to illuminate properly	GOOD. But requires greatest exposed and illuminated area on panel; scale length is limited	FAIR. Offers savings in panel space because only small sections of scale need be exposed and illuminate; long scale is possible

[a] From Van Cott, H. P., and Kinkade, R. G., eds., *Human Engineering Guide to Equipment Design*, rev. ed., Superintendent of Documents, U.S. Government Printing Office, Washington, DC, 1972.

There are many applications for steady or flashing signal lights. They give location, attract attention, indicate status, or give instructions. Size, duration of signal, brightness, flash rate, setting among other lights, and color all affect their performance in the person–machine interface.

Representational displays are pictorial or symbolic displays. Examples are video images, graphs, and maps. These displays may be static or dynamic. Size of display, size of elements displayed, realism, resolution, color, and rate of change affect performance. Symbols were also discussed in Chapter 32.

Display Characters and Elements

Many visual displays have numerical and alphabetical characters. The readability of the characters is important, particularly when there are emergency and adverse conditions. Also important are dial pointers and scale markings. Many factors affect the ability to read displays. Size of characters, aspect ratio, font, stroke width, color of character, background color, and coding all contribute to reading speed and error rate.

The standard viewing distance for displays is assumed to be 28 in from the display to the eye, and recommended dimensions for characters and display elements are based on that distance. For other distances, characters and elements should be adjusted in size. One formula for sizing the height of letters, H, is

$$H = 0.0022D + K_1 + K_2, \tag{33-1}$$

where

H is in inches,
D is viewing distance in inches,
K_1 is a correction factor for illumination and viewing conditions, and
K_2 is a correction factor for importance.

Values for K_1 are as follows:

	Illumination Level	
Reading Conditions	>1.0 fc	<1.0 fc
Favorable	0.06	0.16
Unfavorable	0.16	0.26

K_2 is 0.075 for important items such as emergency labels and 0.0 for all other conditions.

Another estimate of letter height is

$$H = 0.0046D. \tag{33-2}$$

Equation 33-2 does not consider adverse conditions, importance of labels, or vision problems of readers. In some cases, it would be better to adjust sizes upward than to limit them to the values from these empirical equations. This is especially true if one anticipates poor lighting, information critical to safety, and the population of readers that may have vision problems. If characters have long ascenders and descenders, the size of the main body of the characters is the important dimension. Adjustments in overall height are needed to keep readability high.

Sizing and spacing of scale markings are important for accurate reading. Figure 33-3 illustrates recommended dimensions for scale markings and spacing for a 28-in reading distance. The dimensions should be adjusted for other reading distances.

Coding

Codes are representations that have meaning. For example, the inverted triangle shape for traffic signs infers caution and the hexagonal shape is reserved for stop signs. One can use numbers, letters, shapes, colors, and configurations to code visual information. Compound codes use two or more codes at the same time. Codes often are applied for convenience or some practical reason, but they should be used with care. Codes may slow processing, particularly when they are not used regularly, because a person must recall the meaning of the code. In some cases, however, codes may speed up a task.

Auditory Displays

Auditory displays are commonly used for warnings and alarms. They are generally preferred over visual displays for many of the situations listed in Table 33-3. Table 33-4 summarizes characteristics of various audio alarms.

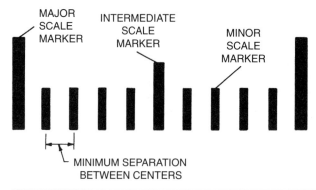

ATTRIBUTE	LOW ILLUMINATION, 0.03 TO 1.0 ft-L	NORMAL VIEWING CONDITIONS, above 1.0 ft-L
Major scale marker height	0.22	0.22
Major scale marker width	0.035	0.0125
Intermediate scale marker height	0.16	0.16
Intermediate scale marker width	0.030	0.0125
Minor scale marker height	0.10	0.09
Minor scale marker width	0.025	0.0125
Minimum separation between scale marker centers *	0.07	0.035
Minimum distance between major scale marker centers	0.5	0.5

Figure 33-3. Recommended minimum scale marking dimensions (inches). The data assume a 28-in reading distance and high contrast between graduation marks and a dial. In the table, the superscript asterisk denotes that the distance should never be less than one stroke width for black marks on a white dial face.

TABLE 33-3   When to Use Auditory and Visual Displays[a]

Use Auditory Presentation if	Use Visual Presentation if
The message is simple	The message is complex
The message is short	The message is long
The message will not be referred to later	The message will be referred to later
The message deals with events in time	The message deals with location in space
The message calls for immediate action	The message does not call for immediate action
The visual system of the person is overburdened	The auditory system of the person is overburdened
The receiving location is too bright or dark-adaptation integrity is necessary	The receiving location is too noisy
The person's job requires moving about continuously	The person's job allows remaining in one position

[a] Van Cott, H. P., and Kinkade, R. G., eds., *Human Engineering Guide to Equipment Design*, rev. ed., Superintendent of Documents, U.S. Government Printing Office, Washington, DC, 1972.

TABLE 33-4   Characteristics and Features of Selected Audio Alarms[a]

Alarm	Intensity	Frequency	Attention-Getting Ability	Noise Penetration Ability
Foghorn	Very high	Very low	Good	Poor in low-frequency noise
Horn	High	Low to high	Good	Good
Whistle	High	Low to high	Good if intermittent	Good if frequency is properly chosen
Siren	High	Low to high	Very good if pitch rises and falls	Very good with rising and falling Frequency
Bell	Medium	Medium to high	Good	Good in low-frequency noise
Buzzer	Low to medium	Low to medium	Good	Fair if spectrum is suited to background noise
Chimes and gong	Low to medium	Low to medium	Fair	Fair if spectrum is suited to background noise
Oscillator	Low to high	Medium to high	Good if intermittent	Good if frequency is properly chosen

[a] Van Cott, H. P., and Kinkade, R. G., eds., *Human Engineering Guide to Equipment Design*, rev. ed., Superintendent of Documents, U.S. Government Printing Office, Washington, DC, 1972.

33-4   CONTROLS AND MOTOR ACTIVITY

Description

After processing information in the person–machine loop, an operator activates some control that modifies what the machine is doing. Most controls require some force to activate them. Controls must be accessed easily, must function within human limits of force, skill, and duration, must operate consistently with expected response of the system, and must have dimensions that fit people. If these and other characteristics are not met, people will make errors in using controls or be unable to operate them.

Compatibility

Compatibility deals with the stimuli and responses that are consistent with human expectations. Spatial, movement, or conceptual relationships with stimuli and responses are all types of compatibility. For example, if a system component moves up and down, a control for that movement should move up and down (movement). If, as in the introductory paragraph on Three Mile Island, there are pairs of controls and displays that go together, they should have the same spatial orientation. Pair A should be on the left and pair B on the right (spatial). If water flows through five pumps in sequence, the controls for the pumps should be in the same sequence as the flow through the pumps (spatial and conceptual). Red means stop or danger; green means go or safe (conceptual).

Many compatibility relationships are learned and some are specific to particular cultures. Compatibility relationships involving control movement are called population stereotypes. For example, turning a steering wheel clockwise infers that the vehicle will turn right.

Compatibility relationships are tendencies, not universal behaviors. There are always some people who will not follow the relationship. Some compatibility relationships are strong and most people will follow them, whereas others are not as strong. In some design situations, more than one compatibility relationship may be involved and a solution may require violating one of the relationships. Use of labels will help those who may not follow the majority in applying a compatibility relationship when operating a control. Table 33-5 lists several compatibility relationships.

Example 33-1   Two workers were setting up a machine. Because vibrations would cause defective parts, it was essential to determine if parts moved together and apart smoothly. Feeling the vibrations was the only way to accomplish the fine adjustment needed. Consequently, the hands of one of the workers were placed in contact with parts of the machine. In this instance, the worker's hands were on a fixed part that was contacted by a moving part called a table.

Because the hands of the first worker were occupied in feeling for the vibration, the second worker was obliged to operate the controls. The control panel was located to the right of the machine parts, and the control for the up and down table movement was a rotary knob. The label for the control indicated that clockwise rotation would lower the table and counterclockwise rotation would raise it. When the first worker said "Drop the table," the second worker turned the control knob counterclockwise and the table moved up, crushing both of the first worker's thumbs.

This accident was probably caused by the lack of compatibility between the motion of the control knob and the movement of the machine part. With the knob on the right of the machine, most people would turn the knob counterclockwise to cause the table to move down.

Another identical machine had controls on the left and for it, the knob required a clockwise rotation to lower the table, just as the label indicated. In this case, the rotation of the control knob was more compatible with the expected up-and-down movement of the table.

Had the control in both cases been an up-and-down toggle switch that defaulted to a neutral position, there would have been an even stronger compatibility relationship between the control and the expected motion of the machine. In addition, the placement to the left or right of the table would have had no significance compared with the rotary knob.

Tracking Controls

Tracking tasks require continuous control. The right amount of movement at the right time is critical. Many kinds of tracking tasks require the operator to keep a system on course. An example is driving a car on a winding road. In other tracking tasks, the operator sees a target on a display and must keep it on course. The display may show vehicle deviation, and it is the operator's task to keep it in the center of the display or within certain bounds.

Two kinds of tracking displays are pursuit and compensatory displays. In pursuit tracking, both the target and a controlled element move. The operator tries to keep the controlled element on the target. The deviation between them is the error. In compensatory displays, either the controlled element or the target is fixed, and the remaining element moves in response to a control. Again, the operator tries to keep the two superimposed. The display may present the deviation between targets as a planar representation or a spatial one. The cursor appearing on a monitor for a computer mouse is an example of a compensatory display.

TABLE 33-5 Compatibility Relationships

Control or Control Movement	Expected System Response
Spatial	
Cooking stove: control-burner units	
Which control on front operates which burner on surface	Left/right controls operate left/right burners
	Front/rear relationships confusing
Aircraft engines	
Four controls, one for each engine should be spatially consistent with engine locations	
Operating engine control on far right	Engine at far right affected
Operating engine control on far left	Engine at far left affected
Location of separate controls for forward movement and backward movement, assuming operator is facing forward	
Operating movement control farthest forward of operator	Affects forward movement
Operating movement control nearest to operator	Affects rearward movement
Movement	
Turning a horizontal steering wheel clockwise	Vehicle turns right
Turning a horizontal steering wheel counterclockwise	Vehicle turns left
Turning a vertical steering wheel clockwise	Vehicle turns right
Turning a vertical steering wheel counterclockwise	Vehicle turns left
Moving a horizontally mounted control lever up	Controlled object moves up
For fixed-scale display and rotary control	
Clockwise rotation of control	Pointer moves clockwise
	Value represented increases
Counterclockwise rotation of control	Pointer moves counterclockwise
	Value represented decreases
Rotary controls and linear controls in the same plane	
Control on left, vertical scale on right	
Clockwise rotation of control knob	Pointer moves down and value decreases
Counterclockwise rotation of control knob	Pointer moves up and value increases
Control on right, vertical scale on left	
Clockwise rotation of control knob	Pointer moves up and value increases
Counterclockwise rotation of control knob	Pointer moves down and value decreases
Rotary controls and display movement in different planes (these may be associated with the similar action of screws and bolts)	
Clockwise rotation of the control	Movement away from control
Counterclockwise rotation of the control	Movement toward the control
Conceptual	
Red, yellow, and green traffic or signal lights or signal lights	Red means stop or dangerous
	Yellow means caution
	Green means go or safe

Control order in tracking tasks is important. Order applies to the way the device or system responds to a movement of the control. Table 33-6 characterizes the terminology and functions of control order.

The movement of a control can be continuous, discrete (step changes), or proportional (ramp change), and the response varies with system or device. Higher-order track-

TABLE 33-6   Control Order for Tracking Tasks

Order	Control Type	Movement of Device Resulting from Movement of Control	Example
Zero	Position control	Movement of device is directly proportional to control movement	Pointing a spotlight at a moving performer
First	Rate control	Rate of change in a device is related to movement of control	Depressing the gas pedal of a car changes the speed (rate of change in position)
Second	Acceleration control	Movement of the control causes a change in the rate of movement of the device	A steering wheel in a car causes the car to change direction proportional to to the angle of the front wheels
Third or greater	Higher-order control	Movement of the control causes a change in the rate of change in the device or higher-order movement	Some ship steering systems approximate a third-order control

ing is very difficult for people to master because a control movement may have too great an effect on the target. Predictor displays and aiding and quickening features in controls help operators manage higher-order control tasks.

Types of Controls

Control devices may be classified as discrete or continuous and linear or rotary. Discrete controls have predetermined positions and transmit discrete information, thereby disallowing a control to be between positions. Conversely, a continuous control can take any position within its range and can transmit values throughout its range of movement. Linear controls essentially move in a line, whereas rotary controls move in a circle or arc. Figure 33-4 illustrates controls for each pair of classifications.

There are many factors that affect the selection and design of controls. Included are control/display (C/D) ratio, direction-of-movement relationships, control resistance, grip, control coding, control function, the control or tracking task, information needs of the operator, space availability, and consequences of inadvertent activation. Some of these factors are reviewed briefly. For further details on control design, refer to more complete discussions in works cited in the bibliography.

Control/Display Ratio

C/D ratio or control/response (C/R) applies to continuous controls and refers to the distance a control moves relative to the movement of the display or system or the response of a system. C/D ratio is important in the ability of an operator to track movement and the time required to move the control and gain the desired response of the system. The significant aspects of C/D ratio vary somewhat with control order. An example of C/D ratio is the difference in steering ratio for a race car compared with a luxury car. The race driver moves the steering wheel a small amount to cause the vehicle to turn sharply. The driver of a luxury car moves the steering wheel much farther to achieve the same degree of turning. The race driver would have difficulty negotiating curves without the proper C/D ratio.

For many continuous controls, the operator moves the display quickly to the approximate desired location (slewing movement) and then uses small movements to adjust the

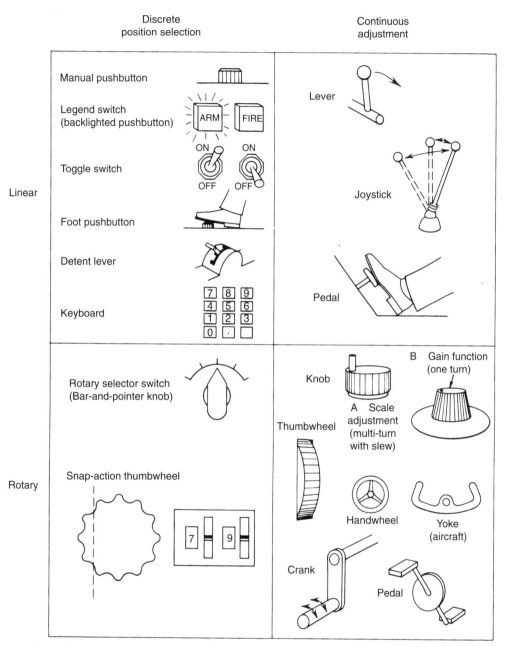

Figure 33-4.   Examples of controls. (Reprinted with permission from Huchingson, R. D., *New Horizons for Human Factors in Design*, McGraw-Hill Book Company, New York, 1981.)

display precisely. This minimizes the time required to move the display on target. C/D ratio affects these two movements: a high C/D ratio increases the time required to move the display on target, whereas a low C/D ratio may cause the operator to overshoot the target and have control problems.

Control Movement

Other factors besides C/D ratio affect response time and errors. If the hand must move a large distance to reach a control, the response time is increased compared with a short movement distance. For movements less than 3 in, vertical movement of the hand is fastest. For larger movements, horizontal movements are faster than vertical. Fore and aft movements are preferred over lateral.

The amount of movement also can impact error rates. If controls have too little movement, operators may actuate them inadvertently. However, compatibility of movement, standardization, and other factors also are important in movement so as to minimize movement time and errors.

Control Resistance

The force required to move a control affects the response time, errors, accidental activation, and operator fatigue. There are different methods for introducing reactive force into a control. Springs, friction, viscous damping, and inertia may be used. Forces that are too low cannot be sensed and increase errors; forces that are too high reduce response rates and may increase fatigue in muscles involved in the control activity and eventually may lead to fatigue errors.

Forces and the variation of force with movement can affect performance. Spring or elastic resistance increases with the distance moved. Near the neutral point of a continuous control movement, the control may have a "sloppy" feel for the operator, and near the center the movements have little resistance. To reduce this feel, a combination of elastic and inertial forces may be desirable. Some rotary controls have a detent for on–off. A high detent force to turn the control on may cause an initial movement that is too large after the control is turned on.

Control Coding

Controls are coded to help an operator identify them. The common types of coding are location, labels, color, shape, and size. Several kinds of coding may apply to a particular control. For example, emergency shutoff push-button controls often are located away from other controls, are larger than other controls, are red, and are labeled in large letters for easy identification and use. In selecting coding, it is important to apply existing standards and to evaluate compatibility with users and other principles.

Preventing Accidental Activation

Some controls that are activated accidentally may have serious consequences. Methods for protecting controls from accidental activation include recessing, location, orientation, covering, locking, operational sequences, and resistance. A combination of methods may be desirable. The application and potential consequences of accidental activation will affect which methods are useful and suitable. If accidental activation leaves a system in a safe condition, rigorous methods are probably not necessary, but if accidental activation may produce serious injury, methods that greatly reduce the likelihood of activation must be applied. Maintenance is also an important consideration. A failure in a protective device may increase the likelihood of accidental activation until the protective device is repaired or replaced. Protective methods should not violate other design principles directed at minimizing errors. Figure 33-5 illustrates several of these methods.

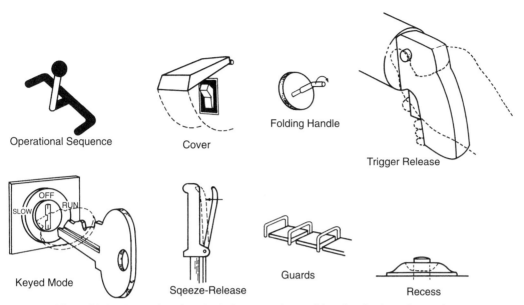

Figure 33-5. Examples of methods for preventing accidental activation of controls.

Coding Coding methods are directed at identification errors. Coding will have little effect on accidentally bumping a control or on population stereotypes when they are inconsistent with them. For example, people often operate a control consistent with expectations, rather than consistent with labeling.

Recessing To prevent activation of a push-button or toggle control, the button should be recessed below the surface of the control panel. The depressed area for the button must be large enough for fingers or activating tools to access the control. Whether an operator wears gloves should be a design consideration, also.

Guards Another technique is to provide raised barriers or guards between push buttons or around them. These are called perimeter guards, and when buttons are adjacent to each other, they prevent activation of two buttons with the same finger. If there is a guard around a push button, the guard will be bumped instead of the button, unless the force is applied inside the guard. Guards require additional spacing between controls.

Another type of control guard is a cover. Covers may be full or partial. A full cover totally encloses the control, and the cover or a portion of it must be moved to gain access to the control. Full or partial covers are common for foot controls where there is a danger of some object falling on the control and activating it. For frequently used controls, covers may introduce inconvenience and may lead to their removal.

Location and Space Locate controls away from other controls and away from movement of people or equipment to prevent inadvertent contact. There should be sufficient space between controls for hand, finger, or tools to operate controls without affecting others. The clearances needed vary with the type of control.

Resistance Another means to prevent inadvertent operation of a control is to introduce a resistance force for fixed positions. Detent wheels and detent on–off rotary switches are examples of resistant forces. Some levers require a breakout force to move them from the neutral position.

Orientation   Controls can be positioned so that normal movement around them will not activate them. If hand movements are lateral, vertical control movement for toggle switches reduce accidental activation. If one can walk by and snag clothing on a control, vertical orientation may provide some protection. If falling objects also can activate a control, both horizontal and vertical orientation may be ineffective protection.

Locking   There are several ways to lock controls into position. One form of lock is a position slot where the control requires two directions of movement to change positions. Another form of lock is a squeeze release on a lever, which requires the operator to squeeze the handle and unlock the position for the lever before it can be moved. A third form of lock is a button on a finger control that must be depressed to release the movement of the finger control. A fourth kind of locking control is a keyed control where a key is inserted into the control to allow the control to be moved to the desired position, after which the key may be removed.

Operational Sequence   Operational sequence involves requiring multiple steps to activate the control. Some of the locking devices have operational sequence. The squeeze release lever and the push button that releases a finger control are both examples. Operational sequences may have interlocks to the next or later steps. Opening a combination padlock is a type of operational sequence. A lock pin through a hole in a hitch pin for a trailer requires an operational sequence to release the pin.

33-5   WORK PHYSIOLOGY

Description

Another area of ergonomics is the ability of people to perform physical work and human physiological factors that limit physical work and related activities. People are limited in the amount and rate of physical work produced, and there are many physiological factors that contribute to such limits. The limits may be local or general. An example of a local limit is a finger or hand that becomes fatigued from operating a hand tool or push button, which results in a declining rate of activity and force generated. The fatigue may involve limitations in muscle activity, circulation or nerve conduction, disease of joints or other tissue, or other limiting factors. Limitations may affect the general ability to perform physical work.

The body must deliver oxygen and food to tissues or the tissues cannot function. The digestive system of the body converts food into useable forms and the excess food is stored as body fat. The circulatory system moves the food to tissues and cells, and the respiratory system transfers oxygen from the air into the blood for transport to the cells.

All these systems in the body have limits. There are limits to the food supply and the rate the body can convert it to the right form, and there are limits to the capabilities of the respiratory system and the circulatory system. Disease, behavior, and environment may limit the normal capacity further. Earlier discussions noted the effect carbon monoxide has on oxygen transport. Another notable example is reduced nerve conduction when potassium levels in the body decline. Reduced nerve conduction affects muscle strength and contraction.

This section does not address all of the factors that affect human physical performance. Discussion is limited to energy expenditure and metabolism. Readers should refer to Astrand and Rodahl[3] or a similar reference for a more complete review of work physiology.

Metabolism and Energy Expenditure

The energy output from muscle activity is related to oxygen consumption during the activity. The body burns fuel through two mechanisms. The first is oxidation of fuels: glucose (carbohydrates), proteins, or lipids (fats). The second involves breaking down glucose and glycogen molecules into two or more fragments, which are then oxidized by other fragments. The first mechanism is aerobic oxidation; the second is anaerobic.

The rate of oxygen uptake in the body increases with exercise and decreases with rest. Figure 33-6 illustrates the oxygen uptake after a change in activity level. It takes the body a few minutes to adjust respiration and heart rate to handle the new activity level. During this period, there is an "oxygen debt," which represents a temporary shift from aerobic to anaerobic oxidation or a supplementation of aerobic activity by anaerobic activity. After several minutes, the body adjusts oxygen uptake for the new activity level, and steady-state conditions denote that the oxygen uptake equals the oxygen requirements of the tissue. Cessation of the activity is followed by a recovery period and a return to a resting level. The oxygen debt is repaid during the recovery period, when the body replaces materials lost in anaerobic oxidation and completes removal of waste.

Humans have energy production limits. Because anaerobic capacity is very limited and of short duration, most activity involves aerobic oxidation. Maximum aerobic power is a function of maximum oxygen uptake. The maximum aerobic capacity, which is determined from physiological tests on treadmills or bicycle ergometers, varies with physical conditioning of an individual. For an individual, activities can be expressed as a percent of maximum aerobic capacity.

The energy produced during an activity is related to the oxygen uptake. Table 33-7 lists the energy cost for various activities for a person weighing 167 lb. Larger people have higher energy expenditures for the same activity because there is more body mass in motion, and smaller people have lower energy costs. Table 33-8 lists a classification of work by severity.

Humans are inefficient at converting oxygen and fuel to useful energy. Efficiencies as high as 30% can be sustained for only 1 or 2 min. For most activities, the energy cost is converted to waste heat.

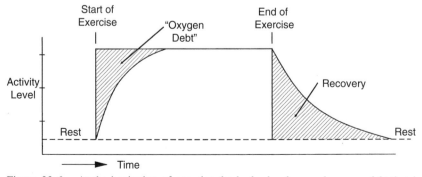

Figure 33-6.   At the beginning of exercise the body develops and *oxygen debt* that is recovered when exercise ends.

TABLE 33-7 Energy Costs for Selected Activities[a,b]

Activity	Energy Cost		
	lb oxygen/hr	kcal/min	Btu/hr
Sleeping	0.04–0.05	1.1–1.2	260–280
Resting, sitting	0.07	1.7	400
Writing	0.07	1.8	430
Typing	0.09	2.3	550
Playing musical instrument	0.11	2.9	690
Golf	0.21	5.4	1,290
Tennis	0.25	6.3	1,500
Swimming at 1 mi/hr	0.26	6.9	1,650
Cycling at 8–11 mi/hr	0.22	5.7	1,360
Walking slowly	0.15	3.8	900
Shoveling sand	0.27–0.30	6.8–7.7	1,620–1,830
Chopping wood	0.29	7.5	1,790
Digging	0.35	8.9	2,120
Climbing stairs	0.47	12.0	2,860
Marching double time	0.52	13.3	3,160

[a] From Webb, P., ed., *Bioastronautics Data Book*, NASA SP 3006, Superintendent of Documents, U.S. Government Printing Office, Washington, DC, 1964.
[b] Data are for a 167-lb person.

TABLE 33-8 Classification of Physical Work by Severity[a]

Grade of Work	Energy Cost			
	lb oxygen/hr	l oxygen/min	kcal/min	Btu/hr
Very light	<0.10	<0.5	<2.5	<595
Light	0.10–0.19	0.5–1.0	2.5–5.0	595–1,190
Moderate	0.19–0.28	1.0–1.5	5.0–7.5	1,190–1,785
Heavy	0.28–0.38	1.5–2.0	7.5–10.0	1,785–2,380
Very heavy	0.38–0.47	2.0–2.5	10.0–12.5	2,380–2,975
Unduly heavy	>0.47	>2.5	>12.5	>2,975

[a] From Christensen, E. H., *Ergonomics Research Society Symposium on Fatigue*, H. K. Lewis, Ltd., London, 1953, p. 93.

Application

Knowledge of energy cost for activities is important in designing physical work for people. Capacity and the physiological limitations that can affect physical work must be known. Energy cost also is important for assessing the dangers of thermal environments because heat gain or loss must be balanced with the heat produced by the activity. Danger increases when activity and heat are combined. Understanding energy costs is important when selecting personal protective suits, particularly full body suits, which must have cooling systems to remove the waste heat from physical work. The movement restrictions of clothing often increase the energy expenditure levels higher than those shown in the tables.

33-6 BIOMECHANICS

Description

Biomechanics is the application of mechanics to biological problems. It builds on anatomy, anthropometry, and kinesiology, the study of human movement. Biomechanics involves kinematics—the geometry and patterns of movement. Kinematic variables are displacement, velocity, and acceleration. Biomechanics also involves kinetics—the forces, energy, power, and work involved in movement.

Use of data on link segment lengths, centers of gravity, moments of inertia, and mass, combined with measurement of linear and angular displacement and forces, allows development of analytical models to describe or evaluate what is going on during various activities. Use of static and dynamic analysis facilitates evaluation of lifting tasks, pushing, pulling, turning, swinging tools, and other motions. Film and video techniques allow viewing of the dangers involved in motion-related activities. Figure 15-1 illustrates a static model. The revised NIOSH lifting equations (Equations 15-1 and 15-2) incorporate knowledge of lifting task biomechanics.

Figure 33-7 diagrams the static loads in the forearm and hand. The reactive loads on the elbow must be created by the muscles acting at the elbow. The weight of the forearm and hand act at some distance (center of mass) from the elbow. For equilibrium, the load created by the segment masses must have a force and a moment at the elbow. The load on the hand creates an additional force and moment at the elbow, transmitted through the wrist joint. If the hand and arm were in motion raising the load, there would be additional inertial forces and moments added to the static components.

Application

There are many applications of biomechanics to safety problems. One application is analysis of resistance forces required for walking on surfaces. The longer the stride, the greater the frictional forces needed to prevent slipping. In a long stride, compared with a short stride, the legs extend forward and rearward further and have greater angles from vertical.

Another application is analysis of repeated motions involved in many jobs. Chapter 13 discussed cumulative trauma disorders (also called musculoskeletal disorders), and Chapter 15 included a discussion of manual materials handling problems and controls.

One can watch for other indicators of task-related biomechanical problems. Extreme joint flexion, unusual postures, large forces, forces not in line with body motion, vibration, and highly repetitive motions are all indicators of potential ergonomic problems. Table 33-9 is a checklist for ergonomic risk factors.

33-7 WORKPLACE AND EQUIPMENT DESIGN

Principles of Workstation Layout

The application of anthropometry, biomechanics, displays, controls, and other ergonomic components are integrated into the design of workstations and equipment. There are workstations for different postures, primarily standing and sitting. The discussion that follows includes only selected concepts, principles, and data for workstation design. Sanders and McCormick[4] identify four general principles for workstation layout: importance, frequency-of-use, functional, and sequence-of-use principles. These principles apply to

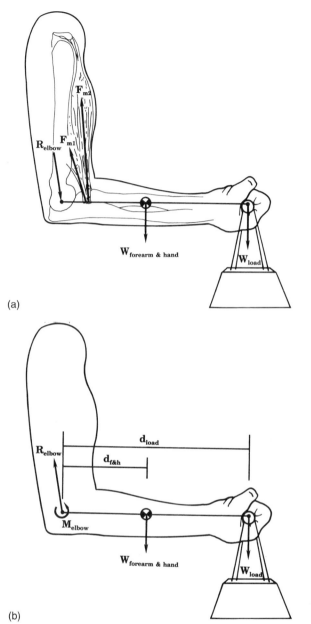

(a)

(b)

Figure 33-7. Example of a biomechanical analysis of the forearm and hand. (a) The reaction force at the elbow must equal the sum of that produced by the forearm and hand. The muscles balance the moment created by the forearm and hand and the weight. (b) The forces and moments are simplified in a free-body diagram.

location and arrangements of components. Items should be placed within reach and view and they should be organized to distribute tasks between hands or between hands and feet.

Importance Principle How important a component is to the overall operation of a system must be considered. Important components should be directly in front of an operator. There are several schemes to judge importance.

TABLE 33-9   Checklist for Ergonomic Risk Factors[a]

Upper extremity	
Shoulder	Elbow held above mid torso
	Hand working above shoulder
	Reaching behind torso
Forearm	Inward rotation with bent wrist
	Outward rotation with bent wrist
	Repetitive twisting
Wrist	Palmar flexion
	More than 30′ extension
	Ulnar or radial deviation
Hand	Pinching
	Cold
	Vibration
High local force concentrations	Anywhere on skin, base of palms, or back of fingers
Low back	
Twisting	With or without load
Bending	More than 15′ with load
Load	Held large distance from body
	Large size or weight
Frequency	Continuous lifting throughout shift

[a] From Armstrong, T. J., "Ergonomics and Cumulative Trauma Disorders," *Hand Clinics*, 2:553–565 (1986).

Frequency-of-Use Principle   Another principle for layout considers how often components are used in operating a system. A display or control that is used very frequently should be given a location that is very convenient for the operator.

Functional Principle   This principle suggests that things that are functionally related should be grouped together. The grouping may be based on conceptual factors, such as flow of fluid through a piping system. Grouping by similarity, for example, keeping all electrical system controls together or all hydraulic system controls together, exhibits another type of functional relationship.

Sequence-of-Use Principle   Another principle for layout considers the order in which one reads displays or operates controls. Those things that are operated in sequence should be located together and in sequential order.

Standing Work Surfaces

Workstation height is very important. Consideration of anthropometrics together with task will determine the preferred height. Table 33-10 lists recommended standing work surface heights for three types of tasks. For lifting tasks, the surface height is somewhat reduced, and when someone must press downward or generate downward forces, lower heights are preferred to make it easier to generate the force and keep the elbow nearly straight. In general, the forearm should be extended slightly beyond horizontal so that the shoulder muscles do not have to carry as much weight of the arms. Work surface edges should be rounded to prevent local forces on the arms, and seated workers at standing height work stations should have foot rests.

TABLE 33-10   Recommended Work Surface Heights[a]

Type Task	Male	Female
Standing		
Precision work, elbows supported	43.0–47.0	40.5–44.5
Light assembly work	39.0–43.0	34.5–38.5
Heavy work	33.5–39.5	31.0–37.0
Seated		
Fine work or assembly	39.0–41.5	35.0–37.5
Precision work (e.g., mechanical assembly)	35.0–37.0	32.5–34.5
Writing or light assembly	29.0–31.0	27.5–29.5
Coarse or medium work	27.0–28.5	26.0–27.5

[a] From Ayoub, M. M., "Work Place Design and Posture," *Human Factors*, 15:265–268 (1973).

Seated Work Surfaces

Preferred work surface heights for seated activities vary with task. Minimal muscle activity to support the arms occurs when the forearms are horizontal or sloping downward a small amount and when upper arms are vertical near the body. Extended and raised arms will fatigue faster. Table 33-10 lists recommended work surface heights for seated work. Although standard desk heights are 30 in, the height should be adjustable where possible. Keyboard heights should be lower than the work surface height. However, the slope of the keyboard can affect the preferred height necessary to minimize the extension of the wrist.

Seating

There are many anthropometric dimensions important for seating. Where possible, seat pan height and back rest height should be adjustable. Most of the weight of the torso bears on the interface between the ischial tuberosities of the pelvis and the seat pan. Part of the leg weight carries to the floor through the feet. Cushioning material and seat pan contour are important in distributing the weight over as large an area of the seat pan as possible so as to minimize concentration of loads and interruption of blood circulation to tissues involved. The seat pan height should be low enough so leg weight does not bear heavily on the front edge of the seat pan and does not interfere with circulation. Foot rests may be necessary when seat pans are not adjustable or do not adjust low enough for short-legged workers. Seat pan depth should be adjustable, often through forward and backward movement of the seat back. The goal is to provide good support for the lower and middle part of the back. Table 33-11 lists recommended dimensions for office chairs. Look for adjustable features to accommodate users.

    Other factors are important in seating. Too often there is no place for factory workers who are seated to place their legs. As a result, they sit with their legs turned and torsos twisted, which leads to back problems. If workers must turn from side to side, seat pans should swivel. Vibration absorption is essential for some seated activities, such as driving a truck.

    To rise from a seated position requires positioning the center of gravity of the body over the feet. Some people, particularly the elderly, have problems getting out of low, soft seating because the problem of moving the center of gravity forward is compounded by

TABLE 33-11 Recommended Dimensions for Office Chairs[a]

Seat	
Height from floor	16–20.5 in[b]
Width (breadth)	17.7 in
Length (depth)	15–17 in[c]
Pan angle	0–10° or adjustable in this range
Seat back-to-pan included angle	90–105°, adjustment preferred
Backrest	
Height	Variable with task and back angle
Width	At least 12.5 in in the lumbar region
Armrest	
Inside distance	At least 17.2 in

[a] From ANSI/HFS 100, *Human Factors Engineering of Visual Display Terminal Workstations.*
[b] Operators with popliteal heights of less than 16 in may need a footrest.
[c] Chairs with seat depths exceeding 16 in shall provide relief to the back side of the knee (such as contouring).

the need to raise it upward to move to the front of the seat. Some chairs have features that assist the user from a seated to a standing position.

Seats in automobiles cannot have the seat pan at the desired height. As a result, the seat pan and seat back are tilted. The tilt increases as the distance between floor and seat pan becomes smaller. Steeply tilted seat backs require the driver to tip the head forward to see, increasing neck muscle activity.

The support system for seats is important for safety. The fewer support legs or casters and the smaller the support area formed by the legs or casters, the greater the chances are of tipping a chair over. Most office chairs today have five support legs and casters to improve stability.

VDT Workstations

Questions about health and various disorders in reference to use of computers rose with expanded use of computers in offices and other operations. Disorders associated with the use of computers and computer terminals have varied origins.

There are eye and vision problems. Having the right eyeglass prescription and correct viewing distance in the prescription is important. Having the right monitor solves many problems: on some screens characters become wiggly from radio interference and circuitry problems; older systems have low-resolution characters; reflective glare from screen surfaces can make viewing of characters difficult; poor color choice can affect character discrimination. Bright background colors can create visual afterimages, which is called the McCollough effect. Adjustable brightness and contrast controls should be included with a monitor.

There are also problems of workstation layout. Putting the screen and keyboard at the right height are important. The correct seating adjustment is also important because the back needs proper support from the chair. Extended use of improperly adjusted heights causes arm, shoulder, and neck muscles to become fatigued, producing pain that can continue well after leaving work. The normal viewing angle is approximately 15° to 20° below horizontal from the eyes because people are not as comfortable looking straight ahead. Copy stands should be close to the screen, and the screen and keyboard should be directly in front of the user. Some monitors now come with bases that permit adjusting the tilt. For people who normally work for extended periods (more than 1 or 2 hr) at a computer workstation, breaks and changes in activity can help reduce problems.

Keyboards have become extremely common control devices with the expansion in the use of computers. There are models that have each half of the keyboard slightly rotated. This feature helps position the keys in line with the arms and hands. Arms and hands cannot be positioned perpendicular to a straight keyboard because of the width of the body. To use a straight keyboard, one must bend the wrists, thus putting some strain on the wrists. Workers should change tasks to reduce the intensity of keyboard data entry and find opportunities to stretch and relax the muscles and tendons involved in typing. Those involved in data entry should look for ways to avoid keyboard data entry, such as through the use of automated systems, optical character recognition, and importing data directly from files. One way to minimize potential trauma from keyboard work is to find ways to avoid keyboard tasks. The other methods are likely to improve productivity in addition to reducing risk of repeated motion disorders.

Maintainability

Too often designers think of normal use and operation of equipment and they forget the tasks related to maintenance and repair.

Designs that incorporate maintainability concepts will reduce errors during maintenance activities and help prevent unnecessary damage to components because workers cannot see or reach into areas where work is carried out. Access points and panels should be convenient and large enough for the work involved. Small holes are needed for hands and arms; large ones for putting the head, shoulders, or entive body into a compartment. There should be viewing ports as well. There are ways to code components so that one does not confuse them. There are ways to design connectors and fittings so that the wrong components are not placed in the wrong locations. References listed in the bibliography provide insight into many techniques for incorporating human factors principles into maintainability. Many involve safety.

33-8   DESIGNING FOR THE WORKFORCE

The Changing Workforce and Population

The demographics of the workforce are changing. The number of women in traditionally male jobs is increasing, and the age of the United States population is increasing and will continue to do so into the twenty-first century. More than in the past, we also look to integrate the disabled into society: most people with disabilities are employable. There are implications for safety related to these changes in the work force. Changes in employment laws reflect these changes and challenge the employer to ensure that the workplace is designed for all people. Manufacturers and designers of products and environments must meet these changes in demographics as well.

Women

Women are capable of most jobs men are, but they have faced problems obtaining protective clothing that fits and workstations designed to accommodate their size. Workstation furnishings need to adjust to fit people of all sizes, regardless of gender. Lifting and bending tasks pose equal dangers for men and women, but statistics indicate that women have higher rates of carpal tunnel syndrome than men. The increasing number of women in previously male-dominated jobs has focused more attention on workstation and task design. Subsequent redesign and reevaluation will reduce risks for everyone.

The Elderly

Typically, the capacities of the elderly are less than those of younger product users and workers. These capacities involve strength, range of motion, and duration of activity as well as diminished vision and hearing. Conditioning can restore some strength and motion for the elderly and treatment may restore vision and improve hearing. Currently, there is a growing sensitivity to the need to design for those with reduced capabilities because of aging. For example, lighting standards now include an adjustment factor for people older than 55 years when selecting illumination levels.

The Disabled

Medicine today extends the life of many people with disabilities. For example, today the survival rate for victims of automobile crashes is much higher than in the past. The unfortunate part is that many of these victims have some permanent disability. Today, people with disabilities from disease and injury are part of the mainstream of society. Federal and state laws require that public buildings, transportation, and certain housing include accessibility for the handicapped. Employment laws prohibit an employer from denying employment to someone who is disabled because the workplace cannot accommodate them. There is even a federal law that mandates that federal agencies ensure that office equipment (including computers) be accessible by disabled individuals.[5]

EXERCISES

1. A warning sign is to be located so it is readable from a distance of 20 ft. Sometimes reading conditions are expected to be unfavorable, but more than 1 fc. The message is important. How high should the letters be?

2. You are to design a VDT workstation that includes use of a keyboard from a seated position. Identify the recommended dimensions or range of dimensions for the following:

   (a) seat height

   (b) seat depth

   (c) seat pan slope

   (d) back rest slope or angle

   (e) clearance between the seat pan and the underside of the work surface

   (f) floor-to-table distance

   (g) keyboard height

   (h) viewing distance

   (i) viewing angle from horizontal

3. You are to design access to the top of a tank trailer. The vehicle operator must climb to the top to check on filling and to seal the hatch. A ladder device must be provided at the midpoint along the trailer length to reach the fill hatch on top. Sketch your design solution. Identify key dimensions and note the source of your data. The basic dimensions of the truck are shown in Figure 33-8.

4. You are to redesign and reconfigure the control/display panel shown in Figure 33-9. The panel is part of a machine that makes sand molds for castings. The panel contains lights, push buttons, selector switches, and selectors that center themselves in

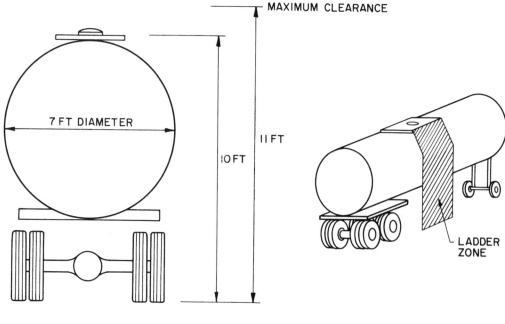

REAR VIEW

Figure 33-8. Diagram for Exercise 3.

a neutral position when not being held. The panel has a five-row and five-column layout. The solution should follow principles of ergonomic design and need not fit in the same panel. The functions of the machine are as shown in Figures 33-10 and 33-11. The automatic and manual modes for the molding machine control components are described in the sequence of functions listed in Table 33-12.

5. Locate the software that adjusts the operation of a computer mouse or game control. Change each setting to see what impact the adjustments have on the ability to use the control device effectively. Identify what control parameter each adjustment affects.

REVIEW QUESTIONS

1. What does the word *ergonomics* mean?
2. Describe what the field of ergonomics is about.
3. How is ergonomics applied?
4. Identify three things that people are good at and three that machines are good at.
5. Name two fundamental principles of ergonomics.
6. What is anthropometry?
7. Much anthropometric data are given in percentiles. Describe how to apply percentile data in design.
8. Name three principles for applying anthropometry in design.
9. Describe the six elements in the people—machine interface.
10. Name two kinds of displays.

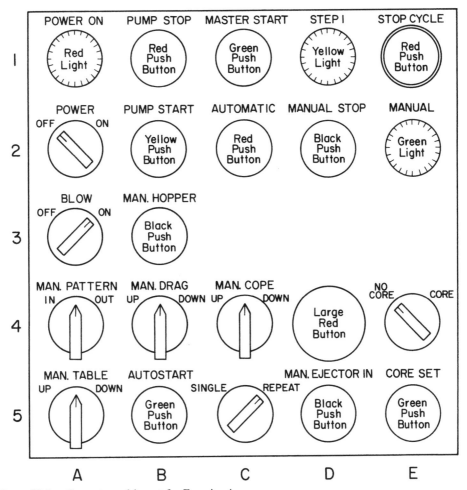

Figure 33-9.   Current panel layout for Exercise 4.

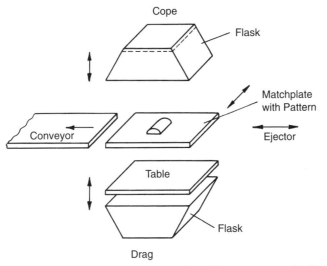

Figure 33-10.   Functional diagram of machine components for Exercise 4.

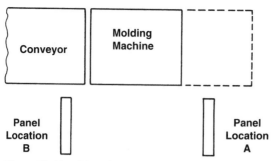

Figure 33-11.   Plan view of machine and panel layout for Exercise 4.

TABLE 33-12   Task Descriptions and Control Use for Exercise 4

Function	Selector Switch	Push Button	Light
		Components Involved	
Energize machine			
Turn power on	A2		A1
Turn sand pump on		B2	
Turn blow valves on	A3		
Select options			
Single or repeat	C5		
Core or no core	E4		
Automatic cycle or manual (default) cycle		C2	
		E2	
Position to start (step 1)			
Manual step the cycle to step 1		C1 & D2 together	
When at step 1			D1
Start cycle (automatic) (may be single or repeating cycle)			
Autostart		C1 & B5 together	
Stop cycle to insert core (see core/no core option) and continue cycle		E5	
Cycle complete			D1
To repeat cycle automatically (see single/repeat option)			
To stop at end of single cycle (see single/repeat option)			
Stop or interrupt automatic cycle			
Stop cycle		E1	
Start actions for manual mode (Default)			
Manual step to step 1		C1 & D2 together	
Manual pattern movement		C1 & A4 together	
Manual drag movement		C1 & B4 together	
Manual cope movement		C1 & C4 together	
Manual table movement		C1 & A5 together	
Manual ejector		D5	
Manual hopper		B3	
Stop manual actions			
Release the buttons			
Deenergize the machine			
Turn blow valves off		A3	
Turn sand pump off		B1	
Turn power off	A2		
Emergency power off		D4	

11. What are three characteristics that display characters should have?

12. What is coding? Give three examples.

13. What is compatibility? Give three examples.

14. What is the difference between pursuit and compensatory tracking tasks?

15. What is the difference between discrete and continuous controls?

16. What is the control/display ratio?

17. Why is control movement important in control tasks?

18. What are the four methods for creating control resistance?

19. Describe five methods for preventing accidental activation of controls.

20. What are the two main ways in which the body converts fuel to energy output?

21. Describe the significance of energy expenditure when wearing personal protective clothing, particularly full suits.

22. What is biomechanics? Kinesiology? Kinematics? Kinetics?

23. What are the four principles for workstation layout?

24. With reference to safety, why is maintainability an important part of ergonomics?

25. Name three changes in the work force and population that place greater emphasis and demands on ergonomics in design of products and workplaces.

NOTES

1 Sheridan, B., "Human Error in Nuclear Power Plants," *Technology Review*, February: 23–33 (1980).

2 Diffrient, N., et al., *Human Scale*, MIT Press, Cambridge, MA, 1981.

3 Astrand, P., and Rodahl, K., *Textbook of Work Physiology*, McGaw-Hill, New York, 1977.

4 Sanders, Mark S., and McCormick, Ernest J., *Human Factors in Engineering and Design*, 7th ed., McGraw-Hill, New York, 1993.

5 Public Law 99-506 (1986), Section 508, Reauthorization of the Rehabilitation Act of 1973.

BIBLIOGRAPHY

ALEXANDER, D. C., *The Practice and Management of Industrial Ergonomics*, Prentice-Hall, Inc., Englewood Cliffs, NJ, 1986.

American National Standards Institute, New York:

ANSI/HFS 100, Human Factors Engineering of Visual Display Terminal Workstations

ANSI A117.1, Specification for Making Buildings and Facilities Accessible to, and Usable by, the Physically Handicapped

ANSI Z53.1, Safety Color Code for Marking Physical Hazards

ANSI Z53.1 Set A, Swatches for Safety Colors Mentioned in ANSI Z53.1, Marking Physical Hazards

ANSI Z53.1 Set B, Swatches for Highway Colors Mentioned in ANSI Z53.1, Marking Physical Hazards

American Society for Testing and Materials, Philadelphia, PA:

ASTM F1168, Standard Practice for Human Engineering Design for Marine Systems, Equipment and Facilities

ASTRAND, P., and RODAHL, K., *Textbook of Work Physiology*, McGraw-Hill, New York, 1977.

AYOUB, M. M., and MITAL, A., *Manual Materials Handling*, Taylor & Francis Ltd., London, 1989.

BALER, I., *Applied Ergonomics Handbook*, 2nd ed., Butterworths, London, 1987.

BONGARRA, J. P., Jr., VANCOTT, H. P., PAIN, R. F., PETERSON, L. R., and WALLACE, R. I., *Human Factors Design Guidlines for Maintainability of Department of Energy Nuclear Facilities*, UCRL-15673, U.S. Department of Energy, Office of Nuclear Safety, Washington, DC, June, 1985.

BROWN, C. M., *Human-Computer Interface Design Guidelines*, Ablex Publishing Corp., Norwood, NJ, 1988.

CAKIR, A., HART, D. J., and STEWART, T. F. M., *Virtual Display Terminals*, Wiley, New York, 1980.

CAREY, J. M., ed., *Human Factors in Management Information Systems*, Ablex Publishing Corp., Norwood, NJ, 1988.

CHAFFIN, D. B., and ANDERSSON, G., *Occupational Biomechanics*, Wiley, New York, 1984.

CHAPANIS, A., ed., *Ethnic Variables in Human Factors Engineering*, Johns Hopkins University Press, Baltimore, MD, 1975.

CONSOLAZIO, C. F., JOHNSON, R. E., and PECORA, L. J., *Physiological Measurements of Metabolic Functions in Man*, McGraw-Hill, New York, 1963.

CORLETT, N., WILSON, J., and MANENICA, I., *The Ergonomics of Working Postures*, Taylor & Francis Ltd., London, 1986.

DAMON, A., STOUDT, H. W., and MCFARLAND, R. A., *The Human Body in Equipment Design*, Harvard University Press, Cambridge, MA, 1966.

DELLEMAN, NICO J., HSLEGRAVE, CHRISTINE M., CHAFFIN, DON B., eds., *Working Postures and Movements: Tools for Evaluating and Engineering*, CRC Press, Boca Raton, FL, 2004.

DIFFRIENT, N., et al., *Human Scale*, MIT Press, Cambridge, MA, 1981.

Ergonomic Design for People at Work, 2 vols., Van Nostrand Reinhold, 1985.

Ergonomic Interventions to Prevent Musculoskeletal Injuries in Industry, Lewis Publishers, Chelsea, MI, 1987.

Ergonomic Principles in Office Automation, Ericsson Information Systems AB, Stockholm, Sweden, 1983.

Ergonomics: A Practical Guide, 2nd ed., National Safety Council, Chicago, IL, 1993.

FISK, ARTHUR, ROGERS, WENDY A., CHARNESS, NEIL, CZAJA, SARA J., and SHARIT, JOSEPH, *Designing for Older Adults: Principles and Creative Human Factors Approaches*, CRC Press, Boca Raton, FL, 2004.

FOLKARD, S., and MONK, T. H., eds., *Hours of Work: Temporal Factors in Work-Scheduling*, Wiley, New York, 1985.

FRASER, T. M., *The Worker at Work*, Taylor & Francis Ltd., London, 1989.

FREIVALDS, ANDRIS, *Biomechanics of the Upper Limbs—Mechanics, Modeling, and Musculoskeletal Injuries*, CRC Press, Boca Raton, FL, 2004.

GRANDJEAN, E., *Ergonomics in Computerized Offices*, Taylor & Francis Ltd., London, 1987.

GRANDJEAN, E., *Fitting the Task to the Man*, 5th ed., Taylor & Francis Ltd., London, 1997.

Health and Ergonomic Considerations of Visual Display Units—Symposium Proceedings, American Industrial Hygiene Association, Akron, OH, 1982.

HENDRICKS, D. E., KILDUFF, P. W., BROOKS, P., MARSHAK, R., and DOYLE, B., *Human Engineering Guidelines for Management Information Systems*, U.S. Army Material Development and Readiness Command, Human Engineering Laboratory, Aberdeen Proving Ground, MD, June, 1983.

HUTCHINGSON, R. D., *New Horizons for Human Factors in Design*, McGraw-Hill, New York, 1981.

HUNT, V., *Work and the Health of Women*, CRC Press, Boca Raton, FL, 1979.

HURRELL, J. J., Jr., MURPHY, L. R., SUTER, S. L., and COOPER, C. L., *Occupational Stress: Issues and Developments in Research*, Taylor & Francis Ltd., London, 1988.

Institute of Electrical and Electronic Engineers, New York:

IEEE 1023, Guide for Application of Human Factors Engineering to Systems, Equipment and Facilities of Nuclear Power Generating Stations

JOHNSON, G. I., and WILSON, J. R., *Ergonomic Matters in Advanced Manufacturing Technology*, Butterworths, London, 1988.

KANTOWITZ, B. H., and SORKIN, D., *Human Factors: Understanding People-System Relationships*, Wiley, New York, 1983.

KARPMAN, V. L., *Cardiovascular System and Physical Exercise*, CRC Press, Boca Raton, FL, 1987.

KONZ, S., *Work Design*, Grid Publishing Company, Columbus, OH, 1979.

KROEMER, K. H. E., *Engineering Physiology: Physiologic Bases of Human Factors/Ergonomics*, Elsevier, New York, 1986.

KUMAR, SHRAWAN, ed., *Muscle Strength*, CRC Press, Boca Raton, FL, 2004.

Making the Job Easier—An Ergonomic Idea Book, National Safety Council, Chicago, IL, 1988.

NASA Anthropometric Source Book, 3 vols., NASA Reference Publication 1024, Washington, DC, July, 1978.

NEMETH, CHRISTOPHER P., *Human Factors Methods for Design: Making Systems Human-Centered,* CRC Press, Boca Raton, FL, 2004.

NICHOLSON, A. S., and RIDD, J. E., *Health, Safety and Ergonomics*, Butterworths, London, 1988.

NORO, K., and IMADA, A. S., *Participatory Ergonomics*, Taylor & Francis Ltd., London, 1990.

PARKER, J. F., and WEST, V. R., eds., *Bioastronautics Data Book*, 2nd ed., National Aeronautics and Space Administration, Washington, DC, 1973.

PHEASANT, S., *Bodyspace: Anthropometry, Ergonomics and Design*, Taylor & Francis Ltd., London, 1986.

PULAT, B. M., and ALEXANDER, D. C., *Industrial Ergonomics*, Institute of Industrial Engineers, Norcross, GA, 1986.

PUTZ-ANDERSON, V., ed., *Cumulative Trauma Disorders: A Mnnual for Musculoskeletal Diseases of the Upper Limbs*, Taylor & Francis Publishers, New York, 1988.

RASMUSSEN, J., *Information Processing and Human-Machine Interaction: An Approach to Cognitive Engineering*, Elsevier Science Publishing, New York, 1986.

RASMUSSEN, J., DUNCAN, K., and LEPLAT, J., *New Technology and Human Error*, Wiley, New York, 1987.

RODAHL, K., *The Physiology of Work*, Taylor & Francis Ltd., London, 1989.

ROEBUCK, J. A., JR., KROEMER, K. H. E., and THOMSON, W. G., *Engineering Anthropometry Methods*, Wiley-Interscience, New York, 1975.

SANDERS, MARK S., and MCCORMICK, ERNEST J., *Human Factors in Engineering and Design*, 7th ed., McGraw-Hill, New York, 1993.

SLAVENDY, G., ed., *Handbook of Human Factors*, Wiley-Interscience, New York, 1986.

Society of Automotive Engineers, Warrendale, PA:

(Note: Other standards on switches, vehicle brake and tail lights, reflector lights, and other items deal with displays and controls. The following standards are selected ones involving various ergonomic topics for vehicles.)

SAE ARP 4032, Human Engineering Considerations in the Application of Color to Electronic Aircraft Displays

SAE J48, Guidelines for Liquid Level Indicators

SAE J89, Dynamic Cushioning Performance Criteria for Snowmobile Seats

SAE J92, Snowmobile Throttle Control Systems

SAE J96, Flashing Warning Lamp for Industrial Equipment

SAE J98, Safety for Industrial Wheeled Equipment

SAE J99, Lighting and Marking of Industrial Equipment on Highways

SAE J107, Operator Controls and Displays on Motorcycles

SAE J115, Safety Signs

SAE J128, Occupant Restraint System Evaluation—Passenger Cars

SAE J137, Lighting and Marking of Agricultural Equipment on Highways

SAE J138, Film Analysis Guides for Dynamic Studies of Test Subjects

SAE J153, Safety Considerations for the Operator

SAE 154a, Operator Enclosures—Human Factor Design Considerations

SAE J185, Access Systems for Off-Road Machines

SAE J209, Instrument Face Design and Location for Construction and Industrial Equipment

SAE J223, Symbols and Color Codes for Maintenance Instructions, Container and Filler Identification

SAE J264, Vision Glossary

SAE J284, Safety Alert Symbol for Agricultural, Construction and Industrial Equipment

SAE J287, Driver Hand Control Reach

SAE J268, Rear View Mirrors—Motorcycles

SAE J297, Operator Controls on Industrial Equipment

SAE J386, Operator Restraint Systems for Off-Road Work Machines

SAE J389b, Universal Symbols for Operator Controls on Agricultural Equipment

SAE J826, Devices for Use in Defining and Measuring Vehicle Seating Accommodation

SAE J833, USA Human Physical Dimensions

SAE J834a, Passenger Car Rear Vision

SAE J841, Operator Controls for Agricultural Wheeled Tractors

SAE J879b, Motor Vehicle Seating Systems

SAE J898, Control Locations for Off-Road Work Machines

SAE J899, Operator's Seat Dimensions for Off-Road Self-Propelled Work Machines

SAE J925, Minimum Service Access Dimensions for Off-Road Machines

SAE J941, Motor Vehicle Driver's Eye Range

SAE J943, Slow-Moving Vehicle Identification Emblem

SAE J974, Flashing Warning Lamp for Agricultural Equipment

SAE J984, Body Forms for Use in Motor Vehicle Passenger Compartment Impact Development

SAE J985, Vision Factors Considerations in Rear View Mirror Design

SAE 1013, Measurement of Whole Body Vibration of the Seated Operator of Off-Highway Work Machines

SAE J1029, Lighting and Marking of Construction and Industrial Machinery

SAE J1038, Recommendations for Children's Snowmobiles

SAE J1048, Symbols for Motor Vehicle Controls, Indicators and Tell-Tales

SAE J1050a, Describing and Measuring the Driver's Field of View

SAE J1051, Deflection of Seat Cushions for Off-Road Work Machines

SAE J1052, Motor Vehicle Driver and Passenger Head Position

SAE J1060, Subjective Rating Scale for Evaluation of Noise and Ride Comfort Characteristics Related to Motor Vehicle Tires

SAE J1062, Snowmobile Passenger Handgrips

SAE J1071, Operator Controls on Graders

SAE J1129, Operator Cab Environment for Heated, Ventilated, and Air Conditioned Construction and Industrial Equipment

SAE J1138, Design Criteria—Driver Hand Controls Location for Passenger Cars, Multi-purpose Passenger Vehicles, and Trucks (10,000 GVW and Under)

SAE J1139, Supplemental Information—Driver Hand Controls Location for Passenger Cars, Multi-Purpose Passenger Vehicles, and Trucks (10,000 GVW and Under)

SAE J1163, Determining Operator Seat Location on Off-Road Work Machines

SAE J1164, Labeling of ROPS and FOPS

SAE 1222, Speed Control Assurance for Snowmobiles

SAE 1257, Rating Chart for Cantilevered Boom Cranes

SAE 1282, Snowmobile Brake Control Systems

SAE 1307, Excavator Hand Signals

SAE 1384, Vibration Performance Evaluation of Operator Seats

SAE 1385, Classification of Earthmoving Machines for Vibration Tests of Operator Seats

SAE 1386, Classification of Agricultural Wheeled Tractors for Vibration Tests of Operator Seats

SAE J1388, Personnel Protection—Skid Steer Loaders

SAE J1441, Subjective Rating Scale for Vehicle Handling

SAE J1460, Human Mechanical Response Characteristics

SAE J1517, Driver Selected Seat Position

SAE J1521, Truck Driver Shin-Knee Position for Clutch and Accelerator

SAE J1522, Truck Driver Stomach Position

STANTON, NEVILLE, HEDGE, ALAN, BOOKHUIS, KAREL, SALAS, EDUARDO, and HENDRICK, HAL W., *Handbook of Human Factors and Ergonomics Methods*, CRC Press, Boca Raton, FL, 2004.

STELLMAN, J. M., *Women's Work, Women's Health—Myths and Realities*, Pantheon Books, New York, 1977.

THOMAS, J., and SCHNEIDER, M. L., *Human Factors in Computer Systems*, Ablex Publishing Corp., Norwood, NJ, 1984.

TICHAUER, E. R., *The Biomechanical Basis of Ergonomics*, Wiley-Interscience, New York, 1978.

VAN COTT, H. P., and Kinkade, R. G., eds., *Human Engineering Guide to Equipment Design*, rev. ed., Superintendent of Documents, U.S. Government Printing Office, Washington, DC, 1972.

WEBB, P., ed., *Bioastronautics Data Book*, NASA SP-3006, National Aeronautics and Space Administration, Washington, DC, 1964.

Weight, Height, and Selected Body Dimensions of Adults: United States, 1960-1962, USPHS Publication 1000, Ser. 11, No. 8, U.S. Public Health Service, Washington, DC, June, 1965.

WICKENS, C. D., *Engineering Psychology and Human Performance*, Charles E. Merrill Publishing Company, Columbus, OH, 1984.

WILSON, J., CORLETT, E. N., HASLEGRAVE, C. M., and MANENICA, I., *Evaluation of Human Work: A Practical Ergonomics Methodology*, Taylor & Francis Ltd., London, 1989.

WINTER, D. A., *Biomechanics of Human Movement*, Riley, New York, 1979.

WOODSON, W. E., *Human Factors Design Handbook*, McGraw-Hill, New York, 1991.

WOODSON, W. E., *Human Factors Reference Guide for Electronics and Computer Professionals*, McGraw-Hill, New York, 1987.

PART V

MANAGING SAFETY AND HEALTH

THIS SECTION of the book deals with management. Engineers need to convert their ideas for managers, they need to talk "management talk." Engineers need to understand how to have safety and health controls implemented in an organization and they need methods for organizing a wide variety of controls and establishing priorities for them. Engineers need to understand the fundamentals of creating an organizational culture that places safety at the top of organizational priorities. The priorities need to be presented in management terms.

Two aspects are covered. The first is incorporating safety and health into an organization. Management of an organization plays an important role in implementing the controls for hazards and getting people in an organization to incorporate safety into their operations. Only a limited amount of discussion addresses this aspect of management of safety.

The second aspect addresses methods and techniques to convert engineering objectives into management actions. Most of this section deals with this aspect of managing safety. These methods help one organize a wide mix of controls for hazards and present it in a form that management understands.

CHAPTER 34

FUNDAMENTALS OF SAFETY MANAGEMENT

January 28, 1986

The O-ring seal in the booster rocket eroded, and blow-by burned a hole in the external fuel tank. Suddenly, mission 51-L exploded before a worldwide audience on live television. The screens in the control room went blank; only a white S remained at the top of each mission control monitor screen. Seventy seconds after launch, the Challenger space shuttle fell in pieces from 50,000 ft to the ocean below.

October 1986

The Presidential Commission on the Space Shuttle Challenger Accident (the Rogers Commission) presented its findings. Chapter 7, titled "The Silent Safety Program," states:

1. Reductions in the safety, reliability, and quality assurance work force at Marshall and NASA Headquarters have seriously limited capability in those vital functions.

2. Organizational structures at Kennedy and Marshall have placed safety, reliability, and quality assurance offices under the supervision of the very organizations and activities whose efforts they are to check.

3. Problem-reporting requirements are not concise and fail to get critical information to the proper levels of management.

4. Little or no trend analysis was performed on O-ring erosion and blow-by problems.

5. As the flight rate increased, the Marshall safety, reliability, and quality assurance work force was decreasing, which adversely affected mission safety.

6. Five weeks after the 51-L accident, the criticality of the Solid Rocket Motor field joint was still not properly documented in the problem reporting system at Marshall.

Another author writes:[1]

the "press on" mentality had taken hold of NASA's management, and they were able to accept conditions that would have surely provoked the scrub of another launch. . . . Rather than demanding that all those supporting the launch prove that conditions were safe, the senior members of the launch team demanded that their subordinates and the contractor representatives prove that it was *not* safe to launch Challenger.

Safety and Health for Engineers, Second Edition, by Roger L. Brauer
Copyright © 2006 John Wiley & Sons, Inc.

February 1, 2003

The space shuttle Columbia came apart and disintegrated during the descent to earth from its orbital mission. The crew of seven astronauts was lost. The Columbia Accident Investigation Board report[2] identified that the physical cause of the loss resulted from a piece of insulating foam from the external fuel tank striking the wing during the second minute after launch. The wing damage manifested itself during re-entry, resulting in structural failure and loss of control and ultimate breakup of the vehicle. The organization's causes of the accident were rooted in cultural traits and organizational practices that were allowed to develop that were detrimental to safety.

There are many stories one could cite that demonstrate the importance management plays in making products and operations safe. As noted early in this book, there are many people who have important roles in safety. Engineers and others who identify hazards that need to be corrected must work with management to be sure that safety is achieved. Implementing safety in an organization requires participation from everyone, from the top of an organization to the bottom. In developing a company culture, top management must incorporate safety as a critical cultural factor.

34-1 ELEMENTS OF MANAGEMENT

What is management? What do managers do? What are the elements of management? The many books and writings on management provide countless definitions of management and what is involved in management. The purpose here is to create a general understanding of management that will provide a framework for addressing how safety can be a part of management and the endeavors of organizations.

Chapter 9 proposed a Goal Accomplishment Model (Figure 9-4) to help identify hazards and controls. The model assumed that organizations have goals that are to be accomplished. There are several elements that contribute to accomplishing the goals:

Activities: knowing what must be done and doing it.

People: having the right number of workers and skills for the activities.

Equipment and Materials: having the right tools, machines, process equipment, and raw materials to extend the activities and to produce the output.

Facilities: having the buildings and systems necessary to make the activities productive.

Physical Environment: keeping the work place and its surround conducive to effective work and healthy workers and community.

Social, Management Environment: providing the leadership, communication, work culture, and motivation for people to work effectively together.

Regulations: having policies, methods, and procedures in place for things to be orderly and safe.

Time: having enough time or organizing time to complete the activities.

Cost: having enough funds to provide the preceding elements necessary to accomplish the organization's goals.

Management involves planning, obtaining, organizing, and orchestrating the elements necessary to achieve the goals.

At a basic level, Grose[3] characterizes the three main elements of management as the three legs of a stool, which represent performance, cost, and schedule. Performance denotes having defined tasks. Cost includes funds and manpower for the tasks. Schedule involves ordering tasks into a sequence and assigning dates for completion. The three-legged stool of management reflects a model quite suitable for management of projects.

Grimaldi and Simonds[4] identify three steps in an orderly pursuit of an objective. The three steps are organizing, administrating, and managing. Organizing is the structuring of authority and activity relationships and using the resources at hand to meet a group's objectives. It includes the arrangement of subtasks into a coordinated effort. Administrating is carrying out the tasks of planning, organizing, coordinating, and measuring performance that move an organization toward its goals. Managing, although similar, adds the dimension of leadership. Leadership sets the tone for the tasks. It involves the use of facts and persuasiveness to strengthen authority and achieve the goals effectively.

Bittel and Ramsey[5] compiled material on 50 vital areas of concern to professional managers. They built the areas out of three basic ones.

1. Primary management functions, such as planning, organizing, activating, controlling, and decision making.

2. Major business activities, such as finance and accounting, operations and production, marketing and sales, and information management.

3. Environmental resources and constraints, such as human resources, materials, funds, equipment and facilities, consumer demand, economic conditions, natural resources, community influences, and government regulations.

The management process assembles resources and converts them into output, such as products and services, working within a framework of various environments. This complex process, illustrated in Figure 34-1, is another way to describe what management is all about.

Management is getting things done through other people. Management is the performance of those functions essential to the success of an organization. Safety is part of those functions.

34-2 SAFETY IN AN ORGANIZATION

As noted in Chapter 32, safety begins in an organization with a policy stating the overriding importance of safety. It assigns responsibility, authority, and accountability. An extension of the policy is procedures for implementation.

There are many ways to structure an organization to make it effective. As a result, there are many ways to assign safety responsibilities. For most organizations, there are two major components: line and staff. Line elements of an organization are those that produce a product or deliver a service; they get the work done. Staff elements take care of business matters, such as finance, accounting, research and development, and sales. Staff elements also take care of special matters, such as legal, security, training, engineering, and maintenance. Staff elements assist and facilitate getting the work done, but they do not have authority over line elements. Figure 34-2 illustrates one organizational structure and where safety specialties could be located.

Safety must be a part of every organizational element. Line elements must be sure that the work is done safely and they depend on the training, procedures, and technical assistance of staff elements to know what is safe. Each level of supervision or management in line elements must keep safety paramount. Otherwise, it is likely to lose its importance for levels below. Safety must be part of the leadership characteristics of every

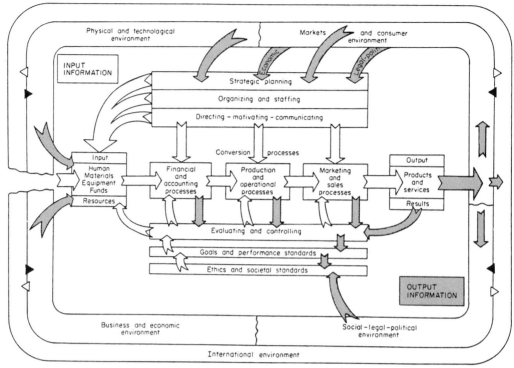

Conceptual rationale for determining scope of subject matter.

Figure 34-1. The management process and its context. (Reprinted with permission from Bittel, L. R., and Ramsey, J. E., eds., *Handbook for Professional Managers*, McGraw-Hill, New York, 1985.)

supervisor and manager. Although safety is important at every level in the line elements, some believe that first-line supervisors are the key. They directly influence the greatest number of workers and the tasks performed. These supervisors have a key role in incorporating safety into the organization.

The safety element of an organization's staff may include several specialties. Beside safety, it may include industrial hygiene, health physics, occupational medicine or nursing, and fire protection. More and more, safety, health, and environmental responsibilities are grouped into one element. Sometimes security may be combined with safety, sometimes all these specialties, with the addition of insurance and workers' compensation functions, are grouped under risk management and sometimes most of these specialties are grouped under a human resources or personnel element.

Where the line element responsible for safety is located in an organization can affect its ability to perform, as seen in the Rogers Commission report on the Challenger accident. It is essential that the safety director or staff person responsible for safety have the ear of top management and serve as the spokesperson for top management in safety matters. Petersen[6] cites four criteria for positioning the staff safety element in an organization:

1. Report to a boss with influence.
2. Report to a boss who wants safety.
3. Have a channel to the top.
4. If possible, install safety under the executive in charge of the major activity.

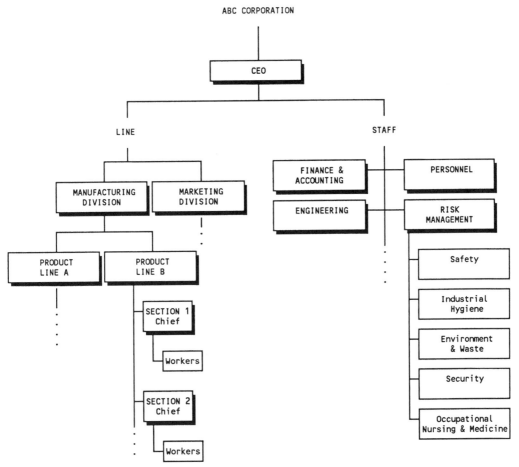

Figure 34-2. Example of one organization and placement of safety functions.

For many small or mid-sized organizations, the staff safety element is only part of one person's job function because having a full-time safety professional as the staff element is not feasible for small organizations. Even part-time specialists should pursue training about hazards, controls, safety management, and legal and regulatory matters. The safety specialist should create access to necessary medical, engineering, and other safety specialists to give the organization support from certified and licensed people. The part-time safety specialists may not qualify for these credentials.

34-3 ACHIEVING SAFETY IN AN ORGANIZATION

There are innumerable theories of management, and managers must apply many theories to the situations they face. Different approaches work better for different situations; different techniques may help or hinder safety. There are many factors that contribute to low accident rates in an organization. These factors are not separate from other objectives for the organization. Achieving safety requires that safety be an integral part of the management process and a component of leadership provided by managers. In this discussion, we consider several management factors that help achieve safety.

Management Style

Different people approach the tasks of management differently. Some are very authoritative, directing virtually every action of their people. Others encourage subordinates to participate in decision making and setting the climate for getting the job done. Management style can affect results.

It is not enough to have safety carefully structured into an organization, nor is having policies, procedures, training, and specialists enough. Having managers at all levels who endorse safety is not enough. The way safety is carried to the workers can significantly influence their performance. Organizational, management, or leadership style strongly influences safety on the job.

One study[7] analyzed the safety records of crews working under particular supervisors. The study found that those supervisors whose crews had a better safety record managed differently from those with poor records. The supervisors with good safety records also had high productivity. The successful supervisors controlled anger and kept stress from the crew, integrated safety into production tasks rather than simply admonishing workers to be safe, and maintained closer contact with their workers.

Accountability

Making safety part of a supervisor's or manager's performance appraisal is one means of achieving safety in an organization. Researchers[8] found that top management can affect safety by knowing the safety records of field managers and using this knowledge for promotions and salary increases. (The knowledge is derived from a cost-accounting system.) The study found that companies using the following techniques had lower accident rates than companies that did not:

1. Use cost accounting to encourage safety of managers.
2. Require managers to do detailed safety planning.
3. Provide new employees with adequate training.

For accountability to work, there must be a way to count success or failure; There must be an accounting system. Many companies use a cost-accounting system (to be discussed further in Section 34-5) for accidents. The cost data must be given to the manager of the relevant department or organization. Data can be organized further by severity and by type of accident, which will help the manager identify what should be corrected.

Other techniques for accountability can be used. One is a safety-by-objectives method. It focuses on actions to be completed, not on costs. For example, a superior can sit with a first-line supervisor, steward, or other manager, and discuss what steps can improve safety in each person's area of responsibility. A safety specialist may participate in the discussion. The outcome is a safety-by-objectives plan that may include employee training, improved discipline, improved housekeeping, motivational programs, and correction of particular hazards. The person responsible for action then lists a few specific objectives to achieve for these areas during a particular period. This one-on-one analysis and planning activity becomes the program for the supervisor. The approach focuses on improvement, regardless of the current record.

Some companies uses completion of a training and certification program for all managers and supervisors in each organizational unit as a "leading" indicator of safety performance. Upper management set quotas for each organizational unit and tracked completion of the required safety training and third-party certification program.[9] The leading indicator does not require the company to wait for accidents, incidents, workers' compensation claims, and associated costs to measure results. Other results from this approach include improvements in the company's safety culture and even productivity enhancements. The productivity enhancements seem to stem from the increased confidence of work units in handling safety issues immediately and directly and not having to use support staff to resolve many of the more straightforward safety matters.

Audits

Another approach is use of audits. An audit is a process of having insiders from other units of the organization or outsiders evaluate management. An audit can vary in depth. A management audit can be applied to an entire organization or a component of it. The purpose of the audit is to challenge existing policies, procedures, and practices and their underlying concepts and principles. It looks at leadership, how well results are achieved, and better ways to achieve them in the future. An audit can be a tool for improvement. Management audits must be handled cautiously so that they do not become a tool for blame or ridicule.

Within a company, a safety specialist or team of management and safety specialists may conduct an audit of a manager's department. Although Table 34-1 is a list of questions directed primarily at chemical processes, the questions are typical of most audits. The findings presented to the manager should include what things need improvement and ways to improve them. The findings also should address policy and procedural problems that need adjustment to make them effective for this part of the operation.

Paying Attention to Details

Paying attention to details and integrating safety into management of activities makes programs successful. These same characteristics lead to productivity. Because workers want to do a good job and are sensitive to what their employers expect of them, motivating workers is important. Motivation involves sending key messages from top management through the management chain to workers. Typical messages are the company cares, do the job right, achieve good productivity, pay attention to details, and integrate safety into the job tasks.

One safety specialist involved in construction for many years noted: "I can tell how well safety is managed on a job by looking at the housekeeping." If materials, equipment, and scrap are scattered, there is little planning and organizing being done by managers and supervisors. Conversely, having things are in order on the job and having work well organized are signs that managers and supervisors attended to the details of work planning and execution. Safety is likely to be an integral part of such management processes.

Enforcement

Enforcement is not just having rules and someone to enforce them. Enforcement involves planning and communication: clearly explaining the need to do a job right, making sure people know how to do it safely, and conveying a clear understanding of what is expected on the job. It includes motivating workers to achieve desired results, including safety and rewarding correct behavior. Having a good accident record should be reinforced by recognizing the effort expended toward safety at all times. Failure to comply and ignoring safety rules must be recognized as well, but should not be the sole component of enforcement.

TABLE 34-1   Questions Typical of a Safety Audit[a]

1. Are written instructions for the operation in place?
2. Are the instructions current? Has the process changed? Do engineering drawings reflect unit modification? Have modifications been reviewed for process safety?
3. Do the operators follow the written instructions?
4. Is the equipment correct for the process?
5. Are reactants handled and stored correctly?
6. Are necessary items of personal protective equipment in the correct locations?
7. Is ventilation sufficient?
8. Is electrical equipment compatible with the area processes (e.g., explosion-proof electrical equipment in an area where flammable vapors or explosive dusts are present)?
9. What hazardous materials are used in the process? How are the materials stored and in what quantity? What emergency procedures and equipment are available to handle accidental events associated with the materials' storage and usage? How are personnel trained to use these procedures and equipment?
10. What errors in operation are possible and within reason? What would be the consequences of such errors? Will the equipment handle these consequences with minimum risk to life and property? Are equipment safeguards in place and operable?
11. Is explosion relief venting adequate? Is the explosion relief properly directed?
12. What are the provisions for uncontrolled reactions?
13. What personal safeguards are advisable (safety showers, fire-resistant clothing, emergency care, protective equipment)?
14. What are the training provisions for new process operators? Is periodic training given to existing operators and what does it consist of? Are operators challenged by means of emergency training exercises? Is a process simulator advisable for training?
15. In addition to the sprinkler system adequacy, is additional fire protection, such as an automatic dry chemical extinguishing system or a other system, warranted? Is an explosion suppression system warranted?
16. What is the manner of receiving, storing, handling, and transferring hazardous materials and combustible liquids and gases? What controls exist to eliminate contamination of product that could create hazardous polymerization as well as accidental release? What are the most likely possibilities of accidental release? What safeguards exist?
17. Are contamination sensitive bulk materials stored above or below ground? If below ground, what steps are taken to prevent contamination and degree of assurance?
18. Are critical spare parts on hand, such as parts for circulation pumps for refrigeration systems on temperature sensitive materials stored in bulk?
19. Is an emergency dump system needed? If already in place, is it adequate?
20. Are critical maintenance procedures identified? Are personnel trained?

[a] Derived from Krivan, S. P., "Avoiding Catastrophic Loss: Technical Safety Audit and Process Safety; Review," *Professional Safety*, February: 21–26 (1986).

Moving Safety Deeper into the Organization

Chapter 3 introduced the concept of errors in management systems and the management principles of Deming and Juran. Their focus is on quality and encouraging participation in the processes of business to minimize errors (or defects) of all kinds up front and the amount of rework that otherwise occurs after things have gone wrong. Motorola applied these principles in achieving the first Malcolm Baldrige Award for quality and used two kinds of metrics in all company departments. The two metrics were errors and process time, with each department establishing their own definitions for each. Both encompass safety, because errors can be incidents and accidents, and process times are longer when recovering from incidents and accidents. The continuous improvements of management processes is driven by listening to the customer and requires the participation of all in finding ways to achieve the desired output with the least amount of errors and the shortest process time. Overall, one can seek to reduce errors to one per million opportunities, approximately six standard deviations (6 sigma, σ) out on a normal distribution curve. Today, the philosophy is captured for some companies through a "Six Sigma" quality and management process improvement program.

In applying these management principles, the old methods that use a top-down approach to managing people in an organization disappear and are replaced with more participatory and collaborative work groups. Under the old approach, managers measured subordinate performance individually, often using the errors and failures of individuals to seek improvement from them. However, as Deming points out, individuals can perform no better than is possible under the management processes given them. Instead, improvement stems from having each employee participating in identifying ways to improve the processes so that there are fewer and fewer errors and the process times are shorter and shorter to benefit the customers. The latter approach requires metrics for the processes of a work group and engaging them in improving their work collectively.

Because incidents and accidents are measured as part of the management process, the work group needs to learn what may have contributed to the opportunity for any incident and accident (one type of error) and how the process can be improved to eliminate them. This requires moving safety knowledge deeper into the work group. The old management approach positioned safety knowledge in the safety department or in upper tiers of positions. The old approach called on the safety people to fix the problem after there was an incident or accident so that it would not happen again. The emphasis on participation in improving the management processes engages the workers and leaders in understanding the processes, including hazards and controls. All participate in identifying what can be done to reduce the errors, to minimize rework, and to reduce rather than lengthen process times. In attending to management process improvement, the workers need to understand the roles that engineering and administrative controls can play. They need to understand how unsafe conditions and unsafe acts impact the processes. They need to learn how to plan for safe work (process improvement) and how to communicate across the work group to ensure that safe practices are implemented from the plan for safe work.

Behavior-Based Safety

The idea of working collaboratively on improving management processes to reduce errors (including incidents and accidents) and to shorten process times to the benefit of external and internal customers of a work group forms much of the foundation for behavior-based safety techniques. There are two main advocates for these techniques.[10] One can refer to their writings for details about applying this approach. The approach can be effective. The approach has also been somewhat controversial.

The general idea is to help workers gain insight into behaviors and to avoid behaviors that may lead to incidents and accidents. There are a number of methods for achieving this objective. The effectiveness is likely to vary with the work management methods in place and within which this approach is applied.

Chapter 3 discussed various theories of incidents and accidents. Included was a discussion about the relationships between unsafe acts and unsafe conditions and their relative roles in causing accidents. The discussion noted that some early studies tried to establish that one or the other was likely to be the cause of an incident or accident. The discussion also cited other studies that established that both were likely to have a role in events leading to an accident or incident.

The management context is likely to impact the effectiveness of engaging workers in understanding and applying information about human behavior in incidents and accidents. In some work environments, workers were assumed to be contractors, and thus, responsible for their own safety. This implies that the worker must be able to control both unsafe conditions and unsafe acts. Management left them on their own. A century ago, this approach was common in mining.

Under the legal philosophies that preceded workers' compensation being a no-fault process, workers were responsible for their own behavior with regard to risk and safety and also had some responsibility for their behavior as it may impact a fellow worker's safety. The theories of assumption of risk, contributory negligence by employers, and the fellow servant rule all placed responsibility for behavior on the individual (see Chapter 6).

Under a top-down management philosophy, a supervisor evaluates a worker's performance based on behaviors exhibited during the evaluation period. Then the supervisor establishes performance goals for the next rating period. Some early approaches to monitoring safe work behavior[11] were formal and some were informal. A more informal approach involved handing out a card to a worker observed performing a task correctly and safely as a means of positive feedback.

In a collaborative environment that focuses on continuously improving the overall process performance, the emphasis is on changing the process through policy, technology, procedures, materials, and other means. In achieving safety, there is a need to analyze many things. Included in the analysis are behaviors associated with the process changes and the risks and dangers that they may pose. There is a need to document how the process is to work and to train everyone engaged in the process about how to perform each task correctly. There is a need to understand what can go wrong and how incorrect task performance may contribute to the resulting errors (including incidents and accidents). The approach may involve simulations of various kinds, encouraging feedback among all participants and even establishing methods for measuring process performance, including individual behavior by participants.

Approaches for changing worker behavior that do not use effective feedback techniques, that operate in work climates with top-down management environments, and do not have clearly defined processes for doing things correctly are less likely to achieve measurable changes in safety performance. Collaborative work environments that engage everyone in the work group in improving processes and the overall performance for customers have a better chance for success with behavior-based safety methods, because work behaviors are simply a component of the processes that are the focus of the group.

Safety Committees

Whether an organization uses traditional management methods or methods that emphasize continuous improvement and quality through highly collaborative work groups, one technique for improving safety is through safety committees. Some state workers' compensation laws recognize safety committees as an effective means of lowering incidents and accidents that lead to workers' compensation claims.

Use of safety committees is one way of expanding participation in safety improvement efforts. The approach falls somewhere between the top-down management methods and the collaborative methods that engage everyone on quality improvement of management processes.

There are several ways to form committees and to select participants. There are many ways to assign roles and responsibilities to committees and their members. One can have one committee for an entire organization with representatives from each major work group. One can also have a committee within each work group with rotating membership from among all members of the work group. There may be a tiered structure for safety committees throughout the organization.

In general, the benefits are expanding participation in achieving safety, increasing the emphasis on safety, and moving safety responsibility from top management and the safety office into work groups. To be effective, the safety committees must have a clearly defined role and members must gain an understanding of hazard recognition, evaluation and control, and how safety can be managed. Members need to know where to obtain assistance from management, specialists, and others and how to impact processes and budgets. They need to know how to impact change to reduce risk.

34-4   SAFETY AND COST

Like any other element of achieving the goals of an organization, safety must be converted to a single common denominator—cost. Accidents and injuries cost money, as do claims, legal fees, lost production time, and compliance with government requirements. Chapter 3 discussed direct, indirect, insured, and uninsured costs, and Table 4-1 listed hidden costs of accidents. Table 34-2 lists uninsured costs. Safety and productivity go hand in hand. The costs of preventing accidents need to show a return on the profit line or at least show avoidance of expenses that would otherwise occur.

Expressing Costs

The accident costs in the United States amount to more than $600 billion per year. This figure is overwhelming to most people, but it has little meaning for the ordinary person, and it certainly has little motivational potential. Costs need to be expressed in terms that are useful. Expressing costs in the right terms can help people understand the importance of safety and its contribution to company profit. Although each level may have a preferred way of expressing cost data, it can help first-line supervisors and workers understand the importance of safety, and if expressions of cost are understood by workers, they are certainly understandable by managers.

TABLE 34-2   Uninsured Costs

Deductible part of insurance policy
Lost wages for those not insured
Wages paid to injured persons not covered by workers' compensation
Overtime work required as a result of an accident
Supervisor time related to an accident
Repairing, replacing, or cleaning up after an accident
Learning period for a new or replacement worker
Accident investigation cost
Costs to prepare and file reports
Uninsured medical costs
Costs of litigation activities
Other

One way to express cost is in dollars per $100 of pay. It is easy to understand that workers' compensation premiums cost $12.85 for each $100 of salary and that lost time costs a given department $8.50 for every $100 paid in salary.

Another way to express cost is in terms of the number of items produced. A company may bake bread or manufacture television sets. It is easy to understand that each $100 loss requires that the company produce and sell an additional 500 loaves of bread or 25 television sets. This is derived from the equation

$$\text{Cost of loss} = P(N)U, \tag{34-1}$$

where

P is the profit margin in percent,

N is the number of its products necessary to cover the loss, and

U is the unit selling price for the products.

The volume of business necessary to cover a loss or expense is another way to express cost. For each $100 loss, a construction company that has a profit margin of 5% will have to bid and be awarded $2,000 in jobs. This is shown mathematically as

$$\text{Cost of loss} = P(V), \tag{34-2}$$

where V is the dollar volume of business. If the company bids on 10 jobs for each 1 it gets, they must bid on $20,000 in jobs to cover the $100 loss. For a claim of $10,000, the company must bid on $2 million and win $200,000 in jobs.

Another way to express cost is in the number of hours a worker must put in to cover the cost of a loss. The cost of a $100 loss for a worker who earns $10.00 per hour is 10 hr. The cost of a $2,000 loss requires 200 hr of work, or approximately 5 weeks.

Another way to present cost, particularly to managers, is to compare actual versus budgeted expenses. A company may use the history of losses or safety costs in the entire company to estimate future costs per unit of time, which may be a month, quarter, or year. The actual costs for each period can be plotted against budgeted costs to demonstrate to managers how well they are doing in controlling safety costs. Presenting similar accident costs for individual departments makes the information more meaningful to particular managers.

Another way to express cost to top managers and company shareholders is cost of accidents or illnesses per share. An alternate is to express the cost of accidents and illness that could have been avoided as additional earnings per share that could have been achieved.

Other expressions for cost are cost-benefit ratios, return on investment (ROI), and risk. The first two are explored further in the following text, and Chapter 36 addresses risk and risk analysis.

Cost and Benefit

A popular way to justify business expenditures is by comparing the cost of some expenditure with the benefit achieved in financial terms. In cost-benefit analysis, the dollar values of all benefits and costs connected with program alternatives are estimated and then compared. Not all cost and benefits can be converted to quantitative terms; some may be expressed only qualitatively. A final decision applies both to quantitative and qualitative factors.

There are several criteria for evaluating costs and benefits:

1. the cost-benefit ratio
2. net benefits, that is, benefits minus costs

TABLE 34-3 Costs and Benefits for Different Parties

Case	Company	Buyer or Public	Opinion
1	Benefits < costs	Benefits < costs	Very unacceptable for both parties
2	Benefits > costs	Benefits > costs	Very acceptable for both parties
3	Benefits < costs	Benefits > costs	Unacceptable unless there is government subsidy for the company
4	Benefits > costs	Benefits < costs	Unacceptable unless: (a) the condition is allowed by government or (b) compensation is made to the public

3. rate of return, such as the annual benefit relative to cost

4. payback period (the time required to recover costs from benefits)

The parties that bear the costs are not always the ones who gain the benefits, which creates some dilemmas for decision makers. Consider the four cases suggested in Table 34-3 for companies and buyers or the public who are involved in costs and benefits. They do not always agree or have the same solutions in mind for differing opinions.

Governments use cost-benefit analysis for public policy. A key question is whether the cost to implement government regulations is worth the benefits derived. The use of cost-benefit analysis for justification of government regulations was challenged in the courts. In June, 1981, the Supreme Court upheld the OSHA cotton dust standards and noted that cost-benefit analysis is not required by the OSHAct, whereas feasibility analysis is. Feasibility means the potential of accomplishing something. Currently, in proposing new or revised regulations. OSHA and other agencies prepare a cost analysis and estimate benefits, including human lives saved.

Another dilemma in evaluating the cost and benefits of safety is assessing an economic value for human life. A related factor is the economic value when human pleasures of activity are interrupted by disability. Is a corporate executive worth more than a laborer? Is an engineer worth more than a housewife? The answers are not easy. Placing a value on human life is not easy. However, in settlements of lawsuits, these decisions are made regularly.

Return on Investment

Return on investment (ROI) is a widely used method for analyzing the performance of investments in a company or investments for an individual. Whether investments are in stocks, bonds, or real estate, an investor wants to know what the annual return will be. The concept suggests that a person must spend money (invest) to earn money on the investment (return). The concept is very similar to a cost-benefit ratio.

There are costs to control hazards, but a company should see a return on this type of investment. The return may be reduced loss rates and reduced insurance premiums, increases in productivity, or better services.

Cost Accounting

The need to use safety costs in management requires keeping track of safety costs. An accounting system must track actual expenditures. There are many kinds of safety costs, as noted in Tables 4-1 and 34-2.

Other Safety Applications for Cost

Cost information can be used in other ways to help achieve safety. For example, accident claim rates and product return and failure costs can be used to select suppliers of parts, assemblies, and components; accident records, insurance rates, and claim records can be used as criteria in selecting contractors; cost of product failures, complaints, accidents, compliance with standards, and claims can be used as criteria for setting performance incentives of suppliers and contractors.

EXERCISES

1. A fast food hamburger restaurant operates on a 4% profit margin.

   (a) If hamburgers sell for $0.60 each, how many hamburgers must the restaurant sell to cover lost profits from a $500 accident?

   (b) If the restaurant sold only hamburgers and on the average, 1,000 were sold each day, how many days' profit are lost because of the accident?

2. A car company decides to reduce cost by not installing an $18 per car safety improvement. If a human life is assumed to be worth $875,000 on the average, how many automobiles would have to be produced to cover 13 death claims per year?

3. Obtain a chart of an organization. Identify where safety elements of the organization are located. Suggest where improvements might be made and what the changes would accomplish.

REVIEW QUESTIONS

1. Describe what managers do.
2. How is safety implemented in an organizational structure?
3. Explain the role of safety specialists in an organization.
4. How can management style affect safety in an operation?
5. Why is accountability important in managing safety?
6. How can accountability be accomplished without using accident costs and accident rates?
7. What is a safety audit? What is its purpose?
8. How is paying attention to details related to successful safety programs?
9. How is enforcement useful in safety management?
10. What are three ways to express cost of safety?
11. Explain how cost and benefit can be used to justify safety programs.
12. Explain return on investment and how it can be used for safety.
13. How can safety cost be used to select and provide incentives for contractors and suppliers?

NOTES

1 McConnell, M., *Challenger—A Major Malfunction*, Doubleday & Co., Inc., Garden City, New York, 1987.

2 *Columbia Accident Investigation Board Report*, National Aeronautics and Space Administration, August, 2003.

3 Grose, V. L., *Managing Risk—Systematic Loss Prevention for Executives*, Prentice-Hall, Englewood Cliffs, NJ, 1987.

4 Grimaldi, J. V., and Simonds, R. H., *Safety Management*, 5th ed., Irwin, Homewood, IL, 1989.

5 Bittel, L. R., and Ramsey, J. E., eds., *Handbook for Professional Managers*, McGraw-Hill, New York, 1985.

6 Petersen, D., *Techniques of Safety Management*, 2nd ed., McGraw-Hill, New York, 1978.

7 Samuelson, N. M., "The Effect of Foremen on Safety in Constuction," Technical Report 219, Department of Civil Engineering, Stanford University, 1977.

8 Levitt, R. E., "The Effect of Top Management on Safety in Construction," Ph.D. dissertation, Civil Engineering, Stanford University, 1976. (Available from University Microfilms, No. 76-5779.)

9 Safety Trained Supervisor Certification, Council on Health, Environmental and Safety Technologists, Savoy, IL (www.cchest.org).

10 Geller, E. Scott, *The Participation Factor—How to Increase Involvement in Occupational Safety*, Prentice Hall, New York, 2001.

Geller, E. Scott, *The Psychology of Safety*, Chilton Book Company, Radnor, PA, 1996.

Krause, Thomas R., *The Behavior-Based Safety Process*, 2nd ed., Van Nostrand Reinhold, New York, 1997.

Krause, Thomas R., *Employee-Driven Systems for Safe Behavior*, Van Nostrand Reinhold, New York, 1995.

Krause, Thomas R., general ed., *Current Issues in Behavior-Based Safety*, Behavioral Science Technology, Inc., Ojai, CA, 1999.

11 Tarrants, W. E., *The Measurement of Safety Performance*, Garland STPM Press, New York, 1979.

BIBLIOGRAPHY

ADAMS, Edward E., *Total Quality Safety Management*, American Society of Safety Engineers, Des Plaines, IL, 2001.

ASFAHL, C. R., *Industrial Safety and Health Management*, 2nd ed., Prentice-Hall, Englewood Cliffs, NJ, 1990.

BIRD, F., and GERMAIN, G. L., *Practical Loss Control Leadership*, Institute Publishing, Loganville, GA, 1986.

BITTEL, L. R., and RAMSEY, J. E., eds., *Handbook for Professional Managers*. McGraw-Hill, New York, 1985.

Case Studies in Safety & Productivity, National Safety Council, Itasca, IL, 2000.

CIANFRANI, CHARLES A., TSIAKALS, JOSEPH J., WEST, JOHN E., and WEST, JACK, *ISO 9001:2000 Explained*, 2nd ed., American Society for Quality, Milwaukee, WI, 2001.

DAVIES, JOHN, ROSS, ALASTAIR, WALLACE, BENDAN, and WRIGHT, LINDA, *Safety Management: A Qualitative Systems Approach*, CRC Press, Boca Raton, FL, 2003.

DEREAMER, R., *Modern Safety and Health Technology*, Wiley, New York, 1980.

ECKENFELDER, D. J., ed., *Readings in Safety Management*, American Society of Safety Engineers, Des Plaines, IL, 1984.

GARNER, CHARLOTTE A., *How Smart Managers Create World-Class Safety, Health, and Environmental Programs*, American Society of Safety Engineers, Des Plaines, IL, 2004.

GRIMALDI, J. V., and SIMONDS, R. H., *Safety Management*, 5th ed., American Society of Safety Engineers, Des Plaines, IL, 1989.

GROSE, V. L., *Managing Risk—Systematic Loss Prevention for Executives*, Prentice-Hall, Englewood Cliffs, NJ, 1987.

JURAN, J. M., and GODFREY, A. BLANTON, *Juran's Quality Handbook*, 5th ed., McGraw-Hill, New York, 1999.

KUHN, R. L., ed., *Handbook for Creative and Innovative Managers*, McGraw-Hill, New York, 1988.

MACKIE, J. B., and KUHLMAN, R. L., *Safety and Health in Purchasing, Procurement and Materials Management*, International Loss Control Institute, Loganville, GA, 1981.

MANUELE, FRED A., *Innovations in Safety Management: Addressing Career Knowledge Needs*, John Wiley & Sons, New York, 2001.

MOTTEL, WILLIAM J., LONG, JOSEPH F., and MORRISON, DAVID E., *Industrial Safety is Good Business: The DuPont Story*, John Wiley & Sons, New York, 1995.

OXENBURGH, MARUICE S., MARLOW, PEPE, and OXENBURGH, ANDREW, *Increasing Productivity and Profit Through Health and Safety*, 2nd ed., CRC Press, Boca Raton, FL, 2004.

PETERSON, D., *Authentic Involvement*, National Safety Council, Itasca, IL, 2001.

PETERSEN, D., *Safety Management—A Human Approach*, 3rd ed., American Society of Safety Engineers, Des Plaines, IL 2001.

PETERSEN, D., *Techniques of Safety Management*, 4th ed., American Society of Safety Engineers, Des Plaines, IL, 1999.

REDINGER, CHARLES F., and LEVINE, STEVEN P., *Occupational Health and Safety Management System Performance Measurement: A Universal Assessment Instrument*, American Industrial Hygiene Association, Fairfax, VA, 1999.

ROUGHTON, JAMES, and MERCURIO, JAMES, *Developing an Effective Safety Culture: A Leadership Approach*, Butterworth-Heinemann, Burlington, MA, 2002.

STEWART, J. M., *Managing for World Class Safety*, John Wiley & Sons, New York, 2001.

SWARTZ, GEORGE, ed., *Safety Culture and Effective Safety Management*, National Safety Council, Itasca, IL, 1999.

Supervisor's Safety Manual, 6th ed., National Safety Council, Chicago, IL, 1987.

WANG, C. K., *How to Manage Workplace Derived Hazards and Avoid Liability*, Noyes Publications, Park Ridge, NJ, 1987.

CHAPTER 35

RISK MANAGEMENT AND ASSESSMENT

35-1   RISK AND LOSSES

In life there are events that result in gains or losses for people and organizations. Most people do not want losses, although they will take a chance at achieving a gain in the face of some potential loss. Risk involves avoidance of losses and unwanted consequences as well as probability and potential for losses.

Rowe[1] defines risk as the potential for realization of unwanted, negative consequences of an event. Risk aversion is action taken to control or reduce risk. There are many definitions for risk. For safety and health, a common definition of risk infers a quantitative concept. Risk is the product of frequency and severity of potential losses. Frequency is the probability of occurrence of an event, such as once per week or once per year or once every 100 years. Severity is the potential loss when an event occurs. The loss may be expressed in human terms, such as loss of life, serious injury, serious illness, number of cancer cases, and so forth. The loss may also be expressed in financial terms, like dollars lost, cost to replace lost equipment, cost of downtime, or cost to replace facilities. Loss may be expressed in legal terms, such as claims, lawsuits, and liability.

Out of these concepts have come formal methods for dealing with risks: formal risk assessment methods and risk management methods. Risk assessment and management applies to general operation of a business and business decisions. The losses and unwanted consequences for a business ultimately are financial. The idea of risk for a business has a broad meaning that implies any kind of detriment to a business. Companies apply risk to financial decisions, security of trade secrets and computer systems, and other potential losses. Risk also is used in dealing with losses associated with accidents, human error, and health exposures. It is the latter aspect of risk that this discussion addresses.

Closely related to risk and risk aversion is loss control, which is the controlling of conditions that can be responsible for a loss. As noted in the previous chapter, accidents and injuries affect the financial picture of a company or its departments. It can affect the financial picture of an individual. The term *loss control* often is associated with insurance. Loss control seeks to reduce the likelihood of an occurrence or reduce its severity.

35-2   RISK MANAGEMENT

The Process

Risk management involves five components:

1. risk identification
2. risk analysis
3. eliminating or reducing risks
4. financing risks
5. administering the risk management process

The objectives of risk management can be divided into two groups; preloss and postloss objectives. Preloss objectives address those things that may happen. Postloss objectives involve application of resources to recover completely and quickly from a loss. Table 35-1 defines preloss and postloss objectives.

Risk Identification

Risk identification is not an easy task because it is easy to overlook something. It requires training and experience to see unsafe conditions and foresee unsafe acts. It is not easy to see how combinations of things and the complexity of operations, equipment, and facilities can lead to undesirable events.

The goal in risk identification is to reduce uncertainty in describing factors that contribute to accidents, injuries, illnesses, and death. Risk identification involves identification of hazards. It improves understanding of risks for particular situations or groups. Risk identification is conducted to determine whether and to what degree effects in one situation apply to another. It involves gathering facts and data. In risk identification, data are analyzed to determine what components contribute to a process that produces injury or

TABLE 35-1 Risk Management Objectives[a]

Preloss objectives	
Economy	Minimizing the economic expenditures consistent with postloss goals for safety programs, risk identification and analysis, insurance premiums, and so forth.
Reduction in anxiety	Reducing the fear and worry over potential losses.
Meeting externally imposed obligations	Satisfying safety, health, and environmental regulations; satisfying employee-benefit plans; acquiring required insurance.
Social responsibility	Meeting the demands for good citizenship to employees, customers, suppliers, and the community. Maintaining public image and social consciousness.
Postloss objectives	
Survival	Being able to resume some operations after a loss.
Continuity of operations	To return to or continue full operations following an interruption. There may be reduction in earnings. Keeping human and material resources available.
Earnings stability	Keeping earnings stable through continued operations with cost control or from funds to replace lost earnings.
Continued growth	Finding ways to expand growth by product development, market expansion, acquisition, and mergers.
Social responsibility	Taking care of employees, customers, suppliers, and the public. Maintaining public relations and public image.

[a] Derived from Mehr, R. I., and Hedges, R. A., *Risk Management: Concepts and Applications*, Richard D. Irwin, Homewood, IL, 1974.

illness and to establish if data from particular cases can be generalized to other situations or populations.

For example, can data from commercial airplane crashes in 1 year be used to characterize the risk of death from flying for all commercial passengers in the future? Is it more accurate to characterize the risk for passengers by aircraft model, by the time since an aircraft has undergone a major overhaul of an engine or airframe, or by the number of takeoffs and landings an aircraft has experienced? Similarly, in assessing the risks of chemicals, one must generalize data from animal and organism studies to humans, characterize the behavior of a chemical within the body, consider interactions with other substances in the body or interactions within cells and organs.

Risks change with time. The process of identifying risks requires a continual and systematic approach. Risk identification involves recognition of hazards and things that can go wrong. It may involve attaching values to potential losses. The values in this step help establish how certain a loss is for general situations.

There are many techniques for identifying risks. Hazard recognition is an important element. One approach is drawing on the past knowledge and history of accidents. Another approach is applying systematic techniques, such as system safety and other analytical methods (see Chapters 36 and 37). It may be necessary to use specialists to help identify risks, because the specialists have unique knowledge and experience and may recognize some important hazards that others may overlook. Checklists of hazards and conditions producing hazards can be developed and used for comparison with the proposed or actual operation, process, equipment, or system. Sometimes energy and energy release analysis are used to identify what failures in a system might occur and what the consequences might be. Sometimes analysis of human behavior and underlying motivating factors helps identify risks.

Frequency and severity data from accidents can help identify risks. A review of accident records and classification of accident data can help. Various statistical methods applied to accident data will help reveal trends in losses and what factors contribute to accidents and injuries. Analyzing claims, such as worker compensation claims[2] or customer claims against products, will help isolate factors associated with losses.

Risk Analysis

Risk analysis is applying qualitative or quantitative techniques to potential risks. It reduces the uncertainties in measuring risks and it usually involves frequency and severity. Frequency deals with the likelihood that an event will occur or that a hazard will be present. Severity is the effect of an event when it occurs. It is measured in deaths, injuries, disease or illnesses, or loss of equipment or property. Severity may also be expressed in financial terms.

Example 35-1   Risk identification may have noted that people traveling in automobiles to and from work face the risk of a vehicle accident. Suppose the data show that there are 29,800,000 automobile accidents per year in the United States, people travel 1.511 billion miles per year by automobile, and the average person travels 4,500 miles each year driving to and from work. Risk analysis would identify the probability of an accident, P, from

$$P = \frac{2.98 \times 10^7 (4,500)}{1.511 \times 10^{12}}$$

$$= 0.089 \text{ accidents per year per person.}$$

RISK ASSESSMENT MATRIX

HAZARD SEVERITY	PROBABILITY				
	Frequent	Probable	Occasional	Remote	Improbable
Catastrophic	▨	▨	▨	▧	
Critical	▨	▨	▧		
Marginal	▧	▧			
Negligible					

▨ Risk reduction required

▧ Written, time-limited waiver endorsed by management required

☐ Operation permissible

Figure 35-1.   Risk reduction decision matrix.

TABLE 35-2   Hazard Severity Classification

Description	Category	Mishap Definition
Catastrophic	I	Death or system loss
Critical	II	Severe injury, severe occupational illness, or major system damage
Marginal	III	Minor injury, minor occupational illness, or minor system damage
Negligible	IV	Less than minor injury, occupational illness, or system damage

It is not possible to quantify all risks. There are many applications where a qualitative risk analysis is more feasible. The risks are classified according to relative frequency and relative severity. One scheme is depicted in Tables 35-2 and 35-3 and in Figure 35-1. The risk or hazard is assigned one of four severity categories from Table 35-2 and one of five probability categories from Table 35-3. These classifications leave out quantities completely. Considerable judgment is needed to apply the categories consistently.

Risk analysis is discussed further in the section on risk assessment.

Eliminating or Reducing Risks

If risks are known, one can attempt to eliminate them. However, it is not possible to eliminate all risks; some can only be reduced. When many risks exist at once or when resources are limited, the problem is what risks to tackle first. This problem requires setting priorities.

There are several methods for setting priorities. Cost-benefit analysis was considered in Chapter 34, and some other methods will be explored in the next two chapters.

TABLE 35-3 Hazard Probability Classification

Description[a]	Level	Specific Individual Item	Fleet[b] or Inventory
Frequent	A	Likely to occur frequently	Continuously experienced
Probable	B	Will occur several times in the life of an item	Will occur frequently
Occasional	C	Likely to occur sometime in the life of an item	Will occur several times
Remote	D	Unlikely but possible to occur in the life of an item	Unlikely but can reasonably be expected to occur
Improbable	E	So unlikely, it can be assumed occurrence may not be experienced	Unlikely to occur, but possible

[a] Definitions of descriptive words may have to be modified based on quantity involved.
[b] The size of the fleet or inventory should be defined.

TABLE 35-4 Example Classification for Costs to Correct Risks

Category	Value
A	<$1,000
B	$1,000–10,000
C	$10,000–100,000
D	>$100,000

Quantitative or qualitative methods can be applied to setting priorities. Alternate ways to achieve elimination or reduction in risks must be explored. Two key factors are the cost of implementing actions and the degree of reduction achieved for each.

The classification of risk severity and probability in Tables 35-2 and 35-3 is extended to a decision matrix in Figure 35-1. Risks or hazards are organized into three categories. Some require elimination or reduction, some are permissible and need no reduction, and others are left for management to wrestle with on an interim basis. Little judgement is necessary in applying the matrix. The greatest judgment involves how to eliminate or reduce catastrophic and critical hazards or finding alternatives when it is impossible or impractical to eliminate or reduce them.

Some methods add a classification table for grouping the correction costs. Table 35-4 gives an example of a cost-to-correct table.

Financing Risks

Managing risks requires decisions regarding how to pay for the risks. Money can be invested in directly eliminating or reducing risk, particularly the most effective reduction alternatives. Developing a cash reserve is one method to pay for risk reduction measures. A second option is to purchase insurance for each risk. The premiums may be lower than the cost to implement a risk-reducing alternative. A third option is to do nothing, which may be logical for events that are not likely to occur and those that have low severity.

Example 35-2 The data from Example 35-1 can be extended to financial aspects. Assume that the cost of vehicle accidents is $39.3 billion per year. The average cost per accident, C, is

$$C = \frac{\$39.3 \times 10^9}{2.98 \times 10^7}$$

$$= \$1,319 \text{ per accident.}$$

Similarly, the financial risk for each person per year, R, is

$$R = 0.089 \times \$1,319$$

$$= \$117.39 \text{ per person per year.}$$

From these data, a decision can be made whether it is better to buy insurance or create a fund to cover the cost of accidents. (In reality, most states require drivers to carry insurance.) In addition, other information may help in deciding if there is any way to reduce the likelihood of an accident or reduce severity if one occurs. Alternatives may reduce the risk if money is spent for driver training, driving a larger car that has lower cost per accident, or incorporating protective features to reduce severity.

Administering the Process

The final step in risk management is administering the process. Part of administration is setting acceptable levels of risk. A company or organization must decide what level of risk it will assume and what level it will transfer. Another aspect of administration is assigning resources to the process. The process may require specialists for risk identification and analysis and financial specialists to help determine the overall costs, benefits, and most economical way to finance risks. Administering the process necessitates monitoring and evaluating if reductions are achieved, if frequency and severity actually resulted as projected, and if expenditures achieve the benefits that were anticipated. Another aspect of administering the process is selecting methods to be used and tracking items analyzed, hazards identified, analysis applied, and decisions made.

Figure 35-2 is an example of a form for tracking risk identification, analysis, and actions taken. Data can also be tracked using computer records. Nomograms have been developed to analyze risks and justify costs. Figure 35-3 gives examples.

35-3   RISK ASSESSMENT

Terminology for risk, risk management, and risk assessment is not fully consistent across sources. In general, risk assessment is a portion of risk management. Risk assessment involves identification, analysis, and evaluation of risk. A diagrammatic representation of risk assessment is given in Figure 35-4, where risk assessment is divided into two components: risk determination and risk evaluation. Risk determination includes identifying risks and risk estimation. There are several approaches that help identify risks. Risk estimation is projecting frequency and severity. Probabilistic risk assessment (PRA) includes risk assessment techniques that evaluate the probability of events occurring and the probability of their severity.

Risk evaluation includes risk aversion and risk acceptance. Risk aversion is estimating how well risk can be reduced or avoided through various alternatives. Risk acceptance involves creating decision tables or standards for deciding what risks are acceptable for individuals, companies, or society. What is acceptable may differ for each group.

Risk assessment is not independent of the process or the personnel involved. The method being used, as well as the participants, can affect the risk assessment. Sometimes,

RISK MANAGEMENT DATA SHEET	
Item or Event	
Risk Identifier	**Method(s) Used**
Risk Ratings Frequency: _____ Severity: _____	
Description and Contributing Factors	
Summary of Supporting Data	
Corrective Action(s) Proposed	
Proposed Action	Effectiveness/Cost

Figure 35-2.   Example of a risk management data sheet.

people who could be affected by a risk are not included as participants, which can create significant biases in the process.

The National Academy of Sciences[3] identified four steps in every complete chemical risk assessment:

1. hazard identification
2. dose-response assessment
3. exposure assessment
4. risk characterization

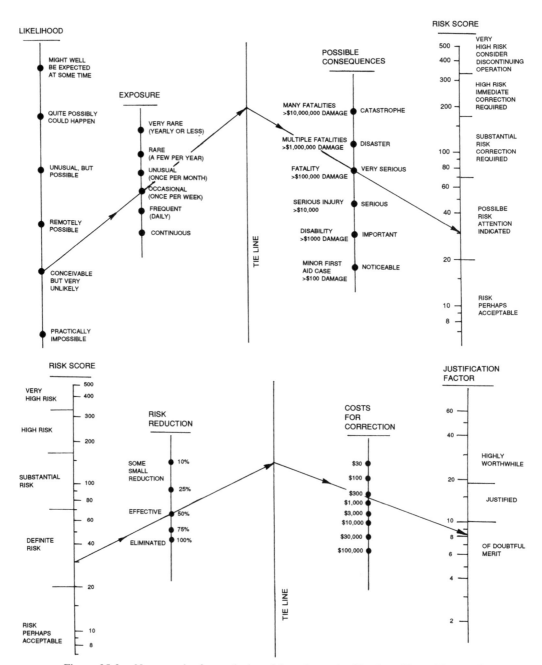

Figure 35-3. Nomographs for analyzing risk and cost justification. (From Kinney, G. F., and Wiruth, A. D., *Practical Risk Analysis for Safety Management*. NWC Technical Publication 5865, Naval Weapons Center, China Lake, CA, 1976.)

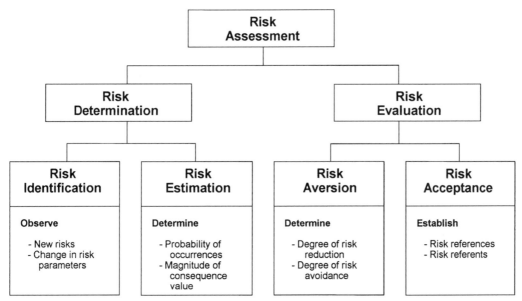

Figure 35-4. Risk assessment concept. (Derived from W. D. Rowe, *An Anatomy of Risk*, Wiley, New York, 1977.)

Hazard Identification

Hazard identification and some methods for it have already been discussed. Hazard identification may include engineering failure assessment, which consists of evaluating the reliability of specific segments of a plant operation and determining probabilistic results. Fault-tree analysis (see Chapter 36) is a common form of engineering failure assessment.

Dose-Response Assessment

For chemicals, dose-response assessment involves describing the quantitative relationship between the amount of exposure and the extent of toxic injury or disease. It requires that the hazard of a material be recognized before the effects are assessed. A dose-response assessment may provide linear equations relating exposure or dose to response or disease. The equations may be derived from regression analysis of dose-response data.

Exposure Assessment

Exposure assessment is describing the nature and size of populations exposed to an agent and the magnitude and duration of the exposures. Exposures include past, present, and future exposures. Exposure assessment may include analysis of toxicants in air, water, or food and it could apply to the prevalence of certain factors in automobile accidents.

Risk Characterization

Risk characterization is integration of data and analysis to determine if people will experience effects of exposure. Risk characterization includes estimating uncertainties associated with the entire process of risk assessment.

Application Issues

Risk assessment often must rely on inadequate scientific information or the lack of data. For example, repair data may not be useful in predicting failures accurately for newly designed equipment because there may be a lack of understanding of phenomena. For example, in toxicological assessment, assumptions must be made about the validity of using animal studies to predict effects for humans. Many people involved in risk assessment will take a conservative approach to avoid overestimating risk, whereas others will use comparison techniques on various options. Then the absolute risk values are not as important as the relative differences between options.

35-4   EXAMPLES OF METHODS

People have applied risk assessment and management in various ways. This section will take a closer look at several approaches.

William Fine

Fine[4] proposed a method for deciding if the cost to correct a hazard is justified and how quickly hazards should be corrected. His method involves the use of risk. A risk score, R, is computed from

$$R = C \times E \times P, \tag{35-1}$$

where

C is the consequence rating value,

E is the exposure value, and

P is the probability value.

The risk score can be used to decide how quickly to act to correct hazards. See Table 35-5 for decision guidance. One can compute a cost justification value, J, from

$$J = \frac{R}{(CF \times DC)}, \tag{35-2}$$

where

CF is a cost factor and

DC is a degree of correction value.

The values for Equations 35-1 and 35-2 are selected from tables (see Table 35-5). Fine suggests that if $J > 10$, the cost is justified and if $J < 10$, the cost is not justified. Fine emphasizes that his method should be used for *guidance* only. The values used in the process and for decision making are somewhat arbitrary. Other definitions could be substituted, other values assigned, and a different value used for J in decision making. However, the approach does provide a simple way to evaluate a variety of hazards and controls and present them to management for approval.

Logical Process Risk Analysis

Frank and Morgan[5] proposed a systematic method for helping managers allocate funds to achieve the greatest risk reduction across several departments within a plant. The method

TABLE 35-5 Values for Fine's Decision Process

Rating	Classification
Consequences, C (most probable result of potential accident)	
100	Catastrophe; numerous fatalities; damage over $1,000,000; major disruption of activities
50	Multiple fatalities; damage $400,000–1,000,000
25	Fatality; damage $100,000–400,000
15	Extremely serious injury (i.e., amputation, permanent disability; damage $1,000–100,000
5	Disabling injury; damage up to $1,000
1	Minor injury or damage
Exposure, E (frequency of occurrence of the hazard event)	
Hazard event occurs	
10	Continuously (or many times daily)
6	Frequently (about once daily)
3	Occasionally (once per week to once per month)
2	Unusually (once per month to once per year)
1	Rarely (it has been known to occur)
0.5	Remotely possible (not known to have occurred)
Probability, P (likelihood that accident sequence will follow to completion)	
Complete accident sequence	
10	Is the most likely and expected result if the hazard event takes place
6	Is quite possible, not unusual, has an even 50–50 chance
3	Would be an unusual sequence or coincidence
0.5	Has never happened after many years of exposure, but is conceivably possible
0.1	Practically impossible sequence(has never happened)
Cost factor, CF (estimated dollar cost of proposed corrective action)	
10	>$50,000
6	$25,000–50,000
4	$10,000–25,000
3	$1,000–10,000
2	$100–1,000
1	$25–100
0.5	Under $25
Degree of correction, DC (degree to which hazard will be reduced)	
1	Hazard positively eliminated 100%
2	Hazard reduced at least 75%
3	Hazard reduced by 50%–75%
4	Hazard reduced by W-50%
6	Slight effect on hazard (<25%)

Risk score summary and actions

Score	Action
200–1,500	Immediate correction required; activity should be discontinued until hazard is reduced
90–199	Urgent; requires attention as soon as possible
0–89	Hazard should be eliminated without delay, but situation is not an emergency

also could be used to compare several plants and to allocate funds to each. The procedure applies risk concepts and was prepared for application to chemical processing plants. The method involves six steps:

1. Compute a risk index for each department.
2. Determine relative risk for each department.
3. Compute the percent risk index for each department.
4. Determine composite exposure dollars for each department.
5. Compute a composite risk for each department.
6. Rank all departments relative to each other based on composite score.

Risk Index   A risk index is developed for a department by evaluating hazards and controls, establishing a hazard score and a control score, and subtracting the hazard score from the control score. The authors developed a hazard checklist (see Table 35-6) and a control checklist (see Table 35-7). The hazard checklist includes six groups of hazards, and there are points associated with each hazard within a group. The points are summed for those hazards that apply within a group, and the product of the sum and a hazard or weighting factor for the group of hazards yields a score for the group. The hazard score for a department is the sum of scores computed for each of the six groups.

The control scores are determined in a similar manner. After identifying what controls apply to a department for each of six groups of controls, the points for each applicable control are summed within a group and multiplied by a group control factor. The control score for a department is the sum of scores for each of the six groups.

The risk index is the control score minus the hazard score. A risk index may have a positive or negative value.

Relative Risk   The goal is to rank departments, not individual hazards. Because the department with the highest risk index (highest positive value) is not likely to need much reduction in hazards (the high risk index value indicates that controls are very effective), it will need funds less than other departments. The authors use the best department risk score as the baseline for all others. All scores are adjusted relative to the score for the best department by subtracting the risk score for the best department from all risk scores. This adjustment makes the relative risk score for the best department zero.

Percent Risk Index   The percent risk index for each department represents the relative contribution each makes toward the total risk of the plant. The relative risk for each department is converted to a percent of all risk by a simple procedure. The total risk for all departments is the sum of absolute values of all relative risk scores. Dividing the absolute value of the relative risk for each department by the total risk gives the percent risk index.

Composite Exposure Dollars   Next, one needs to know what the total dollars-at-risk are for each department. The composite exposure dollars are the sum of the monetary value of three components: property value, business interruption, and personnel exposure. The property value is determined by estimating the replacement cost of all material and equipment at risk in a department. Business interruption is computed as the product of the (1) unit cost of goods produced, (2) the department production capacity per year, and (3) the expected percent of capacity. The personnel exposure is the product of the total number of people in the department during the most populated shift and the monetary value for each person. The authors set the monetary value somewhat high to reflect that loss of life is not acceptable.

TABLE 35-6   Hazard Checklist

Rating Points	Hazard Group and Hazard (Group Hazard Factor in Parentheses)

Fire/explosion potential (10)

2	Large inventory of flammables
2	Flammables generally distributed in the department rather than localized
2	Flammables normally in vapor phase rather than liquid phase
2	Systems opened routinely, allowing flammable/air mix, versus a totally closed system
1	Flammables having low flash points and high sensitivities
1	Flammables heated and processed above flash point

Complexity of process (8)

2	Need for precise reactant addition and control
2	Considerable instrumentation requiring special operator understanding
2	Troubleshooting by supervisor rather than operator
1	Large number of operations and/or equipment monitored by one operator
1	Complex layout of equipment and many control stations
1	Difficult to start up or shut down operations
1	Many critical operations to be maintained

Stability of process (7)

3	Severity of uncontrolled situation
2	Materials that are sensitive to air, shock, heat, water, or other natural contaminants in the process
2	Potential exists for uncontrolled reactions
1	Raw materials and finished goods that require special storage attention
1	Intermediates that are thermally unstable
1	Obnoxious gases present or stored under pressure

Operating pressure involved (6)

3[a]	Process pressure in excess of $110 \, \text{lb/in}^2$ (gauge), or
2[a]	Process pressure above atmosphere but less than $110 \, \text{lb/in}^2$ (gauge), or
1[a]	Process pressure ranges from vacuum to atmospheric
3	Pressures are process rather than utility related
2	High pressure situations are in operator frequented areas
1	Excessive sight glass application
1	Nonmetallic materials of construction in pressure service

Personnel/environmental hazard potential (4)

3	Exposure to process materials pose high potential for severe burns or severe health risks
2	Process materials corrosive to equipment
2	Potential for excursion above threshold limit value (TLV)
1	Spills and/or fumes have high impact on equipment, people, or services
1	High noise levels make communication difficult

High temperatures (2)

1[a]	Equipment temperatures exist in $<100°C$ range (low), or
2[a]	Equipment temperatures exist in $100 <170°C$ range
3[a]	Equipment temperatures exist in $170 <230°C$ range ($350 \, \text{lb/in}^2$ [gauge] steam)
2	High temperature situations are m operator-frequented area
2	Overflows and/or leaks are fairly common
2	Heat stress possibilities from nature of work or ambient air

[a] Maximum of three points from this subgroup.

TABLE 35-7 Control Checklist

Rating Points	Control Group and Control (Group Control Factor in Parentheses)
Fire protection (10)	
4	Automatic sprinkler system capable of meeting demands
2	Supervisors and operators knowledgeable of installed fire protection systems and trained in proper response to fire
1	Adequate distribution of fire extinguishers
1	Fire protection system inspected and tested with regular frequency
1	Building and equipment provided with capability to isolate and control fire
1	Special fire detection and protection provided where indicated
Electrical integrity (8)	
3	Electrical equipment installed to meet National Electrical Code (NEC) area classification
1	Electrical switches labeled to identify equipment served
1	Integrity of installed electrical equipment maintained
1	Class I, division 2 installations provided with sealed devices. Explosion proof equipment provided or purged reliably and good electrical isolation between hazardous and nonhazardous areas.
1	All electrical equipment capable of being locked out
1	Disconnects provided, identified, inspected, and tested regularly
1	Lighting securely installed and facilities properly grounded
Safety devices (7)	
3	Relief devices provided and relieving is to a safe area
2	Confidence that interlocks and alarms are operable
2	Operating instructions are complete and current, and department has continuing training and/or retraining program
1	Safety devices are properly selected to match application
1	Critical safety devices identified and included in regular testing program
1	Fail-safe instrumentation provided
Inerting and dip piping (5)	
2	Vessels handling flammables provided with dip pipes
2	Vessels handling flammables provided with reliable "inerting" system
2	Effectiveness of inerting assured by regular inspection and testing
1	Inerting instruction provided and understood
1	Inerting system designed to cover routine and emergency startup
1	Equipment grounding visible and tested regularly
1	Friction hot spots identified and monitored
Ventilation/open construction (4)	
3	No flammables exist or open air construction is provided
2	Local ventilation provided to prevent unsafe levels of flammable, toxic, or obnoxious vapors
2	Provision made for containing and controlling large spills and leaks of hazardous materials
1	Building design provides for natural ventilation to prevent accumulation of dangerous vapors
1	Sumps, pits, etc., nonexistent or else properly ventilated or monitored
1	Equipment entry prohibited until safe atmosphere assured
Accessibility and/or separation (2)	
2	Critical shutdown devices and/or switches visible and accessible
2	Adjacent operations or services protected from exposure resulting from incident in concerned facility
2	Operating personnel protected from hazards by location
1	Orderly spacing of equipment and materials within the concerned facility
1	Adjacent operations offer no hazard or exposure
1	Hazardous operations within facility well isolated

Composite Risk   The composite risk for a department is the product of the composite exposure dollars and the percent risk index. The composite risk represents the economic value of the relative risk for a department. Units for composite risk are dollars.

Final Ranking   The final step in the process is to rank the departments based on composite risk. Because the goal is to help managers decide where to apply funds to achieve the greatest risk reduction, the departments should be ranked from highest composite score to lowest. The lowest will be zero for the reference department.

Example 35-3   Six departments in a plant are asking for money to improve process safety. The company elects to use logical process risk analysis as a guide for allocating funds. The goal is to reduce potential losses. Analysis is performed for each department to determine hazard scores, control scores, and exposure dollars. Data are presented in Table 35-8.

The data are used to rank departments. The results are presented in Table 35-9.

Risk Analysis with Return on Investment

Another example of risk analysis follows a process of computing risk and cost for controls. Results include return on investment. The items to be analyzed are processes, activities, or equipment. The first step is hazard identification, followed by estimates of

TABLE 35-8   Data for Example 35-3

Exposure Department	Hazard Score	Control Score	Property Value (×$1,000)	Business Interruption Cost (×$1,000)	Composite Personnel Value (×$1,000)	Composite Exposure Dollars (×$1,000)
A	257	304	2,900	1,400	900	5,200
B	71	239	890	1,200	653	2,743
C	181	180	1,700	720	1,610	4,030
D	152	156	290	418	642	1,350
E	156	142	520	890	460	1,870
F	113	336	2,910	3,100	1,860	7,870

TABLE 35-9   Results of Analysis for Example 35-3, Including Final Ranking for Departments

Dept.	Risk Index	Relative Risk	% Risk Index	Composite Exposure Dollars (×$1,000)	Composite Risk (×$1,000)	Rank
A	47	−176	−22.3	5,200	1,160	1
B	168	55	7.0	2,743	192	4
C	−1	−224	28.4	4,030	1,145	2
D	126	−97	12.3	1,350	166	5
E	−14	−237	30.0	1,870	561	3
F	223	0	0.0	7,870	0	6
		−789	100.0			

frequency, and severity of losses for each hazard. Associated with each hazard are one or more controls, the costs and effectiveness of which must be estimated. Effectiveness is reduction in frequency and severity of losses. Implementing a control will reduce risk from what it would be without the control. From this analysis, one can project benefits, return on investment, and payback period.

Example 35-4   A company wants to reduce the hazards of injury for a machine. There are two options that may help: installing a guard or using an interlock device. The goal is to compare each approach and express the solution as a return on investment. Table 35-10 compares the two alternatives.

Chemical Risk Analysis

Although the steps in the risk assessment process seem quite simple, the procedures for applying risk assessment to chemicals is complicated. There are many points in the process at which technical judgement must be applied. As a result, considerable knowledge and experience with the process are necessary to understand the many intricacies that must be accounted for.

Hazard recognition is a starting point in applying risk assessment to hazardous chemicals, and it depends on data from toxicological studies. The amount and type of data vary widely for different chemicals, and there are several difficulties in estimating chemical hazards. For example, animal studies may have considered single doses. As a result, there is little information for estimating chronic exposure problems. Also, animals may have had high dose rates that are not representative of human exposures. There are difficulties in extrapolating animal data to humans, and although most human carcinogens are also animal carcinogens, the converse is not necessarily true.

Additional difficulties arise when making dose-response assessments. First, one must select what measure of dose to use. A common measure is milligrams of chemical per kilogram of body weight per day. Then a scaling factor between species must be applied and an adjustment made for absorption rates, because they are affected by several factors. Extrapolation must be carried out from high to low doses because not all extrapolations are linear or linear over a range of dose rates. Adjustment may be needed for

TABLE 35-10   Risk Analysis for Example 35-4

Step	Option A	Option B
A. Hazard: Becoming caught in a machine		
B. Frequency of occurrence (events per year)	3	3
C. Severity (expected $ loss per event)	$10,000	$10,000
D. Risk, annual (B × C)	$30,000	$30,000
E. Control	Guard	Interlock
F. Control cost (initial, assume no annual recurring costs)	$2,000	$800
G. Control effectiveness (relative to B)	90%	70%
H. Control effectiveness (relative to C)	80%	80%
I. Risk after control is implemented (B × G)(C × H)	$21,600	$16,800
J. Benefit, annual (D − I − F)	$6,400	$12,400
K. Return on investment (100 × J/F)	320%	1,550%
L. Payback period, years (F/J)	4 months	3–4 weeks

threshold effects also. For some substances, there are no-observable-effect levels (NOELs) or lowest-observed-effect levels (LOELs).

The assessment of exposure also has complicating factors. First, the medium of exposure must be considered (air, diet, water, soil). Then the duration of exposure must be accounted for. An exposure may be chronic, covering a major portion of a person's lifetime, or acute, involving one contact or contact for one day or less. Exposures between acute and chronic are called subchronic. What concentration is in an environment must be known and what the intake rate is for a particular environment must be estimated. Some common units for doses in exposure assessment are maximum daily dose (MDD) and lifetime average daily dose (LADD), which is used frequently in carcinogenic risk assessment. LADD is estimated from the product of the average MDD and the fraction of the total lifespan that one is exposed (exposure days per days per lifetime). Furthermore, the individuals exposed need to be characterized because factors such as age, sex, health status, and other exposures (such as cigarette smoke, occupational agents, etc.) may contribute to the risk.

Finally, the risk is estimated from the other data and analysis. From the data, risk can be reported in various ways. One method is establishing allowable dose rates for humans. Typical units are acceptable daily intake (ADI), which are based on NOELs and apply a safety factor to protect sensitive people. If data are available from human studies, a smaller safety factor is applied compared with having data available from only animal studies.

Risk assessment for chemicals is a complicated process of generalizing risk from known, highly controlled situations to situations that were not included in studies. Regardless of its difficulties and uncertainties, risk assessment is a systematic way to organize, analyze, and present information on environmental chemicals and to decide when public protection is needed.

For chemical processes, readers should refer to detailed procedures[6] that go beyond the scope of this book.

EXERCISES

1. A company developed data on injuries resulting from becoming caught in machines. The data show that on the average, 15 workers in company machine shops receive a machine injury each year. The measured severity in financial terms found that the average cost per case is $9,750. If there are 450 machine shop workers in the company, what is the risk (cost per year) for each worker?

2. A company is contemplating changing a process to reduce a hazard. The company has the following data about the change: if there is an accident, there will be multiple fatalities and more than $500,000 in damage. The hazard event occurs once per year. There is a small chance (coincidental) that an accident sequence will follow to completion. The cost to correct the hazard is $60,000. The correction will reduce the hazard by 80%. Use the William Fine method to determine

   (a) risk score and a recommended action to correct the hazard

   (b) cost justification and a decision to implement the change

3. Use the logical process risk analysis method to determine the ranking for applying funds to each of the five plants in a company. Data for the five plants (more were involved in the analysis) are in the following table.

Plant	Relative Risk (%)	Composite Exposure ($)
A	8	822,000
B	15	1,219,000
C	16	759,000
D	7	598,000
E	21	1,021,000

REVIEW QUESTIONS

1. Define the following:

   (a) risk

   (b) risk aversion

   (c) loss control

   (d) risk management

   (e) risk assessment

2. What are the five steps in risk management? Briefly explain each.

3. What are the four steps in risk assessment? Briefly explain each.

4. To what is Fine's risk method applied?

5. To what is logical process risk analysis applied?

6. Discuss some problems in each of the four steps of risk assessment when it is applied to chemical risks.

NOTES

1 Rowe, W. D., *An Anatomy of Risk*, Wiley, New York, 1977.

2 Jensen, R. C., "How to Use Workers' Compensation Data to Identify High-Risk Groups," *Handbook of Occupational Safety and Health*, Slote, L., ed., Wiley, New York, 1987.

3 *Risk Assessment in the Federal Government, Managing the Process*, National Academy of Sciences, National Academy Press, Washington, DC, 1983.

4 Fine, W. T., "Mathematical Evaluation for Controlling Hazards," *J. Safety Res.*, 40:157–166 (1971).

5 Frank, K. H., and Morgan, H. W., "A Logical Process Risk Analysis," *Professional Safety*, June: 23–30 (1979).

6 *Guidelines for Chemical Process Quantitative Risk Analysis*, Center for Chemical Process Safety, American Institute of Chemical Engineers, New York, 1989; *Guidelines for Hazard Evaluation Procedures*, Center for Chemical Process Safety, American Institute of Chemical Engineers, New York, 1985.

BIBLIOGRAPHY

BYRNES, MARK E., KIND, DAVID A., and TIERNO, PHILIP M., JR., *Nuclear, Chemical, and Biological Terrorism: Emergency Response and Public Protection*, CRC Press, Boca Raton, FL, 2003.

CALABRESE, E. J., and KENYON, E., *Air Toxics and Risk Assessment*, Lewis Publishers, Chelsea, MI, 1990.

Chemical-Biological Terrorism: Awareness and Response to Threat, American Industrial Hygiene Association, Fairfax, VA, 2001.

CONWAY, R. A., ed., *Environmental Risk Analysis for Chemicals*, Van Nostrand Reinhold, New York, 1981.

Elements of Technology and Chemical Risk Assessment, Environ Corporation, Washington, DC, 1986.

GREEN, A. E., ed., *High Risk Safety Technology*, Wiley, New York, 1982.

GREENBERG, HARRIS R., and CRAMER, JOSEPH J., *Risk Assessment and Risk Management for the Chemical Process Industry*, Van Nostrand Reinhold, New York, 1991.

GREENWAY, A. ROGER, *Risk Management Planning Handbook*, 2nd ed., Government Institutes, Rockville, MD, 2002.

GROSE, V. L., *Managing Risk: Systematic Loss Prevention for Executives*, Prentice-Hall, Englewood Cliffs, NJ, 1989.

Guidelines for Chemical Process Quantitative Risk Analysis, Center for Chemical Processes Safety, American Institute of Chemical Engineers, New York, 1989.

Guidelines for Hazard Evaluating Procedures, Center for Chemical Process Safety, American Institute of Chemical Engineers, New York, 1985.

HALLENBECK, W. H., and CUNNINGHAM, K. M., *Quantitative Risk Assessment for Environmental and Occupational Health*, 2nd ed., Lewis Publishers, Boca Raton, FL, 1993.

JAYJOCK, MICHAEL, LYNCH, JEREMIAH, and NELSON, DEBORAH IMEL, *Risk Assessment Principles for the Industrial Hygienist*, American Industrial Hygiene Association, Fairfax, VA, 2000.

LOWRANCE, W. W., *Of Acceptable Risk*, William Kaufmann, Los Altos, CA, 1976.

MEHR, R. I., and HEDGES, B. A., *Risk Management: Concepts and Applications*, Richard D. Irwin, Homewood, IL, 1974.

MOLAK, VLASTA, *Fundamentals of Risk Analysis and Risk Management*, Lewis Publishers, Boca Raton, FL, 1997.

OLAJOS, EUGENE J., and STOPFORD, WOODHALL, eds., *Riot Control Agents: Issues in Toxicology, Safety, and Health*, CRC Press, Boca Raton, FL, 2004.

RICCI, P. F., ed., *Principles of Health Risk Assessment*, Prentice-Hall, Englewood Cliffs, NJ, 1985.

"Risk Assessment," Mechanical Engineering, November: 21–59 (1984) (a series of eight articles).

Risk Assessment in the Federal Government. Managing the Process, National Academy of Sciences, National Academy Press, Washington, DC, 1983.

ROWE, W. D., *Evaluation Methods for Environmental Standards*, CRC Press, Boca Raton, FL, 1983.

SANDMAN, PATER M., *Responding to Community Outrage: Strategies for Effective Risk Communication*, American Industrial Hygiene Association, Fairfax, VA, 1993.

THOMSON, J. R., *Engineering Safety Assessment: An Introduction*, Wiley, New York, 1987.

VON LUBITZ, DAG K. J. E., *Bioterrorism: Field Guide to Disease Identification and Initial Patient Management*, CRC Press, Boca Raton, FL, 2004.

WILLIAMS, C. A., Jr., HEAD, G. L., and GLENDENNING, G. W., *Principles of Risk Management and Insurance*, 2 vols., American Institute for Property and Liability Underwriters, Malvern, PA, 1978.

ZIMMERMAN, R., *Governmental Management of Chemical Risk*, Lewis Publishers, Chelsea, MI, 1990.

CHAPTER 36

SYSTEM SAFETY

36-1   INTRODUCTION

System safety is an approach to accident prevention that involves the detection of deficiencies in system components that have a potential for failure or an accident potential. System safety is the application of technical and managerial skills to the systematic, forward-looking identification, and control of hazards throughout the life cycle of a system, project, program, or activity. In this context, a system is an item of equipment or a process. Examples of complex systems are aircraft, weapons, production plants, vehicles, and buildings.

Chapter 3 discussed preventive strategies for accidents. A preventive strategy (Figure 3-4) does not allow accidents to happen before something is done about them, which is contrasted with a reactive strategy (Figure 3-3) that acts after accidents have occurred. The latter is costly and often ineffective, because an existing design and system limit what changes are easy to make. The farther in the development process that changes are made, the greater the costs (see Figure 36-1). Because of cost to change or a preexisting feature of a current system, the changes may not be as comprehensive or as integrally tied into a system as would be desired.

The key element in system safety is hazard analysis. The process identifies, anticipates, and controls hazards. The hazard analysis may consider the entire life cycle of a system. Many kinds of controls extend from the hazard analysis. They may be engineering controls that modify a system to eliminate or reduce the hazards to acceptable levels. Controls include management policy and procedures and identification and implementation of training for system operators, maintainers, and support staff. Controls may include operating procedures, emergency response, and other plans and application of many consensus standards and government standards and regulations for safety.

System safety is not just failure analysis. Hazard analysis may use failure analysis and other analyses to identify hazards, but system safety is a process for safety specialists to identify and deal with safety problems. System safety procedures often include risk assessment, which was discussed in Chapter 35.

The concepts for system safety evolved with aircraft and missile projects in the 1950s and 1960s. There were only a few models built during prototype phases, so design and testing could not afford very many failures or the program was ended. Even production models could not afford many failures, because the aircraft and missiles were very expensive and were a matter of much public attention. Therefore, hazards and failures had to be eliminated or reduced during the design and development phases.

Today, system safety concepts are incorporated in product design, building and facility design, accident prevention management, and other applications. Many system safety

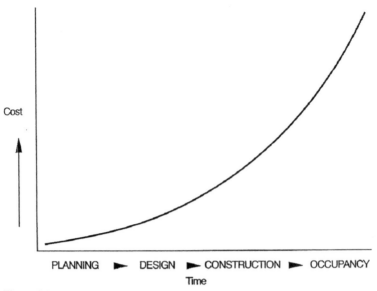

Figure 36-1. Cost for changes increase with stage of development.

techniques are integrated into process safety (see Chapter 30). Some of these applications are discussed further.

36-2   GENERAL PROCEDURES

Presented herein are examples of generally applied system safety procedures.

OSHA Process Safety Standard

The OSHA Process Safety Standard[1] incorporates many system safety concepts. For example, the standard calls for an experienced team to identify and analyze hazards (process hazard analysis, or PHA) using one or more of the following methods:

- What-If
- Checklist
- What-If/Checklist
- Hazard and Operability Study (HAZOP)
- Failure Mode and Effects Analysis (FMEA)
- Fault Tree Analysis
- An appropriate equivalent method

The analysis is then used to address

1. the hazards of the process
2. identification of previous incidents that had a potential for catastrophic consequences in the workplace
3. engineering and administrative controls

4. consequences of failure of engineering and administrative controls

5. facility siting

6. human factors

7. qualitative evaluation of possible safety and health effect of control failures

The final step is establishing a system to address the team's findings and recommendations in a timely manner through an action plan and schedule.

Military Standard 882

There are many variations in system safety procedures as they are applied by different organizations to a variety of systems. Military Standard 882 (MIL-STD 882) addresses an approach for management of environmental, safety, and health mishap risks encountered in the development, test, production, use, and disposal of systems, subsystems, equipment, and facilities. Those engaged in military acquisitions have used the procedures in MIL-STD 882 for a long time to identify, evaluate, and mitigate to an acceptable level mishap risks.

The standard defines a *mishap* as an unplanned event or series of events resulting in death, injury, occupational illness, damage to or loss of equipment or property, or damage to the environment. A *mishap risk* is the possibility and impact of a mishap expressed in terms of potential mishap severity and probability of occurrence. The standard defines *safety* as the freedom from those conditions that can cause death, injury, occupational illness, damage to or loss of equipment or property, or damage to the environment. It also defines a *system* as an integrated composite of people, products, and processes that provide a capability to satisfy a stated need or objective. In addition, *system safety* is the application of engineering and management principles, criteria, and techniques to achieve acceptable mishap risk, within the constraints of operational effectiveness and suitability, time, and cost, throughout all phases of the system life cycle.

The standard outlines requirements for the application of system safety to include:

- Documentation of the system safety approach, including
  - Identification of processes used
  - Integration of the approach into the overall program
  - Defining how hazards and residual mishap risk are communicated to and accepted by authorities and are tracked
- Identification of hazards
- Assessment of mishap risk
- Identification of mishap risk mitigation measures, including (these are often referred to as "design order of preference"; see also Chapter 9)
  - Elimination of hazards through design selection
  - Incorporating safety devices
  - Proving warning devices
  - Developing procedures and training
- Reduction of mishap risk to an acceptable level
- Verification of mishap risk reduction
- Review of hazards and acceptance of residual mishap risk by authorities
- Tracking of hazards, their closures, and residual mishap risk

The standard outlines system safety management methods that may be required in procurements. It references the *System Safety Analysis Handbook*[2] as a publication covering system safety methods. The standard provides suggested classifications for mishap severity categories (see Table 36-1) and for mishap probability levels (see Table 36-2). Classifications from these tables then are used in combination to make decisions. One decision step is to establish mishap risk assessment values, which are used to rank different hazards in terms of mishap risk and to group hazards into mishap risk categories. A second step is to establish risk categories, which help to create specific actions for managing mishap risks. Table 36-3 is an example of a table of mishap risk assessment values, and Table 36-4 is an example of a table of mishap risk categories and mishap risk acceptance levels.

TABLE 36-1 Suggested Mishap Severity Categories

Description	Category	Environmental, Safety, and Health Result Criteria
Catastrophic	I	Could result in death, permanent total disability, loss exceeding $1 million, or irreversible severe environmental damage that violates laws or regulations.
Critical	II	Could result in permanent partial disability, injuries, or occupational illness that may result in hospitalization of at least three personnel, loss exceeding $200,000 but less than $1 million, or reversible environmental damage causing a violation of law or regulation.
Marginal	III	Could result in injury or occupational illness resulting in one or more lost work day(s), loss exceeding $10,000 but less than $200,000, or mitigatible environmental damage without violation of law or regulation where restoration activities can be accomplished.
Negligible	IV	Could result in injury or illness not resulting in a lost work day, loss exceeding $2,000 but less than $10,000, or minimal environmental damage not violating law or regulation.

TABLE 36-2 Suggested Mishap Probability Levels[a]

Description[b]	Level	Specific Individual Item	Fleet or Inventory[c]
Frequent	A	Likely to occur often in the life of an item, with a probability of occurrence greater than 10^{-1} in that life.	Continuously experienced
Probable	B	Will occur several times in the life of an item, with a probability of occurrence less than 10^{-1} but greater than 10^{-2} in that life.	Will occur frequently
Occasional	C	Likely to occur some time in the life of an item, with a probability of occurrence less than 10^{-2} but greater than 10^{-3} in that life.	Will occur several times
Remote	D	Unlikely but possible to occur in the life of an item, with a probability of occurrence less than 10^{-3} but more than 10^{-6} in that life.	Unlikely, but can reasonably be expected to occur
Improbable	E	So unlikely, it can be assumed occurrence may not be experienced, with a probability of occurrence less than 10^{-6} in that life.	Unlikely to occur, but possible

[a] The probability that a mishap will occur during the planned life expectancy of a system, quantified in terms of potential occurrences per unit of time, events, population, items, or activity.
[b] Definitions of descriptive words may be modified based on quantity of items involves.
[c] The expected size of the fleet or inventory should be defined before accomplishing an assessment of the system.

TABLE 36-3 Example Mishap Risk Assessment Values

Probability	Severity			
	Catastrophic	Critical	Marginal	Negligible
Frequent	1	3	7	13
Probable	2	5	9	16
Occasional	4	6	11	18
Remote	8	10	14	19
Improbable	12	15	17	20

TABLE 36-4 Example Mishap Risk Category and Mishap Risk Acceptance Levels

Mishap Risk Assessment Value	Mishap Risk Category	Mishap Risk Acceptance Level
1–5	High	Component acquisition executive
6–9	Serious	Program executive officer
10–17	Medium	Program manager
18–20	Low	As directed

36-3 FAULT TREE ANALYSIS

Fault tree analysis is one system safety method often used for complex systems. Fault tree analysis, which was originated by H. A. Watson at Bell Telephone Laboratories in 1962,[3] is a boolean logic concept that evaluates *events*. The procedure relies on building a tree structure as shown in Figure 36-3. At the top is the principal or top undesired event, which is broken down into contributing factors that are further subdivided into event causes. Fault tree analysis is a deductive process that moves from the general to the specific. Combinations of events are considered in the causal chain. Interactions between events and elements of the system are a vital part of this method.

Fault tree analysis as applied to system safety relies on preliminary hazard analyses (PHA) or other analysis techniques to identify major undesirable events. The tree is developed further from PHA and other analyses. After the tree is constructed, qualitative or quantitative analysis is performed. To perform quantitative analysis, a probability must be assigned to each event cause. Today, computer systems make the procedures of constructing and analyzing fault trees quite easy. Qualitative analysis provides insights into fault paths and critical event causes.

Limitations of Fault Tree Analysis

Analysis of a fault tree can be no better than the events identified for it. A major limitation of fault tree analysis is failure to identify all the events that may lead to a top event. Failure to include an event may simply be oversight, but it may also be lack of experience and knowledge of the system and its behavior or potential behavior. When a system is being developed and analyzed for failures and undesired events, one may not have insight into the kinds of things that may lead to faults and failures in the future or may not be experienced with materials and components used and their potential failure modes.

Another significant difficulty is assigning valid probabilities to event causes. Although considerable data on equipment performance are available from reliability engi-

FREQUENCY OF OCCURANCE	HAZARD CATEGORIES			
	I CATASTROPHIC	II CRITICAL	III MARGINAL	IV NEGLIGIBLE
A FREQUENT	1A	2A	3A	4A
B PROBABLE	1B	2B	3B	4B
C OCCASIONAL	1C	2C	3C	4C
D REMOTE	1D	2D	3D	4D
E IMPROBABLE	1E	2E	3E	4E

Hazard Risk Index	Suggested Criteria
1A, 1B, 1C, 2A, 2B, 3A	Unacceptable
1D, 2C, 2D, 3B, 3C	Undesirable (Management Activity decision required)
1E, 2E, 3D, 3E, 4A, 4B	Acceptable with review by Management Activity
4C, 4D, 4E	Acceptable without review

Figure 36-2. Example hazard risk assessment matrix. (From MIL-STD-882B.)

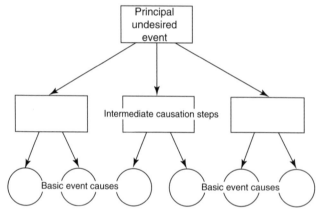

Figure 36-3. Fault tree concept. (Reprinted with permission from Roland, H. E., and Moriarty, B., *System Safety Engineering and Management*, John Wiley & Sons, New York, 1983.)

neering and other sources, placing probabilities on human activities with precision can be quite difficult. Humans may behave very differently under ideal conditions compared with stressful, boring, or distracting conditions. In addition, different people may act quite differently under the same conditions. Data banks on human errors provide reasonable information on simple human errors, but there is little information for estimating mistakes on higher-level tasks involving cognitive functions.

Another limitation on the use of fault tree analysis is cost. Compiling the knowledge for, constructing the fault tree, and assigning probabilities to tree elements can be laborious and costly.

Fault Tree Symbols

Fault tree analysis uses a particular set of symbols. Figure 36-4 illustrates commonly used symbols. There are some variations in symbology among practitioners.

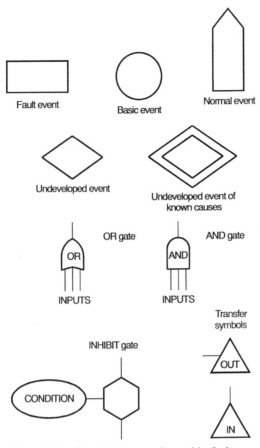

Figure 36-4. Symbols commonly used in fault tree analysis.

Events There are four kinds of events and symbols. A *fault event*, which is represented by a rectangle, is a top or intermediate event that must be described further in the tree. For quantitative analysis, a probability for a fault event is computed from elements below it in the tree.

A *basic event* is an event for which there will be no further analysis. It is represented by a circle and it is the terminus of a branch in the fault tree. Probabilities are assigned to basic events when quantitative analysis is performed.

An *undeveloped event* is represented by a diamond and is an event that an analyst chooses not to analyze. Although it may merit further analysis, an undeveloped event simply may be a curiosity or may not be critical to the problem at hand. Probabilities may be assigned to undeveloped events. Sometimes an undeveloped event of known cause is not developed further, but there is deeper knowledge about that branch of the tree. In diagraming such undeveloped events, some people use a double diamond.

A *normal event* is one that has two states: it occurs or does not occur. Normal events are represented by a house shape and are sometimes called *switch events*. In many cases, analysis of a tree should consider normal events in each of their two states. Frequently, normal events have probabilities of 1.0 or 0.0; sometimes other probabilities are assigned.

Logic Gates Because the elements in a fault tree are related by boolean algebra, symbols are used to depict the kind of relationship among elements. Basic logic relationships are

OR and AND, and are represented by gate symbols. Both AND and OR gate symbols have unique shapes.

An OR gate indicates that any one of the input events can cause an output event. When quantitative analysis is conducted, probabilities for input events attached to an OR gate are summed to compute the probability of the output event.

The other basic logic gate is an AND gate, which indicates that all of the input events must occur to cause the output event. In quantitative analysis, the probability of an output event is the product of all input events.

Special Notations There are other logical relationships that can occur in a fault tree. Various notations to AND and OR symbols indicate that special logical relationships or other symbols are used. For example, two input events for an OR gate may be mutually exclusive; that is, one excludes the other from occurring. An *exclusive* notation attached to the OR gate indicates this condition.

There may be a condition in which at least two of three input events are necessary for an output event to occur at an AND gate. A notation "$A_i \geq 2$" attached to the AND gate would note this special condition.

In another situation, one or more input events may have to occur before a third one has any consequence. This is called a priority modification. A notation "$C \rightarrow R_1, R_2$" would indicate that input event C is not significant unless input events R_1 and R_2 occur first.

Another variation, called a summation gate, is the possibility of having input events that must have certain levels before the output will occur. A summation gate may apply to either an OR or AND gate. A summation sign or note with the gate indicates this special condition.

Sometimes a complex array of conditions determines if an output event will occur at a gate. An "M" notation on a gate indicates that a complex matrix of conditions is processed by this gate.

For some events, certain conditions must be present for the input events to be included in the tree. The input events may inhibit or enable the output event. A hexagon symbol represents an inhibit gate.

When there is not enough space to complete a fault tree, it must be broken into parts. Discontinuities are represented by a transfer symbol that has the shape of a triangle. Identifying numbers or letters on both segments of a drawing indicate where they tie together functionally. A fault tree may have identical branches at more than one location. A transfer symbol reduces the need to completely represent the branches at each location in the tree.

Events

An event describes any element of a fault tree that represents an occurrence. Events may be normal events, failures or faults. Failures are attributes of components that interrupt the function of the component. For example, an electronic relay that sticks open is a failure event.

Fault events are events that contribute to component or system faults. A fault is a condition (not necessarily a failure) of a system, subsystem, or component that contributes to the possible occurrence of an undesired event. For example, failing to act in response to a fire alarm is a fault, but a deaf person not being able to hear an alarm is a failure.

There are four classes of causal events that appear in fault trees. Primary refers to internal attributes or conditions of components; secondary refers to something outside a component.

Primary Failures Primary failures are internal problems with components that make them inoperative. Repairing a primary failure returns a component to full operation. A primary failure also is defined as a failure of a component within the design envelope, such as an inherent characteristic of a component that causes the component to fail. The primary failure of one component cannot contribute to primary failure in another component.

Secondary Failures Secondary failures are external problems that make components inoperative. Repairing a secondary failure does not return a component to operation. A secondary failure is the failure of a component outside the design envelope, such as environmental conditions that affect a component. A primary or secondary failure of one component or a group of components can cause a secondary failure in another component.

Primary Faults Primary faults are events that are abnormal within an operation. They can lead to undesired conditions in a system.

Secondary Faults Secondary faults are event causations that are external causations. One form of secondary fault is a command fault: an inadvertent operation of a component resulting from failure of a control element. An example is accidentally bumping a control switch that energizes a circuit.

Constructing a Fault Tree

Development of a fault tree begins by selecting the top event. Usually, the top event is selected as the most important, most severe or most undesired event. The system to which the top event applies then is clearly defined and the state of the system must also be specified. Then one begins to construct the fault tree.

The first tier of events includes those that are necessary and sufficient causes for the top event. Other tiers are added, and then logical relationships among events are added. It is better to include generic causes at upper levels in a fault tree. This makes it easier to include detailed faults and failures in the tree structure.

Analyzing a Fault Tree

There are several approaches to analyzing a fault tree. Methods involve quantitative and qualitative analysis.

Qualitative Analysis of Fault Trees Creating a fault tree gives analysts insight into the causes of an undesired event and to system behavior. This alone may make the exercise worthwhile.

The elements of a fault tree can be evaluated to gain further insight into the causes of a top event. Causes within the tree can be evaluated and judgments can be made about the likelihood of faults or failures contributing to the top event. Each event sequence can be looked at, and those that are most likely can be considered first.

Another approach is to find the most likely sequences by analyzing the gates using products of input events for AND gates and sums of input events for OR gates. Products of values less than one are smaller than their sums. With this in mind, the most likely event sequence often can be identified quickly by tracing each branch of the tree from the top event to the bottom event. Branches linked by OR gates typically have high probabilities of occurrence, whereas branches linked by AND gates typically have low probabilities of occurrence.

Quantitative Analysis of Fault Trees Quantitative analysis begins at each bottom end of a branch. To perform quantitative analysis on fault trees, a probability must be assigned to each basic and normal event. Probabilities of occurrence may also be assigned to each undeveloped event.

Then boolean algebra is applied to each logic gate to determine the probability of each intermediate event. Ultimately, the analysis calculates the probability for the top event. Example 36-1 illustrates the fundamentals of this process for the fault tree shown in Figure 36-5.

Cut Sets Cut sets are any sequence of events (reading from the bottom of a branch to the top event) that leads to the occurrence of the top event. Each sequence that leads to the top event can be analyzed separately and then compared to the others. The comparison will help identify which sequence is most likely to cause the top event.

Example 36-1 For the fault tree in Figure 36-5, the probabilities for some of the events are as follows:

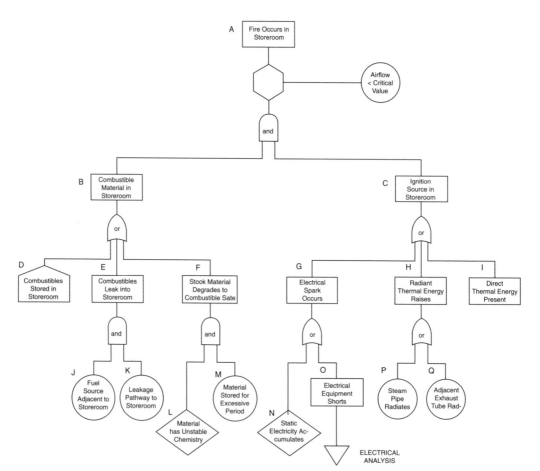

Figure 36-5. Example of a fault tree. (From *Facility System Safety Program Manual*, HNDP 385-3-1, U.S. Army Corps of Engineers, Huntsville Division, Huntsville, AL, October, 1985.)

Event	Probability for Events (Frequency in Days)
D	3.45×10^{-7}
J	6.89×10^{-4}
K	7.33×10^{-3}
L	6.05×10^{-3}
M	1.88×10^{-4}

What is the most likely cause for event B?

The probability for event D is given. The probability for event E is

$$P(J) \times P(K) = (6.89 \times 10^{-4})(7.33 \times 10^{-3})$$
$$= 5.05 \times 10^{-6}.$$

The probability for event F is

$$P(L) \times P(M) = (6.05 \times 10^{-3})(1.88 \times 10^{-4})$$
$$= 1.137 \times 10^{-6}.$$

Event E is the most likely cause. However, event F has a very similar probability and should be given careful consideration in selecting controls.

36-4   FAILURE MODE AND EFFECTS ANALYSIS

Failure mode and effects analysis (FMEA) is an inductive procedure that moves from the specific to the general. Examples of FMEA can be found in the form of diagnostic charts for automobile or appliance repair. The emphasis is not on events, but on conditions. FMEA analyzes *equipment* or *components*; it relates conditions of components to conditions of the system of which they are a part. Failures in components are traced to determine their effects on the system. Of greatest interest are effects that impact safety.

FMEA uses special tables and charts to log data during the analysis. One element of a typical worksheet is a *component description*. The worksheet identifies which individual or combinations of components are analyzed. The worksheet has a column for *failure mode*. Additional columns list *effects on other components* and *effects on the system*. The worksheet also contains a column to identify the hazard category (see Tables 35-2 and 35-3) or risk assessment code (see Figure 35-1). It may also estimate *failure frequency* and *effects probabilities*, which may be qualitative or quantitative. Finally, there is usually a column to identify *control method*, that is, to indicate how to prevent the failure or how to protect against its consequences.

In working across the data columns of a FMEA chart, it is important to recognize that there are many more relationships among data elements than one failure mode for each item, one cause for each failure, one effect for each cause, and so forth.

From a completed FMEA, a critical item list (CIL) can be developed. This list includes failures that exceed the acceptable levels of risk. The CIL may be used for more detailed safety analysis.

Figure 36-6 is an example of a FMEA worksheet.

36-5   SIMULTANEOUS TIMED EVENTS PLOTTING ANALYSIS

Another method for identifying hazards and relating them to systems is simultaneous timed events plotting analysis (STEP), which analyzes *events* from a time or sequence perspec-

FAILURE MODE AND EFFECTS ANALYSIS								
System		Subsystem		Date	Analyst		Page	
Component or Part Name/ Description	Failure Mode	Cause of Failure	Effect on...			Risk or Hazard Category	Probability of Effect	Comments
			Other Components	System	Personnel			

Figure 36-6.   Example format for a Failure Mode and Effects Analysis (FEMA) worksheet.

tive. Sequences of events that occur quickly may require corrective actions different than those that occur more slowly. Event sequences that occur very slowly may be hard to recognize.

STEP procedures involve identifying people or things (called actors) and their actions. An actor plus an action is called an event. For example, "alarm sounds" and "occupant runs" are events. The events for each actor are plotted against a time line and relationships among different actors' events are identified by linking arrows. The resulting chart allows visualization of what events occur when there is a complicated sequence. Figure 36-7 illustrates a chart produced from a STEP analysis of a simple process.

36-6   HAZARD TOTEM POLE

Grose[4] applies system safety principles and techniques in preparing hazard control information for management decisions. His process begins by describing "scenarios" of things that go wrong for each organizational unit or functional element of an operation. The hazards in each scenario are identified and hazards are rated in each for (1) severity, (2) probability of occurrence, and (3) cost to correct. A table for each rating has four categories that are identified by letters instead of numbers. Hazard severity has categories A, B, C, and D, hazard probability has categories J, K, L, and M, and cost to correct has categories R, S, T, and U. Two of the tables have descriptions for categories very similar to those in Tables 35-2 and 35-3.

The combinations of categories from each of the three tables are organized into a decision chart called the hazard totem pole (see Figure 36-8). There are 64 levels in the totem pole (based on a $4 \times 4 \times 4$ matrix = 64 conditions). The totem pole is prepared separately from the evaluation of particular scenarios. At the top of the totem pole is the combination A, J, and R, which represents hazards that are very severe, have a high probability of occurrence, and are very inexpensive to correct. At the bottom of the totem pole is the D, M, and U combination, which represents hazards that are the least severe, are not very likely to occur, and are very expensive to correct. In between are the other combinations in an order acceptable to managers who make decisions about how much to spend on correcting hazards or an order based on risk and criticality, as depicted in Figure 35-1.

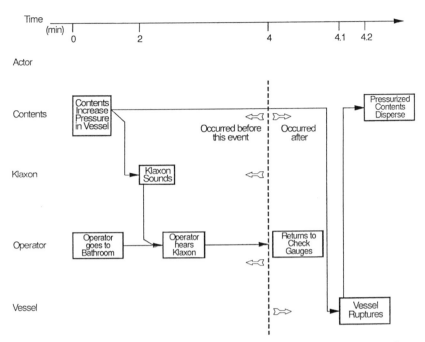

Figure 36-7. Example of a STEP analysis chart for a simple process. (From *Facility System Safety Program Manual*, HNDP 385-3-1, U.S. Army Corps of Engineers, Huntsville Division, Huntsville, AL, October, 1985.)

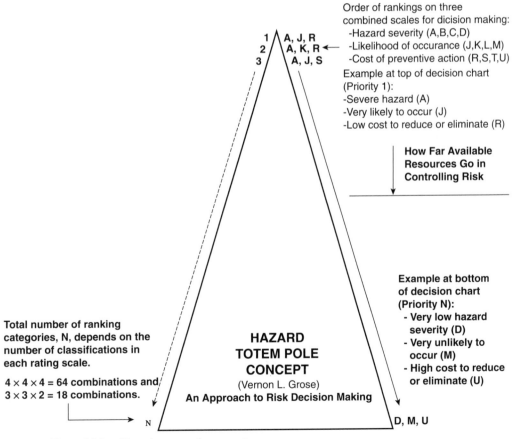

Figure 36-8. Hazard totem pole concept.

Hazards from the analysis of scenarios are organized in a list according to the combination of ratings established in the hazard totem pole. The list forms priorities for allocating funds to correct the hazards. The list can include the cost to correct each hazard and a cumulative cost that sums costs for scenarios starting from the top. Managers have limited resources and are not likely to be able to correct every hazard in the list. They must decide how far down the list they want to go, can afford to go, or both with their budget. At some point there is a cutoff. Funds will not support items below the cutoff line.

The items that are funded are converted to an action plan for implementing the corrections. The process is repeated periodically to identify new scenarios and hazards, and as a result, there is a new funding consideration list for management.

36-7 MANAGEMENT OVERSIGHT AND RISK TREE

Another method that is a derivative of system safety and fault tree analysis techniques is management oversight and risk tree (MORT), which is both a program and a logic diagram. As a program, MORT helps prevent safety-related oversights, errors, and omissions, and it attempts to identify and assess risks associated with an operation and refer them to the proper management level for action. MORT programs help optimize allocation of funds for safety programs and hazard control. MORT incorporates behavioral, organizational, and analytical sciences in dealing with energy transfer, error, change, and risk in a systematic way.

As a diagram, MORT arranges safety program elements in an orderly and logical manner. MORT diagrams structure safety literature and practices into three levels of relationships. At the top level, MORT identifies 98 generic problems or *undesirable events*. At the second level, there are 1,500 possible causes, termed *basic events*, for these problems. At the third level, there are thousands of *criteria* (standards, codes, practices, etc.) to judge whether steps in a safety program are done well or are less than adequate (denoted by LTA).

MORT can be used to investigate accidents or evaluate safety programs. A MORT diagram is an idealized safety system model that uses the logic of fault tree analysis. MORT assumes that in a *perfect* safety system, all components function in a manner that contributes to or complements the achievement of tasks. A safety system or program is a process of eliminating or controlling hazardous events through engineering, design, education, management policy, and supervisory control of conditions and practices.

The general features of a MORT event tree are shown in Figure 36-9. This example gives a general flavor for MORT diagrams and analysis. There are many rules and procedures that cannot be covered here that make MORT an effective tool. In a MORT diagram, generic events are at the top. At the second tier are specific or management oversights and omissions and assumed risks (denoted by R). Specific (denoted by S) refers to events or factors that are specific to an accident. Management (denoted by M) refers to factors in the general management system or context. At the lower tiers, basic events and contributing factors and controls that failed are detailed.

36-8 OTHER ANALYSES AND APPLICATIONS OF SYSTEM SAFETY

There are several other extensions of system safety analysis and techniques that make the system safety approach an effective one. Applications of system safety methods, often with some variance from more formal procedures, have found their way into dealing with many safety problems.

What and How Large Were the Losses?

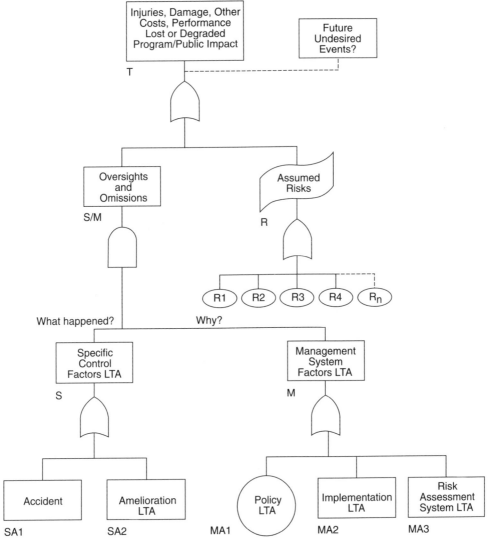

Figure 36-9. An example of the top portion of a MORT diagram. (From *MORT Users' Manual*, Revision 2, DOE 76-45/4, SSDC-4, U.S. Department of Energy, Washington, DC, May, 1983.)

Energy Analysis

For many safety problems, an analysis of energy in various forms, transfer of energy, and release of energy is very useful. Haddon's theories about energy and control of it (see Chapter 3) are a significant part of energy analysis. When analyzing machines, equipment, processes, and operations, an analysis of energy can identify many hazards that need to be controlled. Figure 36-10 is an incomplete diagram of several forms of energy that might be considered during an energy analysis.

For example, a punch press has energy in the moving flywheel. During the machine operation, the energy is transferred to the action of the punch. There are dangers associ-

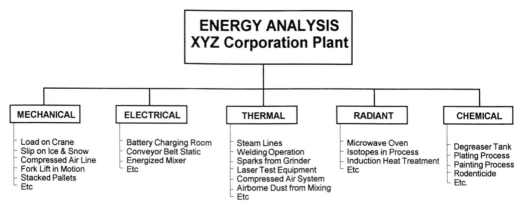

Figure 36-10. Example of an energy analysis chart.

ated with the flywheel and the punch actions that need protection. In addition, there are springs that store energy in the die or elsewhere. Parts of the machine are held in place by brakes during idle phases of the punch cycle. A failure of the brake could release potential energy in the weight of the elevated components and the parts could fall. Similarly, energy could be released while dies are changed. Thus, a die prop is needed to prevent potential energy from becoming kinetic energy. A motor in the press has electrical energy that is converted to mechanical energy. Unprotected electrical energy could lead to injury directly or through its transfer at the wrong time to the machine components.

Energy analysis may be helpful in identifying risks in powered systems and equipment and in establishing engineering and administrative controls as well as lock-out and tag-out procedures.

Buildings

The Department of Defense may require that system safety be applied to construction projects. At minimum, a preliminary hazard analysis must be performed for the facility, its subsystems, and use during its planning and early design. Depending on results, further analysis may be needed to identify further hazards and suitable controls for the building life cycle. The U.S. Army Corps of Engineers developed a procedural guide for applying system safety to building projects.[5]

Fire Safety

The National Fire Protection Association (NFPA) recognized that a building and its fire safety could benefit from a systems analysis. A structure is a system made up of many components. Buildings and structures are modified over time and their conditions often change. Codes and standards for fire safety also change with time. A systems approach for analyzing the fire safety of a building can help identify deficiencies and pinpoint corrective actions.

NFPA has developed a fire safety concepts tree.[6] At the top of the tree is fire safety objective(s), followed by actions to achieve the objectives. Elements of the tree are connected by AND and OR gates, similar to fault tree analysis. Figure 36-11 illustrates the upper levels of the tree. The tree is useful in building analysis and design and can be used for qualitative and quantitative analysis.

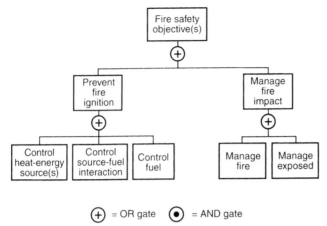

Figure 36-11.   Upper levels of the fire safety concepts tree. (Reprinted with permission from the *Fire Protection Handbook*, 19th ed., 2003, National Fire Protection Association, Quincy, MA, 02169.)

EXERCISES

1. The circuit, fault tree, and probabilities of events in the fault tree are shown in Figure 36-12.

   (a) Compute the failure rates for branches (a) and (b).

   (b) Compute the failure rate for the top event.

   (c) Which branch is more critical (more likely to cause failure)?

   (d) Which failure or fault is most likely to cause the system to fail?

2. Obtain a copy of a fault tree analysis for a system and review the analysis and results. You may wish to look at one of the classic reports on nuclear power plant safety: *Reactor Safety Study: An Assessment of Accident Risks in U.S. Commercial Nuclear Power Plants*, Report WASH-1400, U.S. Nuclear Regulatory Commission, October 1975. (This report is also known as the Rasmussen report.)

REVIEW QUESTIONS

1. What is system safety?

2. Where did system safety originate?

3. What military standard documents the general procedures for system safety?

4. Briefly explain a system safety program and its objectives.

5. List five system safety design requirements.

6. Identify the precedence for meeting system safety requirements.

7. What are the two classes of system safety tasks? Identify five tasks within each class.

8. What does PHA stand for?

9. What is fault tree analysis? For what is this method used to analyze?

10. What are three limitations of fault tree analysis?

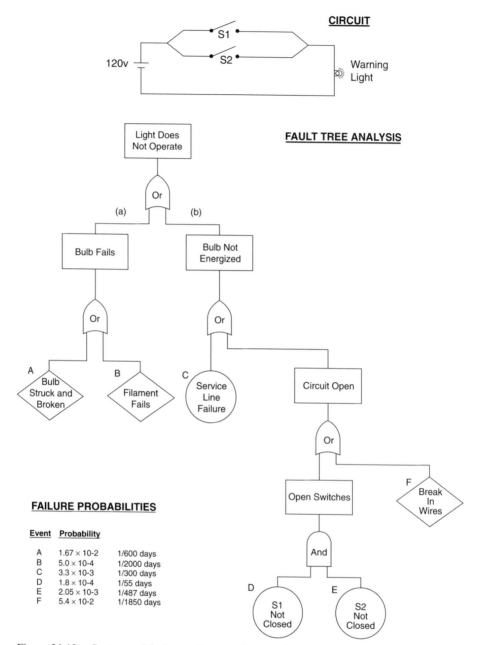

Figure 36-12. System and fault tree diagrams for Exercise 1.

11. What symbols are used in fault tree analysis to represent each of the following factors?

  (a) a basic event

  (b) an undeveloped event

  (c) a normal event

  (d) AND logic gate

  (e) OR logic gate

  (f) inhibit logic gate

  (g) a transfer or discontinuity in a fault tree diagram

12. What is the difference between a fault and a failure?

13. What are the four classes of causal events in fault trees?

14. What kinds of analyses can be performed on a fault tree?

15. What is a cut set?

16. What is FMEA? For what is this method used to analyze?

17. What is STEP analysis? For what is this method used to analyze?

18. What is the hazard totem pole? Explain how is it used.

19. What is MORT? What is this method used for?

NOTES

1 29 CFR 1910.119, Process Safety Management of Highly Hazardous Chemicals.

2 *System Safety Analysis Handbook*, 2nd ed., System Safety Society, Unionville, VA, 1997.

3 Recht, J. L., "Systems Safety Analysis: The Fault Tree," *National Safety News*, 93:37–40 (1966).

4 Grose, V. L., *Managing Risk: Systematic Loss Pre-vention for Executives*, Prentice-Hall, Englewood Cliffs, NJ, 1988.

5 *Facility System Safety Program Manual*, HNDP 385-3-1, U.S. Army Corps of Engineers, Huntsville Division, Huntsville, AL, October, 1985.

6 *Fire Protection Handbook*, 19th ed., National Fire Protection Association, Quincy, MA, 2003.

BIBLIOGRAPHY

BROWN, D. B., *Systems Analysis and Design for Safety*, Prentice-Hall, Englewood Cliffs, NJ, 1976.

DAUGHERTY, E. M., Jr., and Fragola, J. R., *Human Reliability Analysis: A Systems Engineering Approach with Nuclear Power Plant Applications*, Wiley, New York, 1984.

GREEN, A. E., *Safety Systems Reliability*, Wiley, New York, 1984.

HAMMER, W., *Handbook of System and Product Safety*, Prentice-Hall, Englewood Cliffs, NJ, 1972.

HENLEY, E. J., and Kinamoto, H., *Designing for Reliability and Safety Control*, Prentice-Hall, Englewood Cliffs, NJ, 1985.

JOHNSON, W. G., *MORT Safety Assurance Systems*, Marcel Dekker, New York, 1980.

LEVESON, NANCY G., *Safeware—System Safety and Computers*, Addison-Wesley, Reading, MA, 1995.

MALASKY, S. W., *System Safety: Technology and Application*, 2nd ed., Garland STPM Press, 1982.

PERROW, C., *Normal Accidents: Living with High-Risk Technology*, Basic Books, Inc., New York, 1985.

RAHEJA, DEV G., *Assurance Technologies—Principles and Practices*, McGraw-Hill, New York, 1991.

ROLAND, H. E., and Moriarity, B., *System Safety Engineering and Management*, 2nd ed., John Wiley & Sons, New York, 1990.

Standard Practice for System Safety, Military Standard MIL-STD-882D, U.S. Department of Defense, Washington, DC, February 10, 2000.

STEPHANS, RICHARD A., *System Safety for the 21st Century—The Updated and Revised Edition of System Safety 2000*, John Wiley & Sons, New York, 2004.

STEPHANS, RICHARD A., and Talso, Warner W., *System Safety Analysis Handbook*, System Safety Society, Unionville, VA, 1997.

STEPHENSON, JOE, *System Safety 2000—A Practical Guide for Planning, Managing, and Conducting System Safety Programs*, Van Nostrand Reinhold, New York, 1991.

VINCOLI, JEFFREY W., *Basic Guide to System Safety*, Van Nostrand Reinhold, New York, 1993.

CHAPTER *37*

SAFETY ANALYSES AND MANAGEMENT INFORMATION

37-1 PURPOSE OF ANALYSIS

There are several places in this book where analysis is recommended. The preceding two chapters in particular reviewed some of the more prominent analytical procedures frequently used in safety. This chapter considers a few more. Analysis itself can be expensive and can consume many hours of work by many people. Therefore, it is important to consider how much analysis is needed and what the purpose of the analysis is.

One purpose for analysis is to gain understanding. It is not always obvious how things work or what makes things go wrong. Carefully defining the components or elements involved in something and identifying the relationships among them will help provide insights for events, processes, systems, and equipment and will provide understanding of complexities. Analysis makes it possible to discover and observe the intricate relationships between people and the world around them. The need to gain understanding appeals to the human urge to explore.

Another purpose for analysis is to make decisions. Analysis may help managers select the correct action or course of action. It may help managers allocate funds and assign people, equipment, or other resources to ensure that actions are completed in a timely manner.

An additional reason for performing analysis is that it is required. Often laws, regulations, or contracts require people to complete analysis. Someone else may use the information and results in decision making. For example, when someone is involved in a sizeable automobile accident, the vehicle owners must perform enough analysis to prepare a report. In some cases, attending police officers must collect evidence and analyze it to determine if charges should be filed against vehicle drivers or others. Some design contracts require that certain analysis be performed to aid others involved in the design project, production, or use of items being designed.

The dilemma is doing just enough analysis to make decisions with confidence, rather than continuing to analyze something to the point where any further information or results will have no bearing on any more decisions. In safety, analysis helps managers solve the problem "How safe is safe enough?" (see Figure 3-5). Performing analysis for its own sake may be interesting for some, but may not be justified by the usefulness of the results.

From a safety point of view, the purpose of analysis is to prevent accidents. Being able to communicate with managers and explain what hazards exist and what controls should be implemented to eliminate or reduce them is a must, as is the ability to talk in

Safety and Health for Engineers, Second Edition, by Roger L. Brauer
Copyright © 2006 John Wiley & Sons, Inc.

management terms about cost of losses and controls, effectiveness of controls, and benefits derived from allocation of funds and other resources.

37-2 INSPECTIONS

A common form of analysis to prevent accidents is inspections. It is possible to chart on paper many of the possible things that can go wrong, but in operating systems, inspections must be performed to detect actual faults and failures in equipment, activities, and workplaces. Inspections are part of a preventive or proactive approach to accident prevention (see Figure 3-4).

Types of Inspections

There are several kinds of inspections: general and detailed. In a general inspection, individuals or teams look for a range of deficiencies. In a plant, a general inspection is a walkthrough that considers a wide range of safety problems. Figure 37-1 illustrates a checklist that is typical of a general safety inspection. Pilots of aircraft are taught to conduct a general inspection of the aircraft before every flight. They walk slowly around the aircraft looking for any unusual conditions that could affect the airworthiness of the airplane and flight safety. General inspections can cover many topics, but are very precise.

There are also detailed inspections, which are tailored to the activity and equipment involved. In detailed inspection, people look for very specific conditions. For example, a rigging inspector must be able to identify roughly 20 kinds of defects in wire rope, some of which may be easy to recognize and others that are difficult and require special training and considerable experience. Some aircraft inspections are very detailed, requiring several days to complete.

Inspections may be scheduled or unscheduled. Scheduled inspections are those required regularly. The schedule may be governed by the clock, such as a daily inspection of equipment at the beginning of the shift. Other schedules are based on use time or use cycles. Aircraft are inspected by specialists after a certain number of flight hours. The same aircraft are inspected before each flight by pilots and ground crews. Some inspections are structured around events. A person operating a tool crib inspects tools and items at the time of issue and when they are returned. A field engineer may inspect a product while making a trouble-shooting call on a customer.

Unscheduled inspections may be based on random visits by specialists. They may occur when equipment is brought in for maintenance or repair, at which time someone looks for deficiencies and hazardous conditions throughout the item, not just at the components involved in the repair. A good time for a general inspection of a plant is when it is down for maintenance or installation of new equipment or there is downtime because of low product demand. Railroad companies teach workers to inspect passing trains for shifted loads, bearing and wheel failures, and other kinds of problems.

Inspectors

If the goal is to identify hazardous and defective conditions, who should do the inspection? There are several strategies that apply, with some applying to critical conditions. In some cases, detailed training, knowledge, and experience are necessary to be able to recognize a problem. In other cases, nearly anyone can be taught to identify unsafe conditions and activities. Some inspections may require the use of special instruments and tools and someone who knows proper use and procedures for them.

Safety Inspection Report

Company/Division _____

Address/Location _____

Date of Inspection _____

Names of _____
Inspectors

Committee _____
Members

1. Machine Operation. Check to see that guards and devices are in place and in good condition. Look for points of operation where provision of guard could eliminate hazard. Is provision made against accidental starting of machine? Are controls and displays properly color coded and labeled?

Dept	Date Submitted	Date Comp.

2. Transmission and Machine Equipment. Are belts, pulleys, shafts, revolving parts, and set screws guarded? Are machine disconnects easily accessible, in good working order, and capable of being locked out?

Dept	Date Submitted	Date Comp.

3. Hand Tools. Look for broken tools, noting condition of handles and heads. Check to see that non-sparking tools are used where provided. Are portable electric tools properly grounded or double insulated?

Dept	Date Submitted	Date Comp.

4. Housekeeping. Check condition of aisles, work areas, stairs, and outside areas. Are racks and bins provided for small parts and conveniently placed? Are materials in process neatly stored? Are trash receptacles provided and used? Note condition of washroom facilities, ventilation, clothes lockers, etc. Note slip and trip hazards.

Dept	Date Submitted	Date Comp.

5. Handling Materials. Check hand and power trucks. Are materials piled safely? Are weights of manually handled items OK? Are employees taught to lift properly? Is care taken when loading and unloading trucks, elevators, cranes, and conveyors? Is protective clothing provided? Note condition of material, boxes, skids, totes, etc. Note tasks and workstations that should be evaluated for possible ergonomic improvements.

Dept	Date Submitted	Date Comp.

6. Industrial Hygiene. Note work areas that may have excessive levels of dusts, vapors, gases, fumes, noise, etc. Are confined spaces properly identified? Are MSDS sheets accessible for review?

Dept	Date Submitted	Date Comp.

7. Elevators and Other Elevating Equipment. Is every shaft opening gate in good condition? Is elevator pit clear? Is elevator operated by authorized employee and not overloaded? Check light on car and landing, condition of signal system, safety devices, car and shaftway protection, limit switches, and control mechanism. Have cables and hoisting equipment been inspected by and elevator inspector within the past six months? Look for hoistway shear hazards.

Dept	Date Submitted	Date Comp.

8. Floors, Floor Opening, and Hoistway Openings. Are floors free of protruding nails, splinters, holes, slipperiness, unevenness, and loose boards? Are openings properly guarded?

Dept	Date Submitted	Date Comp.

Figure 37-1. Example checklist for a general safety inspection. (Provided by and reprinted with the permission of the Liberty Mutual Group, Boston, MA.)

9. Stairs and Ladders. Note condition of stair treads and supports. Are handrails and lighting adequate? Check condition of ladders. Are they properly stored, or if permanent, firmly fastened in place? Are correct safety feet being used for each job? Note: metal ladders should never be used for electrical repairs!

Dept	Date Submitted	Date Comp.

10. First Aid. Are first aid supplies adequate? Is a trained first aider always available? Are all injuries reported and treated?

Dept	Date Submitted	Date Comp.

11. Electrical Equipment. Note condition of switchboard, transformers, wiring, controlling, and operating apparatus. Is equipment properly protected and isolated? Is area clear of tools or refuse? Report loose or disconnected wires. If wire is in rigid conduit or BX cable, report any point where conduit or BX is broken or separated. Is high voltage equipment locked up and posted with warning signs? Report location of any switch box or open switch where floor is frequently wet or where operator might come in contact with plumbing fixtures. Are rubber gloves, rubber mats, and fuse puller provided.

Dept	Date Submitted	Date Comp.

12. Fire Prevention. Are exits adequate, well located, marked, clear, and well lit? Do fire doors close freely? Are they equipped with fusible links? Is fire fighting equipment conveniently placed and ready for immediate use? Check extinguishers and other equipment. Check to see that materials are not piled too close to sprinkler heads. Is plant kept free of flammable waste? Are flammable, chemical, and explosive materials properly handled and stored? Is power exhaust provided for solvent vapors?

Dept	Date Submitted	Date Comp.

13. Elevated Runways and Platforms. Are they in good condition? Clear of obstructions? Equipped with top and midrails, and toeboards?

Dept	Date Submitted	Date Comp.

14. Boilers, Pressure Apparatus. Note when last inspected. Are safety valves, water gauges, and engine stops checked and tanks drained regularly? Are gas cylinders stored sway from sun and fastened securely in place with separate storage places for different gases? Code requirements met?

Dept	Date Submitted	Date Comp.

15. Safe Practices. Are machines stopped before cleaning and locked out/tagged out during repairs? Is personal protective equipment (eye, face, foot, head, hearing, etc.) in good condition, worn when necessary, and stored properly? Are any unsafe acts observed?

Dept	Date Submitted	Date Comp.

16. Safety Education. Are posters and safety signs displayed? Number? Well located? Is safety literature distributed? How often?

Dept	Date Submitted	Date Comp.

The illustrations, instructions and principles contained in the material are general in scope and, to the best of our knowledge, current at the time of publication. No attempt has been made to interpret any referenced codes, standards or regulations. Please refer to the appropriate code, standard or regulation making authority for interpretation or clarification.

Figure 37-1. *continued*

Some inspections need to be carried out by someone who is not directly involved in performing work on the item or the area being inspected. This results in an independent look at a situation and removes some bias from the inspection task. People may not see their own mistakes or may be too familiar with some equipment to notice that something is amiss.

In some cases, two inspectors are necessary. Consider inspection of completed work. The person who performed the work should be the first inspector, and a coworker, supervisor, or specialist should be the second inspector. The second inspector may be more knowledgeable and experienced. The double inspection provides redundancy to the task.

Each situation dictates who is qualified to conduct an inspection. Inspectors can be workers, coworkers, supervisors, specialists (engineers, safety specialists, and others), or special teams. In some cases, inspectors need to be certified for the inspections they perform. Recertification is also important to make sure that inspection skills are kept high and knowledge is up to date.

Inspection Tools

The most common tool for inspectors is a checklist. Because humans are not good at remembering a long or complex list of items. A checklist helps remove memory errors. Because humans are not good at keeping track of what items in a list have been completed, a checklist provides a means for marking which items have been inspected and what was found. Checklists also provide a place for an inspector's signature, which attests to the fact that a particular person completed the items on a checklist and that the inspector has recorded all significant findings. Signing for someone else is a misuse of a signature block on an inspection form.

When visual inspection is not sufficient, special tools and instruments are needed. Instruments may range from air sampling and analysis devices for confined space entry to special test instruments for pressure and light level assessment and special tools for opening covers on machines.

Inspection Procedures

Here are a few important guidelines for completing inspections.

- Be selective. Look for particular things.
- Know what to look for. Know what failures and abnormal conditions look like.
- Use helps, such as checklists.
- Practice observing.
- Look for facts that are relevant.
- Guard against habit and familiarity. Vary the order of items. Be thorough.
- Be ready to sign that everything has been checked and is in order and that problems are noted.
- Record observations accurately.
- Avoid conversations and distractions that may inhibit careful inspections.
- Obtain additional help to verify a finding if there is any question about something not being in order.
- Continue to report faulty conditions until they are corrected.
- Make sure the inspection report gets to someone who can do something about a deficiency.

37-3 SAFETY AUDITS

Safety audits are systematic procedures for reviewing management procedures and practices implemented to achieve safety. They can include a compliance review, similar to a safety inspection. The audit process also allows for reviewer suggestions and recommendations for improvement. Audits may be part of more general practices to ensure that management systems are in place and applied, such as those covered by ISO standards.[1,2] Some standards detail auditing procedures. Audits can apply to systems, processes, products, programs, or services.

As soon as the scope of an audit is established and it is determined whether internal personnel or contractors or others will conduct the procedures, the audit team is formed and the audit is planned in detail. The members of the audit team should have qualifications suitable to the contents of the audit.

The overall audit process typically involves the following:

- An opening meeting in which the audit team meets with senior management and reviews the process to be followed.

- Information gathering through interviews, examination of documents, observation of activities and operations.

- Documentation of observations and findings, particularly any not in compliance with written standards, policies, and procedures.

- Meetings of the audit team to collaborate on audit activities, progress, and findings.

- A closing meeting by the audit team with senior management to summarize the key results of the audit process.

- Preparation of a written audit report detailing the process, findings, and recommendations and providing the report to senior management.

- Establishing corrective action plan.

- Follow-up on progress in the action plan.

37-4 ACCIDENT INVESTIGATION

Another form of analysis is accident investigation. Although accidents are to be prevented, when they do occur, often valuable lessons are learned. Investigations are part of the reactive approach to accident prevention depicted in Figure 3-3.

Purposes

There are several purposes for investigations of accidents. A primary purpose is to prevent future accidents. Another purpose is to identify causes of accidents and injuries, because such information is necessary to prevent future accidents of a similar nature. A third purpose is to compile legal or liability evidence to be used in the event of claims or lawsuits for losses or injuries. Accident investigations may help assess the degree of damage and the value of losses. For some accidents, data are collected for insurance claims related to injury or property damage.

A general reason for accident investigations is gathering facts. One should use multiple factor accident theories (see Chapter 3) in an investigation so items that may be important are not overlooked. Seldom do accident investigations have the purpose of placing blame or finding fault. When they do, it is for legal purposes and insurance claims in a fault-based claim process.

Types of Investigations

There are two main types of investigations: general and special. Special investigations can be divided into many additional categories.

General   General accident investigations are typical of investigations related to most in-plant accidents. A supervisor may look into causes and prepare a report. In some cases, specialists within the organization may participate in the investigations. Some procedures are generic and do not involve special methods and instruments. General investigations use interviews and visual observations. Investigators interview witnesses of the accident or postaccident events. Investigations may include a meeting of all witnesses to discuss potential causes and corrective actions. Data are limited to company investigation forms. Photographs may be taken and included with a written report.

Special   Special accidents investigations are those that require particular knowledge and skill of investigators. Special investigations are necessary for particular kinds of accidents and incidents. For example, there are fire investigators and methods for fire investigations, and there are special investigators and methods for aviation accidents, train accidents, and automobile accidents. There was a national investigating team during the early 1980s that investigated elevator explosions. Special investigations often require particular tools, instruments, and procedures. For example, fire inspectors need to have kits for taking samples to test for the presence of flammable materials. There are special procedures to avoid contamination of samples and for "reading" what happened in a fire from char and rubble.

Which Accident to Investigate

It may be too expensive to investigate all accidents, and at the very least, it is not possible to bring in specialists for every accident, compile a lot of data, and perform physical, chemical, or computational analyses. Deciding which accidents to investigate can be simplified by using the following criteria.

High Cost and High Severity   Accidents with high property, life, and injury losses are worth investigating. In fact, several parties probably will conduct independent investigations of the same accident to protect liability and other legal interest for different people. When there are high loss rates, there are bound to be arguments over who should pay for the losses.

High Frequency   High frequency accidents should be investigated because accidents that have similar patterns offer the potential for significant preventive actions. A pattern of events and causes may be detected. The trend may appear from accident and incident reports or may arise from customer claims or complaints, from field reports, or from repair or work orders.

High Public Interest   Accidents that gain a great deal of public attention should be investigated. For either a company or public official, there is a need to deal with public concern from a factual basis rather than from a speculative one. If a company has an accident that affects the community around it, an investigation is probably necessary to provide a basis for protecting the company image and dealing honestly with the public.

High Potential Losses   Incidents that may end in large losses of property or human life should be investigated. There is a high potential for preventing such losses in future accidents.

Investigation Tools and Data

Some of the basic tools for accident investigations are rope or security tape to secure the area. Imprinted tape will help keep the curious from damaging or removing evidence. Use of note and sketch pads also is essential. Use of photographic and videotape equipment can be very useful, because some evidence has a short life. A tape recorder is essential for interviewing witnesses. Depending on the kind of accident involved. Special instruments may be important. For radiation releases, Geiger counters or similar instruments are appropriate. For chemical spills, colorimeters, direct reading instruments for particular chemicals, and sampling equipment should be available. Other necessities are uncontaminated containers for specimens of various types, identification tags for marking items of evidence and samples, and tape measures to establish accurately the location of items at an accident scene. In automobile accidents, for example, location data for vehicles and tire marks can be critical.

The data needed may be dictated by law. For example, in certain automobile accidents, police officers are required to acquire information on alcohol and drunkenness. There are standards and guidelines for the data needed in particular kinds of investigations. Photographs and sketches of physical conditions, the surround, and damages are important. Videotape or movie film data may be important, if the occurrence lingers. Statements of witnesses must be compiled accurately. Tape and video recordings, even those from security cameras, and transcripts can help accuracy. Information about where people were, when they were there, and what they saw or heard are important. It may be necessary to collect specimens and samples of various kinds, use recording instruments to collect data on environmental or other conditions, collect damaged equipment or remaining parts, and collect data on injuries. These are some general kinds of data that are collected in an investigation. Particular kinds of accidents require many other items of data.

When to Investigate

The best time to investigate an accident is as soon after the occurrence as possible. Many types of data and evidence disappear or deteriorate with time. Witness recollections also deteriorate with time, and details that may be important can be lost if investigations are delayed.

Investigation Methods and Procedures

There are a few references that detail methods for accident investigation. Figure 37-2 shows the classical steps in accident investigation. In reality, the steps do not occur one at a time; several may overlap and occur at the same time. Figure 37-3 is a general flow chart of accident investigation procedures. Figure 37-4 is an expanded process for the investigative phase.

Analyzing Investigation Data and Report Results

Analysis is a critical part of the investigation process. Many different methods may be used. Some methods are general and valuable to the accident analysis process. Examples are fault tree analysis and management oversight and risk tree (MORT) analysis. For automobile cases, accident reconstruction computer tools are valuable. There are also computer tools for reconstructing some fire behavior in buildings and estimating exiting behavior of people.

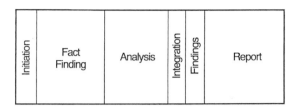

Figure 37-2. Classic steps in accident investigation. (From Accident/Incident Investigation Manual, 2nd ed., DOE/SSDC 76-45/27, SSDC 27, U.S. Department of Energy, Washington, DC, November, 1985.)

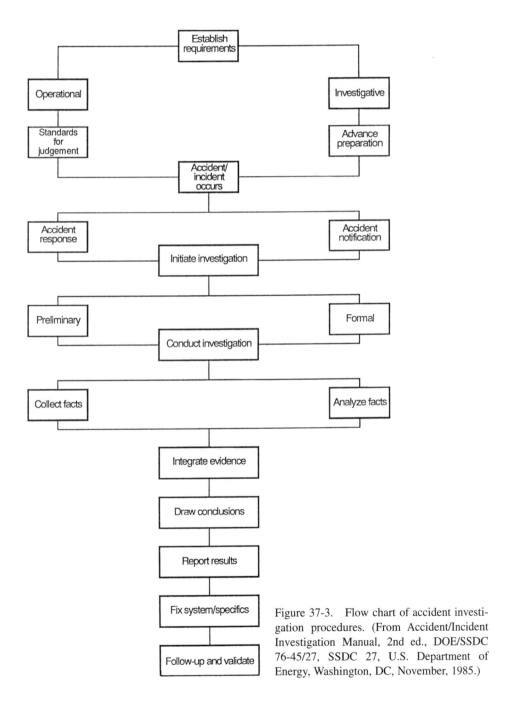

Figure 37-3. Flow chart of accident investigation procedures. (From Accident/Incident Investigation Manual, 2nd ed., DOE/SSDC 76-45/27, SSDC 27, U.S. Department of Energy, Washington, DC, November, 1985.)

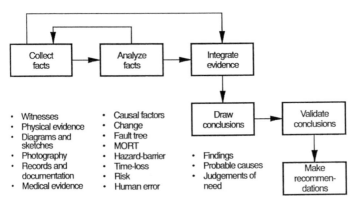

Figure 37-4. Details for the investigation phase of an accident investigation process. (From Accident /Incident Investigation Manual, 2nd ed., DOE/SSDC 76-45/27, SSDC 27, U.S. Department of Energy, Washington, DC, November 1985.)

The analysis leads to findings and conclusions. Written reports should include procedures used. participants, data compiled, analysis and results of analysis, findings, conclusions, and recommendations to prevent such incidents in the future.

Applying Findings

It is not sufficient to end an investigation with a report. The report needs to reach managers, public officials, or others who can do something about the recommendations provided in an investigation report. Recommendations should lead to engineering changes, procedural changes, changes in policy, rules or regulations, improved training, and other kinds of actions.

37-5 SAFETY PERFORMANCE

Management approval for safety programs may be won because the justification analysis may show significant savings or other benefits. Allocation of funds allows implementation of actions to reduce hazards and losses. The question remains whether the real results of corrective actions are similar to those projected in the justification analysis. There are many techniques to determine the effectiveness of safety programs. Some approaches are statistical, some empirical; some are objective, others subjective. Each method has some value for particular situations. Manuele[3] reviews a variety of approaches. A few are discussed here. Others, like inspections, audits, and risk analysis, were discussed previously.

Statistical Approaches

Accident Statistics Some accident statistics were covered in Chapter 8 in a discussion of record keeping. The most common statistics are weekly, monthly, or annual accident or incident occurrences and frequency and severity statistics. These general statistics give an overall picture of an organization's safety performance. For certain industries and types of businesses, performance can be compared with the entire industry group. Frequency rates and some other statistics are available for particular industries from the National Safety Council,[4] from the Bureau of Labor Statistics, and from insurance companies offering workers' compensation insurance.

Although these statistics help measure performance at a global level, they may not be sensitive enough for use in individual departments or small organizations. The accident frequency may be so low per month that individual accidents cause major changes in statistics. Severity rates can shift dramatically because of one long-term disability case. Therefore, it may be necessary to rely on other statistics, such as claims, cost, or others.

A wide range of statistical methods can be applied to determine if a safety program has any effect on accident or incident rates, severity rates, or other measures of performance. Reference to a standard textbook on inferential statistics is helpful. For example, the mean accident rate for 6 months before implementation of a program can be compared with a 6-month period after the program begins. The effect of a program can be determined by comparing the change to a control group or by using multivariate analysis, such as regression analysis and factor analysis, to determine which of several elements of a program has the greatest impact on safety performance. Which statistics apply depends on the information available and the way the program is conducted.

Control Charts   When sampling events, such as accidents and incidents, there is some random variation in events among sampling periods. For example, the number of accidents for a company will vary each month. A problem is knowing whether the variation is strictly random or whether some program or control has produced desired results. Control charts provide a statistical basis for determining whether results of one sampling period are truly an indicator of change or whether the results are the result of random variation.

Figure 37-5 illustrates the idea of a control chart. The number of accidents per month is plotted for a company over a 1-year period. Upper and lower bounds are represented by two horizontal lines, one above and one below the mean and each some distance from the mean number of accidents for the 1-year period. One is the upper control limit and the other is the lower control limit. The limits are based on some statistical level of signifi-

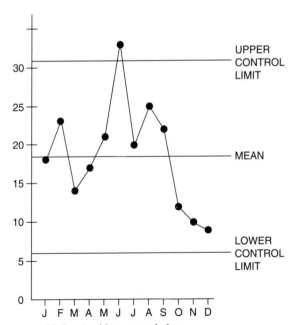

Figure 37-5.   Accident control chart.

cance, such as 95% (two standard deviations from the mean) or 99% (three standard deviation from the mean). If the number of accidents for 1 month falls outside these boundary lines, the variation is most likely the result of something other than random effects. Such a deviation indicates that something is out of control or influenced by some factor. Consider an accident frequency control chart. If both a mean accident frequency and a standard deviation for that data are available, then the upper control limit (UCL) can be computed from

$$\text{UCL} = x + z_{0.025}S, \tag{37-1}$$

where

> x is the mean frequency,
>
> z is the standard normal distribution for the normal curve at the 0.025 probability level, and
>
> S is the standard deviation for the population.

The lower control limit (LCL) can be computed from

$$\text{LCL} = x - z_{0.025}S. \tag{37-2}$$

Example 37-1 An organization compiled accident frequency data for 12 months and wants to establish a control chart for the data. The number of accidents for each of 12 months are: 18, 23, 14, 17, 21, 33, 20, 25, 22, 12, 10, and 9. The mean is 18.67 and the standard deviation is 6.57. From a table of z values for a 0.025 probability level, $z = 1.96$. Then UCL and LCL are

$$\text{UCL} = 18.42 + 1.96(6.34) = 31.55$$

and

$$\text{LCL} = 18.42 - 1.96(6.34) = 5.79.$$

The results are those shown in Figure 37-5. From the control chart, one can see that only one of the months was outside the control limits, even though there was considerable variation in frequencies among the 12 months.

The accurate use of control charts depends on the distribution of data being similar to the statistical normal distribution. Some statistical assumptions are violated unless there is a sufficient sample size and if there are factors that cause a sudden change in data. To ensure that a control chart is valid, there should be somewhere near 60 sampling periods (typically months). In addition, if there is a sudden change in conditions (the number of employees changes, there is a major change in program, etc.), then there should be a new basis for computing a control chart. In Example 37-1, the last 3 months seem to show a significant change in accident frequency. The trend should be looked at further. If a special promotion was underway, the change may be a result of it, in which case the last 3 months may need to begin a new set of data for setting control limits.

Control charts can be used for evaluating safe behavior.[5] The procedures involve use of work-sampling techniques of industrial engineers and work methods. The study could evaluate the effect of a safety program on errors or unsafe behaviors. A series of observations before the program starts could establish a background for the evaluation. Through random sampling of behaviors, one records the portion of observations in a sampling period that were not completed safely. After the program is introduced, the behavior can be tracked similarly. Control limits on the results of the background period form a basis for judging whether the program affected behavior. If, after the program is introduced, the portion of unsafe behaviors decreased and exceeded the control limits, the program had positive effects.

The computation of control limits for proportions is

$$UCL = p + 1.96\left[\frac{p(1-p)}{n}\right]^{0.5} \tag{37-3}$$

and

$$LCL = p - 1.96\left[\frac{p(1-p)}{n}\right]^{0.5}, \tag{37-4}$$

where

p is the mean proportion of observed behaviors that are unsafe or safe for all observation periods and

n is the number of observation periods.

The number of readings, N, required for a certain level of accuracy at a 95% confidence level is

$$N = \frac{4(1-p)}{S^2 p}, \tag{37-5}$$

where

p is the proportion of safe or unsafe acts observed during the study and

S is the desired accuracy (percent per 100 readings).

The accuracy of readings is the proportion observed plus or minus some percent accuracy. For example, the average portion of observed unsafe behaviors across observation periods is 18% ± 10% or 18% ± 2%.

Example 37-2 A study wishes to determine how often workers at 15 workstations involving similar tasks made errors that could lead to accidents. The data will serve as a baseline for a corrective program, if the error rate is excessive. The study must determine the baseline within 15% accuracy of the real behavior. How many work cycles (observations) must be included to achieve this accuracy at a 95% confidence level? A preliminary study found a 24% error rate.

The number of observations required is

$$N = \frac{4(1-0.24)}{(0.15)^2(0.24)} = 563 \text{ observations},$$

which amounts to 563/15 = 38 observations per workstation.

Subjective Methods

Rating and Attitude Scales Another approach used by some organizations is to use validated attitude questionnaires, which may be given to managers and workers. Various scales built into the surveys assess the attitudes and opinions toward safety and safe behavior. The results may be useful in determining if safety programs change management and worker views toward safety and potential behavior.

Judgment During safety audits and other evaluation processes, one or more persons may make subjective judgments about particular aspects of a safety program. Comments logged for items in an inspection checklist may be subjective judgments. The judgments may alert specialists to perform a more detailed evaluation.

Combinations of Methods

Any one method may not be adequate for evaluating the effectiveness of safety programs. If you want to consider the effect on conditions, risks, the attitudes and behavior of personnel, knowledge of workers, management support, compliance with safety standards and company rules, and accident and incident rates, one measure will not provide adequate detail on such a wide range of matters. Some companies have devised their own evaluation system that incorporates many different measures. The problem of assessment is particularly difficult when different departments or divisions of a company have very different operations and hazards.

An example of an evaluation form[6] is shown in Figure 37-6. It is a safety report card of sorts for major organizational unit. It uses a point system distributed over five major areas of evaluation. Most areas have several independent elements. Points for the statistical analysis elements are taken from additional guidance for raters found in Table 37-1.

37-6 PRESENTING RESULTS OF ANALYSES

When results of safety analyses are completed, they are typically presented to managers and others for decision making, training, and other purposes. The way in which results are presented is a key to successful communication. With today's computer tools, it is easy to generate a wide variety of tables and charts. Even three-dimensional graphs can be generated with relative ease. If trends are important, line and bar graphs are very useful. Figure 37-7 gives an example of each. Pie charts are useful for communicating relative portions of a whole. Figure 37-8 gives an example of a pie chart.

To compare organizational units, bar charts can be useful. Various types may be suitable. Stacked bar charts show the contribution of each unit to the whole for different periods or topics. Bar charts that show values in descending or ascending order help compare different groups. Figure 37-9 shows one type of bar chart for communicating safety information. Tables are better than graphs for presenting precise data and complex sets of data.

37-7 COMPUTER SYSTEMS, THE INTERNET, AND SAFETY

Like other professions, the safety specialists today must use computer tools to manage, analyze, and communicate information. This section identifies some of the safety applications for various types of computer software and systems. There are many general computer tools and safety applications that are not covered. The discussion mainly refers to capabilities for microcomputers and software for it. With the emergence of the Internet as a standard resource for everyone, one can access information from everywhere, conduct business and transactions with company customers and clients, and manage information and communications internally within a company. Safety and health applications on the Internet or based on Internet technologies provide new capabilities and resources for a

	Points	
	Possible	**Earned**
1. Statistical Analysis	**20**	
A. Incidence rate for pervious fiscal year	10	
B. Workers compensation claim costs in relation to standard premium	10	
2. Inspection for Compliance with Safety Standards	**15**	

A. Material handling and storage
   B. Machine guarding, interlocks and controls
   C. Health and hygiene
   D. Fire protection
   E. Other standards
	15	

3. Risk Analysis	**25**	
4. Line Management Support	**10**	
A. Top management	5	
B. First Line supervision	5	
5. Safety Program	**30**	
A. Inspections	5	
B. Training	5	
C. Safety coordinator function	10	
D. Safety promotion	5	
E. Safety committee	5	
TOTAL	**100**	

Figure 37-6. An example of a form used in evaluating a safety program. (Reprinted with permission from Brigham, C. J., "Safety Program Evaluation: One Corporation's Approach," *Professional Safety*, May:31–34 (1981), official publication of the American Society of Safety Engineering. Form from Rexnord, Inc.)

wide range of safety functions. They give quick access to information, analysis methods, data collection, and service providers.

Database Management Systems

Database management systems (DBMS) are the workhorses of management information systems. Today the development of applications is quite easy, because DBMS products contain features that eliminate many of the programming tasks. Some DBMS products also produce compiled code for the applications that can be given to other users without additional cost. Most DBMS products allow data entry forms and output reports to be designed on the screen, and all have the ability to query data through some convenient method, such as query by example, by selecting data elements from a list. Data can be selected for standard reports and ad hoc queries, transferred to other programs, or converted to business graphics. Many DBMS products have security features that limit

TABLE 37-1 Guidelines for Scoring Certain Elements of the Safety Analysis and Evaluation (Shown in Figure 37-6)

Incident Rate

Based on No. of OSHA Lost Workday Cases	Points
0.0	10
0.1–0.5	9
0.6–1.0	8
1.1–1.5	7
1.6–2.0	6
2.1–2.5	5
2.6–3.0	4
3.1–3.5	3
3.6–4.0	2
4.1–4.0	1
>4.5	0

Worker's Compensation Cost

Claim Cost Incurred (Dollars per Employee)	Points
0	10
1–17	9
18–33	8
34–50	7
51–66	6
67–83	5
84–100	4
101–116	3
117–133	2
134–149	1
>149	0

Compliance with Safety Standards

1. Two points are deduced for each *serious* deviation from standards that could result in one disabling injury or significant property damage or repeat violation from previous inspection.

2. The formulas below define the points deducted for each *nonserious* deviation from standards. There is a maximum of one point deducted for cumulative deviations from one standard.

Group	No. Employees	Point Formula
I	400 or more	1 for every four items of noncompliance
II	250–399	1 for every four items of noncompliance
III	125–399	1 for every three items of noncompliance
IV	70–124	1 for every three items of noncompliance
V	<70	1 for every two items of noncompliance

Risk Analysis

Based on 23 five-point rating scales on program and procedures. The score is based on the percent of the maximum points possible on the rating scales and that allowed on the rating form.

Line Management Support

Based on structured interviews with top management and line supervisors. Five-point rating scales are used to evaluate the responses to each interview question. Points are based on a percent of maximum points possible.

Safety Program

Based on a review of how safety programs are implemented. A series of questions have five-point rating scales. Points are determined as a percent of maximum points possible.

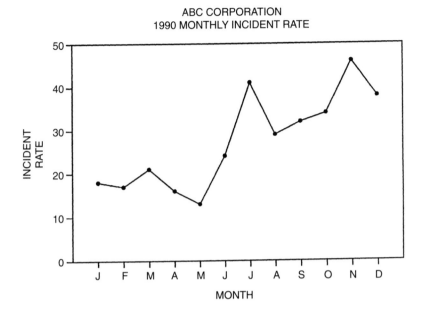

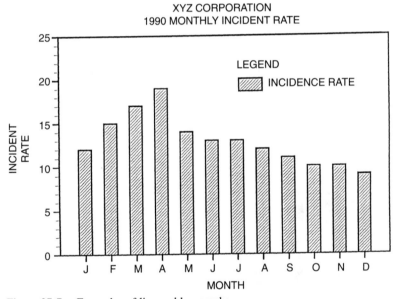

Figure 37-7. Examples of line and bar graphs.

access of certain people to certain data. Having the power of relational database structures gives the user a great deal of information management capability while minimizing data storage.

Over the last decade or so, there have been many commercial DBMS products for managing safety data. Some have survived the rapid rate of change in computer systems and the competition.

There has been a rapid growth in the data required for safety. Some time ago, there was little need for information other than accidents and incidents. Today, it is essential to

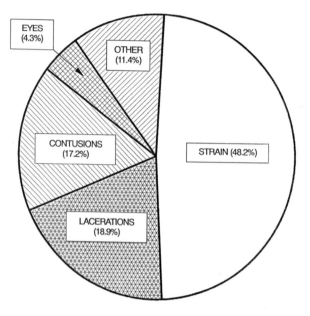

Figure 37-8. Example of a pie chart.

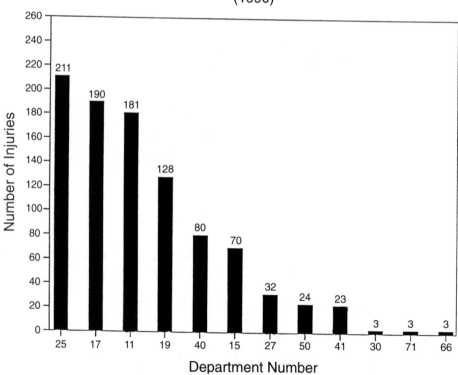

Figure 37-9. Example of a bar chart comparing different departments.

keep track of personal protective equipment and other items issued for safety purposes, environmental exposures of employees, and training given to workers and when training must occur again. There is a need to track hazardous materials data. It may be important to keep track of fire extinguishers, hoses, hydrants, and data about them, such as manufacturer, date of purchase, date of inspection, and tests performed. In safety analysis, hazards and what controls are implemented must be listed, controls must be followed up, and testing and maintenance must be recorded. In addition, manufacturers may track products for recalls, servicing, and modifications.

Relational data bases allow separate but related lists or tables of data to be maintained and help eliminate duplicate data. An example is training people in safety. A person may attend more than one training session in a year, and a single list would contain the course and data about it (title, when given, duration, where given, etc.) and the attendees and data about them (name, department, age, address, phone number, etc.). Building a single list around the training would produce duplicate data for attendees if they attend more than one course, and building the single list around attendees would produce duplicate data about the course a group attended. In a relational data base, data could be divided into at least two tables, one for training and one for attendees. Having one common data element in each list (such as course number or a person's name or identifying number) allows information from each list to be merged.

Figure 37-10 illustrates the structure of tables in a relational data base for an employer who must track many safety matters. Each table contains information about the items included in it, including employees, training, accidents and incidents, equipment issued, exposures, and physical tests (hearing, pulmonary function, biological limit value (BLV) tests, etc.).

Modeling

Another use for computers in safety is modeling the behavior of physical phenomena, people, and other entities. Modeling allows people to anticipate what might happen or what did happen. The accuracy produced by the model depends on the availability of accurate data, the inclusion of factors that can affect the phenomenon, and adequate representation in the mathematical manipulations. There are reasonably successful models in safety for automobile accident reconstruction, for gas dispersion, and for fire behavior in small buildings.

Accident reconstruction models allow analysis of what happened in the sequence of events before, during, and after collision. The models depend on accurate information about particular automobiles and data from crash sites.

Gas dispersion models allow the formation and movement of gases released from point sources to be plotted. The models are useful in planning emergency procedures for accidental releases, but are limited for real time use in deciding what action to take from moment to moment. The limitations arise from difficulties in making sure that actual wind conditions and atmospheric conditions are represented in the model.

The National Institute for Standards and Technology has developed models of fire behavior and exiting behavior of building occupants during a fire. Work in modeling fire behavior in small buildings has been most successful. These systems allow for analysis of the adequacy of building designs and what happened in certain building fires.

Training

Training people about safety must fulfill many requirements. A variety of media are available to accomplish successful training. Computer-based training (CBT) has advanced

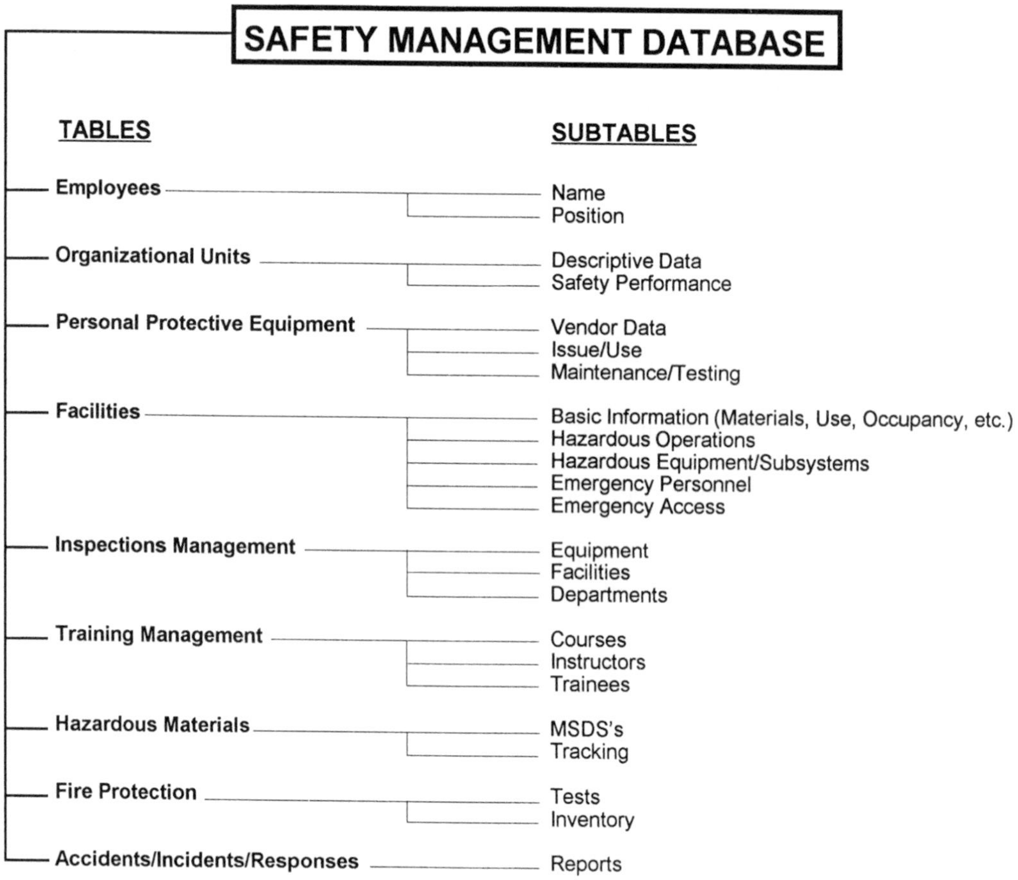

Figure 37-10. Example of a structure for a safety management data base.

rapidly as a tool for training and tracking training accomplished. An important feature that is unique to CBT is the ability to customize training to individuals. A training package may contain remedial or background material that students access if they do not understand the standard material presented. CBT removes peer pressure from the learning process and can record very precisely what learning is achieved in a training session. It allows students to make mistakes without creating accidents and injuries and it can provide immediate feedback on errors.

CBT systems typically include a microcomputer, DVD player, and software. Lessons stored on DVD can have a rich collection of photographic material from field situations. Authoring software for preparing CBT lessons helps design and preparation of text and graphic materials for presentation as well as assists the lesson developer in the preparation and evaluation of materials and examinations. The evaluation items diagnose student deficiencies and guide them to supplemental materials to help them achieve lesson objectives. Presentation software manages the presentation. The student may take several routes to achieve the lesson objectives and may receive extra help when diagnostics indicate the need. Presentation software also can time and score student progress and record completion of a lesson for training management.

Tracking

There is a need to follow the movement of items from location to location. Computers can assist in tracking hazardous materials, documents, and other items from a safety perspective. Bar code devices, optical character readers, magnetic strips, radio frequency transmitters, global positioning systems, and other automated identification technologies are essential. For example, a bar code label can be placed on the containers of a hazardous material. At check points along its normal path of movement and use, the bar code is read and the readings are fed into a common database. Reports from the database can detect time or location deviations and signal an alert.

Another example is tracking safety-related repair and maintenance activities. When a work request comes in, it is assigned an identifying bar code. The maintenance worker logs information about the problem, actions completed, and item repaired on a portable data collection device. The data are loaded daily into a database and subsequent analysis on the database shows what was called in, what was done, what parts were replaced or fixed, and when these tasks were accomplished.

People also can be tracked. For example, to limit access to only authorized personnel, people entering a check point may have to place an identification card in a reader to gain access to the restricted area. Today, security devices can read fingerprints, retinal patterns, voices, and other biological characteristics.

Monitoring

Another important use of computers in safety is monitoring. Sensors and data capture devices are strategically placed to detect the occurrence of unsatisfactory conditions. The data are fed to a computer that analyzes the information to determine if it should act by sounding an alarm. There are monitoring systems for particular gases, for excessive heat or fire, and to detect system failures, such as excessive heat and pressure.

Data Banks

Data banks are another important tool for safety people. Most data banks are now accessed via the Internet. In relying on Internet sources, one must take care to establish the reliability of the source of information. They can help people keep up with rapidly changing laws and regulations and provide convenient knowledge and information for special safety problems. Central organizations may compile information useful to many others and may charge a fee for access and use. Some have bibliographic records, accident and incident records, and other safety-related information. There are data banks on such things as pesticides and their effects, chemical hazards, auto safety, radiation, earthquakes, lightning, tornados, and other topics. There are data bases on legal claims related to products liability and product recalls.

Expert Systems

Expert systems, both inductive and deductive, may prove to be useful computer tools. They help perform some reasoning. One type of application is diagnostics. Expert systems have several uses in safety. They have the potential of assisting someone with little knowledge of a special topic to work through the topic to form and form a decision. There has been some work on identifying hazards in ladders and scaffolding and the use of that equipment, in construction safety and in managing products liability lawsuits and economic decisions about continuing a case, as well as an application to mining safety.

Computer-Aided Design and Drafting

Computer-aided design and drafting (CAD) also is a useful computer capability for safety. It may be useful for charting hazardous areas on site maps or building floor plans and is useful for isolating the safety systems (alarms, fire protection, etc.) in facilities and processes. Separate layers in CAD drawings may be dedicated to these features. Some use a layer in new designs to communicate construction hazards and procedures as reminders to crafts people to ensure that the correct precautions are in place. Process facility drawings may show level of hazard or other characteristics through color or shading. CAD drawings may have special symbols or components of interest to safety. For example, compressed air lines and terminals with $30\,lb/in^2$ or less may have a particular designation. Pressure-relief devices may be specially identified in a list of drawing entities. Safety DBMS systems can be linked to drawings, drawing symbols, and layers and can be managed interactively with drawing changes.

The interface of CAD information with DBMS is valuable. This provides the capability to extend management from safety information even further. For example, there may be a need to associate accident data by department with floor plans to indicate the location and frequency or severity of accidents. In a similar way, risk could be computed and represented by color or shading on a floor plan or site map and hazards could be classified by type or severity and be drawn with each class shown in a different layer in the drawing or represented by color. A list of elements (such as fire extinguishers) in a drawing important for safety could be generated and related with data from DBMS tables about vendors and replacement costs. With the interface between CAD systems and DBMS, one can accomplish all these information management tasks.

Special Computer Devices

There are a number of special computers or computer applications in safety. Some companies have programs and special devices for assessing hearing loss and logging test results into data bases. There are special computers for evaluating pulmonary functions and lifting tasks and there are special computers (some very small and worn on a person) to monitor noise exposure and compute noise dose. There are computer applications for laser hazard analysis, analysis of manual materials handling (computations related to the NIOSH lifting equation), design of ventilation systems, risk analysis, gas monitoring systems, and other safety applications.

EXERCISES

1. A company has the following accident frequency by month over a 2-year period:

Year	J	F	M	A	M	J	J	A	S	O	N	D
1st	19	12	32	24	27	18	26	30	16	20	28	32
2nd	16	36	17	23	29	21	30	18	25	28	13	33

Prepare a control chart for these data and plot the actual data. Calculate and show the upper and lower control limits with the mean frequency. Determine if the frequency for any month extends beyond the control limits.

2. A team was sent to a number of departments to monitor worker behavior. The goal was to determine how safe worker activities were. The team found that 82% of the behavior was safe. The team sampled each department during 20 observation periods. Frequency data suggest that it could be better. A new safety program will be implemented to improve the performance.

  (a) If there are 18 departments and a 95% confidence level is applied, what are the upper and lower control limits for evaluating the behavior in particular departments?

  (b) If the follow-up study is to have an accuracy of 10%, how many observations must be made in each department to achieve this accuracy at a 95% confidence level?

3. Obtain a copy of a report of the National Transportation Safety Board accident investigation of an airplane accident. Find out what procedures were followed, how many investigators participated, and what the findings were.

4. Identify Internet resources for

  (a) regulations and standards

  (b) state and federal laws

  (c) safety equipment and personal protective equipment

  (d) training materials and services

  (e) record keeping systems

  (f) professional development conferences

  (g) accredited safety and health certifications

  (h) safety and related academic programs

REVIEW QUESTIONS

1. What are three purposes for analysis?
2. Where do accidents fit in the process of preventing accidents?
3. What types of inspections are there? Give an example of each applied to safety.
4. Who is qualified to be an inspector?
5. What tools are used in inspections?
6. Where do accident investigations fit in the process of preventing accidents?
7. Why are accident analyses important?
8. What kinds of accident investigations are there? Give an example of each type.
9. Should all accidents be investigated? If not, why?
10. If not all accidents are to be investigated, what strategies can be used to select particular accidents?
11. What tools are used in accident investigations?
12. Describe the classical accident investigation process.
13. What methods can be used to evaluate the safety performance of an organization?

14. Identify four kinds of business graphs that may be useful in presenting results of analysis. When is each more useful than the others? Give an example of each kind of graph.

15. List six kinds of computer tools that are useful for safety? Give an example of how each can be used.

NOTES

1 ISO 9001, *Quality Systems—Model for QA in Design/Development, Production, Installation and Servicing*, International Organization for Standardization, Geneva, Switzerland, 1994.

2 ISO-10011, *Guideline for Auditing Quality Systems*, International Organization for Standardization, Geneva, Switzerland, 1990.

3 Manuele, F. A., "How Do You Know Your Hazard Control Program is Effective?" *Professional Safety*, June:18–24 (1981).

4 *Accident Facts*, National Safety Council, Itasca, IL, annual.

5 Tarrants, W. E., "How to Evaluate Your Occupational Safety and Health Program," *Handbook of Occupational Safety and Health*, Slote, L., ed., Wiley, New York, 1987.

6 Brigham, C. J., "Safety Program Evaluation: One Corporation's Approach," *Professional Safety*, May:31–34 (1981).

BIBLIOGRAPHY

Accident/Incident Investigation Manual, 2nd ed., DOE/SCCD 76-45/27, SSCC 27, U.S. Department of Energy, Washington, DC, November, 1985.

FERRY, T. S., *Elements of Accident Investigation*, Charles C. Thomas, Springfield, IL, 1978.

FERRY, T. S., *Modern Accident Investigation and Analysis*, 2nd ed., Wiley-Interscience, New York, 1988.

FERRY, T. S., ed., *Readings in Accident Investigation*, Charles C. Thomas, Springfield, IL, 1984.

Fire and Life Safety Inspection Manual, National Fire Protection Association, Quincy, MA, 2002.

HENDRICK, K., and BENNER, L., Jr., *Investigating Accidents with STEP*, Marcel Dekker, New York, 1986.

Industrial Hygiene Performance Metrics, American Industrial Hygiene Association, Fairfax, VA, 2001.

JOHNSON, W. G., *MORT Safety Assurance Systems*, Marcel Dekker, New York, 1980.

KALETSKY, RICK, *OSHA Inspections—Preparation and Response*, National Safety Council, Itasca, IL, 2004.

OAKLEY, JEFFREY S., *Accident Investigation Techniques: Basic Theories, Analytical Methods and Applications*, American Society of Safety Engineers, Des Plaines, IL, 2003.

PETERSEN, D., *Analyzing Safety System Effectiveness*, 3rd ed., John Wiley & Sons, New York, 2000.

Root Cause Analysis Handbook: A Guide to Effective Incident Investigation, Government Institutes, Rockville, MD, 1999.

ROSS, C. W., *Computer Systems for Occupational Safety and Health Management*, Marcel Dekker, New York, 1984.

ROUFF, A., and DHILLON B. S., *Safety Assessment—A Quantitative Approach*, Lewis Publishers, Boca Raton, FL, 1994.

STUART, RALPH B. III, *Safety & Health on the Internet*, Government Institutes, Rockville, MD, 1997.

TARRANTS, W. E., *The Measurement of Safety Performance*, Garland STPM Press, 1979.

THOMPSON, J. R., *Engineering Safety Assessment: An Introduction*, Wiley, New York, 1987.

WOODCOCK, HENRY C., and SEIBERT, JOHN, *Investigations: A Handbook for Prevention Professionals*, American Industrial Hygiene Association, Fairfax, VA, 2000.

WRENCH, C. P., *Data Management for Occupational Health and Safety*, Van Nostrand Reinhold, New York, 1989.

CHAPTER 38

SAFETY PLANS AND PROGRAMS

The previous chapters discussed a variety of analysis methods for safety. Recall that the goal is to reduce accidents. Engineers may perform many kinds of analyses to identify hazards and pinpoint controls that will reduce accidents. However, the results of the analysis must be in an appropriate form for managers. Managers want to decide what actions to take. They want to know what the actions will achieve, what the actions will cost, what human resources are necessary to accomplish the actions, and the schedule for completion of the actions.

Although the analyses described in previous chapters develop this information, the key is organizing the information for managers and the decisions they need to make based on the information. One must develop a plan. A plan may begin with hazard identification and analysis methods or it may be the consequence of such methods. A plan includes at least three components: a list of actions or tasks and who will complete them, a schedule for their implementation, and an estimate of costs. Costs may include human resources, as well as monetary resources. Specific plans may have additional components. If a plan is approved, it is set into action.

Plans and programs are much the same thing. Plans are documented strategies for getting things accomplished. They may or may not be implemented. The term *program* infers something larger than a plan. Programs address numerous goals of an organization, whereas plans tend to address a small set of goals. Programs tend to unify several elements of an organization into one strategy for allocation and use of resources.

This chapter looks at examples of plans and programs for achieving safety. It does not cover requirements of laws for particular plans and does not deal with methods for program and project management that ensure successful implementation. Readers should refer to references throughout the book for additional help.

38-1 PRODUCT SAFETY PROGRAM

Many factors contribute to the safety of a product. A manufacturer does not have control over all the factors. However, an important goal is to minimize the contribution a product and its design features make to accidents and injuries. By keeping danger from users, the likelihood of accidents, injuries, and potential losses through claims and lawsuits is minimized. The Consumer Products Safety Commission develops guidelines for manufacturers to help them make products safer,[1] and their guidelines were extended in a publication by the American National Standards Institute (ANSI).[2] Others have developed organizational procedures to reduce product liability claims. Some include system safety and risk analysis procedures in their product safety program.[3] In many cases, product safety programs can be built on existing design and product management processes. Another element is having clear procedures to incorporate safety throughout the design and development process.

Executive Action

Executive actions are those that require support from top management to make them work or work effectively. Executive actions provide the commitment for success, not just support in principle.

Policy The first evidence of top level commitment is a written policy statement. It should state the objective for product safety and cite statutory and voluntary standards to be followed. The policy also should have a commitment to maintaining state-of-the-art safety features in products. The policy should assign responsibilities to particular elements of the organization, which may include internal line and staff organizations and suppliers and distributors. It should state what is to be achieved by effective relations with users and consumers of products. The policy should identify what methods will be used to judge product safety.

Organization and Staffing Internal procedures need to identify what each element of an organization must do to achieve product safety. Responsibilities, authorities, and actions are assigned. Particular procedures are developed for each element of the product safety program, typically including procurement procedures and acceptance methods for components and parts. They would cover design, manufacturing, quality control, distribution, sales, and customer complaints. Procedures would include methods for assessing performance and obtaining assistance with product safety program elements. Detailed procedures would include how each element of the program will be implemented and by whom.

Included in this component of the program are budgets, communication (brochures, technical manuals, etc.), record keeping, and analysis. Financial aspects will include what insurance will be needed to cover potential losses.

Training An important element of a product safety program is training. Participants in the product safety program must know what their responsibilities and roles are and how to perform them properly. Participants need to know that their contributions, no matter how large or small, are essential to the success of the program and the company's products. For some participants in the process, training programs may be formal and may include classroom training or training through other media. For others, they may be informal and involve information in organizational publications.

Technical Requirements

Technical requirements refer to detailed procedures for the life cycle of a product: from planning and design through production and final use by a consumer or user.

Design Review Design review involves the examination of the product itself, its materials and design, and its packaging and labels. The most important part of the review is to identify hazards. The hazards may be inherent in the design or result from use, foreseeable misuse, and the use environments for the product. Comparisons with similar products may be helpful. Tests against existing standards for the same type of product should be completed. Additional tests may be designed and applied. It may be useful to retain test data for evaluation and future reference. If hazards are identified, they must be evaluated for severity of the dangers posed and actions to remove or reduce them are taken. The actions may be based on priorities from the hazard or risk analysis and may include applying warnings and labels to the product or modifications to existing ones to make them more understandable. Procedures should be in place to deal with conflicts among reviewers.

Documentation and Change Control   There must be a method in place to document design changes and the reasons for the changes. The process should pay particular attention to changes intended to reduce hazards. Documentation also should address adequacy of supporting documents for the product, including user manuals, service bulletins, sales brochures, advertising, and maintenance manuals. Any remaining hazards and the actions users must apply to protect themselves from the dangers should be identified clearly in these documents. Information that could be misleading or could lead to improper use should be removed from the documents.

Purchase Control   One cannot depend on materials, parts, or assemblies being free of hazards. It is important to have procedures and clauses in place to be sure that purchased items or assemblies that become part of a product or its packaging do not introduce hazards in the product. Control over purchase decisions and purchase procedures should include steps to prevent hazards in parts or assemblies. The actions to be taken against suppliers who do not comply with purchase requirements, specifications, or tests and certification of tests should be clearly established.

Production   There are many production practices that may affect product safety. One aspect is control of materials and assemblies. Because handling and shipping may introduce defects, checks, inspections, and tests may be needed. Inadequate parts should be identified clearly to prevent them from being introduced into a product inadvertently. Materials that have limited shelf life should be managed so their performance is not impaired and they are not able to contribute to product hazards.

Another aspect is clear work instructions. Production steps should be analyzed for potential introduction of hazards, and instructions to prevent introducing hazards should be written. Workers involved in such steps should be trained about the impact of improper work. Work instructions should include detailed procedures for handling repair or disposal of defective products.

Poor work environments and equipment that does not work properly can add to hazards in a product. The importance of proper conditions and production equipment should be analyzed for its impact on product safety. Environments should have adequate light and other conditions necessary for good worker performance.

The production process should be reviewed for effects on product safety. Inspections, verification of worker qualifications, and other means can help ensure that hazards are not introduced during production.

Quality Control   Procedures during and after production to ensure that parts, assemblies, and the final product meet specifications and standards is an important part of product safety. Various forms of testing, inspection, and analysis may be applicable. When defective items or products are detected, they must be separated so they are not used or shipped inadvertently.

Measurement and Calibration   Many inspection and testing methods depend on quality instruments. It is essential that instruments be properly calibrated at regular intervals.

Distribution Design and selection of packaging and packaging materials is essential to prevent hazards from being introduced during shipping and handling. Packing should be tested. Samples from field points will help determine that packaging provides adequate protection until the product reaches the user.

Consumer Service Consumers may need help with assembly, installation, service, repair, proper repair parts, and other matters. Registration of products and users will help contact users for recalls and special safety notices. Without such user services, users may introduce dangers into products.

Customer Feedback and Records A variety of records can help achieve product safety. The records help identify problems and can assist in notifying product users. Records may include such items as results of tests, inspections, and checks by batch, lot, or serial number. Records of consumer complaints and problems will help. Consumer problems can be compiled from phone calls, correspondence, or field reports of sales and services staff. Records may include location of lots, batches, or serial numbers in the distribution system. For some products, records of ownership can be helpful.

Corrective Action For some products, the manufacturer must notify the federal government of hazards and defects that may affect users. Tracking hazards in a product and the actions taken to eliminate the hazards and prevent defective products from reaching users will help establish compliance with legal requirements.

Audits Through planned and scheduled reviews of all elements of a product safety program will help ensure that the program works effectively. Audits are important to determine that safety standards are met.

38-2 OCCUPATIONAL SAFETY PROGRAM

There are many elements of an occupational safety program that are essential to its success. What elements are essential for an organization will vary. Some elements are essential regardless of the organization, some are required by law, and others vary with the structure and operations of an organization. The emphasis on particular elements will change from time to time as the needs of the organization change. The summary that follows identifies many important elements. Some people would like to add or delete elements or give more emphasis to particular ones, and there is some overlap among elements because they cannot always be independent.

Commitment

No endeavor within an organization can function effectively without strong support from top management. For occupational safety to be an effective program, there must be a commitment at the highest level. That commitment must include a clear statement that safety is important and support for actions that will make safety important.

Policy and Procedures

An occupational safety program begins in an organization with a policy placing paramount importance on occupational safety and protection of employees. That policy certainly encompasses safety while at work and in all matters relating to employment. For some companies, the commitment extends to off-the-job safety as well. For others, it extends to the safety of employees' families. Extending the program beyond the immediate boundaries of employment has several effects. First, it lets employees know that they are valued as individuals. Second, it gets employees involved in safety throughout their lives so that they do not turn safety on when they get to work and off when they leave.

Procedures extend the policy into practice. The procedures assign responsibilities to each unit of the organization. Occupational safety programs must involve everyone at every level in an organization. They include general safety rules that may be part of a book of work rules and may include a compilation of detailed procedures and forms for particular operations, equipment, and materials. The safety manual typically addresses accident reporting, hot work and hot work permits, confined entry practices and permits, and methods and forms for dealing with hazardous materials. The detailed safety manual may extend to particular practices for many activities and equipment and may include emergency responses for fire, weather, spills, leaks, or other conditions as well as information on worker compensation. It may deal with a variety of hazardous operations and use of certain equipment and materials, and it should address safety committees, fire brigades, and other special assignments and how they are operated.

Assigning Responsibility

A safety program must identify clearly who is responsible for what. Assignment of responsibility is normally a part of policy and procedures. Assigning responsibility must include the role for line elements of the organization, staff elements, and special committees or units.

For example, supervisors in line elements are normally responsible for their work areas and the people under them. They may have to manage training, inspections, and reporting of problems that create hazards. Supervisors typically must sign for dangerous operations that will occur in their areas, and they may have safety included in their job description and performance appraisals. Safety performance may be an element of their performance standards.

Staff elements may handle record keeping, company level reports, and other information management functions involving safety. Safety staff will obtain special expertise when needed, if it is not available within the organization, and will see that essential resources are made available. This may include medical services, emergency services and equipment, training materials, and qualified instructors. Other resources may be safety engineering, ergonomics, fire protection, rehabilitation, and similar specialists. The safety staff coordinates the purchase of personal protective equipment. They may review designs of facilities, processes, special activities, and work areas. Typically, they provide measurement and assessment of environments for hazardous conditions and materials, develop and manage the budget for occupational safety matters, perform major inspections or head inspection teams, assist line elements in all areas of safety implementation, and may develop special publications. The safety staff develop and manage policy and procedures for the organization.

Many organizations emphasize creation of a safety culture in which employees at all levels participate in safety. Increased training, validation of knowledge through testing, and recognition all contribute to the culture. Effective leadership by managers and supervisors is essential.

Implementation

Implementation involves setting schedules and tracking activities. For some occupational safety activities, there should be fixed schedules. For example, some equipment or processes should be inspected daily, results of inspections should be recorded, and there should be a regular schedule for activities of special committees and special units, such as a fire brigade. There may be weekly or monthly safety meetings of each organization element to learn about hazards in their operations, identify problems, review safety performance, and recommend controls. In implementation, someone must make sure that all the activities of the occupational safety program are carried out. Data from many of the activities must be collected so that results and status can be evaluated.

Implementation includes compiling information from forms and procedures. For example, the data from accidents and incidents must be recorded. The process of tracking hazardous materials requires entry of data into records. There is a need to maintain hazardous material data sheets and make them available to each element of the organization that needs them, and there is a need to compile information on safety repair and maintenance of equipment and systems.

Implementation includes the issue of personal protective equipment and the inspection of personal protective and other safety equipment. Some items must pass regular tests.

Implementation includes evaluations of work areas for environmental hazards. It may involve collecting and evaluating data from personal monitoring devices or testing and calibration of instrumentation and monitoring systems.

During implementation, the fire protection systems and equipment must be checked. Some components require regular testing and maintenance of inspection and test data. For example, hydrants must be tested for water pressure and flow, extinguishers must be inspected and tested, and alarm systems and sensors must be tested.

There may be a regular schedule for training for emergency procedures. The training may range from fire alarm and evacuation drills to training with fire extinguishing equipment and testing of emergency response plans for a plant and its surrounding communities.

Training

There are many kinds of training associated with an occupational safety program, and many elements were discussed in Chapter 32. Workers must learn about the hazards related to their job and the means for protecting themselves and how to perform particular safety procedures. Users of personal protective equipment must learn how to use properly the equipment issued to them. Special units must learn how to perform functions associated with inspections, responses to emergencies, and other activities they may be required to perform. Someone must track the training required and training completed.

Hazard Recognition and Control

Major elements of an occupational safety program are recognition, evaluation, and control of hazards. These activities may be associated with routine functions. The operations may change, conditions may change, and equipment may change. Inspections, reviews, and other means will help identify the hazards and the likelihood of occurrence. For example, there should be inspections of repair and maintenance work to ensure that guards are replaced or an area is clear of flammable materials and combustibles or sources of heat and fire. Several methods for hazard recognition and control were discussed in previous chapters.

Identification of hazards may be associated with new or modified operations, equipment, or facilities. Someone must ensure that hazards are recognized. Procedures may evaluate hazards for severity and likelihood of accident occurrence. Controls and options for controls must be identified and evaluated for cost and effectiveness.

Hazard recognition and control may include implementation of reviews for operating procedures. One or more people periodically may review operating procedures for each organizational unit or process. The review identifies hazards and whether controls for them are in place and in use.

Budget and Cost

In an implementation program, there should be a budget for the safety program. The budget may be distributed to each organizational unit, be organized centrally, or some combination. The cost of safety activities and purchases should be tracked and should be organized functionally and by organizational unit. This will allow tracking of budgetary performance within organizational units and of what expenses apply to what elements of the safety program.

Program Performance

An important element of an occupational safety program is assessment of performance. This typically includes tracking of frequency and severity of accidents and tracking of program costs. It also should include assessing whether program activities are accomplished. Assessment may be both quantitative and qualitative. For example, an organizational element may complete required safety meetings and training but show no resultant reduction of loss rate or risks.

Performance should include assessment of program objectives and results by organizational unit, program element, and for the program as a whole. Some examples of methods for assessing performance were discussed in previous chapters.

The performance of individuals, organizational units, and program element managers and participants should be rewarded. Part of the assessment should include a safety award program. Competitive awards and performance awards may be within the array of awards. For example, awards should be available to individuals for achieving a safe driver record or safe job record, for having achieved special skills in first aid, for completion of training courses, and to recognize participation on special units or involvement outside the organization in safety activities. Organization units may compete for risk reduction or hazard reduction achievements.

Another part of the assessment element of the safety program is an audit program. The audit should help determine the overall effectiveness of the safety program. Audits help verify compliance with safety standards, company policy, safety plans and programs, and performance on other dimensions.

One means for recognizing effective employer safety and health programs is the Voluntary Protection Program (VPP) program of OSHA. The VPP is intended to encourage employers to partner with OSHA to create safer and healthier workplaces for employees. The program emphasizes developing effective safety and health management systems through incremental steps to improve safety performance. The program design allows for individual employers, corporations, and construction companies to participate. When they meet certain guidelines and achieve certain safety performance standards, participating companies move progressively from a "challenge" level to a "merit" or "star" level. Based on reporting and performance, OSHA recognizes the companies achieving each level. Overall, the VPP program has reduced days away from work to 53% of the average for a typical industry group.

38-3 ERGONOMICS PROGRAM

When an organization or company believes that it has or potentially has losses related to ergonomics, it is useful to establish an ergonomics program. The major elements of the program include:

1. Verify the problem and its scope or potential scope.
2. Gain the support of top management.
3. Develop implementing procedures and organization.
4. Train participants.
5. Track program progress and evaluate its effectiveness.

Verify Ergonomics Problem

In this element, various sources of information are evaluated to establish whether ergonomics problems exist in an organization and are of sufficient frequency or severity to warrant further action. For example, one may review accident records or conduct an analysis of certain jobs. The resulting data may show that there is a significant problem. For some organizations, the driving force behind the analysis is fear of action by inspectors from OSHA. If OSHA were to discover a serious ergonomics problem, large fines could be imposed. An organization that completes its own evaluation could eliminate the ergonomic problems and the possible costs the problems could cause, including potential fines.

Top Management Support

When there is a significant ergonomics problem, it is essential to gain top management support for a preventive program. Like other programs, top management must agree that actions throughout the organization are necessary and valuable, that resources requested are provided, and that actions are completed in a timely manner. With support from top management, a more detailed analysis of the problems and a more detailed plan of action are needed.

Prepare for Implementation

The ergonomics program should gain the participation of many elements of an organization. The goal is to help achieve recognition of particular problems and successful implementation of changes to correct them. Many organizations have ergonomics committees or work groups to oversee the ergonomics program at company-wide levels or within major organizational units. The committees often include someone with expertise in ergonomics, managers, staff specialists, workers, and union representatives. If actions must be taken quickly or management does not fully support the committee recommendations, a committee structure may not be suitable. An alternative to a committee is staff analysis and recommendations implemented by top management.

One major task of the ergonomics group is to identify existing and potential ergonomics problems. Those problems that have already produced accidents and worker compensation claims usually are most easily identified. Additional analysis of job tasks, workstation design, and worker complaints identify additional problems.

Another task of the ergonomics group is determining what procedures will be followed to implement corrections. The procedures may be based on cost, claim reduction potential, and other factors.

The ergonomics group also needs to identify training programs for program participants. The group may have to develop policy and procedures and may need to obtain special assistance with particular problems. One type of procedure involves evaluation. The group needs to establish how the effectiveness of program actions will be determined. Several approaches, such as reduced claims, reduced ergonomics-related accidents, and degree of participation, may be appropriate.

Train Participants

One aspect of training is awareness of causes and contributing factors to repeated motion injuries, lifting injuries, VDT workstation disorders, and other ergonomics problems. Training in proper use of adjustable chairs, tables, and other equipment and proper lifting and bending are important. Training is also important in company policy and commitment to correcting ergonomic problems and preventing ergonomic disorders.

Track Progress and Effectiveness

In this element of the ergonomics program, the progress of various efforts is measured. This may include keeping track of problems identified, actions taken to correct them, and effectiveness of solutions, as well as keeping track of global trends in the entire organization to reduce claims and accidents that are related to ergonomics. Many ergonomic changes also lead to productivity improvements that should be tracked.

38-4 EMERGENCY RESPONSE PLAN

Title 3 of the Superfund Amendments and Reauthorization Act of 1986 is known also as the Emergency Planning and Community Right-to-Know Act of 1986. It is also referred to as SARA Title III. It requires companies that store, use, emit, or move hazardous substances in a community to make details about those substances available to local authorities. Local governmental units, often working closely with such companies, develop plans to respond appropriately should a spill or leak of material occur within the community. This section summarizes many of the components of an emergency response plan. It is based on the guidance of the Federal Emergency Management Agency (FEMA).[4] There are other sources for model plans and planning procedures.

The Process

Step 1: Type of Plan   The process of developing an emergency plan involves at least five steps. The first step is defining the type of plan. This step assumes that a government body should develop an emergency response plan. Given that a plan is needed, it is necessary to determine the depth required. A plan may extend from a simple procedure about who to call and when to a very detailed set of activities for various kinds of emergencies and the responsibilities required of many different organizations and companies.

Step 2: Organizing the Participants The number of participants in an emergency response plan typically increases with the amount of detail covered by the plan. It is important to involve all potential participants in the planning process because they may have information and insight that contributes to completeness and effectiveness of the plan. Participants may involve fire and police departments, medical organizations, companies transporting or using hazardous materials, emergency organizations, and local and state government bodies, as well as companies or organizations with special equipment and materials or specialization in cleanup and disposal services. Participants may involve consultants or specialists in certain fields and may involve the media.

Step 3: Conducting a Hazard Analysis The key step in the process is defining the problems for which emergency responses may be needed. For emergency response plans, the problem definition results from a hazard analysis, identifying hazardous materials that could affect the community. Information about hazardous materials must be obtained from local companies, from users of materials, from transportation organizations and companies, and from waste handlers and disposers. Analysts must determine the danger and the type of hazard for each material and must identify potential quantities present or potentially present in the community. The planning team must know the media in which the material could be distributed (air, water, supply, soil, etc.) and how the community's residents could be affected.

Analysis should include locating sources and impact areas on maps. If materials are shipped, the planners need to know what form of transportation and the type of containers involved. Transportation corridors and zones of release should be mapped to determine what locations are likely to be impacted by leaks and spills.

Analysts need to develop lists of organizations and companies that could be resources for dealing with each type of material. The resources may range from technical advice by telephone to special response equipment and personnel.

Step 4: Writing the Plan The plan should include procedures and assign responsibilities for responding to each type of emergency. An outline of contents for an emergency response plan is given in Table 38-1. Completing the plan is not an easy job. Writers should refer to other plans, and the interfaces to other plans in the community or in the surrounding area should be clearly identified. Because the plan is a dynamic document and must be maintained, it should be developed with computer tools to make the process of updating it easier.

Step 5: Maintaining the Plan The plan should be tested through training and evaluation of incidents. There should be a periodic review process to ensure that data and procedures are up to date. Even simple things must be checked periodically. For example, someone should call telephone numbers on call lists to ensure that they are still correct.

38-5 CONSTRUCTION ACCIDENT PREVENTION PLAN

The U.S. Army Corps of Engineers requires contractors who perform construction work for the federal government to prepare an accident prevention plan.[5] The idea is to analyze

TABLE 38-1 Typical Contents of an Emergency Response Plan

Item	Explanation
Emergency response notification	List of who must be called and what information should be given when there is an emergency.
Record of changes	This is a table of changes and dates for them.
Table of contents introduction	This may include abbreviations, purpose and objective, scope and applicability, policies, and assumptions for the plan.
Emergency response operations	This section details what actions must take place. Typically they include: 1. Notification of spill 2. Required federal notification 3. Initiation of action 4. Coordination of decision making 5. Containment and countermeasures 6. Cleanup and disposal 7. Restoration 8. Recovery of damages 9. Follow-up 10. Special response operations 11. Agent-specific considerations.
Emergency assistance telephone roster	An up-to-date list of people and organizations who may be needed in an emergency.
Legal authority and responsibility	This is a table of references to laws and regulations that provide the authority for the plan.
Response organization structure and responsibilities	This section describes the emergency response groups, their responsibilities, and how they will work together.
Disaster assistance and coordination	This section details where additional assistance may be obtained when the regular response organizations are overburdened.
Procedures for changing or updating the plan	This section details who makes changes and how they are made and implemented.
Plan distribution	This is a list of organizations and individuals who have been given a copy of the plan.
Spill cleanup techniques	This contains detailed information about how response groups should handle cleanups. The methods may be specific for particular materials and kinds of spills.
Cleanup/disposal resources	This is a list of what is available, where it is obtained, and how much is available. The list may include contractors, materials, equipment, protective equipment, qualified personnel, disposal sites, etc.
Laboratory and consultant resource	This is a list of special' facilities and personnel who may be valuable in a response.
Technical library and references	This is a list of libraries and other information sources that may be valuable for those preparing, updating, or implementing the plan.
Hazards analysis	This section details the kinds of emergencies that may be encountered, where they are likely to occur, what areas of the community may be affected, and the probability of occurrence.
Documentation of spill events	These are various incident and investigative reports on spills that have occurred.
Hazardous material information	This is a listing of hazardous materials, their properties, response data, and related information.
Training exercises	This section identifies detailed exercises for testing the adequacy of the plan, training personnel, and introducing changes.

the work to be performed on a specific job and plan accident prevention measures for each phase and component of the work. The plan includes work performed by subcontractors and hazard control measures to be taken by the contractor. The plan must include frequent and regularly scheduled safety inspections of work sites, materials, and equipment by competent persons.

An activity hazard analysis is prepared before beginning each phase of the work. This analysis addresses the hazards for each activity performed in the upcoming phase and presents procedures and safeguards necessary to eliminate the hazards or reduce the risk to acceptable levels.

The content of the accident prevention plan and activity hazard analysis is listed in Table 38-2.

TABLE 38-2 Accident Prevention Plan Format[a]

Administrative section
1. This paragraph lists administrative responsibilities for implementing the plan. Included are identification of contractor personnel responsible for accident prevention.
2. This paragraph lists local requirements and codes that must be compiled with for this job. Examples are noise control and traffic control ordinances.
3. This paragraph details the method the prime contractor proposes to use in controlling and coordinating subcontractor work.
4. Plans for layout of temporary construction buildings and facilities are detailed in this paragraph. Included are control of temporary buildings and facilities of subcontractors.
5. This paragraph explains how the contractor will train employees. Both initial training and continued safety education are addressed.
6. Traffic control is covered in this paragraph. Details include marking of hazards related to haul roads, highway intersections, railroads, utilities, bridges, restricted areas, etc.
7. In this paragraph, the contractor explains how job cleanup, safe access, and egress will be handled.
8. Fire protection and emergencies are covered here. Included are plans for dealing with fires, ambulance service, person overboard on waterways, and similar emergencies.
9. This paragraph details inspection procedures by competent persons and resulting reports and actions taken to correct hazards.
10. This paragraph defines procedures for accident investigations.
11. This paragraph provides details about fall protection systems used in the work.
12. The contractor describes the temporary power distribution system in this paragraph. It includes sketches.
13. This paragraph explains safe clearance procedures. These procedures detail how workers and equipment are protected from energy release from electrical, pressure, thermal, and hazardous material sources.
14. Office trailer anchoring systems are described in this paragraph.
15. This paragraph details contingency plans for severe weather that may affect the work and worksite.

Activity hazard analysis section
    For each phase of the work, the contractor shall develop a plan to identify the activity being done, the sequence of work, specific hazards anticipated, and control measures that must be implemented to minimize or eliminate each hazard. The analysis must be specified to each job.

[a] Derived from EM 385-1-1.

EXERCISES

1. Visit a local company, governmental body, or emergency organization. Find out what emergency plans are in place and analyze them for completeness and for being up to date.

2. Review the safety program for a company. Find out what policies and procedures exist. Find out how the program is maintained and evaluated. Find out what management methods are used to implement it, how effective the methods are, and how program performance is measured.

REVIEW QUESTIONS

1. How are controls for hazards implemented?
2. What is the function of safety plans and programs?
3. List major elements for the following:

   (a) product safety program

   (b) occupational safety program

   (e) ergonomics program

   (d) emergency response plan

   (e) construction accident prevention plan

NOTES

1 *Handbook and Standard for Manufacturing Safer Consumer Products*, rev. ed., U.S. Consumer Products Safety Commission, Washington, DC, May, 1977.

2 *Guidelines for Organizing a Product Safety Program*, ANSI Consumer Council Publication 2, rev. ed., American National Standards Institute, New York, November, 1978.

3 Dudley, R. H., and Heldack, J. M., "Product Liability Planning in the Corporation," Part I, *Professional Safety*, October: 11–18 (1981); Part II, November: 22–27 (1981).

4 *Planning Guide and Checklist for Hazardous Materials Contingency Plans*, FEMA-10, Federal Emergency Management Agency, Washington, DC, July, 1981.

5 *Safety and Health Requirements Manual*, rev. ed., EM 385-1-1, U.S. Army Corps of Engineers, Washington, DC, October, 1987. (Available from the Superintendent of Documents, Washington, DC.)

BIBLIOGRAPHY

Ergonomics: A Practical Guide. National Safety Council, Chicago, IL, 1988.

Guidelines for Organizing a Product Safety Program, rev. ed., ANSI Consumer Council Publication 2, American National Standards Institute, New York, November, 1978.

GARNER, CHARLOTTE A., *How Smart Managers Create World-Class Safety, Health, and Environmental Programs*, American Society of Safety Engineers, Des Plaines, IL, 2004.

Handbook and Standard for Manufacturing Safer Consumer Products, rev. ed., U.S. Consumer Product Safety Commission, Washington, DC, May, 1977.

HEALY, R. J., *Emergency and Disaster Planning*, Wiley, New York, 1969.

JONES, S. E., HOWEIN, H. R., SWALM, G. R., and YABLONSKY, J. F., *Occupational Hygiene Management Guide*, Lewis Publishers, Chelsea, MI, 1989.

MCSWEEN, TERRY E., *The Values Based Safety Process*, 2nd ed., John Wiley & Sons, New York, 2003.

Planning Guide and Checklist for Hazardous Materials Contingency Plans, FEMA-10, Federal Emergency Management Agency, Washington, DC, July, 1991.

ROVINS, C., *Health and Safety in Small Industry: A Practi-* *cal Guide for Managers*, Lewis Publishers, Chelsea, MI, 1989.

SMITH, A. J., *Managing Hazardous Substance Accidents*, McGraw-Hill, New York, 1980.

Appendix A

OSHA PERMISSIBLE EXPOSURE LIMITS

The Occupational Safety and Health Administration publishes information on various toxic and hazardous substances. Some of the information is included in this appendix. The information is often referred to as permissible exposure limits (PELs). The PELs are subject to change at any time. OSHA publishes the justification and technical support for the exposure limits in the *Federal Register*. The tables limit an employee's exposure to various substances.

Below are some notes associated with the three tables included in the OSHA standard.

To achieve compliance with these standards, administrative or engineering controls first must be determined and implemented whenever feasible. When such controls are not feasible to achieve full compliance, protective equipment or any other protective measures shall be used to keep the exposure of employees to air contaminants with the prescribed limits. Any equipment and/or technical measures used for this purpose must be approved for each particular use by a competent industrial hygienist or other technically qualified person. Respirators must comply with OSHA respirator standards.

Table A-1

Only a portion of the chemicals included in OSHA Table Z-1 appear here. Readers should refer to the full table in 29 CFR 1910.1000 for a complete list of regulated chemicals. The PELs are 8-hr time-weighted averages (TWAs) unless otherwise noted. A C designation denotes a ceiling limit. They are to be determined from breathing-zone air samples.

> *Substance with limits preceded by "C": Ceiling Values.* An employee's exposure to any substance shall at no time exceed the exposure limit given for that substance. If instantaneous monitoring is not feasible, then the ceiling shall be assessed as a 15-minute TWA exposure which shall not be exceeded at any time during the working day.

> *Other substances: 8-hr TWAs.* An employee's exposure to any substance that is not preceded by a C shall not exceed the 8-hr TWA given for that substance in any 8-hr work shift of a 40-hr work week.

Table A-2

An employee's exposure to any substance listed in OSHA Table Z-2 shall not exceed the exposure limits specified as follows:

TABLE A-1   Limits of Air Contaminants[a]

Substance	CAS No.[b]	ppm[c]	mg/m^{2d}	Skin Designation
Acetic acid	64-19-7	10	25	
Acetone	67-64-1	1,000	2,400	
Acrolein	107-02-8	0.1	0.25	
Acrylamide	79-27-6	—	0.3	X
Allyl glycidyl ether (AGE)	106-92-3	C 10	C 45	
Ammonia	7664-06-0	50	35	
n-Amyl acetate	628-63-7	100	525	
sec-Amyl acetate	628-38-0	125	650	
Aniline and homologs	62-53-3	5	19	X
Arsine	7784-42-1	0.05	0.2	
Boron oxide	1303-86-2			
Total dust		—	15	
Boron triflouide	7637-07-2	C 1	C 3	
Bromine	7726-95-6	0.1	0.7	
Bromoform	75-25-2	0.5	5	X
2-Butanone (methyl ethyl ketone)	78-93-3	200	590	
2-Butoxyethanol	111-76-2	50	240	X
n-Butyl alcohol	71-36-3	100	300	
Butylamine	109-73-9	C 5	C 15	X
Butyl mercaptan	109-79-5	10	35	
Calcium carbonate	1317-65-3			
Total dust		—	15	
Carbon dioxide	124-38-9	5,000	9,000	
Clorine	7782-50-5	C 1	C 3	
Clorine dioxide	10049-04-4	0.1	0.3	
Chlorine triflouride	7790-91-2	C 0.1	C 0.4	
Chloroacetaldehyde	107-20-0	C 1	C 3	
Chlorobenzene	108-90-7	75	350	
Chlorobromomethane	74-97-5	200	1,050	
Chloroform (trichloromethane)	67-66-3	C 50	C 240	
Cyclohexane	110-82-7	300	1,050	
Cyclohexene	110-83-8	300	1,015	
Dichlorodiphenyltrichloroethane (DDT)	50-29-3	—	1	X
Dichloroethyl ether	111-44-4	C 15	C 90	X
Endrin	72-20-8	—	0.1	X
Ethyl acetate	141-78-6	400	1,400	
Ethyl acrylate	140-88-5	25	100	X
Ethyl ether	60-29-7	400	1,200	
Ethyl mercaptan	75-08-1	C 10	C 25	
Ethylene chlorohydrin	107-07-3	5	16	X
Ethyl acrylate	140-88-5	25	100	X
Heptane (n-heptane)	142-82-5	500	2,000	
Hexachloroethane	67-72-1	1	10	X
n-Hexane	110-54-3	500	1,800	
Hydrazine	302-01-2	1	1.3	X
Hydrogen chloride	7647-01-0	C 5	C 7	
Hydrogen cyanide	74-90-8	10	11	X
Hydrogen peroxide	7722-84-1	1	1.4	
Iodine	7553-56-2	C 0.1	C 1	
Isopropyl alcohol	67-63-0	400	980	

TABLE A-1   continued

Substance	CAS No.[b]	ppm[c]	mg/m[2d]	Skin Designation
Lindane	58-89-9	—	0.5	X
L.P.G. (liquified petroleum gas)	68476-85-7	1,000	1,800	
Magnesite	546-93-0			
Total dust		—	15	
Respirable fraction		—	5	
Magnesium oxide fume	1309-48-4			
Total particulate		—	15	
Manganese compounds (as Mn)	7439-96-5	—	C 5	
Marble	1317-65-3			
Total dust		—	15	
Respirable fraction		—	5	
Methyl alcohol	67-56-1	200	260	
Methyl ethyl ketone (MEK); see 2-butanone				
Methyl formate	107-31-3	100	250	
Methyl hydrazine (monomethyl hydrazine)	60-34-4	C 0.2	C 0.35	X
Naphthalene	91-20-3	10	50	
Nicotine	54-11-5	—	0.5	X
Nitric acid	769737-2	2	5	
Nitric oxide	10102-43-9	25	30	
Nitrobenzene	98-95-3	1	5	X
Nitroglycerin	55-63-0	C 0.2	C 2	X
Paraquat, respirable dust	4685-14-7; 1910-42-5; 2074-50-2	—	0.5	X
Particulates not otherwise regulated				
Total dust		—	15	
Respirable fraction		—	5	
Phosgene (carbonyl chloride)	75-44-5	0.1	0.4	
Phosphoric acid	7664-38-2	—	1	
Phosphorus (yellow)	7723-14-0	—	0.1	
Picric acid	88-89-1	—	0.1	X
Plaster of Paris	26499-65-0			
Total dust		—	15	
Respirable fraction		—	5	
Portland cement	65997-15-1			
Total dust		—	15	
Respirable fraction		—	5	
Propane	74-98-6	1,000	1,800	
Stoddard solvent	8052-41-3	500	2,900	
Sucrose	57-50-1			
Total dust		—	15	
Respirable fraction		—	5	
Sulfur dioxide	7446-09-5	5	15	
Sulfuric acid	7664-93-9	—	1	
Tetraethyl lead (as Pb)	78-00-2	—	0.075	X
Turpentine	8006-64-2	100	560	
Xylenes (o-, m-, p-isomers)	1330-20-7	100	435	
Xylidine	1300-73-8	5	25	X

[a] This listing includes selected contaminants only. Refer to 29 CFR 1910.1000, Table Z-1, for a complete listing.
[b] The CAS number is for information only. Enforcement is based on the substance name. For an entry covering more than one metal compound, measured as the metal, the CAS number for the metal is given—not CAS numbers for the individual compounds.
[c] Parts of vapor or gas per million parts of contaminated air by volume at 25°C and 760 torr.
[d] Milligrams of substance per cubic meter of air. When entry is in this column only, the value is exact; when listed with a ppm entry, it is approximate.

TABLE A-2   Exposure Limits from OSHA Table Z-2 (29 CFR 1910.1000)[a]

Substance	8-hr Time-Weighted Average	Acceptable Ceiling Concentration	Acceptable Maximum Peak Above the Acceptable Ceiling Concentration for an 8-hr Shift	
			Concentration	Maximum Duration
Benzene (Z37.40-1969)[b]	10 ppm	25 ppm	50 ppm	10 minutes
Beryllium and berylllium compounds (Z37.29-1970)	2 µg/m^3	5 µg/m^3	25 µg/m^3	30 minutes
Cadmuim fume[c]	0.1 mg/m^3	0.3 mg/m^3	—	
Cadmium dust (Z37.5-1970)[c]	0.2 mg/m^3	0.6 mg/m^3		
Carbon disulfide (Z37.3-1968)	20 ppm	30 ppm	100 ppm	30 minutes
Carbon tetrachloride (Z37.17-1967)	10 ppm	25 ppm	200 ppm	5 min in any 4 hrs
Chromic acid and chromates (Z37.7-1971)	—	1 mg/10 m^3		
Ethylene dibromide (Z37.31-1970)	20 ppm	30 ppm	50 ppm	5 minutes
Ethylene dichloride (Z37.21-1969)	50 ppm	100 ppm	200 ppm	5 min in any 3 hrs
Flouride as dust (Z37.28-1969)	2.5 mg/m^3	—	—	
Formaldehyde; see 1910.1048	—	—	—	
Hydrogen flouride (Z37.28-1969)	3 ppm	—	—	
Hydrogen sulfide (Z37.2-1966)	—	20 ppm	50 ppm	10 min once, only if no other meas. exp. occurs
Mercury (Z37.8-1971)	—	1 mg/10 m^3	—	
Methyl chloride (Z37.18-1969)	100 ppm	200 ppm	300 ppm	5 min in any 3 hrs
Methylene chloride: See §1919.52				
Organo (alkyl) mercury (Z37.30-1969)	0.1 mg/m^3	0.04 mg/m^3	—	
Styrene (Z37.15-1969)	100 ppm	200 ppm	600 ppm	5 min in any 3 hrs
Tetrachloroethylene (Z37.22-1967)	100 ppm	200 ppm	300 ppm	5 min in any 3 hrs
Toluene (Z37.12-1967)	200 ppm	300 ppm	500 ppm	10 minutes
Trichloroethylene (Z37.19-1967)	100 ppm	200 ppm	300 ppm	5 min in any 2 hrs

[a] References are to specific OSHA standards.

[b] This standard applies to the industry segments exempt from the 1 ppm 8-hr TWA and 5 ppm STEL of the benzene standard at 1910.1028.

[c] This standard applies to any operations or sectors for which the Cadmium standard, 1910.1027, is stayed or otherwise not in effect.

8-hr TWAs. An employee's exposure to any substance listed in OSHA Table Z-2 in any 8-hr work shift of a 40-hr work week shall not exceed the 8-hr TWA limit of a 40-hr work week given for that substance.

Acceptable ceiling concentration. An employee's exposure to a substance listed in Table A-2 shall not exceed at any time during an 8-hr shift the acceptable ceiling concentration limit given for the substance, except for a time period and up to a concentration not exceeding the maximum duration and concentration allowed in the column under "acceptable maximum peak above the acceptable ceiling concentration for an 8-hr shift."

Table A-3

An employee's exposure to any substance listed in OSHA Table Z-3 in any 8-hr work shift of a 40-hr work week, shall not exceed the 8-hr TWA limit given for that substance in the table.

TABLE A-3   Mineral Dusts (From 29 CFR 1910.1000 Table Z-3)

Substance	mppcf[a]	mg/m^3
Silica		
Quartz (respirable)	$\dfrac{250}{\%SiO_2 + 5} - \dfrac{1}{\mu}$ [b]	$\dfrac{10 \text{ mg}/\text{m}^{3\text{e}}}{\%SiO_2 + 2}$
Quartz (total dust)	—	$\dfrac{30 \text{ mg}/\text{m}^3}{\%SiO_2 + 2}$
Cristobalite: use $^1/_2$ the value calculated from the count or mass formulae for quartz		
Tridymite: use $^1/_2$ the value calculated from the formulae for quartz		
Amorphous, including natural diatamaceous earth	20	$\dfrac{80 \text{ mg}/\text{m}^3}{\%SiO_2}$
Silicates (less than 1% crystalline silica):		
Mica	20	
Soapstone	20	
Talc (not containing asbestos)	20^c	
Talc (containing asbestos)—use asbestos limit.		
Tremolite, asbestiform (see 29 CFR 1910.1001)		
Portland cement	50	
Graphite (natural)	15	
Coal dust		
Respirable fraction less than 5% SiO$_2$	—	2.4 mg/m$^{3\text{e}}$
Respirable fraction greater than 5% SiO$_2$	—	$\dfrac{10 \text{ mg}/\text{m}^{3\text{e}}}{\%SiO_2 + 2}$
Insert or nuisance dust^d		
Respirable fraction	15	5 mg/m^3
Total dust	50	15 mg/m^3

[a] Millions of particles per cubic foot of air, based on impinger samples counted by light-field techniques.
[b] The percentage of crystalline silica in the in the formula is the amount determined from airborne samples, except in those instances in which other methods have been shown to be applicable.
[c] Containing less than 1% quartz; if 1% quartz or more, use quartz limit.
[d] All inert or nuisance dusts, whether mineral, inorganic, or organic, not listed specifically by substance name are converted by this limit, which is the same as the particulates not otherwise regulated (PNOR) limit in OSHA Table Z-1.
[e] Bother concentration and percent quartz for the application of this limit are to be determined from the fraction passing a size-selector with the following characteristics:

Aerodynamic diameter (unit density sphere)	Percent passing selector
2	90
2.5	75
3.5	50
5.0	25
10	0

The measurements under this note refer to the use of an AEC (now NRC) instrument. The respirable fraction of coal dust is determined with an MRE; the figure corresponding to that of 2.4 mg/m^3.

Computation Formulae

The computation formula which shall apply to employee exposure to more than one substance for which 8-hr time weighted averages are listed in subpart Z of 29 CFR 1910 to determine whether an employee is exposed over the regulatory limit is as follows:

The cumulative exposure for an 8-hr work shift shall be computed as follows:

$$E = (C_aT_a + C_bT_b + \cdots C_nT_n) \div 8,$$

where

E is the equivalent exposure for the working shift,

C is the concentration during any period of time T where the concentration remains constant, and

T is the duration in hours of the exposure at the concentration C.

The value of E shall not exceed the 8-hr TWA specified in subpart Z of 29 CFR 1910 for the substance involved.

Appendix B

Ergonomics Data

TABLE B-1 Standing Body Dimensions (From MIL-STD-1472D)

| | Percentile Values in Centimeters | | | | | |
| | 5th Percentile | | | 95th Percentile | | |
	Ground Troops	Aviators	Women	Ground Troops	Aviators	Women
Weight (kg)	55.5	60.4	46.4	91.6	96.0	74.5
Standing Body Dimensions						
1 Stature	162.8	164.2	152.4	185.6	187.7	174.1
2 Eye height (standing)	151.1	152.1	140.9	173.3	175.2	162.2
3 Shoulder (acromiale) height	133.6	133.3	123.0	154.2	154.8	143.7
4 Chest (nipple) height[a]	117.9	120.8	109.3	136.5	138.5	127.8
5 Elbow (radiale) height	101.0	104.8	94.9	117.8	120.0	110.7
6 Fingertip (dactylion) height		61.5			73.2	
7 Waist height	96.6	97.6	93.1	115.2	115.1	110.3
8 Crotch height	76.3	74.7	68.1	91.8	92.0	83.9
9 Gluteal furrow height	73.3	74.6	66.4	87.7	88.1	81.0
10 Kneecap height	47.5	46.8	43.8	58.6	57.8	52.5
11 Calf height	31.1	30.9	29.0	40.6	39.3	36.6
12 Functional reach	72.6	73.1	64.0	90.9	87.0	80.4
13 Functional reach, extended	84.2	82.3	73.5	101.2	97.3	92.7

	Percentile Values in Inches					
Weight (lb)	122.4	133.1	102.3	201.9	211.6	164.3
Standing Body Dimensions						
1 Stature	64.1	64.6	60.0	73.1	73.9	68.5
2 Eye height (standing)	59.5	59.9	55.5	68.2	69.0	63.9
3 Shoulder (acromiale) height	52.6	52.5	48.4	60.7	60.9	56.6
4 Chest (nipple) height[a]	46.4	47.5	43.0	53.7	54.5	50.3
5 Elbow (radiale) height	39.8	41.3	37.4	46.4	47.2	43.6
6 Fingertip (dactylion) height		24.2			28.8	
7 Waist height	38.0	38.4	36.6	45.3	45.3	43.4
8 Crotch height	30.0	29.4	26.8	36.1	36.2	33.0
9 Gluteal furrow height	28.8	29.4	26.2	34.5	34.7	31.9
10 Kneecap height	18.7	18.4	17.2	23.1	22.8	20.7
11 Calf height	12.2	12.2	11.4	16.0	15.5	14.4
12 Functional reach	28.6	28.8	25.2	35.8	34.3	31.7
13 Functional reach, extended	33.2	32.4	28.9	39.8	38.3	36.5

[a] Bustpoint height for women.

Safety and Health for Engineers, Second Edition, by Roger L. Brauer
Copyright © 2006 John Wiley & Sons, Inc.

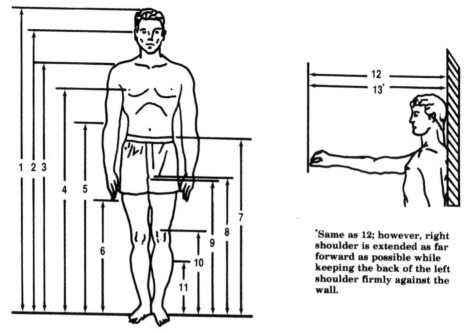

Figure B-1. Standing body dimensions. (From MIL-STD-1472D.)

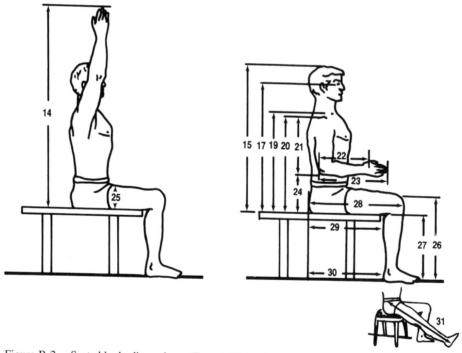

Figure B-2. Seated body dimensions. (From MIL-STD-1472D.)

TABLE B-2 Seated Body Dimensions (From MIL-STD-1472D)

	Percentile Values in Centimeters					
	5th Percentile			95th Percentile		
	Ground Troops	Aviators	Women	Ground Troops	Aviators	Women
Seated Body Dimensions						
14 Vertical arm reach, sitting	128.6	134.0	117.4	147.8	153.2	139.4
15 Sitting height, erect	83.5	85.7	79.0	96.9	98.6	90.9
16 Sitting height, relaxed	81.5	83.6	77.5	94.8	96.5	89.7
17 Eye height, sitting erect	72.0	73.6	67.7	84.6	86.1	79.1
18 Eye height, sitting relaxed	70.0	71.6	66.2	82.5	84.0	77.9
19 Mid-shoulder height	56.6	58.3	53.7	67.7	69.2	62.5
20 Shoulder height, sitting	54.2	54.6	49.9	65.4	65.9	60.3
21 Shoulder–elbow length	33.3	33.2	30.8	40.2	39.7	36.6
22 Elbow–grip length	31.7	32.6	29.6	38.3	37.9	35.4
23 Elbow–fingertip length	43.8	44.7	40.0	52.0	51.7	47.5
24 Elbow rest height	17.5	18.7	16.1	28.0	29.5	26.9
25 Thigh clearance height		12.4	10.4		18.8	17.5
26 Knee height, sitting	49.7	48.9	46.9	60.2	59.9	55.5
27 Popliteal height	39.7	38.4	38.0	50.0	47.7	45.7
28 Buttock–knee length	54.9	55.9	53.1	65.8	65.5	63.2
29 Buttock–popliteal length	45.8	44.9	43.4	54.5	54.6	52.6
30 Buttock–heel length		46.7			56.4	
31 Functional leg length	110.6	103.9	99.6	127.7	120.4	118.6
	Percentile Values in Inches					
Seated Body Dimensions						
14 Vertical arm reach, sitting	50.6	52.8	46.2	58.2	60.3	54.9
15 Sitting height, erect	32.9	33.7	31.1	38.2	38.8	35.8
16 Sitting height, relaxed	32.1	32.9	30.5	37.3	38.0	35.3
17 Eye height, sitting erect	28.3	30.0	26.6	33.3	33.9	31.2
18 Eye height, sitting relaxed	27.6	28.2	26.1	32.5	33.1	30.7
19 Mid-shoulder height	22.3	23.0	21.2	26.7	27.3	24.6
20 Shoulder height, sitting	21.3	21.5	19.6	25.7	25.9	23.7
21 Shoulder–elbow length	13.1	13.1	12.1	15.8	15.6	14.4
22 Elbow–grip length	12.5	12.8	11.6	15.1	14.9	14.0
23 Elbow–fingertip length	17.3	17.6	15.7	20.5	20.4	18.7
24 Elbow rest height	6.9	7.4	6.4	11.0	11.6	10.6
25 Thigh clearance height		4.9	4.1		7.4	6.9
26 Knee height, sitting	19.6	19.3	18.5	23.7	23.6	21.8
27 Popliteal height	15.6	15.1	15.0	19.7	18.8	18.0
28 Buttock–knee length	21.6	22.0	20.9	25.9	25.8	24.9
29 Buttock–popliteal length	17.9	17.7	17.1	21.5	21.5	20.7
30 Buttock–heel length		18.4			22.2	
31 Functional leg length	43.5	40.9	39.2	50.3	47.4	46.7

TABLE B-3   Body Depth and Breadth Dimension (From MIL-STD-1472D)

	Percentile Values in Centimeters					
	5th Percentile			95th Percentile		
	Ground Troops	Aviators	Women	Ground Troops	Aviators	Women
Depth and Breadth Dimensions						
32 Chest depth[a]	18.9	20.4	19.6	26.7	27.8	27.2
33 Buttock depth		20.7	18.4		27.4	24.3
34 Chest breadth	27.3	29.5	25.1	34.4	38.5	31.4
35 Hip breadth, standing	30.2	31.7	31.5	36.7	38.8	39.5
36 Shoulder (bideltoid) breadth	41.5	43.2	38.2	49.8	52.6	45.8
37 Forearm–forearm breadth	39.8	43.2	33.0	53.6	60.7	44.9
38 Hip breadth, sitting	30.7	33.3	33.0	38.4	42.4	43.9
39 Knee-to-knee breadth		19.1			25.5	
	Percentile Values in Inches					
Depth and Breadth Dimensions						
32 Chest depth[a]	7.5	8.0	7.7	10.5	11.0	10.7
33 Buttock depth		8.2	7.2		10.8	9.6
34 Chest breadth	10.8	11.6	9.9	13.5	15.1	12.4
35 Hip breadth, standing	11.9	12.5	12.4	14.5	15.3	15.6
36 Shoulder (bideltoid) breadth	16.3	17.0	15.0	19.6	20.7	18.0
37 Forearm–forearm breadth	15.7	17.0	13.0	21.1	23.9	17.7
38 Hip breadth, sitting	12.1	13.1	13.0	15.1	16.7	17.3
39 Knee-to-knee breadth		7.5			10.0	

[a] Bust depth for women.

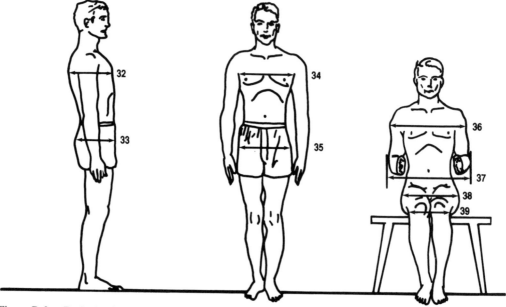

Figure B-3.   Body depth and breadth dimensions. (From MIL-STD-1472D.)

TABLE B-4 Body Circumference and Surface Dimensions (From MIL-STD-1472D)

	Percentile Values in Centimeters					
	5th Percentile			95th Percentile		
	Ground Troops	Aviators	Women	Ground Troops	Aviators	Women
Circumferences						
40 Neck circumference	34.2	34.6	29.9	41.0	41.6	36.7
41 Chest circumference[a]	83.8	87.5	78.4	105.9	109.9	100.2
42 Waist circumference	68.4	73.5	59.5	95.9	101.7	83.5
43 Hip circumference	85.1	87.1	85.5	106.9	108.4	106.1
44 Hip circumference, sitting		97.0	87.7		119.3	110.8
45 Vertical trunk circumference, standing	150.6	156.3	142.2	178.6	181.9	168.3
46 Vertical trunk circumference, sitting		150.4	134.8		175.0	161.0
47 Arm scye circumference	39.6	39.9	33.6	50.3	53.0	41.7
48 Biceps circumference, flexed	27.0	27.8	23.2	37.0	36.9	30.8
49 Elbow circumference, flexed		28.5	23.5		34.2	30.0
50 Forearm circumference, flexed	26.1	26.3	22.2	33.1	33.1	27.5
51 Wrist circumference	15.7	15.3	13.6	18.6	19.2	16.2
52 Upper thigh circumference	48.1	49.6	48.7	63.9	66.9	64.5
53 Calf circumference	31.6	33.3	30.6	41.2	41.3	39.2
54 Ankle circumference	19.3	20.0	18.7	25.2	24.8	23.3
55 Waist back length	39.2	42.4	36.7	50.8	50.9	45.4
56 Waist front length	36.1	35.7	30.5	46.2	44.2	41.4
	Percentile Values in Inches					
Circumferences						
40 Neck circumference	13.5	13.6	11.8	16.1	16.4	14.4
41 Chest circumference[a]	33.0	34.4	30.8	41.7	43.3	39.5
42 Waist circumference	26.9	28.9	23.4	37.8	40.0	32.9
43 Hip circumference	33.5	34.3	33.7	42.1	42.7	41.8
44 Hip circumference, sitting		38.2	34.5		47.0	43.6
45 Vertical trunk circumference, standing	59.3	61.6	56.0	70.3	71.6	65.5
46 Vertical trunk circumference, sitting		59.2	53.1		68.9	63.4
47 Arm scye circumference	15.6	15.7	13.2	19.8	20.9	16.4
48 Biceps circumference, flexed	10.6	11.0	9.1	14.6	14.5	12.1
49 Elbow circumference, flexed		11.2	9.2		13.5	11.8
50 Forearm circumference, flexed	10.3	10.4	8.7	13.0	13.0	10.8
51 Wrist circumference	6.2	6.0	5.4	7.3	7.6	6.4
52 Upper thigh circumference	18.9	19.5	19.2	25.1	26.3	25.4
53 Calf circumference	12.4	13.1	12.0	16.2	16.3	15.4
54 Ankle circumference	7.6	7.9	7.4	9.9	9.7	9.2
55 Waist back length	15.4	16.7	14.4	20.0	20.0	17.9
56 Waist front length	14.2	14.1	12.0	18.2	17.4	16.3

[a] Bust circumference for women.

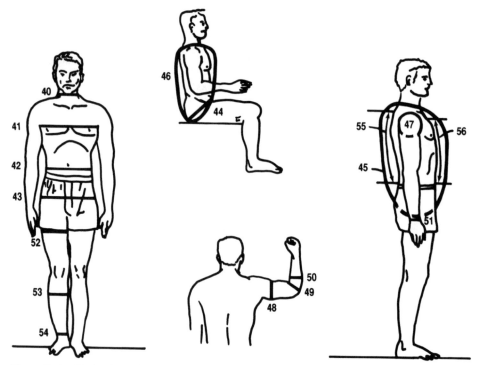

Figure B-4. Body circumferences and surface dimensions. (From MIL-STD-1472D.)

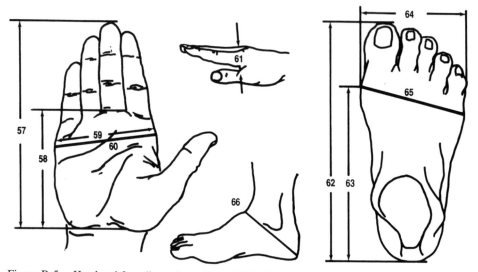

Figure B-5. Hand and foot dimensions. (From MIL-STD-1472D.)

TABLE B-5   Hand and Foot Dimensions (From MIL-STD-1472D)

| | Percentile Values in Centimeters | | | | | |
| | 5th Percentile | | | 95th Percentile | | |
	Ground Troops	Aviators	Women	Ground Troops	Aviators	Women
Hand Dimensions						
57 Hand length	17.4	17.7	16.1	20.7	20.7	20.0
58 Palm length	9.6	10.0	9.0	11.7	11.9	10.8
59 Hand breadth	8.1	8.2	6.9	9.7	9.7	8.5
60 Hand circumference	19.5	19.6	16.8	23.6	23.1	19.9
61 Hand thickness		2.4			3.5	
Foot Dimensions						
62 Foot length	24.5	24.4	22.2	29.0	29.0	26.5
63 Instep length	17.7	17.5	16.3	21.7	21.4	19.6
64 Foot breadth	9.0	9.0	8.0	10.9	11.6	9.8
65 Foot circumference	22.5	22.6	20.8	27.4	27.0	24.5
66 Heel–ankle circumference	31.3	30.7	28.5	37.0	36.3	33.3
	Percentile Values in Inches					
Hand Dimensions						
57 Hand length	6.85	6.98	6.32	8.13	8.14	7.89
58 Palm length	3.77	3.92	3.56	4.61	4.69	4.24
59 Hand breadth	3.20	3.22	2.72	3.83	3.80	3.33
60 Hand circumference	7.68	7.71	6.62	9.28	9.11	7.82
61 Hand thickness		0.95			1.37	
Foot Dimensions						
62 Foot length	9.65	9.62	8.74	11.41	11.42	10.42
63 Instep length	6.97	6.88	6.41	8.54	8.42	7.70
64 Foot breadth	3.53	3.54	3.16	4.29	4.58	3.84
65 Foot circumference	8.86	8.91	8.17	10.79	10.62	9.65
66 Heel–ankle circumference	12.32	12.08	11.21	14.57	14.30	13.11

TABLE B-6   Static Muscle Strength Data (From MIL-STD-1472D)

	Percentile Values in Pounds			
	5th Percentile		95th Percentile	
	Men	Women	Men	Women
Strength Measurements				
A Standing two-handed pull:				
15 in level				
Mean force	166	74	304	184
Peak force	190	89	323	200
B Standing two-handed pull:				
20 in level				
Mean force	170	73	302	189
Peak force	187	84	324	203
C Standing two-handed pull:				
39 in level				
Mean force	100	42	209	100
Peak force	113	49	222	111
D Standing two-handed push:				
59 in level				
Mean force	92	34	229	85
Peak force	106	42	246	97
E Standing one-handed pull:				
39 in level				
Mean force	48	23	141	64
Peak force	58	30	163	72
F Standing one-handed pull:				
Centerline, 18 in level				
Mean force	51	24	152	88
Peak force	61	29	170	101
G Seated one-handed pull:				
Side, 18 in level				
Mean force	54	25	136	76
Peak force	61	30	148	89
H Seated two-handed pull:				
Centerline, 15 in level				
Mean force	134	54	274	173
Peak force	157	64	298	189
I Seated two-handed pull:				
Centerline, 20 in level				
Mean force	118	46	237	142
Peak force	134	53	267	157

TABLE B-7   Arm, Hand and Thumb-Finger Strength (5th Percentile Male Data)

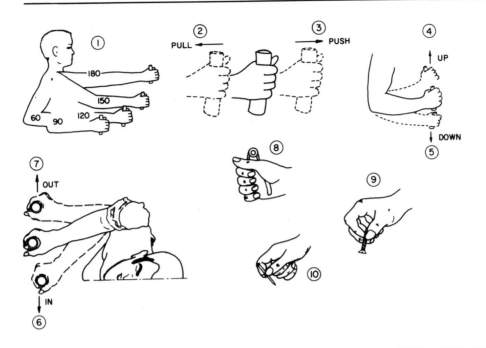

	Arm Strength (lb)												
(1)	(2) Pull		(3) Push		(4) Up		(5) Down		(6) In		(7) Out		
Degree of Elbow Flexion (deg)	L[a]	R[a]	L	R	L	R	L	R	L	R	L	R	
180	50	52	42	50	9	14	13	17	13	20	8	14	
150	42	56	30	42	15	18	18	20	15	20	8	15	
120	34	42	26	36	17	24	21	26	20	22	10	15	
90	32	37	22	36	17	20	21	26	16	18	10	16	
60	26	24	22	34	15	20	18	20	17	20	12	17	

	Hand and Thumb—Finger Strength (N)			
	(8) Hand Grip		(9) Thumb—Finger Grip (Palmer)	(10) Thumb—Finger Grip (Tips)
	L	R		
Momentary hold	56	59	13	13
Sustained hold	33	35	8	8

[a] L = left; R = right.

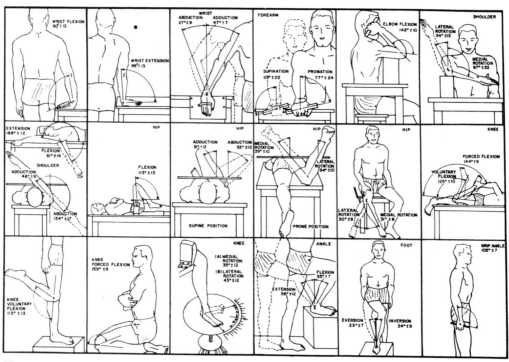

Figure B-6.   Body movements. (From Webb, P., *Bioastronautics Data Book*, NASA SP3006, 1964.)

TABLE B-8   Dimensions for Various Types of Consoles (From MIL-STD-1472D)

Type of Console	Maximum Total Console Height From Standing Surface		Suggested Panel Vertical Dimension (Including Sills)		Writing Surface: Shelf Height From Standing Surface		Seat Height From Standing Surface At Midpoint of G		Maximum Console Width (Not Shown)	
	A		B		C		D			
KEY	m	in	mm	in	mm	in	mm	in	m	in
1. SIT	1.170	46.0	520	20.5	650	25.5	435	17	1.120	44
(with vision	1.335	52.5	520	20.5	810	32.0	595	23.5	1.120	44
over top)	1.435	56.5	520	20.5	910	36.0	695	27.5	1.120	44
2. SIT	1.310	51.5	660	26.0	650	25.5	435	17.0	910	36.0
(without	1.470	58.0	660	26.0	810	32.0	595	23.5	910	36.0
vision over top)	1.570	62.0	660	26.0	910	36.0	695	27.5	910	36.0
3. SIT-STAND (with standing vision over top)	1.535	60.5	620	24.5	910	36.0	695	27.5	910	36.0
4. STAND (with vision over top)	1.535	60.5	620	24.5	910	36.0	NA	NA	1.120	44.0
5. STAND (without vision over top)	1.830	72.0	910	36.0	910	36.0	NA	NA	910	36.0

[a] The range in "A" is provided to allow latitude in the volume of the lower part of the console.
Note the relationship to "C" and "D".

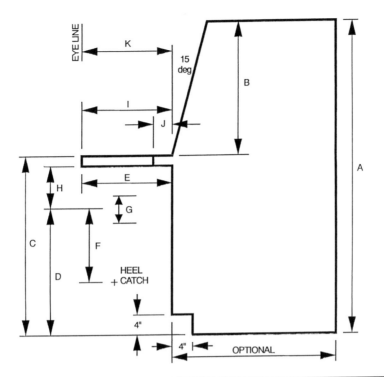

Key	Dimensions	mm	in
A	Maximum total console height from standing surface		
B	Suggested vertical dimension of panel	See Table B-8	
C	Writing surface: shelf height from standing surface		
D	Seat height from standing surface at midpoint of "G"		
E[1]	Minimum knee clearance	460	18.0
F[1]	Foot support to sitting surface[2]	460	18.0
G[1]	Seat adjustability	150	6.0
H[1]	Minimum thigh clearance at midpoint of "G"	190	7.5
I	Writing surface depth including shelf	400	16.0
J	Minimum shelf depth	100	4.0
K	Eye line-to-console front distance	400	16.0

[1] Not applicable to console Types 4 and 5 of Table B-8.
[2] Since this dimension must not be exceeded, a heel catch must be added to the chair if "D" exceeds 460 mm (18.0).
Note: A shelf thickness of 25 mm (1 in) is assumed. For other shelf thicknesses, suitable adjustments should be made.

Figure B-7.   Standard console dimensions. (From MIL-STD-1472D.)

INDEX

A

absorption
  sound, 424–426
  (chemicals) through skin,
    447, 450
acceptable daily intake, 661
accessibility (fire), 295
accident
  and workers' compensation, 58
  cause, 22
  definition, 21, 22, 23
  frequency, 30
  investigation, 690–694
  log, 81, 85
  motor vehicle, 7, 213–216
  prevention, 27, 29, 685
    proactive, 29
    reactive, 29, 79
  reconstruction, 224–226
  recordkeeping, 85
  reports, 81
  severity, 30
  statistics, 694
  theories, 26
    multiple Factor, 26
    four Ms, 26
    energy, 27
    single factor, 29
acclimatization, 346
accountability, 634
acquired immune deficiency
  syndrome, 483
acts
  congressional, 38
  unsafe, 22, 24
aerobic activity, 611
adaptation
  dark, 373
  light, 373
advertising, 70
aerial basket, 156
aerosols, 444
aflatoxin, 492

afterimage, 373
agent orange, 492
air
  bags, 215, 216
  cleaning, 469
    devices, 476–478
  flow, 467–469
    in pipes, 473–474
    pipe entry, 469
  jets, 469
  make up, 469
  recirculating, 470
  speed, 338
    measurement, 355
  supply, 464
  temperature, 338
  terminal, 173
  traffic control system, 227
alarms, 583, 601
alcohol, 213, 219, 570
alpha particle radiation, 400
altitude, 359
American Conference of
  Governmental Industrial
  Hygienists, 106, 346, 386,
  388, 418, 438
American Industrial Hygiene
  Association, 438
American Institute of Architects,
  438
American Institute of Chemical
  Engineers, 438, 442
American National Standards
  Institute, 18, 51, 89, 479
American Petroleum Institute,
  228
American Society for Testing
  and Materials, 51, 228,
  438
American Society of Heating,
  Refrigerating and Air
  Conditioning Engineers,
  345, 438

American Society of
  Mechanical Engineers, 52,
  228, 365
American Table of Distances, 333
Ames test, 452
analysis, reasons for, 685
anemometer, 479
anaerobic activity, 611
angle of repose, 131, 275
anthrax, 484
anthropometry, 596–597
anti-repeat safeguards, 197
aprons, 530
architectural psychology, 594
architecture, 17
arc, arcing, 163, 165, 172
arson, 282
asphyxiant, 445
Association for the
  Advancement of Medical
  Instrumentation, 379
assumption of risk, 55
Atomic Energy Act, 499
attitude, 565
  scales, 697
audiogram, 416
audiology, 416
audition, 416
audits, 635, 689–690
autoignition, 284
automobile, 44, 79, 82
aviation
  safety, 227
awards, safety, 588

B

back
  hoe, 248
  injuries, 239
bacteria, 484
barriers
  explosive, 331
  fall, 144–145

Safety and Health for Engineers, Second Edition, by Roger L. Brauer
Copyright © 2006 John Wiley & Sons, Inc.